Cognitive Linguistics

This book is dedicated to the memory of Larry Trask, 1944–2004, linguist, scholar, teacher, colleague, mentor and friend.

COGNITIVE LINGUISTICS
AN INTRODUCTION

Vyvyan Evans and Melanie Green

EDINBURGH UNIVERSITY PRESS

© Vyvyan Evans and Melanie Green, 2006

Edinburgh University Press Ltd
22 George Square, Edinburgh

Reprinted 2007, 2009, 2011

Typeset in 11/13 Ehrhardt MT and Gill Sans
by Servis Filmsetting Ltd, Manchester, and
printed and bound in Great Britain by
CPI Antony Rowe, Chippenham, Wilts

A CIP record for this book is available from the British Library

ISBN 0 7486 1831 7 (hardback)
ISBN 0 7486 1832 5 (paperback)

The right of Vyvyan Evans and Melanie Green
to be identified as authors of this work
has been asserted in accordance with
the Copyright, Designs and Patents Act 1988.

Contents

Preface   xix
Acknowledgements   xxiii
Abbreviations, symbols and transcription   xxv

Part I Overview of the Cognitive Linguistics Enterprise
Introduction   3

1   **What does it mean to know a language?**   5
    1.1   What is language for?   6
          1.1.1   The symbolic function of language   6
          1.1.2   The interactive function of language 9
    1.2   The systematic structure of language   11
          1.2.1   Evidence for a system   12
          1.2.2   The systematic structure of thought 14
    1.3   What do linguists do?   15
          1.3.1   What?   15
          1.3.2   Why?   16
          1.3.3   How?   16
          1.3.4   Speaker intuitions   16
          1.3.5   Converging evidence   17
    1.4   What it means to know a language   18
    1.5   Summary   20
    Further reading   22
    Exercises   23

2 The nature of cognitive linguistics: assumptions and commitments 27
- 2.1 Two key commitments 27
  - 2.1.1 The 'Generalisation Commitment' 28
  - 2.1.2 The 'Cognitive Commitment' 40
- 2.2 The embodied mind 44
  - 2.2.1 Embodied experience 45
  - 2.2.2 Embodied cognition 46
  - 2.2.3 Experiential realism 47
- 2.3 Cognitive semantics and cognitive approaches to grammar 48
- 2.4 Summary 50
- Further reading 50
- Exercises 52

3 Universals and variation in language, thought and experience 54
- 3.1 Universals in thought and language 55
  - 3.1.1 Typological universals 57
  - 3.1.2 Universals in formal linguistics 60
  - 3.1.3 Universals in cognitive linguistics 63
- 3.2 Cross-linguistic patterns in semantic systems 68
  - 3.2.1 Patterns in the conceptualisation of space 68
  - 3.2.2 Patterns in the conceptualisation of time 75
- 3.3 Cross-linguistic variation in semantic systems 87
  - 3.3.1 Variation in the conceptualisation of space 87
  - 3.3.2 Variation in the conceptualisation of time 92
- 3.4 Linguistic relativity and cognitive linguistics 95
  - 3.4.1 Whorf and the Linguistic Relativity Principle 96
  - 3.4.2 Language as a shaper of thought 98
  - 3.4.3 The cognitive linguistics position 101
- 3.5 Summary 101
- Further reading 102
- Exercises 105

4 Language in use: knowledge of language, language change and language acquisition 108
- 4.1 Language in use 109
  - 4.1.1 A usage event 109
  - 4.1.2 The relationship between usage and linguistic structure 111
  - 4.1.3 Comprehension and production 112
  - 4.1.4 Context 112

		4.1.5	Frequency	114
	4.2	Cognitive Grammar		114
		4.2.1	Abstraction, schematisation and language use	115
		4.2.2	Schemas and their instantiations	115
		4.2.3	Partial sanction	116
		4.2.4	The non-reductive nature of schemas	117
		4.2.5	Frequency in schema formation	118
	4.3	A usage-based approach to language change		120
		4.3.1	Historical linguistics and language change	121
		4.3.2	The Utterance Selection Theory of language change	123
		4.3.3	The Generalised Theory of Selection and the Theory of Utterance Selection	125
		4.3.4	Causal mechanisms for language change	127
	4.4	The usage-based approach to language acquisition		133
		4.4.1	Empirical findings in language acquisition	134
		4.4.2	The cognitive view: socio-cognitive mechanisms in language acquisition	136
		4.4.3	Comparing the generative view of language acquisition	140
	4.5	Summary		146
	Further reading			147
	Exercises			148

Part II Cognitive Semantics

Introduction 153

5 What is cognitive semantics? 156

	5.1	Guiding principles		157
		5.1.1	Conceptual structure is embodied	157
		5.1.2	Semantic structure is conceptual structure	158
		5.1.3	Meaning representation is encyclopaedic	160
		5.1.4	Meaning construction is conceptualisation	162
	5.2	Phenomena investigated within cognitive semantics		163
		5.2.1	The bodily basis of meaning	163
		5.2.2	Conceptual structure	165
		5.2.3	Encyclopaedic semantics	166
		5.2.4	Mappings	167
		5.2.5	Categorisation	168
		5.2.6	Word meaning and polysemy	169
	5.3	Methodology		170
	5.4	Some comparisons with formal approaches to semantics		171

5.5	Summary		172
Further reading			173
Exercises			174

6 Embodiment and conceptual structure — 176

- 6.1 Image schemas — 177
  - 6.1.1 What is an image schema? — 178
  - 6.1.2 Properties of image schemas — 179
  - 6.1.3 Image schemas and linguistic meaning — 189
  - 6.1.4 A provisional list of image schemas — 190
  - 6.1.5 Image schemas and abstract thought — 190
- 6.2 Conceptual structure — 191
  - 6.2.1 Semantic structure — 192
  - 6.2.2 Schematic systems — 194
- 6.3 Summary — 201
- Further reading — 201
- Exercises — 202

7 The encyclopaedic view of meaning — 206

- 7.1 Dictionaries versus encylopaedias — 207
  - 7.1.1 The dictionary view — 207
  - 7.1.2 Problems with the dictionary view — 210
  - 7.1.3 Word meaning versus sentence meaning — 213
  - 7.1.4 The encyclopaedic view — 215
- 7.2 Frame semantics — 222
  - 7.2.1 What is a semantic frame? — 222
  - 7.2.2 Frames in cognitive psychology — 222
  - 7.2.3 The COMMERCIAL EVENT frame — 225
  - 7.2.4 Speech event frames — 228
  - 7.2.5 Consequences of adopting a frame-based model — 229
- 7.3 The theory of domains — 230
  - 7.3.1 What is a domain? — 230
  - 7.3.2 Basic, image-schematic and abstract domains — 232
  - 7.3.3 Other characteristics of domains — 235
  - 7.3.4 Profile/base organisation — 236
  - 7.3.5 Active zones — 238
- 7.4 The perceptual basis of knowledge representation — 240
- 7.5 Summary — 243
- Further reading — 244
- Exercises — 245

8 Categorisation and idealised cognitive models — 248
8.1 Categorisation and cognitive semantics — 249
8.1.1 The classical theory — 251
8.1.2 The definitional problem — 252
8.1.3 The problem of conceptual fuzziness — 253
8.1.4 The problem of prototypicality — 254
8.1.5 Further problems — 254
8.2 Prototype theory — 255
8.2.1 Principles of categorisation — 255
8.2.2 The categorisation system — 256
8.2.3 The vertical dimension — 256
8.2.4 The horizontal dimension — 264
8.2.5 Problems with prototype theory — 268
8.3 The theory of idealised cognitive models — 269
8.3.1 Sources of typicality effects — 270
8.3.2 Radial categories as a further source of typicality effects — 275
8.3.3 Addressing the problems with prototype theory — 278
8.4 The structure of ICMs — 279
8.5 Summary — 281
Further reading — 282
Exercises — 283

9 Metaphor and metonymy — 286
9.1 Literal versus figurative language — 287
9.1.1 Literal and figurative language as complex concepts — 287
9.1.2 Can the distinction be maintained? — 289
9.2 What is metaphor? — 293
9.3 Conceptual Metaphor Theory — 296
9.3.1 The unidirectionality of metaphor — 296
9.3.2 Motivation for target and source — 297
9.3.3 Metaphorical entailments — 298
9.3.4 Metaphor systems — 299
9.3.5 Metaphors and image schemas — 300
9.3.6 Invariance — 301
9.3.7 The conceptual nature of metaphor — 303
9.3.8 Hiding and highlighting — 303
9.4 Primary Metaphor Theory — 304
9.4.1 Primary and compound metaphors — 304
9.4.2 Experiential correlation — 305
9.4.3 Motivating primary metaphors — 306
9.4.4 Distinguishing primary and compound metaphors — 307

	9.5	What is metonymy?	310
	9.6	Conceptual metonymy	314
		9.6.1 Metonymy as an access mechanism	315
		9.6.2 Metonymy-producing relationships	316
		9.6.3 Vehicles for metonymy	317
	9.7	Metaphor–metonymy interaction	318
	9.8	Summary	321
		Further reading	322
		Exercises	325
10	**Word meaning and radial categories**		**328**
	10.1	Polysemy as a conceptual phenomenon	329
	10.2	Words as radial categories	331
	10.3	The full-specification approach	333
		10.3.1 Image schema transformations	337
		10.3.2 Metaphorical extensions	339
	10.4	Problems with the full-specification approach	339
	10.5	The Principled Polysemy approach	342
		10.5.1 Distinguishing between senses	342
		10.5.2 Establishing the prototypical sense	344
		10.5.3 Illustration of a radial category based on Principled Polysemy	347
		10.5.4 Beyond prepositions	348
	10.6	The importance of context for polysemy	352
		10.6.1 Usage context: subsenses	353
		10.6.2 Sentential context: facets	354
		10.6.3 Knowledge context: ways of seeing	355
	10.7	Summary	355
		Further reading	356
		Exercises	359
11	**Meaning construction and mental spaces**		**363**
	11.1	Sentence meaning in formal semantics	364
	11.2	Meaning construction in cognitive semantics	365
	11.3	Towards a cognitive theory of meaning construction	368
	11.4	The architecture of mental space construction	371
		11.4.1 Space builders	371
		11.4.2 Elements	371
		11.4.3 Properties and relations	372
		11.4.4 Mental space lattices	374
		11.4.5 Counterparts and connectors	375
		11.4.6 The Access Principle	376

		11.4.7	Roles and values	381
	11.5	An illustration of mental space construction		382
	11.6	The dynamic nature of meaning construction		386
		11.6.1	Tense and aspect in English	387
		11.6.2	The tense-aspect system in Mental Spaces Theory	389
		11.6.3	Epistemic distance	394
	11.7	Summary		396
	Further reading			397
	Exercises			397
12	**Conceptual blending**			**400**
	12.1	The origins of Blending Theory		401
	12.2	Towards a theory of conceptual integration		403
	12.3	The nature of blending		407
		12.3.1	The elements of conceptual blending	408
		12.3.2	Further linguistic examples	410
		12.3.3	Non-linguistic examples	415
	12.4	Vital relations and compressions		418
		12.4.1	Vital relations	419
		12.4.2	A taxonomy of vital relations and their compressions	420
		12.4.3	Disintegration and decompression	425
	12.5	A taxonomy of integration networks		426
		12.5.1	Simplex networks	426
		12.5.2	Mirror networks	426
		12.5.3	Single-scope networks	427
		12.5.4	Double-scope networks	429
	12.6	Multiple blending		431
	12.7	Constraining Blending Theory		433
	12.8	Comparing Blending Theory with Conceptual Metaphor Theory		435
		12.8.1	Contrasts	435
		12.8.2	When is a metaphor not a blend?	437
		12.8.3	What Blending Theory adds to Conceptual Metaphor Theory	437
	12.9	Summary		439
	Further reading			440
	Exercises			441
13	**Cognitive semantics in context**			**445**
	13.1	Truth-conditional semantics		446
		13.1.1	Meaning, truth and reality	446

		13.1.2	Object language versus metalanguage	446
		13.1.3	The inconsistency of natural language	447
		13.1.4	Sentences and propositions	448
		13.1.5	Truth-conditional semantics and the generative enterprise	449
		13.1.6	Compositionality of meaning	450
		13.1.7	Translating natural language into a metalanguage	451
		13.1.8	Semantic interpretation and matching	452
		13.1.9	Comparison with cognitive semantics	455
	13.2	Relevance Theory		459
		13.2.1	Ostensive communication	459
		13.2.2	Mutual cognitive environment	459
		13.2.3	Relevance	460
		13.2.4	Explicature and implicature	461
		13.2.5	Metaphor	463
		13.2.6	Comparison with cognitive semantics	463
	13.3	Summary		465
	Further reading			466
	Exercises			466

Part III Cognitive Approaches to Grammar

Introduction   471

14 What is a cognitive approach to grammar?   475

	14.1	Guiding assumptions		476
		14.1.1	The symbolic thesis	476
		14.1.2	The usage-based thesis	478
		14.1.3	The architecture of the model	479
	14.2	Distinct cognitive approaches to grammar		480
		14.2.1	The 'Conceptual Structuring System Model'	480
		14.2.2	Cognitive Grammar	480
		14.2.3	Constructional approaches to grammar	481
		14.2.4	Cognitive approaches to grammaticalisation	482
	14.3	Grammatical terminology		483
		14.3.1	Grammar	484
		14.3.2	Units of grammar	484
		14.3.3	Word classes	486
		14.3.4	Syntax	492
		14.3.5	Grammatical functions	494
		14.3.6	Agreement and case	498
	14.4	Characteristics of the cognitive approach to grammar		500

	14.4.1	Grammatical knowledge: a structured inventory of symbolic units	501
	14.4.2	Features of the closed-class subsystem	502
	14.4.3	Schemas and instances	504
	14.4.4	Sanctioning and grammaticality	505
14.5	Summary		506
Further reading			507
Exercises			509

15 The conceptual basis of grammar — 512

- 15.1 The grammatical subsystem: encoding semantic structure — 513
- 15.2 Talmy's 'Conceptual Structuring System Model' — 514
  - 15.2.1 The configuration of SPACE and TIME — 515
  - 15.2.2 Conceptual alternativity — 516
  - 15.2.3 Schematic systems — 517
  - 15.2.4 The 'Configurational Structure System' — 518
  - 15.2.5 The 'Attentional System' — 526
  - 15.2.6 The 'Perspectival System' — 528
  - 15.2.7 The 'Force-Dynamics System' — 531
- 15.3 Langacker's theory of Cognitive Grammar — 533
  - 15.3.1 The conceptual basis of word classes — 533
  - 15.3.2 Attention — 535
  - 15.3.3 Force-dynamics — 544
- 15.4 Categorisation and polysemy in grammar: the network conception — 545
- 15.5 Summary — 548
- Further reading — 549
- Exercises — 550

16 Cognitive Grammar: word classes — 553

- 16.1 Word classes: linguistic categorisation — 554
- 16.2 Nominal predications: nouns — 556
  - 16.2.1 Bounding — 557
  - 16.2.2 Homogeneity versus heterogeneity — 559
  - 16.2.3 Expansibility and contractibility versus replicability — 559
  - 16.2.4 Abstractions — 560
- 16.3 Nominal versus relational predications — 561
- 16.4 Temporal versus atemporal relations — 563
  - 16.4.1 Temporal relations: verbs — 564
  - 16.4.2 Atemporal relations — 565
  - 16.4.3 Class schemas — 570
- 16.5 Nominal grounding predications — 572

		16.5.1	Determiners and quantifiers	572
		16.5.2	Grounding	575
	16.6	Summary		577
	Further reading			577
	Exercises			578
17	**Cognitive Grammar: constructions**			**581**
	17.1	Phrase structure		582
		17.1.1	Valence	583
		17.1.2	Correspondence	584
		17.1.3	Profile determinacy	585
		17.1.4	Conceptual autonomy versus conceptual dependence	585
		17.1.5	Constituency	588
		17.1.6	The prototypical grammatical construction	588
	17.2	Word structure		589
		17.2.1	Phonological autonomy and dependence	590
		17.2.2	Semantic autonomy and dependence	590
		17.2.3	Prototypical stems and affixes	591
		17.2.4	Composite structure	591
		17.2.5	Constructional schemas	592
		17.2.6	Grammatical morphemes and agreement	593
	17.3	Clauses		594
		17.3.1	Valence at the clause level	595
		17.3.2	Grammatical functions and transitivity	601
		17.3.3	Case	606
		17.3.4	Marked coding: the passive construction	609
	17.4	Summary		610
	Further reading			611
	Exercises			612
18	**Cognitive Grammar: tense, aspect, mood and voice**			**615**
	18.1	English verbs: form and function		616
	18.2	The clausal head		617
		18.2.1	The passive construction: [be$_2$ [PERF$_3$ [V]]]	620
		18.2.2	The progressive construction: [be$_1$ [–ing [V]]]	621
		18.2.3	The perfect construction: [have [PERF$_4$ [V]]]	621
	18.3	The grounding predication: mood and tense		624
		18.3.1	Mood	625
		18.3.2	Tense	626
		18.3.3	The epistemic model	627
	18.4	Situation aspect		631

		18.4.1	Situation types	631
		18.4.2	Perfective and imperfective PROCESSES	632
		18.4.3	Aspect and the count/mass distinction	634
	18.5	Summary		637
	Further reading			638
	Exercises			638
19	Motivating a construction grammar			**641**
	19.1	Constructions versus 'words and rules'		642
	19.2	Exploring idiomatic expressions		643
		19.2.1	Typology of idiomatic expressions	643
		19.2.2	Case study I: the *let alone* construction	648
		19.2.3	Case study II: the *what's X doing Y* construction	651
	19.3	Construction Grammar		653
		19.3.1	The Construction Grammar model	653
		19.3.2	Construction Grammar: a 'broadly generative' model	659
		19.3.3	Comparing Construction Grammar with Cognitive Grammar	660
	19.4	The 'Generalisation Commitment'		661
	19.5	Summary		662
	Further reading			662
	Exercises			663
20	The architecture of construction grammars			**666**
	20.1	Goldberg's construction grammar		667
		20.1.1	Assumptions	667
		20.1.2	Advantages of a constructional approach to verb argument structure	669
		20.1.3	The relationship between verbs and constructions	671
		20.1.4	Relationships between constructions	680
		20.1.5	Case studies	684
	20.2	Radical Construction Grammar		692
		20.2.1	Taxonomy of constructions	693
		20.2.2	Emphasis on diversity	693
		20.2.3	Five key features of RCG	693
	20.3	Embodied Construction Grammar		697
		20.3.1	Emphasis on language processing	697
		20.3.2	Analysis and simulation	698
	20.4	Comparing constructional approaches to grammar		699
	20.5	Summary		701

	Further reading	702
	Exercises	703

21 Grammaticalisation 707
- 21.1 The nature of grammaticalisation 708
  - 21.1.1 Form change 710
  - 21.1.2 Meaning change 712
- 21.2 Metaphorical extension approaches 714
  - 21.2.1 Case study: OBJECT-TO-SPACE 718
  - 21.2.2 Case study: SPACE-TO-POSSESSION 719
- 21.3 Invited Inferencing Theory 721
  - 21.3.1 Case study: the evolution of *must* 725
- 21.4 The subjectification approach 728
  - 21.4.1 Case study: *be going to* 730
  - 21.4.2 Case study: the evolution of auxiliaries from verbs of motion or posture 730
- 21.5 Comparison of the three approaches: *be going to* 732
- 21.6 Summary 733
- Further reading 734
- Exercises 736

22 Cognitive approaches to grammar in context 741
- 22.1 Theories of grammar: assumptions, objectives, methodology 741
  - 22.1.1 Cognitive approaches to grammar 743
  - 22.1.2 Generative approaches to grammar 743
  - 22.1.3 Cognitive versus generative models 752
  - 22.1.4 Functional-typological approaches to grammar 758
- 22.2 Core issues in grammar: comparing cognitive and generative accounts 761
  - 22.2.1 Word classes 761
  - 22.2.2 Constituency: heads and dependents 763
  - 22.2.3 The status of tree diagrams 763
  - 22.2.4 Grammatical functions and case 765
  - 22.2.5 The verb string: tense, aspect and mood 767
  - 22.2.6 The passive construction 769
- 22.3 Summary 771
- Further reading 771
- Exercises 773

Part IV Conclusion

23 Assessing the cognitive linguistics enterprise — **777**
    23.1 Achievements — 777
    23.2 Remaining challenges — 779
    23.3 Summary — 782

Appendix: Tables and Figures — 783
References — 792
Index — 812

Preface

The nature of this book

This book represents a general introduction to the area of theoretical linguistics known as cognitive linguistics. It consists of three main parts. Part I provides an overview of some of the main aims, assumptions and commitments of the cognitive linguistics enterprise, and provides an indicative sketch of some of the descriptive analyses and theoretical positions that are representative of cognitive linguistics. The next two parts focus on the two best-developed research frameworks in cognitive linguistics: cognitive semantics (Part II), and cognitive approaches to grammar (Part III). Although some cognitive linguists (notably Langacker) have extended their theories to account for phonology as well as meaning and grammar, we will be mainly concerned with meaning and grammar in this book, and will have little to say about phonology. In part, this reflects the fact that phonology has received relatively little attention within cognitive linguistics (although this situation is changing), and in part this reflects our own interests.

Who is this book for?

Our aim has been to provide a reasonably comprehensive general introduction to cognitive linguistics that is accessible enough for undergraduate students at the university level, while also serving as a work of reference both for linguists and for scholars from neighbouring disciplines. While striving for accessibility, we have also retained considerable detail (including relevant citations in the running text), so that readers (including research students and professional linguists unfamiliar with cognitive linguistics, as well as interested readers from neighbouring disciplines), are provided with a route into the primary literature. In selecting the material presented, and in the presentation itself, we have attempted to provide as balanced a perspective as possible. However, cognitive linguistics represents a collection of approaches rather than a unified theoretical framework, and different authors often take quite distinct positions on similar phenomena, sometimes relying on distinct terminology. It follows that what we present here under the name of 'cognitive linguistics' should be understood as a presentation of the cognitive approach 'as we see it'.

Using the book

We have designed the book so that, in general terms, each chapter builds on preceding chapters. In particular, our decision to present the material on cognitive semantics (Part II) before the material on cognitive approaches to grammar (Part III) reflects the fact that cognitive grammarians assume much of what has been established by cognitive semanticists in developing their approaches. However, because different readers and course tutors will need to use the book in ways tailored to their specific objectives, we have attempted to make Part II and Part III of the book relatively independent so that they can be used for separate courses. The book has sufficient coverage to provide the basis for a number of different courses. We outline below suggestions for 'routes' through the book for three different types of course, assuming 12 teaching weeks at the rate of one chapter per week. Of course, these suggestions can be adjusted depending on teaching time available, level of course and so on. The suggestions made here reflect undergraduate courses taught at the University of Sussex, where this textbook was piloted prior to publication.

Vyvyan Evans and Melanie Green
Linguistics and English Language Department
University of Sussex

March 2005

	Introduction to cognitive linguistics	Cognitive semantics	Cognitive approaches to grammar
Week 1	Ch. 1. What does it mean to know a language?	Ch. 1. What does it mean to know a language?	Ch. 1. What does it mean to know a language?
Week 2	Ch. 2. The nature of cognitive linguistics: assumptions and commitments	Ch. 2. The nature of cognitive linguistics: assumptions and commitments	Ch. 2. The nature of cognitive linguistics: assumptions and commitments
Week 3	Ch. 3. Universals and variation in language, thought and experience	Ch. 5. What is cognitive semantics?	Ch. 14. What is a cognitive approach to grammar?
Week 4	Ch. 4. Language in use: knowledge of language, language change and language acquisition	Ch. 3. Universals and variation in language, thought and experience	Ch. 4. Language in use: knowledge of language, language change and language acquisition
Week 5	Ch 5. What is cognitive semantics?	Ch. 6. Embodiment and conceptual structure	Ch. 15. The conceptual basis of grammar
Week 6	Ch. 6. Embodiment and conceptual structure	Ch. 7. The encyclopaedic view of meaning	Ch. 16. Cognitive grammar: word classes
Week 7	Ch. 7. The encyclopaedic view of meaning	Ch. 8. Categorisation and idealised cognitive models	Ch. 17. Cognitive grammar: constructions
Week 8	Ch. 9. Metaphor and metonymy	Ch. 9. Metaphor and metonymy	Ch. 18. Cognitive grammar: tense, aspect, mood and voice
Week 9	Ch. 14. What is a cognitive approach to grammar?	Ch. 10. Word meaning and radial categories	Ch. 19. Motivating a construction grammar
Week 10	Ch. 19. Motivating a construction grammar	Ch. 11. Meaning construction and mental spaces	Ch. 20. The architecture of construction grammars
Week 11	Ch. 20. The architecture of construction grammars	Ch. 12. Conceptual blending	Ch. 21. Grammaticalisation
Week 12	Ch. 23. Assessing the cognitive linguistics enterprise	Ch. 13. Cognitive semantics in context	Ch. 22. Cognitive approaches to grammar in context

Acknowledgements

In writing this book we have been supported by a large number of people to whom we would like to express our thanks. Firstly, we would like to thank a number of colleagues whose comments and suggestions have helped us to develop and improve the book: Mark Turner and Brigitte Nerlich, our reviewers for Edinburgh University Press, for their advice and encouragement, and Lynne Murphy, Max Wheeler and Jörg Zinken, who read and provided detailed comments on a number of chapters. Of course, these people do not necessarily share our interpretation of the material discussed in this book and any remaining shortcomings are our responsibility.

Secondly, we owe a debt of thanks to our students. Earlier drafts of this textbook were used to teach undergraduate courses at the University of Sussex in 'Cognitive Semantics' during the 2003/4 academic year, and 'Cognitive Grammar', 'Grammar and Mind' and 'Cognitive Semantics' during the 2004/5 academic year. These students engaged fully with the text and provided all sorts of practical suggestions that greatly improved both the presentation and the accessibility of the book. We thank them for their patience and friendly criticism. In particular, we would like to acknowledge the following students from the 2003/4 cohort: Alison Barnes, Nicola Buxton, David Cafferty, Hayley Clark, Steven Cuthbert, Laura Daw, Mary Downham, Mark Hall, Esther Hamon, Shinichiro Harigai, Minna Kirjavainen, Andrew Reid, Kristal Robinson, Claire Simmonds, Rebecca Stubbs, Caroline Veale, Daniel Whaley, Hannah Williams and Susannah Williams; and the following from the 2004/5 cohort: Jonathan Archer, Beatrice Ashmore, Thomas Baker, Mathew Baker, James Brown, Joseph Clark, Tegan French, Simon Harber, Sarah Hayward, Simon Knight, Thomas Kuhn, Alice Lee, Alistair Lockie, Melinda Lowing, Tanja Obradovic, Lucy Meyer, Peter Nichols, Sean Palmer, Melanie Pinet,

Elizabeth Price, Andrew Spratley, Hannah Thomas, Jodi Wenham, Stephanie Wood and Rebecca Wright.

Thirdly, we are grateful to members of the Sussex Cognitive Linguistics Research Group for their constructive comments in response to excerpts from an early draft of the book. In particular we would like to acknowledge Jason Harrison, Basia Golebiowska, Anu Koskela, Shane Lindsay, John Sung and Sarah Witherby. We are also extremely grateful to our colleague Lynne Murphy for her support of the Cognitive Linguistics Research Group, and for her own research, which has informed our thinking on a number of the issues addressed in this book. For financial support in funding the indexing costs we gratefully acknowledge the support of the Department of Linguistics and English Language at Sussex, and the head of department, Richard Coates. Vyvyan Evans gratefully acknowledges the financial support of the University of Sussex, Research Development Fund no. 03R2, which in part supported his participation in this project, and gratefully acknowledges the support of the Dean of the School of Humanities, Stephen Burman, in securing this grant. Melanie Green also gratefully acknowledges her colleagues in the Department of Linguistics and English Language for a period of sabbatical leave that enabled her to complete her part of the project. We also acknowledge our editor at Edinburgh University Press, Sarah Edwards, for her patience and advice, and we remain indebted to Angela Evans for designing the cover image for the book.

Finally, this book is in many ways a product of the unique academic environment at the University of Sussex, most notably the depth and quality of teaching and research in the cognitive sciences, and an atmosphere in which researchers of different theoretical persuasions can share their ideas. It was while the Linguistics and English Language Department was in the School of Cognitive and Computing Sciences (COGS) that the late Professor Larry Trask first proposed and developed undergraduate courses in cognitive linguistics at Sussex, which our students could take alongside courses in formal linguistics and the other cognitive sciences in order to develop a broader and more critical understanding of theoretical models in linguistics. This book emerged from our attempts to teach these courses at a time when appropriate teaching materials were few and far between. Our motivation for writing this book can therefore be directly attributed to the interest in cognitive linguistics, and linguistics in general, that Larry's teaching instilled in others here at Sussex. This book is dedicated to Larry's memory.

Abbreviations, symbols and transcription

A	adjective
ABS	absolutive
ACC	accusative
ANT	anterior
AP	adjective phrase
AUX	auxiliary
BEN	benefactive
DEF	definite
ERG	ergative
F	feminine
HPSG	Head-driven Phrase Structure Grammar
ICM	idealised cognitive model
IMPF	imperfective
INF	infinitive
LF	logical form
M	masculine
MOD	modal
N	noun
NEG	negative
NOM	nominative
NP	noun phrase
OBJ	object
P	preposition
PASS	passive
PERF	perfective
PF	phonological form

PL	plural
PP	preposition phrase
P.PART	past participle
PRED	predicate
PRES	present
PROG	progressive
PRT	particle
S	singular
SFOC	focused subject
SUBJ	subject
TR	trajector
LM	landmark
TAM	tense, aspect, mood
V	verb
VP	verb phrase
1	1st person
2	2nd person
3	3rd person
*	ungrammatical
?	marginal
#	unacceptable
/x/	phoneme
[x]	allophone
v̄	long vowel
v̀	low tone vowel
v̂	falling tone vowel
italics	linguistic form
SMALL CAPS	concept
bold	bold is used for the first mention of key terms introduced in each chapter and for those terms in the summary at the end of each chapter

Part I: Overview of the cognitive linguistics enterprise

Introduction

Cognitive linguistics is a modern school of linguistic thought that originally emerged in the early 1970s out of dissatisfaction with formal approaches to language. Cognitive linguistics is also firmly rooted in the emergence of modern cognitive science in the 1960s and 1970s, particularly in work relating to human categorisation, and in earlier traditions such as Gestalt psychology. Early research was dominated in the 1970s and 1980s by a relatively small number of scholars. By the early 1990s, there was a growing proliferation of research in this area, and of researchers who identified themselves as 'cognitive linguists'. In 1989/90, the International Cognitive Linguistics Society was established, together with the journal *Cognitive Linguistics*. In the words of the eminent cognitive linguist Ronald Langacker ([1991] 2002: xv), this 'marked the birth of cognitive linguistics as a broadly grounded, self conscious intellectual movement'.

Cognitive linguistics is described as a 'movement' or an 'enterprise' because it is not a specific theory. Instead, it is an approach that has adopted a common set of guiding principles, assumptions and perspectives which have led to a diverse range of complementary, overlapping (and sometimes competing) theories. For this reason, Part I of this book is concerned with providing a 'character sketch' of the most fundamental assumptions and commitments that characterise the enterprise as we see it.

In order to accomplish this, we map out the cognitive linguistics enterprise from a number of perspectives, beginning with the most general perspective and gradually focusing in on more specific issues and areas. The aim of Part I is to provide a number of distinct but complementary angles from which the nature and character of cognitive linguistics can be understood. We also draw comparisons with Generative Grammar along the way, in order to set the cognitive approach within a broader context and to identify how it departs from this other well known model of language.

In Chapter 1, we begin by looking at language in general and at linguistics, the scientific study of language. By answering the question 'What does it mean to know a language?' from the perspective of cognitive linguistics, we provide an introductory insight into the enterprise. The second chapter is more specific and explicitly examines the two commitments that guide research in cognitive linguistics: the 'Generalisation Commitment' and the 'Cognitive Commitment'. We also consider the notion of embodied cognition, and the philosophical doctrine of experiential realism, both of which are central to the enterprise. We also introduce the two main approaches to the study of language and the mind adopted by cognitive linguists: cognitive semantics and cognitive (approaches to) grammar, which serve as the focus for Part II and Part III of the book, respectively.

Chapter 3 addresses the issue of linguistic universals and cross-linguistic variation. By examining how cognitive linguists approach such issues, we begin to get a feel for how cognitive linguistics works in practice. We explore the idea of linguistic universals from typological, formal and cognitive perspectives, and look in detail at patterns of similarity and variation in human language, illustrating with an investigation of how language and language-users encode and conceptualise the domains of SPACE and TIME. Finally, we address the Sapir-Whorf hypothesis: the idea that language can influence non-linguistic thought, and examine the status of this idea from the perspective of cognitive linguistics.

In Chapter 4 we focus on the usage-based approach adopted by cognitive linguistic theories. In particular, we examine how representative usage-based theories attempt to explain knowledge of language, language change and child language acquisition. Finally, we explore how the emphasis on situated language use and context gives rise to new theories of human language that, for the first time, provide a significant challenge to formal theories of language.

1

What does it mean to know a language?

Cognitive linguists, like other linguists, study language for its own sake; they attempt to describe and account for its **systematicity**, its **structure**, the **functions** it serves and how these functions are realised by the language system. However, an important reason behind why cognitive linguists study language stems from the assumption that language reflects patterns of thought. Therefore, to study language from this perspective is to study patterns of **conceptualisation**. Language offers a window into cognitive function, providing insights into the nature, structure and organisation of thoughts and ideas. The most important way in which cognitive linguistics differs from other approaches to the study of language, then, is that language is assumed to reflect certain fundamental properties and design features of the human mind. As we will see throughout this book, this assumption has far-reaching implications for the scope, methodology and models developed within the cognitive linguistic enterprise. Not least, an important criterion for judging a model of language is whether the model is psychologically plausible.

Cognitive linguistics is a relatively new school of linguistics, and one of the most innovative and exciting approaches to the study of language and thought that has emerged within the modern field of interdisciplinary study known as cognitive science. In this chapter we will begin to get a feel for the issues and concerns of practising cognitive linguists. We will do so by attempting to answer the following question: what does it mean to know a language? The way we approach the question and the answer we come up with will reveal a lot about the approach, perspective and assumptions of cognitive linguists. Moreover, the view of language that we will finish with is quite different from the view suggested by other linguistic frameworks. As we will see throughout this book, particularly in the comparative chapters at the ends of Part II and Part III, the answer to the title of this chapter will provide a significant challenge to some of these approaches. The cognitive approach also offers exciting glimpses into hitherto hidden aspects of the human mind, human experience and, consequently, what it is to be human.

1.1 What is language for?

We take language for granted, yet we rely upon it throughout our lives in order to perform a range of functions. Imagine how you would accomplish all the things you might do, even in a single day, without language: buying an item in a shop, providing or requesting information, passing the time of day, expressing an opinion, declaring undying love, agreeing or disagreeing, signalling displeasure or happiness, arguing, insulting someone, and so on. Imagine how other forms of behaviour would be accomplished in the absence of language: rituals like marriage, business meetings, using the Internet, the telephone, and so forth. While we could conceivably accomplish some of these things without language (a marriage ceremony, perhaps?), it is less clear how, in the absence of telepathy, making a telephone call or sending an e-mail could be achieved.

In almost all the situations in which we find ourselves, language allows quick and effective expression, and provides a well developed means of **encoding** and **transmitting** complex and subtle ideas. In fact, these notions of encoding and transmitting turn out to be important, as they relate to two key functions associated with language, the **symbolic function** and the **interactive function**.

1.1.1 The symbolic function of language

One crucial function of language is to express thoughts and ideas. That is, language encodes and externalises our thoughts. The way language does this is by using **symbols**. Symbols are 'bits of language'. These might be meaningful subparts of words (for example, *dis-* as in *distaste*), whole words (for example, *cat*, *run*, *tomorrow*), or 'strings' of words (for example, *He couldn't write a pop jingle let alone a whole musical*). These symbols consist of **forms**, which may be spoken, written or signed, and meanings with which the forms are conventionally paired. In fact, a symbol is better referred to as a **symbolic assembly**, as it consists of two parts that are conventionally associated (Langacker 1987). In other words, this symbolic assembly is a **form-meaning pairing**.

A form can be a sound, as in [kæt]. (Here, the speech sounds are represented by symbols from the International Phonetic Alphabet.) A form might be the orthographic representation that we see on the written page: *cat*, or a signed gesture in a sign language. A **meaning** is the conventional ideational or semantic content associated with the symbol. A symbolic assembly of form and meaning is represented in Figure 1.1.

WHAT DOES IT MEAN TO KNOW A LANGUAGE?

Figure 1.1 A symbolic assembly of form and meaning

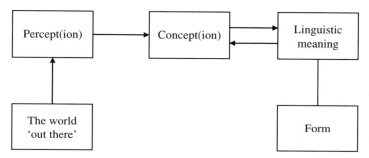

Figure 1.2 Levels of representation

It is important to make it clear that the image of the cat in Figure 1.1 is intended to represent not a particular referent in the world, but the idea of a cat. That is, the image represents the meaning conventionally paired with the form pronounced in English as [kæt]. The meaning associated with a linguistic symbol is linked to a particular mental representation termed a **concept**. Concepts, in turn, derive from **percepts**. For instance, consider a piece of fruit like a pear. Different parts of the brain perceive its shape, colour, texture, taste, smell and so on. This diverse range of perceptual information deriving from the world 'out there' is integrated into a single **mental image** (a representation available to consciousness), which gives rise to the concept of PEAR. When we use language and utter the form *pear*, this symbol corresponds to a conventional meaning, and therefore 'connects' to a concept rather than directly to a physical object in the external world (see Figure 1.2).

Our cognitive abilities integrate raw perceptual information into a coherent and well defined mental image. The meanings encoded by linguistic symbols then, refer to our **projected reality** (Jackendoff 1983): a mental representation of reality, as construed by the human mind, mediated by our unique perceptual and conceptual systems.

We stated above that the symbolic function of language serves to encode and externalise our thoughts. We are now in a position to qualify this view. While our **conceptualisations** are seemingly unlimited in scope, language represents a limited and indeed limiting system for the expression of thought; we've all experienced the frustration of being unable to 'put an idea into words'. There is, after all, a finite number of words, with a delimited set of conventional meanings. From this perspective then, language merely provides **prompts** for the construction of a conceptualisation which is far richer and more elaborate than the minimal meanings provided by language (Fauconnier 1997; Turner 1991). Accordingly, what language encodes is not thought in its complex entirety, but instead rudimentary instructions to the conceptual system to access or create rich and elaborate ideas. To illustrate this point, consider the following illustration adapted from Tyler and Evans (2003):

(1) The cat jumped over the wall.

This sentence describes a jump undertaken by a cat. Before reading on, select the diagram in Figure 1.3 that best captures, in your view, the trajectory of the jump.

We anticipate that you selected the fourth diagram, Figure 1.3(d). After all, the conventional interpretation of the sentence is that the cat begins the jump on one side of the wall, moves through an arc-like trajectory, and lands on the other side of the wall. Figure 1.3(d) best captures this interpretation. On first inspection, this exercise seems straightforward. However, even a simple sentence like (1) raises a number of puzzling issues. After all, how do we know that the trajectory of the cat's jump is of the kind represented in Figure 1.3(d)? What information is there in the sentence that provides this interpretation and excludes the trajectories represented in Figures 1.3(a–c)?

Even though the sentence in (1) would typically be judged as unambiguous, it contains a number of words that have a range of interpretations. The behaviour described by *jump* has the potential to involve a variety of trajectory shapes. For instance, jumping from the ground to the table involves the trajectory represented in Figure 1.3(a). Jumping on a trampoline relates to the trajectory represented in 1.3(b). Bungee jumping involves the trajectory represented in 1.3(c), in which the bungee jumper stops just prior to contact with the surface. Finally, jumping over a puddle, hurdle, wall and so on involves an arc-like trajectory as in 1.3(d).

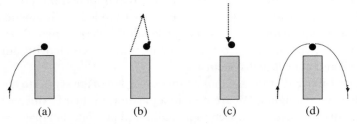

Figure 1.3 Possible trajectories for *The cat jumped over the wall*

If the lexical item *jump* does not in itself specify an arc-like trajectory, but is vague with respect to the shape of the trajectory, then perhaps the preposition *over* is responsible. However, *over* can also have a range of possible interpretations. For instance, it might mean 'across', when we walk *over* a bridge (a horizontal trajectory). It might mean 'above', when an entity like a hummingbird is *over* a flower (higher than but in close proximity to). Equally, *over* could mean 'above' when a plane flies *over* a city (much higher and lacking close proximity). These are just a few of the possibilities. The point to emerge from this brief discussion is that *over* can be used when different kinds or amounts of space are involved, and with a number of different trajectories or paths of motion.

Consider a further complication. Figure 1.3(d) crucially represents the cat's motion ending at a point on the opposite side of the wall relative to the starting position of the jump. Yet no linguistic element in the sentence explicitly provides us with this information.

Example (1) therefore illustrates the following point: even in a mundane sentence, the words themselves, while providing meanings, are only partially responsible for the conceptualisation that these meanings give rise to. Thought relies on a rich array of encyclopaedic knowledge (Langacker 1987). For example, when constructing an interpretation based on the sentence in (1), this involves at the very least the following knowledge: (1) that the kind of jumping cats perform involves traversing obstacles rather than bungee jumping; (2) that if a cat begins a jump at a point on one side of an obstacle, and passes through a point above that obstacle, then gravity will ensure that the cat comes to rest on the other side of the obstacle; (3) that walls are impenetrable barriers to forward motion; (4) that cats know this, and therefore attempt to circumnavigate the obstacle by going over it. We use all this information (and much more), in constructing the rich conceptualisation associated with the sentence in (1). The words themselves are merely prompts for the construction process.

So far, then, we have established that one of the functions of language is to represent or symbolise concepts. Linguistic symbols, or more precisely symbolic assemblies, enable this by serving as prompts for the construction of much richer conceptualisations. Now let's turn to the second function of language.

1.1.2 The interactive function of language

In our everyday social encounters, language serves an **interactive function**. It is not sufficient that language merely pairs forms and meanings. These form–meaning pairings must be recognised by, and be accessible to, others in our community. After all, we use language in order to 'get our ideas across', in other words to **communicate**. This involves a process of transmission by the speaker, and decoding and interpretation by the hearer, processes that involve the construction of rich conceptualisations (see Figure 1.4).

COGNITIVE LINGUISTICS: AN INTRODUCTION

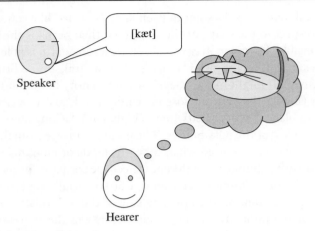

Figure 1.4 The interactive function

The messages we choose to communicate can perform various interactive and **social functions**. For example, we can use language to change the way the world is, or to make things happen:

(2) a. I now pronounce you man and wife.
    b. Shut the door on your way out!

The utterance in (2a), spoken by a suitably qualified person (such as a member of the clergy licensed to perform marriages), in an appropriate setting (like a church), in the presence of two unmarried adults who consent to be joined in matrimony, has the effect of irrevocably altering the social, legal and even spiritual relationship between the two people. That is, language itself can serve as a **speech act** that forever alters an aspect of our reality.

Similarly, in the example in (2b), the utterance represents a command, which is also a type of speech act. Language provides a means of communication, allowing us to share our wishes and desires. Moreover, the way in which these wishes and desires are expressed signals who we are, and what kind of relationship we have with our addressee. We would be unlikely to issue a command like (2b) to the Queen of England, for example.

Another way in which language fulfils the interactive function relates to the notion of **expressivity**. Language is 'loaded', allowing us to express our thoughts and feelings about the world; consider the different mental images evoked by the following expressions, which might be used by different speakers to refer to the same individual:

(3) a. the eminent linguist
    b. the blonde bombshell

While the example in (3a) focuses on the profession of the individual and her relative standing in that profession, the example in (3b) focuses on her physical appearance. Moreover, although both these sentences relate to a female linguist, the person's gender cannot be inferred from the sentence in (3a) while it can from the second sentence due to normative patterns of linguistic behaviour and social stereoptypes. That is, we typically use the expression *blonde bombshell* to describe the physical attributes of women rather than men.

Language also plays a role in how we affect other people in the world, and how we make others feel by our choice of words. That is, language can provide information about **affect** (emotional response):

(4) a. Shut up!
b. I'm terribly sorry to interrupt you, but . . .

These examples also illustrate the way in which we present our public selves through language. The language we choose to use conveys information about our attitudes concerning others, ourselves and the situations in which we find ourselves.

Language can be used to create scenes or **frames** of experience, indexing and even constructing a particular context (Fillmore 1982). In other words, language use can invoke frames that summon rich knowledge structures, which serve to call up and fill in background knowledge.

(5) a. How do you do?
b. Once upon a time . . .

The example in (5a) creates a greeting frame, signalling an acknowledgement of another person and a recognition that this is the first time they have met. It also signals a degree of formality, which expressions like *hey, what's up?* or *hi* would not. Analogously, the utterance in (5b) signals the beginning of a fairytale. In other words, just by hearing or reading the expression in (5b) an entire frame is invoked, which guides how we should respond to what follows, what our expectations should be and so forth.

In summary, we've seen that not only does language encode particular meanings, but also that, by virtue of these meanings and the forms employed to symbolise these meanings which constitute part of shared knowledge in a particular speech community, language can serve an interactive function, facilitating and enriching communication in a number of ways.

1.2 The systematic structure of language

Having seen some examples of what language is used for, let's now consider how language is structured. Language is a system for the expression of meaning and for carrying out its symbolic and interactive functions. So, what evidence is there for the systematicity of language?

1.2.1 Evidence for a system

Language consists of symbolic assemblies that are combined in various ways to perform the functions we described in section 1.1. A symbolic assembly is a conventional **linguistic unit**, which means that it is a piece of language that speakers recognise and 'agree' about in terms of what it means and how it is used. As we will see later in the book, particularly in Part III, one of the prominent concerns in cognitive approaches to grammar is how to model the inventory of linguistic units that make up a language. For example, speakers of Modern English 'agree' that the form *cat* is used to refer to a certain kind of meaning which we illustrated in Figure 1.2. A conventional unit can be a meaningful sub-part of a word, which linguists call a **morpheme** (*anti-dis-establish* . . .), a whole word, a string of words that 'belong' together (a **phrase**) or a whole sentence. Now let's consider another example:

(6) He kicked the bucket

This utterance consists of a sentence that has an **idiomatic meaning** in English. That is, its meaning is not predictable from the integrated meanings of the individual words. A non-native speaker of English who has not learnt the 'special' idiomatic meaning will only be able to interpret example (6) literally. Native speakers of English, on the other hand, while also being able to interpret the sentence literally, often cannot avoid the idiomatic meaning 'he died'. Of course, whether a literal versus an idiomatic interpretation is accessed depends on the situation or **context** in which the utterance occurs.

Focusing for now on the idiomatic interpretation, we can view this utterance as a unit that has a particular meaning associated with it. Therefore, it counts as a symbolic assembly. Another term for symbolic assembly that is employed by some cognitive linguists is **construction** (e.g. Goldberg 1995). We will look in detail at the notion of symbolic assemblies and constructions in Part III of the book.

When we change certain aspects of the sentence in (6), the meaning is affected. For example, if we change the object (the thing being kicked), as in (7), we lose the idiomatic meaning and are left with a **literal** utterance:

(7) He kicked the mop.

For many cognitive linguists, what makes example (7) 'literal' is that this sentence 'as a whole' does not represent a construction. Instead, the meaning of (7) is interpreted by **unifying** the smaller units, the words. In contrast, example (6) is interpreted as a whole single unit: a construction. One way of expressing this idea in more intuitive terms is to use the metaphor of 'storage': suppose we store our knowledge of words, phrases and complex constructions in a mental 'box'. The behaviour of larger constructions, like *kick the bucket*, suggests that these are stored as 'chunks' or single units, just like words. The meanings of sentences like (7) on the other hand are 'built' by unifying the individual words that make them up.

Now consider another example. If we change the structure of example (6) in the following way, we also lose the idiomatic meaning:

(8) The bucket was kicked by him.

This example shows that, in addition to meaning, constructions (form–meaning pairings) have particular formal grammatical patterns associated with them. In other words, the properties of the construction relate not only to the individual words that make it up, as in (6), but also to the grammatical form, or **word order**. The passive construction in (8), in which *the bucket* is placed in subject position, fails to provide the idiomatic meaning associated with the sentence in (6). We can conclude from this that the linear arrangement of the words in the sentence constitutes part of an individual's knowledge of idiomatic constructions like (6).

This point is also illustrated by an **ungrammatical** sentence, a sentence that does not correspond to any of the formal patterns associated with the constructions of English, as in (9), and consequently does not have a conventional meaning associated with it. Ungrammaticality is indicated by an asterisk:

(9) *Bucket kicked he the

As we noted above, the sentence in (6) qualifies as a construction because it consists of particular words arranged in a particular order, and these words are conventionally associated with a particular (idiomatic) meaning. However, we have suggested that constructions can also give rise to 'literal' meanings. To illustrate this, we will examine another sentence that has both idiomatic and literal meanings. For instance, consider the following linguistic joke:

(10) A: Waiter, what is this fly doing in my soup?
B: I think that's the breaststroke, sir!

This joke turns on the ambiguity between the regular interrogative construction, in which a speaker is enquiring after the intention or purpose of something or someone (*What's that seagull doing on the roof? What's that woman doing over there?*), and the 'What's X doing Y construction', studied in detail by cognitive linguists Paul Kay and Charles Fillmore (1999), in which the speaker is indicating that a particular situation is incongruous or unacceptable (*What are you doing wearing those bunny ears? What are those clothes doing on the floor?*). Notice that each of these interpretations requires a different kind of response. For the regular interrogative construction, the response should consist minimally of a piece of information corresponding to the question word (*building a nest; waiting for a bus*). For the 'what's X doing Y' construction, on the other hand, the expected response is typically an explanation, excuse or apology (*I'm going to a fancy-dress party; I've been busy*).

Crucially, for example (10), these two very different meanings are conventionally associated with exactly the same words arranged in the same sequence. The humorous effect of the waiter's reply rests on the fact that he has chosen to respond to the 'wrong' interpretation. While the diner is employing the 'what's X doing Y' construction, the waiter prefers to respond to the interrogative construction.

The examples in this section illustrate the fact that there is a systematic relationship between words, their meanings and how they are arranged in conventional patterns. In other words, language has a systematic structure.

1.2.2 The systematic structure of thought

Does the systematic structure found in language reflect a systematic structure within our conceptual system? Cognitive linguists certainly think so. Cognitive linguists explore the hypothesis that certain kinds of linguistic expressions provide evidence that the structure of our conceptual systems is reflected in the patterns of language. Moreover, as we will see throughout this book, the way the mind is structured can be seen as a reflection, in part, of the way the world (including our sociocultural experience) is structured and organised. Consider the examples in (11).

(11) a. Christmas is fast approaching.
 b. The number of shares we own has gone up.
 c. Those two have a very close friendship.

These examples relate to the abstract **conceptual domains** of TIME (11a), QUANTITY (11b) and AFFECTION (11c). A conceptual domain is a body of knowledge within our conceptual system that contains and organises related ideas and experiences. For example, the conceptual domain of TIME might relate a range of temporal concepts including *Christmas*, which is a temporal event. Notice that in each sentence in (11) the more abstract concepts *Christmas, number (of shares)* and *friendship* are understood in terms of conceptual domains relating to *concrete* physical experience. For instance, *Christmas* is conceptualised in terms of the domain of physical MOTION, which is evident in the use of the word *approaching* in (11a). Clearly *Christmas* (and other temporal concepts) cannot literally be said to undergo motion. Similarly, the notion of *number of shares* is conceptualised in terms of VERTICAL ELEVATION, which is clear from the use of the phrase *gone up* in (11b). Finally, *friendship* is conceptualised in terms of PHYSICAL PROXIMITY in (11c), which is shown by the use of the word *close*.

One of the major findings to have emerged from studies into the human conceptual system is that abstract concepts are systematically structured in terms of conceptual domains deriving from our experience of the behaviour of physical objects, involving properties like motion, vertical elevation and physical proximity (Lakoff and Johnson 1980, 1999). It seems that the language we use to talk about temporal ideas such as *Christmas* provides powerful evidence that our conceptual system 'organises' abstract concepts in terms of more concrete kinds of experiences, which helps to make the abstract concepts more readily accessible.

1.3 What do linguists do?

As we have begun to see, cognitive linguists form hypotheses about the nature of language, and about the conceptual system that it is thought to reflect. These hypotheses are based on observing patterns in the way language is structured and organised. It follows that a theory of language and mind based on linguistic observation must first describe the linguistic facts in a systematic and rigorous manner, and in such a way that the description provides a plausible basis for a speaker's tacit knowledge of language. This foundation for theorising is termed **descriptive adequacy** (Chomsky 1965; Langacker 1987, 1999a). This concern is one that cognitive linguists share with linguists working in other traditions. Below, we provide an outline of what it is that linguists do and how they go about it.

1.3.1 What?

Linguists try to uncover the systems behind language, to describe these systems and to **model** them. Linguistic models consist of theories about language. Linguists can approach the study of language from various perspectives. Linguists may choose to concentrate on exploring the systems within and between sound, meaning and grammar, or to focus on more applied areas, such as the evolution of language, the acquisition of language by children, language disorders, the questions of how and why language changes over time, or the relationship between language, culture and society. For cognitive linguists, the emphasis is upon relating the systematicity exhibited by language directly to the way the mind is patterned and structured, and in particular to conceptual structure and organisation. It follows that there is a close relationship between cognitive linguistics and aspects of cognitive psychology. In addition to this, applied linguistics also informs and is informed by the cognitive linguistics research agenda in various ways (see Chapters 3 and 4 for further discussion of this point).

1.3.2 Why?

Linguists are motivated to explore the issues we outlined above by the drive to understand human cognition, or how the human mind works. Language is a uniquely human capacity. Linguistics is therefore one of the **cognitive sciences**, alongside philosophy, psychology, neuroscience and artificial intelligence. Each of these disciplines seeks to explain different (and frequently overlapping) aspects of human cognition. In particular, as we have begun to see, cognitive linguists view language as a system that directly reflects conceptual organisation.

1.3.3 How?

As linguists, we rely upon what language tells us about itself. In other words, it is ordinary language, spoken every day by ordinary people, that makes up the **'raw data'** that linguists use to build their theories. Linguists describe language, and on the basis of its properties, formulate hypotheses about how language is represented in the mind. These hypotheses can be tested in a number of ways.

1.3.4 Speaker intuitions

Native speakers of any given human language will have strong **intuitions** about what combinations of sounds or words are possible in their language, and which interpretations can be paired with which combinations. For example, native speakers of English will agree that example (6), repeated here, is a well-formed sentence, and that it may have two possible meanings:

(6) He kicked the bucket.

They will also agree that (7) and (8), repeated here, are both well-formed sentences, but that each has only one possible meaning:

(7) He kicked the mop.
(8) The bucket was kicked by him.

Finally, and perhaps most strikingly, speakers will agree that all of the following examples are impossible in English:

(12) a. *bucket kicked he the
     b. *kicked bucket the he
     c. *bucket the kicked he
     d. *kicked he bucket the

Facts like these show that language, and speakers' intuitions about language, can be seen as a 'window' to the underlying system. On the basis of the patterns that emerge from the description of language, linguists can begin to build theoretical 'models' of language. A model of language is a set of statements that is designed to capture everything we know about this hidden cognitive system in a way that is principled, based on empirical evidence and psychologically plausible.

1.3.5 Converging evidence

How do cognitive linguists evaluate the adequacy of their models? One way is to consider **converging evidence** (Langacker 1999a). This means that a model must not only explain linguistic knowledge, but must also be consistent with what cognitive scientists know about other areas of cognition, reflecting the view that linguistic structure and organisation are a relatively imprecise but nevertheless indicative reflection of cognitive structure and organisation. By way of illustration, consider the scene in Figure 1.5.

How might we use language to describe a scene like this? Most English speakers will agree that (13a) is an appropriate description but that (13b) is 'odd':

(13) a. The cat is on the chair.
     b. ?The chair is under the cat.

Figure 1.5 *The cat is on the chair*

Why should (13b) be 'odd'? It's a perfectly grammatical English sentence. From what psychology has revealed about how the human mind works, we know that we have a tendency to focus our attention on certain aspects of a visual scene. The aspect we focus on is something about which we can make certain predictions. For example, in Figure 1.5 we focus on the cat rather than the chair, because our knowledge of the world tells us that the cat is more likely than the chair to move, to make a noise or to perform some other act. We call this prominent entity the **figure** and the remainder of the scene the **ground**, which is another way of saying 'background' (see Chapter 3). Notice that this fact about human psychology provides us with an explanation for why language 'packages' information in certain ways. In (13a) *the cat* has a prominent position in the sentence; any theory of language will tell you that sentence initial position is a 'special' position in many of the world's languages. This accords with the prominence of the corresponding entity in the visual scene. This explanation, based on the figure-ground distinction, also provides us with an explanation for why (13b) is 'odd'. This is an example of how converging evidence works to strengthen or confirm theories of language. Can you think of a situation in which (13b) would not be odd?

1.4 What it means to know a language

Let's look more closely now at some of the claims made by cognitive linguists about how language is represented in the mind. We have established that the linguist's task is to uncover the systematicity behind and within language. What kinds of systems might there be within language? We'll begin to answer this question by introducing one fundamental distinction based on the foundational work of pioneering cognitive linguist Leonard Talmy. Talmy suggests that the **cognitive representation** provided by language can be divided into **lexical** and **grammatical** subsystems. Consider the following example:

(14) **The** hunter track**ed the** tiger**s**.

Notice that certain parts of the sentence in (14) – either whole words (**free morphemes**), or meaningful subparts of words (**bound morphemes**) – have been marked in boldtype. What happens when we alter those parts of the sentence?

(15) a. **Which** hunter track**ed the** tigers?
 b. **The** hunter track**s the** tiger**s**.
 c. **Those** hunter**s** track **a** tiger.

All the sentences in (15) are still about some kind of tracking event involving one or more hunter(s) and one or more tiger(s). What happens when we change the 'little' words like *a*, *the* and *those* and the bound morphemes like *-ed* or *-s* is that we then interpret the event in different ways, relating to information about number (how many hunters or tigers are/were there?), tense (did this event happen before now or is it happening now?), old/new information (does the hearer know which hunters or tigers we're talking about?) and whether the sentence should be interpreted as a statement or a question.

These linguistic elements and morphemes are known as **closed-class** elements and relate to the grammatical subsystem. The term *closed-class* refers to the fact that it is typically more difficult for a language to add new members to this set of elements. This contrasts with the non-boldtype 'lexical' words which are referred to as **open-class**. These relate to the lexical subsystem. The term *open-class* refers to the fact that languages typically find it much easier to add new elements to this subsystem and do so on a regular basis.

In terms of the meaning contributed by each of these two subsystems, while 'lexical' words provide 'rich' meaning and thus have a **content function**, 'grammatical' elements perform a **structuring function** in the sentence. They contribute to the interpretation in important but rather more subtle ways, providing a kind of 'scaffolding' which supports and structures the rich content provided by open-class elements. In other words, the elements associated with the grammatical subsystem are constructions that contribute **schematic meaning** rather than rich contentful meaning. This becomes clearer when we alter the other parts of the sentence. Compare (14) with (16):

(16) a. **The** movie star kiss**ed the** directors.
b. **The** sunbeam illuminat**ed the** rooftops.
c. **The** textbook deligh**ted the** students.

What all the sentences in (16) have in common with (14) is the 'grammatical' elements. In other words, the grammatical structure of all the sentences in (16) is identical to that of (15). We know that both participants in the event can easily be identified by the hearer. We know that the event took place before now. We know that there's only one movie star/sunbeam/textbook, but more than one director/rooftop/student. Notice that the sentences differ in rather a dramatic way, though. They no longer describe the same kind of event at all. This is because the 'lexical' elements prompt for certain kinds of concepts that are richer and less schematic in nature than those prompted for by 'grammatical' elements. The lexical subsystem relates to things, people, places, events, properties of things and so on. The grammatical subsystem on the other hand relates to concepts having to do with number, time reference, whether a piece of information is old or new, whether the speaker is providing information or requesting information, and so on.

Table 1.1 Properties of the lexical and grammatical subsystems

Lexical subsystem	Grammatical subsystem
Open-class words/morphemes	Closed-class words/morphemes
Content function	Structuring function
Larger set; constantly changing	Smaller set; more resistant to change
Prompts for 'rich' concepts, e.g. people, things, places, properties, etc.	Prompts for schematic concepts, e.g. number, time reference, old vs. new, statement vs. question, etc.

A further important distinction between these two subsystems concerns the way that language changes over time. The elements that comprise the lexical (open-class) subsystem make up a large and constantly changing set in any given human language; over a period of time, words that are no longer 'needed' disappear and new ones appear. The 'grammatical' (closed-class) elements that make up the grammatical subsystem, on the other hand, constitute a smaller set, relatively speaking, and are much more stable. Consequently, they tend to be more resistant to change. However, even 'grammatical' elements do change over time. This is a subject we'll come back to in more detail later in the book when we discuss the process known as **grammaticalisation** (see Chapter 21).

Table 1.1 provides a summary of these important differences between the lexical and grammatical subsystems. Together, these two subsystems allow language to present a cognitive representation, encoding and externalising thoughts and ideas.

Having provided a sketch of what it means to know a language from the perspective of cognitive linguistics, we will now begin to examine the cognitive linguistics enterprise in more detail. In particular, we must consider the assumptions and commitments that underlie the cognitive linguistics enterprise, and begin to examine this approach to language in terms of its perspective, assumptions, the cognitive and linguistic phenomena it considers, its methodologies and its approach to theory construction. We turn to these issues in the next chapter.

1.5 Summary

We began this chapter by stating that cognitive linguists, like other linguists, attempt to describe and account for linguistic **systematicity**, **structure** and **function**. However, for cognitive linguists, language reflects patterns of thought; therefore, to study language is to study patterns of **conceptualisation**. In order to explore these ideas in more detail we looked first at the functions of language. Language provides a means of **encoding** and **transmitting** ideas: it has a **symbolic function** and an **interactive function**. Language encodes and externalises our thoughts by using **symbols**. Linguistic symbols consist of **form-meaning pairings** termed **symbolic assemblies**. The **meaning** associated with a linguistic symbol relates to a mental representation termed a **concept**. Concepts derive from **percepts**; the range of perceptual information deriving from the world is integrated into a **mental image**. The meanings encoded by linguistic symbols refer to our **projected reality**: a mental representation of reality as construed by the human mind. While our **conceptualisations** are unlimited in scope, language merely provides **prompts** for the construction of conceptualisations. Language also serves an **interactive function**; we use it to **communicate**. Language allows us to perform **speech acts**, or to exhibit **expressivity** and **affect**. Language can also be used to create scenes or contexts; hence, language has the ability to invoke experiential **frames**.

Secondly, we examined the evidence for a linguistic system, introducing the notion of a conventional **linguistic unit**, which may be a **morpheme**, a **word**, a string of words or a sentence. We introduced the notion of **idiomatic meaning** which is available in certain **contexts** and which can be associated with **constructions**. This contrasts with **literal** meaning, which may be derived by **unifying** smaller constructions like individual words. **Word order** constitutes part of an individual's knowledge of particular constructions, a point illustrated by **ungrammatical** sentences. We also related linguistic structure to the systematic structure of thought. **Conceptual domains** reflected in language contain and organise related ideas and experiences.

Next, we outlined the task of the cognitive linguist: to form hypotheses about the nature of language and about the conceptual system that it reflects. These hypotheses must achieve **descriptive adequacy** by describing linguistic facts in a systematic and rigorous manner. Linguists try to uncover, describe and **model** linguistic systems, motivated by the drive to understand human cognition. Linguistics is therefore one of the **cognitive sciences**. Cognitive linguists carry out this task by examining linguistic **data** and by relying on native speaker **intuitions** and **converging evidence**. As an example of converging evidence, we explored the linguistic reflex of the distinction made in psychology between **figure** and **ground**.

Finally, we looked at what it means to know a language, and introduced an important distinction between kinds of linguistic knowledge: the **cognitive representation** provided by language can be divided into **lexical** and **grammatical** subsystems. The lexical subsystem contains **open-class** elements which perform a **content function**. The grammatical subsystem contains **closed-class** elements, which perform a **structuring function** providing **schematic meaning**.

Further reading

A selection of introductory texts that deal broadly with all aspects of linguistics for those relatively new to the subject

- **Dirven and Verspoor (2004).** This introductory textbook of general linguistics takes a cognitive approach and includes chapters on language and thought, and words, meanings and concepts.
- **Fromkin, Rodman and Hyams (2002).** A very popular introductory textbook of linguistics.
- **Trask (1999).** An accessible introduction to linguistics for the layperson; an entertaining read.

A selection of introductory texts on cognitive science in general

- **Bechtel and Graham (eds) (1999)**
- **Cummins and Cummins (eds) (1999)**
- **Green (ed.) (1996)**

Each of these volumes is an introductory-level collection of papers on various aspects of cognitive science. The Green volume places a particular emphasis on linguistics.

A list of texts that provide an overview of the issues of concern to cognitive linguists

- **Allwood and Gärdenfors (eds) (1999).** A collection of papers on various aspects of cognitive semantics; the paper by Gärdenfors provides a particularly useful overview.
- **Geeraerts (1995).** This article compares cognitive linguistic approaches with cognitive science and generative grammar and provides a very broad survey of work on cognitive linguistics; not as accessible as Radden's chapter.
- **Geeraerts and Cuyckens (2005).** An important reference work featuring articles on a wide range of areas in cognitive linguistics by leading scholars in the field.
- **Goldberg (ed.) (1996).** A collection of conference papers. Provides a representative sample of the range of concerns and issues addressed by cognitive linguists.
- **Janssen and Redeker (1999).** A collection of papers by some of the leading proponents in the field; a good background to cognitive linguistics in general.

- **Lakoff (1987).** Seminal text for cognitive linguistics; lively and accessible.
- **Radden (1992).** Provides a clear and accessible overview of iconicity in language, categorisation, metaphor, cultural models and grammar as a conceptual organising system.
- **Rudzka-Ostyn (1988).** An early collection. Includes seminal papers by, among others, two highly influential scholars, Langacker and Talmy.

A list of texts that relate to the issues dealt with in this chapter

- **Evans (2004a).** Explores the relationship between language and conceptual organisation by focusing on how we think and talk about time and temporal experience.
- **Fillmore, Kay and O'Connor (1988).** Seminal article on the relation between idiomaticity and constructions.
- **Lakoff and Johnson (1980).** An early but hugely influential study which first proposed that language reflects systematic 'mappings' (conceptual metaphors) between abstract and concrete conceptual domains.
- **Langacker (1999a).** A survey article which deals with the notions of the symbolic (in Langacker's terms 'semiotic') and interactive functions associated with language, the notion of converging evidence, and how cognitive linguistics differs from formal and functional approaches to language.
- **Nuyts and Pederson (eds) (1997).** The first chapter provides a good general discussion of the nature of the relationship between language and thought.
- **Talmy (2000).** Chapter 1 deals with the notion of the cognitive representation and the distinction between the lexical (open-class) and grammatical (closed-class) subsystems.
- **Tyler and Evans (2003).** The first chapter addresses the idea that words are merely impoverished 'prompts' for rich conceptualisation. Includes a detailed discussion and illustration of the *The cat jumped over the wall* example.

Exercises

1.1 Linguistic encoding

Consider the following examples in the light of our discussion of example (1). Using the diagrams in Figure 1.3 as a starting point, try to draw similar diagrams that capture the path of motion involved in each example. In each case, how much of this information is explicitly encoded within the meanings of the words themselves? How much seems to depend on what you know about the world?

(a) The baby threw the rattle out of the buggy.
(b) I threw the cat out of the back door.
(c) I tore up the letter and threw it out of the window.
(d) I threw the tennis ball out of the house.
(e) I threw the flowers out of the vase.

1.2 Constructions

The examples below contain idiomatic constructions. If you are a non-native speaker of English, you may need to consult a native speaker or a dictionary of idioms to find out the idiomatic meaning. In the light of our discussion of example (6), try changing certain aspects of each sentence to see whether these examples pattern in the same way. For instance, what happens if you change the subject of the sentence (for example, *the presidential candidate* in the first sentence)? What happens if you change the object (for example, *the towel*)? It's not always possible to make a sentence passive, but what happens to the meaning here if you can?

(a) The presidential candidate threw in the towel.
(b) Before the exam, Mary got cold feet.
(c) She's been giving me the cold shoulder lately.
(d) You are the apple of my eye.
(e) She's banging her head against a brick wall.

What do your findings suggest about an individual's knowledge of such constructions as opposed to sentences containing literal meaning? Do any of these examples also have a literal meaning?

1.3 Word order

Take example (b) from exercise 1.2 above. Believe it or not, a sentence like this with seven words has 5,040 mathematically possible word order permutations! Try to work out how many of these permutations result in a grammatical sentence. What do your findings suggest?

1.4 Concepts and conceptual domains

The examples below contain linguistic expressions that express abstract concepts. In the light of our discussion of the examples in (11), identify the relevant

WHAT DOES IT MEAN TO KNOW A LANGUAGE?

conceptual domain that the concept might relate to. Do these abstract concepts appear to be understood in terms of concrete physical experiences? What is the evidence for your conclusions?

(a) You've just given me a really good idea.
(b) How much time did you spend on this essay?
(c) He fell into a deep depression.
(d) The Stock Market crashed on Black Wednesday.
(e) Unfortunately, your argument lacks a solid foundation.

Now come up with other sentences which illustrate similar patterns for the following conceptual domains:

(f) THEORIES
(g) LOVE
(h) ARGUMENT
(i) ANGER
(j) KNOWING/UNDERSTANDING

1.5 Figure and ground

Consider the scenes in Figure 1.6. For each one, state the sentence that springs first to mind as the most natural way of describing the scene. For example, for the scene in (a), you might come up with *The goldfish is in the bowl*. What happens if you change the sentence around as we did for example (15)? What do your findings suggest about the figure/ground distinction?

1.6 Open-class or closed-class?

Consider the example below in the light of our discussion of examples (15)–(16). First, try to identify the open-class words/morphemes and the closed-class words/morphemes by referring to the properties described in Table 1.1. Next, come up with a set of examples in which only the closed-class words/morphemes have been altered. What kinds of differences do these changes make to the sentence? Finally, try changing the open-class words/morphemes. What kinds of differences do these changes make to the sentence?

The supermodel was putting on her lipstick.

Figure 1.6 Figure and ground

2

The nature of cognitive linguistics: Assumptions and commitments

In this chapter we address the assumptions and commitments that make cognitive linguistics a distinctive enterprise. We begin by outlining two key commitments widely shared by cognitive linguists. These are the **'Generalisation Commitment'** and the **'Cognitive Commitment'**. These two commitments underlie the orientation and approach adopted by practising cognitive linguists, and the assumptions and methodologies employed in the two main branches of the cognitive linguistics enterprise: **cognitive semantics** and **cognitive approaches to grammar**. Once we have outlined the two commitments of cognitive linguistics, we then proceed to address the relationship between language, the mind and experience. The **embodied cognition thesis** is also addressed in some detail as it is at the heart of much research within cognitive linguistics. This thesis holds that the human mind and conceptual organisation are functions of the ways in which our species-specific bodies interact with the environment we inhabit. Finally, we provide a brief overview and introduction to cognitive semantics and cognitive (approaches to) grammar, which are addressed in detail in Parts II and Part III of the book, respectively.

2.1 Two key commitments

In an important 1990 paper, George Lakoff, one of the pioneering figures in cognitive linguistics, argued that the cognitive linguistics enterprise is characterised by two key commitments. These are (1) the 'Generalisation Commitment': a commitment to the characterisation of general principles that are responsible for all aspects of human language, and (2) the Cognitive Commitment: a commitment to providing a characterisation of general principles for language that accords with what is known about the mind and brain from other disciplines. In this section we discuss these two commitments and their implications.

2.1.1 The 'Generalisation Commitment'

One of the assumptions that cognitive linguists make is that there are common structuring principles that hold across different aspects of language, and that an important function of linguistics is to identify these common principles. In modern linguistics, the study of language is often separated into distinct areas such as phonology (sound), semantics (word and sentence meaning), pragmatics (meaning in discourse context), morphology (word structure) syntax (sentence structure) and so on. This is particularly true of **formal approaches**: approaches to modelling language that posit explicit mechanical devices or procedures operating on theoretical **primitives** in order to produce the complete set of linguistic possibilities in a given language. Within formal approaches (such as the Generative Grammar approach developed by Noam Chomsky), it is usually argued that areas such as phonology, semantics and syntax concern significantly different kinds of structuring principles operating over different kinds of primitives. For instance, a syntax 'module' is an area in the mind concerned with structuring words into sentences, whereas a phonology 'module' is concerned with structuring sounds into patterns permitted by the rules of any given language, and by human language in general. This modular view of mind reinforces the idea that modern linguistics is justified in separating the study of language into distinct subdisciplines, not only on grounds of practicality but because the components of language are wholly distinct and, in terms of organisation, incommensurable.

Cognitive linguistics acknowledges that it may often be useful, for practical purposes, to treat areas such as syntax, semantics and phonology as being notionally distinct. The study of syntactic organisation involves, at least in part, the study of slightly different kinds of cognitive and linguistic phenomena than the study of phonological organisation. However, given the 'Generalisation Commitment', cognitive linguists disagree that the 'modules' or 'subsystems' of language are organised in significantly divergent ways, or indeed that distinct modules or subsystems even exist. Below we briefly consider the properties of three areas of language in order to give an idea of how apparently distinct language components can be seen to share fundamental organisational features. The three areas we will look at are (1) categorisation, (2) polysemy and (3) metaphor.

Categorisation

An important recent finding in cognitive psychology is that categorisation is not criterial. This means that it is not an 'all-or-nothing' affair. Instead, human

THE NATURE OF COGNITIVE LINGUISTICS

Figure 2.1 Some members of the category CUP

categories often appear to be **fuzzy** in nature, with some members of a category appearing to be more central and others more peripheral. Moreover, degree of **centrality** is often a function of the way we interact with a particular category at any given time. By way of illustration, consider the images in Figure 2.1. It is likely that speakers of English would select the first image 2.1(a) as being more representative of the category CUP than image 2.1(e). However, when drinking from the container in 2.1(e), a speaker might refer to it as *a cup*. On another occasion, perhaps when using a spoon to eat soup from the same container, the same speaker might describe it as *a bowl*. This illustrates that not only is categorisation fuzzy (for example, when does a cup become a bowl?), but also our interaction with a particular entity can influence how we categorise it.

Although the category members in Figure 2.1 may be rated as being more or less **representative** of the category CUP, each of the members appears to resemble others in a variety of ways, despite the fact that there may not be a single way in which all the members resemble each other. For instance, while the cup in 2.1(a) has a handle and a saucer and is used for drinking beverages like tea or coffee, the 'cup' in 2.1(d) does not have a handle, nor is it likely to be used for hot beverages like tea or coffee; instead, this cup is more likely to contain drinks like wine. Similarly, while the 'cup' in 2.1(e) might be categorised as a 'bowl' when we use a spoon to 'eat' from it, when we hold the 'bowl' to our lips and drink soup from it, we might be more inclined to think of it as a 'cup'. Hence, although the 'cups' in Figure 2.1 vary in terms of how representative they are, they are clearly related to one another. Categories that exhibit degrees of centrality, with some members being more or less like other members of a category rather than sharing a single defining trait, are said to exhibit **family resemblance**.

However, fuzziness and family resemblance are not just features that apply to physical objects like cups; these features apply to linguistic categories like morphemes and words too. Moreover, category-structuring principles of this kind are not restricted to specific kinds of linguistic knowledge but apply across the board. In other words, linguistic categories – whether they relate to phonology, syntax or morphology – all appear to exhibit these phenomena. Formal approaches to linguistics have tended towards the view that a particular category exhibits uniform behaviour which characterises the category. As we will see, however, linguistic categories, despite being related, often do not behave in a uniform way. Instead, they reveal themselves to contain members that exhibit quite divergent behaviour. In this sense, linguistic categories exhibit fuzziness and family resemblance. We illustrate this below – based on discussion in Taylor (2003) – with one example from each of the following areas: morphology, syntax and phonology.

Categorisation in morphology: the diminutive in Italian
In linguistics, the term 'diminutive' refers to an affix added to a word to convey the meaning 'small', and is also used to refer to a word formed by the addition of this affix. In Italian the diminutive suffix has a number of forms such as *-ino*, *-etto*, and *-ello*:

(1) paese → paesino
    'village'   'small village'

While a common meaning associated with this form is 'physically small', as in (1), this is not the only meaning. In the following example the diminutive signals affection rather than small size:

(2) mamma → mammina
    'mum'   'mummy'

When applied to abstract nouns, the diminutive acquires a meaning of short temporal duration, reduced strength or reduced scale:

(3) sinfonia   → sinfonietta
    'symphony'   'sinfonietta' (a shorter symphony, often with fewer instruments)

(4) cena   → cenetta
    'supper'   'light supper'

(5) pioggia   → 'pioggerella
    'rain'   'drizzle'

When the diminutive is suffixed to adjective or adverbs, it serves to reduce intensity or extent:

(6) bello   → bellino
    'beautiful' 'pretty/cute'

(7) bene → benino
    'well'   'quite well'

When the diminutive is added to verbs (the verbal diminutive suffixes are -*icchiare* and -*ucchiare*) a process of intermittent or poor quality is signalled:

(8) dormire → dormicchiare
    'sleep'   'snooze'

(9) lavorare → lavoricciare
    'work'   'work half-heartedly'

(10) parlare → parlucchiare
     'speak'   'speak badly' [e.g. a foreign language]

What these examples illustrate is that the diminutive in Italian doesn't have a single meaning associated with it, but instead constitutes a category of meanings which behave in a variety of distinct ways but nonetheless do appear to be related to one another. The category shares a related form and a related set of meanings: a reduction in size, quantity or quality. Hence, the category exhibits family resemblance.

Categorisation in syntax: 'parts of speech'
The received view in linguistics is that words can be classified into classes such as 'noun' and 'verb', traditionally referred to as **parts of speech**. According to this view, words can be classified according to their morphological and distributional behaviour. For example, a word formed by the addition of a suffix like -*ness* (for example, *happi-ness*) is a noun; a word that can take the plural suffix -*s* (for example, *cat-s*) is a noun; and a word that can fill the gap following a sequence of determiner *the* plus adjective *funny* (for example, *the funny* _____) is a noun. In modern linguistics, the existence of word classes is posited not only for practical purposes (that is, to provide us with a tool of description), but also in an attempt to explain how it is that speakers 'know' how to build new words and how to combine words into grammatical sentences. In other words, many linguists think that these word classes have psychological reality.

However, when we examine the grammatical behaviour of nouns and verbs, there is often significant variation in the nature of the grammatical 'rules' they observe. This suggests that the categories 'noun' and 'verb' are not homogenous, but instead that certain nouns and verbs are 'nounier' or 'verbier' – and hence more representative – than others. In this sense, parts of speech constitute fuzzy categories.

By way of illustration, consider first the **agentive nominalisation** of **transitive** verbs. A transitive verb is a verb that can take an object, such as *import* (e.g. rugs) and *know* (e.g. a fact). However, while transitive verbs can often be nominalised – that is, made into 'agentive' nouns like *driver, singer* and *helper* – some verbs, such as *know*, cannot be:

(11) a. John imports rugs →
John is an importer of rugs

b. John knew that fact →
*John was the knower of that fact

Now consider a second example. While verbs can often be substituted by the 'be V-able' construction, this does not always give rise to a well-formed sentence:

(12) a. His handwriting can be read →
His handwriting is readable

b. The lighthouse can be spotted →
*The lighthouse is spottable

Finally, while most transitive verbs undergo **passivisation**, not all do:

(13) a. John kicked the ball →
The ball was kicked by John

b. John owes two pounds →
*?Two pounds are owed by John

Despite these differences, these verbs do share some common 'verbish' behaviour. For example, they can all take the third person present tense suffix *-s* (*s/he import-s/know-s/read-s/spot-s/kick-s/owe-s* . . .). Therefore, while certain verbs fail to display some aspects of 'typical' verb behaviour, this does not mean that these are not part of the category VERB. In contrast, this variation shows us that there is not a fixed set of criteria that serves to define what it means to be a verb. In other words, the linguistic category VERB contains members that are broadly similar yet exhibit variable behaviour, rather like the physical artefact category CUP.

Now let's consider the linguistic category NOUN. While nouns can be broadly classified according to the morphological and distributional criteria we outlined above, they also show considerable variation. For example, only some nouns can undergo what formal linguists call **double raising**. This term applies to a process whereby a noun phrase 'moves' from an embedded clause to the subject position of the main clause via the subject position of another embedded clause. If you are not familiar with the grammatical terms 'noun phrase', 'subject' or '(embedded) clause', the schematic representation in (14) should help. Noun phrases, which are units built around nouns (but sometimes consist only of nouns (for example in the case of pronouns like *me* or proper names like *George*), are shown in boldtype. Square brackets represent the embedded clauses (sentences inside sentences) and the arrows show the 'movement'. Subject positions are underlined:

(14) a. **It** is likely [___ to be shown [that **John** has cheated]]   →
     b. **John** is likely [___ to be shown [___ to have cheated]]

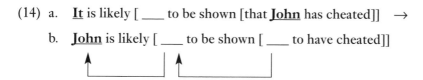

As these examples show, the noun phrase (NP) *John* can only occupy the subject position of a finite or tensed clause: when the verb appears in its '*to* infinitive' form (for example, *to be/to have*), the NP *John* (which we interpret as the 'doer' of the cheating regardless of its position within the sentence) has to 'move up' the sentence until it finds a finite verb like *is*. However, some nouns, like *headway*, do not show the same grammatical behaviour:

(15) a. **It** is likely [___ to be shown [that **no headway** has been made]]
     →
     b. *__No headway__ is likely [___ to be shown [___ to have been made]]

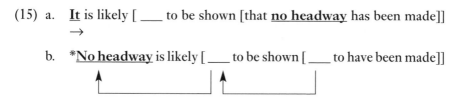

Our next example of variation in the behaviour of nouns concerns question tag formation, a process whereby a **tag question** such as *isn't it?*, *don't you?* or *mustn't he?* can be tagged onto a sentence, where it picks up the reference of some previously mentioned unit. For example, in the sentence *Bond loves blondes, doesn't he?* The pronoun *he* refers back to the subject noun phrase *Bond*. Despite the fact that this grammatical process can apply more or less freely to any subject noun phrase, Taylor (2003: 214) argues that there are nevertheless 'some dubious cases'. For example, the use of a question tag with the noun *heed* is at best marginal:

(16) a. Some headway has been made. →
        Some headway has been made, hasn't it?
     b. Little heed was paid to her. →
        ?*Little heed was paid to her, was it?

As we saw with verbs, examples can always be found that illustrate behaviour that is at odds with the 'typical' behaviour of this category. Although most linguists would not consider this variation sufficient grounds for abandoning the notion of word classes altogether, this variation nevertheless illustrates that categories like NOUN and VERB are not uniform in nature, but are 'graded' in the sense that members of these categories exhibit variable behaviour.

Categorisation in phonology: distinctive features

One of the fundamental concepts in phonology is the **distinctive feature**: an articulatory feature that serves to distinguish speech sounds. For example, the sounds /b/ and /p/ are identical in terms of **place** and **manner** of articulation: both are bilabial sounds (produced by bringing the two lips together) and both are plosives (produced by momentary interruption of the airflow followed by sudden release). However, the two sounds are distinguished by the single feature **voice**: the phenomenon whereby the vocal folds in the larynx are drawn tightly together and vibrate as air passes through them, which affects the quality of the sound. The speech sound /b/ is voiced, whereas /p/ is produced with the vocal folds drawn apart, and is therefore unvoiced. This articulatory feature distinguishes many pairs of consonant sounds that otherwise have a similar manner and place of articulation, for example: /t/ and /d/, as in *tug* versus *dug*; /k/ and /g/, as in *curl* versus *girl*; and /s/ and /z/, as in *Sue* versus *zoo*.

In phonology, these **distinctive features** are traditionally viewed as binary features. In other words, a speech sound can be described in terms of whether it has a positive or a negative value for a certain feature. Binary features are popular in formal linguistics, because they enable linguists to describe units of language by means of a set of properties known as a **feature matrix**. This approach has proven particularly successful in phonology. For example, the sounds /p/ and /b/ can be characterised as follows:

(17) $$\begin{matrix} /p/ \\ \begin{pmatrix} + \text{ bilabial} \\ + \text{ plosive} \\ - \text{ voice} \end{pmatrix} \end{matrix} \quad \begin{matrix} /b/ \\ \begin{pmatrix} + \text{ bilabial} \\ + \text{ plosive} \\ + \text{ voice} \end{pmatrix} \end{matrix}$$

However, Jaeger and Ohala (1984) presented research that questions the assumption that distinctive features are binary in nature. In fact, Jaeger and Ohala found that features like voice are judged by actual users of language as graded or fuzzy categories. Jaeger and Ohala trained naive speakers of English (that is, non-linguists), so that they could identify sounds according to whether they were [+ voice] or [− voice]. They then asked subjects to rate the English plosives, fricatives, nasals and semi-vowels in terms of the voice feature. While plosives involve a sudden release of air from the mouth, fricatives are produced by the gradual release of airflow in the mouth: these are sounds like /f/, /v/, /s/, /z/, and so on. Nasals like /m/ and /n/ involve continuous (uninterrupted) airflow through the nose, and semi-vowels like /w/ and /j/ (which is the IPA symbol for the sound at the start of *yellow*) involve continuous airflow through the mouth.

The researchers found that these sounds were not consistently judged as either voiced or unvoiced. Instead, some sounds were judged as 'more' or 'less' voiced than others. The 'voice continuum' that resulted from Jaeger and Ohala's study is shown in (18a):

(18) a. ← most voiced   least voiced →
     /r,m,n/   /v,ð,z/   /w,j/   /b,d,g/   /f,θ,s,h,ʃ/   /p,t,k/
   b. /r,m,n/   /v,ð,z/   /w,j/   /b,d,g/   /f,θ,s,h,ʃ/   /p,t,k/
     ←————— voiced —————→   ←—— voiceless ——→

The sounds were rated accurately by Jaeger and Ohala's subjects in the sense that voiced and voiceless sounds do not overlap but can be partitioned at a single point on this continuum, as shown in (18b). However, what is striking is that the subjects judged some voiced sounds (like /m/) as 'more voiced' than others (like /z/). These findings suggest that the phonological category VOICED SOUNDS also behaves like a fuzzy category.

Taken together, the examples we have considered from the three 'core' structural areas of human language – morphology, syntax and phonology – suggest that the nature of the linguistic categories we find in each of these areas can be described in rather similar terms. In other words, at least in terms of categorisation, we can generalise across what are often thought of as wholly distinct kinds of linguistic phenomena.

It is worth pointing out at this stage that cognitive linguistics is not unique in seeking to generalise across these 'distinct' areas of human language. Indeed, the quest for binary features in formal linguistics is one example of such an attempt. Encouraged by the relative usefulness of this approach in the area of phonology, formal linguists have, with varying degrees of success, also attempted to characterise word meaning and word classes in terms of binary features. This approach reflects an attempt to capture what are, according to many linguists, the fundamental properties of human language: the 'design features' **discreteness** and **duality of patterning**. Broadly, these features refer to the fact that human language is made of smaller discrete units (like speech sounds, morphemes and words) that can be combined into larger units (like morphemes, words and sentences), and that the capacity for varying the patterns of combination is part of what gives human language its infinite creativity (compare *bin* with *nib*, or *Bond loves blondes* with *blondes love Bond*, for example). Thus different theories of human language are often united in pursuing the same ultimate objectives – here, generalisation – but differ in terms of where and how they seek to reach these objectives.

Polysemy

Polysemy is the phenomenon where a single linguistic unit exhibits multiple distinct yet related meanings. Traditionally, this term is restricted to the area of word meaning (**lexical semantics**), where it is used to describe words like *body* which has a range of distinct meanings that are nevertheless related (for example, the human body; a corpse; the trunk of the human body; the main or central part of something). Polysemy is contrasted with **homonymy**, where two words are pronounced and/or spelt the same way, but have distinct meanings (compare *sole* with *soul*, for example, which are pronounced the same way but which no speaker of English would be likely to judge as having related meanings).

Cognitive linguists argue that polysemy is not restricted to word meaning but is a fundamental feature of human language. According to this view, the 'distinct' areas of language all exhibit polysemy. Cognitive linguists therefore view polysemy as a key to generalisation across a range of 'distinct' phenomena, and argue that polysemy reveals important fundamental commonalities between lexical, morphological and syntactic organisation. Let's look at a few examples.

Polysemy in the lexicon: over
We begin by considering evidence for polysemy at the level of lexical organisation. The word we will consider is the much studied English preposition *over*. Consider the following examples:

(19) a. The picture is over the sofa.   ABOVE
     b. The picture is over the hole.   COVERING
     c. The ball is over the wall.   ON-THE-OTHER-SIDE-OF
     d. The government handed over power.   TRANSFER
     e. She has a strange power over me.   CONTROL

These sentences illustrate various senses of *over*, which are listed in the right-hand column. While each is distinct, they can all be related to one another; they all derive from a central 'above' meaning. We will explore this point in more detail later in the book (see Chapter 10).

Polysemy in morphology: agentive -er *suffix*
Just as words like *over* exhibit polysemy, so do morphological categories. Consider the bound morpheme *-er*, the agentive suffix that was briefly discussed earlier in the chapter:

(20) a. teacher
     b. villager
     c. toaster
     d. best-seller

In each of the examples in (20), the *-er* suffix adds a slightly different meaning. In (20a) it conveys a human AGENT who regularly or by profession carries out the action designated by the verb, in this instance *teach*. In (20b), *-er* relates to a person who lives in a particular place, here a village. In (20c) *-er* relates to an artefact that has the capacity designated by the verb, here *toast*. In (20d) *-er* relates to a particular quality associated with a type of artefact, here the property of selling successfully. Each of these usages is distinct: a teacher is a person who teaches; a toaster is a machine that performs a toasting function; a best-seller is an artefact like a book that has the property of selling well; and a villager is a person who dwells in a village. Despite these differences, these senses are intuitively related in terms of sharing, to a greater or lesser degree, a defining functional ability or attribute: the ability to teach; the 'ability' to toast; the attribute of selling well; and the attribute of dwelling in a specific location. This demonstrates the capacity of morphological categories to exhibit polysemy.

Polysemy in syntax: ditransitive construction
Just as lexical and morphological categories exhibit polysemy, so do syntactic categories. For instance, consider the **ditransitive construction**, discussed by Goldberg (1995). This construction has the following syntax:

(21)   SUBJECT   VERB   OBJECT 1   OBJECT 2

The ditransitive construction also has a range of conventional abstract meanings associated with it, which Goldberg characterises in the terms shown in (22). Note for the time being that terms like AGENT PATIENT and RECIPIENT are labels for 'semantic roles', a topic to which we return in Part III of the book.

(22) a. SENSE 1: AGENT successfully causes recipient to receive PATIENT
INSTANTIATED BY: verbs that inherently signify acts of giving (e.g. *give, pass, hand, serve, feed*)
e.g. [$_{SUBJ}$Mary] [$_{verb}$gave] [$_{OBJ\,1}$ John] [$_{OBJ\,2}$ the cake]

b. SENSE 2: conditions of satisfaction imply that AGENT causes recipient to receive PATIENT
INSTANTIATED BY: verbs of giving with associated satisfaction conditions (e.g. *guarantee, promise, owe*)
e.g. Mary promised John the cake

c. SENSE 3: AGENT causes recipient not to receive PATIENT
INSTANTIATED BY: verbs of refusal (e.g. *refuse, deny*)
e.g. Mary refused John the cake

d. SENSE 4: AGENT acts to cause recipient to receive PATIENT at some future point in time
INSTANTIATED BY: verbs of future transfer (e.g. *leave, bequeath, allocate, reserve, grant*)
e.g. Mary left John the cake

e. SENSE 5: AGENT enables recipient to receive PATIENT
INSTANTIATED BY: verbs of permission (e.g. *permit, allow*)
e.g. Mary permitted John the cake

f. SENSE 6: AGENT intends to cause recipient to receive PATIENT
INSTANTIATED BY: verbs involved in scenes of creation (e.g. *bake, make, build, cook, sew, knit*)
e.g. Mary baked John the cake

While each of the abstract senses associated with 'ditransitive' syntax are distinct, they are clearly related: they all concern volitional transfer, although the nature of the transfer, or the conditions associated with the transfer, vary from sense to sense. We will return to discuss constructions like these in more detail in Part III of the book.

In sum, as we saw for categorisation, cognitive linguists argue that polysemy is a phenomenon common to 'distinct' areas of language. Both 'fuzzy' categories and polysemy, then, are characteristics that unite all areas of human language and thus enable generalisation within the cognitive linguistics framework.

Metaphor

Cognitive linguists also argue that metaphor is a central feature of human language. As we saw in the previous chapter, metaphor is the phenomenon where one conceptual domain is systematically structured in terms of another. One important feature of metaphor is **meaning extension**. That is, metaphor can give rise to new meaning. Cognitive linguists argue that metaphor-based meaning extension can also be identified across a range of 'distinct' linguistic phenomena, and that metaphor therefore provides further evidence in favour of generalising across the 'distinct' areas of language. In this section we'll consider lexicon and syntax.

Metaphor in the lexicon: over *(again)*
In the previous section we observed that the preposition *over* exhibits polysemy. One question that has intrigued cognitive linguists concerns how polysemy is motivated. That is, how does a single lexical item come to have a

multiplicity of distinct yet related meanings associated with it? Lakoff (1987) has argued that an important factor in motivating meaning extension, and hence the existence of polysemy, is metaphor. For instance, he argues that the CONTROL meaning of *over* that we saw in (19e) derives from the ABOVE meaning by virtue of metaphor. This is achieved via application of the metaphor CONTROL IS UP. This metaphor is illustrated by (23):

(23) a. I'm *on top* of the situation.
 b. She's at the *height* of her powers.
 c. His power *rose*.

These examples illustrate that POWER or CONTROL is being understood in terms of greater elevation (UP). In contrast, lack of power or lack of control is conceptualised in terms of occupying a reduced elevation on the vertical axis (DOWN), as shown by (24):

(24) a. Her power is on the *decline*.
 b. He is *under* my control.
 c. He's *low* in the company *hierarchy*.

By virtue of the independent metaphor CONTROL IS UP, the lexical item *over*, which has an ABOVE meaning conventionally associated with it, can be understood metaphorically as indicating greater control. Through frequency of use the meaning of CONTROL becomes conventionally associated with *over* in such a way that *over* can be used in non-spatial contexts like (19e), where it acquires the CONTROL meaning.

Metaphor in the syntax: the ditransitive (again)
One of the observations that Goldberg makes in her analysis of the ditransitive construction is that it typically requires a volitional AGENT in subject position. This is because the meaning associated with the construction is one of *intentional* transfer. Unless there is a sentient AGENT who has the capacity for intention, then one entity cannot be transferred to another. However, we do find examples of this construction where the subject (in square brackets) is not a volitional AGENT:

(25) a. [The rain] gave us some time.
 b. [The missed ball] handed him the victory.

Goldberg argues that examples like these are extensions of the ditransitive construction, and are motivated by the existence of the metaphor CAUSAL EVENTS ARE PHYSICAL TRANSFERS. Evidence for this metaphor comes from examples like the ones in (26), which illustrate that we typically understand abstract causes in terms of physical transfer:

(26) a. David Beckham put a lot of swerve on the ball.
 b. She gave me a headache.

In these examples causal events like causing a soccer ball to swerve, or causing someone to have a headache, are conceptualised as the transfer of a physical entity. Clearly the English soccer star David Beckham, well known for his ability to 'bend' a football around defensive walls, cannot literally put 'swerve' on a football; 'swerve' is not a physical entity that can be 'put' anywhere. However, we have no problem understanding what this sentence means. This is because we 'recognise' the convention within our language system of understanding causal events metaphorically in terms of physical transfer.

Goldberg argues that it is due to this metaphor that the ditransitive construction, which normally requires a volitional AGENT, can sometimes have a non-volitional subject like *a missed ball* or *the rain*. The metaphor licenses the extension of the ditransitive so that it can be used with non-volitional AGENTs.

To conclude the discussion so far, this section has illustrated the view held by cognitive linguists that various areas of human language share certain fundamental organising principles. This illustrates the 'Generalisation Commitment' adopted by cognitive linguists. One area in which this approach has achieved considerable success is in uniting the lexical system with the grammatical system, providing a unified theory of grammatical and lexical structure. As we will see in Part III, cognitive approaches to grammar treat lexicon and syntax not as distinct components of language, but instead as a continuum. However, the relationship between phonology and other areas of human language has only recently begun to be explored from a cognitive perspective. For this reason, while aspects of the foregoing discussion serve to illustrate some similarities between the phonological subsystem and the other areas of the language system, we will have relatively little to say about phonology in the remainder of this book.

2.1.2 The 'Cognitive Commitment'

We turn next to the 'Cognitive Commitment'. We saw above that the 'Generalisation Commitment' leads to the search for principles of language structure that hold across all aspects of language. In a related manner, the 'Cognitive Commitment' represents the view that principles of linguistic structure should reflect what is known about human cognition from other disciplines, particularly the other cognitive sciences (philosophy, psychology, artificial intelligence and neuroscience). In other words, it follows from the 'Cognitive Commitment' that language and linguistic organisation should reflect general cognitive principles rather than cognitive principles that are specific to language. Accordingly, cognitive linguistics rejects the **modular theory** of mind that we mentioned above (section 2.1.1). The modular theory of mind is associated particularly with formal linguistics, but is also explored in other areas of cognitive science such as philosophy and cognitive psychology, and holds that the human mind is organised into distinct 'encapsulated' modules of knowledge, one of which is language, and that these modules serve to 'digest' raw sensory input in such a way that it can then be processed by the central cognitive system (involving deduction, reasoning, memory and so on). Cognitive linguists specifically reject the claim that there is a distinct **language module**, which asserts that linguistic structure and organisation are markedly distinct from other aspects of cognition (see Chapter 4). Below we consider three lines of evidence that, according to cognitive linguists, substantiate the view that linguistic organisation reflects more general cognitive function.

Attention: profiling in language

A very general cognitive ability that human beings have is **attention**, together with the ability to shift attention from one aspect of a scene to another. For instance, when watching a tennis match we can variously attend to the umpire, the flight of the ball back and forth, one or both of the players or parts of the crowd, zooming 'in and out' so to speak. Similarly, language provides ways of directing attention to certain aspects of the scene being linguistically encoded. This general ability, manifest in language, is called **profiling** (Langacker 1987, among others; see also Talmy's (2000) related notion of attentional **windowing**).

One important way in which language exhibits profiling is in the range of grammatical constructions it has at its disposal, each of which serves to profile different aspects of a given scene. For instance, given a scene in which a boy kicks over a vase causing it to smash, different aspects of the scene can be linguistically profiled:

(27) a. The boy kicks over the vase.
 b. The vase is kicked over.
 c. The vase smashes into bits.
 d. The vase is in bits.

In order to discuss the differences between the examples in (27), we'll be relying on some grammatical terminology that may be new to the reader. We will explain these terms briefly as we go along, but grammatical terms are explained in more detail in the grammar tutorial in Chapter 14.

The aspects of the scene profiled by each of these sentences are represented in Figure 2.2. Figure 2.2(a) corresponds to sentence (27a). This is an active sentence in which a relationship holds between the initiator of the action (the boy) and the object that undergoes the action (the vase). In other words, the boy is the AGENT and the vase is the PATIENT. In Figure 2.2(a) both AGENT and PATIENT are represented by circles. The arrow from the AGENT to the PATIENT represents the transfer of energy, reflecting the fact that the AGENT is acting upon the PATIENT. Moreover, both AGENT and PATIENT, as well as the energy transfer, are represented in bold. This captures the fact that the entire **action chain** is being profiled, which is the purpose of the active construction.

Now let's compare sentence (27b). This is a passive sentence, and is represented by Figure 2.2(b). Here, the energy transfer and the PATIENT are being profiled. However, while the AGENT is not mentioned in the sentence, and hence is not in profile, it must be understood as part of the background. After all, an action chain requires an AGENT to instigate the transfer of energy. To represent this fact, the AGENT is included in Figure 2.2(a), but is not featured in bold, reflecting the position that the AGENT is contextually understood but not in profile.

The third sentence, example (27c), profiles the change in the state of the vase: the fact that it smashes into bits. This is achieved via a subject-verb-complement construction. A complement is an obligatory element that is required by another element in a sentence to complete its meaning. In (27c), the complement is the expression *into bits*, which completes the meaning of the expression *smashes*. This is captured by Figure 2.2(c). In figure 2.2(c) it is the internal change of state of the vase that is profiled. The arrow within the circle (the circle depicts the vase) shows that the vase is undergoing an internal change of state. The state the vase is 'moving to' is represented by the box with the letter 'b' inside it. This stands for the state IN BITS. In this diagram the entity, the change of state and the resulting state are all in bold, reflecting the fact that all these aspects of the action chain are being profiled by the corresponding sentence.

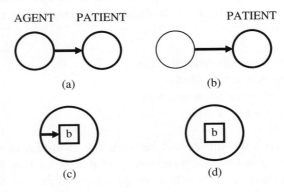

Figure 2.2 Profiling

Finally, consider sentence (27d). The grammatical form of this sentence is the subject–copula–complement construction. The copula is the verb *be*, which is specialised for encoding a particular state. In this case the state is IN BITS, which is captured in Figure 2.2(d).

In sum, each of the constructions ACTIVE, PASSIVE, SUBJECT–VERB–COMPLEMENT and SUBJECT–COPULA–COMPLEMENT is specialised for profiling a particular aspect of an action chain. In this way, linguistic structure reflects our ability to attend to distinct aspects of a scene. These examples demonstrate how linguistic organisation reflects a more general cognitive ability: attention.

It is worth observing at this point that constructions of the kind we have just discussed are not restricted to encoding a canonical action chain (one involving the transfer of energy). For example, the active construction can often be applied in cases where an action is not involved. Consider **stative verbs**, like *own*. A stative verb encodes a relatively stable state that persists over time. This verb can appear in active or passive constructions, even though it describes a state rather than an action:

(28) a. John not Steve owns the shop on Trafalgar Street.   [active]
     b. The shop on Trafalgar Street is owned   [passive]
        by John not Steve.

In Part III of the book, we will return in more detail to the issue of grammatical constructions and the range of meanings associated with them.

Categorisation: fuzzy categories

We saw above that enitites like cups constitute fuzzy categories, which are characterised by the fact that they contain members that are more or less representative of the category. This results in a set of members related by family resemblance rather than a single criterial feature, or a limited set of criterial features possessed by every member of the category. In other words, categories formed by the human mind are rarely 'neat and tidy'. We also saw that fuzzy categories are a feature of language in that members of linguistic categories, despite important similarities, often show quite distinct behaviour. In other words, according to the cognitive framework, the same principles that hold for categorisation in general also hold for linguistic categorisation.

Metaphor

As we began to see in the previous chapter, and as we will see in further detail in Chapter 9, the view adopted in cognitive linguistics is that metaphor is a conceptual rather than a purely linguistic phenomenon. Moreover, the key proponents of the conceptual metaphor approach, George Lakoff and Mark Johnson (1980, 1999), argue that many of the ways in which we think and act are fundamentally metaphorical in nature.

For instance, we conceptualise institutions like governments, universities, and businesses in terms of a hierarchy. Diagrams of such institutions place the person with the highest rank at the top or 'head', while the person with the lowest rank is placed at the lowest point or 'bottom'. In other words, hierarchies are conceptualised and represented non-linguistically in terms of the conceptual metaphor CONTROL/POWER IS UP.

Just as metaphors like CONTROL IS UP show up in a range of modalities, that is different 'dimensions' of expression such as social organisation, pictorial representation or gesture, among others, we have begun to see that they are also manifest in language. The English preposition *over* has a conventional CONTROL meaning associated with it, precisely because of meaning extension due to the conceptual metaphor CONTROL IS UP.

In the foregoing discussion, we have explored three ways in which aspects of general cognition show up in language. Evidence of this kind forms the basis of the cognitive argument that language reflects general cognition.

2.2 The embodied mind

In this section, we turn to **embodiment**, a central idea in cognitive linguistics. Since the seventeenth-century French philosopher René Descartes developed the view that mind and body are distinct entities – the principle of **mind/body dualism** – there has been a common assumption within philosophy and the other more recent cognitive sciences that the mind can be studied without recourse to the body, and hence without recourse to embodiment. In modern linguistics this **rationalist approach** has been most evident in formal approaches such as the **Generative Grammar** approach developed by Noam Chomsky (see Chapter 22) and formal approaches to semantics, such as the framework developed by Richard Montague (see Chapter 13). Proponents of these approaches argue that it is possible to study language as a formal or computational system, without taking into account the nature of human bodies or human experience.

In contrast, cognitive linguistics is not rationalist in this sense, but instead takes its inspiration from traditions in psychology and philosophy that emphasise the importance of human experience, the centrality of the human body, and human-specific cognitive structure and organisation, all of which affect the nature of our experience. According to this **empiricist view**, the human mind – and therefore language – cannot be investigated in isolation from human embodiment.

2.2.1 Embodied experience

The idea that experience is embodied entails that we have a species-specific view of the world due to the unique nature of our physical bodies. In other words, our construal of reality is likely to be mediated in large measure by the nature of our bodies.

One obvious way in which our embodiment affects the nature of experience is in the realm of colour. While the human visual system has three kinds of photoreceptors or colour channels, other organisms often have a different number. For instance, the visual system of squirrels, rabbits and possibly cats, makes use of two colour channels, while other organisms, like goldfish and pigeons, have four colour channels. Having a different range of colour channels affects our experience of colour in terms of the range of colours accessible to us along the colour spectrum. Some organisms can see in the infrared range, like rattlesnakes, which hunt prey at night and can visually detect the heat given off by other organisms. Humans are unable to see in this range. As this simple example demonstrates, the nature of our visual apparatus – one aspect of our physical embodiment – determines the nature and range of our visual experience.

Similarly, the nature of our biological morphology (the kinds of body parts we have), together with the nature of the physical environment with which we interact, determines other aspects of our experience. For instance, while gravity is an objective feature of the world, our experience of gravity is determined by our bodies and by the ecological niche we inhabit. For instance, hummingbirds – which can flap their wings up to a remarkable fifty times per second – respond to gravity in a very different way from humans. In order to overcome gravity, hummingbirds are able to rise directly into the air without pushing off from the ground, due to the rapid movement of their wings. Moreover, due to their small size, their experience of motion is rather different from ours: hummingbirds can stop almost instantaneously, experiencing little momentum. Compare this with the experience of a sprinter at the end of a 100m race: a human cannot stop instantaneously but must take a few paces to come to a standstill.

Now consider organisms that experience gravity in an even more different way. Fish, for example, experience very little gravity, because water reduces its effect. This explains their morphology, which is adapted to the ecological niche they inhabit and enables motion through a reduced-gravity environment. The neuroscientist Ernst Pöppel (1994) has even suggested that different organisms might have different kinds of neural 'timing mechanisms' which underpin abilities such as event perception (see Chapter 3). This is likely to affect their experience of time. The idea that different organisms have different kinds of experiences due to the nature of their embodiment is known as **variable embodiment**.

2.2.2 Embodied cognition

The fact that our experience is embodied – that is, structured in part by the nature of the bodies we have and by our neurological organisation – has consequences for cognition. In other words, the concepts we have access to and the nature of the 'reality' we think and talk about are a function of our embodiment: we can only talk about what we can perceive and conceive, and the things that we can perceive and conceive derive from embodied experience. From this point of view, the human mind must bear the imprint of embodied experience.

In his now classic 1987 book, *The Body in the Mind*, Mark Johnson proposes that one way in which embodied experience manifests itself at the cognitive level is in terms of **image schemas** (see Chapter 6). These are rudimentary concepts like CONTACT, CONTAINER and BALANCE, which are meaningful because they derive from and are linked to human **pre-conceptual experience**: experience of the world directly mediated and structured by the human body. These image-schematic concepts are not disembodied abstractions, but derive their substance, in large measure, from the sensory-perceptual experiences that give rise to them in the first place. Lakoff (1987, 1990, 1993) and Johnson (1987) have argued that embodied concepts of this kind can be systematically extended to provide more abstract concepts and conceptual domains with structure. This process is called **conceptual projection**. For example, they argue that conceptual metaphor (which we discussed briefly above and to which we return in detail in Chapter 9) is a form of conceptual projection. According to this view, the reason we can talk about being *in* states like love or trouble (29) is because abstract concepts like LOVE are structured and therefore understood by virtue of the fundamental concept CONTAINER. In this way, embodied experience serves to structure more complex concepts and ideas.

(29) a. George is in love.
b. Lily is in trouble.
c. The government is in a deep crisis.

The developmental psychologist Jean Mandler (e.g. 1992, 1996, 2004) has made a number of proposals concerning how image schemas might arise from embodied experience. Starting at an early age, and certainly by two months, infants attend to objects and spatial displays in their environment. Mandler suggests that by attending closely to such spatial experiences, children are able to abstract across similar kinds of experiences, finding meaningful patterns in the process. For instance, the CONTAINER image schema is more than simply a spatio-geometric representation. It is a 'theory' about a particular kind of configuration in which one entity is supported by another entity that contains it. In other words, the CONTAINER schema is meaningful because containers are meaningful in our everyday experience. Consider the spatial scene described in (30).

(30) The coffee is in the cup.

Tyler and Evans make the following observations about this spatial scene:

> ... the spatial scene relating to *in* involves a containment function, which encompasses several consequences such as locating and limiting the activities of the contained entity. Being contained in the cup prevents the coffee from spreading out over the table; if we move the cup, the coffee moves with it. (Tyler and Evans 2003: ix)

It is for this reason that the English preposition *in* can be used in scenes that are non-spatial in nature, like the examples in (29). It is precisely because containers constrain activity that it makes sense to conceptualise POWER and all-encompassing states like LOVE or CRISIS in terms of CONTAINMENT. Mandler (2004) describes this process of forming image schemas in terms of a redescription of spatial experience via a process she labels **perceptual meaning analysis**. As she puts it, '[O]ne of the foundations of the conceptualizing capacity is the image schema, in which spatial structure is mapped into conceptual structure' (Mandler 1992: 591). She further suggests that 'Basic, recurrent experiences with the world form the bedrock of the child's semantic architecture, which is already established well before the child begins producing language' (Mandler 1992: 597). In other words, it is experience, meaningful to us by virtue of our embodiment, that forms the basis of many of our most fundamental concepts.

2.2.3 Experiential realism

An important consequence of viewing experience and conceptualisation as embodied is that this affects our view of what reality is. A widely held view in formal semantics is that the role of language is to describe states of affairs in the world. This rests on the assumption that there is an objective world 'out there', which language simply reflects. However, cognitive linguists argue that this **objectivist approach** misses the point that there cannot be an objective reality that language reflects directly, because reality is not objectively given. Instead, reality is in large part constructed by the nature of our unique human embodiment. This is not to say that cognitive linguists deny the existence of an objective physical world independent of human beings. After all, gravity exists, and there is a colour spectrum (resulting from light striking surfaces of different kinds and densities), and some entities give off heat, including body heat, which can only be visually detected in the infrared range. However, the parts of this external reality to which we have access are largely constrained by the ecological niche we have adapted to and the nature of our embodiment. In other words, language does not directly reflect the world. Rather, it reflects our unique human construal of the world: our 'world view' as it appears to us through the lens of our embodiment. In Chapter 1 we referred to human reality as 'projected reality', a term coined by the linguist Ray Jackendoff (1983).

This view of reality has been termed **experientialism** or **experiential realism** by cognitive linguists George Lakoff and Mark Johnson. Experiential realism assumes that there is a reality 'out there'. Indeed, the very purpose of our perceptual and cognitive mechanisms is to provide a representation of this reality, and thus to facilitate our survival as a species. After all, if we were unable to navigate our way around the environment we inhabit and avoid dangerous locations like clifftops and dangerous animals like wild tigers, our cognitive mechanisms would be of little use to us. However, by virtue of being adapted to a particular ecological niche and having a particular form and configuration, our bodies and brains necessarily provide one particular perspective among many possible and equally viable perspectives. Hence, experiential realism acknowledges that there is an external reality that is reflected by concepts and by language. However, this reality is mediated by our uniquely human experience which constrains the nature of this reality 'for us'.

2.3 Cognitive semantics and cognitive approaches to grammar

Having set out some of the fundamental assumptions behind the cognitive approach to language, in this section we briefly map out the field of cognitive linguistics. Cognitive linguistics can be broadly divided into two main areas: **cognitive semantics** and **cognitive (approaches to) grammar**. However, unlike formal approaches to linguistics, which often emphasise the role of grammar, cognitive linguistics emphasises the role of meaning. According to the cognitive view, a model of meaning (a cognitive semantics) has to be delineated before an adequate cognitive model of grammar can be developed. Hence a cognitive grammar assumes a cognitive semantics and is dependent upon it. This is because grammar is viewed within the cognitive framework as a meaningful system in and of itself, which therefore shares important properties with the system of linguistic meaning and cannot be meaningfully separated from it.

The area of study known as cognitive semantics, which is explored in detail in Part II of the book, is concerned with investigating the relationship between experience, the conceptual system and the semantic structure encoded by language. In specific terms, scholars working in cognitive semantics investigate knowledge representation (**conceptual structure**) and meaning construction (**conceptualisation**). Cognitive semanticists have employed language as the lens through which these cognitive phenomena can be investigated. It follows that cognitive semantics is as much a model of mind as it is a model of linguistic meaning.

Cognitive grammarians have also typically adopted one of two foci. Scholars like Ronald Langacker have emphasised the study of the cognitive principles that give rise to linguistic organisation. In his theoretical framework **Cognitive Grammar**, Langacker has attempted to delineate the principles that serve to structure a grammar, and to relate these to aspects of general cognition. Because the term 'Cognitive Grammar' is the name of a specific theory, we use the (rather cumbersome) expression 'cognitive (approaches to) grammar' as the general term for cognitively oriented models of the language system.

The second avenue of investigation, pursued by researchers including Fillmore and Kay (Fillmore *et al.* 1988; Kay and Fillmore 1999), Lakoff (1987), Goldberg (1995) and more recently Bergen and Chang (2005) and Croft (2002), aims to provide a more descriptively detailed account of the units that comprise a particular language. These researchers have attempted to provide an inventory of the units of language. Cognitive grammarians who have pursued this line of investigation are developing a collection of theories that can collectively be called **construction grammars**. This approach takes its name from the view in cognitive linguistics that the basic unit of language is a form-meaning symbolic assembly which, as we saw in Chapter 1, is called a **construction**.

It follows that cognitive approaches to grammar are not restricted to investigating aspects of grammatical structure largely independently of meaning, as is often the case in formal traditions. Instead, cognitive approaches to grammar encompass the entire inventory of linguistic units defined as form-meaning pairings. These run the gamut from skeletal syntactic configurations like the ditransitive construction we considered earlier, to idioms, to bound morphemes like the *-er* suffix, to words. This entails that the received view of clearly distinct 'sub-modules' of language cannot be meaningfully upheld within cognitive linguistics, where the boundary between cognitive semantics and cognitive (approaches to) grammar is less clearly defined. Instead, meaning and grammar are seen as two sides of the same coin: to take a cognitive approach to grammar is to study the units of language and hence the language system itself. To take a cognitive approach to semantics is to attempt to understand how this linguistic system relates to the conceptual system, which in turn relates to embodied experience. The concerns of cognitive semantics and cognitive (approaches to) grammar are thus complementary. This idea is represented in Figure 2.3. The organisation of this book reflects the fact that it is practical to divide up the study of cognitive linguistics into these two areas for purposes of teaching and learning. However, this should not be taken as an indication that these two areas of cognitive linguistics are independent areas of study or research.

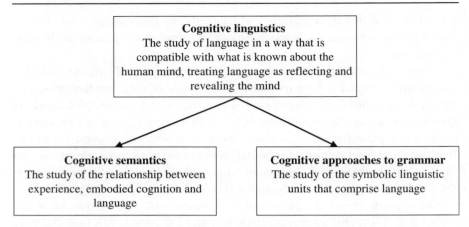

Figure 2.3 The study of meaning and grammar in cognitive linguistics

2.4 Summary

In this chapter, we have provided an overview of the assumptions and commitments that make cognitive linguistics a distinctive enterprise. We have outlined two key commitments widely shared by cognitive linguists. These are the **'Generalisation Commitment'** and the **'Cognitive Commitment'**. These two commitments underlie the orientation and approach adopted by cognitive linguists, and the assumptions and methodologies employed in the two main branches of the cognitive linguistics enterprise, **cognitive semantics** and **cognitive (approaches to) grammar**. We also introduced the **embodied cognition** thesis which is central to much research in cognitive linguistics and addresses the nature of the relationship between language, mind and experience. The view taken in cognitive linguistics is that conceptual organisation within the human mind is a function of the way our species-specific bodies interact with the environment we inhabit. Finally, we provided a brief overview of cognitive semantics and cognitive approaches to grammar which are addressed in detail in Part II and Part III of the book, respectively.

Further reading

Assumptions in cognitive linguistics

The following are all articles by leading cognitive linguists that set out the assumptions and the nature of the cognitive linguistics enterprise:

- **Fauconnier (1999).** A discussion of methodological issues and the nature of the approach adopted in cognitive linguistics, particularly with respect to meaning. Fauconnier, one of the early pioneers in

cognitive linguistics, illustrates with examples from the theory of conceptual blending, which he developed in joint work with Mark Turner.
- **Lakoff (1990).** In the first part of this important article, published in the very first volume of the journal *Cognitive Linguistics*, Lakoff discusses issues relating to the 'Generalisation Commitment' and the 'Cognitive Commitment'. He also explains how cognitive linguistics differs from Generative Grammar.
- **Langacker (1999a).** An important article by another pioneering figure in cognitive linguistics. In this article, Langacker evaluates the approach and methodologies employed in cognitive linguistics and relates this to the formalist and functionalist traditions in linguistics. He illustrates with a discussion from some of the key constructs in his Cognitive Grammar framework.
- **Talmy (2000: Vol. I, 1–18).** In the introduction to his two-volume edifice, *Toward a Cognitive Semantics*, Talmy outlines his view of the cognitive linguistics enterprise and describes how his own work fits in with and has contributed to this endeavour.

Embodied cognition

- **Clark (1997).** Drawing on recent work in robotics, neuroscience, psychology and artificial intelligence, Clark, a leading cognitive scientist, presents a compelling and highly accessible overview of the new science of the embodied mind.
- **Evans (2004a).** This book addresses how time, a fundamental aspect of human experience, is conceptualised. The discussion relates neurological, phenomenological and sensory-perceptual aspects of embodied experience to the experience of temporal cognition as revealed by language. Chapter 4 provides a presentation of some key arguments for the cognitive linguistics perspective on embodied cognition.
- **Lakoff (1987).** This is a classic work by one of the pioneers in cognitive linguistics. Part II of the book is particularly important for the development of experiential realism.
- **Lakoff and Johnson (1980).** This short volume laid the foundations for the approach to embodied cognition in cognitive linguistics.
- **Lakoff and Johnson (1999).** This represents an updated account of experiential realism as developed by Lakoff and Johnson (1980).
- **Mandler (2004).** Influential developmental psychologist Jean Mandler argues for the role of image schemas in the development of conceptual structure and organisation.

- **Varela, Thompson and Rosch (1991).** A highly influential book on embodiment, cognition and human experience by leading cognitive scientists.

Exercises

2.1 Categorisation and family resemblance

The philosopher Wittgenstein famously argued that the category GAME exhibits family resemblance. To test this, first make a list of as many different kinds of games as you can think of. Now see if there is a limited set of conditions that is common to this entire list ('necessary' conditions) and sufficient to distinguish this category from other related categories ('sufficient' conditions) like competitions, amusement activities and so on. Do your conclusions support or refute Wittgenstein's claim?

Now see if you can identify the ways in which the different games you list share family resemblance 'traits'. Try to construct a 'radial' network showing the degrees of family resemblance holding between games of different kinds. A radial network is a diagram in which the most/more prototypical game(s) is/are placed at the centre and less prototypical games are less central, radiating out from the centre.

2.2 Polysemy

Consider the word *head*. Try and come up with as many different meanings for this word as possible. You may find it helpful to collect or create sentences involving the word.

Now consider the closed-class word *you*. Cognitive linguists assume that even closed-class words exhibit polysemy. Collect as many sentences as you can involving *you* and try and identify differences in how this word is used. Do your findings support the view that this word exhibits polysemy?

2.3 Metaphor

Reconsider the different meanings for *head* that you uncovered in the previous exercise. Would you class any of these distinct meanings as metaphorical? Explain your reasoning. Now try and give an account of what motivated the extension from the 'core' meaning of *head* to the metaphoric usage(s).

2.4 Image schemas

The spatial meanings associated with prepositions present a clear case of the way in which image schemas underpin language. In view of this, what sets of image schemas might underpin the semantic distinction between the prepositions *up/down* and *above/under*?

Now consider the metaphoric use of the prepositions *on* and *in* in the following sentences:

(a) The guard is on duty.
(a´) The shoes are on sale.
(b) Munch's painting *The Scream* portrays a figure in despair.
(b´) Sven is in trouble with Nancy.

What might be the experiential basis for the fact that states like SALES and DUTY are described in terms of ON, while states like DESPAIR and TROUBLE are described in terms of IN? We saw in this chapter that the CONTAINER image schema plausibly underpins IN. What might be the image schema underpinning ON?

3

Universals and variation in language, thought and experience

As we saw in Chapter 2, the cognitive linguistics enterprise is characterised by two commitments: (1) the **'Generalisation Commitment'** – a commitment to the characterisation of general principles that are responsible for all aspects of human language; and (2) the **'Cognitive Commitment'** – a commitment to providing a characterisation of general principles for language that accords with what is known about the mind and brain from other disciplines (Lakoff 1990). An important consequence of this approach is the position that language does not result from an encapsulated 'module' of specialised knowledge, separable from general cognition (in contrast with the view developed in formal approaches to linguistics), but instead that language reflects and is informed by non-linguistic aspects of cognition. In particular, given the premise that the principles that inform language reflect general cognitive principles, the language system itself can be seen as a window that enables the direct investigation of **conceptual structure** (knowledge representation, including the structure and organisation of concepts) and **conceptualisation** (the process of meaning construction).

Although cognitive linguists have often been concerned with investigating the general cognitive principles (common to all humans) that govern language, it does not follow from this that all languages are the same, either in terms of grammatical structure or semantic structure. In this chapter, we review some influential cognitively oriented studies that demonstrate that languages can exhibit radically different conceptual organisation and structure. It seems that common cognitive principles do not give rise to uniform linguistic organisation and structure. On the contrary, cross-linguistic variation is widespread. At the same time, the existence of certain common patterns across languages is a matter of empirical fact. These common patterns are known as **linguistic**

universals. For cognitive linguists, these commonalities are explained by the existence of general cognitive principles shared by all humans, in addition to the fundamentally similar experiences of the world also shared by all humans due to embodiment. Nevertheless, given the premise that language reflects cognitive organisation, the existence of cross-linguistic variation entails that speakers of different languages have different underlying conceptual systems. This view has implications for the thesis of **linguistic relativity** or **linguistic determinism** – the view that the language you speak affects or determines how you see the world, most famously expounded in the writings of Benjamin Lee Whorf in the 1930s and 1940s. Hence, once we have developed the cognitive linguistics approach to linguistic universals and cross-linguistic variation as we see it, we will re-examine the Whorfian linguistic relativity principle.

3.1 Universals in thought and language

We begin by considering the issue of linguistic universals. It is important to observe here that the term 'linguistic universal' can be understood in two quite distinct ways. On the one hand, the term can refer to patterns of similarity that are attested in typological studies: these are usually large-scale comparative studies that set out to discover linguistic patterns in relation to a given phenomenon. The existence of the typological universals uncovered by these studies is a matter of empirical fact and is uncontroversial. On the other hand, the term 'universal' can also be used to refer to underlying principles of linguistic organisation and structure that are represented in the human mind. This view is most prominently associated with the generative grammar framework developed by Noam Chomsky, which assumes the existence of a **Universal Grammar**: a set of innate universal principles that equips all humans to acquire their native language and is also held to account for patterns of cross-linguistic similarity. This view is controversial for many linguists, including cognitive linguists. We will briefly set out the assumptions of the Generative Grammar model below (section 3.1.2), and return to these issues in more detail towards the end of the book (Chapter 22), but consider for the time being the following extract from Levinson (1996):

> It may be claimed, the Kantian categories of space, time, cause and so on, form the fundamental ground of our reasoning; they cannot be inferred from experience, but are what we bring to the interpretation of experience from our biological endowment. Thus the conceptual architecture, the essential conceptual parameters, are, as Leibniz would have it, 'innate ideas'. This line of thought dominates current speculations in the cognitive sciences. It is a view reinforced from many quarters: evolutionary biology and neurophysiology stress the

closeness of our neurological equipment to that of our mammalian cousins, studies of human development (following Piaget) assume an unfolding of inborn potential, psychological models of processing are often presumed to be models of 'hardware' properties rather than models of learned or acquired tendencies or 'software', and so on. In linguistics, the adoption of natural science ideals has led to the search for universals without parallel concern for language differences. (Levinson 1996: 133)

As Levinson's comment suggests, the search for linguistic universals (in the sense of universal cognitive principles of language) has preoccupied much of modern linguistics, particularly since the advent of Chomsky's work on generative grammar in the 1950s. However, as Levinson observes, the search for Universal Grammar has prompted some linguists to argue that quite radical cross-linguistic variation has been ignored by formal linguists. To provide just a few examples, languages can range from having between eleven and 141 distinctive speech sounds; some languages lack morphological marking for properties like number (singular or plural) or tense; and some languages appear to lack syntactic constraints on word order, or fail to exhibit familiar word classes such as adjective.

Despite the widespread view within formal linguistics that linguistic structure across languages is broadly similar (and can eventually be stated in terms of a small set of universal principles known as Universal Grammar), studies set within this tradition tend not to be concerned with large-scale cross-linguistic comparison. The branch of linguistics that is concerned with large-scale cross-linguistic comparison, **linguistic typology**, reveals the relative rarity of absolute universals in the sense of patterns of similarity that hold across all languages. Instead, the universals that do emerge are conditional generalisations that can be established to have some statistical validity, as we will see below (section 3.1.1).

As we have already noted, cognitive linguists assume that language reflects conceptual structure and organisation. It follows from this assumption that cross-linguistic differences should point to underlying conceptual differences. Cognitive linguists therefore argue that evidence of variation across languages suggests that languages encode very different kinds of conceptual systems. However, these distinct conceptual systems are thought to emerge from a common **conceptualising capacity**, which derives from fundamental shared aspects of human cognition. Rather than positing universal linguistic principles, then, cognitive linguists posit a common set of cognitive abilities, which serve to both facilitate and constrain the development of our **conceptual systems** (our repository of concepts). Although cross-linguistic analysis reveals that the range of possible conceptual systems found in language is delimited in certain fundamental ways, the languages of the world can and do exhibit a wide range of variation. Cognitive linguists argue that this fact, revealed by typologists, seriously undermines the position that there can be universal principles of language of the kind posited by formal linguists.

3.1.1 Typological universals

According to Croft (2003: 1–2), the term 'linguistic typology' is used in three distinct ways to refer to three different types of approach that fall within the broader discipline of linguistic typology. The first approach, which he calls **typological classification**, involves the assignment of a given language to a single type, based on its properties in a certain area (morphology, word order and so on). The nineteenth- and early twentieth-century typological approach is a representative example, where the emphasis was on developing descriptive taxonomies. For example, in traditional morphological classification, a language is classified as belonging to the 'isolating' type if it lacks grammatical affixes, while a language is classified as belonging to the 'agglutinating' type if it has grammatical affixes that each encode a single grammatical feature.

The second approach within linguistic typology is what Croft calls **typological generalisation**. This involves the search for systematic patterns across languages (linguistic universals), and identifies what patterns of variation can be predicted to exist on the basis of those observed patterns. This approach has its roots in the work begun by Joseph Greenberg in the 1960s, and in emphasising the predictions that emerge from attested patterns about what is a possible human language goes a step further than the essentially taxonomic approach of typological classification.

The third approach within linguistic typology is what Croft calls **functional typology**. This modern approach rests upon typological generalisation, but goes a step further in developing a theoretical framework that seeks to set out explanations for the observed patterns. This approach is called 'functional' typology because it explains these patterns in terms of how language is used for purposes of communication. Functional typology has been developed by typologists such as Bernard Comrie, Talmy Givón, John Haiman, Paul Hopper and William Croft, among others.

Modern linguistic typology adopts large-scale cross-linguistic sampling as its methodology. The size of the sample varies according to the extent to which the phenomenon under investigation is widespread, as well as being constrained by practical considerations; the typical sample size is in the region of 100–200 languages (out of the estimated six thousand living languages in the world). It is important that the generalisations stated by typologists have statistical validity, otherwise they cannot be upheld. The languages that make up these samples are carefully selected, taking into consideration factors that might affect the reliability of the resulting generalisations, such as genetic relationships between languages and contact between neighbouring but genetically unrelated languages.

Linguistic typologists have discovered that, although it is possible to state certain properties that hold for all languages (unrestricted universals), cross-linguistic variation is ubiquitous. However, typologists have also discovered that, while languages can and do vary, cross-linguistic variation is **constrained**, and these constraints can be stated in terms of implicational universals. Indeed, from the perspective of linguistic typology, it is the constraints on variation that make up the universals of language, rather than a set of universal principles that capture the properties that languages have in common (Universal Grammar). Let's look more closely at the distinction between unrestricted universals and implicational universals, which makes this point clearer.

An **unrestricted universal** states that all languages show a particular pattern with respect to some structural feature, while the other logically possible pattern(s) are unattested. Croft (2003: 52) provides the example in (1).

(1) All languages have oral vowels.

This means that the other logical possibility, that there are languages without oral vowels, is not attested. This type of unrestricted universal pinpoints cross-linguistic similarity and is relatively uninteresting to typologists because it does not reveal a pattern in the same way that cross-linguistic differences do.

It is much more common for typologists to state **implicational universals**, which do not state that all languages show the same pattern with respect to a given phenomenon, but instead state the restrictions on the logically possible patterns, usually in the following format: 'If language X has property Y, then it will also have property Z'. As Croft (2003: 54) points out, this type of universal pinpoints patterns in variation rather than similarity, since each implicational universal sets out a set of distinct attested possibilities. Croft provides the example in (2), which was proposed by Hawkins (1983: 84, cited in Croft 2003: 53). This implicational universal rests upon the four logically possible patterns listed in (3).

(2) If a language has noun before demonstrative, then it has noun before relative clause.

(3) a. languages where both demonstratives and relative clauses follow the noun
    b. languages where both demonstratives and relative clauses precede the noun

c. languages where demonstrative precedes and relative clause follows the noun
  d. languages where demonstrative follows and relative clause precedes the noun

Observe that the implicational universal in (2) excludes the possibility described in (3d). In this way, the implicational universal states the limits on cross-linguistic variation by restricting the possibilities to those described in (3a)–(3c), and entails an absolute universal by stating that the pattern in (3d) is unattested. In reality, most of the universals posited by typologists are of this kind, or indeed of a more complex kind. Croft describes the differences between typological and generative approaches as follows:

> One of the major differences between the generative and typological approaches is what direction to generalize first. Given a grammatical phenomenon such as a relative clause structure in English, one could generalize in several directions. One could compare the relative clause structure with other complex sentence structures in English . . . and then generalize over these different structures in English. This is the classic structuralist-generative approach. Alternatively, one could compare relative clause structure in English with relative clause structure in other languages, and then generalize over relative clauses in human languages. This is the classic typological approach . . . the typologist begins with cross-linguistic comparisons, and then compares typological classifications of different structural phenomena, searching for relationships. In contrast, the generative linguist begins with language-internal structural generalizations and searches for correlations of internal structural facts, and only then proceeds to cross-linguistic comparison. (Croft 2003: 285)

A further important difference between functional typology and the generative approach is that functional typologists reject the idea of specialised innate linguistic knowledge (Universal Grammar). Instead, functional typology comes much closer to cognitive linguistics in orientation, in two important ways. Firstly, functional typology emphasises language function and use in developing explanations for linguistic phenomena. Secondly, functional typology appeals to non-linguistic aspects of cognition to explain the properties of language. For example, many typologists adopt some version of a **semantic map model** in accounting for typological patterns (Croft 2003: 133). A semantic map is a language-specific typological pattern, which rests upon a universal **conceptual space** or system of knowledge. We return to look at this idea in more detail at the end of Part III (Chapter 20).

3.1.2 Universals in formal linguistics

We can now look in more detail at the issue of universals from a formal perspective. There are two prominent formal approaches that address this issue: (1) the **Universal Grammar** hypothesis, which relates to grammatical structure; and (2) the **semantic decomposition** approach(es), which relates to semantic structure. What is common to both approaches is the hypothesis that linguistic knowledge has innate pre-specification. From this perspective, while languages may differ 'on the surface' (for example, in terms of the speech sounds they use or in terms of word order), beneath the surface they are broadly similar, and this similarity is explained by the existence of a universal set of primitives together with a universal set of principles that operate over these primitives.

Universal Grammar

The Universal Grammar hypothesis was proposed by Chomsky, and represents an attempt to explain not only why linguistic universals exist, but also how children come to acquire the language(s) they are exposed to so rapidly. The Universal Grammar hypothesis goes hand in hand with the **nativist hypothesis**, which holds that the principles of Universal Grammar are innate rather than learned (see Chapter 4). However, Chomsky does not claim that children are born with a fully specified grammar. Children still have to go through the process of acquiring the grammar of the language(s) they are exposed to. Instead, what is claimed to be universal and innate is the pre-specification, which we can think of as a kind of 'blueprint' that guides what is possible. Chomsky (1965) presented this pre-specification in terms of what he called **formal** and **substantive universals**. Substantive universals are grammatical categories like noun and verb, and grammatical functions like subject and object: what we might think of as the basic 'building blocks' of grammar. Chomsky (1965: 66) suggests that languages select from a universal set of these substantive categories. Formal universals are rules like **phrase structure rules**, which determine how phrases and sentences can be built up from words, and **derivational rules**, which guide the reorganisation of syntactic structures, allowing certain kinds of sentences to be **transformed** into or derived from other kinds of sentences (for example, the transformation of a declarative sentence into an interrogative sentence). In the 1980s, Chomsky developed a more flexible approach to Universal Grammar, called the **Principles and Parameters** approach. According to this model, the innate pre-specification for language is captured in terms of a limited set of principles that can vary according to a small set of parameters of variation. These parameters are 'set' on the basis of the properties of language an individual is exposed to during childhood. For example, given sufficient exposure to spoken language, a child's grammatical system will set the 'head initial/final parameter' at 'initial' for languages like English where verbs precede their objects, but will set this parameter at 'final' for languages like Korean, where verbs follow their objects. The most recent version of Chomsky's theory, the Minimalist Program, also adopts a version of this approach.

Cognitive linguists (and typologists) argue that the fundamental problem with Chomsky's hypothesis is that cross-linguistic comparison reveals there to be little evidence for substantive universals of the kind he assumes. In other words, some typologists argue that categories like adjective or grammatical functions like subject and object are not found in all languages (see Croft 2003: 183–8, for example). Cognitive linguists, among linguists of other theoretical persuasions, also argue that the formal theories of phrase structure proposed by Chomsky in order to account for formal universals are unnecessarily abstract, to the extent that parallels across languages are difficult to ascertain. According to Levinson (1996a: 134) 'it is probably fair to say that the proposals [of Chomsky] need to be taken with a pinch of salt – they are working hypotheses under constant, often drastic, revision.' Indeed, Chomsky himself defines the Minimalist Program as a research programme rather than a fully developed theory, and acknowledges that generative grammar is undergoing constant change

It is important to point out at this point that Universal Grammar is adopted as a working hypothesis by a number of generatively oriented theories of language that depart from Chomsky's transformational approach and adopt a strictly 'monostratal' or **non-derivational approach**. These theories include Head-driven Phrase Structure Grammar (see Borsley 1996, 1999) and Lexical Functional Grammar (see Bresnan 2001). Formal syntacticians view the quest for Universal Grammar as a worthwhile pursuit, not only because it is a hypothesis worth exploring in its own right, whether it turns out to be correct or not, but also because it provides tools that enable precise and careful descriptions of the world's languages as well as close comparisons of languages, both related and unrelated.

For cognitive linguists, the picture of language that emerges from such an approach is artificially narrow, focusing as it does upon **morphosyntax** (word and sentence structure) and having relatively little to say about linguistic meaning or the communicative functions of language.

Semantic universals

The predominant formal approach to semantic universals assumes **semantic primes** or **primitives** and is known as the **semantic decomposition** or **componential analysis** approach. Unlike the Universal Grammar hypothesis, which is associated with generative theories, this approach, or collection of approaches, is not associated with a particular type of theoretical framework. Indeed, semantic decomposition has been advocated, in various guises, by both formal and non-formal theorists, including Jackendoff (1983), Pinker (1994), Li and Gleitman (2002) and Wierzbicka (1996). The intuition behind the semantic decomposition approach is that there is a universal set of primitive semantic concepts, innately given, for which any particular language provides a language-specific label. This idea is expressed by Li and Gleitman in the following way:

> Language has means for making reference to the objects, relations, properties and events that populate our everyday world. It is possible to suppose that these linguistic categories and structures are more or less straightforward mappings from a preexisting conceptual space, programmed into our biological nature. Humans invent words that label their concepts. (Li and Gleitman 2002: 266)

Some linguists who adopt this type of approach argue that words rarely label individual semantic primitives, but combinations or 'bundles' of primitives that combine to create the rather complex concepts that words denote. For instance, Ray Jackendoff, in his pioneering 1983 book *Semantics and Cognition*, argues that conceptual structure consists of a range of **ontological categories**, some of which are primitives. A primitive, in this sense, is an entity that cannot be reduced further, and can be combined with other primitives in order to produce more complex categories. Some of the primitives Jackendoff proposes are [THING], [PLACE], [DIRECTION], [ACTION], [EVENT], [MANNER] and [AMOUNT]. Indeed, these ontological categories can be encoded in language. For instance, each of these corresponds to a *wh*-question word, such as *what*, *who*, *when* and so on. This is illustrated by the question and answer sequences below (drawn or adapted from Jackendoff 1983: 53):

(4) What did you buy?
    A fish   [THING]

(5) Where is my coat?
    On the coat hook   [PLACE]

(6) Where did they go?
    Into the garden   [DIRECTION]

(7) What did you do?
    Went to the cinema   [ACTION]

(8) What happened next?
    The toy fell out of the window   [EVENT]

(9) How did you cook the eggs?
    Slowly   [MANNER]

(10) How long was the fish?
     Over a metre (long)   [AMOUNT]

In addition to primitive ontological categories, the relations that hold between them are also primitives. Consider example (11).

(11) The statue is in the park.

The THEME of the sentence (what the sentence is about) is a particular [THING], lexicalised by the expression *the statue*. Moreover, *the statue* is located with respect to a particular [LOCATION], lexicalised by the expression *in the park*, which consists of the preposition, *in*, and a reference object, *the park*. Given that a [LOCATION] is typically occupied by a [THING], there is a relationship holding between [PLACE] and [THING] in which [THING] is a function of [PLACE]. Jackendoff calls this **thematic relation** [PLACE-FUNCTION].

Jackendoff argues that semantic primitives of this kind derive from the domain of spatial experience and are 'hard wired' or innate. In addition, he posits rules that enable the creation of new combinations as new concepts are acquired. The ontological categories and relations can also be deployed by more abstract concepts. For instance, abstract states can also be structured in terms of the [PLACE-FUNCTION] relation, even though abstract states such as TROUBLE or LOVE cannot be construed as locations:

(12) a. John is in trouble.
     b. John is in love.

According to Jackendoff's theory, the reason that the [PLACE-FUNCTION] relation can be applied to abstract states such as TROUBLE and LOVE is because these more abstract concepts are being structured in terms of more primitive ontological categories.

The semantic decomposition approach faces a number of challenges, as has often been observed by linguists of various theoretical persuasions. In particular, it is difficult to establish empirically what the 'right' semantic primitives might be, or how many there are. Furthermore, 'classical' componential theories, which assume a set of necessary and sufficient conditions, face the problem of accounting for how an entity can still count as an instance of a category in the absence of one or more of these components (for example, a three-legged cat is still described as *cat*). We return to this point in some detail in Chapter 8.

3.1.3 Universals in cognitive linguistics

Cognitive linguists argue against the view that language is pre-specified in the sense that grammatical organisation is mapped out by an innate 'blueprint' for grammar, and semantic organisation by a set of semantic primitives. Instead linguistic organisation is held to reflect embodied cognition, as we discussed in the previous chapter, which is common to all human beings. Instead of seeing language as the output of a set of innate cognitive universals that are specialised for language, cognitive linguists see language as a reflection of embodied cognition, which serves to constrain what it is possible to experience, and thus what it is possible to express in language.

In this section, we discuss some of the ways in which embodied cognition constrains what is possible in language. In subsequent sections we examine how these aspects of human cognition are linguistically manifest in two conceptual domains: SPACE and TIME. The 'Cognitive Commitment' and the 'Generalisation Commitment', together with the embodied cognition thesis, imply a set of constraints that guide the conceptualising capacity as reflected in language. These constraints nevertheless permit a wide range of cross-linguistic variation, as we will see.

Embodiment

Given the fact of human embodiment discussed in Chapter 2, namely that we share similar cognitive and neuro-anatomical architectures (minds, brains and bodies), it follows that the nature of human experience, and the nature of possible conceptual systems that relate to this experience, will be constrained. For instance, as we saw in Chapter 2, the fact that the human visual system lacks access to colour in the infrared range means that humans cannot experience this part of the colour spectrum. This constrains the nature of experience available to us, and the range of concepts we can form based on that experience.

Environment

The nature of the environment humans inhabit has a number of basic commonalities, irrespective of whether one lives in the Arctic or the Kalahari Desert or on a tropical island. Gravity and the other 'physical laws' are experienced by humans in essentially the same way the world over. These 'invariant' features of the environment place important constraints upon what it is possible to experience at the cognitive level.

Experience

There appear to be two broad categories of human experience. The first relates to **sensory experience**. This is experience derived from sensory perception (the 'senses') and concerns perceptual data derived from the external world. Concepts that derive from sensory experience include, among others, those relating to the domains of SPACE, MOTION, TEMPERATURE and so on. The other category of experience is **introspective** or **subjective experience**. Experience of this kind is subjective or internal in nature, and includes emotion, consciousness and experiences of time such as awareness of duration, simultaneity and so on. One of the most fundamental properties of the human conceptualising capacity is its tendency to structure concepts or domains relating to introspective experience in terms of concepts that derive from sensory experience. This is evident in the phenomenon of conceptual metaphor first introduced in Chapter 1 and to which we return in more detail in Chapter 9.

Perception

Sensory experience, discussed above, is received via perceptual mechanisms. These mechanisms are rather sophisticated, however, and provide structure that is not necessarily apparent in the raw perceptual input. In other words, what we perceive is not necessarily the same as what we experience directly. The perceptual mechanisms that facilitate our experience were formalised by the movement known as **Gestalt psychology**, which first emerged at the end of the nineteenth century. Gestalt psychologists such as Max Wertheimer (1880–1943), Wolfgang Köhler (1887–1967) and Kurt Koffka (1886–1941) were interested in the principles that allow unconscious perceptual mechanisms to construct wholes or 'gestalts' out of incomplete perceptual input. For instance, when a smaller object is located in front of a larger one, we perceive the protruding parts of the larger object as part of a larger whole, even though we cannot see the whole because the parts are discontinuous. The **Gestalt principles** therefore provide structure to, and indeed constrain, experience. We briefly survey some of the most important Gestalt principles below, focusing on examples from the domain of visual perception.

Perception: figure-ground segregation
Human perception appears to automatically segregate any given scene into **figure-ground organisation**. A **figure** is an entity that, among other things, possesses a dominant shape, perhaps due to a definite contour or prominent colouring. The figure stands out against the **ground**, the part of a scene that is relegated to 'background'. In the scene depicted in Figure 3.1, the figure is the lighthouse and the ground is made up of the grey horizontal lines against which the figure stands out.

Perception: principle of proximity
This principle holds that elements in a scene that are closer together will be seen as belonging together in a group. This is illustrated in Figure 3.2. The consequence of the greater proximity of the dots on the vertical axis than on

COGNITIVE LINGUISTICS: AN INTRODUCTION

Figure 3.1 Figure–ground segregation

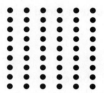

Figure 3.2 Columns of dots

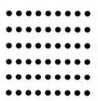

Figure 3.3 Rows of dots

the horizontal axis means that we perceive the dots in this image as being organised into columns rather than rows.

If the scene is altered so that the dots are closer together on the horizontal axis, then we perceive a series of rows, as illustrated in Figure 3.3.

Perception: principle of similarity
This principle holds that entities in a scene that share visual characteristics such as size, shape or colour will be perceived as belonging together in a group. For instance, in Figure 3.4, we perceive columns of shapes (rather than rows). In fact, the shapes are equidistant on both the horizontal and vertical axes. It is due to the principle of similarity that similar shapes (squares or circles) are grouped together and perceived as columns.

Perception: principle of closure
This principle holds that incomplete figures are often completed by the perceptual system, even when part of the perceptual information is missing. For instance, in Figure 3.5, we perceive a white triangle overlaid on three black circles, even though the image could simply represent three incomplete circles.

UNIVERSALS AND VARIATION

Figure 3.4 Columns of shapes

Figure 3.5 A triangle and three black circles

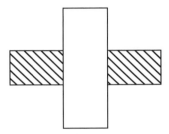

Figure 3.6 Two rectangles

Perception: principle of continuity
This principle holds that human perception has a preference for continuous figures. This is illustrated in Figure 3.6. Here, we perceive two unbroken rectangles, one passing behind another, even though this is not what we actually see. In fact, the shaded rectangle is obscured by the first, so we have no direct evidence that the shaded area represents one continuous rectangle rather than two separate ones.

Perception: principle of smallness
Finally, we consider the principle of smallness. This states that smaller entities tend to be more readily perceived as figures than larger entities. This is illustrated in Figure 3.7. We are more likely to perceive a black cross than a white cross because the black shading occupies a smaller proportion of the image.

Taken together, the Gestalt principles entail that the world is not objectively given. Instead, what we perceive is in part constructed by our cognitive apparatus, and mental representations are thereby constrained by processes

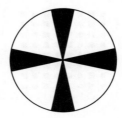

Figure 3.7 A black cross

fundamental to perceptual processing. As we will see below, these facts emerging from the domain of visual perception pattern together with universal constraints in the language of space.

Categorisation

The final constraint we will consider relates to human categorisation. Since the groundbreaking work of the cognitive psychologist Eleanor Rosch in the 1970s, it has been clear that the principles that govern categorisation in the human mind are due in part to the structure of the external world and due in part to innate human abilities. In particular, Rosch found that many human categories are not organised by **necessary and sufficient** conditions, but by **prototypes**. We return to these ideas in more detail in Chapter 8, observing for the time being that, as we saw in Chapter 2, categories are often fuzzy, and that categorisation judgements are made with respect to a prototypical or most representative member of a category.

3.2 Cross-linguistic patterns in semantic systems

In this section we consider cross-linguistic patterns in what cognitive linguists have suggested are arguably the two most fundamental domains of human experience: TIME and SPACE. In section 3.3, we will explore the nature of cross-linguistic variation with respect to the same two domains.

3.2.1 Patterns in the conceptualisation of space

We begin by investigating patterns in the human conceptualisation of space. As we have emphasised, the conceptions we present here are not thought of by cognitive linguists as predetermined semantic universals, but instead represent a set of common patterns in human conceptualisation of space, from which languages appear to elaborate different aspects thereby achieving considerable variation. The discussion presented here on the domain of space is largely based on the work of Leonard Talmy (2000), who proposes that spatial representation in language encodes **spatial scenes**. Spatial scenes are configured according to three parameters:

1. figure–ground segregation;
2. the relative proximity of the figure with respect to the ground; and
3. the location of the figure with respect to the ground. This is achieved by the employment of a particular **reference frame**.

Figure-ground segregation

As we have seen, linguistic representations of spatial scenes reflect a figure-ground asymmetry. While one entity is typically privileged and represents the figure, the second entity is given less prominence and is referred to as the ground or **reference object**. It is a striking fact that language reflects perceptual organisation in the way that spatial scenes are segregated. In English, this is mirrored by the syntax. For instance, in simple sentences like those in (13), the figure (underlined) normally precedes the preposition (*near*), while the reference object (bracketed) follows the preposition. Sentences in which the reference object precedes the preposition, although grammatically well-formed, are semantically odd (indicated by the question mark preceding the sentence):

(13) a. The bike is near [the house].
 b. ?[The house] is near the bike

The semantic 'oddness' of this example can be explained by the fact that the reference object is typically the immovable entity that only serves to locate the figure. Recall that the Gestalt principle of smallness predicts that the smaller entity (the bike) will be perceived as the figure. The criteria for determining figure and reference object, based on linguistic encoding, are listed in Table 3.1.

Primary and secondary reference object

In addition to figure-ground segregation, languages often allow more complex partitioning of spatial scenes. This involves segregating the ground into two reference objects in order to better locate the figure. These are termed **primary reference object** and **secondary reference object**. While the primary reference object is usually explicitly encoded by a lexical item, the secondary reference object need not be, but can instead merely be implied. Consider example (14):

(14) Big Ben is north of the River Thames.

While *the River Thames* is the primary reference object, the secondary reference object, *the Earth*, is implied by the spatial expression *north of*. In other

Table 3.1 Figure-ground segregation, as encoded in language (adapted from Talmy 2000: 183)

Figure	Reference object (or ground)
Has unknown spatial properties, to be determined	Acts as reference entity, having known properties that can characterise the primary object's unknowns
More moveable	More permanently located
Smaller	Larger
Geometrically simpler	Geometrically more complex
More recently on the scene/in awareness	Earlier on the scene/in awareness
Of greater concern/relevance	Of lesser concern/relevance
Less immediately perceivable	More immediately perceivable
More salient, once perceived	More backgrounded, once figure is perceived
More dependent	More independent

words, it is only with respect to the concept THE EARTH that we can process the information that one entity can be 'north of' another. Talmy (2000) identifies two kinds of secondary reference object: **encompassing** and **external**. These are outlined below.

The encompassing secondary reference object is typically asymmetric in orientation and encompasses the primary reference object. This type of reference object provides a frame for locating the primary reference object, which in turn serves to locate the figure. The example in (14) provides an example of this type, where the Earth provides an encompassing secondary reference object containing the primary reference object, *the River Thames*. In addition, it is because the Earth has an asymmetric orientation (the north–south opposition), that it is possible to identify the location of the figure relative to the primary reference object. A similar example is the concept QUEUE, which has asymmetric, front–back orientation:

(15) Jane is ahead of Mary in the queue/line for ice cream.

In example (15), the queue provides an orientational frame that encompasses the primary reference object *Mary*, which in turn locates the figure *Jane*. Observe that it is because of the front–back orientation imposed by the secondary reference object that *Jane's* location with respect to the primary reference object, *Mary*, is established. After all, *Mary* could be facing away from the front of the queue to talk to somebody behind her. Even in this situation, it would still be possible to describe *Jane* as ahead of *Mary* (in *the queue*). We return to the external type of secondary reference object in the next section.

Relative proximity of figure and reference object

The second way in which linguistic variation is constrained with respect to spatial scenes is that languages must encode the relative **proximity** of the figure with respect to the (typically immoveable) ground. At the most schematic level, there are three possibilities relating to proximity: 'contact', 'adjacency' or 'at some distance'. Examples from English that illustrate the linguistic encoding of these distinctions are given below.

Relative proximity: contact
The figure can be in physical contact with the reference object:

(16) a. The mosaic is on the front of the church.
     b. The mosaic is on the back of the church.
     c. The mosaic is on the (right/left-hand) side of the church.

Relative proximity: adjacency
The figure can be adjacent to, but not in contact with, the reference object:

(17) a. The bike is in front of the church.
     b. The bike is behind the church.
     c. The bike is on one side of/beside the church.
     d. The bike is on the right/left of the church.

Relative proximity: at some distance
The figure can be at some remove from the reference object:

(18) a. The bike is to the left/right of the church.
     b. The bike is a way off from the front/rear of the church.

Reference frames

The third parameter for delineating a spatial scene, as evident in the languages of the world, is the **reference frame**. Reference frames represent the means language has at its disposal for using reference objects in order to locate figures. According to Talmy (2000), there is a limited set of reference frames employed by the world's languages. Talmy identifies four kinds, which are illustrated in Figure 3.8. These can be divided into (1) reference frames that involve the primary reference object alone: a **ground-based** reference frame; and (2) reference frames that also involve a secondary reference object. There are three reference frames of this kind: **field-based, guidepost-based** and **projector-based**. In Figure 3.8, primary reference object is abbreviated to PRO, and secondary reference object to SRO.

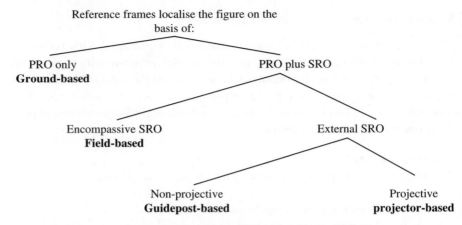

Figure 3.8 Taxonomy of reference frames in the languages of the world (adapted from Talmy 2000: 213)

In order to illustrate each of these reference frames, consider the simple cityscape scene, illustrated in Figure 3.9. Now imagine a situation in which a speaker is directing a hearer to the grocery store. There are a number of ways in which the exact location of the grocery store can be found, in keeping with the four reference frames identified.

Reference frames: ground-based
(19) The grocery store is next to the office building.

This is the simplest kind of reference frame. It involves just a primary reference object, the office building, and employs the intrinsic geometry of this reference object in order to locate the figure: the office building has an intrinsic front, back and sides, to which the speaker appeals in describing the location of the grocery store. Therefore, this type of reference frame is **ground-based**. The example of ground-based reference given in (19) is illustrated in Figure 3.10. The large cross in Figure 3.10, which overlays the office building, indicates that it is the office building that is providing the frame of reference for locating the figure.

Reference frames: field-based
(20) The grocery store is to the west of the office building.

Like the remaining reference frames, the field-based type involves a secondary reference object. Field-based reference is characterised by an encompassing secondary reference object, like the Earth example we discussed earlier. A similar example of field-based reference is given in (20) and illustrated in Figure 3.11.

UNIVERSALS AND VARIATION

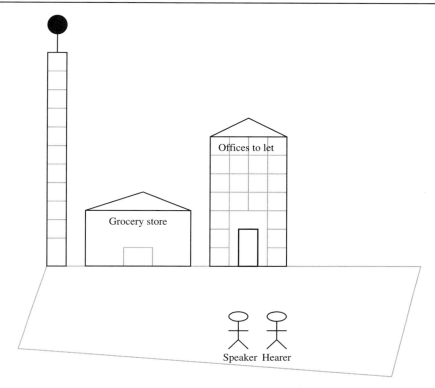

Figure 3.9 Simple cityscape scene

The crossed-lines indicate the **cardinal points** (north, south, east and west) that take their reference from the Earth. It is relative to the cardinal points that the primary reference object (the office building) locates the figure (the grocery store).

Reference frames: guidepost-based
   (21) The grocery store is on the tower side of the office building.

Like the field-based type, guidepost-based reference framing involves a secondary reference object. However, this type involves an external rather than encompassing secondary reference object. In the guidepost-based reference frame, the external secondary reference object is a non-animate entity – the tower in example (21) – which is external to the primary reference object. The example in (21) is represented in Figure 3.12, where it is the tower that identifies that portion of the primary reference object (the office building) with respect to which the grocery store is localised. This explains why this type of reference frame is described as 'guidepost-based'.

COGNITIVE LINGUISTICS: AN INTRODUCTION

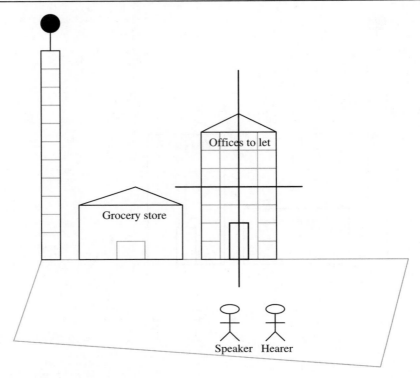

Figure 3.10 Ground-based reference

Reference frames: projector-based
(22) The grocery store is to the left of the office building.

The final kind of reference frame also involves an external secondary reference object. In this type of reference frame, the secondary reference object is an animate entity (here, the speaker), whose location serves as a frame of reference in locating the relevant part of the primary reference object that enables the figure to be located. In example (19), 'left' refers to that side of the office building from the perspective of the speaker. This type of reference frame is called 'projector-based' because the speaker is projecting his or her own location as a frame of reference. Example (22) is illustrated in Figure 3.13.

As the discussion in this section demonstrates, a number of core patterns are evident in the conceptualisation of space as encoded in language. These are (1) figure-ground segregation; (2) the interaction of figure with primary and secondary reference object; and (3) distinct types of reference frame. Moreover, these patterns are independently motivated by psychological principles of perception, which illustrates how the cognitive commitment underlies the statement of linguistic patterns based on evidence from other areas of cognitive science.

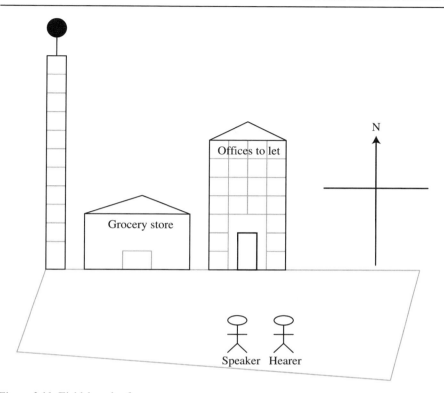

Figure 3.11 Field-based reference

3.2.2 Patterns in the conceptualisation of time

In this section, we address cross-linguistic patterns in the conceptualisation of time. In particular, we focus on how time is encoded in semantic structure. We will not address the grammatical encoding of time by tense systems, to which we will return in Part III of the book (see Chapter 18). Our discussion in this section is based on the 2004 book by Vyvyan Evans, *The Structure of Time*.

Unlike space, time is not a concrete or physical sensory experience. Moreover, unlike the human sensory-perceptual apparatus that is specialised for assessing spatial experience (among others, the visual system), we have no analogous apparatus specifically dedicated to the processing of temporal experience. Despite this, we are aware of the 'passing' of time. This awareness of time appears to be a wholly introspective or subjective experience. According to Evans (2004a), temporal experience can ultimately be related to the same perceptual mechanisms that process sensory experience. That is, perceptual processes are underpinned by temporal intervals, termed **perceptual moments**, which facilitate the integration of sensory experience into perceptual 'windows' or 'time slots'. In other words, perception is a kind of 'windowing' operation, which presents and updates

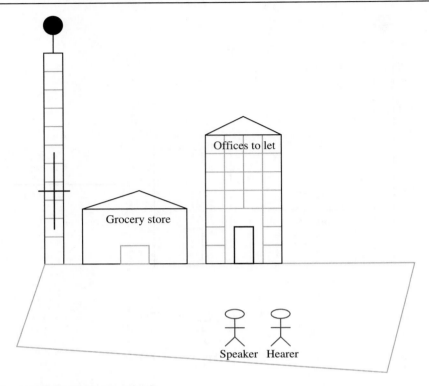

Figure 3.12 Guidepost-based reference

our external environment. The updating occurs as a result of timing mechanisms which hold at all levels of neurological processing and range from fractions of a second in duration to an outer limit of around three seconds.

Evidence for timing mechanisms comes from two sorts of sources. Brain activity can be measured by techniques such as the electroencephalogram (EEG), for instance. The brain produces electrical signals, which are measured by attaching electrodes to the scalp. These read signals and send them to a galvanometer, an instrument that measures small electrical currents. Such techniques allow researchers to observe changes in brain activity over split seconds of time. The brain rhythm assessed by an EEG is measured by the frequency of electrical pulses per second, and is produced on a galvanometer as a series of 'waves' with peaks and troughs (see Figure 3.14)

A second method for assessing timing mechanisms comes from exposing subjects to stimuli of certain kinds at particular points of brain activity. A well known experiment of this kind involves exposing subjects to two flashing lights, and relies on the phenomena known as **apparent simultaneity** and **apparent motion**. If the lights are set to flash with less than 0.1–0.2 seconds between their respective flashes, the lights will be perceived as flashing simultaneously.

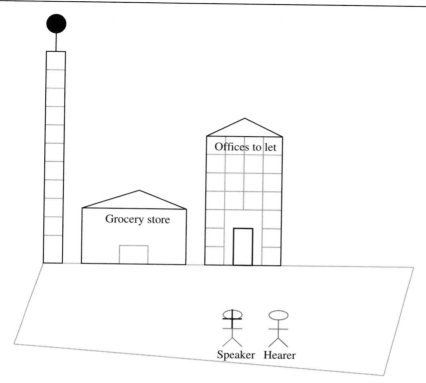

Figure 3.13 Projector-based reference

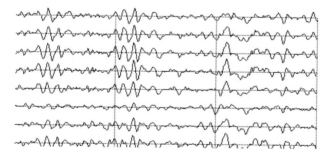

Figure 3.14 Approximately three seconds of data from eight EEG electrodes

This is the phenomenon of apparent simultaneity. If the interval between the two flashing lights is increased slightly, the flashing appears to take place in rapid motion. This is the phenomenon of apparent motion. If the interval between flashes is increased slightly more, the flashing appears to be distinctly sequential. However, when lights are set to flash at an interval close to the transition between apparent simultaneity and apparent motion, and when the flashing is correlated with the brain's own activity, experimenters found that what is perceived depends on when in the subject's own brain rhythm the exposure to the flashing lights takes place.

In the visual cortex, the dominant rhythm, the **alpha rhythm** (named by Hans Berger, who pioneered the EEG technique between 1929 and 1935), has a frequency of around ten pulses per second. It was found that if the lights begin flashing when the alpha rhythm is at a peak, then the subject sees apparent motion. However, when the flashing begins when the alpha rhythm is in a trough, the subject perceives apparent simultaneity. Findings like this provide compelling evidence that it is neurological activity in the brain, innate 'timing mechanisms', which give rise to perceptual moments, and these are in large part responsible for what we perceive.

Evidence that such perceptual moments have an outer limit of around three seconds comes from diverse sources, including language. Language, like other human symbolic behaviours (notably music), appears to manifest rhythmic organisation. For instance, the literary scholar Fred Turner and the neuroscientist Ernst Pöppel, in a (1983) paper entitled *The Neural Lyre*, have shown that the fundamental unit of metered poetry, which they call the **Line**, can contain between four and twenty syllables, depending on the language. This is based on a survey of languages including Latin, Greek, English, Chinese, Japanese, French, German, Ndembu (Zambia), Eipo (New Guinea), Spanish, Italian and Hungarian. Remarkably, however, despite the different numbers of syllables involved, Turner and Pöppel found that the time taken for recitation of the Line among these languages typically ranges from 2.5 to 3.5 seconds. This similarity in the duration of units of meter across such a diverse set of languages suggests that there is a common timing mechanism, or set of mechanisms, that is coordinating rhythmic behaviour.

The discussion so far indicates that, while time is not a physical entity that is objectively given, it is nevertheless a real experience. Our awareness of time emerges from the process of perceiving and from the properties of our perceptual apparatus. It is a consequence, ultimately, of the various 'timing mechanisms' in the brain that give rise to a range of perceptual moments, which in turn underpin perceptual processing. It follows that time enters into all human experience, since it is fundamental to the way in which perceptual processes operate.

One important consequence of this fact is that our subjective experience of time is not a single unitary phenomenon. Instead, it is comprised of a number of experiences that relate to our ability to assess duration, simultaneity and 'points' in time; our sense that sometimes time seems to proceed more slowly or more quickly than usual; our experience of 'now', and so on.

Temporal experience, as it is represented and encoded in language, exhibits two levels of organisation. The first level relates to **lexical concepts**. A lexical concept is the meaning that is represented by a lexical form or word (its **sense**, in traditional terms). Examples of temporal expressions from English include the words *time*, *past*, *present* and *future*, among others. The lexical concepts that underlie words of this kind can be organised in a number of ways at the conceptual level. For instance, the languages of the world appear to structure TIME in terms of MOTION, as we will see below. The second level of organisation relates to **cognitive models** for time. This is a level of organisation in which various lexical concepts are integrated, together with their patterns of conventional imagery. Evans (2004a) calls this process **concept elaboration**. For example, in the expression *a long time*, the lexical concept expressed by the word *time* relates to DURATION, while the imagery that elaborates the lexical concept relates to LENGTH, **lexicalised** or 'put into words' by *long*.

Lexical concepts for TIME

In his discussion of lexical concepts for TIME, Evans (2004a) distinguishes between **primary lexical concepts** and **secondary lexical concepts**. Primary lexical concepts are those that relate to common aspects of human cognitive processing. In other words, they relate to the experiences of time that we mentioned above: duration, simultaneity, temporal 'point' or moment, 'now' and so on. Because experiences of this kind can be traced to underlying perceptual mechanisms and processes, it follows that concepts of this kind are likely to be more common in the languages of the world, and where they occur, to be more similar across languages. In contrast, secondary lexical concepts are **cultural constructs** and thus may often be culture specific. A good example of this is the concept of TIME as a valuable commodity, which can be bought and sold, just like concrete physical merchandise. This concept, while present in the languages of the industrialised world, is entirely absent in the languages of many non-industrialised cultures. Since our focus here is on cross-linguistically robust patterns of lexical concepts for TIME, we limit the discussion in this section to primary lexical concepts.

In order to give an illustration of some of the primary lexical concepts for TIME, we will consider the English lexical item *time*. This form encodes four primary lexical concepts which show up in different contexts. The lexical concepts we will address are DURATION, MOMENT, EVENT and INSTANCE.

Lexical concept: DURATION

The concept of DURATION has two variants that relate to two distinct subjective experiences. The first is called **protracted duration** and relates to the experience that time is proceeding more slowly than usual:

(23) Time drags when you have nothing to do.

(24) My first thought was, 'Where did that car come from?' Then I said to myself, 'Hit the brakes.'. . .I saw her look at me through the open

window, and turn the wheel, hand over hand, toward the right. I also [noticed] that the car was a brown Olds. I heard the screeching sound from my tires and knew . . . that we were going to hit . . . I wondered what my parents were going to say, if they would be mad, where my boyfriend was, and most of all, would it hurt . . . After it was over, I realized what a short time it was to think so many thoughts, but, while it was happening, there was more than enough time. It only took about ten or fifteen seconds for us to hit, but it certainly felt like ten or fifteen minutes. (Flaherty 1999: 52)

Protracted duration is caused by a heightened awareness of a particular stimulus array, either because the interval experienced is 'empty', as in (23), or because the interval is very 'full' due to a great deal being experienced in a short space of time. This is illustrated in (24), which relates a near-death experience involving a car crash.

The second variant of DURATION is called **temporal compression**. This is when we experience time proceeding more quickly than usual, and is most often associated with our experience of routine behaviours which we carry out effortlessly without much attention to the task at hand. Evidence that temporal compression is encoded in language comes from examples like (25)–(27).

(25) The time has sped/whizzed by.

(26) Where has the time gone?

(27) 'Time flies when you're having fun'.

Lexical concept: MOMENT
Another aspect of our temporal experience is the ability to assess time in terms of discrete moments. This experience is also reflected in language. Consider examples (28)–(29).

(28) The time for a decision has come.

(29) Now is the time to address irreversible environmental decay.

Each of the uses of *time* in these examples could be paraphrased by the expression *moment*. In these examples, TIME is conceptualised not in terms of an interval, whose duration can be assessed, but instead as a discrete point.

Lexical concept: EVENT
A third conceptualisation of TIME relates to the notion of an EVENT. This is an occurrence of some kind. Evans (2004a) suggests that events derive, at the perceptual level, from temporal processing, which binds particular occurrences

into a temporally framed unity: a 'window' or 'time slot'. Consider examples (30)–(31).

(30) With the first contraction, the young woman knew her time had come.

(31) The man had every caution given him not a minute before to be careful with the gun, but his time was come as his poor shipmates say and with that they console themselves. (British National Corpus)

In each of these examples a particular event, childbirth and death respectively, is lexicalised by *time*. This suggests that the conceptualisation of an event is closely tied up with temporal experience.

Lexical concept: INSTANCE
The final temporal lexical concept we will consider is INSTANCE. This concept underlies the fact that temporal events can be enumerated, which entails that distinct events can be seen as instances or examples of the 'same' event.

(32) With that 100m race the sprinter had improved for the fourth time in the same season.

In this example, *time* refers not to four distinct moments, but to a fourth instance of the 'improvement' event. This example provides linguistic evidence that separate temporal events can be related to one another and 'counted' as distinct instances of a single event type.

Temporal aspects of an event: Christmas

Now let's consider a word other than *time* which also exhibits these distinct aspects of temporal experience. Consider the word *Christmas*. This relates to a particular kind of temporal event: the kind that is framed (or understood) with respect to the calendar. That is, Christmas is a festival that takes place at the same time each year, traditionally on the 25th of December. While the festival of Christmas is a cultural construct – deriving from the Christian tradition – the expression *Christmas* can be used in contexts that exhibit the same dimensions of temporal experience we described above for the expression *time*: dimensions that appear to derive from our cognitive abilities, and therefore from pre-linguistic experience of time. Consider examples (33)–(36). In example (33), the temporal event Christmas is experienced in terms of protracted duration and thus 'feels' as if it's proceeding more slowly than on previous occasions:

(33) *Protracted* DURATION
Christmas seemed to drag this year.

Equally, Christmas can appear to be proceeding more quickly than usual:

(34) *Temporal compression*
Christmas sped by this year.

Example (35) shows that Christmas can be conceptualised in terms of discrete moments or 'points' of time:

(35) MOMENT
Christmas has finally arrived/is here.

Finally, example (36) shows that instances of Christmas can be counted and compared with one another:

(36) INSTANCE
This Christmas was better than last Christmas.

The elaboration of temporal lexical concepts

One of the most striking ways in which lexical concepts for TIME are elaborated is in terms of **motion**. For example, it is almost impossible to talk about time without using words like *approach, arrive, come, go, pass* and so on. Of course, time is not a physical object that can literally undergo motion. Yet, in languages as diverse as Wolof (a Niger-Congo language spoken in West Africa), Mandarin Chinese, Japanese, Spanish and English, lexical concepts for TIME are systematically structured in terms of motion. Consider examples (37)–(40).

(37) Mandarin (examples from Yu 1998)
    a. Yi   dai   qiu wang libie   luyin   de   shihou
       a   generation ball king   part   green-grass MOD time
       zheng yi   tian   tian chao   women kaojin
       PRT   a   day   day toward us   approach
       'The time when the soccer king of the generation bids farewell to the green is approaching us day by day.'

    b. Liu-shi   de   sui-yue   bu-duan de   chong dan
       flow-pass   MOD   year-month not-break MOD wash faded
       zhe renmen   de   jiyi
       PRT people   MOD memory
       'The (flowing and) passing years are constantly washing away people's memories.'

(38) Japanese (examples from Shinohara 1999)
    a. Toki   ga   nagareru
        time   NOM   flows
        'Time flows'
    b. Toki   ga   sugite   itta
        time   NOM   pass   go.PAST
        'Time passed by'
    c. Kurisumasu   ga   chikazui-teiru
        Christmas   NOM   approach-PROG
        'Christmas is approaching'.

(39) Wolof (example from Moore 2000).
    Tabaski   mungiy   ñów
    Tabaski   3:PRES-IMPF come
    'Tabaski is coming'.   [*Note:* Tabaski is a major holiday.]

(40) Spanish (example from Moore 2000)
    La   Noche   Buena   viene   muy   pronto
    The   night   good   come   very   soon
    'Christmas Eve is coming very soon.'

However, given the specific nature of the lexical concepts we have discussed, it is likely that the range of motion types that languages can rely upon to elaborate specific lexical concepts for TIME will be relatively constrained. For instance, in English, protracted duration can only be elaborated in terms of motion events that involve slow motion or absence of motion:

(41) a. Time seemed to stand still.
     b. The time dragged.

Temporal compression, on the other hand, is elaborated in terms of rapid motion (42a), or motion that is so rapid as to be imperceptible (42b):

(42) a. The time flew/sped/whizzed by.
     b. The time has vanished/disappeared.

Both these kinds of elaboration contrast with the way in which the lexical concepts EVENT and MOMENT are structured. These concepts involve motion directed towards a particular **locus of experience** or **deictic centre** (usually the speaker, from whose perspective the scene is viewed). As examples (43) and (44) show, this is revealed by expressions denoting movement towards the speaker, such as *come, arrive, approach* and so on. Moreover, motion of this kind usually terminates when it reaches the locus of experience.

(43) MOMENT
The time for a decision is approaching/coming/getting closer/has arrived.

(44) EVENT
The young woman's time is approaching/coming/getting closer/has arrived.

Cognitive models for TIME

We now turn to a brief consideration of more complex conceptualisations: cognitive models for TIME. Recall that we defined a cognitive model earlier as a level of organisation in which various lexical concepts are integrated, together with their patterns of conventional imagery. This means that cognitive models are larger-scale knowledge structures than individual lexical concepts. Cross-linguistic evidence suggests that there are three main cognitive models for TIME. While the first two are **ego-based** and typically involve reference to the present or 'now', the third kind is **time-based** and makes no intrinsic reference to the concept of 'now'. The three models are the **moving time model**, the **moving ego model** and the **temporal sequence model** (see Figure 3.15). We briefly discuss each in turn below. These models can be thought of as generalisations over the range of primary (and secondary) lexical concepts for time that are found in the world's languages, including the ways in which these concepts are elaborated.

Cognitive model: moving time
In this model, there is an experiencer, who may either be implicit or linguistically coded by expressions like *I*. The experiencer is called the **ego**, whose location represents the experience of 'now'. In this model, the ego is static. Temporal moments and events are conceptualised as objects in motion. These objects move towards the ego from the future and then beyond the ego into the past. It is by virtue of this motion that the passage of time is understood. In Figure 3.16 the small dark circles represent 'times', and the arrow connecting the 'times'

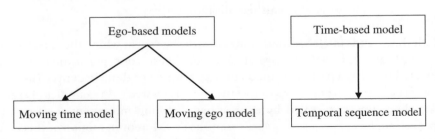

Figure 3.15 Taxonomy of cognitive models for time

UNIVERSALS AND VARIATION

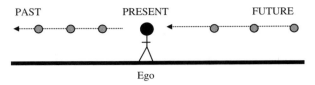

Figure 3.16 The moving time model

indicates motion of the 'times' towards and past the ego. Although present, past and future are marked on the diagram the figure representing the ego is not marked for orientation: while many languages, including English, conceptualise the ego as facing the future with the past behind, there is now good evidence that at least one language, Aymara, spoken in the Andean region of South America, conceptualises the ego as facing the past, with the future behind (Núñez and Sweetser, forthcoming). We illustrate this below (section 3.3.2).

Linguistic evidence for this cognitive model comes from examples like those in (45), in which the passage of time is understood in terms of the motion of a temporal entity towards the ego:

(45) a. Christmas is getting closer.
     b. My favourite part of the piece is coming up.
     c. The deadline has passed.

Cognitive model: moving ego
In this model, TIME is a landscape over which the ego moves, and time is understood by virtue of the motion of the ego across this landscape, towards specific temporal moments and events that are conceptualised as locations. This model is illustrated in Figure 3.17. In this figure, the small circles on the landscape represent future 'times' towards which the ego moves, while 'times' that the ego has already moved beyond now lie in the past. The ego's motion is represented by the direction of the arrow. As with the Figure 3.16, the ego is unmarked for orientation.

Evidence for the moving ego model comes from examples like those in (46):

(46) a. We're moving towards Christmas.
     b. We're approaching my favourite part of the piece.
     c. She's passed the deadline.
     d. We'll have an answer within two weeks.
     e. The meetings were spread out over a month.

In these examples TIME is conceptualised as a stationary location or bounded region in space. It is through the motion of the ego that time's passage is understood.

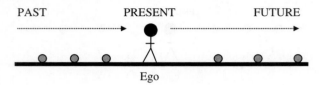

Figure 3.17 The moving ego model

Figure 3.18 The temporal sequence model

Cognitive model: temporal sequence
The third model relates to the concepts EARLIER and LATER. Unlike the previous two models, this one does not involve an ego. Instead, a temporal event is understood relative to another earlier or later temporal event. The model is illustrated in Figure 3.18, and linguistic examples are given in (47).

(47) a. Monday precedes Tuesday.
     b. Tuesday follows Monday.

In these English examples, LATER follows EARLIER: the earlier event, Monday, is understood as being located in front of the later event, Tuesday. In other words, it is relative to Tuesday rather than the ego (the subjective experience of 'now') that Monday is EARLIER. Figure 3.18 captures this as directionality is signalled by the arrow. Earlier events (events are represented by the small circles) are understood as being located in front of later events.

Time-based versus ego-based models

Distinguishing ego-based models from time-based models resolves a puzzling fact in English. Consider the following examples:

(48) a. in the weeks ahead of us
     b. That's all behind us now.

(49) a. in the following weeks
     b. in the preceding weeks

In (48), events relating to the future are conceptualised as being ahead and events relating to the past as being behind. In (49), later events are behind ('following'), and earlier events are ahead ('preceding'). This apparent paradox is reconciled by understanding that each pair of examples rests upon a different cognitive model. The examples in (48) relate to ego-based models for TIME, where time is conceptualised relative to the speaker. In contrast, the examples in (49) relate to the temporal sequence model, which is time-based rather than ego-based: time is conceptualised relative to some other event. As these examples show, the 'location' of a temporal event is interpreted differently depending on what kind of cognitive model is involved. Moreover, the different models relate to different sorts of lexical concepts: PAST/FUTURE in ego-based models and EARLIER/LATER in the time-based model.

3.3 Cross-linguistic variation in semantic systems

In the previous section we discussed some of the patterns of conceptualisation that are shared by languages, due in part to the constraining influence of common experiences and cognitive structures. Nevertheless, while the patterns described above capture some of the broad similarities between languages in the domains of SPACE and TIME, there remains an impressive degree of cross-linguistic variation. The purpose of this section is to provide a glimpse of this diversity.

3.3.1 Variation in the conceptualisation of space

In this section we consider two languages that conceptualise space in very different ways from English: Korean and the Australian language Guugu Yimithirr.

Categorising spatial scenes in English and Korean

One of the ways in which languages diverge is in the kind of spatial relation that holds between the figure and ground, even for objectively similar spatial scenes. A striking illustration of this is the contrast in the ways English and Korean choose to conventionally segregate spatial scenes. This discussion is based on research carried out by Melissa Bowerman and Soonja Choi. Consider the spatial scenes described in (50) and (51), represented in Figure 3.19.

(50) a. put cup on table
 b. put magnet on refrigerator
 c. put hat on
 d. put ring on finger
 e. put top on pen
 f. put lego block on lego stack

Figure 3.19 The division of spatial scenes in English (adapted from Bowerman and Choi 2003: 393)

(51) a. put video cassette in case
   b. put book in case
   c. put piece in puzzle
   d. put apple in bowl
   e. put book in bag

The scenes described in (50) and (51) are lexicalised in English by a verb in conjunction with a spatial particle like *on* or *in*. The expression *put on* suggests placement of the figure in contact with a surface of some kind. The expression *put in* suggests placement of the figure within some bounded landmark or container. The reader familiar only with English might be forgiven for thinking that this is the only way these spatial scenes can be conceptualised. However, the situation in Korean is very different. The English examples in (50), involving the expression *put in*, are categorised into spatial scenes of four different kinds in Korean. This is achieved using the four different Korean verbs in (52):

(52) a. nohta   'put on horizontal surface'
   b. pwuchita   'juxtapose surfaces'
   c. ssuta   'put clothing on head'
   d. kkita   'interlock/fit tightly'

Examples (53)–(56) show which Korean verb corresponds to which of the spatial scenes described using the English expression *put on*.

(53) nohta   'put on horizontal surface'
    e.g.   put cup on table

(54) pwuchita 'juxtapose surfaces'
    e.g.   put magnet on refrigerator

(55) ssuta   'put clothing on head'
    e.g.   put hat on

(56) kkita   'interlock/fit tightly'
    e.g.   a. put ring on finger
                 b. put top on pen
                 c. put lego block on lego stack

Similarly, the English examples in (51), involving the expression *put in*, are categorised into spatial scenes of two different kinds. This is achieved using the two Korean verbs in (57). Observe that the verb *kkita* appears for the second time.

(57) a. kkita   'interlock/fit tightly'
   b. nehta   'put loosely in or around'

The examples in (58) and (59) show which Korean verb corresponds to which of the spatial scenes described using the English expression *put in*.

(58) kkita   'interlock/fit tightly'
    e.g. a.   put video cassette in case
        b.   put book in case
        c.   put piece in puzzle

(59) nehta   'put loosely in or around'
    e.g. a.   put apple in bowl
        b.   put book in bag

The way Korean categorises the scenes we described in (50) and (51) is represented in Figure 3.20, which contrasts with the English model in Figure 3.19.

The psychologist and cognitive linguist Dan Slobin has described phenomena of the kind we have just depicted in terms of **thinking for speaking**: a particular language forces its speakers to pay attention to certain aspects of a scene in order to be able to encode it in language. While English forces speakers to categorise the spatial scenes we have just discussed on the basis of whether the figure is being placed on a surface or in a container, Korean partitions the spatial scenes into different categories. Korean speakers must pay attention to different aspects of the scenes in question, such as what kind of surface is involved (is it horizontal or not?), and what kind of contact is involved (is it simple juxtaposition of surfaces, or does it involve a tight fit or a loose fit?). Clearly, these differences do not arise because people in English-speaking countries experience activities like putting the lid on a pen differently from people in Korea. Instead, these differences reflect the capacity that speakers of different languages have to categorise objectively similar experiences in different ways.

Frames of reference in Guugu Yimithirr

We now turn briefly to Guugu Yimithirr, an indigenous language of North Queensland, Australia, studied extensively by Stephen Levinson and his colleagues. We noted above that the languages of the world provide evidence for a limited number of frames of reference. What is interesting about Guugu Yimithirr is that this language appears to make exclusive use of the field-based reference frame. The field-based terms used in Guugu Yimithirr are shown in Figure 3.21.

Rather than relating strictly to the cardinal points of the compass North, South, East and West (which are marked as N, S, E and W in Figure 3.21), the terms in Guugu Yimithirr actually encompass quadrants, which only roughly correspond to the points of the compass. However, like the points of the compass, the four quadrants are based on the Earth as an absolute frame of reference. In

UNIVERSALS AND VARIATION

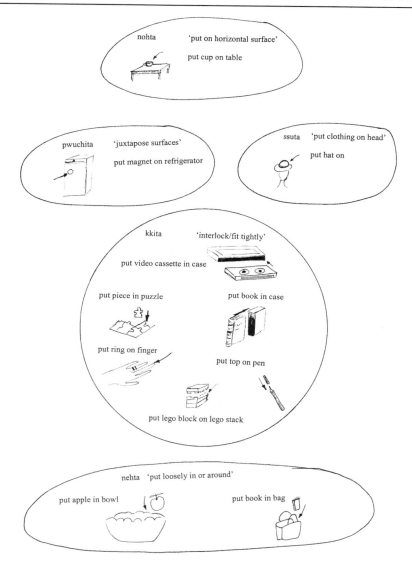

Figure 3.20 The division of spatial scenes in Korean (adapted from Bowerman and Choi 2003: 394)

order to be able to employ a spatial frame of reference for talking about relative locations in space, speakers of Guguu Yimithirr must calculate the location of a particular object with respect to this field-based reference frame. Furthermore, unlike English, which uses field-based terms just for large-scale geographical reference (e.g. *Europe is north of Africa*), Guguu Yimithirr only has access to field-based reference. As the linguistic anthropologist William Foley describes, 'the sun

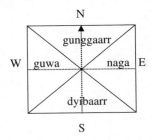

Figure 3.21 The field-based spatial terms of Guugu Yimithirr (Haviland 1993)

doesn't go down, it goes west; the fork isn't at my left, it lies south; the tide doesn't go out, it goes east' (Foley 1997: 217).

3.3.2 Variation in the conceptualisation of time

In this section we consider two languages that conceptualise time in very different ways from English: Aymara and Mandarin.

The past and future in Aymara

Aymara is an indigenous language of South America, spoken in the Andean region of Peru, Chile and Bolivia. There is good linguistic and gestural evidence that while Aymara features variants of both ego-based and time-based cognitive models for time, in the ego-based model, Aymara speakers conceptualise the FUTURE as being located behind the ego, while PAST is conceptualised as being in front of the ego (Núñez and Sweetser, forthcoming). This pattern of elaboration contrasts with the English pattern. Consider example (60).

(60) a. The future lies in front of us.
     b. She has a bright future ahead/in front of her.

These examples show that the lexical concept FUTURE is structured in terms of locations in front of the ego. This is also true of other future-oriented lexical concepts, as (61) illustrates.

(61) a. Old age lies way ahead of him.
     b. Having children is in front of us.
     c. The years ahead of us will be difficult.

Compare the representation of PAST in English:

(62)   The past is behind me.

This example shows that the lexical concept PAST is elaborated in terms of a location behind the ego. This pattern is extended to all past-oriented lexical concepts:

(63) a. My childhood is behind me.
b. Once divorced, she was finally able to put an unhappy marriage behind her.

Now compare the way PAST and FUTURE are conceptualised in Aymara. The Aymaran expressions for PAST and FUTURE are given in (64) and (65), respectively.

(64) mayra   pacha
     front/eye/sight time
     'past time'

(65) q'ipa   pacha
     back/behind time
     'future time'

The expression for the 'past' is literally 'front time', while the expression for 'future' is 'behind time'. This suggests that Aymara has the opposite pattern of elaboration from English. A gestural study of Aymara speakers in which Núñez participated (discussed in Núñez and Sweetser, forthcoming) provides supporting evidence that the past is conceptualised as 'in front' and the future 'behind'. This study reveals that, when speaking about the past, Aymara speakers gesture in front, and when speaking about the future, they gesture behind. A further interesting difference between Aymara and a language like English is that the Aymaran ego-based model for time appears to be 'static'. In other words, there appears to be no evidence that temporal 'events' are conceptualised as moving relative to the ego, nor that the ego moves relative to temporal 'events'. This means that Aymara lacks the 'path-like' ego-based moving time and moving ego cognitive models, but has instead a 'static' ego-based model for time. Aymara speakers also make use of the temporal sequence model. In doing so, however, their gestures relate to temporal events along the left–right axis, rather than the front–back axis.

The pattern of elaboration for PAST and FUTURE in Aymara appears to be motivated by another aspect of the Aymaran language. Aymara is a language that, unlike English, grammatically encodes **evidentiality**: speakers are obliged by the language to grammatically mark the nature of the evidence they rely on in making a particular statement: whether the speaker has witnessed the event described with their own eyes, or whether the event is known to them through hearsay (Mircale and Yapita Moya 1981). It appears likely that the value assigned to visual evidence has consequences for the elaboration of concepts such as PAST and FUTURE. Events that have been experienced are conceptualised as having been seen. Things that are seen are located in front of the ego, due to human physiology. It follows that PAST is conceptualised as being in front of the ego. In contrast, events that have yet to be experienced are conceptualised as being behind the ego, a location that is inaccessible to the human visual apparatus (Mircale and Yapita Moya 1981; Lakoff and Johnson 1999; Evans 2004a; Núñez and Sweetser, forthcoming).

Earlier and later in Mandarin

We now briefly consider how the temporal concepts EARLIER and LATER are conceptualised in Mandarin. Again we find a contrast with the English pattern that we discussed earlier, where concepts relating to the distinction between EARLIER and LATER are elaborated in terms of their relative location on the **horizontal axis**. The following examples illustrate this pattern, where EARLIER is 'before' and LATER is 'after'. Recall Figure 3.18, which shows how LATER follows EARLIER in this model of TIME.

(66) a. Tuesday comes/is before Wednesday.
     b. Wednesday comes/is after Tuesday.

In Mandarin there is a pattern in which the **vertical axis** elaborates the distinction between EARLIER and LATER. Concepts that are earlier (experienced first) are conceptualised as 'higher' or 'upper', while concepts that are later (experienced subsequent to the first) are conceptualised as 'lower'. Examples (67)–(71) from Yu (1998) illustrate this pattern.

(67) a. shang-ban-tian
       upper-half-day
       'morning'
    b. shang-ban-tian
       lower-half-day
       'afternoon'

(68) a. shang-ban-ye
       upper-half-night
       'before midnight'
    b. xia-ban-ye
       lower-half-night
       'after midnight'

UNIVERSALS AND VARIATION

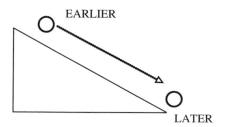

Figure 3.22 The slope model (adapted from Shinohara 2000: 5)

(69) a. shang-ban-yue
upper-half-month
'the first half of the month'
b. xia-ban-yue
lower-half-month
'the second half of the month'

(70) a. shang-ban-nian
upper-half-year
'the first half of the year'
b. xia-ban-nian
lower-half-year
'the second half of the year'

(71) a. shang-bei
upper-generation
'the elder generation'
b. xia-bei
lower-generation
'the younger generation'

According to Shinohara (2000) the motivation for this pattern of elaboration may be due to how we experience slopes. When an object is rolled down a slope, the earlier part of the event is at the top of the slope, while due to the force of gravity the later part of the event is lower down. This idea is represented in Figure 3.22.

3.4 Linguistic relativity and cognitive linguistics

In this final section, we turn to the issue of **linguistic relativity**. Although the nature of the relationship between thought and language has intrigued human beings since the time of the ancient philosophers, within modern linguistics this idea is most frequently associated with the work of Edward Sapir and Benjamin Lee Whorf, and is known as the **Sapir-Whorf hypothesis**. The Sapir-Whorf hypothesis consists of two parts: **linguistic determinism** (the idea that language determines non-linguistic thought) and **linguistic relativity** (the idea that speakers of different languages will therefore think differently). The strong version of this hypothesis holds that language entirely determines thought: a speaker of language X will understand the world in a fundamentally different way from a speaker of language Y, particularly if those two languages have significantly different grammatical systems. In other words, a speaker will only have access to cognitive categories that correspond to the linguistic categories of his or her language. The weak version of this hypothesis, on the other hand, holds that the structure of a language may influence (rather than determine) how the speaker performs certain cognitive processes, because the structure of different languages influences how information is 'packaged'.

Since the rise of the generative model in the 1960s, proponents of formal linguistics have tended to reject the Sapir-Whorf hypothesis altogether, given its incompatibility with the hypothesis that there might exist a universal set of pre-linguistic conceptual primitives, and therefore a universal 'mentalese' or 'language of thought'. The following excerpt from Steven Pinker's book *The Language Instinct* illustrates this position:

> But it is wrong, all wrong. The idea that thought is the same thing as language is an example of . . . a conventional absurdity. . . The thirty-five years of research from the psychology laboratory is distinguished by how little it has shown. Most of the experiments have tested banal 'weak' versions of the Whorfian hypothesis, namely that words can have some effect on memory or categorization. . . Knowing a language, then, is knowing how to translate mentalese into strings of words, and vice versa. (Pinker 1994: 57–82)

While most modern linguists would probably agree that the strong version of the Sapir-Whorf hypothesis is untenable, some interesting findings have emerged in cognitive linguistics and related fields, particularly in linguistic anthropology, cognitive psychology and language acquisition research, which suggest that language can and does influence thought and action. Therefore, a cognitive linguistic approach to the relationship between language, thought and experience, together with the facts of cross-linguistic diversity, is compatible with a weaker form of the linguistic relativity thesis. For this reason, the view we present here might be described as **neo-Whorfian**.

3.4.1 Whorf and the Linguistic Relativity Principle

The most famous proponent of the Linguistic Relativity Principle is Benjamin Lee Whorf (1897–1941), who studied American Indian languages at Yale. However, the tradition of viewing language as providing a distinct world view can be traced back to his teacher at Yale, the anthropologist Edward Sapir (1884–1939), as well as to the linguistic anthropologist Franz Boas (1858–1942), and before that to the German linguist and philosopher Wilhelm Von Humboldt (1767–1835). Whorf was an intriguing and complex writer, who sometimes appeared to take a moderate line, and sometimes expressed a more extreme view of linguistic relativity (Lakoff 1987; Lee 1996). The following much-quoted excerpt states Whorf's position:

> We dissect nature along lines laid down by our native languages. The categories and types that we isolate from the world of phenomena we do not find there because they stare every observer in the face; on the contrary, the world is presented in a kaleidoscopic flux of impressions which has to be organised by our minds – and this means largely by the linguistic systems in our minds. (Whorf 1956: 213)

Setting aside the theoretical objections to the Sapir-Whorf hypothesis put forth by proponents of the generative approach, there is independent empirical evidence against the strong version of the hypothesis. This evidence originally came from work on colour categorisation. It may surprise readers who are only familiar with English to learn that some languages have an extremely small set of **basic colour terms**. These are terms that are morphologically simple (for example, *bluish* is excluded) and are not subsumed under another colour term (for example, *crimson* and *scarlet* are not basic colour terms because they fall within the category denoted by *red*). For instance, the Dani, a tribe from New Guinea, only have two basic colour terms in their vocabulary. The expression *mola*, which means 'light', refers to white and warm colours like red, orange, yellow, pink and purple. The expression *mili*, which means 'dark', refers to black and cool colours like blue and green. Yet, in colour experiments where Dani subjects were shown different kinds of **focal colours** (these are colours that are perceptually salient to the human visual system) they had little difficulty remembering the range of colours they were exposed to (Heider 1972; Rosch 1975, 1978). These experiments involved presenting subjects with a large set of coloured chips, from which they were asked to select the best examples of each colour; in later experiments, they were asked to recall what colours they had selected previously. If language entirely determines thought, then the Dani should not have been able to categorise and remember a complex set of distinct focal colours because they only have two basic colour terms in their language. In another experiment, Rosch taught the Dani subjects sixteen colour names based on words from their own language (clan names). She found that the names for the focal colours were learnt faster than names for non-focal colours. These findings illustrate that humans have common perceptual and conceptualising capacities, as we noted earlier. Due to shared constraints, including environment, experience, embodiment and perceptual apparatus, we can, and often do, conceptualise in fundamentally similar ways, regardless of language. However, this does not entail that variation across languages has no influence on non-linguistic thought.

3.4.2 Language as a shaper of thought

If there is empirical evidence against the hypothesis that language *determines* thought (the strong version of the Sapir-Whorf hypothesis), then the question that naturally arises is whether language can *influence* or *shape* thought in any way. It is this weak version of the Sapir-Whorf hypothesis that underlies much recent research into the nature of the relationship between language and thought, and some of the findings suggest that the answer to this question might be 'yes'. There are two lines of evidence that support a weak version of the Sapir-Whorf hypothesis. These are considered below.

Language facilitates conceptualisation

The first line of evidence relates to linguistic determinism and the idea that language **facilitates** our conceptualising capacity. The assumption in cognitive linguistics is that language reflects patterns of thought, and can be seen as a means of encoding and externalising thought. It follows from this view that patterns of meaning in language represent a **conventional** means (an accepted norm in a given linguistic community) of encoding conceptual structure and organisation for purposes of communication. This is known as the **symbolic function** of language, which we described in Chapter 1. It also follows from this view that different ways of expressing or encoding ideas in language represent different patterns of thought, so that encountering different linguistic 'options' for encoding ideas can influence the way we reason.

A clear example of the influence of language upon thought is the experiment described by Gentner and Gentner (1982) in which they trained different English-speaking subjects in **analogical models** of electricity. An analogical model relies upon a relatively well known scenario or system for understanding a less well known system, where the parts and relations of the well known system stand in a similar relation to those in the less well known system, here electricity. Through **analogy** (comparison based on perceived similarity) subjects can reason about electricity using the well known model. One group was taught that electricity can be represented as a teeming crowd of people, while another group was taught that electricity can be represented as water flowing through a pipe, as in a hydraulic system. The mappings between these two analogical models and an electrical circuit are summarised in Tables 3.2 and 3.3.

Importantly, each analogical model correctly predicted different aspects of the behaviour of an electrical circuit. For example, a circuit with batteries connected serially will produce more current than a circuit with batteries in parallel. This is predicted by the analogy based on the hydraulic system, where serial pumps one after the other will produce a greater flow rate of water. In the moving crowd model, where the battery corresponds simply to the crowd, it is difficult to think of a meaningful contrast between a serial and a parallel connection.

Serial resistors in an electrical circuit reduce current, while parallel resistors increase it. The moving crowd model is better at predicting this aspect of the behaviour of electricity, where resistance is modelled in terms of gates. Parallel gates allow more people through, while serial gates allow fewer people through. Gentner and Gentner hypothesised that if subjects used different analogical models to reason about the circuit, then each group should produce dramatically divergent results, which is exactly what they found. Subjects who were trained in the hydraulic system model were better at correctly predicting the effect of serial versus parallel batteries on current, while subjects who were familiar with the moving crowd model were better at predicting the effect of serial versus parallel resistors on current. This study reveals that different 'choices' of language for representing concepts can indeed affect non-linguistic thought such as reasoning and problem-solving.

Table 3.2 Hydraulic system model (based on Gentner and Gentner 1982: 110)

Hydraulic system	Electric circuit
Pipe	Wire
Pump	Battery
Narrow pipe	Resistor
Water pressure	Voltage
Narrowness of pipe	Resistance
Flow rate of water	Current

Table 3.3 Moving crowd model (based on Gentner and Gentner 1982: 120)

Moving crowd	Electric circuit
Course/passageway	Wire
Crowd	Battery
People	Resistor
Pushing of people	Voltage
Gates	Resistance
Passage rate of people	Current

Cross-linguistic differences and their effect on non-linguistic thought and action

The second thread of evidence in support of a weak version of the Sapir-Whorf hypothesis relates to linguistic relativity: how cross-linguistic differences influence non-linguistic thought and action. We begin by revisiting the domain of SPACE. We noted earlier that Guugu Yimithirr exclusively employs a field-based frame of reference for locating entities in space. An important consequence of this is that speakers of Guguu Yimithirr must be able to dead-reckon their location with respect to the cardinal points of their system, wherever they are in space. Based on a comparative study of Guguu Yimithirr speakers and Dutch speakers, Levinson (1997) found that the ability of Guugu Yimithirr speakers to calculate their location had profound consequences for non-linguistic tasks. It was found that when Guugu Yimithirr speakers were taken to an unfamiliar terrain with restricted visibility, such as a dense rainforest, they were still able to work out their location, identifying particular directions with an error rate of less than 4 per cent. This contrasted with a comparable experiment involving Dutch speakers, who were much less accurate. Like English, Dutch makes extensive use of other non-field-based frames of reference such as ground-based and projector-based reference. According to Levinson, this type of experiment constitutes evidence for a real Whorfian effect, in which the nature of spatial representation in language has consequences for a speaker's non-linguistic abilities. However, it's worth pointing out that experience, as well as language, may play a part in these sorts of experiments. After all, Guugu Yimithirr speakers are likely to have more experience of assessing directions and finding their way around rainforests than the average Dutch speaker.

Next, we consider a study that investigated the influence of the language of time on non-linguistic thought and action. This study was carried out by cognitive psychologist Lera Boroditsky (2001). Boroditsky was interested in investigating whether the different lexical concepts for TIME in English and Mandarin would produce a noticeable effect on reaction time in linguistic experiments. Recall that we observed earlier that a common way of elaborating the concepts EARLIER and LATER in Chinese is by means of positions on the vertical axis: 'upper' and 'lower'. In English, these concepts are elaborated primarily in terms of the horizontal axis: 'before' and 'after'. Boroditsky exposed Mandarin and English speakers to primes like the ones in Figure 3.23, which represented either the vertical or the horizontal axis. A prime is a particular stimulus manipulated by researchers in psycholinguistic experiments. Boroditsky then asked the subjects to answer a series of 'true or false' questions employing the temporal concepts EARLIER or LATER (for example, *March comes earlier than April: true or false?*). Boroditsky found that Mandarin speakers were faster in responding to questions involving the terms *earlier* and *later* when the

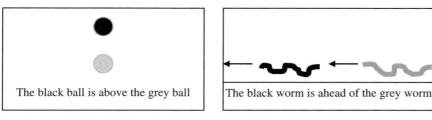

Figure 3.23 Spatial primes (adapted from Boroditsky 2001)

prime related to the vertical axis. In contrast, English speakers were faster when the prime related to the horizontal axis. This remained the case even when both sets of subjects were carrying out the task in English. As Boroditsky puts it, 'it appears that habits in language encourage habits in thought. Since Mandarin speakers showed vertical bias even when thinking for English, it appears that language-encouraged habits in thought can operate regardless of the language that one is currently thinking for' (Boroditsky 2001: 12).

3.4.3 The cognitive linguistics position

The position adopted in cognitive linguistics is that there are commonalities in the ways humans experience and perceive the world and in the ways human think and use language. This means that all humans share a common conceptualising capacity. However, these commonalities are no more than constraints, delimiting a range of possibilities. As we have seen, there is striking diversity in the two domains we have surveyed, which shows that the way English speakers think and speak about space and time by no means represents the only way of thinking and speaking about space and time. According to cognitive linguists, language not only *reflects* conceptual structure, but can also *give rise to* conceptualisation. It appears that the ways in which different languages 'cut up' and 'label' the world can differentially influence non-linguistic thought and action. It follows that the basic commitments of cognitive linguistics are consonant with a weak version of the Sapir-Whorf hypothesis, a position that some linguists argue is gathering increasing empirical support.

3.5 Summary

Linguists of any theoretical persuasion are intrigued by the possible existence of linguistic **universals**, by the form of such universals and by the nature of the relationship between thought and language. In this chapter, we began by comparing the cognitive and formal positions on linguistic universals. While formal linguists have tended to treat universals as resulting from primitive concepts or mechanisms, innately given, cognitive linguists argue instead that there are universal tendencies. We explored the cognitive view in more detail, and outlined a number of constraints on human conceptualisation that go some way to explaining the existence of linguistic universals. These constraints include the nature of human **embodiment**, **Gestalt principles** and the nature of human **categorisation**, all of which collectively constitute a **conceptualising capacity** common to all humans. We then presented some examples of common cross-linguistic patterns in the conceptualisation of the fundamental domains of space and time. In the domain of SPACE we suggested that there are three common cross-linguistic patterns in terms of how languages structure space. These include (1) **figure-ground segregation**; (2) a means of encoding the relative **proximity** of the figure with respect to the ground; and (3) a means of encoding the **location** of the figure with respect to the ground. This is achieved by the employment of a particular **reference frame**. In the domain of TIME, cross-linguistic patterns relate to a small set of **primary lexical concepts** for time, and three large-scale **cognitive models** for time, which integrate these (and other) temporal lexical concepts together with their patterns of **elaboration** (conventional patterns of imagery). We then presented some examples of cross-linguistic variation in the conceptualisation of space and time, which demonstrate that despite some fundamental cross-linguistic similarities in the linguistic representation of space and time, there is nevertheless considerable cross-linguistic variation. Finally, having explored the issue of linguistic universals, we introduced the **Sapir-Whorf hypothesis**: the idea that language might play some role in determining non-linguistic thought, and that speakers of different languages might therefore have different conceptual systems. We concluded that, while the **strong version** of this hypothesis is rejected by most linguists, there is some evidence in favour of the **weak version** of the hypothesis. Although cognitive linguistics makes the case for a common conceptualising capacity, accounting for general cross-linguistic patterns, such a position is nevertheless consistent with and even predicts substantial cross-linguistic variation. Given that the linguistic system both reflects the conceptualising capacity, and in turn influences the nature of knowledge by virtue of the language-specific categories it derives, cognitive linguistics is consistent with a weak version of the Sapir-Whorf hypothesis.

Further reading

Universals, typology and cross-linguistic variation

- **Brown (1991)**. An excellent reanalysis of classic studies in anthropology on human universals and cultural relativity.
- **Comrie (1989); Givón (1991); Greenberg (1990); Haiman (1985); Hopper (1987)**. A selection of representative sources for those interested in learning more about linguistic typology.

- **Croft (2001).** In this provocative study, Croft argues for a cognitive linguistic account of grammatical organisation, taking into account the finding from typology that linguistic categories are not universal, but rather language- and, indeed, construction-specific.
- **Croft (2003).** In this recently revised introductory textbook, Croft presents an excellent introduction to the objectives, methodology and findings of functional typology. Croft also addresses diachronic (historical) explanations of linguistic universals, and compares typological and generative explanations of linguistic universals.

Formal approaches to universals in grammar and meaning

- **Chomsky (2000b).** In this recent collection of articles, Chomsky summarises his ideas about the nature of language as a 'biological object'. He explains why he thinks language can only be meaningfully investigated from an internalist perspective (internal to the mind of the individual) rather than from the (externalist) perspective of language use. Chomsky also considers the development of linguistics in the context of the history of ideas and in the context of the natural sciences.
- **Chomsky (2002).** In this recent collection of essays, Chomsky provides an accessible and up-to-date overview of the generative approach. The editors' introduction provides a useful introduction to key concepts, and the essays by Chomsky focus on the relationship between language and mind, and language and brain. One chapter consists of an interview on the Minimalist Programme, Chomsky's most recent project, and represents the most accessible overview of this framework.
- **Fodor (1998).** In this book, the philosopher Jerry Fodor, the author of the highly influential book *The Modularity of Mind* (1983), presents arguments against semantic decomposition and argues instead that all concepts are atomistic and innate.
- **Jackendoff (1983, 1990, 1992, 1997, 2002).** These books provide an insight into the development of Jackendoff's theory of semantic universals. The 1983 and 1990 books set out this theory in detail, and the 1992 book is a collection of essays that provide short overviews of aspects of his theory, including his arguments in favour of a modular, computational model and his theory of concepts.

Space

- **Bloom, Peterson, Nadel and Garrett (1996).** This edited volume collects together a number of important and influential papers by

leading cognitive scientists who have worked on space and spatial cognition. Particularly relevant papers in this volume include those by Ray Jackendoff, Melissa Bowerman and Jean Mandler.
- **Coventry and Garrod (2004).** This book presents experimental evidence for a perceptual and body-based foundation for spatial prepositions.
- **Levinson (2003).** This book surveys the research conducted by Levinson and his colleagues at the Max Plank Institute at Nijmegen on cross-linguistic diversity in spatial representation.
- **Talmy (2000).** Volume I Chapter 3 presents a revised version of Talmy's pioneering and highly influential study of the way languages structure space. This paper was first published in 1983.
- **Tyler and Evans (2003).** This book explores the semantics and sense networks of English prepositions from a cognitive linguistics perspective.

Time

- **Alverson (1994).** Although some of the claims in this book have been shown to be problematic (see Yu 1998), this represents an important study by a linguistic anthropologist into common cross-linguistic metaphors for time.
- **Evans (2004a).** This book employs the perspective of cognitive linguistics in order to investigate the nature and origin of temporal experience and how we conceptualise time.
- **Evans (2004b).** This paper summarises some of the key ideas from *The Structure of Time* in a single article.
- **Lakoff and Johnson (1999).** Chapter 10 presents a survey of the analysis of TIME in Conceptual Metaphor Theory.
- **Núñez and Sweetser (forthcoming).** This paper presents findings from Aymara, and includes an important discussion on the difference between ego-based and time-based construals of time.
- **Radden (1997; 2003a).** Two articles, summarising the way in which time is often structured conceptually in terms of space, by one of the leaders in the European cognitive linguistics movement. The 2003 paper in particular focuses on cross-linguistic similarities and differences.
- **Turner and Pöppel (1983).** A pioneering article that relates metrical patterns to neurologically instantiated temporal intervals.
- **Yu (1998).** This study contains a chapter on how time is conceptualised in Mandarin.

Linguistic relativity

- **Boroditsky (2001).** In this article, Boroditsky presents experimental evidence for a weak form of the linguistic relativity hypothesis in the domain of TIME.
- **Foley (1997).** Chapter 10 presents a useful overview of the Sapir-Whorf hypothesis from the linguistic anthropology perspective, which is broadly compatible with the concerns of cognitive linguistics.
- **Gentner and Goldin-Meadow (2003).** A recent collection of articles by proponents and opponents of the linguistic relativity hypothesis.
- **Gumperz and Levinson (1996).** A collection of seminal articles, which did much to reopen the debate on linguistic relativity. See in particular articles by Levinson, Slobin and Bowerman.
- **Hunt and Agnoli (1991).** Provides a useful overview as well as insight into one view from cognitive psychology: Hunt and Agnoli argue that language narrowly influences cognition in the sense that 'choices' over language have consequences for processing costs.
- **Lee (1996).** An excellent critical analysis and re-evaluation of Whorf's often complex ideas.
- **Whorf (1956).** This is a collection of many of Whorf's papers.

Exercises

3.1 Cognitive linguistics vs. formal linguistics

How does cognitive linguistics differ from formalist approaches in terms of its approach to universals? Summarise the key points of each position. Is there any shared ground?

3.2 The Sapir-Whorf hypothesis

Summarise the cognitive linguistics position with respect to the Sapir-Whorf hypothesis. What is the evidence for this position?

3.3 Space: reference frames

Classify the following examples based on the taxonomy of reference frames provided in section 3.2. Give your reasoning for each, and provide as much detail as possible.

(a) St Paul's cathedral is to the south of the Thames.
(b) St Paul's is left of the Thames.

(c) St Paul's is on the Bank of England side of the Thames.
(d) St Paul's is in the City of London.
(e) St Paul's is near the London Monument.

3.4 Time

Consider the following examples:

(a) Time passed.
(b) Christmas has vanished.
(c) We've got through the most difficult period of the project.
(d) They have a lot of important decisions coming up.
(e) The general meeting came after we made the decision to liquidate all assets.
(f) The top premiership clubs have three games in the space of five days.

In view of the discussion of the lexical concepts and three cognitive models for TIME presented in this chapter (section 3.2.2), identify which cognitive model each of these utterances is most likely to be motivated by. What problems did you have in identifying the relevant cognitive model? How might these problems be resolved?

3.5 Time: Wolof

Wolof has a number of words that relate to some of the lexical concepts for time found in English. For instance, *dirr* corresponds to the English DURATION concept lexicalised by *time*. In the following examples (drawn from Moore 2000) we'll consider the Wolof word *jot* ('time'). The examples suggest that *jot* is comparable to the English concept of COMMODITY, in which time is conceptualised as a resource that can be possessed, bought or wasted (e.g. *I have all the time in the world*).

(a) Dama   ñàkk   jot   rekk
    SFOC.1   lack   time   only
    'It's just that I don't have time!'

(b) Q: Am   nga   jot?
       have   PERF.2   time
       'Do you have (any) time?'

   A: Fi   ma   tollu   dama   ñàkk   jot
      where   1.SUBJ   be.at.a.point.equivalent.to SFOC.1   lack   time
      'At this point I don't have (any) time.'

(c) Su   ñu   am-ee   jot   ñu   saafal   la
    When we   have-ANT   time   we   roast.BEN   2.OBJ
    'When we have time we will roast [peanuts] for you.'

However, unlike the English concept COMMODITY as lexicalised by *time*, *jot* cannot be transferred to another person (e.g. *can you give/spare me some time?*), nor can it be made, wasted or spent (e.g. *we've made/wasted/spent some time for/on each other*). What does this imply regarding the similarities and differences between the English COMMODITY concept associated with *time*, and the lexical concept for COMMODITY encoded in Wolof by the word *jot*? What might this suggest about how Wolof and English speakers conceptualise time as a resource or commodity? In view of this, is it appropriate to label the meaning associated with *jot* COMMODITY, or can you think of another more appropriate term?

3.6 Kay and Kempton's colour naming experiment

Kay and Kempton (1984) compared English speakers with Tarahumara (Mexican Indian) speakers on naming triads of colour (blue, blue-green, green). Tarahumara has a word for 'blue-green', but not separate words for 'blue' and 'green'. The task was to state whether blue-green colour was closer to blue or green. English speakers sharply distinguished blue and green, but Tarahumara speakers did not. In a subsequent study, English speakers were induced to call the intermediate colours *blue-green*, and the effect disappeared. How might we interpret these findings in the light of the ideas discussed in this chapter?

4

Language in use: knowledge of language, language change and language acquisition

The subject of this chapter is language use and its importance for knowledge of language, for how language evolves over time (language change) and for how we acquire our native language (language acquisition). Some linguistic theories have attempted to separate the mental knowledge of language from language use. For example, in developing the generative framework, Chomsky has argued that language can only be meaningfully investigated from an internalist perspective (internal to the mind of the individual) rather than from the (externalist) perspective of language use. In Chomsky's terms, this is the distinction between **competence** (knowledge) and **performance** (use). Chomksy privileges competence over performance as the subject matter of linguistics. In rejecting the distinction between competence and performance cognitive linguists argue that knowledge of language is derived from patterns of language use, and further, that knowledge of language is knowledge of how language is used. In the words of psychologist and cognitive linguist Michael Tomasello (2003: 5), 'language structure emerges from language use.' This is known as the **usage-based thesis**.

The purpose of this chapter is to provide a sketch of the assumptions and theories that characterise this position in cognitive linguistics. One of the central assumptions is that language use is integral to our knowledge of language, our 'language system' or 'mental grammar'. According to this view, the organisation of our language system is intimately related to, and derives directly from, how language is actually used. It follows from this assumption that language structure cannot be studied without taking into account the nature of language use. This perspective is what characterises cognitive linguistics as a **functionalist** rather than a **formalist** approach to language, a distinction that we explore in more detail in Part III of the book (Chapter 22).

After outlining the main components of a usage-based view of the language system (section 4.1), we focus on three areas of cognitive linguistics that attempt to integrate the usage-based thesis with theoretical models of various linguistic phenomena. The first phenomenon we address is knowledge of language (section 4.2). In this context, the term 'grammar' is used in its broadest sense to refer to the system of linguistic knowledge in the mind of the speaker. In this sense, 'grammar' refers not just to grammatical phenomena like syntax, but also to meaning and sound. As we briefly noted at the end of Chapter 2, the cognitive model of grammar encompasses (1) the units of language (form-meaning pairings variously known as **symbolic assemblies** or **constructions**) which constitute the inventory of a particular language; and (2) the processes that relate and integrate the various constructions in a language system. The specific theory we introduce in this chapter is the framework called **Cognitive Grammar**, developed by Ronald Langacker. This approach explicitly adopts the usage-based thesis; indeed, Langacker was one of the early proponents of the usage-based perspective.

The second phenomenon we consider is language change (section 4.3). Here, we examine William Croft's **Utterance Selection Theory** of language change. This theory views language use as the interface that mediates between the **conventions** of a language (those aspects of use that make a language stable) and mechanisms that result in **deviation** from convention resulting in language change.

The third phenomenon we investigate is language acquisition (section 4.4). We explore how children acquire the grammar of their native language from the perspective of the usage-based model developed by Michael Tomasello, which integrates insights from cognitive linguistics and cognitive psychology into a theory of first language acquisition.

4.1 Language in use

In this section we outline some of the assumptions shared by researchers who have adopted the usage-based thesis in their theoretical accounts of linguistic structure, organisation and behaviour.

4.1.1 A usage event

Perhaps the most important concept underlying usage-based approaches to linguistics is the **usage event**. A usage event is an **utterance**. Consider the following two definitions of the term 'utterance' provided by two of the leading proponents of the usage-based approach:

> [An utterance is] a particular, actual occurrence of the product of human behavior in communicative interaction (i.e., a string of sounds),

as it is pronounced, grammatically structured, and semantically and pragmatically interpreted in its context. (Croft 2001: 26)

An utterance is a linguistic act in which one person expresses towards another, within a single intonation contour, a relatively coherent communicative intention in a communicative context. (Tomasello 2000: 63)

As these statements indicate, an utterance is a situated instance of language use which is culturally and contextually embedded and represents an instance of linguistic behaviour on the part of a **language user**. A language user is a member of a particular linguistic community who, in speaking (and, indeed, in signing or writing), attempts to achieve a particular interactional goal or set of goals using particular **linguistic** and **non-linguistic strategies**. Interactional goals include attempts to elicit information or action on the part of the hearer, to provide information, to establish interpersonal rapport (e.g. when 'passing the time of day') and so on. The linguistic strategies employed to achieve these goals might include the use of speech acts (requesting, informing, promising, thanking and so on), choices over words and grammatical constructions, intonation structures, choices over conforming or not conforming to discourse conventions like turn-taking and so on. Non-linguistic strategies include facial expressions, gesture, orientation of the speaker, proximity of interlocutors in terms of interpersonal space and so on.

As we will define it, a usage event or utterance has a unit-like status in that it represents the expression of a coherent idea, making (at least partial) use of the **conventions of the language** (the 'norms' of linguistic behaviour in a particular linguistic community). In other words, an utterance is a somewhat discrete entity. However, we use the expressions 'unit like' and 'somewhat discrete' because the utterance is not an absolutely discrete or precisely identifiable unit. This is because utterances involve grammatical forms (for example, word order), semantic structures (patterns of meaning), speech sounds, patterns of intonation (for example, pitch contours), slight pauses, and accelerations and decelerations. While these properties converge on discreteness and unity, they do not co-occur in fixed patterns, and therefore do not provide a set of criteria for collectively identifying an utterance. In this respect, utterances differ from the related notion of **sentence**.

A sentence, as defined by linguistics, is an abstract entity. In other words, it is an **idealisation** that has determinate properties, often stated in terms of grammatical structure. For example, one definition of (an English) sentence might consist of the formula in (1):

(1)   S   →   NP VP

In this formula, 'S' stands for sentence, 'NP' for subject noun phrase, and 'VP', for the verb phrase or predicate which provides information about the subject NP. We will look more closely at this idea in Part III of the book (Chapter 14).

The notion of a sentence, while based on prototypical patterns found in utterances, is not the same as an utterance. Utterances typically occur spontaneously, and often do not conform to the grammaticality requirements of a well-formed sentence (recall the discussion of grammaticality in Chapter 1). For example, in terms of structure, an utterance may consist of a single word (*Hi!*), a phrase (*No way!*), an incomplete sentence (*Did you put the . . .?*) or a sentence that contains errors of pronunciation or grammar because the speaker is tired, distracted or excited, and so on. While much of formal linguistics has been concerned with modelling the properties of language that enable us to produce grammatically well-formed sentences, utterances often exhibit **graded grammaticality** (an idea that is discussed in more detail in Chapter 14). This fact is widely recognised by linguists of all theoretical persuasions. As this discussion indicates, while a sentence can be precisely and narrowly defined, an utterance cannot. While sentences represent the structure associated with a prototypical utterance, utterances represent specific and unique instances of language use. Once a sentence is given meaning, context and phonetic realisation, it becomes a (spoken) utterance. Typically, cognitive linguists place little emphasis on the sentence as a theoretical entity. In contrast, the notion of a usage event or utterance is central to the cognitive perspective.

4.1.2 The relationship between usage and linguistic structure

As we indicated above, the generative model separates knowledge of language (competence) from use of language (performance). According to this view, competence determines performance, but performance can also be affected by language-external factors of the type we mentioned above, so that performance often fails to adequately reflect competence. In direct opposition to this view, cognitive linguists argue that knowledge of language is derived from and informed by language use. As we will see below, language acquisition is understood from this usage-based perspective not as the **activation** of an innately pre-specified system of linguistic knowledge (recall the discussion of Universal Grammar in Chapter 3), but instead as the **extraction** of linguistic units or constructions from patterns in the usage events experienced by the child. This process relies upon general cognitive abilities, and the set of units or constructions eventually build up the inventory that represents the speaker's language system or knowledge of language. Furthermore, in usage-based theories of language change, change is seen not as a function of system-internal change, but as a function of interactional and social (usage-based) pressures that motivate changes in the conventions of the language system.

4.1.3 Comprehension and production

Language use involves both the production of language and the comprehension of language. This is because it involves interaction between speakers and hearers. While speakers 'put ideas into words' and utter them, hearers are faced with the task of 'decoding' these utterances and retrieving the ideas behind them. A model of language has to characterise the system that underlies linguistic interaction, regardless of whether it is a model of language knowledge or a model of language processing. However, these two types of model concentrate on explaining somewhat different aspects of this system. Models of language processing, like models of language acquisition, fall within the sub-discipline of psycholinguistics, and seek to explain the 'step-by-step' processes involved in production and comprehension of language. For example, models of language processing seek to discover the principles that govern how speakers match up concepts with words and retrieve those words from the lexicon, how hearers break a string of sounds up into words and find the grammatical patterns in that string, what constraints memory places on these processes, why speech errors happen and so on. In contrast, models of language knowledge concentrate on describing the knowledge system that underlies these processes. Models of language processing usually assume a particular model of language knowledge as a starting point, and place an emphasis on experimental methods. The models we discuss in this book (cognitive and formal models) are models of language knowledge. However, because cognitive linguists adopt the usage-based thesis, the interactional and goal-directed nature of language use is central to the cognitive model.

4.1.4 Context

The context in which an utterance or usage event is situated is central to the cognitive explanation. This is particularly true for word meaning, which is **protean** in nature. This means that word meaning is rather changeable. While words bring with them a conventional meaning, the context in which a word is used has important effects on its meaning. Furthermore, 'context' can mean a number of different things.

One kind of context is **sentential** or **utterance context**. This relates to the other elements in the string. Consider example (2), where we are focusing in particular on the meaning of the preposition *in*:

(2) a. The kitten is in the box.
    b. The flower is in the vase.
    c. The crack is in the vase.

These examples involve spatial scenes of slightly different kinds, where *in* reflects a spatial relationship between the figure and the reference object. In (2a) the figure, *the kitten*, is enclosed by the reference object, *the box*, so that the spatial relationship is one of containment. However, in the other two examples, *in* does not prompt for quite the same kind of meaning. In (2b) the flower is not enclosed by the vase, since it partly protrudes from it. Equally, in (2c) *in* does not prompt for a relationship of containment, because the crack is on the exterior of the vase. As these examples illustrate, the meaning of *in* is not fixed but is derived in part from the elements that surround it.

A second kind of context relates not to the other elements in the utterance itself but to the background knowledge against which the utterance is produced and understood. Consider example (3):

(3) It's dark in here.

If said by one caver to another in an underground cavern, this would be a factual statement relating to the absence of light in the cavern. If uttered by a linguistics professor to a student who happened to be sitting next to the light switch in a poorly lit seminar room, this might be a request to turn the light on. If uttered by one friend to another upon entering a brilliantly lit room, it might be an ironic statement uttered for the purpose of amusement. As this range of possible meanings demonstrates, the context of use interacts with the speaker's intentions and plays a crucial role in how this utterance is interpreted by the hearer. One consequence of the role of context in language use is that **ambiguity** can frequently arise. For example, given the cave scenario we sketched above, example (3) might reasonably be interpreted as an expression of fear, a request for a torch and so on.

In order to distinguish the conventional meaning associated with a particular word or construction, and the meaning that arises from context, we will refer to the former as **coded meaning** and the latter as **pragmatic meaning**. For example, the coded meaning associated with *in* relates to a relationship between a figure and a reference object in which the reference object has properties that enable it to enclose (and contain) the figure. However, because words always occur in context, coded meaning represents an **idealisation** based on the prototypical meaning that emerges from contextualised uses of words. In reality, the meaning associated with words always involves pragmatic meaning, and coded meaning is nothing more than a statement of this prototypical meaning abstracted from the range of pragmatic (situated) interpretations associated with a particular word. According to this view, pragmatic meaning is 'real' meaning, and coded meaning is an abstraction. We explore these ideas in detail in Part II of the book (Chapter 7).

4.1.5 Frequency

The final assumption relating to the usage-based thesis that we introduce in this section is the notion of frequency. If the language system is a function of language use, then it follows that the relative frequency with which particular words or other kinds of constructions are encountered by the speaker will affect the nature of the language system. This is because cognitive linguists assume that linguistic units that are more frequently encountered become more **entrenched** (that is, established as a cognitive pattern or routine) in the language system. According to this view, the most entrenched linguistic units tend to shape the language system in terms of patterns of use, at the expense of less frequent and thus less well entrenched words or constructions. It follows that the language system, while deriving from language use, can also influence language use.

4.2 Cognitive Grammar

In this section, we present an overview of Cognitive Grammar, the model of language developed by Ronald Langacker. The purpose of this section is to illustrate what a **usage-based model** of language looks like, rather than to provide a detailed overview of the theory. We return to the details of Langacker's theory in Part III of the book.

Langacker's model is called 'Cognitive Grammar' because it represents an attempt to understand language not as an outcome of a specialised language module, but as the result of general cognitive mechanisms and processes. According to this view, language follows the same general principles as other aspects of the human cognitive system. In this respect, Cognitive Grammar upholds the generalisation commitment (Chapter 2). It is also important to point out that the term 'grammar' is not used here in its narrow sense, where it refers to a specific subpart of language relating to syntactic and/or morphological knowledge. Instead, the term 'grammar' is used in the broad sense, where it refers to the language system as a whole, incorporating sound, meaning and morphosyntax.

We begin with a brief sketch of the central assumptions of Cognitive Grammar. This approach rejects the modular view adopted by formal models, according to which language is a system of 'words and rules' consisting of a lexicon, a syntactic component containing rules of combination that operate over lexical units, and other components governing sound and sentence meaning. Instead, Cognitive Grammar takes a symbolic or constructional view of language, according to which there is no distinction between syntax and lexicon. Instead, the grammar consists of an inventory of units that are form-meaning pairings: morphemes, words and grammatical constructions. These units, which

Langacker calls **symbolic assemblies,** unite properties of sound, meaning and grammar within a single representation.

4.2.1 Abstraction, schematisation and language use

In Cognitive Grammar, the units that make up the grammar are derived from language use. This takes place by processes of **abstraction** and **schematisation**. Abstraction is the process whereby structure emerges as the result of the generalisation of patterns across instances of language use. For example, a speaker acquiring English will, as the result of frequent exposure, 'discover' recurring words, phrases and sentences in the utterances they hear, together with the range of meanings associated with those units. Schematisation is a special kind of abstraction, which results in representations that are much less detailed than the actual utterances that give rise to them. Instead, schematisation results in **schemas**. These are achieved by setting aside points of difference between actual structures, leaving just the points they have in common. For instance, in example (2), we saw that the three distinct utterances containing the lexical item *in* have slightly different meanings associated with them. These distinct meanings are situated, arising from context. We established that what is common to each of these utterances is the rather abstract notion of enclosure; it is this commonality that establishes the schema for *in*. Moreover, the schema for *in* says very little about the nature of the figure and reference object, only that they must exist, and that they must have the basic properties that enable enclosure. Crucially, symbolic assemblies, the units of the grammar, are nothing more than schemas.

As we saw in Chapter 1, there are various kinds of linguistic units or symbolic assemblies. They can be words like *cat*, consisting of the three sound segments [k], [æ] and [t] that are represented as a unit [kæt], idioms like [*He/she kick*-TENSE *the bucket*], bound morphemes like the plural marker [*-s*] or the agentive suffix [*-er*] in teacher, and syntactic constructions like the ditransitive construction that we met in Chapter 2.

In sum, abstraction and schematisation, fundamental cognitive processes, produce schemas based on usage events or utterances. In this way, Cognitive Grammar makes two claims: (1) general cognitive processes are fundamental to grammar; and (2) the emergence of grammar as a system of linguistic knowledge is grounded in language use.

4.2.2 Schemas and their instantiations

As we mentioned briefly earlier, cognitive linguists argue that grammar not only derives from language use, but also, in part, motivates language use. It does this by licensing or **sanctioning** particular usage patterns. A usage pattern

COGNITIVE LINGUISTICS: AN INTRODUCTION

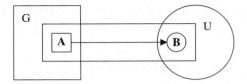

Figure 4.1 An instantiation of a schema (adapted from Langacker 2000: 10)

instantiates its corresponding schema; **instantiations**, therefore, are specific instances of use, arising from a schematic representation. This idea is illustrated in Figure 4.1.

In Figure 4.1, the box labelled G represents the repository of conventional units of language: the grammar. The box labelled U represents a particular usage event: an utterance. The box labelled A in the grammar represents a conventional unit: a symbolic assembly. The circle labelled B represents a specific linguistic element within an utterance. The arrow signals that B instantiates (or 'counts as an instance of') schema A. This means that A sanctions B.

4.2.3 Partial sanction

Of course, language use is not a simple case of language users making use of the finite set of symbolic assemblies represented in their grammar. After all, the richness and variety of situations and contexts in which language users find themselves, and the range of meanings that they need to express, far exceed the conventional range of units a language possesses. Although impressive in its vastness, the inventory of constructions available in a single language is nevertheless finite.

One solution to the restrictions imposed on language use by the finiteness of these resources lies in the use of linguistic units in ways that are only partially sanctioned by the range of constructions available in the language. In other words, language use is often partially **innovative**. For example, consider the word *mouse*. This word has recently acquired a new meaning: it refers not only to a rodent, but also to a computer 'mouse', which has a similar shape. When this new pattern of usage first appeared, it was an innovation, applied by the manufacturers of the computer hardware. This new usage was only partially sanctioned by the existing construction. This is illustrated by the dotted arrow in Figure 4.2. In this diagram, A represents the linguistic unit with the form *mouse* and the meaning RODENT, while the B has the same form but the meaning PIECE OF COMPUTER HARDWARE USED TO CONTROL THE CURSOR.

As we will see when we discuss language change later in the chapter, partial sanction only results in language change when it is diffused through a linguistic community and becomes established as a conventional unit in its own right.

LANGUAGE IN USE

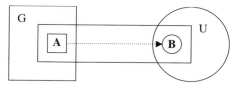

Figure 4.2 Partial sanction by a schema (adapted from Langacker 2000: 10)

4.2.4 The non-reductive nature of schemas

An important feature of Langacker's framework, which results from positing a direct relationship between grammatical organisation and language use, is that the model is **non-reductive**. As we noted above, one of the factors involved in the establishment of constructions is frequency: if a particular linguistic structure recurs sufficiently frequently, it achieves the status of an entrenched unit. As a result of the process of entrenchment, schemas result that have different levels of schematicity. This means that some schemas are **instances** of other, more abstract, schemas. In this way, the grammar acquires an internal hierarchical organisation, where less abstract schemas are instances of more abstract schemas. For example, consider prepositions (P) like *for*, *on* and *in*, which are combined with a complement noun phrase (NP) to form a preposition phrase (PP). In example (4), the NP is bracketed.

(4) a. to [me]
    b. on [the floor]
    c. in [the garage]

The expressions in (4), *to me*, *on the floor* and *in the garage*, are common phrases that probably have unit status for most speakers of English. In other words, they are constructions. However, there is another schema related to these constructions, which has the highly schematic form [P [NP]] and the highly schematic meaning DIRECTION OR LOCATION WITH RESPECT TO SOME PHYSICAL ENTITY. The constructions in (4) are instances of the more abstract schema [P [NP]]. This is illustrated in Figure 4.3.

This view of grammar is non-reductive in the following way. The constructions in (4) can be predicted by the more general schema of which they are instances. However, the fact that they can be predicted does not mean that they can be eliminated from the grammar. On the contrary, the fact that expressions of this kind are frequently occurring ensures that they retain unit status as distinct constructions. Moreover, that fact that they share a similar structure and a common abstract meaning ensures that the more abstract schema also coexists with them in the grammar.

COGNITIVE LINGUISTICS: AN INTRODUCTION

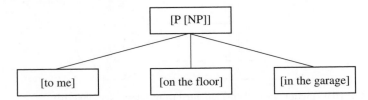

Figure 4.3 Schema-instance relations

This non-reductive model stands in direct opposition to the generative grammar model, which places emphasis on **economy of representation**. This is because the generative model assumes that the rapid acquisition of an infinitely creative system of language can only be plausibly accounted for by a small and efficient set of principles. In particular, the model seeks to eliminate **redundancy**: the same information does not need to be stated in more than one place, as this makes the system cumbersome. According to this view, the fact that the expressions in (4) are predictable from the more abstract schema means that these instances can be eliminated from the grammar and 'built from scratch' each time they are used. In the generative model, the only construction that would be stored in the grammar is the abstract schema. However, this schema would lack schematic meaning and would instead have the status of an 'instruction' about what kinds of forms can be combined to make grammatical units. In the generative model, then, what we are calling a schema is actually a **rule**. While schemas are derived from language use and thus incorporate a meaning element, rules are minimally specified structural representations that predict the greatest amount of information possible in the most economical way possible.

4.2.5 Frequency in schema formation

As we have seen, the central claim of Cognitive Grammar, with respect to the usage-based thesis, is that usage affects grammatical representation in the mind. Furthermore, frequency of use correlates with entrenchment. Two main types of frequency effects have been described in the literature: **token frequency** and **type frequency**. Each of these gives rise to the entrenchment of different kinds of linguistic units. While token frequency gives rise to the entrenchment of instances, type frequency gives rise to the entrenchment of more abstract schemas.

Token frequency refers to the frequency with which specific instances are used in language. For instance, the semantically related nouns *falsehood* and *lie* are differentially frequent. While *lie* is much more commonly used, *falsehood* is much more restricted in use. This gives rise to differential entrenchment of the mental representations of these forms. This is illustrated in the diagrams in

LANGUAGE IN USE

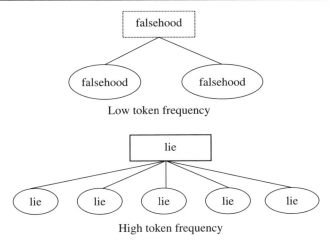

Figure 4.4 Frequency effects and entrenchment of instances

Figure 4.4. The degree of entrenchment of a linguistic unit, whether instance or more abstract schema, is indicated by the degree to which the square box is emboldened.

Now let's consider type frequency. While token frequency gives rise to the entrenchment of instances, type frequency gives rise to the entrenchment of more abstract schemas. For instance, the words *lapped*, *stored*, *wiped*, *signed*, *typed* are all instances of the past tense schema [VERBed]. The past tense forms *flew* and *blew* are instances of the past tense schema [XXew]. As there are fewer usage events involving the distinct lexical items *blew* and *flew* (as there are fewer distinct lexical items of this type relative to past tense forms of the *-ed* type), then it is predicted that the [XXew] type schema will be less entrenched in the grammar than the [VERBed] type schema. This is diagrammed in Figure 4.5.

Recall that, due to the non-reductive nature of the model, the predictability of an instance from a schema does not entail that the instance is not also stored in the grammar. Indeed, a unit with higher token frequency is more likely to be stored. For instance, the form *girls* is predictable from the lexical item *girl*, plus the schema [NOUN-s]. However, due to the high token frequency of the form *girls*, this lexical item is likely to be highly entrenched, in addition to the form *girl* and the plural schema [NOUN-s]. This contrasts with a plural noun like *portcullises* which is unlikely to be entrenched because this expression has low token frequency. Instead, this form would be sanctioned by combination of the plural schema and the singular form *portcullis*.

Bybee and Slobin (1982) provide empirical evidence for the view that frequency correlates with degree of entrenchment. They found that highly frequent irregular forms resist regularisation, while infrequent irregular forms

119

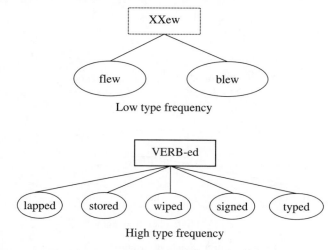

Figure 4.5 Frequency effects and entrenchment of schemas

Table 4.1 Past tense endings of selected verbs in 1982 (based on Bybee and Slobin 1982)

Most frequent	Less frequent	Infrequent
past form *-t* only	past form *-ed* or *-t*	past form *-ed* only
bend – bent	blend – blended/blent	wend – wended
lend – lent	geld – gelded/gelt	
send – sent	gird – girded/girt	
spend – spent	rend – rended/rent	
build – built		

tend to become regularised over time. Bybee and Slobin compared irregular past tense forms of English verbs like *build – built* from Jesperson's (1942) historical grammar of English with their modern forms in the (1982) *American Heritage Dictionary*. They found that more frequently used irregular verbs like *lend* had retained the irregular past tense form (*lent*). In contrast, less frequent forms like *blend* could alternate between the irregular form with *-t* (*blent*) and the regular past tense form with the suffix *-ed* (*blended*). However, highly infrequent forms like *wend* were by (1982) listed only with the regular past tense suffix (*wended*). Table 4.1 lists the past tense endings for these verbs as they appear in the 1982 dictionary.

4.3 A usage-based approach to language change

In this section we examine a usage-based approach to language change, the theory of Utterance Selection developed by William Croft in his (2000) book

Explaining Language Change. Before doing so, we briefly introduce the branch of linguistics concerned with language change, **historical linguistics**.

4.3.1 Historical linguistics and language change

Historical linguistics is concerned with describing how languages change and with attempting to explain why languages change. It concerns the histories and prehistories of languages and relationships between languages. Since the 1960s, explanations in historical linguistics have been revolutionised by the sociolinguistic examination of **language variation**. This is the observation that the language we use (the words and phrases we choose, the way we pronounce them and so on) varies from day to day, from situation to situation and from person to person. Language variation occurs at the level of the individual, in that each speaker employs distinct **registers** of language in different situations (formal, informal, 'motherese' and so on), and at the level of the group, in that speakers can be grouped according to **regional dialect** and **social dialect**. In the process of language change, speakers either consciously or unconsciously target the variation that already exists in the language due to social factors, selecting some variants over others and spreading them through a speech community. Language change can be (and often is) gradual, and in some cases almost imperceptible, but over time the results can be spectacular.

To see how spectacular, let's briefly examine a few changes that have taken place in English. English belongs to the **Germanic** branch of the **Indo-European family** of languages. A language family is a group of 'genetically' related languages, in the sense that they are hypothesised to have emerged from a common 'parent' language. Such relations are established on the basis of systematic correspondences in terms of words, sounds or grammar. Between the years 450 and 550 AD, several Germanic tribes from parts of modern-day Holland, Denmark and Northern Germany arrived and settled in what is now England. In doing so they pushed the native Britons, the Celts, westwards, hence the restriction of the Celtic languages (the ancestors of Cornish and Welsh) to the western peripheries of the country. Within a few centuries, the language spoken by these tribes was sufficiently distinct from the languages of continental Europe to be referred to by a new name. Texts from the period refer to the language as *Englisc*, and from around 1000 AD there is evidence that the country is referred to as *Englaland*, 'land of the Angles', one of the Germanic tribes. In a cruel twist, the displaced inhabitants, the Celts, were labelled *wealas*, meaning 'foreigners', by the invaders, which provides the derivation of the modern forms *Welsh* and *Wales*.

The English spoken in the centuries just after the arrival of the Germanic tribes is called **Old English** (or Anglo-Saxon) by historians of the language.

Old English is spectacularly different from Modern English. To get a sense of some of the differences, consider the sentences in (5) and (6):

(5) sēō   cwēn   geseah   þone   guman
     The   woman   saw   the   man

(6) se   guma   geseah   þā   cwēn
     The   man   saw   the   woman

These sentences illustrate some of the differences between Old and Modern English. Perhaps the most striking difference is the unfamiliar look of some of the words, although some of the sounds are somewhat familiar. For instance, the Old English word for 'woman', *cwēn*, has developed into the modern-day form *queen*. This is an example of a phenomenon called **narrowing**: over time a word develops a more specialised, or narrower, function. Today *queen* can only be applied to a female monarch, whereas in Old English it could be applied to all adult females.

Another striking difference is that Old English had a **case system**. Case is the morphological marking of grammatical relations like subject and object. In example (5), the subject of the sentence features a definite article 'the' marked with nominative (subject) case *sēō*, indicating that what comes next is the subject of the sentence. The definite article *þone* indicates accusative (object) case, indicating that *guman* is the object of the sentence. One consequence of the morphological flagging of subject and object is that word order was not as rigid in Old English as it is in Modern English. In Modern English, we know which expression in a sentence is the subject and which is the object by their position in the sentence: while the subject precedes the verb, the object follows it. One advantage of a case system is that the language is less reliant on word order to provide this kind of information.

Yet another difference illustrated by these sentences also concerns the definite articles: in addition to encoding case, Old English also encoded gender. While *sēō* and *se* in (5) and (6) are both nominative case forms, the former encodes feminine gender and the latter masculine gender. Similarly, while *þa* and *þone* in (5) and (6) both encode accusative case, *þa* encodes masculine gender and *þone* encodes feminine gender. In addition, observe that nouns show **case agreement** with the definite article that precedes them: the distinction between *guman* and *guma* results from case agreement.

Finally, these examples reveal another striking distinction. Some past tense verbs in Old English were marked by the prefix *ge-*, as in *geseah*, which contrasts with the modern past tense equivalent, *saw*. Historical linguistics is concerned, then, with explaining how and why Old English evolved into the version of English that we recognise today.

4.3.2 The Utterance Selection Theory of language change

In this section, we focus on a particular cognitively oriented theory of language change: the Utterance Selection Theory of language change developed by Croft (2000). The key assumption behind this approach is that languages don't change; instead, people change language through their actions. In other words, language is changed by the way people use language. In this respect, Croft's approach takes a usage-based perspective on language change. At first glance, this perspective may seem problematic. Language is a system that people use for communication. Given that humans are not telepathic, then if communication is to succeed, speaker and hearer must share a common **code** (a technical term for a single variety of a language). This means that speaker and hearer follow certain conventions in the way they use language. As we observed earlier, a convention is a regularity in behaviour which all speakers in a particular linguistic community adhere to, either consciously or unconsciously. It follows that a language is a conventional system that allows speakers to express meanings that will be recognised by others in the same linguistic community. For instance, the word *dog* is **arbitrary** in the sense that there is nothing predictable about the sounds that are used to express the lexical concept DOG in English. Other languages use different sounds (e.g. *chien* in French and *Hund* in German). However, a convention of English holds that the word *dog* refers to a particular kind of animal: the word has a conventional meaning. This means that all English speakers can use this word to refer to this animal and in so doing they are following a convention of English. In addition, strings of words can also represent conventions. For example, as we saw in Chapter 1, the idiomatic meaning of the expression *He kicked the bucket*, is 'he died' not 'a male kicked a bucket'. This is a convention of English. Similarly, the phrase: *Can you pass me the salt?* which is literally a question about someone's ability to do something, is actually understood as a request. This is also a convention of English.

If convention is so important to human language and linguistic behaviour, why does language change? If everyone is following the conventions of the language, how do languages change and what causes this change? For this to happen, someone must break a convention and this **innovation** must then undergo **propagation**, which means that the change spreads through the linguistic community and becomes established as a new convention. As we saw above, the conventions of Old English and Modern English are radically different, yet these are two varieties of the same language, separated by time but connected by the process of continuous **replication** (section 4.3.3).

According to Croft, the explanation lies in the fact that 'there cannot be a word or phrase to describe every experience that people wish to communicate' (Croft 2000: 103). In other words, language use has to be partly non-conventional if it is to express all human experience, yet it is also partly conventional in that novel

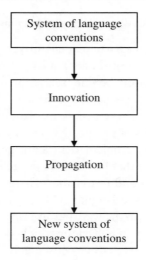

Figure 4.6 The structure of language change

uses rely upon existing aspects of language. One area in which human experience frequently outstrips the conventions of language, and thereby necessitates innovation, is the domain of technological advances. The telephone, the computer, the car and the camcorder are all inventions that have emerged relatively recently. Their emergence has necessitated the coining of new words.

Consider the word *camcorder*. This describes a hand-held camera that records moving pictures. The new word *camcorder* made use of existing conventional forms *camera* and *recorder,* and blended them to create *camcorder*. This is called a **formal blend**. Blending is a productive word formation process in which elements from two existing words are merged to provide a new word, as in the standard textbook example of *smog* from *smoke* and *fog*. Blending relies partly on convention (using existing words), but is also partly innovative, creating a new word.

By assuming the two processes of innovation and propagation, Croft's approach explicitly acknowledges that language change is both a **synchronic** and a **diachronic** phenomenon. A synchronic view of language examines the properties of language at a specific discrete point in time: innovation occurs at a specific point in time. A diachronic view of language considers its properties over a period of time: propagation occurs over a period of time, in that an innovation sometimes requires centuries to become fully **conventionalised**. Figure 4.6 illustrates the structure of language change. A (set of) convention(s) is changed when the convention is first broken: this is innovation. If this innovation is propagated throughout a linguistic community, it can become established as a convention, and this changes the language. The diagram in Figure 4.7 captures the view that language change involves synchronic and

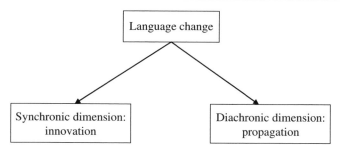

Figure 4.7 The synchronic and diachronic dimensions of language change

diachronic dimensions (in contrast to some theories of language change, which only consider propagation as language change).

4.3.3 The Generalised Theory of Selection and the Theory of Utterance Selection

The theory of Utterance Selection takes its inspiration from neo-Darwinian evolutionary theory, particularly the application of theories of biological evolution to sociocultural constructs like scientific theories. David Hull, a philosopher of science, has attempted to draw out the similarities between various versions of neo-Darwinian evolutionary theory and has developed what he calls a **Generalised Theory of Selection**. Because Croft draws upon Hull's Generalised Theory of Selection in developing his Theory of Utterance Selection, we begin by outlining four key ideas from Hull's theory.

The key concepts in the Generalised Theory of Selection are: (1) replicator; (2) interactor; (3) selection; and (4) lineage. A **replicator** is an entity whose structure can be passed on in successive replications. An example of a replicator from biology is the gene, which contains material that is passed on to offspring through procreation. Crucially, however, the process of **replication** may introduce differences, which result in a slightly different structure from the original replicator. Changes introduced during ongoing replication are cumulative, and result in a replicator that, through successive replications, can have quite different properties from the original replicator. For instance, genes are contained in DNA sequences. Because errors, known as mutations, can occur during the process of replication, new DNA sequences can be replicated. This process is known as **altered replication** and contrasts with **normal replication** which copies the original replicator exactly. An **interactor** is an entity that interacts with its environment in such a way that replication occurs. An example of an interactor from biology is an individual organism.

Selection is the process whereby the extinction or proliferation of interactors results in the differential perpetuation of replicators. For example, if

Table 4.2 Key ideas in the Generalised Theory of Selection (Croft 2000)

Replicator	An entity possessing structure that can be passed on
Replication	The process of copying a replicator
Normal replication	The process of replication resulting in an exact copy
Altered replication	The process of replication whereby the resulting replicator is different from the replicator it copies
Interactor	An entity that interacts with its environment so that replication occurs
Selection	The process whereby replicators are differentially perpetuated (i.e. some replicators are more successful than others)
Lineage	An entity that persists over time due to replication

a particular individual or set of individuals dies out, then the corresponding **gene pool**, the set of replicators, is lost. Finally, **lineage** relates to the persistence of an entity over time, due either to normal or to altered replication. An example of this idea from biology is a species. Table 4.2 summarises these ideas.

Croft's Theory of Utterance Selection applies these notions to language change. However, before looking in detail at what the counterparts of each of these constructs might be in the domain of language, it is important to address the motivations for treating language change in terms of a theory of generalised selection. Recall that cognitive linguists view language change as the result of language use, in particular the result of interaction between interlocutors. As a consequence, there are selectional pressures exerted on linguistic conventions, because language is a system in use that changes as a response to the new uses to which it is put. From this perspective, it makes perfect sense to apply an evolutionary framework to language change.

Next, let's consider what the linguistic counterparts of the constructs illustrated in Table 4.2. might be. We begin with the idea of a replicator. In biology, the gene represents a replicator which is embedded in strands of DNA. In the Theory of Utterance Selection, a replicator is an element of language realised in an utterance. Recall that we defined an utterance as a usage event, each utterance representing a unique speech event bounded in space and time. From this perspective, even if a language user were to repeat an utterance twice, we would still be looking at two distinct utterances. The elements of language that are realised in utterances, and that can therefore count as replicators, include words, morphemes and grammatical constructions. Croft calls these linguistic replicators **linguemes**. Crucially, just as each utterance is a unique event, so is each lingueme.

The linguemes in any given utterance are usually associated with a conventional meaning. Normal replication occurs when linguemes are used in accordance with the conventions of the language. Altered replication, which is essentially innovation, occurs when an utterance provides a meaning that breaks

Table 4.3 Terms for Generalised Theory and linguistic equivalents (Croft 2000)

Replicator		Lingueme	
Interactor	Normal replication	Language user	Conforming to linguistic conventions
	Altered replication		Not conforming to linguistic conventions (innovation)
Selection		Propagation	
Lineage		Etymology	

with the conventions of the language. In other words, altered replication (innovation) occurs when there is a disjunction between the conventional **form-meaning mapping** within an utterance. We discuss this in more detail below.

In the Theory of Utterance Selection, the interactors are the language users. Of course, language change does not depend solely on a group of speakers dying or being more successful at breeding, although language death can be caused by an entire speech community dying out. More commonly, interactors play a role in the selection of utterances by virtue of the various social and communication networks within which they interact, a point to which we return in more detail below.

In terms of language change, just as altered replication can be equated with innovation, so can selection be equated with propagation. The selection and use of a particular utterance containing a particular lingueme or set of linguemes can propagate the altered replication (the innovation), enabling it to diffuse through a linguistic community. In time, this innovation becomes established as a new convention.

Finally, we turn to the concept of lineage. In terms of language change, this relates to **etymology**. Etymology is the study of the history of linguistic units, particularly words; **etymologists** are linguists who study the historical chain of developments affecting word form and meaning. Table 4.3 summarises the notions discussed in the Generalised Theory of Selection and its equivalents in linguistic theory.

4.3.4 Causal mechanisms for language change

In this section, we consider the social mechanisms that give rise to replication, resulting in normal replication, altered replication (innovation), and selection (propagation). Because the Theory of Utterance Selection is usage-based, we are concerned with utterances (usage events), which are embedded in linguistic interaction. For this reason, we require a theory that explains the nature of, and the motivations for, the kinds of interactions that language users engage in.

Recall that the usage-based view of language change assumes that these interactions preserve language stability (by following linguistic conventions), bring about innovation (by breaking linguistic conventions) and give rise to propagation due to the differential selection of certain kinds of linguemes by language users in a sociocultural context, resulting in the establishment of new conventions. In order to account for human behaviour in linguistic interaction, Croft adopts a model proposed by Rudi Keller (1994), which describes linguistic interaction in terms of a number of maxims. The hypermaxims and maxims discussed below are therefore drawn from Keller's work. Note, however, that while we have numbered the maxims for our purposes, these numbers do not derive from Keller's work.

Keller views linguistic behaviour as a form of social action, in keeping with functional approaches to language. He proposes a number of maxims in order to model what language users are doing when they use language. The maxims described here are in service of a more general principle, which Keller (1994) calls a hypermaxim. In Keller's terms, this is the **hypermaxim of linguistic interaction** and can be stated as follows:

(7) **Hypermaxim**: 'Talk in such a way that you are most likely to reach the goals that you set yourself in your communicative enterprise'. (Keller 1994: 106)

Croft argues that by observing the various maxims in the service of fulfilling the hypermaxim of linguistic interaction, speakers facilitate normal replication, altered replication and selection, and thus bring about language change.

Normal replication

As we have seen, a theory of language change must be able to account for the relative stability of language as well as offering an explanation for how and why language changes. Recall that convention is crucial to the success of language as a communicative system. Croft argues that normal replication, which enables stability, arises from speakers following the maxim stated in (8):

(8) **Maxim 1**: 'Talk in such a way that you are understood'. (Keller 1994: 94)

Of course, this maxim states the rather obvious but no less important fact that speakers normally intend to be understood in linguistic interaction. In order to be understood, speakers follow the conventions of the language. Hence, the unintended consequence of observing Maxim 1 is normal replication: stability in language.

Altered replication

Croft argues that innovation arises because, in addition to wanting to be understood, speakers also have a number of other goals. These are summarised by the series of maxims stated in (9)–(12).

(9) **Maxim 2**: 'Talk in such a way that you are noticed'. (Keller 1994: 101)

(10) **Maxim 3**: 'Talk in such a way that you are not recognizable as a member of the group'. (Keller 1994: 101)

(11) **Maxim 4**: 'Talk in an amusing, funny, etc. way'. (Keller 1994: 101)

(12) **Maxim 5**: 'Talk in an especially polite, flattering, charming, etc. way'. (Keller 1994: 101)

These maxims relate to the 'expressive' function of language. In other words, in order to observe the hypermaxim (achieve one's goals in linguistic interaction), speakers might follow Maxims (2)–(5). However, in following these maxims, the speaker may need to break the conventions of the language. As a consequence, innovation or altered replication takes place. We will look at some specific examples below. A further maxim posited by Keller, which may be crucial in altered replication, is stated in (13):

(13) **Maxim 6**: 'Talk in such a way that you do not expend superfluous energy'. (Keller 1994: 101)

This maxim relates to the notion of economy. The fact that frequently used terms in a particular linguistic community are often shortened may be explained by this maxim. Croft provides an example from the community of Californian wine connoisseurs. While in the general English-speaking community wine varieties are known by terms like *Cabernet Sauvignon*, *Zinfandel* and *Chardonnay*, in this speech community, where wine is a frequent topic of conversation, these terms have been shortened to *Cab*, *Zin* and *Chard*. As Croft (2000: 75) observes, 'The energy expended in an utterance becomes superfluous, the more frequently it is used, hence the shorter the expression for it is likely to be(come).' While some theories of language treat economy in terms of mental representation (as a function of psycholinguistic processing costs), Croft argues that Maxim 6, which essentially relates to economy, actually relates to a speaker's interactional goals in a communicative context. In other words, Maxim 6 can only be felicitously followed when it doesn't contravene other maxims, like Maxim 1. It is only in a context involving wine connoisseurs, for instance, that the diminutive forms do not **flout** Maxim 1 and are therefore felicitous.

The observation of the maxims we have considered so far is **intentional**: deliberate on the part of the language user. However, there are a number of mechanisms resulting in altered replication that are **non-intentional**. These processes are nevertheless grounded in usage events. We briefly consider these here.

Altered replication: sound change
The first set of non-intentional mechanisms relates to regular sound change. Sound change occurs when an allophone, the speech sound that realises a phoneme, is replicated in altered form. Because the human articulatory system relies on a highly complex motor system in producing sounds, altered replication can occur through 'errors' in articulation. In other words, the articulatory system can **overshoot** or **undershoot** the sound it is attempting to produce, giving rise to a near (slightly altered) replication. Of course, it seems unlikely that an individual's speech error can give rise to a change that spreads throughout an entire linguistic community, but the famous sociolinguist William Labov (1994) suggests that mechanisms like overshoot or undershoot can give rise to vowel **chain shifts** in languages.

A chain shift involves a series of sound changes that are related to one another. This typically involves the shift of one sound in **phonological space** which gives rise to an elaborate chain reaction of changes. Chain shifts are often likened to a game of musical chairs, in which one sound moves to occupy the place of an adjacent pre-shift sound, which then has to move to occupy the place of another adjacent sound in order to remain distinct, and so on. The net effect is that a series of sounds move, forming a chain of shifts and affecting many of the words in the language.

A well known example of a chain shift is the Great English Vowel Shift, which took effect in the early decades of the fifteenth century and which, by the time of Shakespeare (1564–1616), had transformed the sound pattern of English. The Great Vowel Shift affected the seven long vowels of **Middle English,** the English spoken from roughly the time of the Norman conquest of England (1066) until about half a century after the death of Geoffrey Chaucer (around 1400). What is significant for our purposes is that each of the seven long vowels was raised, which means that they were articulated with the tongue higher in the mouth. This corresponds to a well known tendency in vowel shifts for long vowels to rise, while short vowels fall.

Labov (1994) suggests that chain shifts might be accounted for in purely articulatory terms. In other words, the tendency for long vowels to undergo raising in chain shifts might be due to articulatory pressure for maintaining length, which results in the sound being produced in a higher region of the mouth. This is the phenomenon of overshoot. Undershoot applies to short vowels, but in the opposite direction (lowering). Crucially, this type of mechanism is non-intentional because it does not arise from speaker goals but from purely mechanical system-internal factors.

Another non-intentional process that results in sound change is **assimilation**. Croft, following suggestions made by Ohala (1989), argues that this type of sound change might be accounted for not by articulatory (sound-producing) mechanisms, but by non-intentional auditory (perceptual) mechanisms. Assimilation is the process whereby a sound segment takes on some of the characteristics of a neighbouring sound. For instance, many French vowels before a word-final **nasal** have undergone a process called **nasalisation**. Nasal sounds – like [m] in *mother*, [n] in *naughty* and the sound [ŋ] at the end of *thing* – are produced by the passage of air through the nasal cavity rather than the oral cavity. In the process of nasalisation, the neighbouring vowel takes on this sound quality, and is articulated with nasal as well as oral airflow. For instance, French words like *fin* 'end' and *bon* 'good' feature nasalised vowels. The consequence of this process is that in most contexts the final nasal segment [n] is no longer pronounced in Modern French words, because the presence of a nasalised vowel makes the final nasal sound redundant. Notice that the spelling retains the 'n', reflecting pronunciation at an earlier stage in the language before this process of sound change occurred.

The process that motivates assimilation of this kind is called **hypocorrection**. In our example of hypocorrection, the vowel sound is reanalysed by the language user as incorporating an aspect of the adjacent sound, here the nasal. However, this process of reanalysis is non-intentional: it is a covert process that does not become evident to speakers until the nasalisation of the vowel results in the loss of the nasal sound that **conditioned** the reanalysis in the first place.

Altered replication: form-meaning reanalysis
Altered replication is not restricted to sound change, but can also affect symbolic units. Recall that symbolic units are form-meaning pairings. Language change that affects these units can be called **form-meaning reanalysis** (Croft uses the term **form-function reanalysis**). Form-meaning reanalysis involves a change in the mapping between form and meaning. Consider examples (14) and (15).

(14) I'm going to the library.

(15) I'm going to be an astronaut (when I grow up).

What concerns us here is the meaning of the *be going to* construction. In example (14), this expression describes a physical path of motion, while in (15) it describes future time, which is the more recent meaning associated with this construction. This is an example of a type of form-meaning reanalysis known as **grammaticalisation**, an issue to which we return in detail in Part III of the book (Chapter 21). As we noted above, the term **reanalysis** does not imply a deliberate or intentional process. Instead, the reanalysis is non-intentional, and derives from pragmatic (contextual) factors.

Selection

We now turn to the social mechanisms responsible for selection, and look at how the innovation is propagated through a linguistic community so that it becomes conventionalised. In the Theory of Utterance Selection, mechanisms of selection operate over previously used variants. One such mechanism proposed by Keller is stated in (16).

(16) **Maxim 7**: 'Talk like the others talk'. (Keller 1994: 100)

Croft argues that this maxim is closely related to the theory of **accommodation** (Trudgill 1986). This theory holds that interlocutors often tend to accommodate or 'move towards' the linguistic conventions of those with whom they are interacting in order to achieve greater rapport or **solidarity**. A variant of Maxim 7 posited by Keller is stated in (17).

(17) **Maxim 8**: 'Talk in such a way that you are recognized as a member of the group'. (Keller 1994: 100)

This maxim elaborates Maxim 7 in referring explicitly to group identity. From this perspective, the way we speak is an **act of identity**, as argued by LePage and Tabouret-Keller (1985). In other words, one function of the language we use is to identify ourselves with a particular social group. This means that sometimes utterances are selected that diverge from a particular set of conventions as a result of the desire to identify with others whose language use is divergent from those conventions.

Table 4.4 summarises the various mechanisms for language change and language stability that have been described in this section. Of course, this discussion does not represent an exhaustive list of the mechanisms that are involved in language change, but provides representative examples.

In sum, we have seen that the Theory of Utterance Selection is a usage-based theory of language change because it views language as a system of use governed by convention. Language change results from breaking with convention and selecting some of the new variants created as a result of this departure. While the propagation of new forms can be due to intentional mechanisms relating to the expressive functions associated with language, it also involves non-intentional articulatory and perceptual mechanisms. Finally, the selection

Table 4.4 Causal mechanisms involved in language stability and change (Croft 2000)

Normal replication	Altered replication (innovation)	Selection (propagation)
Follow conventions of the language *Maxim 1*: Talk in such a way that you are understood	**Be expressive** *Maxim 2*: Talk in such a way that you are noticed *Maxim 3*: Talk in such a way that you are not recognizable as a member of the group *Maxim 4*: Talk in an amusing way *Maxim 5*: Talk in an especially polite, flattering or charming way **Be economical** *Maxim 6*: Talk in such a way that you do not expend superfluous energy **Non-intentional mechanisms** (1) Sound change: articulatory factors (over/undershoot) or auditory factors (hypocorrection) (2) Reanalysis of form-meaning mapping	**Accommodation** *Maxim 7*: Talk like the others talk **Act of identity** *Maxim 8*: Talk in such a way that you are recognized as a member of the group **Prestige** Adoption of changes as a result of aspiring to a social group

of variants is due to sociolinguistic processes such as accommodation, identity and prestige.

4.4 The usage-based approach to language acquisition

So far in this chapter, we have seen that a usage-based approach views grammar as a system derived from and grounded in utterances. According to this view, it is from these usage events that the abstracted schemas – the constructions that make up our knowledge of language – arise. We have also explored a usage-based theory of language change. In this section we turn our attention in more detail to the question of *how* linguistic units are derived from patterns of language use by exploring a usage-based account of child language acquisition. In particular, we focus on the acquisition of meaning and grammar rather than phonological acquisition. We base our discussion on the theory proposed by developmental psycholinguist Michael Tomasello in his (2003) book *Constructing a Language*.

A usage-based account of language acquisition posits that language learning involves 'a prodigious amount of actual learning, and tries to minimize the postulation of innate structures specific to language' (Langacker 2000: 2). In this approach to language acquisition, the burden of explanation is placed upon the acquisition of linguistic units rather than upon Universal Grammar. While cognitive linguists do not deny that humans are biologically pre-specified to acquire language, they reject the hypothesis that there exists a specialised and innate cognitive system that equips us for linguistic knowledge. Instead, cognitive linguists argue that humans employ generalised sociocognitive abilities in the acquisition of language.

4.4.1 Empirical findings in language acquisition

The empirical study of first language acquisition is known as **developmental psycholinguistics**. Since the early studies in developmental psycholinguistics such as Braine (1976) and Bowerman (1973), one of the key cross-linguistic findings to have emerged is that infants' earliest language appears to be **item-based** rather than **rule-based**: infants first acquire specific item-based units (words), then more complex item-based units (pairs and then strings of words), before developing more abstract grammatical knowledge (grammatical words and morphemes, complex sentence structures and so on). Cognitive linguists argue that this provides evidence for a usage-based theory of language acquisition, and that more recent empirical findings in developmental psycholinguistics, particularly since the late 1980s and early 1990s, support this view.

Let's look in more detail at what it means to describe early language acquisition as item-based. When a child first produces identifiable units of language at around the age of twelve months (the **one-word stage**), these are individual lexical items. However, these lexical items do not equate with the corresponding adult forms in terms of function. Instead, the child's first words appear to be equivalent to whole phrases and sentences of adult language in terms of communicative intention. For this reason, these early words are known as **holophrases**. These can have a range of goal-directed communicative intentions. In a study of his daughter's early language, Tomasello found that his daughter's holophrases fulfilled a number of distinct functions, which are illustrated in Table 4.5.

Secondly, the item-based nature of first language acquisition is also revealed at the **two-word stage**, which emerges at around eighteen months. After holophrases, children begin to produce multi-word expressions. These are more complex expressions than holophrases in that they contain two or more lexical items. Some of these early multi-word utterances are of the type *ball table*, when a child sees a ball on the table and concatenates two units of equal status (here nouns) in order to produce a more linguistically complex utterance. However, the majority of early multi-word utterances are not like this. Instead, many early multi-word utterances exhibit **functional asymmetry**. This means that the expressions contain a relatively stable element with 'slots' that can be filled by other lexical items. In other words, early multi-word utterances, rather than containing two or more words of equal status, tend to be 'built'

Table 4.5 Holophrases (Tomasello 1992) (adapted from Tomasello 2003: 36–7)

Holophrase	Communicative function
rockin	*First use*: while rocking in a rocking chair
	Second use: as a request to rock in a rocking chair
	Third use: to name the rocking chair
phone	*First use*: in response to hearing the telephone ring
	Second use: to describe activity of 'talking' on the phone
	Third use: to name the phone
	Fourth use: as a request to be picked up in order to talk on the phone
towel	*First use*: using a towel to clean a spill
	Second use: to name the towel
make	*First use*: as a request that a structure be built when playing with blocks
mess	*First use*: to describe the state resulting from knocking down the blocks
	Second use: to indicate the desire to knock down the blocks

Table 4.6 Examples of utterance schemas (based on Tomasello 2003: 66)

Here's the X?	I'm X-ing it
I wanna X	Mommy's X-ing it
More X	Let's X it
It's a X	I X-ed it
There's a X	
Put X here	
Throw X	
X gone	
X here	
X broken	
Sit on the X	
Open X	

around a functionally more salient and stable word. Tomasello calls expressions like these **utterance schemas** (which are also known as **pivot schemas**). Like holophrases, utterance schemas reflect the communicative intention of an equivalent adult utterance, but represent the acquisition of more schematic knowledge, allowing a wider range of lexical items to fill the slots. The obligatory element is known as the **pivot**. Representative examples of utterance schemas are provided in Table 4.6. In this table, X represents the slot that is 'filled in' and corresponds to a word that describes an entity (noun), shown in the left column, or an action (verb), shown in the right column. (There is no significance to the order in which these utterances are listed in the table.) Because most utterance schemas appear to revolve around verb-like elements, Tomasello (1992) labelled these units **verb-island constructions**. Only later do these verb-island constructions develop into the more familiar constructions of adult-like speech.

Tomasello argues that the third way in which early acquisition is item-based rather than rule-based is in its lack of innovation. In other words, early language use is highly specific to the verb-island constructions that the child has already formed and resists innovation. Tomasello argues that this is because early utterance schemas are highly dependent on what children have actually heard rather than emerging from abstract underlying rules. In an experiment carried out by Tomasello and Brooks (1998), two to three year old children were exposed to a nonsense verb *tamming* (meaning 'rolling or spinning') used in an intransitive frame. This is illustrated in example (18).

(18) The sock is tamming.

This usage is intransitive because the verb *tamming* does not have an object. Children were then prompted to use *tamming* in a transitive frame, with an object. One such prompt was a picture in which a dog was causing an object to 'tam'. The question presented to the children was *What is the doggie doing?* However, children were found to be poor at producing *tamming* in a transitive frame (e.g. *He's tamming the car*). Moreover, they were also found in a further study to be poor at understanding the use of *tamming* in a transitive frame. Tomasello draws two conclusions from these findings: (1) two and three year olds were poor at the *creative* use of the novel verb *tamming*; and (2) early utterance schemas are highly dependent on contexts of use in which they have been heard. Tomasello argues that it is only later, as children acquire more complex and more abstract constructions, that they come to be more competent in the creative use of language.

4.4.2 The cognitive view: sociocognitive mechanisms in language acquisition

As we have seen, the fundamental assumption of cognitive approaches to grammar is the **symbolic thesis**: the claim that the language system consists of symbolic assemblies, or conventional pairings, of form and meaning. According to Michael Tomasello and his colleagues, when children acquire a language, what they are actually doing is acquiring constructions: linguistic units of varying sizes and increasing degrees of abstractness. As the complexity and abstractness of the units increases, linguistic creativity begins to emerge. According to this view, the creativity exhibited by young children in their early language happens because they are 'constructing utterances out of various already mastered pieces of language of various shapes and sizes, and degrees of internal structure and abstraction – in ways appropriate to the exigencies of the current usage event' (Tomasello 2003: 307). This view of language acquisition is called **emergentism**, and stands in direct opposition to nativism, the position adopted in generative models. In other words, Tomasello argues that the process of language acquisition involves a huge amount of learning. Recall that cognitive linguists reject the idea that humans have innate cognitive structures that are specialised for language (the Universal Grammar Hypothesis). In light of that fact, we must address the question of what cognitive abilities children bring to this process of language acquisition.

Recent research in cognitive science reveals that children bring a battery of sociocognitive skills to the acquisition process. These cognitive skills are **domain-general**, which means that they are not specific to language but relate to a range of cognitive domains. According to cognitive linguists, these skills facilitate the ability of humans to acquire language. Tomasello argues that there are two kinds of general cognitive ability that facilitate the acquisition of language: (1) **pattern-finding ability**; and (2) **intention-reading ability**.

The pattern-finding ability is a general cognitive skill that enables humans to recognise patterns and perform 'statistical' analysis over sequences of perceptual input, including the auditory stream that constitutes spoken language. Tomasello argues that **pre-linguistic infants** – children under a year old – employ this ability in order to abstract across utterances and find repeated patterns that allow them to construct linguistic units. It is this pattern-finding ability that underlies the abstraction process assumed by Langacker, which we discussed earlier (section 4.2.1).

The evidence for pattern-finding skills is robust and is apparent in pre-linguistic children. For instance, Saffran, Aslin and Newport (1996) found that at the age of eight months infants could recognise patterns in auditory stimuli. This experiment relied on the **preferential looking technique**, which is based on the fact that infants look more at stimuli with which they are familiar. Saffran *et al.* presented infants with two minutes of synthesised speech consisting of the four nonsense words *bidaku*, *padoti*, *golabu* and *tupiro*. These nonsense words were sequenced in different ways so that infants would hear a stream of repeated words such as: *bidakupadotigolabubidakutupiropadoti*..., and so on. Observe that each of these words consisted of three syllables. Infants were then exposed to new streams of synthesised speech, which were presented at the same time, and which were situated to the left and the right of the infant. While one of the new recordings contained 'words' from the original, the second recording contained the same syllables, but in different orders, so that none of the 'words' *bidaku*, *padoti*, *golabu* or *tupiro* featured. The researchers found that the infants consistently preferred to look towards the sound stream that contained some of the same 'words' as the original. This shows that pre-linguistic infants are able to recognise patterns of syllables forming 'words' in an auditory stream and provides evidence for the pattern-finding ability.

Further research (see Tomasello 2003 for a review) demonstrates that infant pattern-finding skills are not limited to language. Researchers have also found

Table 4.7 Human pattern-finding skills (Tomasello 2003)

Human pattern finding abilities

The ability to relate similar objects and events, resulting in the formation of perceptual and conceptual categories for objects and events. Category formation aids recognition of events and objects.
The ability to form sensorimotor schemas based on recurrent perception of action. This is associated with the acquisition of basic sensorimotor skills, and the recognition of actions as events, such as crawling, walking, picking up an object, and so on.
The ability to perform distributional analysis on perceptual and behavioural sequences. This allows infants to identify and recognise recurrent combinations of elements in a sequence and thus identify and recognise sequences.
*The ability to create **analogies** (recognition of similarity) between two or more wholes,* (including utterances), *based on the functional similarity of some of the elements in the wholes.*

that infants demonstrate the same skills when the experiment is repeated with non-linguistic tone sequences and with visual, as opposed to auditory, sequences. Some of the key features associated with the human pattern-finding ability are summarised in Table 4.7.

Finally, this pattern-finding ability appears not to be limited to humans but is also apparent in our primate cousins. For instance, Tamarin monkeys demonstrate the same pattern recognition abilities when exposed to the same kinds of auditory and visual sequencing experiments described above for human infants. Of course, if we share the pattern-finding ability with some of the non-human primates, and if these pattern-finding skills facilitate the acquisition of language, we need to work out why only humans acquire and produce language.

According to Tomasello, the answer lies in the fact that the pattern-finding skills described above are necessary but not sufficient to facilitate language acquisition. In addition, another set of skills are required: **intention-reading abilities**. While pattern-finding skills allow pre-linguistic infants to begin to identify linguistic units, the use of these units requires intention-reading skills, which transform linguistic stimuli from statistical patterns of sound into fully fledged linguistic symbols. In other words, this stage involves 'connecting' the meaning to the form, which gives rise to the form-meaning pairing that make up our knowledge of language. Only then can these linguistic sounds be used for communication. This process takes place when, at around a year old, infants begin to understand that the people around them are **intentional agents**: their actions are deliberate and their actions and states can be influenced. The emergence of this understanding allows infants to 'read' the intentions of others. Some of the features that emerge from this intention-reading ability are summarised in Table 4.8.

Like pattern recognition skills, these intention-reading skills are domain-general. Unlike pattern recognition skills, they are **species-specific**. In other

Table 4.8 Human intention-reading abilities (Tomasello 2003)

The ability to coordinate or share attention, as when an infant and adult both attend to the same object.
The ability to follow attention and gesturing, as when an infant follows an adult's gesture or gaze in order to attend to an object.
The ability to actively direct attention of others, such as drawing attention to a particular object or event, for example by pointing.
The ability of culturally (imitatively) learning the intentional actions of others, such as imitating verbal cues in order to perform intentional actions.

words, only humans possess a complete set of these abilities. The evidence is equivocal as to whether our nearest primate cousins, for instance chimpanzees, recognise conspecifics (members of the same species) as intentional agents. However, Tomasello (1999) argues that the answer is no. Moreover, these intention-reading skills begin to emerge just before the infant's first birthday. Tomasello argues that the emergence of holophrases shortly after the infant's first year is directly correlated with the emergence of these skills.

Tomasello argues that our intention-reading abilities consist of three specific but interrelated phenomena: (1) **joint attention frames**; (2) the understanding of **communicative intentions**; and (3) **role reversal imitation**, which is thought to be the means by which human infants acquire cultural knowledge. According to this view, language acquisition is contextually embedded and is a specific kind of cultural learning.

A joint attention frame is the common ground that facilitates cognition of communicative intention and is established as a consequence of a particular goal-directed activity. When an infant and an adult are both looking at and playing with a toy, for example, the attention frame consists of the infant, the adult and the toy. While other elements that participate in the scene are still perceived (such as the child's clothes or other objects in the vicinity), it is this **triadic relationship** between child, adult and toy that is the joint focus of attention.

The second important aspect of intention-reading involves the recognition of communicative intention. This happens when the child recognises that others are intentional agents and that language represents a special kind of intention: the intention to communicate. For example, when the adult says *teddy bear*, the adult is identifying the toy that is the joint focus of attention and is employing this linguistic symbol to express the intention that the child follow the attention of the adult. This is represented in Figure 4.8, where the unbroken arrow represents the communicative intention expressed by the adult. The dotted arrows represent shared attention.

Finally, Tomasello argues that intention-reading skills also give rise to role reversal imitation. Infants who understand that people manifest intentional

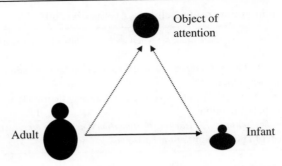

Figure 4.8 The use of a linguistic symbol in a triadic relationship expressing a communicative intention (adapted from Tomasello 2003: 29)

behaviour may attend to and learn (by imitation) the behavioural means that others employ to signal their intentional state. For example, the child may imitate the use of the word *teddy bear* by an adult in directing attention to an object. Tomasello (2003) cites two studies that support the view that infants have a good understanding of the intentional actions of others and can imitate their behaviour. In an experiment reported by Meltzoff (1995), two groups of eighteen-month-old infants were shown two different actions. In one, an adult successfully pulled the two pieces of an object apart. In a second, an adult tried but failed to pull the two pieces apart. However, both sets of infants, when invited to perform the action they had witnessed, successfully pulled the two pieces apart. Meltzoff concludes that even the infants who had not witnessed pieces successfully pulled apart had understood the adult's intention.

In the second experiment, Carpenter, Akhtar and Tomasello (1998) exposed sixteen-month-old infants to intentional and 'accidental' actions. The intentional action was marked vocally by the expression *there!* while the 'accidental' action was marked by *whoops!* The infants were then invited to perform the actions. The children performed the intentional action more frequently than the 'accidental' action. Carpenter *et al.* concluded that this was because the children could distinguish intentional actions from non-intentional ones, and that it is these intentional actions that they attempt to reproduce. In conclusion, Tomasello (2003: 291) claims that language acquisition involves both 'a uniquely cognitive adaptation for things cultural and symbolic (intention reading) and a primate-wide set of skills of cognition and categorization (pattern finding)'.

4.4.3 Comparing the generative view of language acquisition

In this section, we compare the usage-based account of language acquisition with the **nativist** view that is assumed within the generative framework developed by Chomsky. This comparison is important because, in many respects, the usage-based view and the nativist view stand in direct opposition to one another. Furthermore, Chomsky's ideas were influential among developmental psycholinguists, particularly during the 1960s and 1970s, and are sometimes presented as the 'standard' view of language acquisition in many contemporary linguistics textbooks. More recently, cognitive theories of child language acquisition have been developed partly in response to Chomsky's claims. We look in more detail at the nativist hypothesis and the linguistic modularity hypothesis, and at the cognitive response to these hypotheses. We then look at alternative interpretations of empirical findings in language acquisition and, finally, consider localisation of linguistic function in the brain.

The nativist hypothesis

Until the 1960s, the main influence on developmental psychology was the theory of **behaviourism**. This is the doctrine that learning is governed by inductive reasoning based on patterns of association. Perhaps the most famous example of associative learning is the case of Pavlov's dog. In this experiment a dog was trained to associate food with a ringing bell. After repeated association, the dog would salivate upon hearing the bell. This provided evidence, the behaviourists argued, that learning is a type of stimulus–response behaviour. The behaviourist psychologist B. F. Skinner (1904–90), in his 1957 book *Verbal Behavior*, outlined the behaviourist theory of language acquisition. This view held that children learnt language by imitation and that language also has the status of stimulus–response behaviour conditioned by positive reinforcement.

In his famous 1959 review of Skinner's book, Chomsky argued, very persuasively, that some aspects of language were too abstract to be learned through associative patterns of the kind proposed by Skinner. In particular, Chomsky presented his famous argument, known as the **poverty of the stimulus** argument, that language was too complex to be acquired from the impoverished input or stimulus to which children are exposed. He pointed out that the behaviourist theory (which assumes that learning is based on imitation) failed to explain how children produce utterances that they have never heard before, as well as utterances that contain errors that are not present in the language of their adult caregivers. Furthermore, Chomsky argued, children do not produce certain errors that we might expect them to produce if the process of language acquisition were not rule-governed. Chomsky's theory was the first **mentalist** or cognitive theory of human language, in the sense that it attempted to explore the psychological representation of language and to integrate explanations of human language with theories of human mind and cognition. The **poverty of the stimulus** argument led Chomsky to posit that there must be a biologically predetermined ability to acquire language which, as we have seen, later came to be called Universal Grammar.

Tomasello (1995) argues that there are a number of significant problems with this hypothesis. Firstly, Tomasello argues that Chomsky's argument for a Universal Grammar, which was based on his argument from poverty of the stimulus, took the form of a logical 'proof'. In other words, it stemmed from logical reasoning rather than from empirical investigation. Furthermore, Tomasello argues, the poverty of the stimulus argument overlooks aspects of the input children are exposed to that would restrict the kinds of mistakes children might 'logically' make.

For instance, if children were employing the associative or inductive learning strategies proposed by the behaviourists then, as Chomsky pointed out, we might expect them to make mistakes in question formation. For example, based on data like the sentences in (19), children might posit the rule in (20) as part of the inductive process.

(19) a. The man is bald.
　　 b. Is the man bald?

(20) Rule for question formation
　　 Move the verb to the front in the corresponding declarative sentence.

Furthermore, given the data in (21), we might expect children to produce sentences like (22a), which is formed by moving a verb to the front of the sentence. The underscore shows the position of the verb in the corresponding declarative sentence. However, as Chomsky pointed out, children do not make errors like these, despite the absence of any direct evidence that such constructions are not well-formed, and despite the fact that constructions like (22b) are rather rare in 'motherese' or child-directed speech. Despite this, children produce examples like (22b), which rests upon the unconscious knowledge that the first *is* in (21) is 'buried' inside a phrasal unit (bracketed).

(21) [The man who is running] is bald.

(22) a. *Is the man who _____ running is bald?
　　 b. Is the man who is running _____ bald?

According to Chomsky, children must have some innate knowledge that prohibits sentences like (22a) but permits sentences like (22b). According to Tomasello, the problem with this argument is that, in the input children are exposed to, they do not hear the relative pronoun *who* followed by an *-ing* form. In other words, they do have the evidence upon which to make the 'right' decision, and this can be done by means of pattern-finding skills.

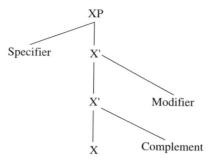

Figure 4.9 The X-bar approach to phrase structure

Tomasello's second argument relates to the nature of the learning skills and abilities children bring with them to the learning process. It has now been established beyond dispute that children bring much more to this task than the inductive learning strategies posited by the behaviourists, which Chomsky demonstrated in 1959 to be woefully inadequate for the task of language acquisition. In the intervening years, research in cognitive science has revealed that infants bring with them an array of cognitive skills, including categorisation and pattern-finding skills, which emerge developmentally and are in place from at least seven months of age. In addition, children also develop an array of sociocognitive (intention-reading) skills, which emerge before the infant's first birthday. On the basis of these facts, there is now a real alternative to the nativist hypothesis.

The third argument that Tomasello raises relates to the notion of language universals. In the 1980s Chomsky proposed a theory of Universal Grammar called the **Principles and Parameters** approach. According to this approach, knowledge of language consists of a set of universal principles, together with a limited set of parameters of variation, which can be set in language-specific ways based on the input received. From this perspective, linguistic differences emerge from parameter setting, while the underlying architecture of all languages is fundamentally similar.

For example, one linguistic universal in the principles and parameters model is the **X-bar schema**. This is a small set of category neutral rules that is argued to underlie the phrase structure of the world's languages. This idea is illustrated in Figure 4.9. In this diagram, X is a variable that can be instantiated by a word of any class, and P stands for phrase. X represents the **head** of the phrase, which projects the 'identity' of the phrase. The **specifier** contains unique elements that occur at one of the 'edges' of the phrase, and the **complement** is another phrasal unit that completes the meaning of the head. A **modifier** adds additional optional information. The name 'X-bar' relates to the levels between head (X) and phrase (XP), which are labelled X' to show that

Table 4.9 Phrase structures in English

Phrase	Specifier	Head	Complement	Modifier
Noun phrase	that	designer	of time machines	in the shed
Verb phrase	Lily	loves	George	distractedly
Adjective phrase	very	fond	of him	
Preposition phrase	right	over	the road	

they have the same categorial status (word class) as X, but are somewhere between word and phrase.

Table 4.9 provides some examples of phrase structures in English that could be built out of this basic structure.

Notice that some of the cells in Table 4.9 are empty. The idea behind the X-bar model is that only the head is obligatory in the phrase, although each individual head may bring with it some requirements of its own for which this structure can be exploited. For example, a transitive verb will require a complement (object), while an intransitive verb will not. Another important feature of this model is that while hierarchical relations between head, specifier, complement and modifier are universal (this means that the phrasal unit underlies the phrase structure of every language), linear relations are not (this means that the parts can occur in different linear orders). This is where the idea of parameter setting comes in. A child exposed to a **head initial** language like English adopts an X-bar structure where the head X precedes the complement. A child exposed to a **head final** language like Korean adopts an X-bar structure where the head follows its complement. Because the X-bar model specifies that the complement always occurs next to the head, only two 'options' are permitted. This illustrates the restricted nature of the parameters of variation in this model.

Tomasello argues, as have many opponents of the generative approach, that the X-bar model does not account for **non-configurational languages** like the native Australian language Dyirbal. A non-configurational language is one in which words are not grouped into obvious phrasal units. The application of X-bar theory to this type of language raises a number of questions about how the Dyirbal child sets his or her head initial/final parameter. Cognitive linguists like Tomasello argue, then, that the 'universals' posited by generative linguists arise from theory-internal considerations rather than appropriately reflecting the diversity and complexity of language.

The linguistic modularity hypothesis

As we have seen, the generative model rests on the hypothesis that there is a specialised and innate cognitive subsystem or 'language faculty': an encapsulated system of specialised knowledge that equips the child for the acquisition of language and gives rise to unconscious knowledge of language or **competence** of the native speaker. This system is often described as a **module** (see Chomsky 1986: 13, 150; Fodor 1983, 2000). Patterns of **selective impairment**, particularly when these illustrate **double dissociation**, are often thought by generative linguists to represent evidence for the encapsulation of such cognitive subsystems. Examples of selective impairment that are frequently cited in relation to the issue of the modularity of language are Williams Syndrome, linguistic savants and Specific Language Impairment. Williams Syndrome is a genetic developmental disorder characterised by a low IQ and severe learning difficulties. Despite this, children with this disorder develop normal or supernormal language skills, characterised by particularly fluent speech and a large and precocious vocabulary. Linguistic savants are individuals who, despite severe learning difficulties, have a normal or supernormal aptitude for language learning. In the case of Specific Language Impairment, a developmental disorder that is probably genetic, individuals perform normally in terms of IQ and learning abilities, but fail to acquire language normally, particularly the grammatical aspects of language. These patterns of impairment constitute a case of double dissociation in the sense that they can be interpreted as evidence that the development of language is not dependent upon general cognitive development and vice versa. This kind of evidence is cited by some generative linguists in support of the modularity hypothesis (see Pinker 1994 for an overview).

Interpretations of empirical findings in child language acquisition

When looking at empirical evidence for or against a particular theory of language, it is important to be aware that the same set of empirical findings has the potential to be interpreted in support of two or more opposing theories at the same time. In other words, while the empirical findings themselves may be indisputable (depending on how well-designed the study is), the interpretation of those findings is rarely indisputable. For example, while Tomasello argues that the one-word and two-word stages in child language provide evidence for item-based learning, generative linguists argue that the existence of these states provides evidence for a 'predetermined path' of language development, and that furthermore the order of units within the two-word expressions provides evidence for the underlying rule-based system that emerges fully later. Moreover, while Tomasello argues that the tendency for infants to attend to familiar linguistic stimuli provides evidence for pattern-finding ability, generative linguists argue that this provides evidence for the existence of a universal 'pool' of speech sounds that the child is equipped to distinguish between, and that parameter setting abilities are evident in the infant. As this brief discussion illustrates, the developmental psycholinguistics literature is fraught with such disputes and represents an extremely complex discipline. The interpretation of such findings should always be approached critically.

Localisation of function in the brain

The final issue we consider here is the localisation of linguistic function in the brain. So far, we have been discussing models of mind rather than brain. Of course, unlike the mind, the brain is a physical object, and neuroscientists have been able to discover much in recent years about what kinds of processes take place in different parts of the brain. In fact, we have known since the nineteenth century that there are parts of the brain that are specialised for linguistic processing, for most if not all people. There is an overwhelming tendency for language processing to take place in the left hemisphere of the brain, and areas responsible for the production of language (Broca's area) and comprehension of language (Wernicke's area) have been shown to occupy distinct parts of the brain. These findings have prompted many linguists to argue that this supports the view that we are biologically predetermined for language. However, this is not an issue about which cognitive linguists and generative linguists disagree. The nature of their disagreement concerns the nature of these biological systems: whether they are domain-general or specialised. The facts concerning localisation of function do not provide evidence for or against either the cognitive or the generative view, given that both are models of mind.

4.5 Summary

In this chapter we have been concerned with the usage-based thesis and how this model accounts for knowledge of language (grammar), for how language evolves over time (language change) and for how we gain or acquire our native language (language acquisition). We began by outlining the main assumptions that characterise the usage-based view of language adopted by cognitive linguists. The first relates to the central importance of the **utterance**, which is a situated instance of language use, culturally and contextually embedded, and represents an instance of linguistic behaviour on the part of a language user. The second key assumption is the idea that **knowledge of language** is derived from and informed by language use. The third key assumption is that human language can only be meaningfully accounted for by emphasising the **interactive** nature of language use. The fourth assumption relates to the central importance of **context** to the usage-based model, particularly in the case of accounting for word meaning. The final assumption is that the relative

frequency of linguistic units affects the nature and organisation of the language system. We then explored these issues by introducing Langacker's usage-based model **Cognitive Grammar**. This model assumes that linguistic units or **symbolic assemblies** are explicitly derived from language use, via a process of **abstraction**, which gives rise to **schemas**. We then introduced the theme of language change, and saw that Croft's model of language change, the **Utterance Selection Theory**, emphasised the importance of **linguistic convention** and interaction in language change. Drawing on ideas from evolutionary theory, Croft argues that language use represents the interface that mediates between linguistic convention, **altered replication** (innovation) of linguistic form-meaning units and **selection** (propagation), giving rise to the adoption of new linguistic conventions (language change). Finally, we examined the work of the developmental psycholinguist Michael Tomasello. Based on empirical findings that early language acquisition is **item-based** rather than **rule-based**, Tomasello argues for a construction-based or symbolic view of language acquisition, which relies upon domain-general **pattern-finding** skills and **intention-reading** skills. Tomasello argues that language use, in the context of **joint attentional frames,** facilitates the imitation of linguistic behaviour, which is a form of cultural learning. We compared Tomasello's usage-based account with Chomsky's **Universal Grammar** model, and found that while cognitive and generative theories stand in direct opposition on the issue of the existence of specialised and innate cognitive systems for language acquisition, they agree that humans are biologically predetermined for language acquisition.

Further reading

Language and use in cognitive linguistics

- **Barlow and Kemmer (2000).** This is a recent collection of papers by leading proponents of the usage-based approach to linguistic theory. The introductory article by Kemmer and Barlow is a particularly useful overview of the main tenets of usage-based approaches.

Langacker's usage-based model

- **Langacker (1987).** Langacker's foundational work, influential in many areas of cognitive linguistics, provides a thorough overview of the usage-based perspective.
- **Langacker (1999b).** Chapter 4 outlines the usage-based model.
- **Langacker (2000).** An article-length overview of the ways in which Cognitive Grammar is usage-based.

- **Langacker ([1991] 2002).** Chapter 10 specifically addresses the usage-based model.

Other usage-based approaches to language change
- **Croft (2000).** In this important book, Croft adopts a usage-based perspective in attempting to develop a new theory of language change.

The usage-based approach to language acquisition
- **Achard and Niemeier (eds) (2000).** A special issue of the journal *Cognitive Linguistics*, devoted to research by cognitively-oriented developmental psycholinguists.
- **Tomasello (1992).** Tomasello's case study of the early linguistic development of his daughter.
- **Tomasello (1995).** A persuasive critique of the Chomskyan perspective on language and language acquisition as presented in Steven Pinker's (1994) book *The Language Instinct*.
- **Tomasello (2000).** In this article, Tomasello presents a succinct overview of some of the ideas developed in his 2003 book (see below).
- **Tomasello (2002).** A collection of articles by leading pioneers in developmental psycholinguists. While not specifically focused on the usage-based perspective, this is an invaluable resource on the state of the art in language acquisition research.
- **Tomasello (2003).** The definitive usage-based account of language acquisition.

Exercises

4.1 A definition of the usage-based approach

In your own words, provide a definition of the usage-based thesis in twenty words or fewer. Make sure you include each of the following expressions in your definition: *utterance, grammar, language change, language acquisition*.

4.2 Grammar and language change

The view advocated by cognitive linguists like Langacker is that a grammar sanctions language use: the conventional symbolic units that make up a language license new and ongoing language use. Adopting this hypothesis, explain how Langacker's usage-based approach allows and explains language change.

4.3 Investigating Language change

During early 2004, the following expressions appeared in the British tabloid press, describing economic migrants coming to Britain from poorer parts of the European Union:

> *Welfare shopping*
> *Benefit tourists*

Explain how you might go about investigating whether, and to what extent, these terms have become conventionalised (propagated) in your English-speaking community.

Now make a list of expressions that you think have entered your speech community recently. Investigate when, where and why they first began to appear, and hypothesise how each expression might have begun to propagate. For each expression, make a prediction as to how conventionalised you think it will become. What is the basis of your prediction?

4.4 Dived vs. dove

In standard British English the past tense of the verb *(to) dive* is *dived*. In many North American varieties, the past tense form is *dove*. Can you explain this difference in terms of the usage-based thesis developed in this chapter? In particular, why might two major English-language speaking communities have evolved different past tense forms? How would you go about investigating and testing the hypotheses you have come up with?

4.5 Holophrases

Consider the early uses of the following holophrases reported by Tomasello (1992, 2003):

(a) *Play-play*: first use, when 'playing' the piano; second use, to name the piano
(b) *Steps*: first use, when climbing or descending stairs (never to name them)
(c) *Bath*: first use, when preparing for a bath; second use, when bathing a baby doll (never to name it)
(d) *Game*: first use, to describe the activity when she plays with a baseball and baseball glove; second use, to describe the activity when others play with a baseball and baseball glove (never to name objects)

Based on these examples and others described in this chapter, what different functions can you discern in the use of holophrases? Is there a pattern that emerges in terms of the order of acquisition in holophrase function? Given that some holophrases come to be used to name an object and others do not, what might this indicate about how a particular holophrase is being analysed by the infant?

4.6 Theories of language acquisition

Summarise the key theoretical and empirical arguments adopted in the usage-based model of child language acquisition. Compare these with the theoretical and empirical arguments adopted in the generative model. Present these arguments as an annotated table. Is there any common ground?

Part II: Cognitive semantics

Introduction

Like the larger enterprise of cognitive linguistics, cognitive semantics is not a unified theory. It represents an approach to the study of mind and its relationship with embodied experience and culture. It proceeds by employing language as a key methodological tool for uncovering conceptual organisation and structure.

In Chapter 5, *What is cognitive semantics?*, we examine the four guiding principles that collectively characterise the collection of approaches that fall within cognitive semantics. These principles can be stated as follows:

1. Conceptual structure is embodied.
2. Semantic structure is conceptual structure.
3. Meaning representation is encyclopaedic.
4. Meaning-construction is conceptualisation.

We examine each of these principles in turn, and provide a preliminary overview of how they are reflected in the concerns addressed by cognitive semanticists. The subsequent chapters address specific theories within cognitive semantics that, to varying degrees, reflect these guiding principles.

Chapter 6, *Embodiment and conceptual structure*, examines the theory of image schemas developed in particular by Mark Johnson and the conceptual structuring system approach developed by Leonard Talmy. The research on image schemas by Johnson and others highlights the embodied basis of conceptual structure while Talmy's research illustrates the ways in which language reflects conceptual structure which in turn reflects embodied experience. Thus these two approaches illustrate the first two of the guiding principles introduced in Chapter 5.

Chapter 7, *The encyclopaedic view of meaning*, is concerned with the third guiding principle of cognitive semantics: the idea that linguistic meaning is encyclopaedic in nature. This issue is explored by presenting, comparing and contrasting the theory of Frame Semantics developed by Charles Fillmore and the theory of domains pioneered by Ronald Langacker.

Chapter 8, *Categorisation and idealised cognitive models*, introduces the research perspective of George Lakoff and discusses his impact on the development of cognitive semantics. In particular, we examine his proposal that experimental research on categorisation and prototype theory from cognitive psychology can be applied and extended in a theoretical account of cognitive representations that he calls 'idealised cognitive models'. Lakoff applied his theory to three distinct aspects of conceptual organisation and language in three influential 'case studies' in his book *Women, Fire and Dangerous Things* (1987). The first two of these, which relate to conceptual metaphor and lexical semantics, are the subjects of the next two chapters.

Chapter 9, *Metaphor and metonymy*, examines the development of Conceptual Metaphor Theory pioneered by George Lakoff in collaboration with Mark Johnson, together with the later development of approaches to conceptual metonymy. According to this model, conceptual metaphor maps structure from one conceptual domain onto another, while metonymy highlights an entity by referring to another entity within the same domain. More recent research suggests that metonymy may be more fundamental to conceptual structure than conceptual metaphor. In the light of this claim, we examine the research of Antonio Barcelona, Zoltán Kövecses and Günter Radden.

In Chapter 10, *Word meaning and radial categories*, we begin by illustrating Lakoff's approach to word meaning. Following influential research by Claudia Brugman, Lakoff argues that words represent categories of meaning or 'senses'. From this perspective, words are conceptual categories like any other, organised with respect to a prototype. However, his approach has been challenged by more recent research in cognitive semantics. In particular, we discuss the 'Principled Polysemy' framework developed by Vyvyan Evans and Andrea Tyler.

In Chapter 11, *Meaning construction and mental spaces*, we examine a model developed by Gilles Fauconnier which is concerned with providing an architecture for modelling meaning construction (sentence meaning) in discourse. Mental spaces are temporary knowledge structures constructed on the basis of ongoing discourse and can form the basis of an account for a range of phenomena including referential ambiguities, tense and aspect, and epistemic distance.

In Chapter 12, *Conceptual blending*, we discuss Blending Theory, the more recent approach that developed from Mental Spaces Theory. Blending Theory was developed by Gilles Fauconnier and Mark Turner and is concerned with generalising key ideas from Mental Spaces Theory and modelling the way that dynamic meaning construction often results in a conceptual representation that is 'more than the sum of its parts'. The approaches discussed in Chapters 11 and 12 illustrate the fourth guiding assumption of the cognitive semantics approach introduced in Chapter 5.

Finally, Chapter 13 compares and contrasts some of the assumptions of cognitive semantics with formal (truth-conditional) semantics and Relevance Theory, a formally-oriented model of communication that presents a view of linguistic meaning that is in certain respects consonant with cognitive approaches, despite directly opposing starting assumptions.

5

What is cognitive semantics?

Cognitive semantics began in the 1970s as a reaction against the **objectivist world-view** assumed by the Anglo-American tradition in philosophy and the related approach, **truth-conditional semantics**, developed within formal linguistics. Eve Sweetser, a leading cognitive linguist, describes the truth-conditional approach in the following terms: 'By viewing meaning as the relationship between words and the world, truth-conditional semantics eliminates cognitive organization from the linguistic system' (Sweetser 1990: 4). In contrast to this view, cognitive semantics sees linguistic meaning as a manifestation of **conceptual structure**: the nature and organisation of mental representation in all its richness and diversity, and this is what makes it a distinctive approach to linguistic meaning. Leonard Talmy, one of the original pioneers of cognitive linguistics in the 1970s, describes cognitive semantics as follows: '[R]esearch on cognitive semantics is research on conceptual content and its organization in language' (Talmy 2000: 4). In this chapter, we will try to give a broad sense of the nature of cognitive semantics as an approach to conceptual structure and linguistic meaning. Cognitive semantics, like the larger enterprise of cognitive linguistics of which it is a part, is not a single unified framework. Those researchers who identify themselves as cognitive semanticists typically have a diverse set of foci and interests. However, there are a number of principles that collectively characterise a **cognitive semantics approach**. In section 5.1 we will identify these guiding principles as we see them. In section 5.2 we will explore some of the major lines of investigation pursued under the 'banner' of cognitive semantics. As we will see, although cognitive semantics began life as a reaction against formal theories of meaning deriving from twentieth-century analytic philosophy and objectivism, the guiding principles adopted within cognitive semantics open up a range of phenomena for direct investigation that transcend the initial point of departure for research in cognitive semantics. In other words, these approaches now go significantly beyond refuting the tradition of truth-conditional semantics. In section 5.3, we will look in more detail at the methodology adopted by cognitive semanticists in investigating these phenomena, and in section 5.4 we will make some explicit comparisons between cognitive approaches and formal approaches to linguistic meaning, setting the scene for some of the more detailed discussions that follow in Part II of the book.

5.1 Guiding principles

In this section we consider four central assumptions of cognitive semantics. These are listed below:

1. Conceptual structure is embodied (the 'embodied cognition thesis').
2. Semantic structure is conceptual structure.
3. Meaning representation is encyclopaedic.
4. Meaning construction is conceptualisation.

These principles can be viewed as outcomes of the two key commitments described in Chapter 2: the 'Generalisation Commitment' and the 'Cognitive Commitment'. The embodied cognition thesis is also one of these assumptions. Let's look at each of these in turn.

5.1.1 Conceptual structure is embodied

A fundamental concern for cognitive semanticists is the nature of the relationship between conceptual structure and the external world of sensory experience. In other words, cognitive semanticists set out to explore the nature of human interaction with and awareness of the external world, and to build a theory of conceptual structure that is consonant with the ways in which we experience the world. One idea that has emerged in an attempt to explain the nature of conceptual organisation on the basis of interaction with the physical world is the **embodied cognition thesis**, which we introduced in Chapter 2. As we saw, this thesis holds that the nature of conceptual organisation arises from bodily experience, so part of what makes conceptual structure meaningful is the bodily experience with which it is associated.

Let's illustrate this idea with an example. Imagine a man in a locked room. A room has the structural properties associated with a **bounded landmark**: it has enclosed sides, an interior, a boundary and an exterior. As a consequence of these properties, the bounded landmark has the additional functional property of **containment**: the man is unable to leave the room. Although this seems rather obvious, observe that this instance of containment is partly a consequence of the properties of the bounded landmark and partly a consequence of the properties of the human body. Humans cannot pass through minute crevices like gas can, or crawl through the gaps under doors like ants can. In other words, containment is a meaningful consequence of a particular type of physical relationship that we have experienced in interaction with the external world.

The concept associated with containment is an instance of what cognitive linguists call an **image schema**. In the cognitive model, the image-schematic concept represents one of the ways in which bodily experience gives rise to meaningful concepts. While the concept CONTAINER is grounded in the directly embodied experience of interacting with bounded landmarks, image-schematic conceptual structure can also give rise to more abstract kinds of meaning. For example, consider the following examples from Lakoff and Johnson (1980: 32):

(1) a. He's *in* love.
 b. We're *out of* trouble now.
 c. He's *coming out of* the coma.
 d. I'm *slowly getting into* shape.
 e. He *entered* a state of euphoria.
 f. He *fell into* a depression.

Lakoff (1987) and Johnson (1987) both argue that examples like the ones in (1) are licensed by the **metaphorical projection** of the CONTAINER image schema onto the abstract conceptual domain of STATES, to which concepts like LOVE, TROUBLE and HEALTH belong. This results in the conceptual metaphor STATES ARE CONTAINERS. The idea behind metaphorical projection is that meaningful structure from bodily experience gives rise to concrete concepts like the CONTAINER image schema, which in turn serves to structure more abstract conceptual domains like STATES. In this way, conceptual structure is embodied. We will look in detail at image schemas in Chapter 6.

5.1.2 Semantic structure is conceptual structure

This principle asserts that language refers to concepts in the mind of the speaker rather than to objects in the external world. In other words, **semantic structure** (the meanings conventionally associated with words and other linguistic units) can be equated with concepts. As we saw in Chapter 3, these conventional meanings associated with words are **linguistic concepts** or **lexical concepts**: the conventional form that conceptual structure requires in order to be encoded in language.

However, the claim that semantic structure can be equated with conceptual structure does not mean that the two are identical. Instead, cognitive semanticists claim that the meanings associated with words, for example, form only a subset of possible concepts. After all, we have many more thoughts, ideas and feelings than we can conventionally encode in language. For example, we have a concept for the place on our faces below our nose and above our mouth where moustaches go. We must have a concept for this part of the face in order to understand that the hair that grows there is called a *moustache*. However, as Langacker (1987) points out, there is no English word that conventionally encodes this concept (at least not in the non-specialist vocabulary of everyday language). It follows that the set of lexical concepts is only a subset of the entire set of concepts in the mind of the speaker.

For a theory of language, this principle is of greater significance than we might think. Recall that semantic structure relates not just to words but to all **linguistic units**. A linguistic unit might be a word like *cat*, a **bound morpheme** such as *-er*, as in *driver* or *teacher*, or indeed a larger conventional pattern, like the structure of an active sentence (2) or a passive sentence (3):

(2) William Shakespeare wrote *Romeo and Juliet*. [active]

(3) *Romeo and Juliet* was written by William Shakespeare. [passive]

Because active and passive constructions are conventionally associated with a functional distinction, namely the point of view we are adopting with respect to the subject of the sentence, cognitive linguists claim that the active and passive structures are themselves meaningful: in active sentences we are focusing on the active participant in an event by placing this unit at the front of the construction. In passive sentences, we are focusing on the participant that undergoes the action. The conventional meanings associated with these grammatical constructions are admittedly schematic, but they are nevertheless meaningful. According to the view adopted in cognitive semantics, the same holds for smaller grammatical units as well, including words like *the* and tense morphemes like *-ed* in *wondered*. This is an idea that we discuss in more detail in Part III of the book.

For present purposes, the idea that grammatical categories or constructions are essentially conceptual in nature entails that closed-class elements as well as open-class elements fall within the purview of semantic analysis. Indeed, Talmy (2000) explicitly focuses upon **closed-class semantics**. One of the properties that makes cognitive semantics different from other approaches to language, then, is that it seeks to provide a unified account of lexical and grammatical organisation rather than viewing these as distinct subsystems.

There are two important caveats that follow from the principle that semantic structure represents a subpart of conceptual structure. Firstly, it is important to point out that cognitive semanticists are not claiming that language relates to concepts internal to the mind of the speaker and nothing else. This would lead to an extreme form of **subjectivism**, in which concepts are divorced from the world that they relate to (see Sinha 1999). Indeed, we have concepts in the first place either because they are useful ways of understanding the external world, or because they are inevitable ways of understanding the world, given our cognitive architecture and our physiology. Cognitive semantics therefore steers a path between the opposing extremes of subjectivism and the objectivism encapsulated in traditional truth-conditional semantics (section 5.4) by claiming that concepts relate to lived experience.

Let's look at an example. Consider the concept BACHELOR. This is a much-discussed example in the semantics literature. This concept, which is traditionally defined as an 'unmarried adult male', is not isolated from ordinary experience because we cannot in fact apply it to all unmarried adult males. We understand that some adult males are ineligible for marriage due either to vocation or to sexual preference (at least while marriage is restricted to occurring between members of the opposite sex). It is for this reason that we would find it odd to apply the term *bachelor* to either the Pope or a homosexual male, even though they both, strictly speaking, meet the 'definition' of BACHELOR.

The second caveat concerns the notion of semantic structure. We have assumed so far that the meanings associated with words can be defined: for example, BACHELOR means 'unmarried adult male'. However, we have already begun to see that word meanings, which we are calling lexical concepts, cannot straightforwardly be defined. Indeed, strict definitions like 'unmarried adult male' fail to adequately capture the range and diversity of meaning associated with any given lexical concept. For this reason, cognitive semanticists reject the definitional or dictionary view of word meaning in favour of an encyclopaedic view. We will elaborate this idea in more detail below (section 5.1.3).

5.1.3 Meaning representation is encyclopaedic

The third central principle of cognitive semantics holds that semantic structure is **encyclopaedic** in nature. This means that words do not represent neatly packaged bundles of meaning (the **dictionary** view), but serve as 'points of access' to vast repositories of knowledge relating to a particular concept or conceptual domain (e.g. Langacker 1987). We illustrated this idea above in relation to the concept BACHELOR. Indeed, not only do we know that certain kinds of unmarried adult males would not normally be described as bachelors, we also have cultural knowledge regarding the behaviour associated with stereotypical bachelors. It is 'encyclopaedic' knowledge of this kind that allows us to interpret this otherwise contradictory sentence:

(4) 'Watch out Jane, your husband's a right bachelor!'

On the face of it, identifying Jane's husband (a married man) as a bachelor would appear to be contradictory. However, given our cultural stereotype of bachelors, which represents them as sexual predators, we understand the utterance in (4) as a warning issued to Jane concerning her husband's fidelity. As this example illustrates, the meanings associated with words often draw upon complex and sophisticated bodies of knowledge. We will look in detail at the encyclopaedic view of meaning in Chapter 7.

Of course, to claim that words are 'points of access' to encyclopaedic meaning is not to deny that words have conventional meanings associated with them. The fact that example (5) means something different from example (6) is a consequence of the conventional range of meanings associated with *safe* and *happy*.

(5) John is safe.

(6) John is happy.

However, cognitive semanticists argue that the conventional meaning associated with a particular word is just a 'prompt' for the process of **meaning construction**: the 'selection' of an appropriate interpretation against the context of the utterance. For example, the word *safe* has a range of meanings, and the meaning that we select emerges as a consequence of the context in which the word occurs. To illustrate this point, consider the examples in (7) against the context of a child playing on the beach.

(7) a. The child is safe.
    b. The beach is safe.
    c. The shovel is safe.

In this context, the interpretation of (7a) is that the child will not come to any harm. However, (7b) does not mean that the beach will not come to harm. Instead, it means that the beach is an environment in which the risk of the child coming to harm is minimised. Similarly, (7c) does not mean that the shovel will not come to harm, but that it will not cause harm to the child. These examples illustrate that there is no single fixed property that *safe* assigns to the words *child*, *beach* and *shovel*. In order to understand what the speaker means, we draw upon our encyclopaedic knowledge relating to children, beaches and shovels, and our knowledge relating to what it means to be safe. We then 'construct' a meaning by 'selecting' a meaning that is appropriate in the context of the utterance.

Just to give a few examples, the sentence in (7b) could be interpreted in any of the following ways, given an appropriate context. Some of these meanings can be paraphrased as 'safe from harm', and others as 'unlikely to cause harm': (1) this beach has avoided the impact of a recent oil spill; (2) this beach is not going to be dug up by property developers; (3) due to its location in a temperate climate, you will not suffer from sunburn on this beach; (4) this beach, which is prone to crowding, is free of pickpockets; (5) there are no jellyfish in the sea; (6) the miniature model beach with accompanying model luxury hotels, designed by an architect, which was inadvertently dropped before an important meeting, has not been damaged.

5.1.4 Meaning construction is conceptualisation

In this section, we explore the process of meaning construction in more detail. The fourth principle associated with cognitive semantics is that language itself does not encode meaning. Instead, as we have seen, words (and other linguistic units) are only 'prompts' for the construction of meaning. According to this view, meaning is constructed at the conceptual level: meaning construction is equated with **conceptualisation**, a dynamic process whereby linguistic units serve as prompts for an array of conceptual operations and the recruitment of background knowledge. It follows from this view that meaning is a process rather than a discrete 'thing' that can be 'packaged' by language. Meaning construction draws upon encyclopaedic knowledge, as we saw above, and involves **inferencing strategies** that relate to different aspects of conceptual structure, organisation and packaging (Sweetser 1999). The dynamic quality of meaning construction has been most extensively modelled by Gilles Fauconnier (e.g. 1994, 1997), who emphasises the role of **mappings**: local connections between distinct **mental spaces**, conceptual 'packets' of information, which are built up during the 'on-line' process of meaning construction.

Let's look at an example that illustrates the conceptual nature of meaning construction. Consider the following example from Taylor (2002: 530):

(8) In France, Bill Clinton wouldn't have been harmed by his relationship with Monica Lewinsky.

Sentences of this kind are called **counterfactuals**, because they describe a scenario that is counter to fact. This sentence prompts us to imagine a scenario in which Bill Clinton, the former US President, is actually the President of France, and that the scandal that surrounded him and the former Whitehouse intern, Monica Lewinsky, took place not in the United States but in France. In the context of this scenario, it is suggested that Bill Clinton would not have been politically harmed by his extramarital affair with Lewinsky. According to Gilles Fauconnier and Mark Turner (e.g. 2002), we actually have to engage in conceptual feats of breathtaking complexity in order to access this kind of meaning. These conceptual feats are performed on a second-by-second basis in the ongoing construction of meaning in discourse, and without conscious awareness.

According to this view, which is called **Conceptual Blending Theory**, the sentence in (8) prompts us to set up one mental space, a 'reality space', in which Clinton is the US President, Lewinsky is his intern, they have an affair, they are found out and scandal ensues. We also set up a second 'reality space', which contains the President of France together with knowledge about French culture which deems it permissible for French presidents to have extra-marital relations, and 'public' and 'private' families. In a third **blended space**, Clinton is the President of France, he has an affair with Lewinsky, they are found out, but there is no scandal. Because of the conceptual mappings that relate the first two spaces to the third blended space, we come to understand something additional about the original 'input' or reality spaces. We learn that the cultural and moral sensitivities regarding extramarital affairs between politicians and members of their staff are radically different in the United States and France. This meaning is constructed on the basis of complex mapping operations between distinct reality-based scenarios, which combine to create a new counterfactual scenario. The blended space, then, gives rise to a new meaning, albeit counterfactual, which is not available from encyclopaedic knowledge. This new meaning rests upon Clinton as French President escaping scandal despite his affair with Lewinsky. We will look in detail at mental spaces and the idea of conceptual blending in Chapters 11–12. Table 5.1 summarises the four key assumptions of cognitive semantics that we have discussed in this section.

5.2 Phenomena investigated within cognitive semantics

Having established the guiding principles that underpin cognitive semantics, we turn in this section to a brief overview of some of the phenomena investigated within this approach. This provides some elaboration on issues addressed in the previous section, and gives a flavour of the nature and scope of cognitive semantics.

5.2.1 The bodily basis of meaning

Given the thesis of embodied cognition that we discussed earlier (section 5.1.2), a key area of investigation within cognitive semantics concerns the

Table 5.1 The guiding principles of cognitive semantics

Conceptual structure is embodied	The nature of conceptual organisation arises from bodily experience
Semantic structure is conceptual structure	Semantic structure (the meanings conventionally associated with words and other linguistic units) is equated with concepts
Meaning representation is encyclopaedic	Words (and other linguistic units) are treated as 'points of access' to vast repositories of knowledge relating to a particular concept
Meaning construction is conceptualisation	Meaning construction is equated with conceptualisation, a dynamic process whereby linguistic units serve as prompts for an array of conceptual operations and the recruitment of background knowledge

bodily basis of meaning (see Chapter 6). Given the assumption that conceptual structure is meaningful by virtue of being tied to directly meaningful pre-conceptual (bodily) experience, much research within the cognitive semantics tradition has been directed at investigating conceptual metaphors. According to this approach, conceptual metaphors give rise to systems of conventional conceptual mappings, held in long-term memory, which may be motivated by image-schematic structure. If image schemas arise from bodily experience, then we may be able to explain conceptual metaphor on the basis that it maps rich and detailed structure from concrete domains of experience onto more abstract concepts and conceptual domains. We have seen several examples of this phenomenon already. Consider again example (9), which was first presented in Chapter 1.

(9) The number of shares has gone up.

According to Lakoff and Johnson, examples like this are motivated by a highly productive conceptual metaphor that is also evident in (10).

(10) a. John got the highest score on the test.
b. Mortgage rates have fallen.
c. Inflation is on the way up.

This metaphor appears to relate the domains of QUANTITY and VERTICAL ELEVATION. In other words, we understand greater quantity in terms of increased height, and decreased quantity in terms of lesser height. Conceptual metaphor scholars like Lakoff and Johnson argue that this conventional pattern of conceptual mapping is directly grounded in ubiquitous everyday experience. For example, when we pour a liquid into a glass, there is a simultaneous increase in the height and quantity of the fluid. This is a typical example of the correlation between height and quantity. Similarly, if we put items onto a pile, an increase in height correlates with an increase in quantity. This **experiential correlation** between height and quantity, which we experience from an early age, has been claimed to motivate the conceptual metaphor MORE IS UP, also known as QUANTITY IS VERTICAL ELEVATION (see Chapter 9).

5.2.2 Conceptual structure

As we have seen, an important line of investigation within cognitive semantics focuses on how language encodes (and reflects) conceptual structure. This line of investigation concerns the conceptual structuring mechanisms apparent in linguistic structure. One way of uncovering conceptual structure in language is by investigating the distinct functions associated with open-class and closed-class semantic systems. Talmy (2000) argues that these two systems encode our **Cognitive Representation** (CR) in language. The closed-class semantic system (the system of meaning associated with grammatical constructions, bound morphemes and grammatical words like *and* and *the*) provides scene-structuring representation. The open-class semantic system (the system of meaning associated with content words and morphemes) provides the substantive content relating to a particular scene. In Chapter 1, we illustrated the distinction between the open-class and closed-class subsystems with the following example:

(11) **The** hunter track**ed the** tigers

The elements marked in bold, as well as the **declarative** word order (as opposed to the **interrogative** *Did the hunter track the tigers?* for example) form part of the system of closed-class semantics. They provide the 'concept structuring' elements of the meaning described in this scene, and provide information about when the event occurred, how many participants were involved, whether the participants are familiar to the speaker and hearer in the current discourse, whether the speaker asserts the information (rather than, say, asking a question about it) and so on. We can think of these closed-class elements as providing a kind of frame or scaffolding, which forms the foundations of the meaning in this sentence. The open-class semantic system relates to words like *hunter*, *track* and *tiger*, which impose rich contentful meaning upon this frame: who the participants are and the nature of event described in the scene. We look at these ideas in more detail in Chapter 6.

5.2.3 Encyclopaedic semantics

Research into the encyclopaedic nature of meaning has mainly focused on the way semantic structure is organised relative to conceptual knowledge structures. One proposal concerning the organisation of word meaning is based on the notion of a **frame** against which word-meanings are understood. This idea has been developed in linguistics by Charles Fillmore (1975, 1977, 1982, 1985a). Frames are detailed knowledge structures or schemas emerging from everyday experiences. According to this perspective, knowledge of word meaning is, in part, knowledge of the individual frames with which a word is associated. A theory of **frame semantics** therefore reveals the rich network of meaning that makes up our knowledge of words (see Chapter 7).

By way of illustration, consider the verbs *rob* and *steal*. On first inspection it might appear that these verbs both relate to a THEFT frame, which includes the following roles: (1) THIEF; (2) TARGET (the person or a place that is robbed); and (3) GOODS (to be) stolen. However, there is an important difference between the two verbs: while *rob* profiles THIEF and TARGET, *steal* profiles THIEF and GOODS. The examples in (12) are from Goldberg (1995: 45).

(12) a. [Jesse] robbed [the rich] (of their money).   <THIEF TARGET GOODS>
     b. [Jesse] stole [money] (from the rich).   <THIEF TARGET GOODS>

In other words, while both verbs can occur in sentences with all three participants, each verb has different requirements concerning which two participants it needs. This is illustrated by following examples (although it's worth observing that (13a) is acceptable in some British English dialects):

(13) a. *Jesse robbed the money.
     b. *Jesse stole the rich.

As these examples illustrate, our knowledge of word meaning involves complex networks of knowledge.

A related approach is the theory of **domains**, developed by Langacker (e.g. 1987). In his theory of domains (also discussed in Chapter 7), Langacker argues that knowledge representation can be described in terms of **profile-base organisation**. A linguistic unit's **profile** is the part of its semantic structure upon which that word focuses attention: this part is explicitly mentioned. The aspect of semantic structure that is not in focus, but is necessary in order to understand the profile, is called the **base**. For instance, the lexical item *hunter* profiles a particular participant in an activity in which an animal is pursued with a view to it being killed. The meaning of *hunter* is only understood in the context of this activity. The hunting process is therefore the base against which the participant *hunter* is profiled.

5.2.4 Mappings

Another prominent theme in cognitive semantics is the idea of conceptual **mappings**. Fauconnier (1997) has identified three kinds of mapping operations: (1) projection mappings; (2) pragmatic function mappings; and (3) schema mappings.

A **projection mapping** projects structure from one domain (**source**) onto another (**target**). We mentioned this kind of mapping earlier in relation to conceptual metaphor. Another example is the metaphor TIME IS THE MOTION OF OBJECTS, where TIME is conceptualised in terms of MOTION (recall the discussion of the 'moving time' model in Chapter 3). Consider the examples in (14).

(14) a. Summer has just zoomed by.
 b. The end of term is approaching.
 c. The time for a decision has come.

In these sentences, temporally framed concepts corresponding to the expressions *summer, the end of term* and *the time for a decision* are structured in terms of MOTION. Of course, temporal concepts cannot undergo literal motion because they are not physical entities. However, these conventional metaphoric mappings allow us to understand abstract concepts like TIME in terms of MOTION. We explore conceptual metaphor in detail in Chapter 9.

Pragmatic function mappings are established between two entities by virtue of a shared frame of experience. For example, **metonymy**, which depends upon an association between two entities so that one entity can stand for the other, is an instance of a pragmatic function mapping. Consider example (15).

(15) The ham sandwich has wandering hands.

Imagine the sentence in (15) uttered by one waitress to another in a restaurant. In this context, the salient association between a particular customer and the food he orders establishes a pragmatic function mapping. We also look in detail at metonymy in Chapter 9.

Schema mappings relate to the projection of a schema (another term for frame) onto particular utterances. As intimated in section 5.2.1, a frame is a relatively detailed knowledge structure derived from everyday patterns of interaction. For instance, we have an abstract frame for PURCHASING GOODS, which represents an abstraction over specific instances of purchasing goods, such as buying a stamp in a post office, buying groceries in a supermarket, ordering a book through an on-line retailer, and so on. Each instance of PURCHASING GOODS involves a purchaser, a vendor, merchandise, money (or credit card) and so on. Consider example (16):

(16) The Ministry of Defence purchased twenty new helicopters from Westland.

We make sense of this sentence by mapping its various components onto the roles in the PURCHASING GOODS frame. This frame enables us to understand the role assumed by each of the participants in this example: that the Ministry of Defence is the PURCHASER, the contractor Westland is the VENDOR and the helicopters are the MERCHANDISE. We look in more detail at schema mappings in Chapters 11 and 12, where we address two theories that rely upon this idea: Mental Spaces Theory and Conceptual Blending Theory.

5.2.5 Categorisation

Another phenomenon that has received considerable attention within cognitive semantics is categorisation: our ability to identify entities as members of groups. Of course, the words we use to refer to entities rest upon categorisation: there are good reasons why we call a cat '*cat*' and not, say, '*fish*'. One of the reasons behind the interest in this area stems from the 'Cognitive Commitment': the position adopted by cognitive linguists that language is a function of generalised cognition (Chapter 2). The ability to categorise is central to human cognition; given the 'Cognitive Commitment', we expect this ability to be reflected in linguistic organisation. The other reason behind the interest in this area relates to a question that has challenged philosophers (and, more recently, linguists) since ancient times: can word meaning be defined?

In the 1970s, pioneering research by cognitive psychologist Eleanor Rosch and her colleagues presented a serious challenge to the classical view of categorisation that had dominated Western thought since the time of Aristotle. According to this classical model, category membership is defined according to a set of necessary and sufficient conditions, which entails that category membership is an 'all-or-nothing' affair. For example, as we observed in Chapter 2, the artefacts depicted in Figure 5.1 can, depending on the situation and the way the artefact is being used, be identified as members of the category CUP. However, these are not all 'equal' members of that category.

WHAT IS COGNITIVE SEMANTICS?

Figure 5.1 Some members of the category CUP

The findings of Eleanor Rosch and her team revealed that categorisation is not an all or nothing affair, but that many categorisation judgements seemed to exhibit **prototype** or **typicality effects**. For example, when we categorise birds, certain types of bird (like robins or sparrows) are judged as 'better' examples of the category than others (like penguins).

In his famous book *Women, Fire and Dangerous Things*, George Lakoff (1987) explored some of the consequences of the observations made by Rosch and her colleagues for a theory of conceptual structure as manifested in language. An important idea that emerged from Lakoff's study is the theory of **idealised cognitive models (ICMs)**, which are highly abstract frames. These can account for certain kinds of typicality effects in categorisation.

For example, let's consider once more the concept BACHELOR. This is understood with respect to a relatively schematic ICM MARRIAGE. The MARRIAGE ICM includes the knowledge that bachelors are unmarried adult males. As we have observed, the category BACHELOR exhibits typicality effects. In other words, some members of the category BACHELOR (like eligible young men) are 'better' or more typical examples than others (like the Pope). The knowledge associated with the MARRIAGE ICM stipulates that bachelors can marry. However, our knowledge relating to CATHOLICISM stipulates that the Pope cannot marry. It is because of this mismatch between the MARRIAGE ICM (with respect to which BACHELOR is understood) and the CATHOLICISM ICM (with respect to which the Pope is understood) that this particular typicality effect arises.

5.2.6 Word meaning and polysemy

Another area in which Lakoff's work on ICMs has been highly influential is lexical semantics. As we have begun to see (recall example (7)), **lexical items** (words) typically have more than one meaning associated with them. When the meanings are related, this is called **polysemy**. Polysemy appears to be the norm rather than the exception in language. Lakoff proposed that lexical units like words should be treated as conceptual categories, organised with respect to an ICM or prototype. According to this point of view, polysemy arises because words are linked to a network of lexical concepts rather than to a single such concept. However, there is usually a central or 'typical' meaning that relates the others. In this respect, word meanings are a bit like the category BIRD. We look in more detail at word meaning in Chapter 10.

5.3 Methodology

In this section, we briefly comment on issues relating to methodology in cognitive semantics. First of all, it is important to explain how cognitive semantics is different from cognitive approaches to grammar, which we explore in Part III of the book. Cognitive semantics is primarily concerned with investigating conceptual structure and processes of conceptualisation, as we have seen. This means that cognitive semanticists are not primarily concerned with studying linguistic meaning for its own sake, but rather for what it can reveal about the nature of the human conceptual system. Their focus on language is motivated by the assumption that linguistic organisation will reflect, at least partially, the nature and organisation of the conceptual system; this does not mean that language directly mirrors the conceptual system, as we were careful to point out earlier in this chapter. For cognitive semanticists, then, language is a tool for investigating conceptual organisation.

In contrast, cognitive approaches to grammar are primarily concerned with studying the language system itself, and with describing that system, and our knowledge of that system, on the basis of the properties of the conceptual system. It follows that cognitive semantics and cognitive approaches to grammar are 'two sides of the same coin': cognitive semanticists rely on language to help them understand how the conceptual system works, while cognitive grammarians rely on what is known about the conceptual system to help them understand how language works.

In employing language for the purposes of investigating patterns of conceptual organisation, cognitive semanticists rely upon the methodology of seeking **converging evidence**, an idea that we introduced in Chapter 2. This means that when patterns in language suggest corresponding patterns in conceptual structure, cognitive semanticists look for related evidence of these patterns in other areas of investigation. For example, linguistic patterns suggest conceptual patterns relating to time, where PAST is 'behind' and FUTURE is 'in front'. Evidence from gesture studies provides independent support for the existence of this conceptual pattern: while English speakers gesture behind themselves while talking about the past, they gesture in front of themselves when talking about the future. Converging evidence from two distinct forms of communication (language and gesture) suggests that a common conceptual pattern underlies those two different forms. This explains why cognitive semanticists rely upon evidence from other disciplines, particularly cognitive psychology and neuroscience, in building a theory of the human conceptual system.

5.4 Some comparisons with formal approaches to semantics

In this section, we sketch out some of the differences between cognitive semantics and formal approaches to meaning. These different points are developed at relevant points throughout Part II of the book, and in Chapter 13 cognitive semantics is compared with two influential formal theories of meaning: Formal Semantics and Relevance Theory. To begin with, formal approaches to meaning such as truth-conditional semantics, which aim to be broadly compatible with the generative model, assume a dictionary model of linguistic meaning, rather than an encyclopaedic model. According to this view, linguistic meaning is separate from 'world knowledge', and can be modelled according to precise and formally stated definitions. Often, formal models of meaning rely on semantic decomposition along the lines we outlined in Chapter 3. One consequence of the strict separation of linguistic knowledge from world knowledge is the separation of semantics from pragmatics. While semantic meaning relates to the meaning 'packaged' inside words, regardless of their context of use, pragmatic meaning relates to how speakers make use of contextual information to retrieve speaker meaning by constructing inferences and so on. Of course, both semantic and pragmatic meaning interact to give rise to the interpretation of an utterance, but the formal model holds that only semantic meaning, being 'purely linguistic', belongs in the lexicon. As we will discover, cognitive semantics rejects this sharp division between semantics and pragmatics. Furthermore, in assuming a prototype model of word meaning, cognitive semantics also rejects the idea that word meaning can be modelled by strict definitions based on semantic decomposition.

A related issue concerns the assumption of compositionality that is assumed within formal models Not only is word meaning composed from semantic primitives, but sentence meaning is composed from word meaning, together with the structure imposed on those words by the grammar. While this view might work well enough for some sentences, it fails to account for 'non-compositional' expressions: those expressions whose meaning cannot be predicted from the meanings of the parts. These include idioms and metaphors (recall our discussion of the idiomatic expression *kick the bucket* in Chapter 1). This view implies that non-compositional expressions are the exception rather than the norm. As we will see, cognitive linguists also reject this view, adopting a constructional rather than compositional view of sentence meaning. Furthermore, cognitive semanticists argue that figurative language is in fact central to our way of thinking as well as to the way language works.

The final difference that we mention here relates to the model of truth-conditional semantics that is adopted by most formal models of linguistic meaning. This approach assumes an objectivist position, which means that it assumes an objective external reality against which descriptions in language can be judged true or false. In this way, it builds a model of semantic meaning that can be made explicit by means of a logical metalanguage. For example, the sentences *Lily devoured the cake* and *The cake was devoured by Lily* stand in a sentence meaning relation of paraphrase. The truth-conditional model characterises this meaning relation by describing the two sentences, or rather the propositions they express, as both holding true of the same state of affairs in the world. The appeal of this model is that it allows for precise statements that can be modelled by logic (a point to which we return in Chapter 13). One of the main disadvantages is that it can only account for propositions (roughly, descriptions of states of affairs). Of course, many utterances do not express propositions, such as questions, commands, greetings and so on, so that the truth-conditional model can only account for the meaning of a subset of sentence or utterance types. This view stands in direct opposition to the experientialist view adopted within cognitive semantics, which describes meaning in terms of human construal of reality.

Of course, there are many different formal models of linguistic meaning, and we cannot do justice to them all here. For purposes of comparison in this book, we refer to the 'standard' truth-conditional approach that is set out in most textbooks of semantics, while drawing the reader's attention to the fact that more recent formal approaches, notably the Conceptual Semantics model developed by Ray Jackendoff (1983, 1990, 1992, 1997), are consonant with the cognitive view in a number of important ways. For example, like cognitive semanticists, Jackendoff assumes a non-objective representational rather than denotational view of meaning: a mentalist model, which treats meaning as a relationship between language and world that is mediated by the human mind. Jackendoff also rejects the truth-conditional approach. However, as we saw in Chapter 3, Jackendoff adopts the semantic decomposition approach, and aims to build a model that is compatible with generative assumptions, including the nativist hypothesis and the modularity hypothesis.

5.5 Summary

In this chapter we have presented the four fundamental principles that characterise the approach to linguistic meaning known as cognitive semantics. In contrast to **objectivist semantics**, cognitive semantics adopts the position that language refers not to an objective reality, but to concepts: the conventional meanings associated with words and other linguistic units are seen as relating to thoughts and ideas. Hence, the first main assumption of cognitive semantics concerns the nature of the relationship between conceptual structure and human interaction with, and awareness of, the external world of sensory experience. Cognitive semanticists posit the **embodied cognition thesis**: the idea that the nature of conceptual organisation arises from bodily experience. In other words, conceptual structure is meaningful in part because of the bodily experiences with which it is associated. The second assumption is that **semantic structure is conceptual structure**. The third assumption associated with cognitive semantics holds that **meaning representation is encyclopaedic**: words (and other linguistic units) are 'points of access' to vast repositories of knowledge concerning a particular lexical concept. The fourth assumption holds that language itself does not encode meaning. Instead, words (and other linguistic units) serve as 'prompts' for the construction of meaning. This gives rise to the thesis that **meaning construction is conceptualisation**, a dynamic process whereby linguistic units serve as prompts for an array of conceptual operations and the recruitment of background knowledge.

Further reading

Introductory texts

- **Croft and Cruse (2004)**
- **Lee (2001)**
- **Saeed (2002)**
- **Ungerer and Schmid (1996)**

These are all textbooks that provide good coverage of cognitive semantics. The Lee book is the most accessible. The Croft and Cruse book is the most advanced. The Saeed book is an excellent general introduction to the study of linguistic meaning, addressing both formal and non-formal perspectives, and includes one chapter focusing on cognitive semantics as well as a chapter on Jackendoff's conceptual semantics framework.

Foundational texts

The following are among the foundational book-length texts in cognitive semantics, providing an insight into issues explored, phenomena investigated and the kinds of methodologies employed. We will look in detail at all these theories in subsequent chapters.

- **Fauconnier (1994)**. Mental Spaces Theory.
- **Fauconnier and Turner (2002)**. Conceptual Blending Theory.
- **Johnson (1987)**. Image schemas.
- **Lakoff (1987)**. Addresses categorisation and provides a theory of mental models. Also addresses the philosophical basis of cognitive semantics.
- **Lakoff and Johnson (1980)**. The earliest sketch of Conceptual Metaphor Theory.

- **Lakoff and Johnson (1999)**. An updated and detailed treatment of Conceptual Metaphor Theory.
- **Langacker (1987)**. Part II presents an overview of the nature of semantic structure necessary in order to support grammatical representation in language.
- **Sweetser (1990)**. Addresses the metaphorical basis of meaning extension.
- **Talmy (2000)**. A compendium of Talmy's now classic papers detailing his work on the schematic systems that underpin linguistic organisation.

Theoretical and philosophical overviews

- **Johnson (1992)**
- **Lakoff (1987**: chapter 17)
- **Sinha (1999)**
- **Turner (1992)**

These are all article-length contributions by leading figures in cognitive semantics. They address both theoretical and philosophical issues relating to cognitive semantics.

Exercises

5.1 Defining cognitive semantics

'Cognitive semantics is an approach not a theory.' Discuss this statement. What does it mean? Do you agree?

5.2 Experience and conceptual structure

In example (1) in the main text, abstract states are conceptualised in terms of containers, which is shown by the use of the preposition 'in'. Now consider the following examples:

(a) The guard is on duty.
(b) The blouse is on sale.
(c) We're on red alert.

Can you think of a reason why states like these might be lexicalised using *on* rather than *in*? What does this reveal about the relationship between experience and conceptual structure?

5.3 Meaning construction and conceptualisation

Consider the following exchange at a dinner party, and answer the questions that follow.

> *Guest:* Where shall I sit?
> *Host:* Can you sit in the apple juice seat?

(i) If you were the guest what would you make of this? Make a list of all the possible interpretations of 'apple juice seat'.
(ii) What is the most likely meaning, from those you've listed, given the context of a dinner party?
(iii) Now imagine that the guest is teetotal and the rest of the guests are drinking wine with their dinner. What does this tell you about the meaning of 'apple juice seat'?
(iv) Finally, what does this example illustrate in light of our discussion of the role of language in meaning construction (section 5.1.4)?

5.4 Word meaning

Consider the following examples.

(a) That parked BMW over there is a fast car.
(b) They were travelling in the fast lane on the motorway.
(c) That car is travelling fast.
(d) He can think through a problem fast.
(e) Christmas went by fast this year.

Each of these uses of *fast* means something slightly different. Identify the meaning of *fast* in each sentence. What do these different readings reveal about the nature of word meaning?

5.5 Mappings

Consider the following exchange which takes place in a library:

> *Librarian:* Yes?
> *Elderly man:* I can't reach Shakespeare on the top shelf.

What does the sentence uttered by the elderly man mean? In light of the discussion of the three types of mapping proposed by Fauconnier (section 5.2.4), identify the type of mapping that accounts for the meaning of this sentence.

6

Embodiment and conceptual structure

This chapter explores in more detail two of the central principles of cognitive semantics introduced in Chapter 5. These are: (1) the thesis that conceptual structure derives from embodiment, also known as the **embodied cognition** thesis; and (2) the thesis that semantic structure reflects **conceptual structure**. The reason for exploring these two principles together in a single chapter is because they are inextricably linked: once we have established that conceptual structure is embodied, in the sense that the nature of our embodiment determines and delimits the range and nature of concepts that can be represented, we can then examine how these concepts are encoded and externalised via language by looking at how the language system provides meaning based on concepts derived from embodiment.

We address the thesis of embodied cognition by presenting the theory of **image schemas** developed by Johnson (1987), among others. As we began to see in the previous chapter, image schemas are relatively abstract conceptual representations that arise directly from our everyday interaction with and observation of the world around us. That is, they are concepts arising from embodied experience. Once we have described the research on image schemas, and how they derive from embodiment, we then address the second principle. This is the thesis that embodiment, as the basis of conceptual organisation, should be evident in semantic structure: the meanings associated with words and other linguistic elements. In order to explore this thesis, we examine Leonard Talmy's theory of conceptual structure. In his influential work, Talmy has argued that one of the ways that language encodes conceptual representation is by providing **structural meaning**, also known as **schematic meaning**. This kind of meaning relates to structural properties of **referents** (the entities that language describes: objects, people, and so on) and **scenes**

EMBODIMENT AND CONCEPTUAL STRUCTURE

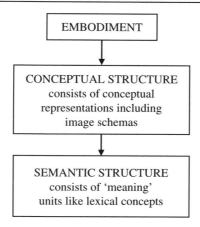

Figure 6.1 From embodiment to linguistic meaning

(the situations and events that language describes). Talmy argues that schematic meaning is directly related to fundamental aspects of embodied cognition, and can be divided into a number of distinct **schematic systems**, each of which provides a distinct type of meaning that is closely associated with a particular kind of embodied experience. Talmy's work presents compelling evidence from language that semantic structure reflects conceptual structure, and that conceptual structure arises from embodied experience.

The reader should bear in mind that Johnson's theory of image schemas and Talmy's work on the conceptual system represent two highly influential yet independent lines of research within cognitive semantics. However, we treat them together in this chapter because they relate to two of the most basic guiding principles of cognitive semantics: (1) that conceptual structure reflects embodied experience, which Johnson's theory addresses; and (2) that semantic structure reflects this conceptual structure, which Talmy's theory addresses. The relationship between these areas of investigation is represented in Figure 6.1.

6.1 Image schemas

In this section we consider the theory of image schemas, which was first developed within cognitive semantics and has come to be highly influential in neighbouring areas of study such as cognitive and developmental psychology. The notion of an image schema is closely associated with the development of the embodied cognition thesis, proposed by early researchers in cognitive semantics, notably George Lakoff and Mark Johnson. One of the central questions raised by Lakoff and Johnson in their (1980) book *Metaphors We Live By* can be stated as follows: Where does the complexity associated with our conceptual representation come from? The answer they offered was that this complexity is, in large measure, due to a tight correlation between the kinds of concepts human beings are capable of forming and the nature of the physical bodies we have. From this perspective, our embodiment is directly responsible for structuring concepts. In this section, therefore, we address the idea central to the thesis of embodied cognition: the image schema.

6.1.1 What is an image schema?

In his (1987) book *The Body in the Mind*, Mark Johnson proposed that embodied experience gives rise to image schemas within the conceptual system. Image schemas derive from sensory and perceptual experience as we interact with and move about in the world. For example, given that humans walk upright, and because we have a head at the top of our bodies and feet at the bottom, and given the presence of gravity which attracts unsupported objects, the vertical axis of the human body is functionally asymmetrical. This means that the vertical axis is characterised by an up-down or top-bottom asymmetry: the top and bottom parts of our bodies are different.

Cognitive semanticists argue that the asymmetry of the body's vertical axis is meaningful for us because of the way we interact with our environment. For example, gravity ensures that unsupported objects fall to the ground; given the asymmetry of the human vertical axis, we have to stoop to pick up fallen objects and look in one direction (downwards) for fallen objects and in another (upwards) for rising objects. In other words, our physiology ensures that our vertical axis, which interacts with gravity, gives rise to meaning as a result of how we interact with our environment.

According to Johnson, this aspect of our experience gives rise to an image schema: the UP-DOWN schema. Moreover, as shown by the developmental psychologist Jean Mandler, image schemas are **emergent**. This means that because this experience is a function of our bodies and of our interaction in the world, this type of experience arises in conjunction with our physical and psychological development during early childhood. In other words, image schemas are not claimed to be innate knowledge structures. For example, we know from work in developmental psychology that in the early stages of development infants learn to orient themselves in the physical world: they follow the motion of moving objects with their eyes, and later reach out their hands intentionally to grasp those moving objects and so on (Mandler 2004).

The term 'image' in 'image schema' is equivalent to the use of this term in psychology, where **imagistic** experience relates to and derives from our experience of the external world. Another term for this type of experience is **sensory experience**, because it comes from sensory-perceptual mechanisms that include, but are not restricted to, the visual system. Some of these sensory-perceptual mechanisms are summarised in Table 6.1. It is therefore important

Table 6.1 Some sensory-perceptual systems

System	Sensory experience	Physical location
Visual system	Vision	Eye, optic nerve
Haptic system	Touch	Beneath the skin
Auditory system	Hearing	Ear/auditory canal
Vestibular system	Movement/balance	Ear/auditory canal

to emphasise that although the term 'image' is restricted to visual perception in everyday language, it has a broader application in psychology and in cognitive linguistics, where it encompasses all types of sensory-perceptual experience.

Imagistic experience is contrasted with what psychologists call **introspective experience**: internal subjective experience such as feelings or emotions. The term 'schema' in 'image schema' is also very important: it means that image schemas are not rich or detailed concepts, but rather abstract concepts consisting of patterns emerging from repeated instances of embodied experience. If we take a parallel example from language, words like *thing* or *container* have rather more schematic meanings than words like *pencil* or *teacup*. This use of the term 'schema' is therefore consistent with the range of ways in which the term is used elsewhere in cognitive linguistics.

By way of illustration, the image schema CONTAINER results from our recurrent and ubiquitous experiences with containers as revealed by this extract from Johnson's (1987) book, which describes the start of an ordinary day:

> You wake *out of* a deep sleep and peer *out from* beneath the covers *into* your room. You gradually emerge *out of* your stupor, pull yourself *out from* under the covers, climb *into* your robe, stretch *out* your limbs, and walk *in* a daze *out* of the bedroom and *into* the bathroom. You look *in* the mirror and see your face staring *out* at you. You reach *into* the medicine cabinet, take *out* the toothpaste, squeeze *out* some toothpaste, put the toothbrush *into* your mouth, brush your teeth *in* a hurry, and rinse *out* your mouth. (Johnson 1987: 331; our italics differ from the original)

As this example reveals by the recurrent use of the expressions *in* and *out*, a great number of everyday objects and experiences are categorised as specific instances of the schematic concept CONTAINER: not only obvious containers like bathroom cabinets and toothpaste tubes or less obvious 'containers' like bed-covers, clothing and rooms, but also states like sleep, stupor and daze.

6.1.2 Properties of image schemas

In this section, we further develop the notion of image schema by outlining a number of properties associated with this aspect of the conceptual system.

Image schemas are pre-conceptual in origin

According to Johnson, image schemas like the CONTAINER schema are directly grounded in embodied experience: they relate to and derive from sensory experience. This means that they are pre-conceptual in origin. Mandler (2004) argues that they arise from sensory experiences in the early stages of human development that precede the formation of concepts. However, once the recurrent patterns of sensory information have been extracted and stored as an image schema, sensory experience gives rise to a conceptual representation. This means that image schemas are **concepts**, but of a special kind: they are the foundations of the conceptual system, because they are the first concepts to emerge in the human mind, and precisely because they relate to sensory-perceptual experience, they are particularly schematic. Sometimes it is more difficult to grasp the idea of an image-schematic concept than it is to grasp the idea of a very specific concept like CAT or BOOK. This is because these specific concepts relate to ideas that we are aware of 'knowing about'. In contrast, image schemas are so fundamental to our way of thinking that we are not consciously aware of them: we take our awareness of what it means to be a physical being in a physical world very much for granted because we acquire this knowledge so early in life, certainly before the emergence of language.

An image schema can give rise to more specific concepts

As we have already seen, the concepts lexicalised by the prepositions *in, into, out, out of* and *out from* in the passage cited above are all thought to relate to the CONTAINER schema: an abstract image-schematic concept that underlies all these much more specific **lexical concepts**. As we have seen in previous chapters, a lexical concept is a concept specifically encoded and externalised by a specific lexical form.

Of course, cognitive semanticists face the same problems that semanticists of any theoretical persuasion face in attempting to describe linguistic meaning in an economical and memorable way. There are a limited number of options available to us. Most semanticists, including cognitive semanticists, use words from natural language to represent pre-linguistic elements of meaning. Our use of words in small capitals to represent concepts is an example of this strategy. As we have already mentioned, some semanticists use a formal metalanguage, usually logic, to represent the meaning of larger units like sentences or propositions. Cognitive linguists often attempt to support their formal representations of meaning elements by using diagrams. Although concepts are labelled with ordinary words, the advantage of a diagram is that it can represent a concept independently of language.

EMBODIMENT AND CONCEPTUAL STRUCTURE

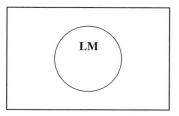

Figure 6.2 CONTAINER image schema

For example, the CONTAINER schema is diagrammed in Figure 6.2. This image schema consists of the structural elements interior, boundary and exterior: these are the minimum requirements for a CONTAINER (Lakoff 1987). The **landmark** (LM), represented by the circle, consists of two structural elements, the interior – the area within the boundary – and the boundary itself. The exterior is the area outside the landmark, contained within the square. The container is represented as the landmark because the boundary and the exterior together possess sufficient Gestalt properties (e.g. closure and continuity) to make it the figure, while the exterior is the ground (recall our discussion of Gestalt principles in Chapter 3).

Of course, the reason why this diagram does not resemble any specific type of container (like a teacup, a house or a bad mood) is precisely because of its schematic meaning. The idea behind this type of diagram is that it 'boils down' the image-schematic meaning to its bare essence, representing only those properties that are shared by all instances of the conceptual category CONTAINER.

Although Figure 6.2 represents the basic CONTAINER schema, there are a number of other image schemas that are related to this schema which give rise to distinct concepts related to containment. For instance, let's consider just two variants of the CONTAINER schema lexicalised by *out*. These image schemas are diagrammed in Figures 6.3 and 6.4, and are illustrated with linguistic examples. The diagram in Figure 6.3 corresponds to example (1). The **trajector** (TR) *John*, which is the entity that undergoes motion, moves from a position inside the LM to occupy a location outside the LM. The terms 'TR' and 'LM' are closely related to the notions of figure and reference object or ground that we discussed in Chapter 3. The terms 'TR' and 'LM' derive from the work of Langacker (e.g. 1987), and have been widely employed in cognitive semantics by scholars including Lakoff and Johnson, among others.

(1) John went out of the room.   OUT1

The image schema in Figure 6.4 corresponds to example (2). In this example, the meaning of *out* is 'reflexive', which is a technical way of saying that something refers to itself: we could paraphrase example (2), albeit redundantly, as

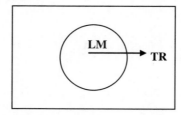

Figure 6.3 Image schema for OUT1

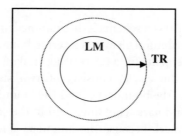

Figure 6.4 Image-schema for OUT2

The honey spread itself out. In other words, liquid substances like *honey*, because of their physical properties, can simultaneously be the LM and the TR. The LM is the original area occupied by the honey, while the honey is also the TR because it spreads beyond the boundary of its original location.

(2) The honey spread out.   OUT2

The image schemas shown in Figures 6.3 and 6.4 represent two concepts that are more specific and detailed than the image schema diagrammed in Figure 6.2, because they involve motion as well as containment. This shows that image schemas can possess varying degrees of schematicity, where more specific image schemas arise from more fundamental or schematic ones.

Image schemas derive from interaction with and observation of the world

As we have seen, because image schemas derive from embodied experience, they derive from the way in which we interact with the world. To illustrate this idea, consider the image schema for FORCE. This image schema arises from our experience of acting upon other entities, or being acted upon by other entities, resulting in the transfer of motion energy. Johnson illustrates the **interactional derivation** of this image schema (in other words, how it arises from experience) as follows:

> [F]orce is always experienced through interaction. We become aware of force as it affects us or some object in our perceptual field. When

you enter an unfamiliar dark room and bump into the edge of the table, you are experiencing the interactional character of force. When you eat too much the ingested food presses outwards on your taughtly stretched stomach. There is no schema for force that does not involve interaction or potential interaction. (Johnson 1987: 43).

The idea of FORCE is also central to Talmy's theory of conceptual structure, as we will see later in the chapter (section 6.2.2).

Image schemas are inherently meaningful

Because image schemas derive from interaction with the world, they are inherently meaningful. Embodied experience is inherently meaningful in the sense that embodied experiences have predictable consequences. Let's illustrate this point with another example. Imagine a cup of coffee in your hand. If you move the cup slowly up and down, or from side to side, you expect the coffee to move with it. This is because a consequence of containment, given that it is defined by boundaries, is that it constrains the location of any entity within these boundaries. In other words, the cup exerts force-dynamic control over the coffee. Of course, this seems rather obvious, but this kind of knowledge, which we take for granted, is acquired as a consequence of our interaction with our physical environment. For example, walking across a room holding a cup of coffee without spilling it actually involves highly sophisticated motor control that we also acquire from experience: we would be unlikely to ask a two-year-old to perform the same task. This experience gives rise to knowledge structures that enable us to make predictions: if we tip the coffee cup upside-down, the coffee will pour out.

The force-dynamic properties just described for the CONTAINER schema also show up in linguistic meaning, as illustrated by the meaning of the preposition *in*. Consider the diagram in Figure 6.5, drawn from the work of Claude Vandeloise (1994).

Vandeloise observes that the image depicted in Figure 6.5 could either represent a bottle or a lightbulb. Observe from example (3) that we can use the preposition *in* to describe the relation between *the lightbulb* (TR) and *the socket* (LM).

Figure 6.5 A bottle or a lightbulb? (Adapted from Vandeloise 1994)

(3) The bulb is in the socket.

In contrast, we cannot use the preposition *in* to describe the relation between a bottle and its cap, as example (4) shows. (The symbol preceding this example indicates that the sentence is semantically 'odd'.)

(4) The bottle is in the cap

Vandeloise points out that the spatial relation holding between the TR and LM in each of these sentences is identical, and yet while (3) is a perfectly acceptable sentence, (4) is semantically odd. Vandeloise suggests that it is not the spatial relation holding between the TR and LM that accounts for the acceptability or otherwise of *in*. He argues that the relevant factor is one of force-dynamics: '[W]hile the socket exerts a force on the bulb and determines its position, the opposite occurs with the cap and the bottle' (Vandeloise 1994: 173). In other words, not only is the position and the successful function of the bulb contingent on being *in* (contained by) the socket, but the socket also prevents the bulb from succumbing to the force of gravity and falling to the ground. In contrast, the position and successful function of the bottle is not contingent on being *in* the cap. This suggests that our knowledge of the functional consequences associated with the CONTAINER image schema affects the contextual acceptability of a preposition like *in*.

Image schemas are analogue representations

Image schemas are **analogue** representations deriving from experience. In this context, the term 'analogue' means image schemas take a form in the conceptual system that mirrors the sensory experience being represented. In other words, although we can try to describe image schemas using words and pictures, they are not represented in the mind in these kinds of **symbolic** forms. Instead, image-schematic concepts are represented in the mind in terms of holistic sensory experiences, rather like the memory of a physical experience. Let's illustrate this idea with an analogy: learning to drive a car properly cannot simply be achieved by reading a driving manual, or even by listening to a driving instructor explain the 'rules' of driving. At best, these provide very rough clues. Instead, we have to 'learn' how it 'feels' to drive a car by experiencing it at first hand. This learning is a complex process, during which we master an array of interrelated sensori-motor routines. Because image schemas derive from sensory experience, they are represented as summaries of **perceptual states** which are recorded in memory. However, what makes them conceptual rather than purely perceptual in nature is that they give rise to concepts that are consciously **accessible** (Mandler 2004). In other words, image schemas structure (more complex) lexical concepts.

EMBODIMENT AND CONCEPTUAL STRUCTURE

Figure 6.6 The PATH image schema

Image schemas can be internally complex

Image schemas are often, perhaps typically, comprised of more complex aspects that can be analysed separately. For example, the CONTAINER schema is a concept that consists of interior, boundary and exterior elements. Another example of a complex image schema is the SOURCE-PATH-GOAL or simply PATH schema, represented in Figure 6.6. Because a path is a means of moving from one location to another, it consists of a starting point or SOURCE, a destination or GOAL and a series of contiguous locations in between which relate the source and goal. Like all complex image schemas, the PATH schema constitutes an **experiential Gestalt**: it has internal structure but emerges as a coherent whole.

One consequence of internal complexity is that different components of the PATH schema can be referred to. This is illustrated in example (5), where the relevant linguistic units are bracketed. In each of these examples, different components of the path are profiled by the use of different lexical items.

(5) a. SOURCE
John left [England].
b. GOAL
John travelled [to France].
c. SOURCE-GOAL
John travelled [from England] [to France].
d. PATH-GOAL
John travelled [through the Chunnel] [to France].
e. SOURCE-PATH-GOAL
John travelled [from England] [through the Chunnel] [to France].

Image schemas are not the same as mental images

Close your eyes and imagine the face of your mother or father, child or close friend. This is a mental image, relatively rich in detail. Image schemas are not the same as mental images. Mental images are detailed and result from an effortful and partly conscious cognitive process that involves recalling visual memory. Image schemas are schematic and therefore more abstract in nature, emerging from ongoing embodied experience. This means that you can't close your eyes and 'think up' an image schema in the same way that you can 'think up' the sight of someone's face or the feeling of a particular object in your hand.

Image schemas are multi-modal

One of the reasons why we are not able to close our eyes and 'think up' an image schema is because image schemas derive from experiences across different modalities (different types of sensory experience) and hence are not specific to a particular sense. In other words, image schemas are buried 'deeper' within the cognitive system, being abstract patterns arising from a vast range of perceptual experiences and as such are not available to conscious introspection. For instance, blind people have access to image schemas for CONTAINERS, PATHS and so on precisely because the kinds of experiences that give rise to these image schemas rely on a range of sensory-perceptual experiences in addition to vision, including hearing, touch and our experience of movement and balance, to name but a few.

Image schemas are subject to transformations

Because image schemas arise from embodied experience, which is ongoing, they can undergo **transformations** from one image schema into another. In order to get a sense of what this means, consider the following example from Lakoff (1987):

> Imagine a herd of cows up close – close enough to pick out the individual cows. Now imagine yourself moving back until you can no longer pick out individual cows. What you perceive is a mass. There is a point at which you cease making out individuals and start perceiving a mass. (Lakoff 1987: 428)

According to Lakoff, perceptual experiences of this kind mediate a transformation between the COUNT image schema, which relates to a grouping of individual entities that can be individuated and counted, and the MASS image schema, which relates to an entity that is perceived as internally homogenous. The COUNT and MASS schemas are reflected in the grammatical behaviour of nouns, relating to the distinction between **count** and **mass nouns**. Count but not mass nouns can be determined by the indefinite article:

(6) a. He gave me a pen/crayon/ruler/glass of water.
    b. *He gave me a sand/money/gold

However, count nouns can be transformed into mass nouns and vice versa, providing linguistic evidence for the count-mass image-schematic transformation. If a count noun, like *tomato* in example (7), is conceived as a mass, it takes on the grammatical properties of a mass noun, as shown in (8).

(7) Count noun
   a. I have a tomato.
   b. *I have tomato

(8) Mass noun
   a. After my fall there was tomato all over my face.
   b. *After my fall there was a tomato all over my face

In essence, the grammatical transformation from count to mass, which Talmy (2000) calls **debounding**, and the transformation from mass to count, which he calls **excerpting**, is held to be motivated by an image-schematic transformation that underpins our ability to grammatically encode entities in terms of count or mass. As we will see, this distinction is also important in Lakoff's theory of word meaning, which we examine in Chapter 10.

Image schemas can occur in clusters

Image schemas can occur in clusters or networks of related image schemas. To illustrate this, consider again the FORCE schema, which actually consists of a series of related schemas. Force schemas share a number of properties (proposed by Johnson 1987) which are summarised in Table 6.2.

Johnson identifies no fewer than seven force schemas that share the properties detailed in Table 6.2. These schemas are shown in Figures 6.7 to 6.13 (after Johnson 1987: 45–8). The small dark circle represents the source of the force, while the square represents an obstruction of some kind. An unbroken arrow represents the force vector (the course taken by the force), while a broken arrow represents a potential force vector.

The first FORCE schema is the COMPULSION schema (Figure 6.7). This emerges from the experience of being moved by an external force, for example being pushed along helplessly in a large dense crowd, being blown along in a very strong wind and so on.

The second force-related image schema is the BLOCKAGE schema (Figure 6.8). This image schema derives from encounters in which obstacles resist force, for example when a car crashes into an obstacle like a tree.

Table 6.2 Shared characteristics of FORCE schemas

Force schemas are always experienced through interaction
Force schemas involve a force vector, i.e. a directionality
Force schemas typically involve a single path of motion
Force schemas have sources for the force and targets that are acted upon
Forces involve degrees of intensity
Forces involve a chain of causality, a consequence of having a source, target, force vector and path of motion, e.g. a child throwing a ball at a coconut

Figure 6.7 The COMPULSION image schema

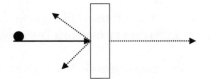

Figure 6.8 The BLOCKAGE image schema

Figure 6.9 The COUNTERFORCE image schema

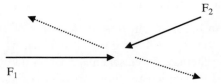

Figure 6.10 The DIVERSION image schema

The third force-related image schema is the CONTERFORCE schema (Figure 6.9). This derives from the experience of two entities meeting with equal force, like when we bump into someone in the street. F_1 and F_2 represent the two counterforces.

The fourth force-related image schema is the DIVERSION schema (Figure 6.10). This occurs when one entity in motion meets another entity and this results in diversion. Examples include a swimmer swimming against a strong current so that she is gradually pushed along the shoreline, or the ricochet of a bullet.

The fifth force-related image schema is the REMOVAL OF RESTRAINT schema (Figure 6.11). This captures a situation in which an obstruction to force is removed, allowing the energy to be released. This describes a situation like leaning on a door that suddenly opens.

The sixth force-related image schema is the ENABLEMENT schema (Figure 6.12). This image schema derives from our sense of potential energy, or lack of it, in relation to the performance of a specific task. While most people who are fit and well feel able to pick up a bag of grocery shopping, for example, few people feel able to lift up a car. It is important to observe that while this image schema does not involve an actual force vector, it does involve a poten-

EMBODIMENT AND CONCEPTUAL STRUCTURE

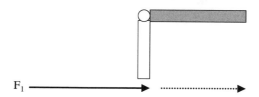

Figure 6.11 The REMOVAL OF RESTRAINT image schema

Figure 6.12 The ENABLEMENT image schema

Figure 6.13 The ATTRACTION image schema

tial force vector. According to Johnson, it is this property that marks the ENABLEMENT schema as a distinct image schema.

Finally, the ATTRACTION schema (Figure 6.13) derives from experiences in which one entity is drawn towards another entity due to the force exerted upon it. Examples include magnets, vacuum cleaners and gravity.

6.1.3 Image schemas and linguistic meaning

As we have begun to see in our discussions of the preposition *in* (recall examples (3)–(4)) and the distinction between count and mass nouns (recall examples (6)–(8)), image schemas can serve as the conceptual representation that underpins lexical items. In this section, we briefly examine the relationship between the FORCE schemas we have just considered and the English **modal auxiliary verbs** (e.g. *must, may, can*). Johnson suggests that certain FORCE schemas underlie the basic or **root** meanings of these verbs: these meanings relate to sociophysical experience, as illustrated in the following sentences:

(9) a. You **must** move your foot or the car will crush it.
 [physical necessity]
 b. You **may** now kiss the bride.
 [no parental, social or institutional barrier now prevents the bride from being kissed by the groom]
 c. John **can** throw a javelin over 20 metres.
 [he is physically capable of doing this]

Johnson argues that the root meaning of *must* (physical necessity) derives from the COMPULSION schema, while the root meaning of *may* (permission) to relates to the REMOVAL OF RESTRAINT schema and the root meaning of *can* (physical capacity) derives from the ENABLEMENT schema. Thus his claim is that the meanings associated with the modal verbs have an image-schematic basis which arises from embodied experience.

6.1.4 A provisional list of image schemas

To consolidate the discussion of image schemas presented in this section, we provide in Table 6.3 a list of image schemas compiled from Cienki (1998), Gibbs and Colston (1995), Johnson (1987), Lakoff (1987) and Lakoff and Turner (1989). While far from exhaustive, this list provides an idea of the range of image schemas that have been proposed so far in the literature. Following suggestions by Clausner and Croft (1999), we group the image schemas according to the nature of their experiential grounding, although our listing is arranged slightly differently.

6.1.5 Image schemas and abstract thought

One of the most striking claims made by cognitive semanticists is that abstract thought has a bodily basis. In their influential research on **conceptual metaphors**, George Lakoff and Mark Johnson (1980) have argued that conceptual structure is in part organised in terms of a **metaphor system**, which is characterised by related sets of conventional associations or **mappings** between concrete and abstract **domains**. A domain *in Conceptual Metaphor Theory* is a body of knowledge that organises related concepts. The importance of image schemas is that they can provide the concrete basis for these metaphoric mappings. We have seen some examples like this in earlier

Table 6.3 A partial list of image schemas

SPACE	UP-DOWN, FRONT-BACK, LEFT-RIGHT, NEAR-FAR, CENTRE-PERIPHERY, CONTACT, STRAIGHT, VERTICALITY
CONTAINMENT	CONTAINER, IN-OUT, SURFACE, FULL-EMPTY, CONTENT
LOCOMOTION	MOMENTUM, SOURCE-PATH-GOAL
BALANCE	AXIS BALANCE, TWIN-PAN BALANCE, POINT BALANCE, EQUILIBRIUM
FORCE	COMPULSION, BLOCKAGE, COUNTERFORCE, DIVERSION, REMOVAL OF RESTRAINT, ENABLEMENT, ATTRACTION, RESISTANCE
UNITY/MULTIPLICITY	MERGING, COLLECTION, SPLITTING, ITERATION, PART-WHOLE, COUNT-MASS, LINK(AGE)
IDENTITY	MATCHING, SUPERIMPOSITION
EXISTENCE	REMOVAL, BOUNDED SPACE, CYCLE, OBJECT, PROCESS

chapters: for example, recall our discussion in Chapter 5 of the conceptual metaphor STATES ARE CONTAINERS. Let's consider one more example.

Consider the image schema OBJECT. This image schema is based on our everyday interaction with concrete objects like desks, chairs, tables, cars and so on. The image schema is a schematic representation emerging from embodied experience, which generalises over what is common to objects: for example, that they have physical attributes such as colour, weight and shape, that they occupy a particular bounded region of space, and so forth. This image schema can be 'mapped onto' an abstract entity like 'inflation', which lacks these physical properties. The consequence of this metaphoric mapping is that we now understand an abstract entity like 'inflation' in terms of a physical object. This is illustrated by the examples in (10).

(10) a. If there's much more inflation we'll never survive.
b. Inflation is giving the government a headache.
c. Inflation makes me sick.
d. Lowering interest rates may help to reduce the effects of inflation.

Notice that it is only by understanding 'inflation' in terms of something with physical attributes that we can quantify it and talk about its effects. Thus image schemas which relate to and derive ultimately from pre-conceptual embodied experience can serve to structure more abstract entities such as inflation. We return to a detailed investigation of conceptual metaphor in Chapter 9.

6.2 Conceptual structure

In this section, we explore the thesis that **semantic structure** encodes and externalises conceptual structure. As we explained in the introduction to this chapter, this issue follows on from our investigation of the embodied cognition thesis: once we have uncovered evidence for the idea that embodied experience determines and delimits the range and nature of concepts that can be represented, we can then examine how these concepts are encoded and externalised in language. We do this by looking at how the language system provides meaning based on concepts derived from embodiment.

As we also mentioned in the introduction to this chapter, Talmy has argued that one of the ways that language reflects conceptual representation is by providing **structural meaning**, also known as **schematic meaning**. This kind of meaning relates to structural properties of **referents** (the entities that language describes) and **scenes** (the situations that these entities are involved in). Talmy also argues that this schematic meaning is directly related to fundamental aspects of embodiment.

6.2.1 Semantic structure

Linguistic expressions refer to entities or describe situations or scenes. Entities and scenes can be relatively concrete objects or events, or they can relate to more subjective experiences, such as feeling remorse or joy or experiencing unrequited love. According to Talmy, the way language conveys entities and scenes is by reflecting or encoding the language user's **Cognitive Representation (CR)** or conceptual system. In other words, although the conceptual system is not open to direct investigation, the properties of language allow us to reconstruct the properties of the conceptual system and to build a model of that system that, among other things, explains the observable properties of language. Talmy suggests that the CR, as manifested in language, is made up of two systems, each of which brings equally important but very different dimensions to the scene that they construct together. These systems are the **conceptual structuring system** and the **conceptual content system**. While the conceptual structuring system, as its name suggests, provides the structure, skeleton or 'scaffolding' for a given scene, the content system provides the majority of rich substantive detail. It follows from this view that the meaning associated with the conceptual structuring system is highly schematic in nature, while the meaning associated with the conceptual content system is rich and highly detailed. This distinction is captured in Figure 6.14.

It is important to emphasise that the system represented in Figure 6.14 relates to the conceptual system as it is encoded in semantic structure. In other words, semantic structure represents the conventional means of encoding conceptual structure for expression in language. The bifurcation shown in Figure 6.14 reflects the way language conventionally encodes the conceptual structure that humans externalise in language. Nevertheless, we reiterate a point here that we made in Chapter 5: while lexical concepts are conceptual in nature, in the sense

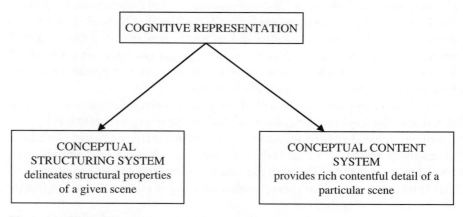

Figure 6.14 The bifurcation in the cognitive representation (CR)

that they prompt for conceptual structures of various kinds, the range of lexical concepts conventionally encoded in language must represent only a small fraction of the range and complexity of conceptual structure in the mind of any given human being. Indeed, as we will see in various chapters throughout Part II of the book, the range of concepts available in the conceptual system and the meaning potential associated with these concepts is vast. This means that while semantic structure must, to some extent at least, reflect conceptual structure, and while semantic structure can be thought of as a subset of conceptual structure – a system of lexical concepts specialised for expression in language – the relationship between conceptual structure and semantic structure is nevertheless complex and indirect. (As we will see later in this part of the book, the conceptual structure associated with linguistic units such as words are prompts for complex processes of conceptualisation, what Gilles Fauconnier refers to as **backstage cognition**.)

Given the hypothesis that semantic structure reflects conceptual structure, the system of semantic structure is also divided into two subsystems, reflecting the bifurcation in the CR. These two systems are the **open-class semantic system** and the **closed-class semantic system** that have already been introduced in previous chapters. These semantic subsystems correspond to the formal distinction between **open-class elements** (for example, nouns like *man*, *cat*, *table*, verbs like *kick*, *run*, *eat*, and adjectives like *happy*, *sad*) and **closed-class elements** (idioms like *kick the bucket*, grammatical patterns like declarative or interrogative constructions, grammatical relations like subject or object, word classes like the category verb, grammatical words like *in* or *the*, and bound morphemes like *-er* in *singer*).

As we have seen, the crucial difference between open-class and closed-class semantics is that while open-class semantics provides rich content, closed-class semantics contributes primarily to the structural content. However, a caveat is in order here. Given the view within cognitive linguistics that meaning and grammar cannot be divorced, the division of semantic structure into two subsystems sets up a somewhat artificial boundary (as we will see in Part III of the book). After all, free morphemes like prepositions (*in*, *on*, *under* and so on) which belong to the closed-class system exhibit relatively rich meaning distinctions. Therefore the distinction between the closed-class and open-class semantic subsystems might be more insightfully viewed in terms of distinct points on a continuum rather than in terms of a clear dividing line. We will elaborate this position in Part III by presenting the arguments put forward by cognitive grammarian Ronald Langacker, who suggests that while there is no principled distinction between the lexicon and the grammar, there are nevertheless qualitatively distinct kinds of phenomena that can be identified at the two ends of the continuum. The idea of a lexicon-grammar continuum is represented in Figure 6.15. We might place a lexical concept like FLUFFY at the open-class end,

⟵ Open-class elements   Closed-class elements ⟶

Figure 6.15 The lexicon–grammar continuum

and the concept PAST relating to a grammatical morpheme like *-ed* at the closed-class end, while the lexical concept relating to *in* might be somewhere in the middle of the continuum.

Talmy's research has examined the way in which both the open-class and closed-class semantic systems encode the CR. However, he has been primarily concerned with elaborating the semantics of the closed-class subsystem, the part of semantic structure that is at the grammar 'end' of the continuum shown in Figure 6.15. We defer a detailed presentation of this aspect of Talmy's theory until Part III of the book which explicitly focuses on grammar (Chapter 15). However, Talmy's work is important for our investigation of cognitive semantics for at least two reasons: (1) Talmy's theory illustrates that the closed-class or grammatical subsystem is meaningful (albeit schematic); (2) Talmy's findings suggest that the grammatical subsystem encodes meaning that relates to key aspects of embodied experience, such as the way SPACE and TIME are configured in language, and the way that the closed-class system encodes experiential meaning arising from phenomena such as attention, perspective and force-dynamics. For these reasons, Talmy's research both illustrates and supports the position adopted in cognitive semantics that semantic structure reflects conceptual structure which in turn reflects embodied experience. We turn next to Talmy's proposals concerning the schematic systems that comprise the CR.

6.2.2 Schematic systems

According to Talmy the conceptual structuring system is based upon a limited number of large-scale **schematic systems**. These provide the basic organisation of the CR upon which the rich content meaning encoded by open-class elements can be organised and supported. The basic architecture of these schematic systems has been described in a series of highly influential papers by Leonard Talmy, which are collected in his two-volume set *Toward a Cognitive Semantics* (2000).

Talmy proposes that various schematic systems collaborate to structure a scene that is expressed via language. Each schematic system contributes different structural aspects of the scene, resulting in the overall delineation of the scene's skeletal framework. There are four key schematic systems identified by Talmy: (1) the **'Configurational System'**; (2) the **'Perspectival System'**; (3) the **'Attentional System'**; and (d) the **'Force-Dynamics System'** (see Figure 6.16). We provide a brief overview of each of these systems in turn.

EMBODIMENT AND CONCEPTUAL STRUCTURE

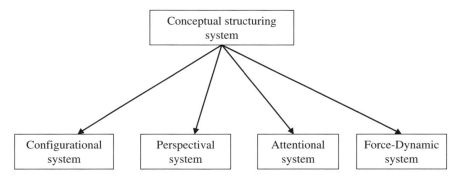

Figure 6.16 The key schematic systems within the 'Conceptual Structuring System'

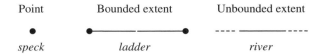

Figure 6.17 Degree of extension for matter (adapted from Talmy 2000: 61)

The 'Configurational System'

The 'Configurational System' structures the temporal and spatial properties associated with a scene, such as the division of a scene into parts and participants. Schematic systems like the 'Configurational System' can be further divided into **schematic categories**. In order to see how both the open-class and closed-class semantic systems encode configurational structure, we will consider one example of a schematic category within this system: the category **degree of extension**. Degree of extension relates to the degree to which matter (space) or action (time) are extended. Consider the open-class words *speck, ladder* and *river*, which exemplify this category as it relates to matter. The degree of extension of each of these is illustrated in Figure 6.17.

Lexical items like these include in their semantic specification information relating to degree of extension. For example, part of the meaning of *river* is schematic, relating to the degree of extension associated with rivers. The rich encyclopaedic meaning associated with the lexical item *river* relates to its specific properties as an entity involving water, which occupies a channel of certain dimensions, and which flows under the force of gravity from higher ground sometimes over many miles to the sea, and so on. In contrast to this rich and detailed specific meaning, its schematic meaning concerns the degree of extension associated with this entity. The schematic category 'degree of extension' has three values: a point, a bounded extent or an unbounded extent. Rivers are typically unbounded within the perceptual

195

field of a human experiencer. In other words, while we may know from looking at maps that rivers have beginnings and ends and are thus bounded, our 'real' experience of rivers is usually that they are unbounded because we cannot see the beginning and end.

The examples in (11)–(13) relate to action rather than matter, and employ closed-class elements in order to specify the degree of extension involved. (Note that 'NP' stands for noun phrase; the relevant NP is bracketed.)

(11) Point $\quad$ at + $NP_{\text{point-of-time}}$
The train passed through at [noon].

(12) Bounded extent $\quad$ in + $NP_{\text{extent-of-time}}$
She went through the training circuit in [five minutes flat].

(13) Unbounded extent $\quad$ '$keep$ -ing' + '-er and -er'
The plane kept going higher and higher.

As these examples illustrate, some closed-class elements encode a particular degree of extension. For instance, in (11) the preposition *at* together with an NP that encodes a temporal point encodes a point-like degree of extension. The NP does not achieve this meaning by itself: if we substitute a different preposition, a construction containing the same NP *noon* can encode a bounded extent (e.g. *The train arrives between noon and 1 pm*). The punctual nature of the temporal experience in example (11) forms part of the conceptual structuring system and is conveyed in this example by the closed-class system. The nature of the punctual event, that is the passage of a train through a station rather than, say, the flight of a flock of birds overhead, relates to the conceptual content system.

In the example in (12), the preposition *in* together with an NP that encodes a bounded extent encodes a bounded degree of extension. In (13) the closed-class elements *keep* -*ing* + -*er* and -*er* encodes an unbounded degree of extension. Each of these closed-class constructions provides a grammatical 'skeleton' specialised for encoding a particular value within the schematic category 'degree of extension'. The conceptual content system can add dramatically different content meaning to this frame (e.g. *keep singing louder and louder; keep swimming faster and faster; keep getting weaker and weaker*), but the schematic meaning contributed by the structuring system remains constant (in all these examples, time has an unbounded degree of extension).

The 'Perspectival System'

In contrast to the 'Configurational System' which partitions a scene into actions and participants with certain properties, the 'Perspectival System' specifies the perspective from which one 'views' a scene. This system includes schematic categories that relate to the spatial or temporal **perspective point** from which a scene is viewed, the distance of the perspective point from the entity viewed, the change of perspective point over time and so on. To illustrate this system, we will consider one schematic category subsumed by this system, namely **perspectival location** (traditionally called **deixis**). This relates to the position of a **perspective point** or **deictic centre** from which a scene is 'viewed'. In intuitive terms, the deictic centre corresponds to the 'narrator', from whose perspective you can imagine the scene being described. In spoken language, the 'narrator' is the speaker. In each of the following two examples, the perspective point from which the scene is described is different. In (14), the perspective point is located inside the room, while in (15) the perspective point is located outside the room.

(14) Interior perspective point
      The door slowly opened and two men walked in.

(15) Exterior perspective point
      Two men slowly opened the door and walked in.
      (Talmy 2000: 69)

Examples like these raise the following question: how do we know where the perspective point is located? After all, there does not appear to be anything in these sentences that explicitly tells us where it is. However, it is not the case that there is no explicit encoding that conveys the perspective point. It is simply that the perspective point is encoded by the grammatical or closed-class system: here, by the grammatical construction of the sentence. In example (14), the subject of the sentence is *the door*, which is the THEME: a passive entity whose location or state is described. In this example, *open* is an intransitive verb: it requires no object. In example (15), the subject of the sentence is *two men*, which is the AGENT: the entity that intentionally performs the action of opening the door. In this example, *open* is transitive (it requires an object: *the door*).

Why does changing the grammatical structure of the sentence, and thus the subject, affect our understanding of the perspective point? The reason is that what comes first in the sentence (the subject) corresponds to what is viewed first by the speaker/narrator, and this provides us with clues for reconstructing the perspective point. In the first clause of example (14), the initiator(s) of the action are not mentioned, so we deduce that the initiators of the action are not visible. From this we conclude that the perspective point must be inside the room. In example (15) the initiators of the event are mentioned first, so we deduce that the perspective point is exterior to the room. The way in which grammatical organisation mirrors experience is called **iconicity**. This features prominently in explanations offered by functional typologists (see Croft 2002), and has also influenced the cognitive semantics framework. These examples illustrate that the grammatical organisation of the sentence provides schematic information that enables us to determine where the perspective point is located.

The 'Attentional System'

This system specifies how the speaker intends the hearer to direct his or her attention towards the entities that participate in a particular scene. For instance, this system can direct attention to just one part of a scene. By way of illustration, consider the pattern of distributing attention that is called the **windowing of attention**:

(16) a. Initial and final windowing
 The crate fell out of the plane into the ocean.
 b. Initial, medial and final windowing
 The crate fell out of the plane, through the air and into the sea.

The examples in (16) relate to **path windowing**. Path windowing is a way of focusing attention on a particular subpart of a path of motion. Consider the path of motion represented in Figure 6.18, where the line between point A and point B represents the path of motion followed by a crate that falls from an airborne plane travelling over water. Point A represents the initial location of the crate, the line represents the trajectory of descent and point B represents the final location of the crate once it hits the water.

Path windowing allows language users to **window** (focus attention on) subparts of the trajectory associated with the motion of an object. In principle, windowing can operate over the initial portion of the path, the medial portion or the final portion. The examples in (17) illustrate some more of the ways in which language can encode the windowing of attention. Recall from our discussion of example (5) that it is the internal complexity of the PATH image

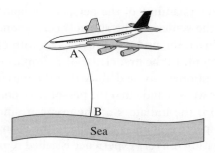

Figure 6.18 The path associated with an object falling out of a plane

schema that enables attention to be focused on distinct subparts of the path of motion. The initial, medial and final windows therefore correspond to the SOURCE, PATH and GOAL of the image schema, respectively.

(17) a. Medial and final windowing
   The crate fell [through the air] and [into the ocean].
   b. Initial windowing
   The crate fell [out of the plane].
   c. Medial windowing
   The crate fell [through the air].
   d. Final windowing
   The crate fell [into the ocean].

The 'Force-Dynamics System'

Talmy argues that this system, as it is manifested in semantic structure, relates to the way in which objects are conceived relative to the exertion of force. It is worth pointing out that while the other schematic systems we have discussed so far relate primarily to information derived from visual perception, the 'Force-Dynamics System' derives from **kinaesthesia** (our bodily experience of muscular effort or motion) and **somesthesia** (our bodily experience of sensations such as pressure and pain). To illustrate this system and the linguistic devices that give rise to force-dynamics distinctions, consider the following examples drawn or adapted from Talmy (2000: 412).

(18) Physical force
   a. The ball <u>was rolling</u> along the beach
   b. The ball <u>kept rolling</u> along the beach

The examples in (18) highlight a contrast in physical force. The expression in (18a) depicts a scene that is neutral with respect to force, in the sense that, while encyclopaedic knowledge tells us that something or someone must have caused the motion of the ball, the sentence does not refer to this knowledge. In contrast, the use of the *keep* V-*ing* construction in (18b) conveys a scene in which we understand that the ball's natural tendency towards rest is overcome by some external force, perhaps the wind, which ensures that the ball remains in a state of motion. Again, the only difference between these two examples is in the grammatical constructions: specifically, the auxiliary verb *be* versus the quasi-auxiliary *keep*, together with the progressive participle V-*ing*. According to Talmy, FORCE forms part of the conceptual structure associated with our CR, the 'Force-Dynamics System', and can be encoded via closed-class elements like grammatical constructions.

The 'Force-Dynamics System' does not just relate to physical force, but can also relate to 'psychological' force. Consider example (19).

(19) Psychological force
    a. He <u>didn't close the door</u>.
    b. He <u>refrained from closing the door</u>.

In this example, the contrast is between an AGENT's non-action, as in (19a), and the AGENT's resistance of the urge to act, as in (19b). In other words, the construction *not VP* in (19a) is, like (19a), neutral with respect to force. In contrast, the construction *refrain from* VP*ing* encodes a force-dynamics conflict internal to the agent.

Finally, consider example (20), which illustrates social force.

(20) Social force
    a. She'<u>s got to go to the park</u>.
    b. She <u>gets to go to the park</u>.

The *have (got) to* VP construction in (20a) encodes a scene in which the subject's desire not to act is overcome by an external force so that she is forced to act. Our encyclopaedic knowledge tells us that the force that obliges someone to go to the park is likely to be of a social rather than a physical nature: this construction therefore expresses obligation. The *get to VP* construction in (20b), on the other hand, encodes a scene in which the subject's desire to act is unimpeded by any external inhibiting force so that she is able to act. This construction therefore expresses permission. Both scenes depict the same end result, but the grammatical constructions encode different force-dynamics of a social nature that lead to this result.

The discussion in this section has provided only the briefest introduction to a number of extremely complex schematic systems proposed by Talmy, each of which consists of a number of schematic categories. It is important to point out that the systems described here do not, in all likelihood, represent an exhaustive list of the subsystems that make up the conceptual structuring system, as Talmy himself acknowledges. However, even this brief discussion reveals that systematic patterns in language, both in the open-class and the closed-class semantic systems, represent evidence for a conceptual system that structures knowledge according to embodied experience. As this discussion indicates, Talmy's theory requires a significant grammatical vocabulary in order to be fully understood. For this reason, we defer a more detailed investigation of this model until Part III of the book (Chapter 15), where our focus is on cognitive approaches to grammar.

6.3 Summary

This chapter has explored two guiding principles of cognitive semantics: (1) the thesis that conceptual structure derives from **embodied experience**; and (2) the thesis that semantic structure reflects **conceptual structure**. Conceptual structure is the cognitive system that represents and organises experience in a form that can serve as the input for processes like reasoning and expression in language. Semantic structure is the system wherein concepts are conventionally encoded in a form in which they can be externalised by language. The first part of the chapter focused on the relationship between embodied experience and conceptual structure, and introduced the theory of **image schemas**. Image schemas are relatively abstract representations that derive from our everyday interaction with and observation of the world around us. These experiences give rise to embodied representations that, in part, underpin conceptual structure. The second part of the chapter addressed the relationship between conceptual structure and semantic structure, and introduced Talmy's theory of the conceptual system. On the basis of evidence from linguistic representation, conceptual structure can be divided into two systems, the **conceptual structuring system** and the **conceptual content system**. While the conceptual structuring system provides structural or schematic information relating to a particular scene, the conceptual content system provides the rich content or detail. Talmy argues that the conceptual structuring system can be divided into a number of **schematic systems** which together serve to provide the structure or 'scaffolding' for the rich content provided by the conceptual content system. Crucially, the nature of these schematic systems relates to fundamental aspects of embodied sensory-perceptual experience, such as how referents and scenes encoded in language are structured, the perspective taken with respect to such scenes, how attention is directed within scenes and force-dynamics properties. In sum, both the open-class and closed-class semantic systems reflect and encode fundamental aspects of embodied experience, mediated by conceptual structure.

Further reading

Image schemas: theory and description

- **Cienki (1998).** An in-depth analysis of the single image schema STRAIGHT, its experiential basis and its metaphoric extensions, with data from English, Japanese and Russian.
- **Hampe (forthcoming).** This excellent collected volume constitutes an up-to-date review by leading authors of the state of the art in image schema research. Of particular importance are the papers by Grady,

Johnson and Rohrer, and Zlatev, who develops the notion of what he refers to as the 'mimetic schema'.

- **Johnson (1987).** Mark Johnson's book represents the original statement on image schemas; now a classic.
- **Lakoff (1987).** Lakoff discusses image schemas in the development of his theory of cognitive models. See in particular his influential study of *over*.
- **Lakoff (1990).** Lakoff explores the thesis that metaphoric thought is due to image schemas and their extensions to abstract domains.

Applications of image schema theory

- **Gibbs and Colston (1995).** This paper reviews findings from psycholinguistics and cognitive and developmental psychology that support the position that image schemas are psychologically real.
- **Mandler (2004).** Jean Mandler is a developmental psychologist. She argues that image schemas may form the basis of early conceptual development in infants.
- **Turner (1996).** Mark Turner, an influential figure in cognitive linguistics, applies the notion of image schemas to literary and poetic thought and language.

Schematic systems

- **Talmy (2000).** Chapter 1 of the first volume provides an influential discussion of the Cognitive Respresentation system (CR), and how it relates to the concept and content structuring systems and closed-class and open-class semantics. This volume also collects together Talmy's influential papers on the schematic systems.

Exercises

6.1 Image schemas

A number of image schemas are listed below. We have seen that image schemas derive from embodied experience. Make a list of the kinds of situations that are likely to give rise to these image schemas and the sensory-perceptual modalities to which these experiences relate (you may wish to consult Table 6.1). The first example has been done for you.

(a) COMPULSION *situations*: being moved by external forces like wind, water, physical objects and other people *sensory-perceptual modalities*: haptic system (touch, pressure on skin); vestibular system (balance, orientation); kinaesthesia (awareness of motion, other-initiated motion, inability to stop oneself from moving, directionality of motion, and so on)

(b) CONTAINER
(c) MATERIAL OBJECT
(d) PROCESS
(e) CENTRE–PERIPHERY
(f) CONTACT
(g) NEAR–FAR
(h) SCALE

6.2 Image schemas and metaphor

Consider the following sentences. Identify the image schemas that serve as source domains in these sentences.

(a) We need to weigh up the arguments.
(b) They're in trouble.
(c) The logic of her argument compelled me to change my mind.
(d) Interest rates have gone up again.
(e) The current rate of borrowing on credit will prove to be a heavy burden for the nation.

6.3 Cognitive Representation

List the main differences between the conceptual structuring and conceptual content systems. How are these systems reflected in language? Can you provide some examples of your own to illustrate your answer?

6.4 Schematic category: degree of extension

In view of the discussion of the schematic category 'degree of extension', consider the following examples. Identify the sentences that relate to point, bounded extent and unbounded extent. Some of the sentences relate to matter (SPACE) and action (TIME). Identify which is which. You may wish to refer to Figure 6.17.

(a) When the sheep all died, we moved out of the farm.
(b) The house is (exactly) 10 metres away from the farm.
(c) The sheep kept dying.
(d) The house seems to go on and on.

(e) I read that book twenty years ago.
(f) The house is 10 metres wide.
(g) The sheep all died in six weeks.
(h) She read the book in two days.
(i) She kept reading the book.

6.5 The intersection of schematic categories

Consider two new schematic categories that relate to the configurational system: 'plexity' and 'state of boundedness'. The category 'plexity' relates to the division of matter or action into equal elements. In the domain of matter, plexity relates to the grammatical category 'number' with its member notions 'singular' and 'plural'. In the domain of action it relates to the traditional aspectual distinction between 'semelfactive' and 'iterative' (the distinction between one and more than one instance of a point-like event, respectively). This category and its member notions of 'uniplex' and 'multiplex' are illustrated below:

	Matter	*Action*
Uniplex	A bird flew in.	He sighed (once).
Multiplex	Birds flew in.	He kept sighing.

Now consider the schematic category 'state of boundedness'. This relates to the categories count noun and mass noun, and to the distinction between perfective and imperfective verbs (these describe events that change through time or remain constant through time, respectively). This category has two member notions, 'bounded' and 'unbounded' as illustrated below:

	Matter	*Action*
Unbounded	Water makes up three-quarters of the planet.	The Eiffel Tower stands across from the Trocadero.
Bounded	We came across a small lake.	She kicked the ball.

These schematic categories intersect. For instance, the lexical item *timber* is both unbounded (consisting of the set of all trees) and multiplex (consisting of more than one element). Place the following lexical items in the appropriate place in the table provided below:

(a) furniture
(b) (a) grove
(c) (a) cat
(d) (to) snore
(e) (to) moult, e.g. *The dog moulted*
(f) (a) tree
(g) (to) breathe
(h) (a) family

	Uniplex	*Multiplex*
Bounded		
Unbounded		

Now consider the lexical item *trees*. Where would you place this? Did you have any difficulties in deciding? What does this illustrate?

Finally, state which of the lexical items relates to matter and which to action. Is there a distinction in terms of word class ('part of speech')?

7

The encyclopaedic view of meaning

In this chapter we explore the thesis that meaning is encyclopaedic in nature. This thesis, which we introduced in Chapter 5, is one of the central assumptions of cognitive semantics. The thesis has two parts associated with it. The first part holds that semantic structure (the meaning associated with linguistic units like words) provides access to a large inventory of structured knowledge (the conceptual system). According to this view, word meaning cannot be understood independently of the vast repository of **encyclopaedic knowledge** to which it is linked. The second part of the thesis holds that this encyclopaedic knowledge is grounded in human interaction with others (social experience) and the world around us (physical experience). We will look in detail at the two parts of this thesis, and at the end of the chapter we also briefly consider the view that encyclopaedic knowledge, accessed via language, provides **simulations** of perceptual experience. This relates to recent research in cognitive psychology that suggests that knowledge is represented in the mind as **perceptual symbols**.

In order to investigate the nature of encyclopaedic knowledge, we explore two theories of semantics that have given rise to this approach to meaning. These are (1) the theory of **Frame Semantics**, developed in the 1970s and 1980s by Charles Fillmore; and (2) the theory of **domains**, developed by Ronald Langacker (1987). In fact, these two theories were originally developed for different purposes: Fillmore's theory derived from his research on **Case Grammar** in the 1960s, and continued to be developed in association with his (and others') work on **Construction Grammar** (see Part III). Langacker's theory of domains provides part of the semantic basis for his theory of **Cognitive Grammar** (also discussed in Part III). However, despite these different starting points, both theories address related phenomena. For this reason, we suggest that together they form the basis for a theory of **encyclopaedic semantics**. We will see that Langacker argues that **basic domains**, knowledge structures derived from pre-conceptual sensory-perceptual experience, form the basis of more complex **abstract domains** which correspond to the **semantic frames** proposed by Fillmore. Together, these two types of knowledge structure make up encyclopaedic knowledge. Indeed, this perspective is presupposed by much current work on word meaning and conceptual structure in cognitive semantics.

At this point, it is worth explaining why this chapter focuses on encyclopaedic knowledge, while a later chapter (Chapter 10) focuses on word meaning. After all, when we introduced the idea of encyclopaedic knowledge in Chapter 5, we illustrated it with the proposition that words provide a 'point of access' to this system of knowledge, and indeed we will have quite a bit to say about word meaning in this chapter. However, the focus of this chapter is to explore in detail the **system of conceptual knowledge** that lies behind lexical concepts and their associated linguistic units, while the focus of Chapter 10 is to explore in detail the nature and organisation of those lexical concepts themselves.

7.1 Dictionaries versus encyclopaedias

We begin by considering the traditional view of linguistic meaning, which is often called the **dictionary view**. By explaining how this traditional model works, we will establish a basis for exploring how the encyclopaedic view adopted and developed within cognitive semantics is different. The theoretical distinction between dictionaries and encyclopaedias has traditionally been an issue of central importance for **lexicologists** (linguists who study word meaning) and **lexicographers** (dictionary writers). Since the emergence of the **mentalist** approach to language in the 1960s, it has also been widely assumed that a distinction parallel to the dictionary/encyclopaedia distinction exists at the level of the mental representation of words. This view has been widely adopted, particularly by formal linguists who assume a **componential view** of word meaning (recall our discussion of Universal Grammar and semantic universals in Chapter 3). More recently, however, linguists have begun to argue that the distinction traditionally drawn between 'dictionary knowledge' (word meaning) and 'encyclopaedic knowledge' (non-linguistic or 'world knowledge') is artificial. If this can be established, the alternative view emerges that dictionary knowledge is a subset of more general encyclopaedic knowledge. This is the position adopted by cognitive semanticists.

7.1.1 The dictionary view

The traditional view in semantic theory holds that meaning can be divided into a dictionary component and an encyclopaedic component. According to this view, it is only the dictionary component that properly constitutes the study of **lexical semantics**: the branch of semantics concerned with the study of word meaning. In contrast, encyclopaedic knowledge is external to linguistic knowledge, falling within the domain of 'world knowledge'. Of course, this view is consistent with the modularity hypothesis adopted within formal linguistics, which asserts that linguistic knowledge (e.g. knowing the meaning of a word like *shoelaces*) is specialised to language, and distinct in nature from other kinds of 'world' or 'non-linguistic' knowledge (like knowing how to tie your shoelaces, or that you can usually buy them in the supermarket). From this perspective, then, dictionary knowledge relates to knowing what words mean, and this knowledge represents a specialised component, the 'mental dictionary' or **lexicon**. While this component is mainly concerned with word meaning, formal theories differ quite considerably on the issue of what other kinds of information might also be represented in the lexicon, such as grammatical information relating to word class and so on. However, a common assumption within formal theories is that the word meanings stored in our minds can be defined, much as they appear in a dictionary.

In the **componential analysis** or **semantic decomposition** approach, which is one version of the dictionary model, word meaning is modelled in terms of semantic features or **primitives**. For instance *bachelor* is represented as [+MALE, + ADULT, −MARRIED], where each of these binary features represents a conceptual primitive that can also contribute to defining other words, such as *man* [+MALE, + ADULT], *girl* [−MALE, −ADULT], *wife* [−MALE, +ADULT, +MARRIED], and so on. Early examples of this approach are presented in Katz and Postal (1964) and Katz (1972). Another more recent variant of this approach is represented in the work of Anna Wierzbicka (1996), who takes the position that words are comprised of universal innate semantic primitives or **primes**, in terms of which other words can be defined. We consider these componential approaches in more detail below.

According to the dictionary view, the core meaning of a word is the information contained in the word's definition (for example that *bachelor* means 'unmarried adult male'), and this is the proper domain of lexical semantics. Encyclopaedic knowledge (for example, stereotypical connotations relating to bachelor pads, sexual conquests and dirty laundry) is considered non-linguistic knowledge. In this way, the dictionary model enables lexical semanticists to restrict their domain of investigation to intrinsic or non-contextual word meaning, while questions concerning how the outside world interacts with linguistic meaning are considered to fall within the domain of **pragmatics**, an area that some linguists consider to be external to the concerns of linguistics proper.

A number of dichotomies follow from the dictionary view of word meaning. Firstly, the core meaning of a word (**sense**), which is contained in the mental dictionary, stands in sharp contradistinction to what that word refers to in the outside world (**reference**). This distinction is inherited from referential theories of meaning dating back to Plato's (fourth century BC) *Cratylus Dialogue: The Realm of Ideas and Truth*. Referential theories hold that word meaning arises from a direct link between words and the objects in the world that they refer to. As the philosopher Frege (1892 [1975]) argued, however, it is possible for a word to have meaning (sense) without referring to a real object in the world (e.g. *dragon, unicorn*), hence the distinction between sense and reference.

The second dichotomy that arises from the dictionary view of meaning is the distinction between **semantics** and **pragmatics**. As we saw above, the dictionary view assumes a sharp distinction between knowledge of word meaning (semantics), and knowledge about how contextual factors influence linguistic meaning (pragmatics).

Thirdly, the dictionary view treats knowledge of word meaning as distinct from cultural knowledge, social knowledge (our experience of and interaction with others) and physical knowledge (our experience of interaction with the world). As we have seen, a consequence of this view is that semantic knowledge is autonomous from other kinds of knowledge, and is stored in its own mental repository, the **mental lexicon**. Other kinds of knowledge belong outside the language component, represented in terms of **principles of language use** (such as Grice's 1975 Cooperative Principle and its associated maxims, which represent a series of statements summarising the assumptions that speakers and hearers make in order to communicate successfully). This dichotomy between knowledge of language and use of language, where only the former is modelled within the language component, is consistent with the emphasis within formal approaches on the mental representation of linguistic knowledge rather than situated language use. Table 7.1 summarises the dictionary view.

It is worth mentioning here that word meaning is only 'half' of what traditional semantics is about. While lexical semantics is concerned with describing the meanings of individual words as well as the relationships between them: **lexical relations** or **sense relations** such as **synonymy**, **antonymy** and **homonymy** (see Murphy 2003 for an overview), the other 'half' of semantics involves sentence meaning or **compositional semantics**. This relates to the

Table 7.1 The dictionary view of key distinctions in the study and representation of meaning

Dictionary (linguistic) knowledge	Encyclopaedic (non-linguistic) knowledge
Concerns **sense** (what words mean)	Concerns **reference** (what speakers do with words)
Relates to the discipline **semantics**	Relates to the discipline **pragmatics**
Is stored in the **mental lexicon**	Is governed by **principles of language use**

study of the ways in which individual lexical items combine in order to produce sentence meaning. While the two areas are related (words, after all, contribute to the meaning of sentences), the two 'halves' of traditional semantics are often seen as separate subdisciplines, with many linguists specialising in one area or the other. We return to a discussion of the formal approach to sentence meaning in Chapter 13. In cognitive semantics, the distinction between lexical and compositional semantics is not seen as a useful division. There are a number of reasons for this, which we will return to shortly (section 7.1.3).

7.1.2 Problems with the dictionary view

According to the perspective adopted in cognitive semantics, the strict separation of lexical knowledge from 'world' knowledge is problematic in a number of ways. To begin with, the dictionary view assumes that word meanings have a semantic 'core', the 'essential' aspect of a word's meaning. This semantic core is distinguished from other non-essential aspects of the word's meaning, such as the associations that a word brings with it (recall our discussion of *bachelor*). Indeed, this distinction is axiomatic for many semanticists, who distinguish between a word's **denotation** (the set of entities in the world that a word can refer to) and its **connotation** (the associations evoked by the word). For example, the denotation of *bachelor* is the set of all unmarried adult males, while the connotations evoked by *bachelor* relate to cultural stereotypes concerning sexual and domestic habits and so on. Let's consider another example. Most speakers would agree that the words *bucket* and *pail* share the same denotation: the set of all cylindrical vessels with handles that can be used to carry water. These words share the same denotation because they are **synonyms**. Thus either of these lexical items could refer to the entity depicted in Figure 7.1.

However, while *bucket* and *pail* have the same (or at least very similar) denotations, for speakers who have both these words in their dialects they have very different connotations. For these speakers, a pail can be metal or wooden but not plastic, and it is associated with vessels of a certain size (for example, a child's small bucket used for making sandcastles on the beach could not be

Figure 7.1 Bucket or pail?

described as a pail). It follows from this that *pail* also shows a different linguistic distribution from its synonym. For example, it does not participate in the same collocational expressions as *bucket*: we can say *bucket and spade* but not *pail and spade*. Given these observations, cognitive linguists argue that the decision to exclude certain kinds of information from the 'core' meaning or denotation of a word, while including other kinds information, is arbitrary: on what basis is it decided that a particular piece of information is 'core' or 'non-core'?

The second way in which cognitive linguists argue that the dictionary view is problematic relates to background knowledge. The dictionary view assumes that words, although related to other words by lexical relations like synonymy and so on, can nevertheless be defined in a context-independent way. In contrast, a number of scholars, such as Fillmore (1975, 1977, 1982, 1985a and Fillmore and Atkins 1992) and Langacker (1987) have presented persuasive arguments for the view that words in human language are never represented independently of context. Instead, these linguists argue that words are always understood with respect to **frames** or **domains** of experience.

As we will see in detail below, a frame or domain represents a schematisation of experience (a knowledge structure), which is represented at the conceptual level and held in long-term memory, and which relates elements and entities associated with a particular culturally-embedded scene, situation or event from human experience. According to Fillmore and Langacker, words (and grammatical constructions) are relativised to frames and domains so that the 'meaning' associated with a particular word (or grammatical construction) cannot be understood independently of the frame with which it is associated. For example, the word *aorta* relates to a particular lexical concept, but this lexical concept cannot be understood without the frame of the MAMMALIAN CIRCULATORY SYSTEM. We explore these ideas in detail below (section 7.2–7.3).

The third problem that cognitive linguists identify with the dictionary view is the dichotomy between sense and reference. As we have seen, this view restricts linguistic meaning to a word's sense. From the perspective of the usage-based approach adopted in cognitive linguistics (recall Chapter 4), this dichotomy is problematic because a word's sense, what we have called **coded meaning**, is a function of language use or **pragmatic meaning**. In other words, the usage-based view holds that a word only comes to be meaningful as a consequence of use. This view stands in direct opposition to the dictionary view, which holds that a word's meaning or sense is primary and determines how it can be used.

Cognitive semanticists argue that the division of linguistic meaning into semantics (context-independent meaning) and pragmatics (context-dependent meaning) is also problematic. This dichotomy arises for historical as well as theoretical reasons. The discipline of semantics originated with the ancient Greek philosophers and was only recognised as a subdiscipline of linguistics as recently as the nineteenth century. Until this point linguists had concerned themselves mainly with describing the observable structural properties of language (grammar and phonology). Indeed, as recently as the twentieth century the famous American linguist Leonard Bloomfield (1933: 140) described the study of semantics as 'the weak point in language study'. The 'mentalist' approach to linguistics pioneered by Chomsky gave rise to a new interest in linguistic meaning as part of the competence of the native speaker, but due to the historical development of the discipline within the philosophical tradition, the resulting formal models tended to emphasise only those aspects of meaning that could be 'neatly packaged' and modelled within the truth-conditional paradigm (see Chapter 13), hence the predominance of the dictionary view. Meanwhile, in the 1950s and 1960s, the **natural language philosophers** such as Austin and Grice, who argued that the truth-conditional model was artificially limiting the study of linguistic meaning, began to focus attention on the principles that governed the use of language in interactive contexts. For this reason, pragmatics emerged as a largely independent approach, and has often been seen as peripheral with respect to the concerns of formal linguistics, which relate to modelling knowledge of language rather than use of language, or competence rather than performance. An important exception to this generalisation is the Relevance Theory model, developed by Sperber and Wilson (1995). We will consider this approach in Chapter 13.

As many linguists have argued, imposing a principled distinction between semantics and pragmatics results in a rather artificial boundary between the two types of meaning. After all, context of use is often critical to the meaning associated with words, and some linguistic phenomena cannot be fully explained by either a semantic or a pragmatic account in isolation. For example, Saeed (2003) makes this point in relation to **deictic** expressions: words like *bring* and *take*, and *today* and *tomorrow*. These expressions clearly have 'semantic' content, yet their meaning cannot be fully determined in isolation from context. Levinson (1983: 55) provides a revealing example. Imagine you are on a desert island and you find this message in a bottle washed up on the beach. The message reads *Meet me here a week from now with a stick about this big*. This example illustrates the dependence of deictic expressions on contextual information. Without knowing the person who wrote the message, where the note was written or the time at which it was written, you cannot fully interpret *me*, *here* or *a week from now*. Observe that we also rely upon visual signals to interpret expressions like *this big*, where the speaker would hold his or her hands a certain distance apart to indicate the size of the object being described. Such expressions are not fully meaningful in the absence of this visual information. It is the deictic or context-dependent properties of expressions like these that also explain why it is less than helpful for a shopkeeper to go out for lunch and leave a sign on the door reading *Back in an hour!*

In view of these observations, cognitive semanticists argue that the dichotomy between semantics and pragmatics represents an arbitrary distinction: linguistic knowledge cannot be separated in a principled way from 'world' knowledge, nor can 'semantic' knowledge be separated from 'pragmatic' knowledge. From the cognitive perspective, the kinds of knowledge subsumed under these headings constitute a continuum. The encyclopaedic view adopted within cognitive semantics assumes that there are no principled distinctions of the kind discussed here, but that any apparent distinctions are simply a matter of degree. In other words, while there are conventional meanings associated with words (the **coded meanings** we discussed in Chapter 4), these are abstracted from the range of contexts of use associated with any given lexical item. Furthermore, words are sometimes used in ways that are only partially sanctioned by these coded meanings: language use is often partly innovative, for the reasons laid out in Chapter 4. Moreover, the degree to which any given usage of a coded meaning is innovative varies according to contextual factors.

7.1.3 Word meaning versus sentence meaning

Before elaborating the encyclopaedic view of meaning, we first briefly return to the traditional distinction between word meaning (lexical semantics) and sentence meaning (compositional semantics). As noted above, cognitive semanticists also view this distinction as artificial. There are a number of reasons for this position, which we briefly review here.

Word meaning is protean in nature

The traditional distinction between lexical and compositional semantics is based on the assumption that word meanings combine, together with the grammatical structure of the sentence, to produce sentence meaning. This is known as the **principle of compositionality**. The way the 'division of labour' works in most formal approaches is that lexical semanticists work out how to represent the meanings of words, while compositional semanticists work out the principles governing the combination of words into larger units of meaning and the relationships between words within those larger units.

From the perspective of cognitive semantics, the problem with the compositional view of sentence meaning is that word meanings cannot be precisely defined in the way that is required by this approach. Instead, cognitive semanticists argue that, while words do have relatively well-entrenched meanings stored in long-term memory (the coded meaning), word meaning in language is 'protean' in nature. This means that the meaning associated with a single word is prone to shift depending on the exact context of use. Thus cognitive semanticists argue that the meaning of any given word is constructed 'on line' in the context in which it is being used. We saw an example illustrating this when we discussed various uses of the word *safe* in Chapter 5. One problem with the compositional view of sentence meaning, then, is that it relies upon the assumption that the context-independent meanings associated with words can be straightforwardly identified.

The conceptual nature of meaning construction

The second problem with dividing semantics into the study of word meaning on the one hand and sentence meaning on the other relates to **meaning construction**, which has traditionally been regarded as the remit of compositional semantics. Meaning construction is the process whereby language encodes or represents complex units of meaning; therefore this area relates to sentence meaning rather than word meaning. The principle of compositionality assumes that words 'carry' meaning in neatly packaged self-contained units, and that meaning construction results from the combination of these smaller units of meaning into larger units of meaning within a given grammatical structure. However, as we have begun to see, cognitive semanticists argue that words are **prompts** for meaning construction rather than 'containers' that carry meaning. Furthermore, according to this view, language actually represents highly underspecified and impoverished prompts relative to the richness of conceptual structure that is encoded in semantic structure: these prompts serve as 'instructions' for conceptual processes that result in meaning construction. In other words, cognitive linguists argue that meaning construction is primarily conceptual rather than linguistic in nature. From this perspective, if meaning construction is conceptual rather than linguistic in nature, and if words themselves do not 'carry' meaning, then the idea that sentence meaning is built straightforwardly out of word meanings is largely vacuous. We will explore these ideas further in Chapters 11 and 12 where we address meaning construction in detail.

Grammatical constructions are independently meaningful

Finally, as we saw in Part I of the book and as will see in detail in Part III, cognitive linguistics adopts the **symbolic thesis** with respect to linguistic structure and organisation. This thesis holds that linguistic units are form-meaning pairings. This idea is not new in linguistics: indeed, it has its roots in the influential work of the Swiss linguist Ferdinand de Saussure (1857–1913) and is widely accepted by linguists of all theoretical persuasions. The innovation in cognitive linguistics is that this idea is extended beyond words to larger constructions including phrases and whole sentences. According to this view, it is not just words that bring meaning to sentences, but the grammatical properties of the sentence are also meaningful in their own right. In one sense, this does not appear significantly different from the compositional view: all linguists recognise that *George loves Lily* means something different from *Lily loves George*, for example, and this is usually explained in terms of grammatical functions like subject and object which are positionally identified in a language like English. However, the claim made in cognitive linguistics is stronger than the claim that grammatical structure contributes to meaning via the structural identification of grammatical functions like subject and object. The cognitive claim is that grammatical constructions and grammatical functions are themselves inherently meaningful, independently of the content words that fill them. From this perspective, the idea that sentence meaning arises purely from the composition of smaller units of meaning into larger ones is misleading. We look in detail at the idea that grammatical constructions are meaningful in Part III of the book.

7.1.4 The encyclopaedic view

For the reasons outlined in the previous section, cognitive semanticists reject the 'dictionary view' of word meaning in favour of the 'encyclopaedic view'. Before we proceed with our investigation of the encyclopaedic view, it is worth emphasising the point that, while the dictionary view represents a model of the knowledge of linguistic meaning, the encyclopaedic view represents a model of the system of conceptual knowledge that underlies linguistic meaning. It follows that this model takes into account a far broader range of phenomena than purely linguistic phenomena, in keeping with the 'Cognitive Commitment'. This will become evident when we look at Fillmore's theory of frames (section 7.2) and Langacker's theory of domains (section 7.3). There are a number of characteristics associated with this model of the knowledge system, which we outline in this section:

1. There is no principled distinction between semantics and pragmatics.
2. Encyclopaedic knowledge is structured.
3. There is a distinction between encyclopaedic meaning and contextual meaning.
4. Lexical items are points of access to encyclopaedic knowledge.
5. Encyclopaedic knowledge is dynamic.

There is no principled distinction between semantics and pragmatics

Firstly, cognitive semanticists reject the idea that there is a principled distinction between 'core' meaning on the one hand, and pragmatic, social or cultural meaning on the other. This means that, among other things, cognitive semanticists do not make a sharp distinction between semantic and pragmatic knowledge. Knowledge of what words mean and knowledge about how words are used are both types of 'semantic' knowledge, according to this view. This is why cognitive semanticists study such a broad range of (linguistic and non-linguistic) phenomena in comparison to traditional or formal semanticists, and this also explains why there is no chapter in this book called 'cognitive pragmatics'. This is not to say that the existence of pragmatic knowledge is denied. Instead, cognitive linguists claim that semantic and pragmatic knowledge cannot be clearly distinguished. As with the lexicon-grammar continuum, semantic and pragmatic knowledge can be thought of in terms of a continuum. While there may be qualitative distinctions at the extremes, it is often difficult in practice to draw a sharp distinction.

Cognitive semanticists do not posit an autonomous mental lexicon that contains semantic knowledge separately from other kinds of (linguistic or non-linguistic) knowledge. It follows that there is no distinction between dictionary knowledge and encyclopaedic knowledge: there is only encyclopaedic knowledge, which subsumes what we might think of as dictionary knowledge.

The reason for adopting this position follows, in part, from the usage-based perspective developed in Chapter 4. The usage-based thesis holds, among other things, that context of use guides meaning construction. It follows from this position that word meaning is a consequence of language use, and that pragmatic meaning, rather than coded meaning, is 'real' meaning. Coded meaning, the stored mental representation of a lexical concept, is a **schema**: a skeletal representation of meaning abstracted from recurrent experience of language use. If meaning construction cannot be divorced from language use, then meaning is fundamentally pragmatic in nature because language in use is situated, and thus contextualised, by definition. As we have seen, this view is in direct opposition to the traditional view, which holds that definitional meaning is the proper subject of semantic investigation while pragmatic meaning relies upon non-linguistic knowledge.

Encyclopaedic knowledge is structured

The view that there is only encyclopaedic knowledge does not entail that the knowledge we have connected to any given word is a disorganised chaos. Cognitive semanticists view encyclopaedic knowledge as a structured system of knowledge, organised as a network, and not all aspects of the knowledge that is, in principle, accessible by a single word has equal standing. For example, what we know about the word *banana* includes information concerning its shape, colour, smell, texture and taste; whether we like or hate bananas; perhaps information about how and where bananas are grown and harvested; details relating to funny cartoons involving banana skins; and so on. However, certain aspects of this knowledge are more **central** than others to the meaning of *banana*.

According to Langacker (1987), centrality relates to how **salient** certain aspects of the encyclopaedic knowledge associated with a word are to the meaning of that word. Langacker divides the types of knowledge that make up the **encyclopaedic network** into four types: (1) **conventional**; (2) **generic**; (3) **intrinsic**; and (4) **characteristic**. While these types of knowledge are in principle distinct, they frequently overlap, as we will show. Moreover, each of these kinds of knowledge can contribute to the relative salience of particular aspects of the meaning of a word.

The conventional knowledge associated with a particular word concerns the extent to which a particular facet of knowledge is shared within a linguistic community. Generic knowledge concerns the degree of generality (as opposed to specificity) associated with a particular word. Intrinsic knowledge is that aspect of a word's meaning that makes no reference to entities external to the referent. Finally, characteristic knowledge concerns aspects of the encyclopaedic information that are characteristic of or unique to the class of entities that the word designates. Each of these kinds of knowledge can be thought of as operating along a continuum: certain aspects of a word's meaning are more or less conventional, or more or less generic, and so on, rather than having a fixed positive or negative value for these properties.

Conventional knowledge
Conventional knowledge is information that is widely known and shared between members of a speech community, and is thus likely to be more central to the mental representation of a particular lexical concept. The idea of conventional knowledge is not new in linguistics. Indeed, the early twentieth-century linguist Ferdinand de Saussure (1916), who we mentioned earlier in relation to the symbolic thesis, also observed that conventionality is an important aspect of word meaning: given the **arbitrary** nature of the sound–meaning pairing (in other words, the fact that there is nothing intrinsically meaningful about individual speech sounds, and therefore nothing predictable about why a certain set of sounds and not others should convey a particular meaning), it is only because members of a speech community 'agree' that a certain word has a particular meaning that we can communicate successfully using language. Of course, in reality this 'agreement' is not a matter of choice but of learning, but it is this 'agreement' that represents conventionality in the linguistic sense.

For instance, conventional knowledge relating to the lexical concept BANANA might include the knowledge that some people in our culture have bananas with their lunch or that a banana can serve as a snack between meals. An example of non-conventional knowledge concerning a banana might be that the one you ate this morning gave you indigestion.

Generic knowledge
Generic knowledge applies to many instances of a particular category and therefore has a good chance of being conventional. Generic knowledge might include our knowledge that yellow bananas taste better than green bananas. This knowledge applies to bananas in general and is therefore generic. Generic knowledge contrasts with specific knowledge, which concerns individual instances of a category. For example, the knowledge that the banana you peeled this morning was unripe is specific knowledge, because it is specific to this particular banana. However, it is possible for large communities to share specific (non-generic) knowledge that has become conventional. For instance, generic knowledge relating to US presidents is that they serve a term of four years before either retiring or seeking re-election. This is generic knowledge, because it applies to US presidents in general. However, a few presidents have served shorter terms. For instance, John F. Kennedy served less than three years in office. This is specific knowledge, because it relates to one president in particular, yet it is widely known and therefore conventional. In the same way that specific knowledge can be conventional, generic knowledge can also be non-conventional, even though these may not be the patterns we expect. For example, while scientists have uncovered the structure of the atom and know that all atoms share a certain structure (generic knowledge), the details of atomic structure are not widely known by the general population.

Intrinsic knowledge
Intrinsic knowledge relates to the internal properties of an entity that are not due to external influence. Shape is a good example of intrinsic knowledge relating to objects. For example, we know that bananas tend to have a characteristic curved shape. Because intrinsic knowledge is likely to be generic, it has a good chance of being conventional. However, not all intrinsic properties (for example, that bananas contain potassium) are readily identifiable and may not therefore be conventional. Intrinsic knowledge contrasts with extrinsic knowledge. Extrinsic knowledge relates to knowledge that is external to the entity: for example, the knowledge that still-life artists often paint bananas in bowls with other pieces of fruit relates to aspects of human culture and artistic convention rather than being intrinsic to bananas.

Characteristic knowledge
This relates to the degree to which knowledge is unique to a particular class of entities. For example, shape and colour may be more or less characteristic of an entity: the colour yellow is more characteristic of bananas than the colour red is characteristic of tomatoes, because fewer types of fruit are yellow than red (at least, in the average British supermarket). The fact that we can eat bananas is not characteristic, because we eat lots of other kinds of fruit.

THE ENCYCLOPAEDIC VIEW OF MEANING

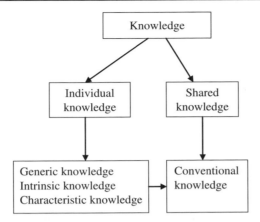

Figure 7.2 Identifying knowledge types which give rise to centrality

The four types of knowledge we have discussed thus far relate to four continua, which are listed below. Knowledge can fall at any point on these continua, so that something can be known by only one person (wholly non-conventional) known by the entire discourse community (wholly conventional) or somewhere in between (for example, known by two people, a few people or many but not all people.

1. Conventional ⟷ Non-conventional
2. Generic ⟷ Specific
3. Intrinsic ⟷ Extrinsic
4. Characteristic ⟷ Non-characteristic

Of course, conventionality versus non-conventionality stands out in this classification of knowledge types because it relates to how widely something is known whereas the other knowledge types relate to the nature of the lexical concepts themselves. Thus it might seem that conventional knowledge is the most 'important' or 'relevant' kind when in fact it is only one 'dimension' of encyclopaedic knowledge. Figure 7.2 represents the interaction between the knowledge types discussed here. As this diagram illustrates, while generic, intrinsic and characteristic knowledge can be conventional (represented by the arrow going from the box containing these types of knowledge to the box containing conventional knowledge) they need not be. Conventional knowledge, on the other hand, is, by definition, knowledge that is shared.

Finally, let's turn to the question of how these distinct knowledge types influence **centrality**. The centrality of a particular aspect of knowledge for a linguistic expression will always be dependent on the precise context in which the expression is embedded and on how well established the knowledge

Table 7.2 Four kinds of knowledge that relate to the centrality of encyclopaedic knowledge of word meaning

Conventional knowledge	Knowledge that is widely known
Generic knowledge	Knowledge that is general rather than specific in nature
Intrinsic knowledge	Knowledge deriving from the form of the entity or relation in question
Characteristic knowledge	Knowledge that is (relatively) unique to the entity or relation in question

element is in memory. Moreover, the closer knowledge is to the left-hand side of the continua we listed above, the more salient that knowledge is and the more central that knowledge is to the meaning of a lexical concept. For example, for Joe Bloggs, the knowledge that bananas have a distinctive curved shape is conventional, generic, intrinsic and characteristic, and is therefore highly salient and therefore central to his knowledge about bananas and to the meaning of the lexical concept BANANA. The knowledge that Joe Bloggs has that he once peeled a banana and found a maggot inside is non-conventional, specific, extrinsic and non-characteristic, and hence is much less salient and less central to his knowledge about bananas. We summarise the four categories of encyclopaedic knowledge in Table 7.2.

There is a distinction between encyclopaedic meaning and contextual meaning

The third issue concerning the encyclopaedic view relates to the distinction between **encyclopaedic meaning** and **contextual meaning** (or situated meaning). Encyclopaedic meaning arises from the interaction of the four kinds of knowledge discussed above. However, encyclopaedic meaning arises in the context of use, so that the 'selection' of encyclopaedic meaning is informed by contextual factors. For example, recall our discussion of *safe* in Chapter 5. We saw that this word can have different meanings depending on the particular context of use: *safe* can mean 'unlikely to cause harm' when used in the context of a child playing with a spade, or *safe* can mean 'unlikely to come to harm' when used in the context of a beach that has been saved from development as a tourist resort. Similarly, the phenomenon of frame-dependent meaning briefly mentioned earlier suggests that the discourse context actually guides the nature of the encyclopaedic information that a lexical item prompts for. For instance, the kind of information evoked by use of the word *foot* will depend upon whether we are talking about rabbits, humans, tables or mountains. This phenomenon of **contextual modulation** (Cruse 1986) arises when a particular aspect of the encyclopaedic knowledge associated with a lexical item is privileged due to the discourse context.

Compared with the dictionary view of meaning, which separates core meaning (semantics) from non-core meaning (pragmatics), the encyclopaedic view makes very different claims. Not only does semantics include encyclopaedic knowledge, but meaning is fundamentally 'guided' by context. Furthermore, the meaning of a word is 'constructed' on line as a result of contextual information. From this perspective, fully-specified pre-assembled word meanings do not exist, but are selected and formed from encyclopaedic knowledge, which is called the **meaning potential** (Allwood 2003) or **purport** (Cruse 2000) of a lexical item. As a result of adopting the usage-based approach, then, cognitive linguists do not uphold a meaningful distinction between semantics and pragmatics, because word meaning is always a function of context (pragmatic meaning).

From this perspective, there are a number of different kinds of **context** that collectively serve to modulate any given instance of a lexical item as it occurs in a particular **usage event**. These types of context include (but are not necessarily limited to): (1) the **encyclopaedic information** accessed (the lexical concept's context within a network of stored knowledge); (2) **sentential context** (the resulting sentence or utterance meaning); (3) **prosodic context** (the intonation pattern that accompanies the utterance, such as rising pitch to indicate a question); (4) **situational context** (the physical location in which the sentence is uttered); and (5) **interpersonal context** (the relationship holding at the time of utterance between the interlocutors). Each of these different kinds of context can contribute to the contextual modulation of a particular lexical item.

Lexical items are points of access to encyclopaedic knowledge

The encyclopaedic model views lexical items as **points of access** to encyclopaedic knowledge. According to this view, words are not containers that present neat pre-packaged bundles of information. Instead, they provide access to a vast network of encyclopaedic knowledge.

Encyclopaedic knowledge is dynamic

Finally, it is important to note that while the central meaning associated with a word is relatively stable, the encyclopaedic knowledge that each word provides access to, its encylopaedic network, is dynamic. Consider the lexical concept CAT. Our knowledge of cats continues to be modified as a result of our ongoing interaction with cats, our acquisition of knowledge regarding cats, and so on. For example, imagine that your cat comes home looking extremely unwell, suffering from muscle spasms and vomits a bright blue substance. After four days in and out of the animal hospital (and an extremely large vet's bill) you will have acquired the knowledge that metaldehyde (the chemical used in slug pellets) is potentially fatal to cats. This information now forms part of your encyclopaedic knowledge prompted by the word *cat*, alongside the central knowledge that cats are small fluffy four-legged creatures with pointy ears and a tail.

7.2 Frame semantics

Having provided an overview of what an encyclopaedic view of word meaning entails, we now present the theory of Frame Semantics, one theory that has influenced the encyclopaedic model adopted within cognitive semantics. This approach, developed by Charles Fillmore (1975, 1977, 1982, 1985a; Fillmore and Atkins 1992), attempts to uncover the properties of the structured inventory of knowledge associated with words, and to consider what consequences the properties of this knowledge system might have for a model of semantics.

7.2.1 What is a semantic frame?

As we saw in Chapter 5, Fillmore proposes that a **frame** is a schematisation of experience (a knowledge structure), which is represented at the conceptual level and held in long-term memory. The frame relates the elements and entities associated with a particular culturally embedded scene from human experience. According to Fillmore, words and grammatical constructions are relativised to frames, which means that the 'meaning' associated with a particular word (or grammatical construction) cannot be understood independently of the frame with which it is associated. In his 1985a article, Fillmore adopts the terms **figure** and **ground** from Gestalt psychology in order to distinguish between a particular lexical concept (the specific meaning designated by a lexical item) and the background frame against which it is understood. The specific meaning designated by a lexical item is represented by the figure, and is a salient subpart of a larger frame, which represents the ground relative to which the figure is understood. Frames thus represent a complex knowledge structure that allows us to understand, for example, a group of related words and that also plays a role in licensing their grammatical behaviour in sentences.

7.2.2 Frames in cognitive psychology

Before developing Fillmore's theory of semantic frames in more detail, we begin by exploring the development of this idea in cognitive psychology. This will enable us to obtain a richer picture of the kind of conceptual entity that Fillmore assumes as the basis of his theory. In psychology, the basic unit of knowledge is the **concept**. Theories of **knowledge representation** attempt to model the kinds of concepts that people appear to have access to, including the relationships holding between concepts and the kinds of operations that people use concepts for such as categorisation judgements (explored in more detail in the next chapter) and conceptualisation or meaning construction (explored in Chapters 11 and 12).

A common system for modelling knowledge representation is the **feature list approach**. This entails listing the range of distinct features or attributes associated with a particular concept. From this perspective, we might hypothesise that the concept of CAR, for instance, has a range of features or attributes associated with it that relate to its parts (wheel, tyre, windscreen, bonnet, boot, steering wheel, engine and so on), as well as the fact that cars require petrol or diesel in order to function, are driven by humans who must first obtain a driving licence and so on. However, one of the problems associated with modelling knowledge solely in terms of feature lists is that people's knowledge regarding conceptual entities is relational. For example, we know that cars have engines which provide the mechanism for moving the vehicle. We also know that this motion is effected by the engine causing the axles to turn which then causes the wheels to turn. Moreover, we know that unless a driver is operating the vehicle, which involves turning on the ignition, the engine will not start in the first place. Thus a serious problem with viewing a concept as a straightforward list of features is that there is no obvious way of modelling how the relationships between the components of the list might be represented. The theory of frames represents an attempt to overcome this shortcoming.

Since Bartlett's (1932) theory of **schemata**, there has been a tradition in cognitive psychology of modelling knowledge representation in terms of frames. We will base our discussion of frames on a recent version of this theory proposed by Lawrence Barsalou (1992a, 1992b), who defines frames as complex conceptual structures that are used to 'represent all types of categories, including categories for animates, objects, locations, physical events, mental events and so forth' (Barsalou 1992a: 29). According to this view, frames are the basic mode of knowledge representation. They are continually updated and modified due to ongoing human experience, and are used in reasoning in order to generate new inferences. Below, we describe two basic components of frames: **attribute-value** sets and **structural invariants**. In order to illustrate these notions, we present a vastly simplified frame for CAR. This is illustrated in Figure 7.3.

Attributes and values

We begin by examining the ideas of **attribute** and **value**. Barsalou (1992a: 30) defines an attribute as 'a concept that describes an aspect of at least some

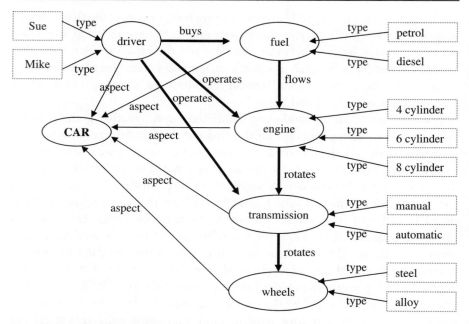

Figure 7.3 A partial frame for CAR (adapted from Barsalou 1992a: 30)

category members'. For instance, ENGINE represents one aspect of the members of the category CAR, as do DRIVER, FUEL, TRANSMISSION and WHEELS. An attribute is therefore a concept that represents one aspect of a larger whole. Attributes are represented in Figure 7.3 as ovals. Values are subordinate concepts which represent subtypes of an attribute. For instance, SUE and MIKE are types of DRIVER; PETROL and DIESEL are types of FUEL; MANUAL and AUTOMATIC are types of TRANSMISSION, and so on. Values are represented as dotted rectangles in Figure 7.3. Crucially, while values are more specific than attributes, a value can also be an attribute because it can also have subtypes. For instance, PETROL is an attribute to the more specific concepts UNLEADED PETROL and LEADED PETROL which are values of PETROL. Attributes and values are therefore superordinate and subordinate concepts within a taxonomy: subordinate concepts, or values, which are more specific inherit properties from the superordinate concepts, or attributes, which are more general.

Structural invariants

As Barsalou observes, 'Attributes in a frame are not independent slots but are often related correlationally and conceptually . . . a frame's core attributes correlate highly, often appearing together across contexts' (Barsalou 1992a: 35). In other words, attributes within a frame are related to one another in consistent ways across **exemplars**: individual members of a particular category. For example, in most exemplars of the category CAR it is the driver who controls the speed of the ENGINE. This relation holds across most instances of cars, irrespective of the values involved, and is therefore represented in the frame as a **structural invariant**: a more or less invariant relation between attributes DRIVER and ENGINE. In Figure 7.3, structural invariants are indicated by bold arrows.

Simulations

The final issue that remains to be addressed is the dynamic quality associated with frames. Humans have the ability to imagine or **simulate** a conceptual entity, such as an action involving a particular object, based on a particular frame. For example, we can mentally simulate the stages involved in filling a car up with petrol, including mentally rehearsing the actions involved in taking the petrol cap off, removing the petrol nozzle from the pump, placing it in the petrol tank, pressing the lever so that the petrol flows into the tank, and so on. The most recent theories of knowledge representation attempt to account for this ability. This is an issue we will return to later in the chapter, once we have investigated two theories that are specifically concerned with semantic knowledge representation: conceptual structure as it is encoded in language.

7.2.3 The COMMERCIAL EVENT frame

We now return to our discussion of Fillmore's theory of semantic frames. The semantic frame is a knowledge structure required in order to understand a particular word or related set of words. Consider the related group of words *buy, sell, pay, spend, cost, charge, tender, change*, and so on. Fillmore argues that in order to understand these words, we need access to a COMMERCIAL EVENT frame which provides 'the background and motivation for the categories which these words represent' (Fillmore 1982: 116–17). Recall the PURCHASING GOODS frame that we discussed in Chapter 5; this is a subpart of the COMMERCIAL EVENT frame. The COMMERCIAL EVENT frame includes a number of attributes called **participant roles** which must, at the very least, include BUYER, SELLER, GOODS and MONEY. This skeletal frame is represented in Figure 7.4.

According to Fillmore, **valence** is one of the consequences of a frame like this. Valence concerns the ways in which lexical items like verbs can be combined with other words to make grammatical sentences. More precisely, the valence (or **argument structure**) of a verb concerns the number of participants or **arguments** required, as well as the nature of the arguments, that is the **semantic roles** assumed by those participants. For example, *buy* is typically

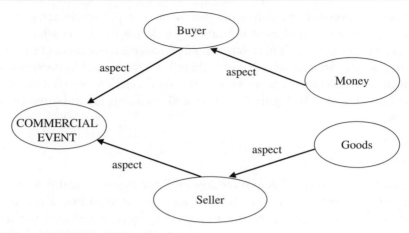

Figure 7.4 Partial COMMERCIAL EVENT frame

'divalent' which means that it requires two participants, the BUYER and the GOODS. *Pay*, on the other hand, is typically 'trivalent', which means that it requires three participants: the BUYER, the SELLER and the GOODS. Observe that valence is not a stable feature of verbs, however. *Pay* could also occur in a sentence with two participants (*I paid five hundred pounds*) or with four participants (*I paid John five pounds for that pile of junk*). While *buy* and *pay* relate to the actions of the BUYER, *buy* relates to the interaction between the BUYER and the GOODS, while *pay* relates to the interaction between the BUYER and the SELLER. This knowledge, which is a consequence of the COMMERCIAL EVENT frame, has consequences for grammatical organisation (recall our discussion of *rob* and *steal* in Chapter 5). Consider the following sentences:

(1) a. John bought the car (from the salesperson).
    b. *John bought the salesperson

(2) a. John paid the salesperson (for the car).
    b. *John paid the car

The sentences in (1) demonstrate that *bought* and *paid* take the same number of arguments. These are realised as subject and object, and optionally as **oblique object**: an object like *from the salesperson* which is introduced by a preposition. The verb *bought* profiles a relation between the participant roles BUYER and GOODS, not a relation between BUYER and SELLER. This explains why the sentence in (1b) is ungrammatical. Of course, if we invoke a SLAVE TRADE frame then (1b) might be acceptable on the interpretation that *the salesperson* represents the GOODS role. Example (2) shows that the verb *pay* relates the BUYER role with the SELLER role rather than the GOODS role. In addition, *pay* can also prompt for a relation between BUYER and AMOUNT PAID, or between BUYER, SELLER and AMOUNT PAID, as illustrated by examples (3) and (4), respectively.

(3) John paid £2,000 (for the car).

(4) John paid the salesperson £1,000 (for the car).

These examples demonstrate that *pay* relates to that aspect of the COMMERCIAL EVENT frame involving the transfer of money from BUYER to SELLER in order to receive the GOODS. The frame thus provides a structured set of relationships that define how lexical items like *pay* and *buy* are understood and how they can be used. As we have seen, this has consequences for the grammatical behaviour of these lexical items. Indeed, frames of this kind have played a central role in the development of Construction Grammar (e.g. Goldberg 1995), to which we return in Part III.

One way of interpreting the structured set of linguistic relationships licensed by the frame is to analyse the frame as a knowledge representation system that provides a potentially wide range of event sequences. According to this view, the frame provides **event-sequence potential**. Given that verbs such as *buy* and *sell* encode particular kinds of dynamic processes, we can analyse these verbs as designating particular configurations of events. According to this view, the verb selected by the speaker (for example, *buy* vs. *sell* vs. *pay*) designates a particular 'route' through the frame: a way of relating the various participant roles in order to highlight certain aspects of the frame. While some 'routes' include obligatory relationships (invariant structure), others are optional. For instance, *pay* designates a relation between BUYER and the SELLER, which has the potential to make optional reference to GOODS and MONEY. However, not all these participant roles need to be mentioned in any given sentence, and when they are not mentioned, they are 'understood' as part of the background. For example, in the sentence *I paid five pounds*, we understand that this event must also have involved a SELLER and some GOODS, even though these are not explicitly mentioned in the sentence. This knowledge derives from our knowledge of the event frame. Table 7.3 summarises the 'routes' connecting the participants encoded by verbs that are understood with respect to the COMMERCIAL EVENT frame. Brackets indicate that an element is optional and can therefore be omitted (that is, not explicitly mentioned in the sentence). The symbol Ø indicates that an element cannot be included in the sentence, for example **I spent John five hundred pounds for that pile of junk*. 'I-object' indicates that an element is the indirect object: the first element in a double object construction like *I paid John five hundred pounds for that pile of junk*. 'Oblique' indicates that an element is introduced by a preposition, like *for that pile of junk*.

Table 7.3 The valence of the verbs relating to the COMMERCIAL EVENT frame (adapted from Fillmore and Atkins 1992: 79)

	BUYER	SELLER	GOODS	MONEY
buy	subject	(oblique)	object	(oblique)
	e.g. John bought the car (from the salesperson) (for £10,000)			
sell	(oblique)	subject	object	(oblique)
	e.g. Susan sold the car (to John) (for £10,000)			
charge	(I-object)	subject	(oblique)	object
	e.g. Susan charged (John) £10,000 (for the car)			
spend	subject	Ø	(oblique)	object
	e.g. John spent £10,000 (on the car)			
pay	subject	(I-object)	(oblique)	object
	e.g. John paid (Susan) £10,000 (for the car)			
pay	subject	(oblique)	(oblique)	object
	e.g. John paid £10,000 (to Susan) (for the car)			
cost	(I-object)	Ø	subject	object
	e.g. The car cost (John) £10,000			

7.2.4 Speech event frames

While semantic frames like the COMMERCIAL EVENT frame describe a knowledge inventory independent of the speech event, a second kind of frame provides a means of framing the discourse or communication context. This type of frame is called the **speech event frame**. These frames schematise knowledge about types of interactional context which contribute to the interpretation and licensing of particular lexical items and grammatical constructions. For example, we have speech event frames for fairytales, academic lectures, spoken conversations, obituaries, newspaper reports, horoscopes and business letters, among others. In other words, these speech event frames contain schematic knowledge about **styles** or **registers** of language use. It is important to point out that while these frames are described as 'speech event frames', they encompass not only events relating to spoken language, but also events relating to written language. Each of these provides a means of framing a particular type of linguistic interaction, with respect to which choices about language and style (including choices about vocabulary and grammatical constructions) can be made and understood. Indeed, many lexical items explicitly index a specific speech event frame, like the English expression *once upon a time*, which indexes the generic FAIRYTALE frame, bringing with it certain expectations. Speech event frames, then, are organised knowledge structures that are culturally embedded.

7.2.5 Consequences of adopting a frame-based model

In this section, we briefly explore some of the consequences that arise from adopting a frame-based model of encyclopaedic knowledge.

Words and categories are dependent on frames

A theory based on semantic frames asserts that word meanings can only be understood with respect to frames. Fillmore (1982) provides an example of this, which relates to language change. According to semantic frame theory, words disappear from language once the frame with respect to which they are understood is superseded by a different frame. As Fillmore observes, the word *phlogiston* (meaning 'a substance without colour, odour or weight, believed to be given off in burning by all flammable materials') has now disappeared from the English language. This is because the frame against which the corresponding lexical concept was understood, a theory of combustion developed in the late seventeenth century, had, by the end of the eighteenth century, been shown to be empirically inaccurate. As the frame disappeared, so did the word.

Frames provide a particular perspective

The words *coast* and *shore*, while both relating to the strip of land adjacent to the sea, do so with respect to different frames: LAND DWELLING versus SEAFARING. While *coast* describes the land adjacent to the sea from the perspective of a person on land, *shore* describes the same strip of land from the perspective of a person out at sea. It follows that a trip from 'coast to coast' is an overland trip, while a trip from 'shore to shore' entails a journey across the sea or some other body of water. In this way, lexical choice brings with it a particular background frame that provides its own perspective. Fillmore calls this perspective a particular **envisionment of the world.**

Scene-structuring frames

From the frame semantics perspective, both closed-class and open-class units of language are understood with respect to semantic frames. As Fillmore observes, and as we saw in the previous chapter, cognitive semanticists view open-class semantics as 'providing the "content" upon which grammatical structure performs a "configuring" function. Thinking in this way, we can see that any grammatical category or pattern imposes its own "frame" on the material it structures' (Fillmore 1982: 123). For instance, the distinction between active and passive constructions is that they provide access to distinct scene-structuring frames. While the active takes the perspective of the AGENT in a sentence, the passive takes the perspective of the PATIENT. This is an idea that we will explore further in Part III of the book when we address conventional schematic meanings associated with closed-class constructions of this kind.

Alternate framing of a single situation

The same situation can be viewed, and therefore linguistically encoded, in multiple ways. For example, someone who is not easily parted from his money could be described either as *stingy* or as *thrifty*. Each of these words is understood with respect to a different background frame which provides a distinct set of evaluations. While *stingy* represents a negative assessment against an evaluative frame of GIVING AND SHARING, *thrifty* relates to a frame of HUSBANDRY (management of resources), against which it represents a positive assessment. In this way, lexical choice provides a different way of framing a situation, giving rise to a different construal. In other words, language is rarely 'neutral', but usually represents a particular perspective, even when we are not consciously aware of this as language users.

7.3 The theory of domains

Langacker's theory of domains, like Fillmore's theory of Frame Semantics, is based on the assumption that meaning is encyclopaedic, and that lexical concepts cannot be understood independently of larger knowledge structures. Langacker calls these knowledge structures **domains**. Langacker's theory of domains complements Fillmore's theory of Frame Semantics in a number of ways.

7.3.1 What is a domain?

According to Langacker, 'Domains are necessarily cognitive entities: mental experiences, representational spaces, concepts, or conceptual complexes' (Langacker 1987: 147). In other words, domains are conceptual entities of varying levels of complexity and organisation. The only prerequisite that a knowledge structure has for counting as a domain is that it provides background information against which lexical concepts can be understood and used in language. For instance, expressions like *hot*, *cold* and *lukewarm* designate lexical concepts in the domain of TEMPERATURE: without understanding the temperature system, we would not be able to use these terms. In this respect, the theory of domains is very much like Fillmore's theory of frames.

However, the theory of domains adds to the theory of Frame Semantics in four important respects. Firstly, while Fillmore acknowledges that concepts can be structured in terms of multiple frames (or domains), Langacker argues that this is actually the typical arrangement. The range of domains that structure a single lexical concept is called the **domain matrix** of that concept. Clausner and Croft illustrate this idea in the following way:

> Our commonsense knowledge about birds for example includes their shape, the fact that they are made of physical material, their activities such as flying and eating, the avian lifecycle from egg to death, etc. These aspects of the concept *bird* are specified in a variety of different domains such as SPACE, PHYSICAL OBJECTS, LIFE, TIME, and so on. (Clausner and Croft 1999: 7)

Secondly, Langacker addresses an additional level of conceptual organisation that, although implicit in Fillmore's work, was not explicitly worked out within the theory of Frame Semantics. This relates to the distinction between **basic domains** and **abstract domains**. This distinction rests upon the notion of **experiential grounding** or **embodiment** which we discussed in Chapter 6. While some basic domains like SPACE and TIME derive directly from the nature of our embodied experience, other domains like MARRIAGE, LOVE or MEDIEVAL MUSICOLOGY are more abstract, in the sense that, although they are ultimately derived from embodied experience, they are more complex in nature. For instance, our knowledge of LOVE may involve knowledge relating to basic domains, such as directly embodied experiences like touch, sexual relations and physical proximity, and may also involve knowledge relating to abstract domains, such as experience of complex social activities like marriage ceremonies, hosting dinner parties and so on. While Fillmore's theory primarily addresses abstract domains, Langacker's theory addresses both basic and abstract domains.

Thirdly, as we will see in the next section, domains are organised in a hierarchical fashion in Langacker's model. This means that a particular lexical concept can simultaneously presuppose a domain lower down the hierarchy and represent a subdomain for a lexical concept further up the hierarchy (see Figure 7.5). For example, while the concept ELBOW is understood with respect to the domain ARM, the concept ARM is understood with respect to the domain BODY. In this way, the relationship between domains reflects meronymic (part–whole) relations.

Finally, Fillmore's emphasis in developing a theory of Frame Semantics is somewhat different from Langacker's emphasis in developing a theory of domains. While Fillmore, particularly in more recent work (e.g. Fillmore and Atkins 1992), views frames as a means of accounting for grammatical behaviour like valence relations (recall examples (1)–(2)), Langacker's theory of domains is more concerned with **conceptual ontology**: the structure and organisation of knowledge, and the way in which concepts are related to and understood in terms of others.

COGNITIVE LINGUISTICS: AN INTRODUCTION

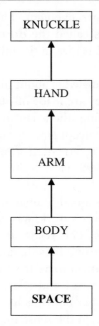

Figure 7.5 Location of the lexical concept KNUCKLE in a hierarchy of domain complexity

7.3.2 Basic, image-schematic and abstract domains

If concepts presuppose the domains against which they are understood, it follows that there is a **hierarchy of complexity** leading ultimately to domains that do not presuppose anything else. In other words, conceptual structure must ultimately be based on knowledge that is not dependent upon other aspects of conceptual organisation, otherwise the system would suffer from the problem of circularity. Domains that are not understood in terms of other domains are the basic domains we introduced above. However, given that cognitive linguists reject the idea that concepts are innately given, since this view runs counter to the cognitive theses of experientialism and emergentism, it is important to establish the origins of these basic domains. Of course, Langacker argues that basic domains derive from pre-conceptual experience, such as sensory-perceptual experience, which forms the basis of more complex knowledge domains.

In order to illustrate the theory of domains and look at how they are related, let's consider a specific example of a hierarchy of complexity. Consider the word *knuckle*. This relates to a lexical concept that is understood with respect to the domain HAND. In turn, the lexical concept HAND is understood with respect to the domain ARM. The lexical concept ARM is understood with respect to the domain BODY, and the lexical concept BODY is understood more generally in terms of (three-dimensional) SPACE. However, it is difficult to envisage another domain in terms of which we understand SPACE. After all, SPACE is a domain that derives directly from sensory experience of the world, such as visual perception and our experience of motion and touch. Therefore SPACE appears not to be understood in terms of a further conceptual domain but in terms of fundamental pre-conceptual experience. This hierarchy of complexity is illustrated in Figure 7.5. Because SPACE is presupposed by all the concepts above it, it is situated at the lowest point in the hierarchy; because KNUCKLE requires knowledge of a greater number of domains, it is placed at the highest point in this hierarchy.

According to Langacker, then, basic domains derive from directly embodied experiences that are pre-conceptual in nature. This means that such experiences derive either from subjective or 'internal' embodied experiences like emotion, consciousness or awareness of the passage of time, or from sensory-perceptual experiences which relate to information derived from the external world. Subjective experiences and sensory-perceptual experiences are both directly embodied pre-conceptual experiences; once experienced, they are represented as concepts at the conceptual level. Of course, the reader will have noticed that this discussion is reminiscent of the discussion of image schemas that was presented in Chapter 6. Let's consider, then, how image schemas relate to Langacker's theory of domains.

Firstly, we consider in more detail what might count as basic domains and what kinds of subjective and sensory experiences might give rise to these domains. We begin with the sensory experiences that relate to the external world. Vision contributes to at least two basic domains: COLOUR and SPACE. The word 'contribute' is important here, particularly as it relates to the domain of SPACE. After all, people who are blind or partially sighted still develop concepts relating to SPACE. This means that other sensory capacities also contribute to this domain, including touch, and **kinaesthetic perception** (the ability to perceive self-motion). Other basic domains include PITCH (arising from hearing experience) and TEMPERATURE, PRESSURE and PAIN (arising from touch experience). All these domains are directly tied to sensory experience and do not presuppose other conceptual domains.

Experiences that are subjective in nature give rise to a basic domain (or domains) relating to EMOTION and TIME, among others. A (non-exhaustive) inventory of basic domains is shown in Table 7.4.

Based on our discussion so far, we can identify three attributes associated with basic domains. These are summarised in Table 7.5.

Let's now consider how basic domains relate to **image schemas**. As we saw in the previous chapter, image schemas, like basic domains, are conceptual representations that are directly tied to pre-conceptual experience. Moreover, a large number of lexical concepts appear to presuppose image schemas, also a

Table 7.4 Partial inventory of basic domains

Basic domain	Pre-conceptual basis
SPACE	Visual system; motion and position (proprioceptive) sensors in skin, muscles and joints; vestibular system (located in the auditory canal – detects motion and balance)
COLOUR	Visual system
PITCH	Auditory system
TEMPERATURE	Tactile (touch) system
PRESSURE	Pressure sensors in the skin, muscles and joints
PAIN	Detection of tissue damage by nerves under the skin
ODOUR	Olfactory (smell) system
TIME	Temporal awareness
EMOTION	Affective (emotion) system

Table 7.5 Attributes of basic domains

Basic domains:

Provide the least amount of complexity in a complexity hierarchy, where 'complexity' relates to level of detail
Are directly tied to pre-conceptual embodied experience
Provide a 'range of conceptual potential' in terms of which other concepts and domains can be understood.

characteristic of domains. For example, the CONTAINER image schema appears to underlie a number of lexical concepts that we have discussed so far throughout this book. This suggests that the CONTAINER schema might be equivalent to a domain. However, Clausner and Croft (1999) argue that image schemas, while deriving from sensory experience, are not quite the same thing as basic domains. For example, they argue that the CONTAINER image schema is a relatively complex knowledge structure, which is based on the basic domain SPACE and another image schema MATERIAL OBJECT. Therefore the CONTAINER schema does not relate to a level of **least complexity** and, according to this criterion, is not equivalent to a basic domain.

A second distinction between basic domains and image schemas relates to the idea that image schemas are abstracted from recurrent patterns of experience. It follows that image schemas are likely to contribute to the domain matrices of a wide range of concepts (a domain matrix is the network of domains that underlies a concept). In contrast, basic domains need not occur in a wide range of domain matrices. For example, compare the image schema MATERIAL OBJECT with the basic domain TEMPERATURE. Because MATERIAL OBJECT derives from

Table 7.6 Distinctions between basic domains and image schemas

Basic domain	Image schema
Occupies lowest position in the hierarchy of complexity, e.g. SPACE, TIME, TEMPERATURE, PITCH	Need not occupy lowest position in the hierarchy of complexity, e.g. UP-DOWN, FRONT-BACK, CONTAINMENT, PATH
Need not occur in a wide range of domain matrices, e.g. TEMPERATURE, ODOUR	Occurs in the widest range of domain matrices, e.g. SCALE, PROCESS, OBJECT, CONTAINMENT
Derived from subjective experience, e.g. TIME, EMOTION, or sensory-perceptual experience, e.g. SPACE, TEMPERATURE	Derived from sensory-perceptual experience only, e.g. UP-DOWN, FRONT-BACK, CONTAINMENT, SURFACE

experience of material objects, it will contribute to the domain matrix of all material objects: CAR, DESK, TABLE, CHAIR, VASE, TREE, BUILDING and so on. However, TEMPERATURE contributes to the domain matrices of a more restricted set of concepts: THERMOMETER, HOT, COLD and so on. Therefore, basic domains can have a narrower distribution within the conceptual system than image schemas.

A third distinction between basic domains and image schemas concerns the idea that all image schemas are **imagistic** in nature: they derive from sensory experience and therefore have **image content**. However, while some basic domains like SPACE and TEMPERATURE also have image content because they are based on pre-conceptual sensory experience, other basic domains like TIME are ultimately derived from subjective (introspective) experience and are not intrinsically imagistic in nature. This does not mean, however, that basic domains that arise from subjective experience cannot be conceptualised in terms of image content. For example, as we have seen, various emotional STATES can be structured in terms of the CONTAINER schema, as a result of conceptual metaphor. We will explore this idea further in Chapter 9. The distinctions between basic domains and image schemas are summarised in Table 7.6.

In sum, an assumption central to cognitive semantics is that all human thought is ultimately grounded in basic domains and image schemas. However, as Langacker observes, 'for the most part this grounding is indirect, being mediated by chains of intermediate concepts' (Langacker 1987: 149–50). These intermediate concepts, which correspond to the non-bold type domains in Figure 7.5, are abstract domains. As we have seen, an abstract domain is one that presupposes other domains ranked lower on the complexity hierarchy.

7.3.3 Other characteristics of domains

Langacker's proposal that encyclopaedic knowledge consists of an inventory of basic and more abstract domains is only one step in developing a theory of the architecture of human conceptual organisation. In addition, Langacker sets out a number of characteristics that identify domains.

Dimensionality

The first characteristic is **dimensionality**: some domains are organised relative to one or more dimension. For example, the basic domains TIME, TEMPERATURE and PITCH are organised along a single dimension and are thus one-dimensional: TEMPERATURE is structured in terms of a series of points that are conceptualised as an ordinal sequence. In contrast, SPACE is organised with respect to two or three dimensions (a drawing of a triangle on a page is two-dimensional, while a flesh-and-blood human is three-dimensional), and COLOUR is organised with respect to three dimensions (BRIGHTNESS, HUE and SATURATION). These dimensions of colour relate to distinct neuro-perceptual mechanisms, which allow us to detect differences along these three dimensions, affecting our perception of colour. Abstract domains can also be organised with respect to a particular dimension or set of dimensions. For example, CARDINAL NUMBERS (1, 2, 3, 4 . . .) represent a domain ordered along a single dimension. However, some domains cannot be characterised in terms of dimensionality; it is not clear how we might describe the domain of EMOTION in this way, for example.

Locational versus configurational domains

A further characteristic of domains is that they can be distinguished on the basis of whether they are **configurational** or **locational**. This distinction relates to whether a particular domain is **calibrated** with respect to a given dimension. For example, COLOUR is a locational domain because each point along each of its dimensions (for example, HUE) is calibrated with respect to the point adjacent to it. In other words, each colour sensation occupies a distinct 'point' on the HUE dimension, so that a different point along the dimension represents a different colour experience. This contrasts with the domain of SPACE, which is not calibrated in this way: SPACE is not locational but configurational. For example, regardless of its position with respect to the dimension of SPACE, the shape TRIANGLE remains a triangle rather than, say, a SQUARE.

7.3.4 Profile/base organisation

We noted earlier that lexical concepts (the meanings associated with words) are understood with respect to a domain matrix. In other words, lexical concepts are typically understood with respect to a number of domains, organised in a network. One consequence of this claim is that, as we have already seen, a word

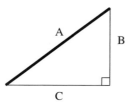

Figure 7.6 Scope for the concept HYPOTENUSE

provides a point of access to the entire knowledge inventory associated with a particular lexical concept. However, if we assume that a domain matrix underlies each lexical concept, then we need to explain why different facets of the encyclopaedic knowledge network are differentially important in the understanding of that concept. For example, consider the word *hypotenuse*. The lexical concept behind this word relates to the longest side of a right-angled triangle, which is illustrated in Figure 7.6. In this diagram, the hypotenuse is the side of the triangle in bold type labelled A.

While *hypotenuse* provides a point of access to a potentially infinite knowledge inventory, relating to RIGHT-ANGLED TRIANGLES, TRIANGLES in general, GEOMETRIC FIGURES, GEOMETRIC CALCULATION, SPACE and so on, only part of this knowledge network is essential for an understanding of the meaning of the lexical concept. Langacker suggests an explanation for this in terms of scope, profile and base. The essential part of the knowledge network is called the **scope** of a lexical concept. The scope of a lexical concept is subdivided into two aspects, both of which are indispensable for understanding what the word means. These are the **profile** and its **base**, which we first introduced in Chapter 5. The profile is the entity or relation designated by the word, and the base is the essential part of the domain matrix necessary for understanding the profile. In the case of our example *hypotenuse*, this word profiles or **designates** the longest side in a right angled-triangle, while the base is the entire triangle, including all three of its sides. Without the base, the profile would be meaningless: there is no hypotenuse without a right-angled triangle. Hence, the word *hypotenuse* designates a particular substructure within a larger conceptual structure. As Langacker explains it, 'The semantic value of an expression resides in neither the base nor the profile alone, but only in their combination' (Langacker 1987: 183).

One consequence of the profile/base relation is that the same base can provide different profiles. Consider Figure 7.7, which depicts a CIRCLE. This base can give rise to numerous profiles, including ARC (Figure 7.7(a)), RADIUS (Figure 7.7(b)), DIAMETER (Figure 7.7(c)), CIRCUMFERENCE (Figure 7.7(d)), and so on.

Now let's consider a more complex example. The word *uncle* profiles an entity with a complex domain matrix. This includes at least the following abstract domains: GENEALOGY, PERSON, GENDER, SEXUAL INTERCOURSE, BIRTH, LIFE

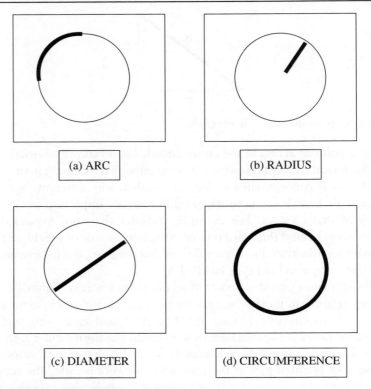

Figure 7.7 Different profiles derived from the same base

CYCLE, PARENT/CHILD RELATIONSHIP, SIBLING RELATIONSHIP, EGO. The base for the lexical concept UNCLE is the conceived network of FAMILIAL RELATIONS represented in Figure 7.8. Against this base, *uncle* profiles an entity related to the EGO by virtue of being a MALE SIBLING of EGO's mother or father.

7.3.5 Active zones

As we have seen, the encyclopaedic view of meaning recognises that, in ordinary speech, the meaning associated with a lexical item undergoes 'modulation' as a result of the context in which it is used. This means that typically only part of an entity's profile is relevant or active within a particular utterance. This part of the profile is called the **active zone**. Consider the examples in (5).

(5) a. The footballer headed the ball.
    b. The footballer kicked the ball.
    c. The footballer frowned at the referee.
    d. The footballer waved at the crowd.

THE ENCYCLOPAEDIC VIEW OF MEANING

Figure 7.8 Familial network in which UNCLE is profiled

While *the footballer* is profiled in each of these examples, a different active zone is evident in each example. For instance, in (5a) the active zone is the footballer's forehead (Figure 7.9(a)); in (5b) the active zone is the footballer's foot (Figure 7.9(b)); in (5c) the active zone is the footballer's face (Figure 7.9(c)); and in (5d) the active zone is the footballer's hands and arms (figure 7.9(d)).

Let's now illustrate how the phenomenon of active zones is evident in language use. Consider the example in (6).

(6) This red pen isn't red.

The idea of active zones helps to explain why this apparently contradictory sentence can give rise to a non-contradictory interpretation. If we interpret the sentence in (6) to mean that a pen whose ink is red is not coloured red, or indeed that a pen that is coloured red does not contain red ink, then we do so by assigning each instance of *red* a different active zone. One active zone relates to the contents of the pen that result in coloured marks on paper while the other active zone corresponds to the outer body of the pen. This example shows how active zone phenomena are at work in discourse, enabling speakers and hearers to 'search through' the inventory of knowledge associated with each word and to 'select' an interpretation licensed by the context.

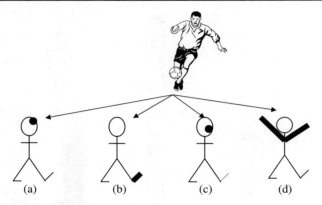

Figure 7.9 Active zones for the sentences in (5)

7.4 The perceptual basis of knowledge representation

In this section, we return to the issue of how cognitive psychologists characterise conceptual structure. In particular, we return to the issue of simulations, which we introduced briefly in section 7.2.2, and attempt to see how these can be incorporated into a theory of frames. Of course, this relates to the more general question we have been pursuing in this chapter: what do the mental representations that underpin language 'look like'? For cognitive linguists, the answer lies in the thesis of embodied cognition which gives concepts a fundamentally perceptual character. As Langacker argues, for instance, concepts are ultimately grounded in terms of basic domains which represent knowledge arising from foundational aspects of experience relating either to sensory experience of the external world or to subjective (or introspective) states. Our objective in this section, then, is to provide a sense of how the models of knowledge representation being developed in cognitive semantics are increasingly consonant with theories being developed in cognitive psychology. In particular, we address some of the more recent ideas that have been proposed by cognitive psychologist Lawrence Barsalou.

In his (1999) paper *Perceptual Symbol Systems*, Barsalou argues that there is a common representational system that underlies both **perception** (our ability to process sensory input from the external world and from internal body states such as consciousness or experience of pain) and **cognition** (our ability to make this experience accessible to the conceptual system by representing it as concepts, together with the information processing that operates over those concepts). One property of cognition that distinguishes it from perception is that cognition operates **off-line**. In other words, cognitive processing employs mental representations (concepts) that are stored in memory, and thereby frees itself from the process of experiencing a particular phenomenon every time that experience is accessed and manipulated. For instance, when planning a long car journey, we can predict roughly at what points in the journey we will need to stop and refuel. In other words, we can make predictions based on our concept – or frame – for CAR. We can make these predictions on the basis of past experiences, which come to form part of the mental representation associated with our mental knowledge of cars. This means we can make predictions about fuel consumption on a forthcoming journey rather than just getting into the car and waiting to see when the petrol runs out.

According to Barsalou, perceptual symbols (concepts) are neural representations stored in sensory-motor areas of the brain. He describes perceptual symbols as 'records of the neural states that underlie perception. During perception, systems of neurons in sensory-motor regions of the brain capture information about perceived events in the environment and in the body' (Barsalou 1999: 9). For example, consider the concept HAMMER. The perceptual symbol for this concept will consist of information relating to its shape, weight, texture, colour, size and so on, as well as sensory-motor patterns consistent with the experience of using a hammer (derived from our experience of banging a nail into a piece of wood, for example). It follows that perceptual symbols are **multi-modal**, drawing information from different sensory-perceptual and introspective (subjective) input 'streams'.

However, perceptual symbols do not exist independently of one another. Instead, they are integrated into systems called **simulators**. A simulator is a mental representation that integrates and unifies related perceptual symbols (for example, all our experiences with hammers). Two kinds of information are extracted from simulators. The first is a frame, which we discussed earlier in the chapter (section 7.2.2). A frame is **schematic** in nature, abstracting across a range of different perceptual symbols for hammers. Hence, it provides a relatively stable representation (a concept) of HAMMER, drawing together what is uniform about our experience with tools of this kind.

The second kind of information extracted from a simulator is a **simulation**. A simulation is an 'enactment' of a series of perceptual experiences, although in attenuated (weakened) form. For instance, if we say 'imagine you're using a hammer . . .', this utterance allows you to construct a simulation in which you can imagine the hammer, feel a sense of its weight and texture in your hand, and sense how you might swing it to strike another object. Therefore, part of our knowledge of the concept HAMMER includes a schematic frame relating to the kinds of knowledge we associate with hammers, as well as simulations that provide representations of our perceptual experience of hammers. Crucially, both frames and simulations derive from perceptual experience.

Evidence for the view that conceptual structure has a perceptual basis, and for the view that concepts (represented in terms of frames) can give rise to simulations, comes from a range of findings from **neuroscience**, the interdisciplinary study of brain function. This area of investigation has begun to provide support for the thesis that cognition is grounded in perceptual symbol systems of the kind proposed by Barsalou. For example, it is now clear that damage to parts of the brain responsible for particular kinds of perception also impairs our ability to think and talk about concepts that relate to those areas of perceptual experience. For example, damage to motor and somatosensory (touch) areas affects our ability to think about and identify conceptual categories like tools which relate to motor and somatosensory experience. Similarly, damage to areas of the brain that process visual perception affects our ability to access or manipulate conceptual categories that relate to visual experience. Evidence from experiments based on descriptive tasks also suggests that conceptual representation is perceptual in nature. For example, when a subject sitting in a lab without a perceptual stimulus is asked to describe a car, he or she will typically describe the car from a particular 'perspective': subjects tend not to list attributes in a random order, but to describe the parts of the car that are near each other first. Moreover, when a context is provided, this can influence the simulated perspective: subjects who are told to imagine that they are standing outside the car will describe different attributes of a car, and in a different order, compared with subjects who are told to imagine that they are sitting inside the car. This type of experiment suggests that the CAR frame, together with its associated simulations, is based on sensory-motor experience of cars.

Before concluding, let's briefly compare models that assume a perceptual basis for mental representation with the type of model adopted in formal linguistics. Since the emergence of the Chomskyan mentalist model of language in the mid-twentieth century which firmly focused attention on language as a cognitive phenomenon and the simultaneous rise of cognitive science, theories of mental representation have adopted a **non-perceptual** view. This is sometimes called an **amodal view**, because it views conceptual structure as based not on perceptual (modal) states, but on a distinct kind of representational system. According to Barsalou, cognitive science was influenced in this respect by formalisms that emerged from branches of philosophy and mathematics (such as logic), and from the development of computer languages in computer science and artificial intelligence. Moreover, the prevalence of the modular theory of mind, not only in linguistics but also in cognitive psychology, represented a widespread view of perception and cognition as separable systems, operating according to different principles. This view is inherent in Fodor's theory of mind, for example, which is outlined in his book *The Modularity of Mind* (1983). According to this theory, there are three distinct kinds of mental mechanisms: **transducers** (which receive 'raw' sensory-perceptual input and 'translate' it into a form that can be manipulated by the other cognitive systems), **central systems** (which do the 'general' cognitive work such as reasoning, inference and memory) and **modules** (specialised and encapsulated systems of knowledge that mediate between the transducers and the central systems).

In non-perceptual systems for mental representation, words assume primary importance as symbols for mental representations. For example, in early approaches to lexical semantics, feature lists employed words to stand for semantic features:

(7)   Bachelor
$$\begin{pmatrix} + \text{ MALE} \\ - \text{ MARRIED} \\ + \text{ ADULT} \end{pmatrix}$$

In formal semantics, the language of predicate calculus was adopted, which also based semantic features on words. While semanticists who rely upon componential and formal methods do not assume that words literally make up the content of the mental representations they stand for, they do rely upon items of natural language as a metalanguage for describing natural language, an approach that entails obvious difficulties. For example, if we rely on real words to express concepts, this limits the set of concepts to the set of real words. As we have seen, recent developments in cognitive psychology suggest that conceptual structure actually has a perceptual basis. These ideas, together with the empirical evidence that is beginning to be gathered, is consonant with the claims of cognitive semantics, particularly the thesis of embodied cognition.

7.5 Summary

In this chapter, we have explored one of the central theses of cognitive linguistics: that meaning is **encyclopaedic** in nature. This view relates to the open-class semantic system and holds that word meaning cannot be understood independently of the vast system of encyclopaedic knowledge to which it is linked. In addition, cognitive semanticists argue that semantic knowledge is grounded in human interaction with others (social experience) and with the world around us (physical experience). The thesis of **embodied cognition** central to cognitive linguistics entails that mental representations are perceptual in nature. We briefly considered recent perspectives from cognitive psychology that also suggest that knowledge is represented in the mind as **perceptual symbols**: representations that are fundamentally perceptual in nature. In order to elaborate the notion of **encyclopaedic semantics**, we explored two theories of semantics that have been particularly influential in developing this approach to meaning: (1) the theory of **Frame Semantics** developed by Charles Fillmore, and (2) the theory of **domains** developed by Ronald Langacker. While these two theories were developed for different purposes, together they provide the basis for a theory of encyclopaedic semantics that is presupposed by much current work on lexical semantics and conceptual structure in cognitive semantics, and in cognitive linguistics more generally.

Further reading

The encyclopaedic view of meaning

- **Haiman (1980).** Haiman (a typologist) considers and rejects arguments for assuming a dictionary view of word meaning. Haiman argues in favour of an encyclopaedic account.
- **Langacker (1987).** The first volume in Langacker's two-volume overview of Cognitive Grammar provides a detailed case for an encyclopaedic approach to linguistic meaning. See Chapter 4 in particular.
- **Tyler and Evans (2003).** Tyler and Evans also make the case for an encyclopaedic account of word meaning, applying this approach to a single and highly complex lexical class: the English prepositions.

Frame semantics

- **Fillmore (1975)**
- **Fillmore (1977)**
- **Fillmore (1982)**
- **Fillmore (1985a)**
- **Fillmore and Atkins (1992)**

Listed above are the key papers that have given rise to the Frame Semantics approach. The paper by Fillmore and Atkins (1992) presents a detailed analysis of the semantic frame for RISK. The words in this set include: *risk, danger, peril, hazard* and neighbouring words such *gamble, invest* and *expose*. More recently, Fillmore has been leading the FrameNet project. This project applies the theory of Frame Semantics with a view to developing an electronic frame-based dictionary. For further details and references see the FrameNet website: www.icsi.berkeley.edu/framenet/.

The theory of domains

- **Langacker (1987).** This is the key source for the theory of domains. See Part II of the book in particular.

- **Taylor (2002).** This introduction to Langacker's theory has a number of very good chapters on the theory of domains. See in particular chapters 10, 11, 22 and 23.

Frames and perceptual symbol systems

- **Barsalou (1992a).** This paper provides a comprehensive and yet concise introduction to an influential theory of frames and framing by a leading researcher in this area.
- **Barsalou (1992b).** An excellent and very accessible overview of key ideas in cognitive psychology. Chapter 7 is a particularly good introduction to knowledge representation, concepts and frames.
- **Barsalou (1999).** This paper provides points of entry into the literature on perceptual symbol systems and simulation in mental representation. In particular it develops Barsalou's own theory of the percepetual basis of conceptual structure.
- **Barsalou (2003).** This paper summarises and reviews the empirical evidence that supports the perspective presented in Barsalou's 1999 paper.

Exercises

7.1 Examining the dictionary view

What distinctions are central to the dictionary view of word meaning? Outline the advantages and disadvantages of this account.

7.2 Centrality

In view of the distinction between conventional, generic, intrinsic and characteristic knowledge (section 7.1.4), provide a characterisation for the following lexical items: *apple*, *diamond*, *crocodile*.

7.3 Fillmore's Frame Semantics versus Langacker's theory of domains

What are the key similarities and differences, as you see them, between Fillmore's Frame Semantics and Langacker's theory of domains?

7.4 Frames

Identify the frames associated with the following lexical items:

(a) Saturday
(b) breakfast

(c) widow
(d) celibacy
(e) (to) lend

7.5 Frames and participant roles

Provide a Frame Semantics analysis of the distinction between the verbs (*to*) *borrow* and (*to*) *lend*. You will need to say what participant role(s) each verb is associated with and provide evidence with example sentences.

7.6 Framing and culture

Now consider the lexical item *Prime Minister*. Say what frame this belongs to, giving as much detail as possible in terms of other elements. In what way is this frame culture-dependent?

7.7 Base, domain and domain matrix

What is the distinction between a base, a domain and a domain matrix? Provide examples to illustrate.

7.8 Domains and hierarchies of complexity

Provide hierarchies of complexity for the following lexical items:

(a) toe
(b) spark plug
(c) (a) second [= unit of time]
(d) Prime Minister

Did you have any difficulties establishing a hierarchy of complexity for *Prime Minister*? Comment on why this might be.

7.9 Domain matrix

Provide a domain matrix for *Prime Minister*. Does this shed any light on why you may have had difficulties in exercise 7.8(d)? Now consider the domain matrices for *President* and *Monarch* respectively. What are your assumptions in terms of political systems?

7.10 Profile-base organisation

Give a characterisation of *Prime Minister* in terms of profile-base organisation. How is this distinct from profile-base organisation for *President*?

7.11 Image schemas versus basic domains

Consider the following lexical items. Based on the discussion in this chapter, which aspects of the meaning associated with these lexical items would you model in terms of image schemas and which in terms of (basic) domains? Explain how you reached your conclusions.

(a) cup
(b) container
(c) (to) push

8

Categorisation and idealised cognitive models

In this chapter, we continue our exploration of the human conceptual system by focusing on categorisation: our ability to identify perceived similarities (and differences) between entities and thus group them together. Categorisation both relies upon and gives rise to concepts. Thus categorisation is central to the conceptual system, because it accounts, in part, for the organisation of concepts within the network of encyclopaedic knowledge. Categorisation is of fundamental importance for both cognitive psychologists and semanticists, since both disciplines require a theory of categorisation in order to account for knowledge representation and indeed for linguistic meaning. Central to this chapter is the discussion of findings that emerged from the work of cognitive psychologist Eleanor Rosch and her colleagues in the 1970s, and the impact of these findings on the development of cognitive semantics. In particular, we will be concerned with the work of George Lakoff, who addressed findings relating to **prototype structure** and **basic level categories** revealed by research in cognitive psychology, and who developed a cognitive semantic theory of **idealised cognitive models** (ICMs) in order to account for these phenomena. The influence of Lakoff's research, and of his book *Women, Fire and Dangerous Things* (1987), was important for the development of cognitive semantics. In particular, this book set the scene for cognitive semantics approaches to conceptual metaphor and metonymy, lexical semantics (word meaning) and grammatical structure. In this chapter, then, we set out the theoretical background of Chapters 9 and 10 where we will address Lakoff's theory of conceptual metaphor and metonymy and his theory of word meaning in detail.

We begin the chapter by explaining how Rosch's research on categorisation was important in the development of cognitive semantics, setting this discussion against the context of the classical view of categorisation that was superseded by Rosch's findings. We then look in detail at the findings to emerge from Rosch's research (section 8.2) and explore the development of Lakoff's theory of cognitive models that was developed in response to this research (section 8.3). Finally, we briefly explore the issue of linguistic categorisation in the light of the empirical findings and theoretical explanations presented in this chapter (section 8.4).

8.1 Categorisation and cognitive semantics

In the 1970s the **definitional** or **classical theory** of human categorisation – so called because it had endured since the time of the ancient Greek philosophers over 2,000 years ago – was finally called into question. The new ideas that contributed most significantly to this development are grouped together under the term **prototype theory**, which emerged from the research of Eleanor Rosch and her colleagues. In fact, 'Prototype Theory' was less a theory of knowledge representation than a series of findings that provided startling new insights into human categorisation. In so far as the findings led to a theory, Rosch proposed in her early work that humans categorise not by means of the **necessary and sufficient conditions** assumed by the classical theory (to which we return below), but with reference to a **prototype**: a relatively abstract mental representation that assembles the key attributes or features that best represent instances of a given category. The prototype was therefore conceived as a schematic representation of the most salient or central characteristics associated with members of the category in question.

A problem that later emerged was that the view of prototypes as mental representations failed to model the **relational knowledge** that humans appear to have access to (recall from the last chapter that relational knowledge is one of the properties of encyclopaedic knowledge addressed by Frame Semantics). These criticisms led to further developments in prototype theory. Some scholars argued for a revised view of the prototype, suggesting that the mental representation might correspond to an **exemplar**: a specific category member or 'best example' of a category, rather than a schematic group of attributes that characterise the category as a whole. However, these **exemplar-based models** of knowledge representation were also problematic because they failed to represent the **generic information** that humans have access to when they use concepts in order to perform a host of conceptual operations, including categorisation. Indeed, the most recent theories of categorisation assert that a key aspect of knowledge representation is the **dynamic** ability to form **simulations**, an idea that was introduced in the previous chapter. Thus, in a number of respects, prototype theory has been superseded by more recent empirical findings and theories. Despite this, there are a number of reasons why a chapter on categorisation in general, and prototype theory in particular, is essential for a thorough understanding of cognitive semantics.

Firstly, an investigation of prototype theory provides a picture of the historical context against which cognitive linguistics emerged as a discipline. The development of prototype theory in the 1970s resonated in important ways with linguists whose research would eventually contribute to defining the field of cognitive semantics. Charles Fillmore and George Lakoff were both members of faculty at the University of California at Berkeley where Eleanor Rosch was also conducting her research, and both were influenced by this new approach to categorisation. For Lakoff in particular, Rosch's discovery that psychological categories did not have clearly definable boundaries but could instead be described as having 'fuzzy' boundaries reflected his own views about language: Lakoff thought that lexical and grammatical categories might also be most insightfully conceived as categories with rather fluid membership. This led Lakoff to apply this new view of psychological categories to linguistic categories (such as word meanings). In this way, 'Prototype Theory' inspired some of the early research in cognitive semantics.

Secondly, and perhaps more importantly, although it now seems that prototype theory cannot be straightforwardly interpreted as a theory of knowledge representation, the empirical findings that emerged from this research demand to be accounted for by any theory of categorisation. In other words, the **prototype effects** or **typicality effects** that Rosch discovered are psychologically real, even if the early theories of knowledge representation that were proposed to account for these effects have been shown to be problematic. Indeed, a central concern in Lakoff's (1987) book was to address the problems that early prototype theory entailed, and to propose in its place a theory of cognitive models.

Thirdly, as we mentioned above, Lakoff's (1987) book set the scene for the development of three important strands of research within cognitive linguistics: (1) **Conceptual Metaphor Theory** (Chapter 9); (2) **cognitive lexical semantics** (Chapter 10); and (3) a **cognitive approach to grammar** that influenced the well-known constructional approach developed by his student Adele Goldberg (to which we return in Part III of this book).

Finally, *Women, Fire and Dangerous Things*, despite its rather meandering presentation, in many ways defines the two key commitments of cognitive linguistics: the 'Generalisation Commitment' and the 'Cognitive Commitment'. Lakoff's book took what was then a relatively new set of findings from cognitive psychology and sought to develop a model of language that was compatible with these findings. In attempting to model principles of language in terms of findings from cognitive psychology, Lakoff found himself devising and applying principles that were common both to linguistic and conceptual phenomena, which thus laid important foundations for the cognitive approach to language.

8.1.1 The classical theory

Before presenting Rosch's findings concerning categorisation, it is important to set her research in some historical context. The 'classical theory' of categorisation was the prevalent model since the time of Aristotle and holds that conceptual and linguistic categories have **definitional structure**. This means that an entity represents a category member by virtue of fulfilling a set of **necessary and (jointly) sufficient conditions** for category membership. These conditions are called 'necessary and sufficient' because they are individually necessary but only collectively sufficient to define a category. Traditionally, the conditions were thought to be sensory or perceptual in nature. To illustrate, consider once more the familiar lexical concept BACHELOR. For an entity to belong to this category, it must adhere to the following conditions: 'is not married'; 'is male'; 'is an adult'. Each of these conditions is necessary for defining the category, but none of them is individually sufficient because 'is not married' could equally hold for SPINSTER, while 'is male' could equally hold for HUSBAND, and so on. In theories of linguistic meaning, necessary and sufficient conditions have taken the form of **semantic primitives** or **componential features**, an idea that we have mentioned in previous chapters (recall our discussion of semantic universals in Chapter 3 and our discussion of the dictionary view of linguistic meaning in Chapter 7). As we have seen, the idea of semantic primitives has been influential in semantic theories that adopt the formal 'mentalist' view proposed by Chomsky, which is primarily concerned with modelling an innate and specialised system of linguistic knowledge. This is because, in principle at least, semantic primitives suggest the possibility of a set of universal semantic features that can be combined and recombined in order to give rise to an infinite number of complex units (word meanings). This approach is reminiscent of the characterisation of human speech sounds in phonetics and phonology, where a bundle of articulatory features makes up each speech sound. It is also reminiscent of the characterisation of sentence structure in terms of strings of words that combine to make phrases, which then combine to make sentences. In other words, the influence of the semantic decomposition approach reflects the influence of structural approaches to sound and grammar upon the development of theories of word meaning. This kind of approach is attractive for a formal theory because it enables the formulation of precise statements which are crucial to the workings of the 'algorithmic' or 'computational' model favoured by these approaches. For example, Katz (1972) argued that the English noun *chair* names a category that can be decomposed into the set of semantic features or markers shown in Table 8.1.

However, while many (usually formal) linguists would argue that 'decompositional' approaches have worked rather well for modelling the structural aspects of language such as phonology or syntax, many linguists (both formal

Table 8.1 Semantic features or markers for the category CHAIR

OBJECT
PHYSICAL
NON-LIVING
ARTEFACT
FURNITURE
PORTABLE
SOMETHING WITH LEGS
SOMETHING WITH A BACK
SOMETHING WITH A SEAT
SEAT FOR ONE

and cognitive) also recognise that the classical decompositional theory of word meaning suffers from a number of problems. We discuss here three of the most serious problems with this approach.

8.1.2 The definitional problem

While the classical theory holds that categories have definitional structure, in practice it is remarkably difficult to identify a precise set of conditions that are necessary and sufficient to define a category. This requires the identification of all those features that are shared by all members of a category (necessary features) and that together are sufficient to define that category (no more features are required). The following famous passage from the philosopher Wittgenstein's discussion of the category GAME illustrates the difficulty inherent in this approach:

> Consider for example the proceedings that we call 'games'. I mean board-games, card-games, ball-games, Olympic games and so on. What is common to them all? – Don't say: 'There must be something common, or they would not be called "games"' – but look and see whether there is anything common to all. – For if you look at them you will not see something that is common to all, but similarities, relationships, and a whole series of them at that. To repeat: don't think, but look! – For example at board-games, with their multifarious relationships. Now pass to card-games; here you find many correspondences with the first group, but many common features drop out, and others appear. When we pass next to ball-games, much that is common is retained, but much is lost. – Are they all 'amusing'? Compare chess with noughts and crosses. Or is there always winning and losing, or competition between players? Think of patience. In ball-games there is winning and losing; but when a child throws his ball at the wall and catches it again, this feature has disappeared. Look at the parts played

by skill and luck; and at the difference between skill in chess and skill in tennis. Think now of games like ring-a-ring-a-roses; here is the element of amusement, but how many other characteristic features have disappeared! And we can go through the many, many other groups in the same way; we see how similarities crop up and disappear. (Wittgenstein 1958: 66)

This passage reveals that there is no single set of conditions that is shared by every member of the category GAME. While some games are characterised by AMUSEMENT, like tiddlywinks, others are characterised by LUCK, like dice games, still others by SKILL or by COMPETITION, like chess. In other words, it appears to be impossible to identify a definitional structure that neatly defines this category. To present a simpler example, consider the category CAT. We might define this category as follows: 'is a mammal'; 'has four legs'; 'is furry'; 'has a long tail'; 'has pointy ears'. What happens if your cat gets into a fight and loses an ear? Or gets ill and loses its fur? Does it then stop being a member of the category CAT? The definitional approach therefore suffers not only from the problem that the definitions are often impossible to identify in the first place, but also from the problem that definitions are, in reality, subject to **exceptions**. A three-legged one-eared hairless cat is still a cat. It seems, then, that a category need not have a set of conditions shared by all members in order to 'count' as a meaningful category in the human mind. It is important to emphasise here that we are not dealing with scientific categories, but with the everyday process of categorisation that takes place in the human mind on the basis of perceptual features. While a biologist could explain why a three-legged one-eared hairless cat still 'counts' as a member of that species from a scientific perspective, what cognitive psychologists and linguists want to explain is how the human mind goes about making these kinds of everyday judgements in the absence of scientific knowledge.

8.1.3 The problem of conceptual fuzziness

A second problem with the classical view is that definitional structure entails that categories have definite and distinct boundaries. In other words, an entity either will or will not possess the 'right' properties for category membership. Indeed, this appears to be the case for many categories. Consider the category ODD NUMBER. As we learn at school, members of this category are all those numbers that cannot be divided by 2 without leaving a remainder: 1, 3, 5, 7, 9 and so on. This category has clearly defined boundaries, because number is either odd or even: there is no point in between. However, many categories are not so clearly defined but instead have 'fuzzy' boundaries. Consider the category FURNITURE. While TABLE and CHAIR are clearly instances of this category, it is less clear whether CARPET should be considered a member. Consider the

Table 8.2 Problems for the classical theory of categorisation

Definitional problem: difficult or impossible to identify the set of necessary and sufficient conditions to define a category
The problem of conceptual fuzziness: not all categories have clear boundaries
The problem of typicality: many categories, including some with clear boundaries, exhibit typicality effects

category BIRD. While it is obvious that birds like ROBIN and SPARROW belong to this category, it is less obvious that animals like PENGUINS and OSTRICHES do, neither of which can fly. The difficulty in deciding to set the boundary for certain categories is the problem of conceptual 'fuzziness'. If the classical theory of categorisation is correct, this problem should not arise.

8.1.4 The problem of prototypicality

The third problem with the definitional view of categories is related to the problem of conceptual fuzziness, but while the problem of conceptual fuzziness concerns what happens at the **boundaries** of a category, the problem of prototypicality concerns what happens at the **centre** of a category. As we will see in the next section, findings from experimental cognitive psychology reveal that categories give rise to prototype or **typicality effects**. For example, while people judge TABLE or CHAIR as 'good examples' or 'typical examples' of the category FURNITURE, CARPET is judged as a less good example. These asymmetries between category members are called typicality effects. While we might expect this to happen in the case of categories that have fuzzy boundaries, experiments have revealed that categories with distinct boundaries also show typicality effects. For example, Armstrong *et al.* (1983) found that the category EVEN NUMBERS exhibits typicality effects: participants in their experiments consistently rated certain members of the category including '2', '4', '6', and '8' as 'better' examples of the category than, say, '98' or '10,002'. Categories that exhibit typicality effects are called graded categories. Typicality effects represent a serious challenge for the classical theory, because if each member of a category shares the same definitional structure, then each member should be equally 'typical'. These problems with the classical theory of categorisation are summarised in Table 8.2.

8.1.5 Further problems

Laurence and Margolis (1999) discuss further problems with this approach which we mention only briefly here. These are what they call **the problem of psychological reality** and **the problem of ignorance and error**.

The problem of psychological reality relates to the fact that there is no evidence for definitional structure in psychological experiments. For example, we might expect words with a relatively 'simple' definitional structure or small set of features (like, say, *man*) to be recognised more rapidly in word-recognition experiments than words with a more 'complex' definitional structure or greater number of features (like, say, *cousin*). This expectation is not borne out by experimental evidence. The problem of ignorance and error relates to the fact that it is possible to possess a concept without knowing what its properties are. In other words, possessing a concept is not dependent upon knowing its definition. For example, it is possible to have the concept WHALE while mistakenly believing that it belongs to the category FISH rather than the category MAMMAL.

8.2 Prototype theory

Prototype theory is most closely associated with the experimental research of cognitive psychologist Eleanor Rosch and her colleagues. In this section, we present an overview and discussion of Rosch's research, which is largely based on experimental findings.

8.2.1 Principles of categorisation

Prototype theory posits that there are two basic principles that guide the formation of categories in the human mind: (1) the **principle of cognitive economy**, and (2) the **principle of perceived world structure**. These principles together give rise to the human **categorisation system**.

Principle of cognitive economy

This principle states that an organism, like a human being, attempts to gain as much information as possible about its environment while minimising cognitive effort and resources. This cost-benefit balance drives **category formation**. In other words, rather than storing separate information about every individual stimulus experienced, humans can group similar stimuli into categories, which maintains **economy** in cognitive representation.

Principle of perceived world structure

The world around us has **correlational structure**. For instance, it is a fact about the world that wings most frequently co-occur with feathers and the ability to fly (as in birds), rather than with fur or the ability to breathe underwater. This principle states that humans rely upon correlational structure of this kind in order to form and organise categories.

8.2.2 The categorisation system

These two principles give rise to the human categorisation system. While the principle of cognitive economy has implications for the level of detail or **level of inclusiveness** with which categories are formed, the principle of correlational structure has implications for the **representativeness** or **prototype structure** of the categories formed (Rosch 1977, 1978). Rosch (1978) suggests that this gives rise to a categorisation system that has two dimensions: a horizontal and a vertical dimension. This idea is represented in Figure 8.1.

The vertical dimension relates to the level of inclusiveness of a particular category: the higher up the vertical axis a particular category is, the more inclusive it is. Consider the category DOG in Figure 8.1. Relative to this category, the category MAMMAL is higher up the vertical axis and includes more members than the category DOG. The category MAMMAL is therefore more inclusive than the category DOG. The category COLLIE, however, is lower on the vertical axis and has fewer members; this category is less inclusive than the category DOG. In contrast, the horizontal dimension relates to the category distinctions at the same level of inclusiveness. Hence, while DOG and CAR are distinct categories, they operate at the same level of detail. In the next two subsections, we look in more detail at the evidence for these two dimensions of categorisation.

8.2.3 The vertical dimension

The vertical dimension derives from the discovery by Rosch and her colleagues (Rosch *et al.* 1976) that categories can be distinguished according to **level of inclusiveness**. Inclusiveness relates to what is subsumed within a particular category. As we have seen, the category FURNITURE is more inclusive than the category CHAIR because it includes entities like DESK and TABLE in addition to CHAIR. In turn, CHAIR is more inclusive than ROCKING CHAIR because it includes other types of chairs in addition to rocking chairs. The category ROCKING CHAIR

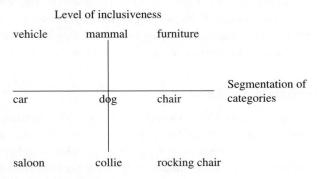

Figure 8.1 The human categorisation system

Table 8.3 Example of a taxonomy used by Rosch *et al.* (1976) in basic-level category research

Superordinate level	Basic level	Subordinate level
	CHAIR	KITCHEN CHAIR
		LIVING-ROOM CHAIR
FURNITURE	TABLE	KITCHEN TABLE
		DINING-ROOM TABLE
	LAMP	FLOOR LAMP
		DESK LAMP

only includes rocking chairs, and therefore represents the least inclusive level of this category. Rosch and her colleagues found that there is a level of inclusiveness that is **optimal** for human beings in terms of providing optimum cognitive economy. This level of inclusiveness was found to be at the mid-level of detail, between the most inclusive and least inclusive levels: the level associated with categories like CAR, DOG and CHAIR. This level of inclusiveness is called the **basic level**, and categories at this level are called **basic-level categories**. Categories higher up the vertical axis, which provide less detail, are called **superordinate** categories. Those lower down the vertical axis, which provide more detail, are called **subordinate** categories. This is illustrated in Table 8.3.

In a remarkable series of experiments, Rosch found that basic-level categories provided the most inclusive level of detail at which members of a particular category share features in common. In other words, while the superordinate level (e.g. MAMMAL) is the most inclusive level, members of categories at this level of inclusiveness share relatively little in common when compared to members of categories located at the basic level of inclusiveness (e.g. DOG).

Attributes

Rosch *et al.* (1976) found that the basic level is the level at which humans are best able to list a cluster of **common attributes** for a category. To investigate this, Rosch and her colleagues gave subjects 90 seconds to list all the attributes they could think of for each of the individual items listed in a particular taxonomy. Six of the taxonomies used by Rosch *et al.* are presented in Table 8.4. (It is worth pointing out to British English readers that because Rosch's experiments were carried out in the United States, some of the American English expressions may be unfamiliar.)

Table 8.5 lists common attributes found for three of these taxonomies. In the table, lower levels are assumed to have all the attributes listed for higher levels and are therefore not repeated. Table 8.5 illustrates the fact that subjects were only able to provide a minimal number of shared attributes for superordinate categories. In contrast, a large number of attributes were listed as being shared

Table 8.4 Six of the taxonomies used by Rosch et al. (1976) as stimuli

Superordinate	Basic level	Subordinates	
MUSICAL INSTRUMENT	GUITAR	FOLK GUITAR	CLASSICAL GUITAR
	PIANO	GRAND PIANO	UPRIGHT PIANO
	DRUM	KETTLE DRUM	BASE DRUM
FRUIT	APPLE	DELICIOUS APPLE	MACKINTOSH APPLE
	PEACH	FREESTONE PEACH	CLING PEACH
	GRAPES	CONCORD GRAPES	GREEN SEEDLESS GRAPES
TOOL	HAMMER	BALL-PEEN HAMMER	CLAW HAMMER
	SAW	HACK HAND SAW	CROSS-CUTTING HAND SAW
	SCREWDRIVER	PHILLIPS SCREWDRIVER	REGULAR SCREWDRIVER
CLOTHING	PANTS	LEVIS	DOUBLE KNIT PANTS
	SOCKS	KNEE SOCKS	ANKLE SOCKS
	SHIRT	DRESS SHIRT	KNIT SHIRT
FURNITURE	TABLE	KITCHEN TABLE	DINING-ROOM TABLE
	LAMP	FLOOR LAMP	DESK LAMP
	CHAIR	KITCHEN CHAIR	LIVING ROOM CHAIR
VEHICLE	CAR	SPORTS CAR	FOUR-DOOR SEDAN CAR
	BUS	CITY BUS	CROSS-COUNTRY BUS
	TRUCK	PICK-UP TRUCK	TRACTOR-TRAILER TRUCK

Table 8.5 Examples of attribute lists (based on Rosch et al. 1976: appendix I)

tool	clothing	furniture
make things	you wear it	no attributes
fix things	keeps you warm	CHAIR
metal	PANTS	legs
SAW	legs	seat
handle	buttons	back
teeth	belt loops	arms
blade	pockets	comfortable
sharp	cloth	four legs
cuts	two legs	wood
edge	LEVIS	holds people – you sit on it
wooden handle	blue	KITCHEN CHAIR
CROSS-CUTTING	DOUBLE-KNIT	no additional
HAND SAW	PANTS	LIVING-ROOM CHAIR
used in construction	comfortable	large
HACK HAND SAW	stretchy	soft
no additional		cushion

by basic-level categories, while just one or two more specific attributes were added for subordinate categories. Hence, while subordinate categories have slightly more attributes, the basic level is the most inclusive level at which there is a cluster of shared attributes.

Motor movements

In this experiment, Rosch *et al.* set out to establish the most inclusive level at which properties of human physical interaction with a category are found to cluster. This experiment also revealed that basic level categories were the most inclusive level at which members of categories share **motor movements**. To demonstrate this, subjects were asked to describe the nature of their physical interaction with the objects listed. It was found that while there are few motor movements common to members of a superordinate category, there are several specific motor movements listed for entities at the basic level, while entities at the subordinate level make use of essentially the same motor movements. This provides further evidence that the basic level is the most inclusive level, this time with respect to common interactional experiences. This is illustrated in Table 8.6.

Similarity of shapes

For this experiment, Rosch *et al.* sought to establish the most inclusive level of categorisation at which shapes of objects in a given category are most similar. In order to investigate this, the researchers collected around 100 images from sources like magazines and books representing each object at each level in the taxonomies listed in Table 8.4. The shapes were scaled to the same size and then superimposed upon one another. Areas of overlap ratios were then measured, which allowed the experimenters to determine the degree of similarity in shape. While objects at the superordinate level are not very similar in terms of shape (compare the outline shapes of car, bus and motorcycle, for example, as instances

Table 8.6 Motor movements for categories at three levels of inclusiveness (based on Rosch *et al.* 1976: appendix II)

Movement for superordinate categories	**FURNITURE**	
	Eyes:	scan
Additional movements for basic-level categories	**CHAIR**	
	Head:	turn
	Body:	turn, move back position
	Knees:	bend
	Arm:	extend-touch
	Waist:	bend
	Butt:	touch
	Body-legs:	release weight
	Back-torso:	straighten, lean back
Additional movements for subordinate categories	**LIVING-ROOM CHAIR**	
	Body:	sink

of the category VEHICLE), and while objects at the subordinate level are extremely similar, the basic level was shown to the most inclusive level at which object shapes are similar. In other words, the basic level includes a much greater number of instances of a category than the superordinate level (for example, DOG versus COLLIE) that can be identified on the basis of shape similarity.

Identification based on averaged shapes

In a fourth experiment, Rosch and her team devised averaged shapes of particular objects. They did this by overlapping outlines of entities belonging to a particular category. For all points where the two outlines did not coincide, the central point between the two lines was taken. Subjects were then shown the shapes and provided with superordinate, basic-level and subordinate terms to which they were asked to match the shapes. The success rate of matching shapes with superordinate terms was no better than chance, while subjects proved to be equally successful in matching averaged shapes with basic-level and subordinate terms. For example, the superordinate category VEHICLE consisted of overlapped shapes for car, bus and motorcycle, which are significantly different in shape and therefore less recognisable. On the other hand, the basic-level category CAR, represented by overlapping shapes of different types of cars, did not involve significant differences in shape, and was easily identifiable. Again, although there is a greater degree of similarity at the subordinate level, the basic level is more inclusive. The absence of shape similarity at the superordinate level compared to the evident shape similarity at the basic level goes some way towards explaining why the basic level is the optimum categorisation level for the human categorisation system, which is based, among other things, on perceptual similarity.

Cognitive economy versus level of detail

The major finding to emerge from Rosch's research on basic-level categorisation is that this level of categorisation is the most important level for human categorisation because it is the most inclusive and thus most informative level. It is worth emphasising why this should be the case. After all, Rosch et al.'s findings seem to show that the subordinate level is at least as informative as the basic level, if not more so, given that it provides more detailed information in addition to the information represented at the basic level. Recall that, when asked to list attributes of CAR and SPORTS CAR, subjects typically listed more attributes for SPORTS CAR than for CAR. This is because the subordinate category SPORTS CAR is likely to be identified with the same attributes as CAR, plus some extra attributes specific to SPORTS CAR.

The reason why the basic level is the most salient level of categorisation relates to the tension between similarity of members of a category and the principle of cognitive economy. While entities at the subordinate level are most alike (rocking chairs have most in common with other rocking chairs), different categories at the subordinate level are also very similar (rocking chairs are pretty similar to kitchen chairs). At the basic level, on the other hand, while there are also similarities within a particular category (all chairs are pretty similar to one another), there are far fewer between-category similarities (a chair is not that similar to a table). To illustrate this point, let's compare and contrast the basic-level and subordinate level categories given in Table 8.7.

Crucially, for a category to achieve cognitive economy (to provide the greatest amount of information at the lowest processing cost), it must share as many common within-category attributes as possible, while maintaining the highest possible level of between-category difference. In intuitive terms, it is easier to spot the differences between a chair and a lamp than between a desk lamp and a floor lamp. This demonstrates why the basic level of categorisation is 'special': it is the level which best reconciles the conflicting demands of cognitive economy. Therefore the basic level is the most informative level of categorisation.

This notion of cognitive economy has been described in terms of **cue validity**. According to Rosch (1977: 29) 'cue validity is a probabilistic concept' which predicts that a particular cue – or attribute – becomes more valid or relevant to a given category the more frequently it is associated with members of that category. Conversely, a particular attribute becomes less valid or relevant to a category the more frequently it is associated with members of other categories. Thus 'is used for sitting on' has 'high cue validity' for the category CHAIR, but 'is found in the home' has low cue validity for the category CHAIR because many other different categories of object can be found in the home in addition to chairs.

Cue validity is maximised at the basic level, because basic level categories share the largest number of attributes possible while minimising the extent to which these features are shared by other categories. This means that basic-level categories simultaneously maximise their **inclusiveness** (the vertical dimension) and their **distinctiveness** (the horizontal dimension) which results in optimal cognitive economy by providing a maximally efficient way of representing information about frequently encountered objects.

Table 8.7 Comparison between levels of categorisation

Basic level	Subordinate level
TABLE	DINING TABLE
	KITCHEN TABLE
CHAIR	DINING CHAIR
	LOUNGE CHAIR

Perceptual salience

It is clear from Rosch's findings that categorisation arises from perceptual stimuli. When we categorise objects, we do so according to various types of sensory-perceptual input, including shape, size, colour and texture, as well as kinaesthetic input representing how we interact physically with objects. Another way of describing the importance of the basic level, then, is by relating it to **perceptual salience**. There are a number of additional lines of evidence that support the position that the basic level represents the most salient level of categorisation.

The basic level appears to be the most abstract (that is, the most inclusive and thus the least specific) level at which it is possible to form a mental **image**. After all, we are unable to form an image of the category FURNITURE without imagining a specific item like a chair or a table: a basic-level object. This is consistent with the finding that averaged shapes cannot be identified at the superordinate level as there are insufficient similarities between entities at this very high level of inclusiveness. This is also consistent with the fact that Rosch's subjects often struggled to list attributes for the superordinate level. You can try this experiment yourself: if you ask a friend to draw you a picture of 'fruit' or 'furniture' they will draw you apples and bananas or tables and chairs. These are all basic-level categories. There is no recognisable or meaningful shape that represents the superordinate level of categorisation.

Based on a picture verification task, Rosch *et al.* (1976) also found that objects are **perceived** as members of basic-level categories more rapidly than as members of superordinate or subordinate categories. In this experiment, subjects heard a word like *chair*. Immediately afterwards, they were presented with a visual image. If the word matched the image, subjects pressed a 'match' response key. If the word did not match the image, they pressed a different response key. This enabled experimenters to measure the reaction times of the subjects. It emerged that subjects were consistently faster at identifying whether an object matched or failed to match a basic level word than they were when verifying images against a superordinate or subordinate level word. This suggests that in terms of perceptual verification, objects are recognised more rapidly as members of basic-level categories than other sorts of categories.

Language acquisition

Rosch *et al.* (1976) found that basic-level terms are among the first concrete nouns to emerge in child language. This investigation was based on a case study of a single child, consisting of weekly two-hour recordings dating from the initial period of language production. All relevant utterances were independently rated by two assessors in order to determine whether they were superordinate, basic or subordinate level terms. The study revealed that the individual noun-like utterances were overwhelmingly situated at the basic level. Rosch *et al.* argued that this finding provided further support for the primacy of the basic level of categorisation.

Basic-level terms in language

The language system itself also reveals the primacy of the basic level in a number of ways. Firstly, basic-level terms are typically **monolexemic**: comprised of a single word-like unit. This contrasts with terms for subordinate level categories which are often comprised of two or more lexemes – compare *chair* (basic-level object) with *rocking chair* (subordinate-level object). Secondly, basic-level terms appear to occur more frequently in language use than superordinate or subordinate level expressions. More speculatively, Rosch (1978) has even suggested basic-level terms may have emerged prior to superordinate- and subordinate-level terms in the process of language evolution. Of course, given that evidence for the primacy of the basic level is so overwhelming, we might wonder why we need the other levels of categorisation at all. In fact, the superordinate and subordinate levels, while they may not be cognitively salient, have extremely useful functions. As Ungerer and Schmid (1996) explain, the superordinate level (for example, VEHICLE) highlights the **functional attributes** of the category (vehicles are for moving people around), while also performing a **collecting function** (grouping together categories that are closely linked in our knowledge representation system). Subordinate categories, on the other hand, fulfil a **specificity function**.

Are basic-level categories universal?

Of course, if we can find evidence for basic-level categories among English speakers, two questions naturally arise. Firstly, do members of all cultures or speech communities categorise in this way? Given that all humans share the same cognitive apparatus, it would be surprising if the answer to this question were 'no'. This being so, the second question that arises is whether the same basic-level categories are evident in all cultures or speech communities. Clearly, this question relates to 'the extent to which structure is "given" by the world versus created by the perceiving organism' (Rosch *et al.* 1976: 429). Put another way:

> [B]asic objects for an individual, subculture, or culture must result from *interaction* between potential structure provided by the world and the particular emphases and state of knowledge of the people who are categorizing. However, the environment places constraints on categorizations. (Rosch *et al.* 1976: 430)

It follows that while the environment partly delimits and thus determines the nature of the categories we create, these categories are also partly determined by the nature of the interaction between human experiencers and their environment. This finding, of course, is consonant with the thesis of embodied cognition.

This view of categorisation entails that while the organisation of conceptual categories into basic, superordinate and subordinate levels may be universal, the level at which particular categories appear may not be. This relates not only to cross-linguistic or cross-cultural variation in the broader sense, but is also reflected within a single speech community or culture where acquired specialist knowledge may influence an individual's taxonomy of categories. For instance, Rosch *et al.* (1976) found that for most of their North American subjects the category AIRPLANE was situated at the basic level. However, for one of their subjects, a former aircraft mechanic, this category was situated at the superordinate level, with specific models of aircraft being situated at the basic level. This reveals how specialist knowledge in a particular field may influence an individual's categorisation system. At the cross-cultural level, the cultural salience of certain objects may result in taxonomic differences. For example, the anthropologist Berlin and his colleagues (1974) investigated plant naming within the Mayan-speaking Tzeltal community in Southern Mexico. They found that in basic naming tasks members of this community most frequently named plants and trees at the (scientific) level of genus or kind (for example, *pine* versus *willow*) rather than at the (scientific) level of class (for example, *tree* versus *grass*). When Rosch *et al.* (1976) asked their North American students to list attributes for TREE, FISH and BIRD as well as subordinate instances of these categories, they found that, on average, the same number of attributes were listed for TREE, FISH and BIRD as for the subordinate examples, suggesting that for many speakers TREE, FISH and BIRD may be recognised as a basic-level category. The differences between the Tzeltal and North American speakers indicates that aspects of culture (for example, familiarity with the natural environment) can affect what 'counts' as the basic level of categorisation from one speech community to another. However, it does not follow from this kind of variation that any category can be located at any level. While our interaction with the world is one determinant of level of categorisation, the world itself provides structure that also partly determines categorisation, an issue to which we now turn.

8.2.4 The horizontal dimension

The horizontal dimension of the categorisation system (recall Figure 8.1) relates in particular to the principle of perceived world structure which we introduced earlier. This principle states that the world is not unstructured, but possesses correlational structure. As Rosch points out, 'wings correlate with feathers more than fur' (Rosch 1978: 253). In other words, the world does not consist of sets of attributes with an equally probable chance of co-occurring. Instead, the world itself has structure, which provides constraints on the kinds of categories that humans represent within the cognitive system.

One consequence of the existence of correlational structure in the world is that cognitive categories themselves reflect this structure: the category **prototype** reflects the greater number of correlational features. Recall that categories often exhibit typicality effects, where certain members of the category are judged as 'better' or more representative examples of that category than other members. Members of a category that are judged as highly **prototypical** (most representative of that category) can be described as category prototypes. This feature of category structure was investigated in a series of experiments reported in Rosch (1975), which established that prototypical members of a category were found to exhibit a large number of attributes common to many members in the category, while less prototypical members were found to exhibit fewer attributes common to other members of the category. In other words, not only do categories exhibit typicality effects (having more or less prototypical members), category members also exhibit **family resemblance** relations. While for many categories there are no attributes common to all members (not all members of a family are identical in appearance), there is sufficient similarity between members that they can be said to resemble one another to varying degrees (each having some, but not all, features in common).

Goodness-of-example ratings

In order to investigate the prototype structure of categories, Rosch (1975) conducted a series of experiments in which subjects were asked to provide **goodness-of-example** ratings for between fifty and sixty members of each category, based on the extent to which each member was representative of the category. Typically, subjects were provided with a seven-point scale. They were asked to rate a particular member of the category along this scale, with a rating of 1 indicating that the member is highly representative, and a rating of 7 indicating that the entity was not very representative. Presented in Table 8.8 are the highest- and lowest-ranked ten examples for some of the categories rated by American undergraduate students. It is worth observing that the experiments Rosch employed in order to obtain goodness-of-example rating were 'linguistic' experiments. That is, subjects were presented with word lists rather than visual images.

Family resemblance

Rosch argues that prototype structure, as exhibited by goodness-of-example ratings, serves to maximise shared information contained within a category. As

Table 8.8 A selection of goodness-of-example ratings (based on Rosch 1975: appendix)

Rank	BIRD	FRUIT	VEHICLE	FURNITURE	WEAPON
Top eight (from more to less representative)					
1	Robin	Orange	Automobile	Chair	Gun
2	Sparrow	Apple	Station wagon	Sofa	Pistol
3	Bluejay	Banana	Truck	Couch	Revolver
4	Bluebird	Peach	Car	Table	Machine gun
5	Canary	Pear	Bus	Easy chair	Rifle
6	Blackbird	Apricot	Taxi	Dresser	Switchblade
7	Dove	Tangerine	Jeep	Rocking chair	Knife
8	Lark	Plum	Ambulance	Coffee table	Dagger
9	Swallow	Grapes	Motorcycle	Rocker	Shotgun
10	Parakeet	Nectarine	Streetcar	Love seat	Sword
Bottom ten (from more to less representative)					
10	Duck	Pawpaw	Rocket	Counter	Words
9	Peacock	Coconut	Blimp	Clock	Hand
8	Egret	Avocado	Skates	Drapes	Pipe
7	Chicken	Pumpkin	Camel	Refrigerator	Rope
6	Turkey	Tomato	Feet	Picture	Airplane
5	Ostrich	Nut	Skis	Closet	Foot
4	Titmouse	Gourd	Skateboard	Vase	Car
3	Emu	Olive	Wheelbarrow	Ashtray	Screwdriver
2	Penguin	Pickle	Surfboard	Fan	Glass
1	Bat	Squash	Elevator	Telephone	Shoes

Rosch puts it, 'prototypes appear to be those members of a category that most reflect the redundancy structure of the category as a whole' (Rosch 1978: 260). In other words, the more frequent a particular attribute is among members of a particular category, the more representative it is. The prototype structure of the category reflects this 'redundancy' in terms of repeated attributes across distinct members, or exemplars. This entails that another way of assessing prototype structure is by establishing the set of attributes that a particular entity has (Rosch and Mervis 1975). The more category-relevant attributes a particular entity has, the more representative it is.

In order to investigate this idea, Rosch and Mervis (1975) presented twenty subjects with six categories: FURNITURE, VEHICLE, FRUIT, WEAPON, VEGETABLE and CLOTHING. For each category, the experimenters collected twenty items that were selected to represent the full goodness-of-example scale for each category, from most to least representative. The subjects were each given six items from each category and asked to list all the attributes they could think of for each item. Each attribute then received a score on a scale of 1–20, depending on how many items in a category that attribute had been listed for: the attributes that were listed most frequently were allocated more points than those listed less frequently. The degree of family resemblance of a particular item (for example, CHAIR in the category FURNITURE) was the sum of the score for each of the attributes listed for that item: the higher the total score, the greater the family resemblance. Rosch and Mervis's findings showed a high degree of correlation between items that received a high score and their goodness-of-example ratings. Table 8.9 illustrates these ideas by comparing some of the attributes common across the category BIRD against two members of the category: ROBIN (judged to be highly representative) and OSTRICH (judged to be much less representative).

This table illustrates that the number of relevant attributes possessed by a particular category member correlates with how representative that member is judged to be. Robins are judged to be highly prototypical: they possess a large number of attributes found across other members of the BIRD category. Conversely, ostriches, which are judged not to be very good examples of the category BIRD, are found to have considerably fewer of the common attributes found among members of the category. Therefore, while OSTRICH and ROBIN are representative to different degrees, they nonetheless share a number of attributes and thus exhibit a degree of family resemblance. The claim that category members are related by family resemblance relations rather than by necessary and sufficient conditions entails that categories are predicted to have fuzzy boundaries. In other words, we expect to reach a point at which, due to the absence of a significant number of shared characteristics, it becomes unclear whether a given entity can be judged as a member of a given category or not.

Table 8.9 Comparison of some attributes for ROBIN and OSTRICH

Attributes	ROBIN	OSTRICH
lays eggs	yes	yes
beak	yes	yes
two wings	yes	yes
two legs	yes	yes
feathers	yes	yes
small	yes	no
can fly	yes	no
chirps/sings	yes	no
thin/short legs	yes	no
short tail	yes	no
short neck	yes	no
moves on the ground by hopping	yes	no

8.2.5 Problems with prototype theory

As we noted at the outset of this chapter, it has been argued that prototype theory is inadequate as a theory of knowledge representation. In this section, we briefly review some of the objections, as well as consider whether Rosch and her colleagues intended their findings to be interpreted directly as a model of knowledge representation.

We begin with a number of criticisms discussed by Laurence and Margolis (1999), who present a survey of the criticisms that have been levelled against prototype theory in the literature. The first criticism, which Laurence and Margolis describe as **the problem of prototypical primes**, concerns the study of ODD NUMBERS that we discussed earlier (Amstrong et al. 1983). Recall that this study found that even a 'classical category' of this nature exhibits typicality effects. Armstrong et al. argue that this poses potentially serious problems for Prototype Theory since such effects are not predicted for classical categories.

The second criticism that Laurence and Margolis identify is that, like the classical theory, prototype theory also suffers from **the problem of ignorance and error**: it fails to explain how we can possess a concept while not knowing or being mistaken about its properties. The basis of this criticism is that a concept with prototype structure might incorrectly include an instance that is not in fact a member of that category. The example that Laurence and Margolis use to illustrate this point is that of a prototypical GRANDMOTHER, who is elderly with grey hair and glasses. According to this model, any elderly grey-haired woman with glasses might be incorrectly predicted to be a member of this category. Conversely, concepts with a prototype structure may incorrectly exclude instances that fail to display any of the attributes that characterise the prototype (for example, a cat is still a cat without having any of the prototypical attributes of a cat).

The third criticism that Laurence and Margolis discuss is called **the missing prototypes problem**: the fact that it is not possible to describe a prototype for some categories. These categories include 'unsubstantiated' (non-existent) categories like US MONARCH and heterogeneous categories like OBJECTS THAT WEIGH MORE THAN A GRAM. In other words, the fact that we can describe and understand such categories suggests that they have meaning, yet prototype theory as a model of knowledge representation fails to account for such categories.

Finally, Laurence and Margolis describe **the problem of compositionality**, which was put forward by Fodor and Lepore (1996). This is the criticism that prototype theory provides no adequate explanation for the fact that complex categories do not reflect prototypical features of the concepts that contribute to them. To illustrate this point, Laurence and Margolis cite Fodor and Lepore's example of PET FISH. If a prototypical PET is fluffy and affectionate and a prototypical FISH is grey in colour and medium-sized (like a mackerel), this does not predict that a prototypical PET FISH is small and orange rather than medium, grey, fluffy and affectionate.

As this brief discussion of the criticisms levelled against prototype theory indicates, Rosch's findings have often been interpreted directly as a theory of knowledge representation (a theory about the structure of categories as they are represented in our minds). Indeed, Rosch explored this idea in her early work (albeit rather speculatively). Consider the following passage:

> [A prototype can be thought of] as the abstract representation of a category, or as those category members to which subjects compare items when judging category membership, or as the internal structure of the category defined by subjects' judgments of the degree to which members fit their 'idea' or 'image' of the category. (Rosch and Mervis 1975: 575)

Rosch retreats from this position in her later writings. As she later makes explicit, 'The fact that prototypicality is reliably rated and is correlated with category structure does not have clear implications for particular processing models nor for a theory of cognitive representations of categories' (Rosch 1978: 261). In other words, while typicality effects are 'real' in the sense that they are empirical findings, it does not follow that these findings can be directly 'translated' into a theory of how categories are represented in the human mind. In other words, experiments that investigate typicality effects only investigate the categorisation judgements that people make rather than the cognitive representations that give rise to these judgements.

This point is central to Lakoff's (1987) discussion of Rosch's findings. Lakoff argues that it is mistaken to equate prototype or typicality effects with cognitive representations. Rather, typicality effects are 'surface phenomena'. This means that they are a consequence of complex mental models that combine to give rise to typicality effects in a number of ways. Typicality effects might therefore be described in intuitive terms as a superficial 'symptom' of the way our minds work, rather than a direct reflection of cognitive organisation. Lakoff (1987) therefore attempts to develop a theory of cognitive models that might plausibly explain the typicality effects uncovered by Rosch and her colleagues. As we will see in the next section, Lakoff's theory of cognitive models avoids the problems that we summarised above which follow from assuming Prototype Theory as a model of knowledge representation.

8.3 The theory of idealised cognitive models

In his book, *Women, Fire And Dangerous Things* (1987), George Lakoff set out to develop a theory of category structure at the cognitive level that could account for the empirical findings presented by Rosch and her colleagues. This theory was called the **theory of idealised cognitive models**, and represented one of the early frameworks that helped define cognitive semantics as a research programme.

Lakoff argued that categories relate to **idealised cognitive models (ICMs)**. These are relatively stable mental representations that represent **theories** about the world. In this respect, ICMs are similar to Fillmore's notion of frames, since both relate to relatively complex knowledge structures. While ICMs are rich in detail, they are 'idealised' because they abstract across a range of experiences rather than representing specific instances of a given experience. In Lakoff's theory, ICMs guide cognitive processes like categorisation and reasoning. For example, Barsalou (1983) argues that 'ad hoc' categories like WHAT TO TAKE FROM ONE'S HOME DURING A FIRE also exhibit typicality effects. Lakoff argues that categories of this kind, which are constructed 'online' for local reasoning, are constructed on the basis of pre-existing ICMs. In other words, faced with a house fire, our ability to construct a category of items to be saved relies on pre-existing knowledge relating to the monetary and sentimental value attached to various entities, together with knowledge of the whereabouts in the house they are, the amount of time likely to be available and so on. In the next two subsections, we look in more detail at the properties of ICMs.

8.3.1 Sources of typicality effects

Lakoff argues that typicality effects can arise in a range of ways from a number of different sources. In this section, we present some of the ICMs proposed by Lakoff, and show how these are argued to give rise to typicality effects.

The simplest type of typicality effects

Typicality effects can arise due to mismatches between ICMs against which particular concepts are understood. To illustrate, consider the ICM to which the concept BACHELOR relates. This ICM is likely to include information relating to a monogamous society, the institution of marriage and a standard marriageable age. It is with respect to this ICM, Lakoff argues, that the notion of BACHELOR is understood. Furthermore, because the background frame defined by an ICM is idealised, it may only partially match up with other cognitive models, and this is what gives rise to typicality effects. Consider the Pope, who is judged to be a poor example of the category BACHELOR. An individual's status as a bachelor is an 'all or nothing' affair, because this notion is understood with respect to the legal institution of MARRIAGE: the moment the marriage vows have been taken, a bachelor ceases to be a bachelor. The concept POPE, on the other hand, is primarily understood with respect to the ICM of the CATHOLIC CHURCH whose clergy are unable to marry. Clearly, there is a mismatch between these two cognitive models: in the ICM against which BACHELOR is understood, the Pope is 'strictly speaking' a bachelor because he is unmarried. However, the Pope is not a prototypical bachelor precisely because the Pope is understood with respect to a CATHOLIC CHURCH ICM in which marriage of Catholic clergy is prohibited.

Typicality effects due to cluster models

According to Lakoff, there is a second way in which typicality effects can arise. This relates to **cluster models**, which are models consisting of a number of converging ICMs. The converging models collectively give rise to a complex cluster, which 'is psychologically more complex than the models taken individually' (Lakoff 1987: 74). Lakoff illustrates this type of cognitive model with the example of the category MOTHER, which he suggests is structured by a cluster model consisting of a number of different MOTHER **subcategories**. These are listed below.

1. THE BIRTH MODEL: a mother is the person who gives birth to the child.
2. THE GENETIC MODEL: a mother is the person who provides the genetic material for the child.
3. THE NURTURANCE MODEL: a mother is the person who brings up and looks after the child.
4. THE MARITAL MODEL: a mother is married to the child's father.
5. THE GENEALOGICAL MODEL: a mother is a particular female ancestor.

While the category MOTHER is a composite of these distinct sub-models, Lakoff argues that we can, and often do, invoke the individual models that contribute to the larger cluster model. The following examples reveal that we can employ different models for MOTHER in stipulating what counts as a 'real mother' (Lakoff 1987: 75).

(1) a. BIRTH MODEL
I was adopted and I don't know who my real mother is.
b. NURTURANCE MODEL
I am not a nurturant person, so I don't think I could ever be a real mother to my child.
c. GENETIC MODEL
My real mother died when I was an embryo, and I was later frozen and implanted in the womb of the woman who gave birth to me.

d. BIRTH MODEL
I had a genetic mother who contributed the egg that was planted in the womb of my real mother, who gave birth to me and raised me.

e. BIRTH MODEL
By genetic engineering, the genes in the egg my father's sperm fertilised were spliced together from genes in the eggs of twenty different women. I wouldn't call any of them my real mother. My real mother is the woman who bore me, even though I don't have any single genetic mother.

Lakoff argues that cluster models give rise to typicality effects when one of the ICMs that contributes to the cluster is viewed as primary. This results in the other subcategories being ranked as less important: 'When the cluster of models that jointly characterize a concept diverge, there is still a strong pull to view one as the most important' (Lakoff 1987: 75). This is reflected in dictionary definitions, for example, which often privilege one of the MOTHER sub-models over the others. Although many dictionaries treat the BIRTH MODEL as primary, Lakoff found that *Funk and Wagnall's Standard Dictionary* selected the NURTURANCE MODEL while the *American College Dictionary* chose the GENEALOGICAL MODEL.

Typicality effects due to metonymy

Lakoff argues that a third kind of typicality effect arises when an **exemplar** (an individual instance) stands for an entire category. The phenomenon whereby one conceptual entity stands for another is called **metonymy** and is explored in much more detail in the next chapter. To illustrate metonymy consider example (2):

(2) Downing Street refused comment.

In this example, the official residence of the British Prime Minister stands for the Prime Minister. In other words, it is the Prime Minister (or his or her press officer) who refuses to comment. Similarly, in example (3) it is the vehicle owner who is standing for the car.

(3) I'm parked out the back.

A metonymic ICM can be a subcategory, as in the case of one of the subcategories of a cluster model, or an individual member of a category that comes to stand for the category as a whole. An important consequence of this is that the metonymic model, by standing for the whole category, serves as a **cognitive**

reference point, setting up norms and expectations against which other members of the category are evaluated and assessed. It follows that metonymic ICMs give rise to typicality effects, as other members of the category are judged as atypical relative to the metonymic model.

An example of a metonymic ICM is the cultural stereotype HOUSEWIFE-MOTHER, in which a married woman does not have paid work but stays at home and looks after the house and family. The HOUSEWIFE-MOTHER stereotype can give rise to typicality effects when it stands for, or represents, the category MOTHER as a whole. Typicality effects arise from resulting expectations associated with members of the category MOTHER. According to the HOUSEWIFE-MOTHER stereotype, mothers nurture their children, and in order to do this they stay at home and take care of them. A WORKING MOTHER, by contrast, is not simply a mother who has a job, but also one who does not stay at home to look after her children. Hence the HOUSEWIFE-MOTHER model, by metonymically representing the category MOTHER as a whole, serves in part to define other instances of the category such as WORKING MOTHER, which thus emerges as a non-prototypical member of the category.

Lakoff proposes a number of different kinds of metonymic models, any of which can in principle serve as a cognitive reference point and can thus give rise to typicality effects. We briefly outline some of these below.

Social stereotypes
The HOUSEWIFE-MOTHER model is an example of a **social stereotype**. These are conscious ICMs which emerge from public discussion. Against this background, we can re-evaluate the category BACHELOR. The stereotypical bachelor in our culture is a womaniser who lacks domestic skills. Typicality effects can arise if a particular bachelor contrasts with this stereotype. For instance, an unmarried man with one sexual partner who enjoys staying at home cooking and takes pride in his housework may be judged atypical with respect to the social stereotype for bachelors. This shows how the social stereotype BACHELOR, which represents one element in the category BACHELOR, can stand for the category as a whole thus giving rise to typicality effects.

Typical examples
Typicality effects can also arise in relation to **typical examples** of a particular category. For instance, in some cultures ROBIN and SPARROW are typical members of the category BIRD. This is because in some parts of the world these birds are very common. In this respect, our environment has consequences for what we judge as good examples of a category. Furthermore, Lakoff argues that we may evaluate a member of the category bird with respect to a typical example. In this way, typicality effects arise when the typical example stands for the entire category.

Ideals
Lakoff suggests that some categories are understood in terms of **ideals**, which may contrast with typical or stereotypical instances. For example, we might have an ideal for the category POLITICIAN: someone who is public-spirited, altruistic, hardworking and so on. This may contrast with our stereotype of politicians as egotistical, power-hungry and obsessed with 'spin'. Once more, typicality effects occur when the ideal stands metonymically for the entire category. For instance, with respect to our ideal the utterance *He's a great politician* might be interpreted as a positive evaluation. However, with respect to our social stereotype, the same utterance would be interpreted as a negative evaluation.

Paragons
Individual category members that represent ideals are **paragons**. For instance, David Beckham, arguably the world's best-known soccer star, is good-looking, a committed father, glamorous, married to a pop star and captain of the England team, as well as being one of the world's most successful footballers. For many people around the world, Beckham represents a FOOTBALL paragon. Similarly, Rolls-Royce represents a paragon in terms of LUXURY CARS, Nelson Mandela represents a paragon in terms of POLITICAL LEADERS, Winston Churchill in terms of WAR LEADERS, Noam Chomsky in terms of GENERATIVE LINGUISTS, and so on. Because paragons stand for an entire category, they set up norms and expectations against which other members of the category may be evaluated. For instance, the comment, 'He's no Nelson Mandela' about a particular political leader may represent a negative assessment as to the leader's altruism and so forth. In this way, paragons give rise to typicality effects.

Generators
According to Lakoff, members of some categories are 'generated' by a core subset of category members called **generators**. These generators are judged to be more prototypical than the other category members that they generate. For example, the natural numbers are represented by the set of integers between zero and nine, which are combined in various ways in order to produce higher natural numbers. For instance, the number 10 combines the integers 1 and 0. Thus the entire category NATURAL NUMBERS is generated from a small subset of single-digit integers. Lakoff argues that this is why the numbers 1 to 9 are judged as prototypical members of the category NATURAL NUMBERS than much larger numbers. Another example of a set of generators is Morse Code. In this system the generators are the 'dot' and the 'dash'. While the 'dot' represents the letter 'E', the 'dash' represents the letter 'T'. Because all other letters represent combinations of dots and/or dashes, the 'letters' 'E' and 'T' are likely to be more prototypical than the others for regular Morse Code users.

Table 8.10 Summary of some metonymic ICMs

Stereotypes	represent cultural norms and expectations regarding instances of the category
Typical examples	represent the most frequent or commonly encountered instances of the category
Ideals	combine the ideal properties of the category
Paragons	represent actual instances of an ideal
Generators	members of a category are 'generated' by a core subset of members
Salient examples	represent memorable or well-known actual instances of a category

Salient examples
Finally, memorable or **salient examples** can also give rise to a type of metonymic ICM. For instance, Oxford University is a salient example of a university, in part due to its history (it received its royal charter in the thirteenth century), in part due to the esteem in which its teaching and scholarship have traditionally been held and in part due to the nature of the colleges that make up the university, both in terms of the structure of the institution and its architecture. Although in many ways atypical in terms of British and other international higher education institutions, people, particularly in the United Kingdom, often rely upon Oxford as a point of comparison for other universities. Typicality effects occur when Oxford serves to establish a means of evaluating and assessing another university.

In other words, salient examples, like prototypes in general, provide cognitive reference points that not only structure a category metonymically, but can influence the decisions we make, for instance whether we decide to go to a particular university based on how similar it is to a salient example like Oxford. Table 8.10 provides a summary of some of the types of metonymic ICMs proposed by Lakoff.

In sum, Lakoff argues that cluster models and metonymic ICMs can give rise to typicality effects in different ways. While the cluster model provides a converging cluster of cognitive models which gives rise to typicality effects by ranking one of the subcategories as more important than the others in the cluster, a metonymic model can stand for the category as a whole and gives rise to typicality effects by defining cultural expectations relating to this category. We will look in more detail at metonymy in Chapter 9.

8.3.2 Radial categories as a further source of typicality effects

Lakoff proposes that the cluster model for MOTHER and the metonymic HOUSEWIFE-MOTHER stereotype taken together contribute to a **composite prototype** for MOTHER: a prototype derived from two models. This prototype provides **representative structure** for the category. For example, the composite prototype for the category MOTHER includes a female who gave birth to the child, was supplier of 50 per cent of the genetic material, stayed at home in order to nurture the child, is married to the child's father, is one generation older than the child and is also the child's legal guardian. In other words, the composite prototype draws upon information from the BIRTH MODEL, the GENETIC MODEL, the NURTURANCE MODEL, the MARITAL MODEL, the GENEALOGICAL MODEL and the HOUSEWIFE MODEL, which is a social stereotype. This type of prototype is an idealisation which provides schematic information. Importantly, further models can be derived from this composite prototype. These models include ADOPTIVE MOTHER, FOSTER MOTHER, BIRTH MOTHER and SURROGATE MOTHER. As Lakoff points out:

> These variants are not generated from the central model by general rules; instead, they are extended by convention and must be learned one by one. But the extensions are by no means random. The central model determines the possibilities for extensions, together with the possible relations between the central model and the extension models. (Lakoff 1987: 91)

A composite prototype and extensions of this kind are modelled in terms of a radiating lattice structure. The composite prototype is positioned centrally with other subcategories represented as extending from the **central case** (see Figure 8.2).

Crucially, the non-central cases in such **radial categories** are not strictly predictable from the central case but are cultural products. For instance, the

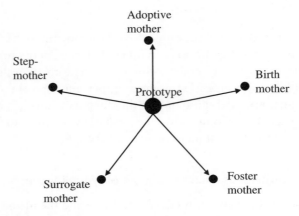

Figure 8.2 Radial network for the category MOTHER

subcategories of MOTHER listed below are all understood in terms of how they diverge from the central case.

1. STEPMOTHER – married to the father but didn't supply genetic material or give birth.
2. ADOPTIVE MOTHER – provides nurturance and is the legal guardian.
3. BIRTH MOTHER – gave birth and supplied genetic material but put the child up for adoption hence does not nurture the child and has no legal responsibilities.
4. FOSTER MOTHER – charged by the state to nurture the child but is not the child's legal guardian.
5. SURROGATE MOTHER – gives birth to the child, typically does not supply the genetic material and has no other obligations to the child.

Thus radial categories of this kind provide a fourth way in which typicality effects can arise. These effects occur when the subcategories are seen to deviate from the composite prototype. Moreover, as particular categories can become more conventionalised than others, different subcategories in a radial category can develop different degrees of prototypicality.

Importantly, radial categories are not 'generators'. The central case does not productively generate new subcategories of the MOTHER category. While the subcategories are **motivated** in the sense that they are licensed by the prototype, this is a consequence of our cultural experience. For instance, the subcategory SURROGATE MOTHER is a consequence of recent achievements in medicine and cultural trends and has appeared in the second half of the twentieth century. In sum, radial categories are motivated, but knowing a prototype does not predict what subcategories will become conventionally adopted in the culture. We will have more to say about radial categories and how they apply to word meaning in Chapter 11.

To summarise this section, we have seen that there are four ways in which Lakoff accounts for typicality effects. The first kind of typicality effect arises from mismatches between ICMs. The second kind of typicality effect arises from more complex cognitive models which Lakoff calls cluster models. These consist of a number of distinct subcategory models. Typicality effects occur when one subcategory is deemed to be more salient than the others. The third kind of typicality effect relates to metonymic ICMs. These are essentially exemplar-based cognitive models in which a particular member of a given category stands for the category as a whole. Assessed with respect to the metonymic models, other members of a category may be evaluated as being atypical. The fourth kind of typicality effect arises from radial categories, in which members of a radial category exhibit degrees of typicality depending on how close to the composite prototype they are.

8.3.3 Addressing the problems with prototype theory

In section 8.2.5, we reviewed a number of problems that have been claimed to undermine the validity of prototype theory as a model of knowledge representation. In this section, we look at how Lakoff's theory of ICMs addresses these problems.

The first problem we saw was the problem of prototypical primes, which relates to the unexpected typicality effects exhibited by 'classical' categories. Lakoff argues that this finding is not problematic for a prototype-based theory of cognitive models, because these effects can be explained by the nature of the cognitive model that underlies them. Recall that the integers 0–9 are generators: they have a privileged place in the category REAL NUMBER precisely because they form the basis of the category. Within this set, there is a submodel EVEN NUMBERS, which consists of numbers that can be divided by 2, and a submodel ODD NUMBERS for those that cannot. Lakoff argues that because a set of generators can metonymically stand for the category or model as a whole, then the generators included in the submodel ODD NUMBERS (the numbers 1, 3, 5, 7, 9) can stand for the entire category. Against this metonymic model, other odd numbers appear to be less representative of the category, resulting in typicality effects. Although the category ODD NUMBER remains a 'classical' category in the sense that it has definite rather than fuzzy boundaries, it still exhibits typicality effects, which Lakoff argues can be accounted for by the theory of cognitive models. Of course, if typicality effects were interpreted as a direct reflection of cognitive representation of categories, the findings of Armstrong *et al.*'s study would certainly be unexpected. This example goes some way towards explaining why prototype theory cannot be straightforwardly translated into a model of cognitive representation.

The second problem we saw was the problem of ignorance and error. This relates to the idea that it is possible to possess a concept while not knowing or being mistaken about its properties. For example, a concept with prototype structure might incorrectly include an instance that is not in fact a member of that category, or incorrectly exclude instances that are a member of the category but fail to display any of the attributes that characterise the prototype. However, this problem only arises on the assumption that typicality effects are equivalent to cognitive representation. In other words, tendencies to categorise elderly women with grey hair and spectacles as members of the category GRANDMOTHER (when they might not be) or the failure to categorise sprightly blonde women as members of the category GRANDMOTHER (when they might be) arise from the social stereotype for GRANDMOTHER which can stand for the category as a whole. In Lakoff's model, this is only one ICM among several for the category GRANDMOTHER, which means that both 'correct' and 'incorrect' instances of categorisation can be accounted for. Equally, it is possible to possess the concept WHALE while believing it is an instance of the category FISH rather than MAMMAL. Again, this can be accounted for on the basis of metonymic models. A typical property of fish is that they have fins and live in the sea while a typical property of mammals is that they have legs and live on land. Thus, based on the typicality of attributes within the ICM, a whale might be 'miscategorised' as a fish.

The third problem we saw relates to 'missing prototypes'. According to this criticism, it should be possible to describe a prototype for any category we can conceive, yet it is not possible to describe a prototype for 'unsubstantiated' (non-existent) categories like US MONARCH and heterogeneous categories like OBJECTS THAT WEIGH MORE THAN A GRAM. Once more, this problem only arises on the assumption that typicality effects equate to cognitive representation. According to the theory of idealised cognitive models, categories like these are constructed 'on-line' from pre-existing cognitive models, like the 'ad hoc' categories we discussed earlier. Recall that ICMs are relatively stable knowledge structures that are built up on the basis of repeated experience: it is the non-conventional status of non-existent and heterogeneous categories that predicts that such categories would be unlikely to exhibit typicality effects.

The final problem we saw related to compositionality: the criticism that prototype theory fails to provide an adequate explanation for the fact that complex categories do not reflect prototypical features of the concepts that contribute to them. For example, we saw that the category PET FISH does not represent prototypical attributes of the categories PET and FISH. Observe, however, that this criticism assumes that PET FISH is a straightforward composite of the meanings of the two conceptual categories PET and FISH. According to the cognitive model this concept has category structure independently of the two categories to which it is related. In other words, although a pet fish is a type of pet and a type of fish, experience of pet fish gives rise to an independently structured cognitive model in which the prototypical pet fish is the goldfish. The experiential basis of the cognitive model therefore explains why the attributes of this category differ from those of PET and FISH.

8.4 The structure of ICMs

In this section, we explore in more detail the structure of ICMs. So far, we have likened the ICM to Fillmore's notion of a frame and have shown how ICMs can give rise to typicality effects of various kinds. However, we will show that Lakoff's ICMs encompass a wider range of conceptual phenomena than frames and that frames are just one kind of ICM. In Lakoff's theory, ICMs are complex structured systems of knowledge. ICMs structure **mental spaces**: conceptual 'packets' of knowledge constructed during ongoing meaning construction (see Chapter 12). As Lakoff observes, '[a] mental space is a medium for conceptualization and thought. Thus any fixed or ongoing state of affairs as we conceptualize it is represented by a mental space' (Lakoff 1987: 281). Examples include our understanding of our immediate reality, a hypothetical situation or a past event. In particular, language prompts for the construction of mental spaces in ongoing discourse. The role of ICMs is to provide the background knowledge that can be recruited in order to structure mental spaces. We referred to this process as **schema mapping** in Chapter 5, a process that is also called **schema induction**. According to Lakoff, ICMs depend upon (at least) five sorts of structuring principles for their composition: (1) image schemas; (2) propositions; (3) metaphor; (4) metonymy; and (5) symbolism. We briefly consider each of these structuring principles in turn.

Image schematic ICMs

For Lakoff, a fundamental 'building-block' of conceptual structure is the image schema (recall Chapter 6). Lakoff argues that, in many respects, image schemas serve as the foundation for conceptual structure. He argues that our experience and concepts of SPACE are structured in large part by image schemas like CONTAINER, SOURCE–PATH–GOAL, PART–WHOLE, UP–DOWN, FRONT–BACK and so on. This means that image schemas like these structure our ICM (or mental model) for SPACE.

Propositional ICMs

Lakoff uses the term 'propositional' in the sense that ICMs of this kind are not structured by 'imaginative devices' (1987: 285) like metaphor and metonymy. Instead, propositional ICMs consist of elements with properties and relations that hold between those elements. An ICM of this kind consists of propositional (or factual) knowledge. For example, our knowledge of the 'rules' involved in requesting a table and ordering food in a restaurant emerges from a propositional ICM. Another sort of propositional ICM might be a taxonomic classification system, for example the biological systems that classify plants and animals.

Metaphoric ICMs

Metaphoric ICMs are structured by the projection or mapping of structure from a source domain to a target domain. For example, when the domain or ICM of LOVE is metaphorically structured in terms of a JOURNEY, as illustrated by expressions like *Their relationship has come a long way*, the ICM for LOVE is metaphorically structured. We return to this subject in more detail in the next chapter.

Metonymic ICMs

We have already examined metonymic ICMs in some detail. As we saw above, ICMs like stereotypes, paragons and ideals are metonymic in the sense that a single type or individual stands for the entire category. We also examine metonymy in more detail in the next chapter.

Symbolic ICMs

ICMs of this kind represent the knowledge structures that Fillmore described in terms of semantic frames. Semantic frames involve lexical items (and grammatical constructions), which cannot be understood independently of the other lexical items relative to which they are understood. Recall the examples of *buy*, *sell* and so on which are understood with respect to the COMMERCIAL EVENT frame that we discussed in the previous chapter. Because this kind of ICM (or semantic frame) is explicitly structured by language (rather than providing a purely conceptual structure that underlies language), its structure contains symbolic units; this is why Lakoff describes it as symbolic.

8.5 Summary

In this chapter we outlined the classical theory of categorisation, which assumes necessary and sufficient conditions, and identified the problems inherent in this approach. We then looked in some detail at **prototype theory**, the model of categorisation that emerged from research carried out by cognitive psychologist Eleanor Rosch and her colleagues. This research revealed that many categories have **prototype structure** rather than **definitional structure**. In addition, Rosch found that categories for concrete objects are most informative at the **basic level**. However, we saw that assumptions concerning the direct 'translation' of Rosch's findings into a model of knowledge representation gave rise to a number of problems. We then looked at how the empirical findings from this research inspired the development of Lakoff's theory of **idealised cognitive models (ICMs)**. The main claim to emerge from this research was that typicality effects are surface phenomena, arising from underlying ICMs of various kinds. Lakoff argues that prototype structure is not to be directly equated with conceptual structure and organisation, but that typicality effects emerge from three sources: mismatches between ICMs; one subcategory becoming primary in a **cluster model**; and **metonymic ICMs**. The latter two types of ICM additionally give rise to **radial categories** which give rise to a fourth source of typicality effect. Finally, we examined the nature of ICMs in more detail and looked at the various ways in which they are structured. Lakoff argues that ICMs structure **mental spaces** (entities that serve as the locus for on-line conceptualisation) by providing the background knowledge that structures these mental spaces. ICMs can be structured in a range of ways. We considered **image schematic ICMs, propositional ICMs, metaphoric ICMs**, metonymic ICMs and **symbolic ICMs**. We will return immediately to metaphor and metonymy in the next chapter. We return to radial categories in Chapter 10 and to mental spaces in Chapter 11.

Further reading

Prototypes and basic-level categories

- **Rosch (1975)**
- **Rosch (1977)**
- **Rosch (1978)**
- **Rosch and Mervis (1975)**
- **Rosch** *et al.* **(1976)**

These are among the key articles by Rosch and her collaborators which present their findings concerning prototypes and basic-level categories. The two 1975 papers deal with experimental evidence for prototype effects. The 1976 paper is concerned with basic level categories. The 1977 and 1978 papers provide summaries and overviews of key developments based on the earlier findings. The 1978 paper is particularly important because Rosch explicitly distances herself from earlier suggestions that experimental findings can be considered a direct reflection of cognitive organisation of category structure.

The theory of idealised cognitive models

- **Lakoff (1987)**. While long and sometimes meandering, this book is one of the seminal volumes that sets out the cognitive semantics framework. It introduces and develops the theory of ICMs.
- **Taylor (2003)**. Taylor's book, first published in 1989 and now in its third edition, is an excellent introduction to Rosch's research and the interpretation of these findings within cognitive semantics. Moreover, Taylor elaborates on and extends many of the issues first addressed by Lakoff, particularly as they apply to language.

Other views of categorisation and conceptual organisation

- **Komatsu (1992); Laurence and Margolis (1999)**. Both these articles provide overviews of different approaches to categorization, including prototype theory. These articles are of particular interest

CATEGORISATION AND IDEALISED COGNITIVE MODELS

because prototype theory is compared and contrasted with other approaches. The Komatsu article is shorter and more accessible. The Laurence and Margolis volume consists of collected papers by the foremost researchers in the field, including cognitive linguists, formal linguists, philosophers and psychologists.

Exercises

8.1 The classical theory

What are the main claims associated with the classical theory of categorisation? What kinds of problems are inherent in this approach?

8.2 Prototype theory

How is the theory of prototypes and basic level categories different from the classical theory? What do the principles of cognitive economy and perceived world structure contribute to this theory?

8.3 Prototype structure

Try Rosch's experiments for yourself.

(i) List as many attributes as you can for each level of the following taxonomy. What do your findings show?

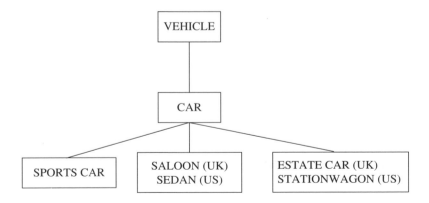

(ii) List all the motor movements relating to each level of the following taxonomy. What does this experiment reveal?

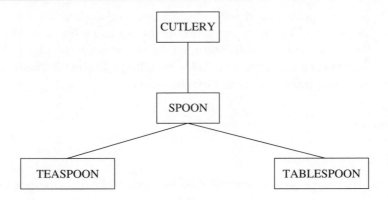

(iii) Collect judgements from three non-linguists for the following members of the category KITCHEN UTENSIL. Ask them to rank the members on a 1 (good example) to 7 (bad example) scale. Discuss your findings in the light of Rosch's claims.

bread-bin *pepper-mill*
blender *plate*
bowl *sink-plunger*
cafetiere *rolling-pin*
chopping board *salad spinner*
fork *saucepan*
frying pan *saucer*
grater *scales*
juicer *spatula*
knife *spoon*
microwave *teacup*
mixer *teapot*
mug *toaster*
nutcracker *whisk*
oven *wooden spoon*
peeler *sink plug*

8.4 Idealised cognitive models (ICMs)

What are the ICMs against which the following terms are understood: *bachelor, spinster, boy, girl*? How do these distinct ICMs contribute to the quite different connotations associated with the pairs *bachelor–spinster* and *boy–girl*? (You will need to state first what the common connotations associated with each of these words are.)

8.5 The theory of ICMs

In view of the theory of ICMs, give a detailed account of why the following concepts might be judged as non-prototypical with respect to their corresponding categories. You will first need to state your assumptions about the prototypical attributes associated with the categories in question.

(a) STEPFATHER   [category: FATHER]
(b) 977   [category: CARDINAL NUMBERS]
(c) OSTRICH   [category: BIRD]
(d) TARZAN   [category: BACHELOR]
(e) NORTH KOREA   [category: NATION]

8.6 Radial categories

Consider the category KNIFE. What are the various subcategories associated with this category? What is the prototype? Explain your reasoning.

9

Metaphor and metonymy

In this chapter, we will examine the central claims associated with **Conceptual Metaphor Theory**. This framework was first proposed by George Lakoff and Mark Johnson in their 1980 book *Metaphors We Live By* and has been developed in a number of subsequent publications. Conceptual Metaphor Theory was one of the earliest theoretical frameworks identified as part of the cognitive semantics enterprise and provided much of the early theoretical impetus for the cognitive approach. The basic premise of Conceptual Metaphor Theory is that metaphor is not simply a stylistic feature of language, but that thought itself is fundamentally metaphorical in nature. According to this view, conceptual structure is organised according to **cross-domain mappings** or correspondences between conceptual domains. Some of these mappings are due to pre-conceptual embodied experiences while others build on these experiences in order to form more complex conceptual structures. For instance, we can think about and talk about QUANTITY in terms of VERTICAL ELEVATION, as in *She got a really high mark in the test*, where *high* relates not literally to physical height but to a good mark. According to Conceptual Metaphor Theory, this is because the **conceptual domain** QUANTITY is conventionally structured and therefore understood in terms of the conceptual domain VERTICAL ELEVATION. Conceptual operations involving mappings, such as conceptual metaphor, are known more generally as **conceptual projection**. The claims made by conceptual metaphor theorists like Lakoff and Johnson and their collaborators directly relate to two of the central assumptions associated with cognitive semantics which we identified in Chapter 5. The first is the embodied cognition thesis, which holds that conceptual structure is grounded in embodied experience, and the second is the thesis that semantic structure reflects conceptual structure.

Recent work, particularly since Gibbs (1994), has also begun to emphasise the importance of a cognitive operation called **conceptual metonymy**. Research since the early 1990s has begun to suggest that this operation may be as least as important as conceptual metaphor in terms of providing conceptual structure (Kövecses and Radden 1998; Radden and Panther 1999). For this reason, both conceptual metaphor and conceptual metonymy are discussed in this chapter.

9.1 Literal versus figurative language

In this section we begin our examination of metaphor and metonymy by considering whether there really is a distinction to be made between **literal language** and **figurative language**. The traditional position, both in philosophy and in linguistics – and indeed the everyday view – is that (1) there is a stable and unambiguous notion of **literality**, and (2) that there is a sharp distinction to be made between literal language, on the one hand, and non-literal or figurative language on the other. According to this view, while literal language is precise and lucid, figurative language is imprecise, and is largely the domain of poets and novelists. In his 1994 book *The Poetics of Mind*, cognitive psychologist and cognitive linguist Raymond Gibbs examined this issue. Based on a close examination of the key features that are held to distinguish literal and figurative language, and based on a wide-ranging survey of different kinds of psycholinguistic experiments aimed at uncovering such a distinction, Gibbs found that there is no evidence for a principled distinction between literal and figurative language. In the following section, we begin by considering the two main claims associated with the traditional view.

9.1.1 Literal and figurative language as complex concepts

The basic assumption made by the traditional view is there are two kinds of meaning that can be straightforwardly distinguished: literal and figurative meaning. However, as Gibbs shows, there are many different kinds of literal and figurative meaning.

Definitions of literal language

Gibbs identifies a number of different definitions of literal meaning assumed within the cognitive science literature, four of which are presented in the following excerpt (Gibbs 1994: 75):

> *Conventional literality*, in which literal usage is contrasted with poetic usage, exaggeration, embellishment, indirectness, and so on.

Nonmetaphorical literality, or directly meaningful language, in which one word (concept) is never understood in terms of a second word (or concept).

Truth conditional literality, or language that is capable of 'fitting the world' (that is, referring to objectively existing objects or of being objectively true or false).

Context-free literality, in which the literal meaning of an expression is its meaning [independent of any communicative situation].

We return below to examine each of these in turn, observing for the time being that there is more than one idea about what defines literality in language.

Definitions of non-literal language

Not only have different scholars assumed different definitions of literal language, there are many definitions of non-literal language. Here, we consider just a few categories of 'non-literal' language use: irony, zeugma and metonymy.

An expression is ironic when what is meant is the opposite of what is said. This is illustrated by the response of 'Teenage son' to his mother in example (1).

(1) Mother: Time for bed . . . You have a BIG exam in the morning!
    Teenage son: I can't wait (uttered without enthusiasm).

Zeugma is a kind of ellipsis, in which a lexical item is understood, but 'left out' in subsequent clauses within a sentence, and where this lexical item has a different semantic or grammatical status in each case. One consequence is that when a lexical item has more than one meaning, a different meaning can be invoked in each clause. This can result in a humorous effect, as in example (2), where two different meanings of *expire* are invoked:

(2) On the same day my old Dad expired, so did my driving licence.

Metonymy depends upon an association between two entities so that one entity can stand for the other. Consider example (3):

(3) a. My wheels are parked out (the) back.
    b. My motor is parked out (the) back.

In this example, a salient component of a car, namely the wheels or the motor, can be used to refer to the car as a whole.

This brief survey reveals that both 'literal language' and 'non-literal (or figurative) language' are complex concepts. We must therefore question the assumption that there are two distinct and discrete kinds of language use that can be unambiguously identified. In the next section, we focus in more detail on the question of whether literal and non-literal language are fully discrete.

9.1.2 Can the distinction be maintained?

Recall from above that the traditional view holds that literal language is markedly distinct from non-literal or figurative language. In this section, we investigate whether the various categories of literal language can actually be meaningfully distinguished from non-literal language.

Conventional versus non-conventional language use

This distinction relies upon the idea that while literal language is the conventional 'ordinary' or 'everyday' way we have of talking about things, figurative language is 'exotic' or 'literary' and only need concern creative writers. According to this view, most ordinary language is literal. However, on closer inspection, much of our ordinary everyday language turns out to be figurative in nature. Consider the following examples, in which the figurative expressions are highlighted:

(4) Things are *going smoothly* in the operating theatre.

(5) He was *in* a state of shock after the election result.

(6) The economy *is going from* bad *to* worse.

These sentences are representative of 'ordinary', 'everyday' ways of talking about events like operations, emotional or psychological states, and changes in the economy. However, each sentence makes use of language that relates to motion, physical location or change in location in order to describe non-physical entities. Consider sentence (4): while sailing boats can 'go smoothly' across a lake or an ocean, abstract entities like operations are not physical objects that can undergo motion. Similarly, in sentence (5), while we can be physically located within bounded landmarks like rooms or buildings, we cannot be literally located within a state of shock, because shock is not a physical entity. Finally, in example (6) a change of state is understood in terms of a physical change in location. From this perspective, the italicised expressions in examples (4)–(6) have non-literal meanings in these sentences. Despite this, these expressions represent conventional means of talking about events, states and changes. This observation presents a serious challenge to the view that literal language provides the conventional means for talking about everyday events and situations.

Metaphorical versus non-metaphorical language use

Another definition of literality identified by Gibbs is non-metaphorical literality. According to this view, literal language is language that directly expresses meaning rather than relying upon metaphor. This view entails that we should always be able to express our 'true' meaning without recourse to metaphorical language, which involves expressing one idea in terms of another. For example, while the sentence in (7) has literal meaning, the sentence in (8) does not because it employs a metaphor: Achilles is understood in terms of a lion, which conveys the idea that Achilles has some quality understood as typical of lions such as fearlessness. This interpretation arises from our folk knowledge of lions, which stipulates that they are brave.

(7) Achilles is brave.

(8) Achilles is a lion.

However, it is difficult to find a non-metaphorical way of thinking and talking about certain concepts. For example, try talking about TIME without recourse to expressions relating to SPACE or MOTION. Consider example (9).

(9) a. Christmas is *approaching*.
    b. We're *moving towards* Christmas.
    c. Christmas is not very *far away*.

Each of these expressions relies upon language relating to motion or space in order to convey the idea that the temporal concept CHRISTMAS is imminent. These expressions represent ordinary everyday ways of talking about time. Indeed, it turns out to be more difficult to find ways of describing temporal concepts that do not rely on metaphorical language (see Evans 2004a). If certain concepts are wholly or mainly understood in metaphorical terms, then the non-metaphorical definition of literality entails that concepts like CHRISTMAS or TIME somehow lack meaning in their own right. Indeed, some scholars have actually claimed that time is not a 'real' experience. However, many everyday concepts appear to be understood in metaphorical terms. Consider the concept ANGER. Emotions like anger are, in developmental terms, among the earliest human experiences. Despite this, the way we conceptualise and describe this concept is highly metaphorical in nature, as the following examples illustrate.

(10) a. You make my *blood boil*.
   b. He was *red with* anger.
   c. She's just *letting off steam*.
   d. Don't *fly off the handle*.
   e. Try to *get a grip on yourself*.
   f. He almost *burst a blood vessel*.

Consider another example. We typically think and talk about ARGUMENT in terms of WAR. The examples in (11) are from Lakoff and Johnson (1980: 4).

(11) a. Your claims are *indefensible*.
   b. He *attacked every weak point* in my argument.
   c. His criticisms were *right on target*.
   d. I *demolished* his argument.
   e. I've never *won* an argument with him.
   f. You disagree? Okay, *shoot!*
   g. If you use that strategy, he'll *wipe you out*.
   h. He *shot down* all of my arguments.

As these examples demonstrate, the non-metaphorical definition of literality, which entails that we should always be able to express ourselves without recourse to metaphoric language, does not appear to present an accurate picture of the facts.

Literal truth versus literal falsity in language use

The truth-conditional view of literality rests upon the assumption that the basic function of language is to describe an objective external reality, and that this relationship between language and the world can be modelled in terms of truth or falsity (this idea was introduced in Chapter 5). The intuition behind this approach is that an important function of language is to describe states of affairs. Consider example (12).

(12) It's raining in London.

This sentence describes a state of affairs in the world and can be assessed as either true or false of a given situation, real or hypothetical. According to the truth-conditional definition of literality, example (12) represents literal language because it can either be literally true or false of a given situation. In contrast, expressions like *It's raining in my heart* or *You are my sunshine* can only be literally false and are therefore figurative. However, many linguistic expressions do not describe situations at all, and cannot therefore be meaningfully evaluated as true or false. Consider the examples in (13).

(13) a. Get well soon!
     b. Can you pass the salt please?
     c. I now pronounce you man and wife.

These examples represent speech acts. For instance, the function of the example in (13c) is not to describe a situation, but to change some aspect of the world (this idea was introduced in Chapter 1). If we adopt the truth-conditional view of literality, which rests upon the idea of literal truth, expressions like those in (13) are neither literal nor figurative since they cannot be evaluated as true (or false) with respect to a given situation.

Context-free versus context-dependent language use

The truth-conditional view also holds that literal meaning is context-independent. This means that literal meaning does not require a context in order to be fully interpreted. Consider example (14).

(14) a. The cat sat on the mat.
     b. My cat is a greedy pig.

According to this view, (14a) is fully interpretable independent of any context and the meaning we retrieve from (14a) is literal. In contrast, example (14b), which contains a metaphor, relies upon a context in which a cat habitually eats a lot in order to be fully understood. If this example were interpreted literally it would result in contradiction, since a cat cannot literally be a pig.

However, according to the encyclopaedic view of meaning assumed by cognitive semanticists (see Chapter 7) even the sentence in (14a) is not context-independent because it is interpreted against the background of rich encyclopaedic knowledge. Cultural associations, for instance, dictate what kind of cat we have in mind, and our experience of the world entails the assumption that gravity and normal force-dynamics apply so that we do not envisage the cat in (14a) on a flying carpet. In other words, a considerable number of background assumptions are brought to bear even on the interpretation of a relatively simple sentence. This brief discussion illustrates that it is difficult to pin down what aspects of meaning might be fully context-independent, which in turn calls into question the context-independent definition of literality.

In sum, we have examined a number of different definitions of literality identified by Gibbs in the cognitive science literature. We have seen that each of these definitions is problematic in certain respects. In particular, it seems that it is difficult to establish a neat dividing line between literal and figurative meaning. In the remainder of this chapter, we examine metaphor and metonymy: two phenomena that have traditionally been described as categories of figurative language use. As we will see, cognitive semanticists view metaphor and metonymy as phenomena fundamental to the structure of the conceptual system rather than superficial linguistic 'devices'.

9.2 What is metaphor?

For over 2,000 years, metaphor was studied within the discipline known as **rhetoric**. This discipline was first established in ancient Greece, and was focused on practical instruction in how to persuade others of a particular point of view by the use of rhetorical devices. Metaphor was one of these devices, which were called **tropes** by rhetoricians. Due to its central importance, metaphor came to be known as the **master trope**. Within this approach, metaphor was characterised by the schematic form: A is B, as in *Achilles is a lion*. As a consequence, metaphor has been identified since the time of Aristotle with implicit **comparison**. In other words, while metaphor is based on the comparison of two categories, the comparison is not explicitly marked. This contrasts with **simile**, where the comparison is overtly signalled by the use of *as* or *like*: *Achilles is as brave as a lion*; *Achilles is brave, like a lion*.

Clearly, examples of metaphor like *Achilles is a lion* are based on comparison. Following Grady (1997a, 1999) we will use the term **perceived resemblance** to describe this comparison. In this case, the resemblance is not physical: Achilles does not actually look like a lion. Instead, due to cultural knowledge which holds that lions are courageous, by describing Achilles as a lion we associate him with the lion's qualities of courage and ferocity. Metaphors of this kind are called **resemblance metaphors** (Grady 1999).

Resemblance metaphors based on physical resemblance have been called **image metaphors** (Lakoff and Turner 1989). In other words, image metaphors are one subset of resemblance-based metaphors. For instance, consider the following translation of the beginning of André Breton's surrealist poem 'Free Union', cited in Lakoff and Turner (1989: 93):

> My wife whose hair is a brush fire
> Whose thoughts are summer lightning
> Whose waist is an hourglass
> Whose waist is the waist of an otter caught in the teeth of a tiger
> Whose mouth is a bright cockade with the fragrance of a star of the first magnitude
> Whose teeth leave prints like the tracks of white mice over snow

Several of these lines represent image metaphors. For example, in the third line the poet is establishing a visual resemblance between the shape of his wife's waist and the shape of an hourglass.

Resemblance metaphors have received considerable attention within conceptual metaphor theory, particularly within the approach now known as **Cognitive Poetics** (see Lakoff and Turner 1989 for a seminal study; see also Stockwell 2002, and Gavins and Steen 2003). However, for the most part, research in the conceptual metaphor tradition has not been primarily concerned with metaphors of this kind. Instead, research in this tradition has focused on the kind of everyday language illustrated in the following examples. These examples represent common ways of referring to particular experiences of relationships like marriage. The examples in (15) are from Lakoff and Johnson (1980: 44–5).

(15) a. Look *how far* we've *come*.
b. We're at *a crossroads*.
c. We'll just have to *go our separate ways*.
d. We can't *turn back* now.
e. I don't think this relationship is *going anywhere*.
f. *Where* are we?
g. We're *stuck*.
h. It's been *a long, bumpy road*.
i. This relationship is *a dead-end street*.
j. We're just *spinning our wheels*.
k. Our marriage is *on the rocks*.
l. This relationship *is foundering*.

What is striking about these examples is that they represent ordinary everyday ways of talking about relationships: there is nothing stylised or overtly poetic about these expressions. Moreover, for the most part, they do not make use of the linguistic formula A is B, which is typical of resemblance metaphors. However, these expressions are clearly non-literal: a relationship cannot literally spin its wheels, nor stand at the crossroads.

Although a slim volume, Lakoff and Johnson's 1980 book *Metaphors We Live By* changed the way linguists thought about metaphor for two important reasons. Firstly, Lakoff and Johnson observed that metaphorical language appears to relate to an underlying **metaphor system**, a 'system of thought'. In other words, they noticed that we cannot choose any conceptual domain at random in order to describe relationships like marriage. Observe that the expressions in (15) have something in common: in addition to describing experiences of relationships, they also rely upon expressions that relate to the conceptual domain JOURNEYS. Indeed, our ability to describe relationships in terms of journeys appears to be highly productive.

This pattern led Lakoff and Johnson to hypothesise a conventional link at the conceptual level between the domain of LOVE RELATIONSHIPS and the domain of JOURNEYS. According to this view, LOVE, which is the **target** (the domain being described), is conventionally structured in terms of JOURNEYS, which is the **source** (the domain in terms of which the target is described). This association is called a **conceptual metaphor**. According to Lakoff and Johnson, what makes it a metaphor is the conventional association of one domain with another. What makes it conceptual (rather than purely linguistic) is the idea that the motivation for the metaphor resides at the level of conceptual domains. In other words, Lakoff and Johnson proposed that we not only speak in metaphorical terms, but also think in metaphorical terms. From this perspective, linguistic expressions that are metaphorical in nature are simply reflections of an underlying conceptual association.

Lakoff and Johnson also observed that there are a number of distinct roles that populate the source and target domains. For example, JOURNEYS include TRAVELLERS, a MEANS OF TRANSPORT, a ROUTE followed, OBSTACLES along the route and so on. Similarly, the target domain LOVE RELATIONSHIP includes LOVERS, EVENTS in the relationship and so on. The metaphor works by mapping roles from the source onto the target: LOVERS become TRAVELLERS (*We're at a crossroads*), who travel by a particular MEANS OF TRANSPORT (*We're spinning our wheels*), proceeding along a particular ROUTE (*Our relationship went off course*), impeded by obstacles (*Our marriage is on the rocks*). As these examples demonstrate, a metaphorical link between two domains consists of a number of distinct correspondences or **mappings**. These mappings are illustrated in Table 9.1.

It is conventional in the conceptual metaphor literature, following Lakoff and Johnson, to make use of the 'A is B' formula to describe conceptual metaphor: for example, LOVE IS A JOURNEY. However, this is simply a convenient shorthand for a series of discrete conceptual mappings which license a range of linguistic examples.

The second important claim to emerge from *Metaphors We Live By* was that conceptual metaphors are grounded in the nature of our everyday interaction with the world. That is, conceptual metaphor has an **experiential basis**.

Table 9.1 Mappings for LOVE IS A JOURNEY

Source: JOURNEY	Mappings	Target: LOVE
TRAVELLERS	→	LOVERS
VEHICLE	→	LOVE RELATIONSHIP
JOURNEY	→	EVENTS IN THE RELATIONSHIP
DISTANCE COVERED	→	PROGRESS MADE
OBSTACLES ENCOUNTERED	→	DIFFICULTIES EXPERIENCED
DECISIONS ABOUT DIRECTION	→	CHOICES ABOUT WHAT TO DO
DESTINATION OF THE JOURNEY	→	GOALS OF THE RELATIONSHIP

Consider the following linguistic evidence for the metaphor QUANTITY IS VERTICAL ELEVATION:

(16) a. The price of shares is *going up*.
b. She got a *high* score in her exam.

In these sentences there is a conventional reading related to QUANTITY. In (16a) the sentence refers to an increase in share prices. In (16b) it refers to an exam result that represents a numerical quantity. Although each of these readings is perfectly conventional, the lexical items that provide these readings, *going up* and *high*, refer literally to the concept of VERTICAL ELEVATION. Examples like these suggest that QUANTITY and VERTICAL ELEVATION are associated in some way at the conceptual level. The question is, what motivates these associations?

QUANTITY and VERTICAL ELEVATION are often correlated and these correlations are ubiquitous in our everyday experience. For instance, when we increase the height of something there is typically more of it. If an orange farmer puts more oranges on a pile, thereby increasing the height of the pile, there is a correlative increase in quantity. Similarly, water poured into a glass results in a correlative increase in both height (vertical elevation) and quantity of water. According to Lakoff and Johnson, this kind of correlation, experienced in our everyday lives, gives rise to the formation of an association at the conceptual level which is reflected in the linguistic examples. According to this view, conceptual metaphors are always at least partially motivated by and grounded in experience. As we have seen, then, cognitive semanticists define metaphor as a conceptual mapping between source and target domain. In the next section, we look in more detail at the claims made by Conceptual Metaphor Theory.

9.3 Conceptual Metaphor Theory

Conceptual Metaphor Theory has been highly influential both within cognitive linguistics and within the cognitive and social sciences, particularly in neighbouring disciplines like cognitive psychology and anthropology. In this section we summarise and outline some of the key aspects of Conceptual Metaphor Theory as they emerged between the publication of *Metaphors We Live By* and the mid-1990s.

9.3.1 The unidirectionality of metaphor

An important observation made by conceptual metaphor theorists is that conceptual metaphors are **unidirectional**. This means that metaphors map structure from a source domain to a target domain but not vice versa. For example, while we conceptualise LOVE in terms of JOURNEYS, we cannot conventionally structure JOURNEYS in terms of LOVE: travellers are not conventionally described as 'lovers', or car crashes in terms of 'heartbreak', and so on. Hence, the terms 'target' and 'source' encode the unidirectional nature of the mapping.

Lakoff and Turner (1989) observed that unidirectionality holds even when two different metaphors share the same domains. For example, they identified the two metaphors PEOPLE ARE MACHINES and MACHINES ARE PEOPLE, which are illustrated in examples (17) and (18), respectively.

(17) PEOPLE ARE MACHINES
    a. John always gets the highest scores in maths; he's a human calculator.
    b. He's so efficient; he's just a machine!
    c. He's had a nervous breakdown.

(18) MACHINES ARE PEOPLE
    a. I think my computer hates me; it keeps deleting my data.
    b. This car has a will of its own!
    c. I don't think my car wants to start this morning.

Although these two metaphors appear to be the mirror image of one another, close inspection reveals that each metaphor involves distinct mappings: in the PEOPLE ARE MACHINES metaphor, the mechanical and functional attributes associated with computers are mapped onto people, such as their speed and efficiency, their part-whole structure and the fact that they break down. In the MACHINES ARE PEOPLE metaphor, it is the notion of desire and volition that is mapped onto the machine. This shows that even when two metaphors share the same two domains, each metaphor is distinct in nature because it relies upon different mappings.

9.3.2 Motivation for target and source

Given that metaphorical mappings are unidirectional, two points of interest arise. The first relates to whether there is a pattern in terms of which conceptual domains typically function as source domains and which function as targets. The second point relates to what might motivate such a pattern. Based on an extensive survey, Kövecses (2002) found that the most common source domains for metaphorical mappings include domains relating to the HUMAN BODY (*the heart of the problem*), ANIMALS (*a sly fox*), PLANTS (*the fruit of her labour*), FOOD (*he cooked up a story*) and FORCES (*don't push me!*). The most common target domains included conceptual categories like EMOTION (*she was deeply moved*), MORALITY (*she resisted the temptation*), THOUGHT (*I see your point*), HUMAN RELATIONSHIPS (*they built a strong marriage*) and TIME (*time flies*).

Turning to the second point, the prevalent explanation until the mid-1990s was that target concepts tended to be more abstract, lacking physical characteristics and therefore more difficult to understand and talk about in their own terms. In contrast, source domains tended to be more concrete and therefore more readily 'graspable'. As Kövecses (2002: 20) puts it, 'Target domains are abstract, diffuse and lack clear delineation; as a result they 'cry out' for metaphorical conceptualization.' The intuition behind this view was that target concepts were often 'higher-order concepts': although grounded in more basic embodied experiences, these concepts relate to more complex and abstract experiential knowledge structures. Consider the conceptual domain TIME, an abstract domain *par excellence*. Time is primarily conceptualised in terms of SPACE, and MOTION through space, as illustrated by the examples in (19).

(19) a. Christmas is *coming*.
 b. The relationship lasted a *long time*.
 c. The time for a decision *has come*.
 d. *We're approaching* my favourite time of the year.

Lakoff and Johnson (1999) argue that TIME is structured in terms of MOTION because our understanding of TIME emerges from our experience and awareness of CHANGE, a salient aspect of which involves MOTION. For instance, whenever we travel from place A to place B, we experience CHANGE in location. This type of event also corresponds to a temporal span of a certain duration. From this perspective, our experience of time – that is, our awareness of change – is grounded in more basic experiences like motion events. Lakoff and Johnson argue that this comparison of location at the beginning and end points of a journey, gives rise to our experience of time: embodied experiences like MOTION partially structure the more abstract domain TIME. This gives rise to the general metaphor TIME IS MOTION.

9.3.3 Metaphorical entailments

In addition to the individual mappings that conceptual metaphors bring with them, they also provide additional, sometimes quite detailed knowledge. This is because aspects of the source domain that are not explicitly stated in the mappings can be inferred. In this way, metaphoric mappings carry **entailments** or rich **inferences**. Consider the examples in (20), which relate to the conceptual metaphor AN ARGUMENT IS A JOURNEY:

(20) a. We will proceed in a *step-by-step fashion*.
 b. We have *covered a lot of ground*.

In this metaphor, PARTICIPANTS in the argument correspond to TRAVELLERS, the ARGUMENT itself corresponds to a JOURNEY and the PROGRESS of the argument corresponds to the ROUTE taken. However, in the source domain JOURNEY, travellers can get lost, they can stray from the path, they can fail to reach their destination, and so on. The association between source and target gives rise to the entailment (the rich inference) that these events can also occur in the target domain ARGUMENT. This is illustrated by the examples in (21) which show that structure that holds in the source domain can be inferred as holding in the target domain.

(21) a. I *got lost in* the argument.
 b. We *digressed from* the main point.
 c. He failed *to reach* the conclusion.
 d. I *couldn't follow* the argument.

9.3.4 Metaphor systems

An early finding by Lakoff and Johnson (1980) was that conceptual metaphors interact with each other and can give rise to relatively complex **metaphor systems**. These systems are collections of more schematic metaphorical mappings that structure a range of more specific metaphors like LIFE IS A JOURNEY. Lakoff (1993) outlines a particularly intricate example of a metaphor system which he calls the **event structure metaphor**. This is actually a series of metaphors that interact in the interpretation of utterances. The individual metaphors that make up the event structure metaphor, together with linguistic examples, are shown in table 9.2.

In order to illustrate how the event structure metaphor applies, consider the specific metaphor LIFE IS A JOURNEY. This is illustrated by the examples in (22).

(22) a. STATES ARE LOCATIONS
 He's *at a crossroads* in his life.
 b. CHANGE IS MOTION
 He went *from* his forties *to* his fifties without a hint of a mid-life crisis.
 c. CAUSES ARE FORCES
 He *got a head start* in life.
 d. PURPOSES ARE DESTINATIONS
 I can't ever seem to *get to where I want to be* in life.
 e. MEANS ARE PATHS
 He followed *an unconventional course* during his life.

Table 9.2 The event structure metaphor

Metaphor:	STATES ARE LOCATIONS (BOUNDED REGIONS IN SPACE)
Example:	*John is in love*
Metaphor:	CHANGE IS MOTION (FROM ONE LOCATION TO ANOTHER)
Example:	*Things went from bad to worse*
Metaphor:	CAUSES ARE FORCES
Example:	*Her argument forced me to change my mind*
Metaphor:	ACTIONS ARE SELF-PROPELLED MOVEMENTS
Example:	*We are moving forward with the new project*
Metaphor:	PURPOSES ARE DESTINATIONS
Example:	*We've finally reached the end of the project*
Metaphor:	MEANS ARE PATHS (TO DESTINATIONS)
Example:	*We completed the project via an unconventional route*
Metaphor:	DIFFICULTIES ARE IMPEDIMENTS TO MOTION
Example:	*It's been uphill all the way on this project*
Metaphor:	EVENTS ARE MOVING OBJECTS
Example:	*Things are going smoothly in the operating theatre*
Metaphor:	LONG-TERM PURPOSEFUL ACTIVITIES ARE JOURNEYS
Example:	*The government is without direction*

f. DIFFICULTIES ARE IMPEDIMENTS TO MOTION
   Throughout his working life problematic professional relationships had somehow always *got in his way*.
g. PURPOSEFUL ACTIVITIES ARE JOURNEYS
   His life had been *a rather strange journey*.

The target domain for this metaphor is LIFE, while the source domain is JOURNEY. The EVENTS that comprise this metaphor are life events, while the PURPOSES are life goals. However, because this metaphor is structured by the event structure metaphor, LIFE IS A JOURNEY turns out to be a highly complex metaphor that represents a composite mapping drawing from a range of related and mutually coherent metaphors: each of the examples in (22) **inherits** structure from a specific metaphor within the event structure complex. Similarly, other complex metaphors including AN ARGUMENT IS A JOURNEY, LOVE IS A JOURNEY and A CAREER IS A JOURNEY also inherit structure from the Event Structure Metaphor.

9.3.5 Metaphors and image schemas

Subsequent to the development of image schema theory (Chapter 6), the idea that certain concepts were image-schematic in nature was exploited by Conceptual Metaphor Theory (e.g. Lakoff 1987, 1990, 1993). Lakoff and Johnson both argued that image schemas could serve as source domains for metaphoric mapping. The rationale for this view can be summarised as follows: image schemas appear to be knowledge structures that emerge directly from pre-conceptual embodied experience. These structures are meaningful at the conceptual level precisely because they derive from the level of bodily experience, which is directly meaningful. For example, our image-schematic concept COUNTERFORCE arises from the experience of being unable to proceed because some opposing force is resisting our attempt to move forward. Image schemas relating to FORCES metaphorically structure more abstract domains like CAUSES by serving as source domains for these abstract concepts. This is illustrated by the event structure metaphor, where the image-schematic concept BOUNDED LOCATIONS structures the abstract concept STATES, while the image-schematic concept OBJECTS structures the abstract concept EVENTS, and so on.

The striking consequence to emerge from this application of image schema theory to Conceptual Metaphor Theory is that abstract thought and reasoning, facilitated by metaphor, are seen as having an image-schematic and hence an embodied basis (e.g. Lakoff 1990). Clearly, highly abstract concepts are unlikely to be directly structured in terms of simple image schemas but are more likely to be structured in complex ways by **inheritance relations**: a network of intermediate mappings. It also seems likely that certain concepts must relate in part to subjective experiences like emotions (a point we return to below). Despite these caveats, Conceptual Metaphor Theory holds that abstract concepts can, at least in part, be traced back to image schemas.

9.3.6 Invariance

As a result of the emergence of these ideas, a preoccupation for conceptual metaphor theorists in the late 1980s and early 1990s centred on how metaphoric mappings could be constrained (Brugman 1990; Lakoff 1990, 1993; Lakoff and Turner 1989; Turner 1990, 1991). After all, if metaphor is ultimately based on image schemas, with chains of inheritance relations giving rise to highly abstract and specific metaphors like LOVE IS A JOURNEY, ARGUMENT IS WAR and so on, it is important to establish what licenses the selection of particular image schemas by particular target domains and why unattested mappings are not licensed.

There appear to be certain restrictions in terms of which source domains can serve particular target domains, as well as constraints on metaphorical entailments that can apply to particular target domains. For example, Lakoff and Turner (1989) observed that the concept of DEATH is **personified** in a number of ways (which means that a concept has human-like properties attributed to it, such as intentionality and volition). However, the human-like qualities that can be associated with DEATH are restricted: DEATH can 'devour', 'destroy' or 'reap', but as Lakoff (1993: 233) observes, 'death is not metaphorized in terms of teaching, or filling the bathtub, or sitting on the sofa.' In order to account for these restrictions, Lakoff posited the **Invariance Principle**:

> Metaphorical mappings preserve the cognitive topology (that is, the image schema structure) of the source domain, in a way consistent with the inherent structure of the target domain. (Lakoff 1993: 215)

There are a number of specific death personification metaphors, including DEATH IS A DEVOURER, DEATH IS A REAPER and DEATH IS A DESTROYER, which inherit structures from a more schematic metaphor, which Lakoff and Turner (1989) call a **generic-level metaphor**: EVENTS ARE ACTIONS (or INANIMATE PHENOMENA ARE HUMAN AGENTS). What the invariance principle does is guarantee that image-schematic organisation is invariant across metaphoric mappings. This means that the structure of the source domain must be preserved by the mapping in a way consistent with the target domain. This constrains potentially incompatible mappings.

Let's elaborate this idea in relation to the DEATH metaphors mentioned above. While DEATH can be structured in terms of the kinds of agents we have noted (DEVOURER, REAPER or DESTROYER), it cannot be structured in terms of any kind of agent at random. For example, it would not be appropriate to describe DEATH as KNITTER, TEACHER or BABYSITTER. Agents that devour, reap or destroy bring about a sudden change in the physical state of an entity. This corresponds exactly to the nature of the concept DEATH, whose 'cognitive topology' or 'inherent' conceptual structure is preserved by the attested mappings like DEATH IS A DESTROYER but not the unattested mapping *DEATH IS A KNITTER.

The Invariance Principle also predicts that metaphoric entailments that are incompatible with the target domain will fail to map. Consider the examples in (23), which relate to the metaphor CAUSATION IS TRANSFER (OF AN OBJECT):

(23) a. She gave him a headache.   STATE
     b. She gave him a kiss.   EVENT

While the source domain for both of these examples is TRANSFER, the first example relates to a STATE and the second to an EVENT. The source domain TRANSFER entails that the recipient is in possession of the transferred entity. However, while this entailment is in keeping with STATES because they are temporally unbounded, the same entailment is incompatible with EVENTS because they are temporally bounded and cannot therefore 'stretch' across time. This is illustrated by (24).

(24) a. She gave him a headache and he still has it.   STATE
     b. *She gave him a kiss and he still has it.   EVENT

The process that prevents entailments from projecting to the target domain is called **target domain override** (Lakoff 1993).

9.3.7 The conceptual nature of metaphor

A consequence of the claim that conceptual organisation is in large part metaphorical is that thought itself is metaphorical. In other words, metaphor is not simply a matter of language, but reflects 'deep' correspondences in the way our conceptual system is organised. This being so, we expect to find evidence of metaphor in human systems other than language. Indeed, this view comes from studies that have investigated the metaphorical basis of a diverse range of phenomena and constructs, including social organisation and practice, myths, dreams, gesture, morality, politics and foreign policy, advertisements and mathematical theory. For example, the organisation of a business institution is often represented in terms of a diagram that represents a hierarchical structure, in which the CEO is at the highest point and other officers and personnel of the company are placed at lower points; relative positions upwards on the vertical axis correspond to relative increases in importance or influence. This type of diagram reflects the conceptual metaphor SOCIAL INSTITUTIONS ARE HIERARCHICAL STRUCTURES. Conceptual metaphor theorists argue that this metaphor is in turn grounded in more basic kinds of experience, such as the correlation between height or size and influence, or the fact that the head (which controls the body) is the uppermost part of the body.

To provide a second example, linguistic theories themselves can have a metaphorical basis. The dominant metaphor in Generative Grammar, for example, could be described in terms of SENTENCE STRUCTURE IS A HIERARCHY. This explains why a proliferation of terminology emerged from this theory that reflected hierarchical relationships, including terms like *dominate*, *govern*, *control*, *bind* and so on. Moreover, sentence structure is visually represented in a number of syntactic theories by 'tree diagrams', structures that are hierarchically organised so that the sentence 'dominates' or 'contains' phrases, which in turn 'dominate' or 'contain' words. Equally, Mental Spaces Theory (Chapter 11) is a model of meaning construction that relies upon the metaphor COGNITIVE REPRESENTATIONS ARE CONTAINERS to describe the process of on-line meaning construction. According to cognitive semanticists, examples illustrate the central importance of metaphor in human thinking.

9.3.8 Hiding and highlighting

An important idea in Conceptual Metaphor Theory relates to **hiding** and **highlighting**: when a target is structured in terms of a particular source, this highlights certain aspects of the target while simultaneously hiding other aspects. For example, invoking the metaphor ARGUMENT IS WAR highlights the adversarial nature of argument but hides the fact that argument often involves an ordered and organised development of a particular topic (*He won the argument, I couldn't defend that point*, and so on). In contrast, the metaphor AN ARGUMENT IS A JOURNEY highlights the progressive and organisational aspects of arguments while hiding the confrontational aspects (*We'll proceed in step-by-step fashion; We've covered a lot of ground*). In this way, metaphors can **perspectivise** a concept or conceptual domain.

9.4 Primary Metaphor Theory

As observed by Murphy (1996), among others, one problem with Conceptual Metaphor Theory, as formalised by the Invariance Principle, is the potential contradiction inherent in the claim that a target domain possesses an invariant 'inherent structure' that limits the metaphorical mappings and entailments that can apply, and at the same time that the target domain is abstract in the sense that it is not clearly delineated. According to Conceptual Metaphor Theory, the purpose of metaphor is to map structure onto abstract domains; if a target already has its own invariant structure, why should it require metaphoric structuring?

9.4.1 Primary and compound metaphors

In an influential study, Joseph Grady (1997a) addresses this problem by proposing that there are two kinds of metaphor: **primary metaphor** and **compound metaphor**. While primary metaphors are foundational, compound metaphors are constructed from the unification of primary metaphors. Grady's central claim, which marks his approach as distinct from earlier work in Conceptual Metaphor Theory, is that primary metaphors conventionally associate concepts that are equally 'basic', in the sense that they are both directly experienced and perceived. This means that Grady rejects the view that the distinction between the target and source of a metaphoric mapping relates to abstract versus concrete concepts. Instead, Grady argues that the distinction between target and source relates to **degree of subjectivity** rather than how clearly delineated or how abstract a concept is. This view means that the Invariance Principle is redundant because the foundational primary metaphors, upon which more complex metaphor systems are based, are not viewed as providing an 'abstract' target with 'missing' structure. Consider the following examples of primary metaphors proposed by Grady, together with example sentences.

(25) SIMILARITY IS NEARNESS
That colour is quite close to the one on our dining-room wall.

(26) IMPORTANCE IS SIZE
    We've got a big week coming up at work.

(27) QUANTITY IS VERTICAL ELEVATION
    The price of shares has gone up.

(28) CAUSES ARE FORCES
    Vanity drove me to have the operation.

(29) CHANGE IS MOTION
    Things have shifted a little since you were last here.

(30) DESIRE IS HUNGER
    We're hungry for a victory.

Grady accounts for these metaphors in the following terms (small capitals added):

> ... the target concepts [e.g. SIMILARITY, IMPORTANCE, QUANTITY, CAUSES, CHANGE and DESIRE] lack the kind of perceptual basis which characterises the source concepts ... CHANGE, for instance, can be detected in any number of domains, including non-physical ones (e.g. a change in the emotional tone of a conversation), whereas the detection of physical MOTION is directly based on physical perception. DESIRE is an affective state while HUNGER is a physical sensation. QUANTITY is a parameter in any realm, while VERTICAL ELEVATION is a physical variable, perceived by the senses. (Grady n.d.: 5/14–15)

In other words, primary target concepts reflect subjective responses to sensory perception, and represent 'judgements, assessments, evaluations and inferences' (Grady n.d.: 5/15). From this perspective, target concepts like SIMILARITY, QUANTITY and DESIRE are not dismissed as 'abstract' but are recognised as being among the most fundamental and direct experiences we have as human beings. This explains why Grady describes them as 'primary'. The key distinction between target and source in Grady's theory is that primary source concepts relate to sensory-perceptual experience, while primary target concepts relate to *subjective responses* to sensory-perceptual experience. This is reminiscent of the distinction between imagistic experience and introspective experience that we introduced in Chapter 6.

9.4.2 Experiential correlation

If primary target and primary source concepts are equally 'basic' which renders the Invariance Principle redundant, what motivates their association? Grady maintains the assumption fundamental to Conceptual Metaphor Theory that there is an experiential basis for primary metaphor formation. However, in Grady's theory there must be a clear and direct experiential basis: an **experiential correlation**. Consider again the examples in (16), repeated here:

(16) a. The price of shares is *going up*.
b. She got a *high* score on her exam.

In our earlier discussion of these examples, we observed that QUANTITY and HEIGHT correlate in experiential terms. This experience provides the basis for the conventional association between the concepts QUANTITY and VERTICAL ELEVATION. In this respect, Grady provides a more principled theory of the experiential basis of conceptual metaphor, linking this directly to the licensing of metaphorical mappings.

9.4.3 Motivating primary metaphors

Like the more general framework of Conceptual Metaphor Theory, Primary Metaphor Theory assumes that primary metaphors are unidirectional. However, because primary metaphors involve the association of a target and a source that are equally basic and are derived from real and directly apprehended experiences, there must be a different explanation for the unidirectionality: for what makes a source a source and a target a target. Recall that the earlier view in Conceptual Metaphor Theory was that target concepts (or domains) were more abstract than the source concept (or domain), and that the source provided the target with structure that made it possible to think and talk about these abstract concepts.

In Primary Metaphor Theory, the mapping from source to target is explained in the following terms: because primary target concepts relate to subjective responses, they operate at a level of cognitive processing to which we have low conscious access. Primary target concepts are responses and evaluations, which derive from background operations (an idea that we illustrate below). According to this view, the function of primary metaphor is to structure primary target concepts in terms of sensory images in order to foreground otherwise backgrounded cognitive operations. This is achieved by employing source concepts that are more accessible because they relate to sensory rather than subjective experience. Primary source concepts, which derive from external sensory experience, are said to have **image content** while primary target concepts, which are more evaluative and hence subjective in nature, are said to have **response content**.

Recall example (25), which illustrates the primary metaphor SIMILARITY IS NEARNESS. The target concept SIMILARITY relates to a covert (background) process of evaluation that is intrinsic to judgement. For instance, when we look at two people's faces and judge that they have similar appearances and might therefore be members of the same family, the cognitive operations that allow us to identify these similarities are part of the background. What is important or salient to us are the faces themselves and our resulting judgement of their similarity. While the concept NEARNESS is derived from sensory experience, the concept SIMILARITY relates to a subjective evaluation produced by mechanisms that are typically covert, or at least operate at a relatively low level of conscious access.

9.4.4 Distinguishing primary and compound metaphors

Recall that Grady proposes that there are two types of conceptual metaphor: primary metaphor and compound metaphor. In this section, we examine how primary metaphor and compound metaphor are distinguished in Grady's theory and how the two interact. This discussion is based on Grady's (1997b) investigation of the conceptual metaphor THEORIES ARE BUILDINGS, originally proposed by Lakoff and Johnson (1980). The following examples are used by Lakoff and Johnson as evidence for the metaphor:

> Is that the *foundation* for your theory? The theory needs more *support*. The argument is *shaky*. We need some more facts or the argument will *fall apart*. We need to *construct* a *strong* argument for that. I haven't figured out yet what the *form* of the argument will be. Here are some more facts to *shore up* the theory. We need to *buttress* the theory with *solid* arguments. The theory will *stand* or *fall* on the *strength* of that argument. The argument *collapsed*. They *exploded* his latest theory. We will show that theory to be without *foundation*. So far we have put together only the *framework* of the theory. (Lakoff and Johnson 1980: 46)

According to Grady, THEORIES ARE BUILDINGS fails as an instance of primary metaphor according to three criteria, and must therefore be considered an example of compound metaphor. We consider each of these criteria below.

Association of complex domains

Primary metaphors are simple. As Grady (n.d. 5/30) puts it, 'they refer to simple aspects or dimensions of subjective experience, not confined to any particular, rich domain, but crosscutting these domains; not associated with particular, rich, scenarios but inhering within broad categories of scenarios.' In other words, primary metaphors relate two 'simple' concepts from distinct domains. In contrast, compound metaphors relate entire complex domains of experience, like THEORIES ARE BUILDINGS. Figure 9.1, in which the small

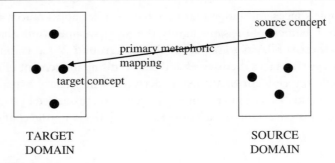

TARGET　　　　　　　　　　　　SOURCE
DOMAIN　　　　　　　　　　　　DOMAIN

Figure 9.1 Primary metaphor

circles represent distinct concepts, illustrates the idea that primary metaphors link distinct concepts from distinct domains rather than linking entire domains. Since both THEORIES and ARGUMENTS are relatively complex and rich in detail, they do not qualify as primary target and source concepts, respectively. A consequence of the view that primary source and target concepts are associated by virtue of experiential correlations arising from human physiology and a shared environment is that primary metaphors are likely to represent **cross-linguistic universals**. In contrast, because compound metaphors arise from more detailed and specific knowledge structure, they are more likely to be culture-dependent. This theory predicts that communities with a significantly different material culture from that of the West (for example, nomadic tent-dwellers or cave-dwellers) would be unlikely to employ the metaphor THEORIES ARE BUILDINGS, but might instead structure the concept THEORIES in terms of some other culturally salient concept.

Poverty of mapping

Further evidence that the THEORIES ARE BUILDINGS metaphor does not qualify as a primary metaphor relates to what Grady calls poverty of mapping. Because primary metaphors relate to relatively simple knowledge structures – in other words, concepts rather than conceptual domains – they are expected to contain no **mapping gaps**. In other words, because a primary metaphor maps one single concept onto another, there is no part of either concept that is 'missing' from the mapping. Indeed, it is difficult to imagine how primary source concepts like MOTION, FORCE and SIZE could be broken down into component parts in the first place.

In contrast, the compound metaphor THEORIES ARE BUILDINGS relies upon two complex conceptual domains, each of which can be can be broken down into component parts. For example, BUILDINGS have WINDOWS, TENANTS and RENT, among other associated concepts, yet these components fail to map onto the target concept, as the examples in (31) illustrate (Grady 1997b: 270).

(31) a. ?This theory has French windows.
b. ?The tenants of her theory are behind in their rent.

The occurrence of 'mapping gaps' reveals that THEORIES and BUILDINGS do not qualify as the basic or simple concepts that are associated in primary metaphors.

Lack of clear experiential basis

Finally, as we have seen, Grady argues that primary metaphors emerge from a clear experiential basis. Clearly, the metaphorical association between THEORIES and BUILDINGS lacks this experiential basis: we can hardly claim that theories and buildings are closely correlated with one another in our everyday experience of the world. Although we often discuss theories in buildings, buildings are only incidentally associated with theories: we might just as easily discuss theories outdoors, in a tent or on a boat.

In conclusion, since THEORIES ARE BUILDINGS lacks the characteristics of primary metaphor, Grady concludes that it represents an instance of compound metaphor. Grady suggests that this particular compound metaphor derives from the unification of two primary metaphors. This is illustrated in Figure 9.2.

According to Grady, this unification combines two independently motivated primary metaphors: PERSISTING IS REMAINING UPRIGHT and ORGANISATION IS PHYSICAL STRUCTURE. Their unification licenses the complex metaphor THEORIES ARE BUILDINGS. The salient characteristics of THEORIES are that they have relatively complex organisation, based on models, hypotheses, premises, evidence and conclusions. Moreover, a good

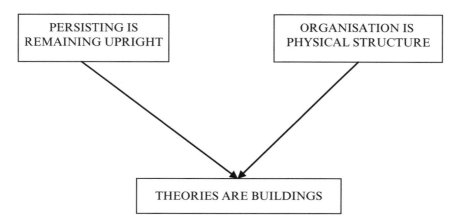

Figure 9.2 Compound metaphor

Table 9.3 Mappings for AN ABSTRACT ORGANISED ENTITY IS AN UPRIGHT PHYSICAL OBJECT

Target: ABSTRACT ORGANISED ENTITY	mappings	Source: UPRIGHT PHYSICAL OBJECT
Complex abstract entity	→	Complex physical object
Abstract constituents of the entity	→	Physical parts
Logical relations among constituents	→	Physical arrangement of parts
Persistence	→	Verticality
Asymmetrical dependence	→	Support

theory is one that stands the test of time. Two salient characteristics associated with BUILDINGS are they remain upright for a long time and have complex physical structure. In other words, the salient characteristics that unite THEORIES and BUILDINGS are exactly those found as target and source in the two more foundational primary metaphors PERSISTING IS REMAINING UPRIGHT and ORGANISATION IS PHYSICAL STRUCTURE. Grady argues that we conceptualise THEORIES in terms of buildings because, in our culture, buildings are a particularly salient – indeed prototypical – form of physical structure that is both upright and complex in structure. Furthermore, Grady accounts for 'mapping gaps' on the basis that only salient parts of the physical structure of buildings are licensed to map onto the target: although we know that BUILDINGS have WINDOWS and OCCUPANTS, these do not perform a supporting function within the physical structure of the building and are therefore unlicensed to map onto the target. Table 9.3 lists the licensed mappings that Grady provides for the unified compound metaphor THEORIES ARE BUILDINGS, which might more generally be called AN ABSTRACT ORGANISED ENTITY IS AN UPRIGHT PHYSICAL OBJECT.

Finally, the ability to construct compound metaphors has been argued to facilitate the process of **concept elaboration** (Evans 2004a), an idea that we discussed in Chapter 3. According to this perspective, the nature and scope of concepts can be developed and extended through the conventional association between (lexical) concepts and imagery. In other words, when the concept THEORY is elaborated via mechanisms like conceptual metaphor, the conceptual metaphor serves as a vehicle for **conceptual evolution** (Musolff 2004). This explanation for why concepts like THEORY are associated with metaphor provides an alternative to the argument that it is the abstract nature of concepts that motivates metaphor.

9.5 What is metonymy?

In *Metaphors We Live By*, Lakoff and Johnson pointed out that, in addition to metaphor, there is a related conceptual mechanism that is also central to human thought and language: **conceptual metonymy**. Like metaphor, metonymy has traditionally been analysed as a trope: a purely linguistic device. However, Lakoff and Johnson argued that metonymy, like metaphor, was conceptual in nature. In recent years, a considerable amount of research has been devoted to metonymy. Indeed, some scholars have begun to suggest that metonymy may be more fundamental to conceptual organisation than metaphor, and some have gone so far as to claim that metaphor itself has a metonymic basis, as we will see. Here, we present an overview of the research in cognitive semantics that has been devoted to this topic.

The earliest approach to conceptual metonymy in cognitive semantics was developed by Lakoff and Johnson (1980). They argued that, like metaphor, metonymy is a conceptual phenomenon, but one that has quite a distinct basis. Consider example (32).

(32) The ham sandwich has wandering hands.

Imagine that the sentence in (32) is uttered by one waitress to another in a café. This use of the expression *ham sandwich* represents an instance of metonymy: two entities are associated so that one entity (the item the customer ordered) stands for the other (the customer). As this example demonstrates, linguistic metonymy is **referential** in nature: it relates to the use of expressions to 'pinpoint' entities in order to talk about them. This shows that metonymy functions differently from metaphor. For example (32) to be metaphorical we would need to understand *ham sandwich* not as an expression referring to the customer who ordered it, but in terms of a food item with human qualities. Imagine a cartoon, for example, in which a ham sandwich sits at a café table. On this interpretation, we would be attributing human qualities to a ham sandwich, motivated by the metaphor AN INANIMATE ENTITY IS AN AGENT. As these two quite distinct interpretations show, while metonymy is the conceptual relation 'X stands for Y', metaphor is the conceptual relation 'X understood in terms of Y'.

A further defining feature of metonymy pointed out by Lakoff and Johnson is that it is motivated by physical or causal associations. Traditionally, this was expressed in terms of **contiguity**: a close or direct relationship between two entites. This explains why the waitress can use the expression *the ham sandwich* to refer to the customer: there is a direct experiential relationship between the ham sandwich and the customer who ordered it.

A related way of viewing metonymy is that metonymy is often **contingent** on a specific context. Within a specific discourse context, a salient vehicle **activates** and thus highlights a particular target. Hence, while correlation-based (as opposed to resemblance-based) metaphors are pre-conceptual in origin and are therefore in some sense inevitable associations (motivated by the nature of our bodies and our environment), conceptual metonymies are motivated by communicative and referential requirements.

Finally, Lakoff and Turner (1989) added a further component to the cognitive semantic view of metonymy. They pointed out that metonymy, unlike metaphor, is not a cross-domain mapping, but instead allows one entity to stand for another because both concepts coexist within the same domain. This explains why a metonymic relationship is based on contiguity or conceptual 'proximity'. The reason *ham sandwich* in (32) represents an instance of metonymy is because both the **target** (the customer) and the **vehicle** (the ham sandwich) belong to the same CAFÉ domain. Kövecses and Radden summarise this view of metonymy as follows:

> Metonymy is a cognitive process in which one conceptual entity, the vehicle, provides mental access to another conceptual entity, the target, within the same domain, or ICM. (Kövecses and Radden 1998: 39)

Observe that Kövecses and Radden frame the notion of metonymy in terms of **access** rather than mapping. Indeed, other scholars have suggested that metonymy might be usefully considered in terms of a mapping process that **activates** or **highlights** a certain aspect of a domain (for discussion see Barcelona 2003b; Croft 1993). From this perspective, metonymy provides a 'route' of access for a particular target within a single domain. For example, while it is not usual to describe a human in terms of food, from the perspective of a waitress, the food ordered may be more salient than the customer. For this reason, the food ordered 'activates' the customer sitting at a particular table in the café.

Metonymies are represented by the formula 'B for A', where 'B' is the vehicle and 'A' is the target, e.g. PLACE FOR INSTITUTION. This contrasts with the 'A is B' formula that represents conceptual metaphor. For instance, in example (33) *Buckingham Palace* is the vehicle (PLACE) which stands for the BRITISH MONARCHY, the target (INSTITUTION):

(33) Buckingham Palace denied the rumours.

This utterance is an example of the metonymy PLACE FOR INSTITUTION. Figure 9.3 illustrates the distinction between conceptual metaphor and conceptual metonymy.

There are a number of distinct kinds of metonymy that have been identified in the cognitive semantics literature. We briefly illustrate some of these below. In each of the following examples, the vehicle is italicised.

(34) PRODUCER FOR PRODUCT
    a. I've just bought a new *Citröen*.

METAPHOR AND METONYMY

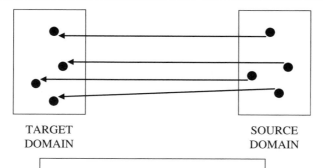

TARGET DOMAIN   SOURCE DOMAIN

Conceptual metaphor (compound): cross-domain mapping between source and target

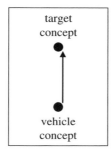

target concept

vehicle concept

Conceptual metonymy: mapping within a single domain between a vehicle concept and a target concept

Figure 9.3 Comparison between metaphor and metonymy

    b. Pass me the *Shakespeare* on the top shelf.
    c. She likes eating *Burger King*.

(35) PLACE FOR EVENT
    a. *Iraq* nearly cost Tony Blair the premiership.
    b. American public opinion fears another *Vietnam*.
    c. Let's hope that *Beijing* will be as successful an Olympics as *Athens*.

(36) PLACE FOR INSTITUTION
    a. *Downing street* refused comment.
    b. *Paris* and *Washington* are having a spat.
    c. *Europe* has upped the stakes in the trade war with the *United States*.

(37) PART FOR WHOLE
    a. My *wheels* are parked out the back.

b. Lend me *a hand*.
c. She's not just a *pretty face*.

(38) WHOLE FOR PART
a. *England* beat *Australia* in the 2003 rugby World Cup final.
b. *The European Union* has just passed new human rights legislation.
c. *My car* has developed a mechanical fault.

(39) EFFECT FOR CAUSE
a. He has *a long face*.
b. He has *a spring in his step* today.
c. Her *face is beaming*.

While most of the examples of metonymy we have considered so far relate to noun phrases, metonymic vehicles are not restricted to individual lexical items. For instance, Panther and Thornburg (2003) have argued that indirect speech acts represent instances of metonymy. Consider example (40):

(40) Can you pass the salt?

Recall from Chapter 1 that a speech act is an utterance that performs a (linguistic) action. The example in (40) is 'indirect' because it counts as a conventional way of making a request, but does so 'via' a question about the ability of the addressee to carry out the action (signalled by the interrogative form of the clause), rather than making the request directly (by using an imperative clause like *Pass me the salt*). Panther and Thornburg argue that indirect speech acts are metonymic, in that the question stands for the request. In other words, the ability to perform the action is a necessary prerequisite (or 'felicity condition') for a request to be carried out (Searle 1969), and a question about this ability stands for the request itself.

9.6 Conceptual metonymy

As we have seen, cognitive semanticists argue that metonymy, like conceptual metaphor, is not a purely linguistic device but is central to human thought. Indeed, we have already seen some non-linguistic instances of metonymy; these were illustrated in the previous chapter, where we discussed Lakoff's claims concerning the metonymic function of idealised cognitive models (ICMs) which give rise to prototype effects. According to Lakoff's theory of cognitive models, ideals, stereotypes and salient examples can metonymically represent an entire category. In this section, we look in more detail at the explanations that cognitive linguists have proposed in order to account for metonymy as a conceptual phenomenon.

9.6.1 Metonymy as an access mechanism

We noted above that Kövecses and Radden define metonymy in terms of the conceptual access it affords. This idea is based on proposals made by Langacker (1993: 30) who argues that 'the entity that is normally designated by a metonymic expression serves as a reference point affording mental access to the desired target (that is, the entity actually being referred to)'. In other words, metonymy serves as point of access to a particular aspect of a domain and thus provides access to the target concept. Furthermore, each vehicle provides a different route into the relevant conceptual domain.

According to Croft (1993), a target is accessed within a domain as a result of **domain highlighting**. Croft takes as his starting point the encyclopaedic view of meaning and adopts Langacker's theory of domains (see Chapter 7). Recall that Langacker's theory holds that a concept profile is understood with respect to a domain matrix: the range of domains that contribute to our ultimate understanding of the concept. This accounts for the fact that lexical items relate to potentially huge knowledge structures. Croft's proposal is that, from the perspective of encyclopaedic semantics, metonymy functions by highlighting one domain within a concept's domain matrix. Thus a particular usage of a lexical concept can highlight distinct domains within the concept's domain matrix on different occasions. Consider the following examples drawn from Croft (1993):

(41) a. Proust spent most of his time in bed.
 b. Proust is tough to read.

Part of the domain matrix associated with Marcel Proust is that he was a man known for particular habits relating to how much time he spent in bed. This is knowledge about Proust the man. Another aspect of the domain matrix relates to Proust's literary work and his career as a writer. While the expression *Proust* in (41a) highlights the domain for Proust (Proust the man), the expression *Proust* in (41b) highlights the literary work of Proust. Thus, from the perspective of domain matrices, a particular expression can metonymically highlight distinct, albeit related, aspects of our encyclopaedic knowledge relating to Proust.

The claim that metonymy relates to a highlighted domain in a domain matrix does not amount to the claim that metonymy is a cross-domain relationship in the sense intended by metaphor theorists. Clearly, the example in (41b) is still an 'X stands for Y' relation (a metonym) rather than an 'X understood in terms of Y' relation (a metaphor). Croft argues that while metaphor requires an association across two wholly distinct sets of domain matrices, as we have seen, metonymy highlights a particular aspect of a single domain matrix.

9.6.2 Metonymy-producing relationships

The idea that metonymy provides access to (or highlights a particular aspect of) a domain matrix leads to two closely related questions. Firstly, what common patterns of access are there? Secondly, what are good vehicles for access? We address the first of these questions in this section, and the second of these questions in section 9.6.3. Our discussion is based on the study by Kövecses and Radden (1998).

In their paper, Kövecses and Radden examine the kinds of relationships that give rise to the metonymies that occur frequently in language. They observe that there appear to be two main kinds of motivating relationships: (1) those relating to the part-whole organisation of a given domain (or domain matrix) so that parts (or substructures) of a domain represent the entire domain; (2) those involving parts of a domain that stand for other parts. These are illustrated below with just a few examples taken from the extensive taxonomy provided by Kövecses and Radden.

Part-whole, whole-part relationships

(42) WHOLE THING FOR PART OF A THING
America for 'United States'

(43) PART OF A THING FOR THE WHOLE THING
England for 'United Kingdom' [Kövecses and Radden 1998: 50]

(44) A CATEGORY FOR A MEMBER OF THE CATEGORY
The pill for 'birth control pill'

(45) A MEMBER OF A CATEGORY FOR THE CATEGORY
Aspirin for 'any pain-relieving tablet' [Kövecses and Radden 1998: 53]

These examples illustrate that the part-whole structure of a domain provides a 'route' of access via metonymy. A whole entity can be accessed by a part, or a part can be accessed by the entire domain.

Domain part-part relationships

This type of metonymic relationship is illustrated here as it relates to the domain of ACTION which involves INSTRUMENTS, an AGENT, a PATIENT, an end RESULT and so on. These 'parts' or substructures within the domain of ACTION can be metonymically related, as the following examples from Kövecses and Radden (1998: 54–5) illustrate:

(46) INSTRUMENT FOR ACTION
to ski, to shampoo one's hair

(47) AGENT FOR ACTION
to butcher the cow, to author a book

(48) ACTION FOR AGENT
snitch (slang: 'to inform' and 'informer')

(49) OBJECT INVOLVED IN THE ACTION FOR THE ACTION
to blanket the bed

(50) ACTION FOR OBJECT INVOLVED IN THE ACTION
Give me one bite

(51) RESULT FOR ACTION
a screw-up (slang: 'to blunder' and 'blunder')

(52) ACTION FOR RESULT
a deep cut

(53) MEANS FOR ACTION
He sneezed the tissue off the table.

(54) MANNER OF ACTION FOR THE ACTION
She tiptoed to her bed.

(55) TIME PERIOD OF ACTION FOR THE ACTION
to summer in Paris

(56) DESTINATION FOR MOTION
to porch the newspaper

(57) TIME OF MOTION FOR AN ENTITY INVOLVED IN THE MOTION
the 8.40 just arrived

These examples from the domain of ACTION illustrate that a part of the domain can metonymically provide access to another part. Thus, together with the examples relating to part-whole structure of domains, these two sets of examples illustrate the ways in which metonymy provides access within a domain (or domain matrix).

9.6.3 Vehicles for metonymy

Kövecses and Radden (1998) propose a number of cognitive and communicative principles in order to account for the selection of a vehicle for metonymic relationships. In this section, we briefly present two of the cognitive principles:

(1) HUMAN OVER NON-HUMAN; and (2) CONCRETE OVER ABSTRACT. A central aspect of their explanation is that our anthropocentric perspective entails our tendency to privilege human and other humanly relevant entities and attributes for metonymic vehicles. The HUMAN OVER NON-HUMAN principle holds that human vehicles are preferred over non-human vehicles. Examples of metonymy that illustrate this principle include the following:

(58) CONTROLLER FOR CONTROLLED
Schwarzkopf defeated Iraq.

(59) PRODUCER FOR PRODUCT
He's reading Shakespeare.

The CONCRETE OVER ABSTRACT principle holds that concrete vehicles are preferred over abstract vehicles. This principle is illustrated by the following metonymic relationships:

(60) BODILY OVER ACTIONAL
hold your tongue (for 'stop speaking')

(61) BODILY FOR EMOTIONAL
heart (for 'kindness'), e.g. *He's heartless*

(62) BODILY OVER PERCEPTUAL
ear (for 'hearing'), e.g. *lend me your ear*

(63) VISIBLE OVER INVISIBLE
to save one's skin (for 'to save one's life')

The purpose of these principles is to provide generalisations that account for the vehicles that provide a basis for metonymy in language. Although we do not elaborate further, Table 9.4 summarises the principles proposed by Kövecses and Radden.

9.7 Metaphor-metonymy interaction

We have seen that metaphor and metonymy are viewed by cognitive linguists as conceptual processes that contribute to providing structure to the human conceptual system. According to this view, metaphor and metonymy as they appear in language are reflections of the organisation of the underlying conceptual system. Given that metaphor and metonymy are both conceptual phenomena, and given that they may in principle both relate to the same conceptual domains, questions arise concerning the interaction of metaphor and metonymy within the conceptual system. We therefore conclude this

Table 9.4 Constraints on possible vehicles in metonymy (Kövecses and Radden 1998)

Cognitive principles

Human experience
HUMAN OVER NON-HUMAN
CONCRETE OVER ABSTRACT
INTERACTIONAL OVER NON-INTERACTIONAL
FUNCTIONAL OVER NON-FUNCTIONAL

Perceptual selectivity
IMMEDIATE OVER NON-IMMEDIATE
OCCURRENT OVER NON-OCCURRENT
MORE OVER LESS
DOMINANT OVER LESS DOMINANT
GOOD GESTALT OVER POOR GESTALT
BOUNDED OVER UNBOUNDED
SPECIFIC OVER GENERIC

Cultural preferences
STEREOTYPICAL OVER NON-STEREOTYPICAL
IDEAL OVER NON-IDEAL
TYPICAL OVER NON-TYPICAL
CENTRAL OVER PERIPHERAL
BASIC OVER NON-BASIC
IMPORTANT OVER LESS IMPORTANT
COMMON OVER LESS COMMON
RARE OVER LESS RARE

Communicative principles
CLEAR OVER LESS CLEAR
RELEVANT OVER IRRELEVANT

chapter with a brief discussion of the ways in which metaphor and metonymy interact.

Metaphtonymy

In an important article, Goossens (1990) presented an analysis of the way in which metaphor and metonymy interact. He calls this phenomenon **metaphtonymy**. Goossens identified a number of logically possible ways in which metaphor and metonymy could potentially interact; however, he found that only two of these logically possible interactions were commonly attested.

The first way in which metaphor and metonymy interact is called **metaphor from metonymy**. In this form of interaction, a metaphor is grounded in a metonymic relationship. For example, the expression *close-lipped* can mean 'silent', which follows from metonymy: when one has one's lips closed, one is (usually) silent, therefore to describe someone as *close-lipped* can stand metonymically for silence. However, *close-lipped* can also mean 'speaking but giving little away'. This interpretation is metaphoric, because we understand the absence of meaningful information in terms of silence. Goossens argues that the metaphoric interpretation has a metonymic basis in that it is only because being closed-lipped can stand for silence that the metaphoric reading is possible: thus metaphor from metonymy.

The second common form of interaction is called **metonymy within metaphor**. Consider the following example adapted from Goossens (1990):

(64) She caught the Prime Minister's ear and persuaded him to accept her plan

This example is licensed by the metaphor ATTENTION IS A MOVING PHYSICAL ENTITY, according to which ATTENTION is understood as a MOVING ENTITY that has to be 'caught' (the minister's ear). However, within this metaphor there is also the metonymy EAR FOR ATTENTION, in which EAR is the body part that functions as the vehicle for the concept of ATTENTION in the metaphor. In this example, the metonym is 'inside' the metaphor.

The metonymic basis of metaphor

According to some cognitive semanticists (e.g. Barcelona 2003c; Taylor 2003), metonymy is an operation that may be more fundamental to the human conceptual system than metaphor. Barcelona (2003c: 31) goes so far as to suggest that 'every metaphorical mapping presupposes a prior metonymic mapping.' One obvious way in which metaphor might have a metonymic basis relates to the idea of experiential correlation that we discussed earlier. As we saw, primary metaphors are argued to be motivated by experiential correlation. Yet, as Radden (2003b) and Taylor (2003) have pointed out, correlation is fundamentally metonymic in nature. For example, when height correlates with quantity, as when fluid is poured into a glass, greater height literally corresponds to an increase in quantity. When this correlation is applied to more abstract domains, such as HIGH PRICES, we have a metaphor from metonymy, in the sense of Goossens. Indeed, as Barcelona argues, given the claim that primary metaphors underpin more complex compound metaphors and the claim that primary metaphors have a metonymic basis, it follows that all metaphor is ultimately motivated by metonymy.

However, although Taylor (1995: 139) has observed that 'It is tempting to see all metaphorical associations as being grounded in metonymy', he observes some counter-examples to this thesis. These include so-called

synaesthetic metaphors, in which one sensory domain is understood in terms of another, as in *loud colour*. Examples like these are problematic for the thesis that all metaphor is grounded in metonymy because there does not appear to be a tight correlation in experience between LOUDNESS and COLOUR that motivates the metaphor. Barcelona (2003c) argues that even metaphors like these can be shown to have a metonymic basis. He suggests that the metaphor that licenses expressions like *loud colour* relate not to the entire domain of SOUND as the source domain, but to a SUBDOMAIN which he calls DEVIANT SOUNDS. In this respect, Barcelona's treatment of metonymy is consonant with Croft's. According to Barcelona, these sounds are deviant because they deviate from a norm and thus attract involuntary attention. This provides the metonymic basis of the metaphor: there is a tight correlation in experience between deviant (or loud) sounds and the attraction of attention, so that a deviant sound can metonymycally represent attraction of involuntary attention. For this reason, the subdomain of deviant sounds can be metaphorically employed to understand deviant colours which also attract involuntary attention.

9.8 Summary

In this chapter we discussed two kinds of conceptual projection, **conceptual metaphor** and **conceptual metonymy**, both introduced by Lakoff and Johnson (1980) in their development of **Conceptual Metaphor Theory**. As we have seen, cognitive linguists view metaphor and metonymy as more than superficial linguistic 'devices'. According to the cognitive view, both these operations are conceptual in nature. While metaphor **maps** structure from one domain onto another, metonymy is a mapping operation that **highlights** one entity by referring to another entity within the same domain (or domain matrix). In earlier versions of Conceptual Metaphor Theory, metaphor was thought to be motivated by the need to provide relatively abstract target domains with structure derived from more concrete source domains. More recently, the theory of **primary metaphor** has challenged this view, arguing that a foundational subset of conventional metaphors – primary metaphors – serve to link equally basic concepts at the cognitive level. According to this theory, primary target concepts are no less experiential than primary source concepts, since both primary target concepts and primary source concepts are directly experienced. However, primary target concepts are less consciously accessible than primary source concepts because they relate to background cognitive operations and processes. Due to correlations in experience, primary source concepts come to be associated pre-linguistically with primary target concepts in predictable ways. The cognitive function of metaphor, according to this theory, is to **foreground** otherwise background operations. Moreover, primary metaphors can be unified in order to provide more complex conceptual mappings called **compound metaphors**. In contrast to metaphor, metonymy appears to be the result of contextually motivated patterns of **activation** that map **vehicle** and **target** within a single source domain. Within a specific discourse context, a salient vehicle activates and thus highlights a particular **target**. Hence, while **correlation-based** (as opposed to **resemblance-based**) metaphors are pre-conceptual in origin and are thus in some sense inevitable associations (motivated by the nature of our bodies and our environment), conceptual metonymies are motivated by communicative and referential requirements and the 'routes' of access that they provide to a particular target within a single domain.

Further reading

As noted in the text, Conceptual Metaphor Theory was one of the earliest coherent frameworks to have emerged in Cognitive Semantics. Consequently, there is a vast literature devoted to this topic, as reflected in the nature and breadth of the sources listed here.

Introductory textbook

- **Kövecses (2001)**. A useful introductory overview of Conceptual Metaphor Theory by one of its leading proponents.

Key texts in the development of Conceptual Metaphor Theory

- **Gibbs (1994)**
- **Gibbs and Steen (1999)**
- **Lakoff (1990)**
- **Lakoff (1993)**
- **Lakoff and Johnson (1980)**
- **Lakoff and Johnson (1999)**

The foundational text is the extremely accessible 1980 book by Lakoff and Johnson. An updated and more extended version is presented in their 1999 book. The 1994 book by Gibbs provides an excellent review of the relevant literature relating to experimental evidence for Conceptual Metaphor Theory. The 1999 Gibbs and Steen book provides a collection of articles representing contemporary metaphor research. There is also a 1994 list of metaphors, 'The Master Metaphor List', compiled by Lakoff and his students, available on the Internet: http://cogsci.berkeley.edu/Metaphor Home.html.

Applications of metaphor theory

- Chilton and Lakoff (1995)
- Cienki (1999)
- Johnson (1994)
- Kövecses (2000)
- Lakoff (1991)
- Lakoff (2002)
- Lakoff and Núñez (2000)
- Nerlich, Johnson and Clarke (2003)
- Sweetser (1990)

This (non-exhaustive) list provides a flavour of the range and diversity of applications to which Conceptual Metaphor Theory has been put. Lakoff has applied metaphor theory to politics (1991, 2002), as have Chilton and Lakoff in their 1995 paper. Metaphor theory has also been applied to gesture (Cienki), semantic change (Sweetser), morality (Johnson), mathematics (Lakoff and Núñez) and media discourse (Nerlich *et al.*).

Conceptual metaphor and literature

- Freeman (2003)
- Lakoff and Turner (1989)
- Turner (1991)
- Turner (1996)

Mark Turner has been a leading pioneer both in the development of Conceptual Metaphor Theory and in its application to literature, giving rise to the related areas of Cognitive Poetics and cognitive stylistics. His 1996 book is an accessible presentation of the cognitive basis of literature, and the 1989 Lakoff and Turner book develops a theory of and methodology for the investigation of poetic metaphor. Cognitive Poetics, which has its roots in Conceptual Metaphor Theory, is introduced in two companion volumes: Stockwell (2002) and Gavins and Steen (2003). An excellent overview is presented in the volume edited by Semino and Culpeper (2003), which provides a collection of articles by leading literary scholars who apply insights from cognitive linguistics in general, including Conceptual Metaphor Theory, to literary and stylistic analysis.

Primary metaphor theory

- Grady (1997a)
- Grady (1997b)

- Grady (1998)
- Grady (1999)
- Grady and Johnson (2000)
- Grady, Taub and Morgan (1996)

A good place to begin is Grady's (1997b) paper. A more detailed treatment is offered in his (1997a) doctoral thesis.

Other views on metaphor

This section lists some sources that address many of the concerns associated with 'classic' Conceptual Metaphor Theory but either take issue with aspects of the approach and/or present competing accounts.

- **Evans (2004a).** This study investigates how we experience and conceptualise time. Evans argues that TIME represents a more complex conceptual system than is typically assumed by conceptual metaphor theorists, particularly within Primary Metaphor Theory.
- **Haser (2005).** In this important and compelling book-length review of Lakoff and Johnson's work, Haser provides a close reading and examination of the philosophical underpinnings of Conceptual Metaphor Theory. She concludes that much of the philosophical basis is extremely shaky and the theory itself is, in certain key respects, not convincing.
- **Leezenberg (2001).** In this book-length treatment, Leezenberg emphasises the context-dependent nature of metaphoric interpretations, a point which plays little part in the Lakoff and Johnson account.
- **Murphy (1996).** Presents an influential critique of early metaphor theory, including problems with the Invariance Principle.
- **Ortony (1993).** This volume, which includes an essay by George Lakoff, presents an excellent overview of the diverse traditions and approaches that have investigated metaphor.
- **Stern (2000).** Presents a critique of Conceptual Metaphor Theory that focuses on its lack of attention to the context-sensitive nature of metaphor.
- **Zinken, Hellsten and Nerlich (forthcoming).** This paper argues that Conceptual Metaphor Theory has traditionally paid little attention to the situatedness of metaphor. In introducing the notion of **discourse metaphor**, the authors argue that culture-specific discourse-based metaphors may not derive from 'more basic' experientially-grounded primary metaphors but may co-evolve with the cultures in which they are used.

Conceptual metonymy

- **Kövecses and Radden (1998).** One of the first serious attempts to provide a detailed and carefully articulated theory of metonymy within cognitive semantics.
- **Panther and Thornburg (2003).** This edited volume brings together a number of important papers on the relationship between metonymy and inferencing, including articles by Panther and Thornburg, Coulson and Oakley, and Barcelona.
- **Radden and Panther (1999).** This book is an edited volume that brings together leading scholars in the field of conceptual metonymy.

Comparing metaphor and metonymy

- **Barcelona (2000); Dirven and Pörings (2002).** Both these volumes compare and contrast conceptual metaphor and conceptual metonymy. The Dirven and Pörings volume reproduces influential articles on the topic of metaphor and metonymy; see in particular the articles by Croft, and by Grady and Johnson. The Barcelona volume includes an excellent introduction by Barcelona, together with his own article in the volume which claims that all metaphors have a metonymic basis.

Exercises

9.1 Conceptual Metaphor Theory

Summarise the key claims of Conceptual Metaphor Theory.

9.2 Identifying mappings

The following sentences are motivated by the metaphor TIME IS (MOTION ALONG) A PATH, which relates to the moving ego model that we introduced in Chapter 3. Following the model provided in Table 9.1, identify the set of mappings underlying these examples.

(a) We're approaching Christmas.
(b) Graduation is still a long way away.
(c) Easter is ahead of us.
(d) We've left the summer behind us.
(e) When he was a boy he used to play football over the summer vacation. Now he has to work.

9.3 Identifying metaphors

Identify the metaphors that underlie these examples. Identify possible source and target domains, and state the metaphor in the form 'A is B'.

(a) That marriage is on the rocks.
(b) This once great country has become weaker over the years.
(c) In defending her point of view she took no prisoners.
(d) Those two are still quite close.
(e) We've got a big day ahead of us tomorrow.
(f) A different species is going extinct everyday.

9.4 Primary vs. compound metaphors

For the metaphors you identified in exercise 9.3, determine whether these are likely to be examples of primary or compound metaphor. In view of the discussion in section 9.4, explain your reasoning for each example.

9.5. Correlation vs. resemblance-based metaphors

Consider the following examples. Explain how the metaphors that underlie them illustrate the distinction between metaphors motivated by **correlation** versus metaphors motivated by **perceived resemblance**:

(a) My boss is a real pussycat.
(b) So far, things are going smoothly for the Liberal Democrats in the election campaign.

9.6 Metaphor vs. metonymy

Describe the main differences between conceptual metaphor and conceptual metonymy, and explain how the function of each type of conceptual projection differs.

9.7 Identifying metonymies

Identify the conceptual metonymies that underlie each of the following examples. For each example, identify the vehicle and the target, and explain how you reached your conclusions.

(a) George Bush arrested Saddam Hussein.
(b) The White House is refusing to talk to the Elysée Palace these days while the Kremlin is talking to everyone.

(c) Watergate continues to have a lasting impact on American politics.
(d) She loves Picasso.
(e) The restaurant refused to serve the couple as they weren't properly dressed.
(f) She xeroxed the page.
(g) Jane has a long face.
(h) She's not just a pretty face.
(i) All hands on deck!

9.8. Textual analysis

Select an excerpt from a newspaper or magazine article. Analyse the excerpt with respect to conceptual metaphor and metonymy. Identify the source/vehicle and target in each case, and explain your reasoning. Below are some examples of the sorts of texts you might consider selecting:

(a) an article from a woman's interest magazine relating to make-up and beauty products;
(b) an example from a men's magazine dealing with health and/or fitness;
(c) an article from a newspaper relating to sports coverage, such as rivalry between football teams or their managers;
(d) an article from a newspaper's 'opinion/comment' page(s), dealing with a current political controversy;
(e) an excerpt from an agony-aunt column dealing with relationships;
(f) a pop-song lyric dealing with love;
(g) slogans or text from advertisements that appear in newspapers or magazines.

10

Word meaning and radial categories

In this chapter we build on insights developed in the previous two chapters in order to develop the approach taken to word meaning in cognitive semantics. This approach is known as **cognitive lexical semantics**. Pioneered by Claudia Brugman and George Lakoff, cognitive lexical semantics built upon Lakoff's work on categorisation and idealised cognitive models which we presented in Chapter 8. This approach to word meaning also incorporated ideas from Conceptual Metaphor Theory which we explored in Chapter 9. Cognitive lexical semantics takes the position that **lexical items** (words) are **conceptual categories**: a word represents a category of distinct yet related meanings that exhibit typicality effects. Thus, Lakoff argued, words are categories that can be modelled and investigated using the theory of idealised cognitive (ICMs) that we presented in Chapter 8. In particular, Lakoff argued that lexical items represent the type of complex categories he calls **radial categories**: recall that a radial category is structured with respect to a composite prototype, and the various category members are related to the prototype by convention rather than being 'generated' by predictable rules. As such, word meanings are stored in the mental lexicon as highly complex structured categories of meanings or **senses**. This chapter presents an overview of Lakoff's approach to the lexical item as a category of senses by illustrating how he modelled the lexical item *over* in the second of his famous 'case studies' from *Women, Fire and Dangerous Things* (sections 10.1–10.3). Lakoff's approach has been highly influential and has given rise to a significant body of subsequent work, some of which has been critical of certain aspects of his approach. In particular, Lakoff's model has been criticised for taking an excessively fine-grained approach to word meaning which results in a very large number of distinct senses conventionally associated with individual lexical items (section 10.4). Hence we will consider a more recent development by Vyvyan Evans and Andrea Tyler and their theory of **Principled Polysemy** which provides a methodology for constraining the number of distinct senses associated with an individual word (section 10.5). Finally, having developed a detailed account of approaches to word meaning within cognitive semantics, we revisit the role of context in word meaning (section 10.6).

10.1 Polysemy as a conceptual phenomenon

We begin by comparing and contrasting **polysemy** with **homonymy**. While both polysemy and homonymy give rise to **lexical ambiguity** (two or more meanings associated with a word), the nature of the ambiguity is different in each case. Polysemy is the phenomenon whereby a lexical item is commonly associated with two or more meanings that appear to be related in some way. Consider the following examples containing the English preposition *over*.

(1) a. The picture is over the sofa.   ABOVE
    b. The ball landed over the wall. ON THE OTHER SIDE
    c. The car drove over the bridge. ACROSS

Each of these instances of *over* is associated with a slightly different meaning or sense (listed on the right), but these senses are nevertheless relatively closely related. This shows that *over* exhibits polysemy.

Polysemy contrasts with homonymy, which relates to two distinct words that happen to share the same form in sound (homophones) and/or in writing (homographs). For example, the form *bank* relates to two different words with unrelated meanings, 'financial institution' and 'bank of a river'. These two senses are not only synchronically unrelated (unrelated in current usage) but also historically unrelated. The word *bank* meaning 'side of river' has been in the English language for much longer, and is related to the Old Icelandic word for 'hill', while the word *bank* meaning 'financial institution' was borrowed from Italian *banca*, meaning 'money changer's table' (*Collins English Dictionary*).

While formal linguists have long recognised the existence of polysemy, it has generally been viewed as a surface phenomenon, in the sense that **lexical entries** are underspecified (abstract and lacking in detail) and are 'filled in' either by context (Ruhl 1989) or by the application of certain kinds of lexical generative devices (Pustejovsky 1995). According to this view, polysemy is epiphenomenal, emerging from **monosemy**: a single relatively abstract meaning from which other senses (like the range of meanings associated with *over*) are derived on the basis of context, speaker intention, recognition of that intention by the hearer, and so on. A monosemy account is plausible in principle when accounting for senses like those in example (1), which are all spatial in nature and could therefore be accounted for in terms of a single abstract spatial sense. However, *over* also exhibits non-spatial senses. Consider example (2).

(2) Jane has a strange power over him   CONTROL

While the meaning of *over* in (2) might be characterised as a 'control' sense, it is difficult to see how a single abstract meaning could derive the three spatial senses in (1) as well as this non-spatial 'control' sense. After all, the sentence in (2) does not describe a spatial scene (*Jane* is not located above *him* in space), but has an abstract sense relating to a power relationship between two people.

One way of analysing the meaning of *over* in (2) would be to treat it as a distinct sense of *over* from the spatial senses in (1). This would amount to the claim that *over* in (2) is a homonym: a distinct word. A second possible analysis, which preserves the monosemy position, might claim that a single abstract underlying sense licenses both the spatial and non-spatial senses, but that while the spatial senses are literal, the non-spatial sense is metaphorical and is interpreted by applying pragmatic principles to retrieve the speaker's intended meaning. As we develop the cognitive semantic position on polysemy, we will see why these lines of analysis are both rejected in favour of a radial category model of polysemy.

In their work on cognitive lexical semantics Claudia Brugman (1981; Brugman and Lakoff 1988) and George Lakoff (1987) claimed that *over* is stored as a category of distinct polysemous senses rather than a single abstract monosemous sense. It follows from this position that polysemy reflects conceptual organisation and exists at the level of mental representation rather than being a purely surface phenomenon. In this respect, cognitive lexical semantics approaches diverged both from traditional and from more recent formal approaches to word meaning, in particular in developing the position that polysemy is a fundamentally conceptual phenomenon and that lexical organisation at the mental level determines polysemy as it is manifested in language use. Thus, in the same way that units of language are conceived as being stored in an inventory-like grammar (as we will see in Part III), the meanings associated with each linguistic unit are conceived as being stored as distinct, yet related, semantic entities, which we have referred to in previous chapters as **lexical concepts**. In addition, the cognitive approach, which posits highly detailed and fine-grained lexical structure, is at odds with the monosemy position, which posits highly abstract word meanings. Indeed, the monosemy view is widely held in formal lexical semantics and is adopted in order to ensure that lexical representation exhibits economy, an important concern for formal lexical semanticists.

The position originally proposed by Claudia Brugman was that polysemy as a conceptual phenomenon should form the basis of a theory of word meaning.

This idea was developed within the theory of ICMs and radial categories developed by Lakoff, and integrated with the theory of conceptual metaphor developed by Lakoff and Johnson. Having explored these approaches in the previous two chapters, we are now in a position to approach word meaning from the perspective of cognitive semantics.

10.2 Words as radial categories

In this section, we present Lakoff's account of the semantics of *over*, which has been highly influential in the development of cognitive lexical semantics. Lakoff's account was based on ideas proposed in a master's thesis by Claudia Brugman, his former student. As we have already indicated, the idea underpinning Lakoff's approach was that a lexical item like *over* constitutes a conceptual category of distinct but related (polysemous) senses. Furthermore, these senses, as part of single category, can be judged as more prototypical (central) or less prototypical (peripheral). This means that word senses exhibit typicality effects, like the cognitive categories that we saw in Chapter 8. For instance the ABOVE sense of *over* in example (1a) would be judged by most native speakers of English as a 'better' example of *over* than the CONTROL sense in example (2). While the prototypical ABOVE sense of *over* relates to a spatial configuration, the CONTROL sense does not. The intuition that the spatial meanings are somehow prototypical led Brugman and Lakoff (1988) and Lakoff (1987) to argue that the CONTROL sense of *over* is derived metaphorically from the more prototypical spatial meaning of *over*. However, this approach departs in important ways from the monosemy account that we sketched above, as we will see.

Lakoff (1987) proposed that words represent radial categories. As we saw in Chapter 8, a radial category is a conceptual category in which the range of concepts are organised relative to a central or prototypical concept. The radial category representing lexical concepts has the same structure, with the range of lexical concepts (or **senses**) organised with respect to a prototypical lexical concept or sense. This means that lexical conceptual categories have structure: more prototypical senses are 'closer' to the central prototype, while less prototypical senses are 'further from' the prototype (peripheral senses). In cognitive semantics, radial categories are modelled in terms of a radiating lattice configuration, as shown in Figure 10.1. In this diagram, each distinct sense is represented by a node (indicated by a black circle). While all senses are related by virtue of belonging to the same conceptual category, arrows between nodes indicate a close relationship between senses.

Central to this approach is the assumption that radial categories of senses are represented or **instantiated** in long-term **semantic memory**. (In cognitive semantics, the term 'semantic memory' is used interchangeably with the more traditional term '(mental) **lexicon**'.) According to this view, the reason we are

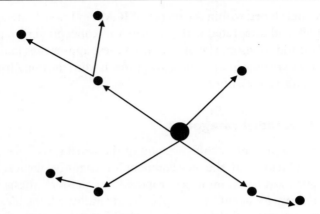

Figure 10.1 A radiating lattice diagram ('semantic network') for modelling radial categories

able to use *over* with a CONTROL meaning is because this sense of *over* is instantiated in long-term memory. This means that the range of senses associated with *over* are **conventionalised** (Chapter 4). In other words, most native speakers of English simply 'know' the range of senses associated with *over*. From this perspective, a radial category is not a device for *generating* distinct meanings from the central or prototypical sense. Instead, it is a model of how distinct but related meanings are stored in semantic memory. In this important respect, the cognitive account of word meaning departs from the monosemy account, which holds that a single abstract sense is stored which is 'filled in' by context on each occasion of use.

An important concern for cognitive semanticists has been to explain how polysemy arises. Because cognitive semanticists assume that linguistic categories are no different, in principle, from other kinds of conceptual categories, it follows that linguistic categories are structured by the same general cognitive mechanisms that structure non-linguistic conceptual categories. According to this view, less prototypical senses are derived from more prototypical senses by cognitive mechanisms that facilitate **meaning extension**, including conceptual metaphor and image schema transformations (Chapter 6). These mechanisms result in the systematic extension of lexical categories resulting in **meaning chains**. This gives rise to polysemy: a **semantic network** for a single lexical item that consists of multiple related senses. It follows that the radial category in Figure 10.1 also represents a semantic network. A semantic network might consist of a number of distinct senses that are peripheral and hence not strictly predictable with respect to the prototype, but which are nevertheless motivated by the application of general cognitive mechanisms. In addition, this model predicts the emergence of senses that are intermediate with respect to the prototype and the peripheral senses. The process that connects these central and peripheral senses is called **chaining**. In the next section, we explore in more detail how this process works. Table 10.1 summarises the main assumptions that characterise the cognitive approach to lexical semantics.

10.3 The full-specification approach

Lakoff's analysis of the English preposition *over* is sometimes described as the **full-specification approach** to lexical semantics. Central to Lakoff's account is the view that the senses associated with prepositions like *over*, which are grounded in spatial experience, are structured in terms of image schemas. As we noted above, the spatial senses of *over* are judged to be more prototypical by native speakers than non-spatial meanings, as illustrated by the fact that spatial senses are listed as primary senses by lexicographers. Lakoff argued that the prototypical sense of *over* is an image schema combining elements of both ABOVE and ACROSS. The distinct senses associated with *over* are structured with respect to this image schema which provides the category with its prototype structure. Recall from Chapter 6 that image schemas are relatively abstract schematic representations derived from embodied experience. The central image schema for *over*, proposed by Lakoff, is shown in Figure 10.2.

Table 10.1 Assumptions of cognitive lexical semantics

Words and their senses represent conceptual categories, which have much in common with non-linguistic conceptual categories. It follows that linguistic categories have prototype structure.
Word meanings are typically polysemous, being structured with respect to a central prototype (or prototypes). Lexical categories therefore form radial categories which can be modelled as a radiating lattice structure.
Radial categories, particularly meaning extensions from the prototype, are motivated by general cognitive mechanisms including metaphor and image schema transformation.
The senses that constitute radial categories are stored rather than generated.

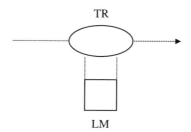

Figure 10.2 The central schema for *over* (adapted from Lakoff 1987: 419)

Lakoff argues that the schema depicted in Figure 10.2 underlies examples like (3):

(3) The plane flew over.

As we have seen, the abbreviations **TR** and **LM** are derived from Langacker's theory of Cognitive Grammar (e.g. 1987), which is discussed in detail in Part III of the book. TR stands for **trajector** and relates to the entity in the scene that is smaller and that is typically capable of motion. LM stands for **landmark** and relates to the entity with respect to which the TR moves. TR and LM are therefore Langacker's terms for figure and ground (or reference object), respectively, which we introduced in Chapter 3. In the central schema for *over* the LM is unspecified. The oval represents the TR and the arrow represents its direction of motion. The TR and its path of motion are located above the LM. According to Lakoff, this central image schema is highly schematic, lacking detail not only about the nature of the LM but also about whether there is contact between the TR and the LM.

Lakoff proposes a number of further more detailed image schemas related to this central schema. These are developed by the addition of further information that specifies properties of the landmark and the existence and nature of any contact between the TR and LM. For example, landmarks can be 'horizontally extended', which means that they can extend across the horizontal plane of the LM. This is illustrated in example (4), where the bird's flight (TR) extends across the yard (LM).

(4) The bird flew over the yard.

Lakoff annotates this property with the symbol X (horizontally eXtended). For contexts in which there is no contact between the TR and LM, which is also illustrated by example (4), Lakoff uses the abbreviation NC (No Contact). According to Lakoff, examples like (4) therefore relate to a distinct sense of *over* arising from a distinct image schema. This image schema is represented in Figure 10.3. Like the central image schema (labelled schema 1) in Figure 10.2, the moving entity is designated by TR, but this schema contains an overt

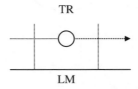

Figure 10.3 The bird flew over the yard (Schema 1.X.NC) (adapted from Lakoff 1987: 421)

horizontal landmark, represented by LM. This LM corresponds to *the yard* in example (4).

Some landmarks are simultaneously vertically and horizontally extended, like *hill* in example (5).

(5) The plane flew over the hill.

Lakoff annotates landmarks that are vertically extended with V. Therefore, a landmark that is both vertically and horizontally extended is represented by VX. According to Lakoff, the schema in Figure 10.4, which corresponds to example (5), represents a distinct sense for *over*, which counts as an instance of the central schema with the additional features VX.NC.

While the previous two schemas involve landmarks that are horizontally extended, example (6) designates a landmark (*the wall*) that is vertically extended but not horizontally extended. Lakoff's image schema for examples of this kind is depicted in Figure 10.5.

(6) The bird flew over the wall.

In addition to the variants of schema 1 represented in Figures 10.3, 10.4 and 10.5, none of which involve contact between the TR and LM, Lakoff also proposes instances of this schema that involve contact between the TR and LM. These are annotated as 'C' rather than 'NC'. These are illustrated in Figures 10.6, 10.7 and 10.8.

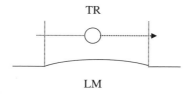

Figure 10.4 The plane flew over the hill (Schema 1.VX.NC) (adapted from Lakoff 1987: 421)

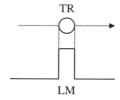

Figure 10.5 The bird flew over the wall (Schema 1.V.NC) (adapted from Lakoff 1987: 421)

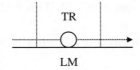

Figure 10.6 John walked over the bridge (Schema 1.X.C) (adapted from Lakoff 1987: 422)

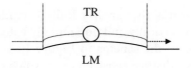

Figure 10.7 John walked over the hill (Schema 1.VX.C) (adapted from Lakoff 1987: 422)

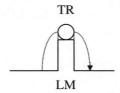

Figure 10.8 Sam climbed over the wall (Schema 1.V.C) (adapted from Lakoff 1987: 422)

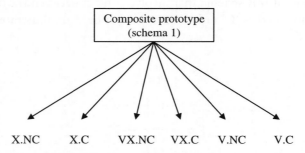

Figure 10.9 Instances of schema 1, the central image schema (adapted from Lakoff 1987: 423)

In sum, Lakoff claims that each of the schemas considered so far represent distinct senses associated with *over*. According to this model of word meaning, the central schema for *over* in Figure 10.2 has at least six distinct and closely related variants (see Figure 10.9), each of which is stored in semantic memory.

It should now be clear why Lakoff's approach is described as the 'full-specification approach'. Given the range of senses *over* is associated with in addition to the ABOVE-ACROSS sense (summarised in Table 10.2), this model results in a potentially vast proliferation of senses for each lexical item. As we will see in section 10.4, some cognitive semanticists argue that the level of detail

Table 10.2 Schemas proposed by Lakoff (1987) for *over* in addition to the central schema

Schema type	Basic meaning	Example
ABOVE schema	The TR is located above the LM	*The helicopter is hovering over the hill*
COVERING schema	The TR is covering the LM	*The board is over the hole*
REFLEXIVE schema	The TR is reflexive: the TR is simultaneously the TR and the LM. The final location of the TR is understood with respect to its starting position	*The fence fell over*
EXCESS schema	When *over* is employed as a prefix it can indicate 'excess' of TR relative to LM	*The bath overflowed*
REPETITION schema	*Over* is used as an adverb to indicate a process that is repeated	*After receiving a poor grade, the student started the assignment over (again)*

or **granularity** that characterises the full specification approach is problematic for a number of reasons.

10.3.1 Image schema transformations

As we have seen, some of the distinct senses posited by Lakoff are reflections of individual schemas, which are stored image-schematic representations that specify the central schema in more detail. However, Lakoff argues that distinct senses can also be derived by virtue of **image schema transformations**. In Chapter 6, we saw that image schemas are dynamic representations that emerge from embodied experience, and that one image schema can be transformed into another (for example, when we understand the relationship between a SOURCE and a GOAL in terms of a PATH, and vice versa). One consequence of a shift in focus from PATH to GOAL is that we achieve **endpoint focus**: the end of a path takes on particular prominence. In other words, image schema transformations relate to the **construal** of a scene according to a particular **perspective**.

Lakoff has argued that the transformation from a SOURCE schema to an endpoint focus or GOAL schema gives rise to two distinct senses associated with the ABOVE–ACROSS schema (schema 1) that we discussed above. Consider once more the senses depicted in Figures 10.6 and 10.7, illustrated by examples (7) and (8).

(7) John walked over the bridge. [1.X.C: represented in Figure 10.6]

(8) John walked over the hill. [1.VX.C: represented in Figure 10.7]

As a result of image schema transformation, an endpoint focus can be added to these senses. This is illustrated by examples (9) and (10):

(9) St Paul's Cathedral is over the bridge.

(10) John lives over the hill.

By following a mental path, a process that Langacker (1987) refers to as **subjective motion**, attention is focused on the location of St Paul's in example (9) and on where John lives in example (10). In other words, the meaning of *over* in these examples is focused not on the path itself, but on the endpoint of the path. Lakoff argues that sentences like these relate to the image schemas shown in Figures 10.10 and 10.11. Observe that the TR is located at the endpoint of the path.

Lakoff argues that endpoint focus is not supplied by the subject *John*, nor by the verb, nor by the landmark; it follows that this 'additional' meaning is supplied by *over*. Lakoff annotates this aspect of meaning by adding E (endpoint focus) to the representations in (9) and (10), resulting in 1.X.C.E and 1.VX.C.E respectively. As these annotations indicate, Lakoff argues that *over* has two distinct endpoint focus senses, one relating to horizontally extended landmarks, illustrated by sentence (9), and the other relating to vertically extended landmarks, illustrated by sentence (10). In sum, these endpoint focus senses are the result of image schema transformation. Moreover, Lakoff claims that image schema transformations like these result in addition of 'endpoint focus' senses to the semantic network for *over*. In other words, they represent distinct lexical concepts or senses instantiated in semantic memory. According to Lakoff, the fact that senses of this kind exist provides further evidence for the cognitive reality of image schemas and illustrates their important role in meaning extension.

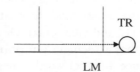

Figure 10.10 St Paul's Cathedral is over the bridge (Schema 1.X.C.E) (adapted from Lakoff 1987: 424)

Figure 10.11 John lives over the hill (Schema 1.VX.C.E) (adapted from Lakoff 1987: 423)

10.3.2 Metaphorical extensions

As we indicated earlier, conceptual metaphor also has a central role in Lakoff's theory of radial categories. Consider the following example (Lakoff 1987: 435).

(11) She has a strange power over me

In this example, *over* is understood metaphorically, which results in a CONTROL sense. In other words, this sentence does not literally mean that the TR (*she*) is literally moving above and across the LM (*me*), nor that the TR is located in a static position above the LM. This CONTROL sense of *over* is peripheral rather than central and is licensed by the metaphor CONTROL IS UP. Because *over* has a conventional ABOVE schema associated with it (see Table 10.2), this metaphor allows the ABOVE schema to be extended metaphorically, providing a new meaning for *over*: the CONTROL sense. Furthermore, Lakoff argues that just as schemas can be extended via metaphor, some schemas are derived via metaphor in the first place. Consider the REPETITION schema, which is illustrated in (12).

(12) The student started the assignment over

According to Lakoff, this schema is derived from the X.C variant of Schema 1 (recall Figure 10.6). However, the REPETITION meaning is derived via two metaphors. Firstly, this sense relies upon the metaphor A PURPOSEFUL ACTIVITY IS A JOURNEY: because purposeful activities like university assignments can be understood as journeys, the X.C instance of the ABOVE-ACROSS schema associated with *over* is licensed. Secondly, the REPETITION sense relies upon the metaphor EVENTS ARE OBJECTS: the LM is metaphorically understood as an earlier performance of the activity, where each performance event is understood as an object. According to this theory, REPETITION is understood in terms of movement ACROSS an earlier performance of the activity, which gives rise to the repetition sense. As with meanings which derive from image schema transformations, meanings derived by metaphor can be instantiated in semantic memory as distinct lexical concepts. Table 10.3 provides a summary of the main claims to emerge from Lakoff's full specification approach.

10.4 Problems with the full-specification approach

While Lakoff's theory of lexical semantics has been hugely influential, there nevertheless remain a number of outstanding problems that have attracted a fair degree of attention in the literature. As we mentioned earlier, Lakoff's full-specification view had been criticised as it entails a potentially vast proliferation of distinct senses for each lexical item. For example, Lakoff's approach

Table 10.3 The main findings of the full-specification approach (Lakoff 1987)

Words represent radial categories: related senses organised with respect to a central sense.
A radial category consists of abstract schemas, which may also consist of more detailed instances.
Radial categories are highly granular in nature, ranging from relatively schematic senses to very detailed senses. The lexicon (semantic memory) fully specifies the majority of the senses associated with a lexical item.
Senses may derive from image schema transformations and/or metaphorical extension.
Because radial categories have prototype structure, they exhibit polysemy; while some senses are closely related, others are more peripheral (e.g. metaphorical extensions).

entails that *over* has, at the very least, several dozen distinct senses. A proliferation of senses is not problematic *per se* because cognitive linguists are not concerned with the issue of economy of representation. However, the absence of clear methodological principles for establishing the distinct senses is problematic. In this section, we focus on two main problems that have been pointed out in relation to this issue.

Polysemy and vagueness: the role of context

The first problem concerns a failure to distinguish between polysemy and **vagueness.** A linguistic expression is vague rather than polysemous if context rather than information stored in semantic memory provides the meaningful detail about the entity in question. Consider the word *thing*. This expression could be used to refer to almost any entity or event, yet it seems unlikely that semantic memory links this expression to all the possible entities that it could refer to. Instead, the meaning of this expression is fully specified by context. A less extreme example is the expression *aunt*, which can refer either to a maternal or a paternal aunt. While our knowledge associated with this expression contains this information, the distinction between these senses is fully dependent upon non-linguistic context. Therefore, while a polysemous expression relates to a range of conventional senses, a vague expression is characterised by a lack of conventional sense distinctions.

Based on proposals by Tuggy (1993), the distinction between polysemy and vagueness is illustrated in Figure 10.12. Polysemy is illustrated in Figure 10.12(a) and vagueness is illustrated in Figure 10.12(b). In the case of polysemy, A represents the central sense and other senses are represented by the boxes marked B and C. All the boxes are marked with bold lines which represent the idea that all three representations have equal degrees of **entrenchment** in memory (Chapter 4). The lines between the boxes indicate that the senses are related. In the case of vagueness, on the other hand, A **licenses** the interpretations designated by B and C: the arrows represent the idea that inter-

WORD MEANING AND RADIAL CATEGORIES

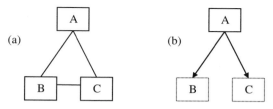

Figure 10.12 The distinction between polysemy and vagueness

pretations B and C are 'computed' from the basic meaning A plus context. The dashed lines represent the idea that meanings B and C are not stored in semantic memory as distinct senses, but emerge from 'on-line' processing.

Given this distinction, it becomes clear that one of the reasons Lakoff's full-specification model results in such a large number of distinct senses is that the model fails to distinguish between polysemy (distinct senses stored in memory) and vagueness (meaning 'filled in' by context). Recall that Lakoff argued for at least six distinct senses associated with the ABOVE-ACROSS schema alone. This number rises to eight if we include the two image schema transformations resulting in endpoint focus. A number of cognitive semanticists have argued that this proliferation of senses results from a failure to take into account the role of context in determining meaning ('filling in' information). From this perspective, Lakoff's full-specification model represents the opposite extreme of the monosemy approach by denying the role of context in meaning altogether. Some cognitive linguists have argued for a position somewhere between these two extremes. For example, Tyler and Evans (2003) argue that the examples in (13) do not represent distinct senses of *over* (one specifying contact and one specifying lack of contact):

(13) a. The bird flew over the wall.
     b. Sam climbed over the wall.

Instead, Tyler and Evans argue that the interpretation of *over* with respect to contact or lack of contact derives from the integration of *over* with the other elements in the sentence. Our knowledge about birds (they can fly) and people (they cannot), provides us with the inference that birds do not come into contact with walls when crossing over them while people do. In other words, the linguistic context together with encyclopaedic knowledge provides the details relating to the presence or absence of contact. According to Tyler and Evans, *over* in (13) is **vague** with respect to contact. Tyler and Evans argue that while Lakoff's position on polysemy as a conceptual phenomenon is correct, it is also important to take into account the crucial role of context in word meaning (recall the discussion in Chapter 7).

The polysemy fallacy: unconstrained methodology

The full-specification approach has also been criticised for a lack of methodological constraints. In other words, Lakoff provides no principled criteria for determining what counts as a distinct sense. This means that the polysemy account presented for *over* (or whatever lexical item we might apply the approach to) results purely from the intuitions (and perhaps also the imagination) of the analyst rather than actually representing the way a particular category is represented in the mind of the language user. This problem has been discussed in some detail by Sandra and Rice (1995) and by Sandra (1998). Sandra argues that to view all context bound usages of a particular lexical item as instances of polysemy is to commit what he calls the **polysemy fallacy**: just because lexical items *can* exhibit polysemy, it does not follow that all or even many distinct senses associated with a lexical item *are* instances of polysemy. Indeed, Sandra has even suggested that the lack of clear methodological principles underpinning Lakoff's semantic network analysis undermines its status as a true linguistic theory. As he puts it, 'what is lacking from the exercise is a set of *scientifically valid* [decision] principles' (Sandra 1998: 371; original emphasis).

10.5 The Principled Polysemy approach

The Principled Polysemy approach proposed by Vyvyan Evans and Andrea Tyler (e.g. Evans 2004a; Tyler and Evans 2003) takes up Sandra's challenge to develop clear decision principles that make semantic network analyses objective and verifiable. These decision principles should achieve two goals: (1) they should serve to determine what counts as a distinct sense and thus distinguish between senses stored in semantic memory (polysemy) and context-dependent meanings constructed 'on-line' (vagueness); (2) they should establish the prototypical or central sense associated with a particular radial category. The second point is important because cognitive semanticists have not always agreed about the central senses of categories. For example, while Lakoff argued that the central sense for *over* is the ABOVE-ACROSS meaning, Kreitzer (1997) has argued more recently that it is an ABOVE meaning. In their (2003) book *The Semantics of English Prepositions*, Tyler and Evans sought to provide decision principles that could be applied to the entire class of English prepositions. In the remainder of this section, we look in detail at how this approach works.

10.5.1 Distinguishing between senses

Tyler and Evans provide two criteria for determining whether a particular sense of a preposition counts as a distinct sense and can therefore be established as a case of polysemy:

1. for a sense to count as distinct, it must involve a meaning that is not purely spatial in nature, and/or a spatial configuration holding between the TR and LM that is distinct from the other senses conventionally associated with that preposition; and
2. there must also be instances of the sense that are context-independent: instances in which the distinct sense could not be inferred from another sense and the context in which it occurs.

To see how these criteria are applied, consider the sentences in (14) and (15):

(14) The hummingbird is hovering over the flower

(15) The helicopter is hovering over the city

In (14), *over* designates a spatial relation in which the TR, coded by *the hummingbird*, is located higher than the LM, coded by *the flower*. In (15), *over* also designates a spatial relationship in which the TR, *the helicopter*, is located higher than the LM. In these examples, neither instance of *over* involves a non-spatial interpretation and both senses encode the same spatial relation. According to Tyler and Evans's first criterion, then, the two instances do not encode distinct senses so the second criterion does not apply. The sense of *over* that is represented in both these examples is what Tyler and Evans call the ABOVE sense. According to Tyler and Evans, this is the central sense, a point to which we return below. Now compare the example in (16) with (14) and (15).

(16) Joan nailed a board over the hole in the ceiling

In (16), the spatial configuration between the TR and LM is not consistent with the ABOVE meaning in (14) and (15): in (16) the board is actually below the hole in the ceiling. In addition, there is a non-spatial aspect to this sense: part of the meaning associated with *over* in (16) relates to COVERING, because the LM (*the hole*) is obscured from view by the TR. This COVERING meaning is not apparent in examples (14) and (15). The presence of this non-spatial aspect in the sense of *over* in (16) meets the first assessment criterion stated by Tyler and Evans, which means we can now consider the second criterion. In doing so, we must establish whether the COVERING meaning is context-independent. Recall that if the meaning is 'computed' on-line, based on the central ABOVE meaning of *over* plus contextual and/or encyclopaedic knowledge, then this sense qualifies as vagueness rather than polysemy. Tyler and Evans argue that the meaning of *over* in (16) cannot be computed on-line, and is therefore context-independent. In other words, the knowledge that *over* in (15) has an ABOVE meaning does not allow us to infer a COVERING meaning from the context supplied by (16).

To elaborate this point, Tyler and Evans provide a different example in which the COVERING meaning is derivable from context. Consider example (17).

(17) The tablecloth is over the table.

In (17), the TR (*the tablecloth*) is above (and in contact with) the LM (*the table*). The interpretation that the table is covered or obscured by the tablecloth can be inferred from the fact that the tablecloth is above the table, together with our encyclopaedic knowledge that tablecloths are larger than tables and the fact that we typically view tables from a vantage point higher than the top of the table. This means that the sense of *over* in (17) can be inferred from the central ABOVE sense together with encyclopaedic knowledge. This type of inference is not possible in (16) because the spatial relation holding between the TR and the LM is one that would normally be coded by the expression *below* (*The board is below the hole in the ceiling*), given our typical vantage point in relation to ceilings. The COVERING meaning of *over* in (16) must therefore be stored as a conventional sense associated with *over*, which means that we can conclude that this is an instance of polysemy.

It is worth observing that Tyler and Evans argue that examples like (17) – which give rise to a 'covering' inference while conventionally encoding the ABOVE meaning of *over* – represent the means by which new senses are added to a lexical category. According to this view, when context-dependent inferences are reanalysed as distinct meanings (a process called **pragmatic strengthening**) a lexical item develops new senses. This perspective is somewhat at odds with Lakoff's view that conceptual metaphor and image schema transformations hold a central place in meaning extension. By arguing that contextual factors can give rise to new senses, Tyler and Evans emphasise the usage-based nature of semantic change, adopting a position that owes much to the **Invited Inferencing Theory** of semantic change (Chapter 21).

10.5.2 Establishing the prototypical sense

Recall that Tyler and Evans argue that the central sense of *over* is the ABOVE sense. In this section, we look at the criteria Tyler and Evans provide for establishing the central sense of a polysemous lexical item. These relate to four types of linguistic evidence (listed below) that Tyler and Evans suggest can be relied upon to provide a more objective means of selecting a central sense. Taken together, these criteria form a substantial body of evidence pointing to one sense from which other senses may have been extended.

1. earliest attested meaning;
2. predominance in the semantic network;

3. relations to other prepositions;
4. ease of predicting sense extensions.

We examine each of these criteria in turn. To begin with, given the very stable nature of spatial expressions within a language (prepositions represent a closed class), one likely candidate for the central sense is the historically earliest sense. Moreover, unlike other word classes, the earliest attested sense for many prepositions is still an active component of the **synchronic semantic network**. For example, *over* is related to the Sanskrit *upan* 'higher' as well as the Old Teutonic comparative form *ufa* 'above', representing in both cases a spatial configuration in which the TR is higher than the LM. This historical evidence points to the ABOVE meaning as the central sense.

The second criterion relates to predominance within a semantic network. This criterion holds that the central sense will be the one most frequently involved in or related to the other distinct senses. For example, by applying the two criteria discussed in the previous section, Tyler and Evans (2003) identified fifteen distinct senses associated with *over*. Of these, eight directly involve the location of the TR ABOVE the LM; four involve a TR located ON THE OTHER SIDE OF the LM relative to the vantage point; one involves occlusion (COVERING); two (REFLEXIVE and REPETITION) involve a **multiple TR-LM configuration**: a situation in which there is more than one TR and/or LM; and one involves temporal 'passage'. The criterion of predominance therefore suggests that the central sense for *over* is the ABOVE sense.

The third criterion concerns relations to other prepositions. Within the entire group of English prepositions, certain clusters of prepositions appear to form **contrast sets** that divide up various spatial dimensions. For example, *above*, *over*, *under* and *below* form a compositional set that divides the vertical dimension into four related subspaces, as illustrated in Figure 10.13. As this

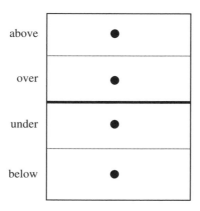

Figure 10.13 Division of the vertical axis into subspaces by prepositions

diagram shows, *over* and *under* tend to refer to those subspaces along the vertical axis that are physically closer to the LM, while *above* and *below* tend to designate relations in which the TR is further away from the LM. In Figure 10.13, the bold horizontal line refers to the LM while the dotted lines refer to areas of vertical space higher and lower than the LM which count as proximal. The dark circles represent TRs in each subspace corresponding to the prepositions listed on the left of the diagram.

Evidence for the proximity distinction comes from the fact that sentences relating to an unspecified region higher than the LM appear to be less natural with *over* but more natural with *above* (for example, compare *The birds are somewhere above us?* with *The birds are somewhere over us*). To a large extent, the lexical item assigned to designate a particular TR–LM configuration is determined by how it contrasts with other members of the set. For example, what we label as *over* is partially determined by what we label as *under*. The sense used in the formation of such a contrast set would thus seem a likely candidate for a primary sense. For *over*, the sense that distinguishes this preposition from *above, under* and *below* relates to a TR located ABOVE but in proximity to the LM. This criterion therefore also suggests that the ABOVE sense is central.

Finally, the fourth criterion relates to the ease with which sense extensions can be predicted from a given sense: the greater the ease of prediction, the more central the sense. Because the central sense is likely to be the sense from which the other senses in the semantic network have derived diachronically, it seems likely that the central sense should be the best predictor of other senses in the network.

The approach to establishing the central sense proposed by Tyler and Evans differs markedly from the approach proposed by Lakoff. Rather than assuming an idealised composite image schema as Lakoff does, Tyler and Evans provide a number of distinct criteria that can be applied to other prepositions, providing empirically testable predictions and a methodology that can be replicated.

Finally, it is important to point out that in Tyler and Evans's theory, the central sense for a preposition such as *over* is directly grounded in a specific kind of recurring spatial scene. This spatial scene, which relates a TR and an LM in a particular spatio-geometric configuration, is called the **proto-scene**. While the proto-scene is a type of image schema, it is distinct from the central image schema proposed by Lakoff becuase it relates to a distinct and discrete spatial scene. The proto-scene for *over* is illustrated in Figure 10.14. The small circle represents the TR and the unbroken line the LM. The fact that the TR is located above the LM indicates that the spatio-geometric relation involves a 'higher-than' or ABOVE relation. The dashed line indicates that the TR must be within a region proximal to the LM.

WORD MEANING AND RADIAL CATEGORIES

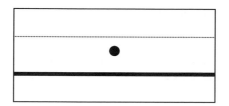

Figure 10.14 The proto-scene for *over* (Tyler and Evans 2003)

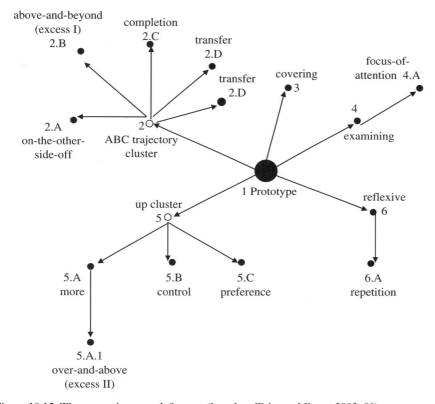

Figure 10.15 The semantic network for *over* (based on Tyler and Evans 2003: 80)

10.5.3 Illustration of a radial category based on Principled Polysemy

On the basis of the Principled Polysemy approach, Tyler and Evans (2003) propose that *over* can be modelled in terms of a semantic network consisting of fifteen distinct senses, as shown in Figure 10.15. Each distinct sense is shown as a dark circle which represents a node in the network. The central sense occupies a central position, indicating its status as the prototypical sense. Some senses within the radial category appear to be more closely related to one another. These

347

Table 10.4 Distinct senses for *over* identified in Tyler and Evans (2003)

	Sense	Example
1	ABOVE (central sense)	The picture is over the sofa
2.A	ON-THE-OTHER-SIDE-OF	St Paul's is over the river from Southwark
2.B	ABOVE-AND-BEYOND (excess I)	Your article is over the page limit
2.C	COMPLETION	The movie is over
2.D	TRANSFER	The discredited government hand power over to an interim authority
2.E	TEMPORAL	The relationship had altered over the years
3	COVERING	The clouds are over the sun
4	EXAMINING	Mary looked over the document quite carefully
4.A	FOCUS-OF-ATTENTION	The committee agonised over the decision
5.A	MORE	Jerome found over forty kinds of shells on the beach
5.A.1	OVER-AND-ABOVE (excess II)	The heavy rains caused the river to flow over its banks
5.B	CONTROL	She has a strange power over me
5.C	PREFERENCE	I would prefer tea over coffee
6	REFLEXIVE	The fence fell over
6.A	REPETITION	After the false start, they started the race over

senses are represented in clusters, arranged with respect to an unshaded circle. Distance from the prototype reflects intuitions about degree of centrality. Direction of arrows represents possible paths of derivation, discussed in a little more detail below. A key to the distinct senses is given in Table 10.4.

10.5.4 Beyond prepositions

The attraction for cognitive semanticists in studying prepositions like *over* has been their direct grounding in spatial experience. In this respect, prepositions provide a transparent illustration of the thesis of embodied cognition, particularly in terms of how concepts in the spatio-physical realm are extended to concepts that are less clearly grounded in spatio-physical experience such as COMPLETION, CONTROL and TIME. However, the approach developed by Tyler and Evans is, in principle, applicable to all lexical classes. We illustrate this point with a discussion of two lexical items from other word classes: the noun *time* and the verb *fly*.

Noun: time

Evans (2004a) further developed the Principled Polysemy approach in order to investigate the polysemy associated with the abstract noun *time*. Evans proposes three criteria for establishing distinct senses associated with *time*:

1. The meaning criterion:
   For a sense to count as distinct, it must contain additional meaning not apparent in any other senses associated with *time*.
2. The concept elaboration criterion:
   A distinct sense will feature unique or highly distinct patterns of **concept elaboration**. Concept elaboration relates to **semantic selection restrictions** which determine how the lexical concept can be metaphorically structured and thus elaborated at the linguistic level. Concept elaboration may relate to how the noun is modified (*a short time*), to the verb phase that forms a sentence with the noun phrase (*The time sped by*), or to an adverbial element (*The time went by **quickly***).
3. The grammatical criterion:
   A distinct sense may manifest unique or highly distinct **structural dependencies**. That is, it may occur in specific kinds of grammatical constructions. Hence, for a sense to be distinct it must exhibit distinctive grammatical behaviour.

In order to illustrate how these criteria apply, consider examples (18) and (19).

(18) a. Time flies when you're having fun.
     b. Last night at the fair the time seemed to whiz by.

(19) a. The time has arrived to finally tackle environmental pollution.
     b. A time will come when we'll have to say no to further deforestation of the Amazon region.

In (18), the examples relate to one aspect of our experience of DURATION in which time appears to be proceeding more quickly than usual. As we saw in Chapter 3, this psychologically real phenomenon is called **temporal compression**. In contrast, the examples in (19) do not relate to our experience of duration but our experience of discrete points in time, without regard for their duration (MOMENT). Hence, the expression *time* has quite distinct meanings associated with it in each set of examples. This means that the two senses are distinct according to the meaning criterion.

In terms of the second criterion, the examples in (18) and (19) have distinct patterns of concept elaboration (metaphorical structuring) associated with them. The TEMPORAL COMPRESSION meaning associated with *time* can be elaborated in terms of motion, which is either rapid as in example (18) or imperceptible as in example (20).

(20) a. The time has vanished.
     b. The time seems to have sneaked by.

On the other hand, the MOMENT meaning in (19) has to be elaborated in terms of motion that is terminal in nature which is therefore oriented with respect to a specific reference point (e.g. NOW). In other words, elaborating the MOMENT sense of *time* in terms of rapid or imperceptible motion results in extremely unnatural sentences that are difficult to interpret. This is illustrated by example (21a), which can be explained on the basis that rapid or imperceptible motion is incompatible with a reading involving the imminent occurrence of a discrete temporal MOMENT:

(21) a. ??The time has vanished/whizzed by to finally tackle environmental pollution
     b. ??A time will whizz by/vanish when we'll have to say no to further deforestation of the Amazon region

Equally, elaborating the TEMPORAL COMPRESSION sense of *time* in terms of terminal motion cancels the TEMPORAL COMPRESSION reading and forces a MOMENT reading as illustrated by example (22).

(22) The time has arrived.
     [Intended reading: the experience of duration is abnormally compressed; that is, time *feels* as if it's proceeding more 'quickly' than usual]

This fact that these two senses of *time* respond differently to concept elaboration satisfies the second criterion, suggesting that these readings qualify as distinct senses.

In terms of the third criterion which relates to the grammatical realisation of distinct senses, observe that the TEMPORAL COMPRESSION sense is encoded by a mass noun, one diagnostic of which is that *time* cannot take the singular indefinite article (*a*), as shown in (23).

(23) *A time raced by

In contrast, the MOMENT sense is encoded by a count noun and can co-occur with the indefinite article:

(24) A time will come when we'll finally have to address global warming.

The fact that the two senses of *time* pattern differently in terms of grammatical behaviour means that they are also distinct senses according to the third criterion. Taken together, these three criteria provide persuasive evidence for the view that we are dealing with two distinct lexical concepts or senses of *time*.

Verb: fly

Although they were originally developed for the analysis of the single lexical item *time* which relates to a relatively narrow subset of one lexical class (abstract nouns), the criteria discussed above provide a promising direction for the analysis of concrete nouns and other lexical classes including adjectives and verbs. For example, consider how these criteria might serve to provide a lexical semantic analysis of the motion verb *fly*, illustrated by the examples in (25):

(25) a. The plane/bird is flying (in the sky).
     b. The pilot is flying the plane (in the sky).
     c. The child is flying the kite (in the breeze).
     d. The flag is flying (in the breeze).

In terms of the meaning criterion, each instance of *fly* in (25) represents a distinct sense. The meaning in (25a), which we will identify as sense 1, can be represented as SELF-PROPELLED AERODYNAMIC MOTION and entails absence of contact with the ground. The meaning in (25b), sense 2, can be represented as OPERATION BY AGENT OF ENTITY CAPABLE OF AERODYNAMIC MOTION. The meaning in (25c), sense 3, can be represented as CONTROL OF LIGHTWEIGHT ENTITY BY AGENT (for example, using an attachment like a piece of string, with the result that it remains airborne). The meaning in (25d), sense 4, can be represented as SUSPENSION OF LIGHTWEIGHT OBJECT (like a flag, with the result that it remains extended and visible).

In terms of the second criterion, which relates to concept elaboration and resulting semantic selectional restrictions, there are a number of distinct patterns in evidence. For example, the different senses of *fly* appear to require distinct kinds of **semantic arguments**. For instance, sense 1 can only apply to entities that are capable of self-propelled aerodynamic motion. Entities that are not self-propelled, like tennis balls, cannot be used in this sense (**the tennis ball is flying in the sky*).

Sense 2 is restricted to the operation by an AGENT of entities that can undergo self-propelled aerodynamic motion and the entity must therefore be able to accommodate the AGENT and thereby serve as a means of transport. This explains why planes and hot air balloons are compatible with this sense but entities unable to accommodate an AGENT are not. This is illustrated by example (26).

(26) ??He flew the sparrow across the English Channel

In the case of sense 3, this sense is restricted to entities that are capable of becoming airborne by turbulence and can be controlled by an AGENT on the

Table 10.5 Some senses of *fly*

	Sense	Example
Sense 1	SELF-PROPELLED AERODYNAMIC MOTION (no contact with the ground)	The bird is flying
Sense 2	OPERATION BY AGENT OF ENTITY CAPABLE OF AERODYNAMIC MOTION	The pilot is flying the plane
Sense 3	CONTROL OF LIGHTWEIGHT ENTITY BY AGENT (so that it remains airborne)	The child is flying the kite
Sense 4	SUSPENSION OF LIGHTWEIGHT OBJECT (which is thus extended and visible)	The flag is flying

ground. This sense appears to be specialised for objects like kites and model aeroplanes.

Sense 4 relates to entities that can be horizontally extended by virtue of air turbulence yet retain contact with the ground by virtue of remaining physically attached to another (non-agentive) fixed entity. This sense can be applied to flags as well as hair and scarves, which can 'fly in the wind'. In sum, each of the four senses discussed here appear to restrict the kind of entities to which the verb can be applied and are therefore distinct senses according to the second criterion.

In terms of the third criterion, there are also grammatical differences associated with the senses presented which are most clearly manifested in terms of transitivity. For instance, while senses 1 and 4 are **intransitive** (they cannot take a direct object), senses 2 and 3 are **transitive** (they either can (sense 2) or must (sense 3) take a direct object). Hence, it appears that the three lines of evidence developed in Evans (2004a, 2005) provide the basis for a methodology for distinguishing distinct senses across a wider range of lexical classes. The senses of *fly* discussed in this section are summarised in Table 10.5.

10.6 The importance of context for polysemy

In the foregoing discussion, we have assumed that it is possible to provide criteria for establishing word senses, and that it is therefore possible to determine where sense boundaries occur. However, in practice, it is not always a straightforward matter to determine whether a particular sense of a word counts as a distinct sense and thus establishes polysemy. This is because word meanings, while relatively stable, are always subject to **context** (recall the discussion in Chapter 7). The consequence of this fact is that while polysemy as a conceptual phenomenon entails a number of wholly distinct yet demonstrably related senses, the reality is that some word senses, while appearing to be distinct in certain contexts, appear not to be in others. In other words, polysemy is often a matter of degree and exhibits **gradability** due to contextual influence. In a number of studies, Alan Cruse has identified a number of ways in which context affects the nature of polysemy. We discuss three of these, which we refer to as **usage context, sentential context** and **knowledge context**. We briefly consider both of these below.

10.6.1 Usage context: subsenses

A subsense is a distinct word meaning that appears to be motivated by usage context: the specific situational context in which the word (and the utterance in which the word is embedded) occurs. However, the distinct sense disappears in other contexts. This suggests that subsenses (also known as **micro-senses**; Croft and Cruse 2004) lack what Cruse calls full **autonomy**: the degree of conventionalisation that secures relative context-independence and thus identifies distinct senses. Example (27), taken from Cruse (2000: 36), illustrates a context-specific subsense of the lexical item *knife*:

(27) Mother: Haven't you got a knife, Billy?
     Billy: (at table, fingering his meat: has penknife in his pocket, but no knife of the appropriate type) No.

Although Billy does have a knife (a penknife), the context (sitting at the meal table) stipulates that it is not a knife of the appropriate kind. In other words, the usage context narrows down the meaning of knife to CUTLERY KNIFE.

At this point, we might pause to consider whether the notion of subsenses can be subsumed under vagueness: could it be that the expression *knife* is vague rather than polysemous like the expression *aunt*? Cruse argues that this is not the case based on evidence such as the **identity constraint**. Consider the following examples adapted from Croft and Cruse (2004: 129):

(28) John needs a knife; so does Sarah.

(29) John has an aunt; so does Sarah.

In the first sentence, we are likely to interpret the second conjunct as referring to the same sense of *knife* (e.g. they both need a table knife): this illustrates the identity constraint. However, in (29), there is no such constraint. The second conjunct could refer to either a maternal or paternal aunt. These examples illustrate Cruse's claim that, while subsenses adhere to the identity constraint, lexical items that are vague do not.

Now let's consider why subsenses are not fully conventional senses. Cruse observes that in certain situations the distinct subsenses CUTLERY KNIFE and PENKNIFE disappear:

(30) The drawer was filled with knives of various sorts.

This sentence could appropriately be used to describe a drawer that contained a cutlery knife, a penknife, a surgeon's knife, a flick knife, a soldier's knife and so on. In other words, the example in (30) appeals to a unified meaning of *knife* in which the contextually induced subsenses disappear. This demonstrates that subsenses do not qualify as fully distinct senses because they require specific kinds of context in order to induce them. Hence, the polysemy associated with the lexical item appears to be heavily dependent upon usage context.

10.6.2 Sentential context: facets

A **facet** is a sense that is due to the part-whole structure of an entity, and is selected by a specific sentential context. As with subsenses, facets are context-dependent because the distinctions between facets only arise in certain sentential contexts. For example, consider the lexical item *book*. By virtue of its structure, the concept BOOK consists of both TEXT (the informational content of a book) and TOME (the physical entity consisting of pages and binding). These two meanings are facets rather than subsenses because they relate to the intrinsic structure and organisation of books in general rather than relating to contexts of use. However, these facets only become apparent in certain sentential contexts. Consider the examples in (31).

(31) a. That book is really thick.
b. That book is really interesting.

The example in (31a) refers to the TOME facet of *book* while (31b) refers to the TEXT facet. Observe that it is sentential context (the presence of the expressions *thick* versus *interesting*) rather than context of use that induces a particular facet. However, just as with subsenses, the distinction between facets can disappear in certain contexts:

(32) Although it's an expensive book, it's well worth reading.

In this example, while price (*it's an expensive book*) relates to the TOME facet, the fact that the book is interesting (*it's well worth reading*) relates to the TEXT facet. The fact that the example in (32) coordinates these two facets without the difference in meaning being marked in any way suggests that the distinction between the facets disappears in this context. In this example, the facets combine to form a unified meaning of *book* that includes both TEXT and TOME. The example in (32) contrasts with examples of **zeugma**, illustrated by example (33), which we presented in the previous chapter. In (33), the two coordinated meanings of *expire* are striking. This suggests that while *expire* exhibits 'full' polysemy, *book* does not.

(33) On the day that my dad expired, so did my driving licence.

10.6.3 Knowledge context: ways of seeing

The third and final kind of context that we will consider is knowledge context. This relates to encyclopaedic knowledge rather than context of use or sentential context. The fact that each individual has different experiences entails that each individual also has different mental representations relating to their experience of particular entities. This creates an encyclopaedic knowledge context that can influence how words are interpreted. Cruse (2000; Croft and Cruse 2004) calls this phenomenon **ways of seeing**. For example, Croft and Cruse (2004: 138) show that the expression *an expensive hotel* can be interpreted in (at least) three different ways depending upon 'ways of seeing':

(34) *an expensive hotel*
'Kind' way of seeing: 'a hotel that is/was expensive to buy'
'Functional' way of seeing: 'a hotel that is expensive to stay at'
'Life-history' way of seeing: 'a hotel that is/was expensive to build'

In sum, Cruse (1986) refers to contextual effects upon interpretation that we have discussed in this section as **contextual modulation**. This idea is in keeping with the encyclopaedic view of meaning that we discussed in Chapter 7. Contextual factors **modulate** or conceptually **highlight** (in Croft's (1993) terms), different aspects of our knowledge associated with a particular entity. As we have seen, these contextual factors might include the situation in which an expression is used, the sentence in which an expression occurs and/or the encyclopaedic knowledge that particular individuals bring to bear upon the interpretation of an expression. The idea of contextual modulation is reminiscent of Barsalou's theory of **background dependent framing**, which we introduced in Chapter 7. For example, the way we interpret the expression *shoe* depends in large measure on the frame we activate (HUMAN versus HORSE, for example). As the discussion in this section has demonstrated, language use involves a complex interaction between polysemy, contextual factors and encyclopaedic knowledge.

10.7 Summary

In this chapter we have introduced the approach to **lexical semantics** that has been developed by cognitive semanticists: **cognitive lexical semantics**.

This approach treats the **polysemy** exhibited by **lexical items** (words) as a psychologically real conceptual phenomenon. Lexical items are viewed as **conceptual categories**, structured with respect to a prototype. A consequence of treating word senses as conceptual categories is that the theory of word meaning assumed within cognitive semantics is motivated by independent evidence from psychology. Lakoff (1987) proposes a **radial category** model for the representation of word meaning which reflects empirical facts relating to word meaning, particularly with respect to polysemy and prototype structure. While Lakoff's approach has been extremely influential, not least because he was one of the first scholars to propose that lexical items should be modelled as conceptual categories, his approach has also faced criticism. In particular, his highly 'granular' model may not be psychologically valid and may underplay the role of context in determining word meaning. Furthermore, Lakoff's approach has been criticised for lacking a rigorous methodology for determining when a sense is **conventionalised** (stored in semantic memory) and when a meaning is inferred on-line as a result of contextual information. A recent approach that has addressed these concerns is the **Principled Polysemy** framework proposed by Evans and Tyler. Finally, we saw in more detail how contextual factors of various kinds, as described by Cruse, serve to **modulate** word meaning. Thus word meaning involves a complex interaction between polysemy, context and encyclopaedic knowledge.

Further reading

Introductory text

- **Aitchison (1996).** This is a popular introductory textbook to lexical semantics which addresses many of the concerns of cognitive semantics.

The distinction between polysemy, homonymy and vagueness.

- **Dunbar (2001)**
- **Geeraerts (1993)**
- **Tuggy (1993)**

These articles provide a cognitive perspective on the traditional problem of distinguishing between polysemy, homonymy and vagueness. The Geeraerts paper provides a comprehensive consideration of problems for traditional distinctions between these notions. Such difficulties have given rise to the view that these notions constitute a continuum.

Analysis of theoretical developments

- **Croft (1998)**
- **Sandra (1998)**
- **Tuggy (1999)**

These articles appeared in the journal *Cognitive Linguistics* in the late 1990s and provide an interesting and insightful commentary on some of the debates relating to theoretical models for lexical semantics.

The development of the radial categories model of word meaning

- **Geeraerts (1994)**
- **Lakoff (1987)**

Lakoff's book introduced and developed the notion of radial categories and prototype semantics. Geeraerts develops a model of prototype semantics that can be applied to historical semantic change.

The Principled Polysemy approach

- **Evans (2004a)**
- **Tyler and Evans (2003)**

These are two book-length treatments that introduce and develop different aspects of the Principled Polysemy approach. See also the papers by Evans and Tyler/Tyler and Evans listed in the next section.

The polysemy of spatial particles

- **Brugman and Lakoff (1988)**
- **Deane (forthcoming)**
- **Dewell (1994)**
- **Coventry and Garrod (2004)**
- **Evans and Tyler (2004a)**
- **Evans and Tyler (2004b)**
- **Herskovits (1986)**
- **Kreitzer (1997)**
- **Lindner (1981)**
- **Lindstromberg (1997)**
- **Sinha and Kuteva (1995)**

- Tyler and Evans (2001b)
- Vandeloise (1991)
- Vandeloise (1994)
- Zelinsky-Wibbelt (1993)

There is a vast literature in cognitive semantics that addresses the polysemy of spatial particles in English and other languages. The references listed here provide a flavour of the nature and extent of this research.

The psycholinguistics of polysemy

- **Cuyckens, Sandra and Rice (2001)**
- **Gibbs and Matlock (2001)**
- **Rice, Sandra and Vanrespaille (1999)**
- **Sandra and Rice (1995)**

Increasingly, cognitive linguists have turned to experimental techniques for testing theoretical models of polysemy. The sources listed here provide some examples of this experimental research.

Corpus linguistics and cognitive lexical semantics

Recent work by scholars such as Stefan Gries and Anatol Stefanowitsch has made a compelling case for incorporating new techniques from corpus linguistics into cognitive linguistics. Nowhere is the utility and benefit of such techniques clearer than in cognitive lexical semantics.

- **Gries (2005).** This paper makes a compelling case for the use of techniques from corpus linguistics in shedding light on many of the issues explored in this chapter.
- **Gries and Stefanowitsch (2005).** This is an edited collection of important papers on topics relating to the application of corpus linguistics to cognitive linguistics.

The frame semantic approach to lexical semantics

- Fillmore and Atkins (1992)
- Fillmore and Atkins (2000)

Although we have not explicitly discussed a frame semantics approach to word meaning in this chapter (see Chapter 7 for an overview), this has been a prominent tradition within cognitive semantics.

10. The role of context in polysemy

- **Cruse (1986)**
- **Cruse (2000)**
- **Cruse (2002)**
- **Croft and Cruse (2004)**

Cruse's contribution to the cognitive framework has been important not least for his work on the role of context in sense-delimitation. Perhaps the most accessible introduction to some of the issues he addresses is Chapter 6 of his 2002 textbook *Meaning in Language*.

Surveys and edited volumes

- **Cuyckens and Zawada (2001)**
- **Cuyckens, Dirven and Taylor (2003)**
- **Nerlich, Todd, Herman and Clarke (2003)**
- **Ravin and Leacock (2002)**

The volumes listed here provide recent collections of articles that address many of the issues considered in this chapter. The first two listed are recent collections that contain papers by leading cognitive lexical semanticists. The second two also include papers by scholars working outside cognitive semantics. For an introduction to some of the recent concerns in cognitive lexical semantics the volume by Cuyckens, Dirven and Taylor is a good place to start.

Exercises

10.1 Comparing cognitive and traditional models of word meaning

What criticisms are levelled against the traditional approach to lexical semantics? How does the cognitive lexical semantics approach address these concerns?

10.2 Comparing Principled Polysemy with the full-specification approach

How does Tyler and Evans's (2003) Principled Polysemy approach differ from Lakoff's 'full-specification' approach? In what respects are the two theories in agreement?

10.3 Prepositions and image schemas

In this chapter we saw that prepositions can be modelled in terms of image schemas. Consider the following examples.

(a) i. The lifejacket is kept under the seat.
    ii. The nurse deftly slipped the pillow under the patient's head.
    iii. ??The valley is far under the tallest peak

(b) i. The water level fell below 10 metres.
    ii. ??The nurse deftly slipped the pillow below the patient's head
    iii. The valley is far below the tallest peak.

Based on these examples, propose and diagram image schemas for *under* and *below* that take account, where necessary, of (i) the spatio-geometric properties of their respective TRs and LMs, and (ii) the spatio-geometric character of the TR/LM relationship. Then state in informal terms what the meaning difference is between *under* and *below*.

10.4 Preposition meaning and context

In view of your response to exercise 10.3, how would this allow you to account for the following example?

(a) Passenger: I can't find my life jacket.
    Flight attendant: You'll find the life jacket **below** the seat.

10.5 The semantic network for through

Consider the following sentences featuring the English preposition *through*:

(a) The relationship is through.
(b) The tunnel through Vale Mountain was completed in the 1980s.
(c) She did it through love.
(d) The trip abroad was funded through the miscellaneous fund.
(e) The ball whizzed through the hole in the net.
(f) He looked through the window.
(g) The relationship seemed to have evolved through the years.
(h) The dog jumped through the hoop.
(i) The skewer is through the meat.
(j) The stream runs through the pasture.
(k) The jogger ran through the tunnel.

Based on these examples provide an analysis of the semantic network for *through*. Your analysis should provide:

(i) labels for each distinct sense you posit (not all the examples may represent distinct senses); categorise the examples by sense;

(ii) 'decision principles' for determining what counts as a distinct sense;
(iii) the prototypical sense and the reasons for your decision;
(iv) a semantic network (radial category) showing how the senses might be related.

10.6 The semantic network for by

Following the methodology outlined in exercise 10.5, consider the following sentences that contain the preposition *by*. Using appropriate criteria, identify distinct senses for *by* and identify a central sense. Include in your answer a semantic network accounting for all the senses you identify and comment on the nature and arrangement of the semantic network.

(a) The man is by the tree.
(b) He will arrive by 3 o'clock.
(c) Paris is beautiful by night.
(d) Day by day her condition worsened.
(e) John put his work by until later.
(f) The frame measures 6 metres by 4 metres.
(g) We purchase the beer by the barrel.
(h) She did well by her children.
(i) I have put money by.
(j) Are you paid by the hour?

10.7 Data collection

Consider the words below. For each word, collect examples of how it is used from written texts, from conversations or from a good dictionary. Carry out a semantic network analysis of each word.

(a) (to) run
(b) (to) crawl
(c) (a) foot

10.8 Like

Now consider the lexical item *like*. Based on the sorts of sources you used in the previous exercise, identify the range of meanings associated with *like*. How do these meanings relate to different grammatical functions? Now attempt to provide a semantic network of the range of meanings exhibited by this form. Check the order in which the various meanings of *like* entered the language by consulting the OED.

10.9 Fly

We noted in this chapter (section 10.5.4) that one of the senses of the verb *fly* is OPERATION BY AGENT OF ENTITY CAPABLE OF AERODYNAMIC MOTION, which is illustrated by example (a). We further noted that entities that are unable to accommodate the AGENT and thus serve as a mode of transport are incompatible with this sense, which is illustrated by example (b). In the light of these observations, how would you account for example (c)?

   (a) The pilot flew the plane to France.
   (b) ??He flew the sparrow across the English Channel.
   (c) She decided to fly her large spotted homing pigeon in the competition.

10.10 Subsenses versus vagueness

In view of the discussion between subsenses and vagueness (section 10.6.1), consider the lexical items below. Based on the discussion in the chapter, determine which of these items exhibit distinct subsenses and which are vague.

   (a) equipment
   (b) cousin
   (c) card
   (d) car
   (e) best friend

11

Meaning construction and mental spaces

This chapter explores the view of **meaning construction** developed in cognitive semantics. In the previous chapter, we were concerned with the meaning of words. In this chapter, we consider how larger units of language like sentences and texts (units of discourse larger than the sentence) are meaningful. It is to this level of linguistic organisation that the term 'meaning construction' applies. Recall from Chapter 7 that cognitive semanticists see linguistic expressions as 'points of access' to the vast repository of encyclopaedic knowledge that we have at our disposal. According to this view, language underdetermines the content of the conceptual system. **Meaning construction** is the process whereby language 'prompts for' novel cognitive representations of varying degrees of complexity. These representations relate to conceived scenes and aspects of scenes, such as states of affairs in the world, emotion and affect, subjective experiences, and so on.

Cognitive semanticists treat meaning construction as a process that is fundamentally conceptual in nature. From this perspective, sentences work as 'partial instructions' for the construction of complex but temporary conceptual domains, assembled as a result of ongoing discourse. These domains, which are called **mental spaces**, are linked to one another in various ways, allowing speakers to 'link back' to mental spaces constructed earlier in the ongoing linguistic exchange. From this perspective, meaning is not a property of individual sentences, nor simply a matter of their interpretation relative to the external world. Instead, meaning arises from a dynamic process of meaning construction, which we call **conceptualisation**.

This chapter is primarily concerned with presenting **Mental Spaces Theory**, developed by Gilles Fauconnier ([1985] 1994, 1997). This approach holds that language guides meaning construction directly in context.

According to this view, sentences cannot be analysed in isolation from ongoing discourse. In other words, **semantics** (traditionally, the context-independent meaning of a sentence) cannot be meaningfully separated from **pragmatics** (traditionally, the context-dependent meaning of sentences). This is because meaning construction is guided by context and is therefore subject to situation-specific information. Moreover, because meaning construction is viewed as a fundamentally conceptual process, this approach also takes account of general cognitive processes and principles that contribute to meaning construction. In particular, meaning construction relies on some of the mechanisms of **conceptual projection** that we have already explored, such as metaphor and metonymy.

11.1 Sentence meaning in formal semantics

Because Fauconnier's Mental Spaces Theory represents a reaction to the **truth-conditional** model of sentence meaning adopted in **formal semantics**, we begin with a very brief overview of this approach. The truth-conditional model works by establishing 'truth conditions' of a sentence: the state of affairs that would have to exist in the world, real or hypothetical, for a given sentence to be true. For example, relative to a situation or 'state of affairs' in which the cat stole my breakfast, the sentence *The cat stole my breakfast* is true, while the sentence *The cat did not steal my breakfast* is false. The truth-conditional approach is not concerned with empirical truth but rather with establishing a model of meaning based on 'what the world would have to be like' for a given sentence to be true. In other words, it is not important to find out whether the cat stole my breakfast or not, nor indeed whether I even have a cat. What is important is the fact that speakers know 'what the world would have to be like' for such a sentence to be true. Establishing the truth conditions of a sentence then enables sentences to be compared, and the comparison of their truth conditions gives rise to a model of (some aspect of) their meaning. For example, if the sentence *The cat stole my breakfast* is true of a given situation, the sentence *My breakfast was stolen by the cat* is also true of that situation. These sentences stand in a relation of **paraphrase**. According to the truth-conditional model, they 'mean the same thing' (at least in semantic or context-independent terms) because they share the same truth conditions: they can both be true of the same state of affairs. Compare the two sentences we saw earlier: *The cat stole my breakfast* and *The cat did not steal my breakfast*. These two sentences stand in a relation of **contradiction**: they cannot both be true of the same state of affairs. If one is true, the other must be false, and vice versa. These examples illustrate how truth conditions can be used to model meaning relationships between sentences, like paraphrase (if A is true B is true, and vice versa) and contradiction (if A is true B is false, and vice versa). This very brief description of the truth-conditional model will be elaborated in Chapter 13. For the time being, we observe that although this model does not rely on empirical truth – you don't have to witness your cat stealing your breakfast before you can understand that the sentences discussed above stand in the kinds of meaning relationships described – the model nevertheless relies on the **objectivist thesis**.

The objectivist thesis holds that the 'job' of language is to represent an objectively defined external world. In modern truth-conditional approaches, this objective external reality may be mediated by mental representation (external reality as it is construed by the human mind), but in order for a formal truth-conditional model to work, it requires certain objectively defined primitives and values. Furthermore, as we saw in Chapter 7, this kind of approach to linguistic meaning assumes the **principle of compositionality**: the meaning of a sentence is built up from the meaning of the words in the sentence together with the way in which the words are arranged by the grammar. According to this view, then, the semantic meaning of a sentence is the output of this compositional process and is limited to what can be predicted from the context-independent meanings of individual words and from the properties of the grammar. Any additional meaning, such as the inferences a hearer can draw from the utterance of a particular sentence within a particular context, falls outside the immediate concerns of semantic theory into the domain of pragmatics. From this perspective, semantics is concerned with what words and sentences mean, while pragmatics is concerned with what speakers mean when they use words and sentences in situated language use, and how hearers retrieve this intended meaning. From the formal perspective, these two areas of investigation can be meaningfully separated.

11.2 Meaning construction in cognitive semantics

In contrast to formal semantics which relies on the objectivist thesis, cognitive semantics adopts an **experientialist perspective**. According to this view, external reality exists, but the way in which we mentally represent the world is a function of embodied experience (recall the discussion of embodied cognition in Chapter 2). Thus meaning construction proceeds not by 'matching up' sentences with objectively defined 'states of affairs', but on the basis of linguistic expressions 'prompting' for highly complex conceptual processes which construct meaning based on sophisticated encyclopaedic knowledge.

In one important respect then, the view of 'meaning' developed in earlier chapters oversimplifies the picture. Throughout the book, we have used terms like 'encode' and 'externalise' in order to describe the function of language in relation to concepts. According to this view, semantic structure is the conventional form that conceptual structure takes when encoded in language, and represents a body of stored knowledge that language simply reflects. However, the expression 'encode' oversimplifies the relationship between language and cognition and requires some qualification.

Firstly, the meanings 'encoded' in language (the semantic representations associated with linguistic units) are partial and incomplete representations of conceptual structure. For example, we saw in Chapter 7 that conceptual structure is underpinned by information derived from perceptual processes, including sensory and introspective (or subjective) experience. While the representations of this experience that make up our conceptual system (including frames, domains, ICMs, conceptual metaphors and so on) are less rich in detail than perceptual experience itself, the representations encoded by semantic structure are still further reduced in detail. Moreover, conceptual representation is thought to be ultimately perceptual in nature, a view that is suggested by the perceptual simulations that conceptual structure can provide. For example, one can mentally simulate (that is, mentally rehearse or imagine) the stages involved in taking a penalty kick in a football match. In contrast, semantic representation is specialised for expression via a **symbolic system**. This means that the linguistic system, which consists of spoken, written or signed symbols, 'loses' much of the richness associated with the multimodal character of conceptual representation. By way of analogy, if we were to take the six-stream digital sound reproduction available in modern cinema multiplexes and compress this through a single speaker, not only would some of the sounds be lost (for example, the bass track, background sounds and the experience of 'moving' sounds), but the nature and detail of the remaining sounds would also be significantly impoverished: the mono sound becomes a very partial and incomplete clue to what the original sounds might have been like.

In a similar way, although semantic structure 'encodes' conceptual structure, the format of semantic structure ensures that language can only ever provide minimal clues to the precise mental representation intended by the speaker. In other words, language does encode 'meaning', but this meaning is impoverished and functions as **prompts** for the construction of richer patterns of conceptualisation by the hearer. The cognitive semanticist Mark Turner has expressed this idea in the following way:

> Expressions do not mean; they are prompts for us to construct meanings by working with processes we already know. In no sense is the meaning of [an]. . .utterance 'right there in the words.' When we understand an utterance, we in no sense are understanding 'just what the words say'; the words themselves say nothing independent of the richly detailed knowledge and powerful cognitive processes we bring to bear. (Turner 1991: 206)

Secondly, the cognitive view holds that conceptualisation emerges from language use in context. It follows that there is no principled distinction between semantics and pragmatics. Formal approaches often assume that assigning meaning to an utterance is a two-stage process. In the first stage, context-independent word meanings are decoded by the hearer and composed into the context-independent semantic representation of a sentence. In the second stage, the utterance undergoes pragmatic processing which brings to bear information relating to context, background knowledge and inferences made by the hearer regarding speaker intentions. In contrast, Mental Spaces Theory assumes that conceptualisation is guided by discourse context, which forms an integral part of the meaning construction process. According to this view, meaning construction is localised and situated, which entails that pragmatic (context-dependent) information and knowledge inform and guide the meaning construction process. Thus, while pragmatic knowledge may be qualitatively distinct from semantic knowledge (the impoverished information encoded by linguistic prompts), semantic knowledge is only meaningful in context. As we saw in Chapter 7, cognitive semanticists therefore reject the assumption that there are distinct 'semantic' and 'pragmatic' stages in meaning construction, together with the assumption that there exists some meaningful boundary between these two kinds of knowledge: both are aspects of encyclopaedic knowledge.

Finally, conceptualisation is held to rely upon complex conceptual processing, which involves conceptual projections of the kind that have been discussed so far in this book. These include conceptual metaphors, conceptual metonymies and the process of **schema induction** that was first introduced in Chapter 5. This is the process whereby our conceptualisations are elaborated and enriched by the application of large-scale and pre-assembled knowledge structures which serve a contextualising function. Schema induction is of central importance for meaning construction, as we will see in this chapter. Conceptual projection mechanisms like metaphor, metonymy and schema induction establish **mappings**. As we have already established (Chapter 9), a mapping connects entities in one conceptual region with another. These mappings can be highly conventionalised, as in the case of primary conceptual metaphors, or they can be constructed 'on-line' for purposes of local understanding. Gilles Fauconnier summarises this position as follows:

> Language, as we use it, is but the tip of the iceberg of cognitive construction. As discourse unfolds, much is going on behind the scenes: New domains appear, links are forged, abstract meanings operate, internal structure emerges and spreads, viewpoint and focus keep shifting. Everyday talk and commonsense reasoning are supported by

invisible, highly abstract, mental creations, which ... [language] ... helps to guide, but does not by itself define. (Fauconnier 1994: xxii–xxiii)

In sum, meaning is not simply pre-existing stored knowledge encoded by language. Cognitive semanticists argue that the naive view, which views words as 'containers' for meaning and language as a conduit for the transfer or externalisation of pre-existing meaning, is erroneous (see Reddy [1979] 1993). Instead, meaning construction is seen as a complex process that takes place at the conceptual level. Words and grammatical constructions are merely partial and impoverished prompts upon which highly complex cognitive processes work giving rise to rich and detailed conceptualisation.

In his pioneering work on meaning construction, Fauconnier demonstrates that much of what goes on in the construction of meaning occurs 'behind the scenes'. He argues that language does not encode thought in its complex entirety, but encodes rather rudimentary instructions for the creation of rich and elaborate ideas. It is because the principles and strategies that guide this conceptualisation process are largely unseen that the rather simplistic view has arisen that meaning construction is achieved by simply 'decoding' the meaning inherent 'in' language. Fauconnier calls the unseen conceptualisation processes that are involved in meaning construction **backstage cognition**.

11.3 Towards a cognitive theory of meaning construction

Gilles Fauconnier is the leading proponent of Mental Spaces Theory, a highly influential cognitive theory of meaning construction. Fauconnier develops this approach in his two landmark books *Mental Spaces* ([1985] 1994) and *Mappings in Thought and Language* (1997). More recently, Fauconnier and Turner have extended this theory, which has given rise to a new framework called **Conceptual Blending Theory**. We outline Mental Spaces Theory in the present chapter and explore its more recent development into Conceptual Blending Theory in the next chapter.

According to Fauconnier, meaning construction involves two processes: (1) the building of **mental spaces**; and (2) the establishment of **mappings** between those mental spaces. Moreover, the mapping relations are guided by the local discourse context, which means that meaning construction is always **situated** or context-bound. Fauconnier defines mental spaces as 'partial structures that proliferate when we think and talk, allowing a fine-grained partitioning of our discourse and knowledge structures' (Fauconnier 1997: 11). As we will see, the fundamental insight that this theory provides is that mental spaces partition meaning into distinct conceptual regions or 'packets'.

We begin here by providing a general overview of Mental Spaces Theory before exploring its architecture in more detail.

Mental spaces are regions of conceptual space that contain specific kinds of information. They are constructed on the basis of generalised linguistic, pragmatic and cultural strategies for recruiting information. However, because mental spaces are constructed 'on-line', they result in unique and temporary 'packets' of conceptual structure, constructed for purposes specific to the ongoing discourse. The principles of mental space formation and the relations or mappings established between mental spaces have the potential to yield unlimited meanings. For example, consider the following utterance similar to one discussed by Fauconnier (1997):

(1) If I were your father I would smack you.

This utterance gives rise to a **counterfactual** conceptualisation. That is, it sets up a scenario that runs counter to a presupposed reality. This scenario represents a mental space. Intuitively, you can think of a mental space as a 'thought bubble', rather like the strategy cartoonists use to reveal the inner thoughts of their characters. Crucially, Mental Spaces Theory holds that you can have many 'thought bubbles' working simultaneously.

Depending on the context, the utterance in (1) can give rise to different counterfactual scenarios. This is because the context guides mapping operations between the state of affairs that holds in reality and the states of affairs that are set up in different versions of the counterfactual scenario. Imagine that a childminder, Mary, utters the sentence in (1) after the child in her care, James, is particularly unruly. We consider here three distinct possible interpretations of (1) and see how Mental Spaces Theory accounts for them.

The lenient father interpretation ('your father should be stricter')

In this interpretation, the childminder Mary thinks that the unruly child's father should demonstrate more authority and punish the child by smacking him. In terms of mapping operations between reality and the counterfactual scenario, this interpretation is derived by Mary with her stricter disposition 'replacing' the father with his more lenient disposition. This mapping is partial in the sense that the child's father remains the same in all other respects: he has a beard, rides a bike, gets home at the same time in the evening and so on. What changes in this counterfactual scenario is that the father is now less tolerant of the child's unruly behaviour and smacks the child. A consequence of this interpretation is that in the reality scenario, which is presupposed by the counterfactual scenario, the father is being critically compared to the speaker Mary. Because the childminder would smack the child, by implication the failure of the father to smack the child is interpreted as a fault on his part. In this way, the counterfactual scenario entails consequences for how we view the father and his approach to parenting in reality.

The stern father interpretation ('you're lucky I'm not as strict as your father')

In this interpretation, it is the father, who has a stricter disposition, who is replacing the childminder Mary. In other words, Mary is advising the child that he is lucky that she is looking after him rather than his father, because otherwise the child would have been smacked. In this interpretation, it is the father who is strict and Mary who is lenient in reality, and it is the father who assumes Mary's place in the counterfactual scenario. The implication of this counterfactual scenario for reality might be that where the father would smack the child, Mary exhibits greater restraint. This interpretation might therefore imply a positive assessment of Mary in her role as childminder.

The role interpretation ('the only reason I'm not smacking you is because I'm not allowed to')

In this interpretation, Mary is saying that if she could assume the role of the child's father then she would smack the child. This interpretation assumes nothing about the child's father who may (or may not) smack the child in reality. Instead, this counterfactual scenario replaces the father role with Mary. In this counterfactual scenario, Mary-as-father would smack the child. The implication of this interpretation for reality is that it comments on Mary's role and the limitations that it entails: in her role as childminder, she is legally prohibited from smacking the child.

Several important points emerge from the discussion of example (1). Firstly, the same utterance can prompt for a number of different interpretations, each of which arises from different mappings between reality and the counterfactual scenario that is constructed. Secondly, each of these mappings brings with it different implications for how we view the participants in reality (for example, criticism versus a positive assessment and so on). Finally, this example illustrates that meaning is not 'there in the words' but relies on the conceptual processes that make connections between real and hypothetical situations. These processes result in representations that are consistent with, but only partially specified by, the prompts in the linguistic utterance. Of course, the precise interpretation constructed will depend upon the precise details of the context in which it is uttered, upon the speaker's intentions and upon how these intentions are interpreted by the hearer. For example, if James has a father who is far stricter than his childminder in reality, he might be most likely to construct the second of these possible interpretations.

11.4 The architecture of mental space construction

As we saw above, linguistic expressions are seen as underdetermined prompts for processes of rich meaning construction: linguistic expressions have **meaning potential**. Rather than 'encoding' meaning, linguistic expressions represent partial **building instructions**, according to which mental spaces are constructed. Of course, the actual meaning prompted for by a given sentence will always be a function of the discourse context in which it occurs, which entails that the meaning potential of any given sentence will always be exploited in different ways dependent upon the discourse context. In this section, we consider in detail the cognitive architecture that underlies this process of meaning construction.

11.4.1 Space builders

According to this theory, when we think and speak we set up mental spaces. Mental spaces are set up by **space builders**, which are linguistic units that either prompt for the construction of a new mental space or shift attention back and forth between previously constructed mental spaces. Space builders can be expressions like prepositional phrases (*in 1966, at the shop, in Fred's mind's eye, from their point of view*), adverbs (*really, probably, possibly, theoretically*), connectives (*if . . . then . . .; either . . . or . . .*), and subject-verb combinations that are followed by an embedded sentence (*Fred believes* [*Mary likes bananas*], *Mary hopes . . ., Susan states . . .*), to name but a few. What is 'special' about space builders is that they require the hearer to 'set up' a scenario beyond the 'here and now', whether this scenario reflects past or future reality, reality in some other location, hypothetical situations, situations that reflect ideas and beliefs, and so on.

11.4.2 Elements

Mental spaces are temporary conceptual domains constructed during ongoing discourse. These spaces contain **elements**, which are either entities constructed on-line or pre-existing entities in the conceptual system. The linguistic expressions that represent elements are noun phrases (NPs). These include linguistic expressions like names (*Fred, Elvis, Madonna, Elizabeth Windsor, Tony Blair, James Bond*), descriptions (*the Queen, the Prime Minister, a green emerald, a Whitehouse intern, an African elephant*), and pronouns (*she, he, they, it*).

NPs can have a **definite interpretation** or an **indefinite interpretation**. Briefly, NPs that have a definite interpretation include those that occur with the definite article *the*, (*the sleepy koala*) and names (*Margaret Thatcher, James Bond*). NPs that have indefinite interpretation include those occurring with the indefinite article *a* (*a sleepy koala*) and 'bare plurals' (*koalas*). NPs with indefinite interpretation typically introduce new elements into the discourse: elements that are unfamiliar or have not already been mentioned in the conversation (*I've bought a new sofa!*). NPs with definite interpretation are said to function in the **presuppositional mode**, because they presuppose existing knowledge. This means that they refer to elements that are already accessible: elements familiar to speaker and hearer, or already part of the conversation (*The new sofa clashes with the curtains*). In Mental Spaces Theory, elements introduced in the presuppositional mode are said to be **propagated**, which means that they spread to neighbouring spaces. This process of propagation is governed by the **Optimisation Principle**. This principle allows elements, together with their properties and relations, to spread through the network or **lattice** of mental spaces, unless the information being propagated is explicitly contradicted by some new information that emerges as the discourse proceeds. This principle enables mental space configurations to build complex structures with a minimum of explicit instructions.

11.4.3 Properties and relations

In addition to constructing mental spaces and setting up new or existing elements within those spaces, meaning construction also processes information about how the elements contained within mental spaces are related. Space builders specify the **properties** assigned to elements and the **relations** that hold between elements within a single space. Consider example (2).

(2) In that play, Othello is jealous.

The space builder in example (2) is the phrase *in that play*, which sets up a mental space. In Figure 11.1 we diagram the mental space using a circle and label this mental space PLAY to show that the mental space represents the 'world' inside the play. The name *Othello* introduces an element into the mental space, which we

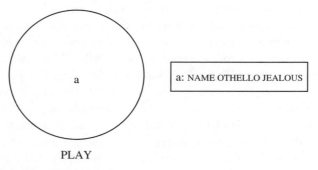

Figure 11.1 *In that play, Othello is jealous*

label a, and the expression *jealous* assigns a property to the element (JEALOUS). This information is captured in the 'dialogue box' next to the mental space.

Now consider example (3).

(3) In the picture, a witch is riding a unicorn.

Again, the prepositional phrase (PP) *in the picture* is a space builder that sets up a mental space which we label PICTURE in Figure 11.2. This shows that the mental space relates to the 'world' inside the picture. Two new elements are introduced: a witch and a unicorn. These are introduced as 'new' in the discourse because they have indefinite interpretation. In Figure 11.2, a represents the element prompted for by the expression *witch*, and b the element prompted for by the expression *unicorn*.

So far, the mental space in Figure 11.2 is only a partial representation of the sentence, because while it tells us that the picture contains a witch and a unicorn, it does not tell us whether a relation holds between them nor does it describe the nature of that relation. Mental spaces are **internally structured** by existing knowledge structures: frames and idealised cognitive models. The space builders, the elements introduced into a mental space and the properties and relations prompted for **recruit** this pre-existing knowledge structure, a process that we identified above as schema induction. For example, the space builder in sentence (3) prompts for the recruitment of a frame for PICTURES. The elements introduced prompt for the recruitment of frames relating to WITCHES AND WITCHCRAFT and MYTHICAL CREATURES such as UNICORNS. Finally, the expression *is riding* expresses a **relation** between the two elements and prompts for the RIDE frame. The RIDE frame brings with it two participant roles, one for a RIDER and one for the ENTITY RIDDEN. The RIDER role is mapped onto element a, introduced by the expression *witch*, and the ENTITY RIDDEN role is mapped onto element b, introduced by the expression *unicorn*. This establishes a relation between the two elements in the mental space. The completed

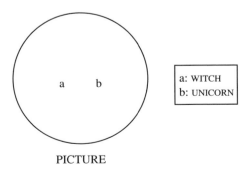

PICTURE

Figure 11.2 *In the picture, a witch is riding a unicorn*

mental space for example (3) with the additional structure resulting from schema induction is illustrated in Figure 11.3.

11.4.4 Mental space lattices

Once a mental space has been constructed, it is linked to the other mental spaces established during discourse. At any given point in the discourse, one of the spaces is the **base**: the space that remains accessible for the construction of a new mental space, a point that we elaborate below. As discourse proceeds, mental spaces proliferate within a network or lattice as more schemas are induced and links between the resulting spaces are created. This is illustrated in Figure 11.4. The circles represent the mental spaces and the dotted lines indicate links between spaces. The base is the space at the top of the lattice.

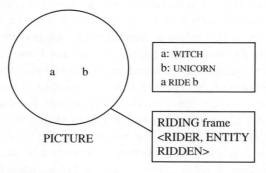

Figure 11.3 Schema induction

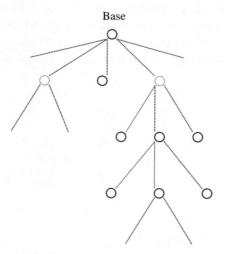

Figure 11.4 A lattice of mental spaces

11.4.5 Counterparts and connectors

In order to explain how different mental spaces are linked to one another, we begin by exploring the idea that elements within different mental spaces can be linked. Elements in different spaces are linked by **connectors** which set up mappings between **counterpart** elements. Counterparts are established on the basis of **pragmatic function**: when two (or more) elements in different mental spaces have a related pragmatic function, they are counterparts. One salient type of pragmatic function is **identity**. For instance, in Ian Fleming's novels, *James Bond* is the name of the fictional British spy character and *007* is the code name used by the British Secret Service (MI6) to identify this spy. The pragmatic function relating the entities referred to as *James Bond* and *007* is co-reference or identity. In other words, both expressions refer to the same individual and together form a **chain of reference**. Elements in different mental spaces that are co-referential (counterparts related by identity) are linked by an **identity connector**. To illustrate the linking of counterparts in two separate mental spaces by an identity connector, consider example (4).

(4) James Bond is a top British spy. In the war, he was an officer in the Royal Navy.

Each sentence in (4) sets up its own mental space, although it is not always the case that every sentence gives rise to its own mental space. We only need to set up a new mental space if the utterance contains a new space builder. As this example illustrates, not every mental space is introduced by an explicit space builder. For example, the base space introduced by the first sentence in (4) is established by our background knowledge that James Bond is a fictional character in the book or movie being described. The expression *James Bond* induces the schema that is associated with this knowledge. This shows that background knowledge can function as an implicit space builder. If this space builder were made explicit, the sentence might begin *In the book.* . . . When a mental space lacks an explicit space builder, it does not receive a label like PLAY or BOOK because this information is implicit.

In the first sentence in (4), the first mental space is set up by the introduction of the element corresponding to the name *James Bond*. This entity is assigned the property introduced by the indefinite NP *a top British spy*, which describes James Bond rather than introducing a separate entity because the two expressions are connected by *is*. This mental space is the base space. In the second sentence, the PP *in the war* is a space builder which constructs a new WAR space. This mental space also features an element, introduced by *he*, which also has a property assigned to it, *an officer in the Royal Navy*. Notice that *he* refers to the same person as *James Bond*. In linguistics, the process whereby one expression relies on another for full interpretation is called **anaphora**. The dependent expression (*he*) is called an **anaphor** and the expression it relies upon for its meaning (*James Bond*) is called the **antecedent**. The establishment of a link between an anaphor and an antecedent is a type of **inference**, an interpretation we 'work out' on the basis of establishing coreference between the two expressions. Anaphora relies on inference because an expression like *he*, unlike the name *James Bond*, lacks the semantic properties to uniquely define its referent: it could in principle refer to any male entity. This means that the hearer has to 'work out' which entity it refers to by searching the context for a likely candidate.

11.4.6 The Access Principle

In an example like (4) an identity connector is set up between the anaphor *he* and the antecedent *James Bond*. The elements a_1 and a_2 in Figure 11.5 are counterparts and are linked by an identity connector. This connector provides **access** to a counterpart in a different mental space. It is important to point out that the identity connector (which is represented as a line linking a_1 and a_2 in Figure 11.5) is not overtly introduced into the representation by any linguistic expression. Instead, the identity connector represents a **mapping**, a conceptual 'linking' operation established by the inference.

Fauconnier formalises this structuring property of mental space configurations in terms of the **Access Principle**, which states that 'an expression that

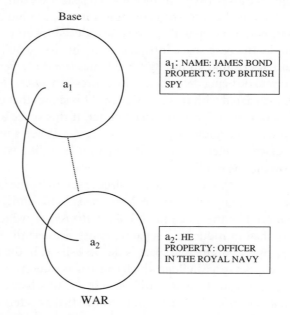

Figure 11.5 Linking counterparts

names or describes an element in one mental space can be used to access a counterpart of that element in another mental space' (Fauconnier 1997: 41). This means that connectors are a type of conceptual projection: like the conceptual metaphors and conceptual metonymies described in the previous chapter, connectors establish relationships or mappings across regions of conceptual structure.

One consequence of the Access Principle is that expressions referring to a particular counterpart can typically provide access to entities in mental spaces in either direction. In other words, connectors can 'link upwards' or 'link downwards' between spaces. When this occurs, the connector is said to be **open**. For example, the element corresponding to the anaphor *he* in example (4) serves as the **trigger** to access the element corresponding to the element a (*James Bond*), the **target**, in the base. In this example, the connector 'links upwards' to a previously established space. Access can also 'link downwards' from one mental space to a subsequently established space. Suppose we add example (5) to the text in (4):

(5) James Bond served on HMS *Espionage*.

This sentence adds structure to the WAR space by prompting for a new frame to be added containing information regarding WARSHIPS and the relationship between naval officers and the ships they serve on. Because the expression *James Bond* is used, which corresponds to element a in the base space, the counterpart of element a (labelled a_1) in the WAR space is accessed. New information can then be added with respect to element a_1. In this example, element a in the base space, which is identified by *James Bond*, is the trigger for element a_1, the target, which is in the WAR space. In this way, a_1 in the WAR space is accessed via the base space. Another way of thinking about this is to say that the space that is in 'focus', the WAR space, which is the space where structure is being added, is accessed from the perspective of the base space. This additional structure and the direction of the connector is represented in Figure 11.6.

Another consequence of the Access Principle is that multiple counterparts can be accessed. This is illustrated in the next example, discussed by Fauconnier (1994), which relates to a fictitious movie about the life of the famous film director Alfred Hitchcock. In his movies, Hitchcock invariably made a cameo appearance as a minor character. In the fictitious movie, Hitchcock is played by Orson Welles:

(6) In the movie Orson Welles played Hitchcock, who played a man at the bus stop.

This sentence contains the space builder *in the movie*. This sets up a MOVIE space containing the characters *Hitchcock* and *the man at the bus stop*. As we have

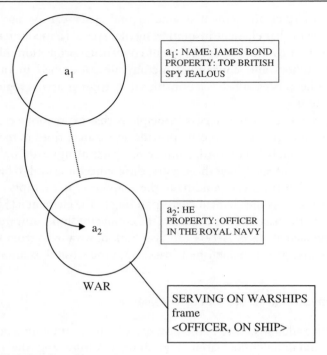

Figure 11.6 Directionality of connectors

seen, a mental space either represents the base space or is constructed relative to a base space; the base space contains **default information** currently available to the discourse context, including contextually relevant background frames. The base space for example (6) relates to the film set, which includes the director, the actors and so on. This information is not provided by specific linguistic expressions in example (6), but is supplied by schema induction arising from our knowledge of the MOVIE frame which also sets up connectors between actors and the characters they play.

In the base, which represents the reality space, both the element introduced by *Orson Welles* and the element introduced by *Hitchcock* are present. This is default information: both individuals exist as actors in the reality space. In the MOVIE space, based on our knowledge of the MOVIE frame, the information provided by *played* instructs us to link Orson Welles the actor (in the base) with Hitchcock the character (in the MOVIE space) as counterparts, linked by an **actor-character connector**. This is represented by connector 1 in Figure 11.7. In addition, while Hitchcock is identified as a character in the MOVIE space (by virtue of the actor-character connector), he is also identified as an actor by the subsequent part of the sentence: *who played a man at the bus stop*. This relation between Hitchcock-as-character (established in the MOVIE

MEANING CONSTRUCTION AND MENTAL SPACES

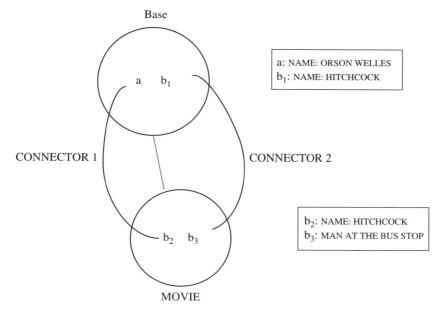

Figure 11.7 Hitchcock and the movie

space) and Hitchcock–as–actor (established in the base space) is set up by the expression *who*, which is an instruction to set up a connector between these two counterparts. This is represented by connector 2 in Figure 11.7.

Now suppose we add example (7) to the information established in (6).

(7) Hitchcock liked himself in that movie.

This sentence is ambiguous. It could mean either that (the real) Hitchcock liked the character played by Orson Welles (Hitchcock-as-actor), or that he liked the man at the bus stop (Hitchcock-as-character). That is, from the perspective of the base, b_1 (the real) *Hitchcock* can be linked either to counterpart b_2 in the MOVIE space (Hitchcock–as actor, introduced by *who*) or to counterpart b_3 in the MOVIE space (*a man at the bus stop*). This is illustrated in Figure 11.8, which shows that the ambiguity in the sentence arises from the fact that b_1 (the real) *Hitchcock* has two potential connectors which link it to two counterparts in the MOVIE space. In other words, b_1 (*Hitchcock*) is a trigger with two targets established by pragmatic function: (1) the connector linking b_1 with b_2 (Hitchcock-as-actor, introduced by *who*), which is established by virtue of an identity connector; and (2) the connector linking b_1 (*Hitchcock*) with b_3 (*the man at the bus stop*), which is established by an actor-character connector. Crucially, the ambiguity is a function of the mapping possibilities across mental spaces.

379

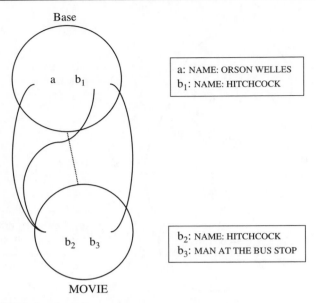

Figure 11.8 Two connectors to one element

As this discussion reveals, one appeal of Mental Spaces Theory is that it offers a plausible account of how language prompts for different referential possibilities. It is precisely because we partition discourse into distinct mental spaces, with mappings holding between elements in different mental spaces, that we are able to construct the complex patterns of reference illustrated here.

One of the challenges for truth-conditional theories of sentence meaning is that referential ambiguities cannot be straightforwardly accounted for. This is because truth-conditional models rely upon the idea that each sentence has a truth value that can be assessed relative to a stable and objectively defined 'state of affairs', as we discussed earlier. A truth-conditional approach would be forced to claim that each interpretation arising from example (7) has a different set of truth conditions, which is inconsistent with the view that the meaning of a sentence can be modelled in terms of its truth or falsity relative to a given state of affairs. In other words, given a state of affairs in which Hitchcock liked the character Hitchcock-as-actor in the movie, the sentence in (7) would be simultaneously true (on the corresponding interpretation) and false (on the interpretation that Hitchcock liked the man at the bus stop). This gives rise to a logical inconsistency, because this model holds that a sentence cannot simultaneously be true and false in relation to the same state of affairs. In contrast to this view, because Mental Spaces theory holds that elements are set up in mental spaces rather than in some objectively defined 'state of affairs', no inconsistency arises in a single element having two distinct counterparts: it is possible, and even likely, that two or more distinct interpretations of a single sentence may coexist simultaneously.

11.4.7 Roles and values

An important aspect of Mental Spaces Theory is its treatment of NPs with definite interpretation, an issue that also relates to potential ambiguity. As we have seen, NPs of this kind include common nouns co-occurring with the definite article (*the President*) or proper nouns (*James Bond*). Mental Spaces Theory claims that NPs with definite interpretation do not have **rigid reference**, which means that they may or may not refer to a unique referent. This is illustrated by the following examples from Fauconnier (1994: 39):

(8) a. The president changes every seven years.
    b. Your car is always different.

The sentences in (8) are ambiguous. Example (8a) could mean that every seven years the person who is president changes in some way, for instance goes bald, becomes insane, grows a moustache and so on. Alternatively, (8a) could mean that every seven years the person who serves as president changes. Similarly, (8b) could mean that every time we see your car, some aspect of the car has changed; it might have had a respray, acquired some new hubcaps and so on. Alternatively, this sentence could mean that you have a new car every time we see you.

Ambiguities like these illustrate that NPs with definite interpretation can either have what Fauconnier calls a **role** reading or a **value** reading. For example, the role reading of *the President* relates to the position of president, regardless of who fills it (our second interpretation of (8a)). The value reading relates to the individual who fills the role (our first interpretation of (8a)). Roles and values both introduce elements into mental spaces, but each gives rise to different mapping possibilities. This is illustrated by example (9):

(9) Tony Blair is the Prime Minister. Margaret Thatcher thinks she is still the Prime Minister and Tony Blair is the Leader of the Opposition.

In the base, the elements *Tony Blair, Prime Minister* and *Margaret Thatcher* are all present. These are default elements established by the discourse or by encyclopaedic knowledge. This is indicated by the fact that they have definite reference, which shows that they are not set up as new elements but are pre-existing. In this base, *Tony Blair* is a value element linked to the role element *Prime Minister*. In other words, there is a role-value relationship holding between the two elements, which are co-referential. This relationship could be established on the basis of background knowledge, but in (9) it is explicitly

COGNITIVE LINGUISTICS: AN INTRODUCTION

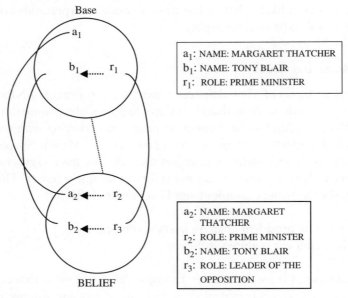

Figure 11.9 Roles and values

signalled by the first sentence. This relationship is captured in Figure 11.9 by the dotted arrows between the value element *Tony Blair* and the role element *the Prime Minister*. The second sentence sets up a new space, because it contains the space builder *Margaret Thatcher thinks*. . . . In Margaret Thatcher's BELIEF space, *she* (which is linked to *Margaret Thatcher* by an identity connector) corresponds to the value element linked to the role element *the Prime Minister*, while *Tony Blair* corresponds to the value element linked to the role element *the Leader of the Opposition*. Figure 11.9 illustrates the interpretation of roles and values in example (9).

11.5 An illustration of mental space construction

In this section, we analyse a short text so that we can apply some of the aspects of mental space construction that have been introduced so far. Although this text is very simple, it nevertheless involves meaning construction processes of considerable complexity.

(10) Fido sees a tortoise. He chases it. He thinks that the tortoise is slow. But it is fast. Maybe the tortoise is really a cat.

As we have seen, mental space construction always proceeds by the establishment of a base that represents the starting point for any particular stage in the discourse. We can think of 'stages' in discourse as topics of conversation. Elements are introduced into the base by indefinite descriptions or are identified as pre-existing by definite descriptions or by non-linguistic factors such as contextual salience. Salience can arise in a number of ways, for example if the speaker is referring to something that is visible or familiar to both speaker and hearer (*Pass me the scissors*) or something they have been discussing previously (*I found the book*). The first sentence in (10) provides a definite description, *Fido*. This is in presuppositional mode, which signals that the element *Fido* is present in the discourse context. Observe that we can make this assumption regardless of whether we have access to the previous discourse context. If (10) is part of a spoken story, for example, we probably already know who or what Fido is. But if (10) begins a written story, we 'construct' this background context. This element is therefore set up in the base space as part of the background. Moreover, Fido is a name, and background knowledge tells us that it is a name typically associated with a male dog. We can therefore deduce that the expression refers to a dog. There is also an indefinite description in this sentence: *a tortoise*. The indefinite description introduces a new element to the discourse, and this is set up in the base space. The verb *see* introduces a relation between the two elements based on a SEE frame which involves at least two participant roles: SEER and SEEN. This frame is projected to the base space by means of schema induction, and the SEER role is mapped onto *Fido* (element a_1) while the SEEN role is mapped onto *a tortoise* (element b_1). This is illustrated in Figure 11.10.

The second sentence employs the anaphors *he* and *it*. Because we already know from background knowledge that the name *Fido* refers to a male animal, *he* identifies a_1 in the base space and *it* refers to the animal whose sex has not been identified: element b_1. The verb *chase* prompts for further structure to be added to the base space: the projection of the CHASE frame via schema induction. Like the SEE frame, CHASE also has two participant roles: CHASER and

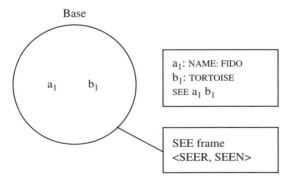

Figure 11.10 *Fido sees a tortoise*

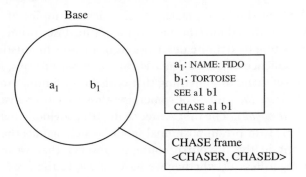

Figure 11.11 *He chases it*

CHASED. These are mapped onto a_1 and b_1, respectively. This is illustrated by Figure 11.11.

The third sentence contains the space builder, *he thinks that*. This sets up a new BELIEF space which is established relative to the base. *He* prompts for a_2, a counterpart of a_1 (*Fido*), while *the tortoise* introduces an element in the presuppositional mode because this element has already been introduced into the discourse by the indefinite expression *a tortoise*. This prompts for a counterpart in the base: *the tortoise* introduces element b_2, counterpart of b_1 (*a tortoise*). In both cases, the pragmatic function that links the counterparts is the identity relation. The Access Principle entails that connectors are established between the counterparts and the Optimisation Principle ensures that information in the base space is automatically transferred to the new belief space. This means that the properties and relations holding for the counterparts of a_1 and b_1 – namely a_2 and b_2 – are set up in the belief space. This includes the participant roles that follow from the SEE and CHASE frames. In addition, the property SLOW is associated with b_2 (*the tortoise*) in Fido's BELIEF space. This is represented by Figure 11.12.

In the fourth sentence, new information is added which states that the tortoise is fast. Because this information relates to reality, it is added to the base space rather than to Fido's BELIEF space. The use of *but*, which introduces a **counter-expectational** interpretation, overtly signals that the Optimisation Principle does not apply to this information, which means that the information that the tortoise is fast is limited to the base space. This is because information in the BELIEF space, namely that the tortoise is slow, contradicts information in the base. In this way, the Optimisation Principle prevents contradictory information (that the tortoise is fast) from spreading to the BELIEF space: Fido cannot simultaneously think that the tortoise is slow and that the tortoise is fast. This is illustrated in Figure 11.13.

The final sentence includes the space builder *maybe*. This sets up a POSSIBILITY space. In this space, the counterpart of the tortoise (b_1) is a cat (b_3).

MEANING CONSTRUCTION AND MENTAL SPACES

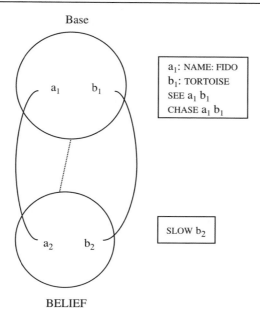

Figure 11.12 *He thinks that the tortoise is slow*

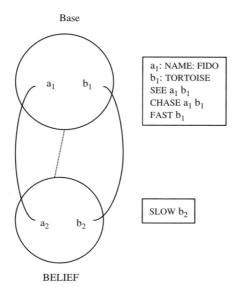

Figure 11.13 *But it is fast*

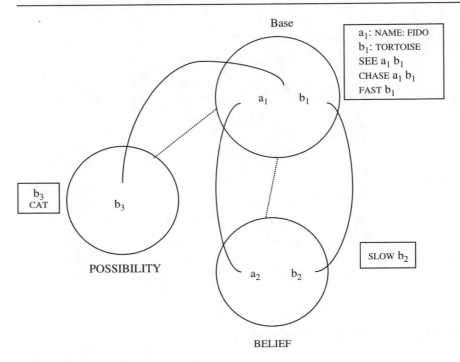

Figure 11.14 *Maybe the tortoise is really a cat*

The expression *really* signals that this POSSIBILITY space is set up from the perspective of the base space rather than from the perspective of Fido's BELIEF space, because the base space is the reality space (see Figure 11.14).

As this relatively simple example demonstrates, even a short piece of discourse involves active participation on the part of the hearer/reader in terms of the construction of a number of different mental spaces in which linked but potentially contradictory information can be held. This model goes some way towards explaining the complex cognitive operations that go on 'in the background' during meaning construction, and shows how language prompts for knowledge within the conceptual system. In the next section, we look at how Mental Spaces Theory can account for two other aspects of linguistic meaning: counterfactual *if* . . . *then* . . . constructions and the tense-aspect-modality (TAM) system.

11.6 The dynamic nature of meaning construction

In this section we focus on the dynamic aspect of meaning construction. This relates to the way in which **interlocutors** (discourse participants) keep track of the spaces that have been set up during ongoing discourse, including the content of the various spaces, the links between them and their sequence. Language assists in this process in two main ways: (1) the grammatical tense-aspect system signals **time reference** (the location in time of one space relative to another); and (2) the grammatical system of **epistemic modality** signals **epistemic distance**. Epistemic modality is a type of grammatical marking that reflects the speaker's knowledge or opinion concerning the likelihood, possibility or certainty of the proposition expressed by a sentence. Epistemic modality therefore concerns the reality status of one space with respect to another. Because tense, aspect and modality are often closely interwoven within the grammatical systems of languages, this area is often abbreviated to the 'TAM' system. We explore the Mental Spaces Theory approach to these two aspects of the TAM system in the following sections.

11.6.1 Tense and aspect in English

We begin by looking at how the English tense-aspect system prompts for information relating to the timing of events. To begin with the fundamentals, tense is a feature of the closed-class system, usually marked morphologically on verbs or independent inflection words. Tense marks a sentence with information concerning the time of the event described relative to the moment of speaking. Present tense signals that the time referred to and the time of speaking are equivalent. Past tense signals that the time referred to precedes the time of speaking. Future tense signals that the time referred to follows the time of speaking. Linguists often use a relatively simple representational system to capture the relationship between event time and time of speaking called the **SER (Speech-Event-Reference) system** (Reichenbach 1947). In this system, S stands for 'moment of speaking' and R stands for 'reference time' (the time referred to in the utterance).

(11) Past tense: $R < S$
     Present tense: $S = R$
     Future tense: $S < R$

In English, present and past tense are marked on the verb with suffixes, but in the present tense this suffix is only marked on the third person singular *he/she/it* form in the case of most verbs (for example, *I/you/we/they sing* vs. *she sing-s*). However, the 'irregular' verb *be* shows a wider range of present tense forms (*I am, you/we/they are, he/she/it is*). Past tense is marked on many verbs by the suffix *-ed* (for example, *I played*). Strictly speaking, English lacks a future tense, because there is no bound morpheme indicating future time that forms part of the same grammatical system as present and past tense. However, English has a number of ways of referring to future time, including the use of the modal verb *will*, for example *I will sing*, which we can loosely refer to as future tense.

Tense interacts with grammatical aspect (see Chapter 18 for the distinction between grammatical and lexical aspect). Unlike tense, aspect does not refer to the time of the event described relative to the moment of speaking, but instead describes whether the event is viewed as 'completed' or 'ongoing'. The traditional term for a 'completed' event is **perfect aspect** and traditional terms for an 'ongoing' event include the terms **imperfect** or **progressive aspect**. In English, perfect aspect is introduced by the **auxiliary verb** *have* (for example, *I have finished*) and progressive aspect is introduced by the auxiliary verb *be* (for example, *I am singing*). For novice linguists, this is a difficult system to get to grips with, not least because the verbs *have* and *be* do not always function as auxiliary verbs. They can also function as **lexical verbs**. The easiest way to tell the difference between auxiliary and lexical verbs is that the former are followed by another verb form called a **participle** (*I am singing; You have finished*), while the latter are not (*I am hungry; You have green eyes*). In the SER system, aspect is represented as the interaction between R (reference time) and E (event). In the case of perfect aspect, the whole completed event is located prior to the reference time, indicating that, relative to the time referred to in the utterance, the event is viewed as 'completed':

(12) Perfect aspect: $E < R$

Progressive aspect is represented in the SER system as B . . . F (which stand for 'beginning' and 'finish', respectively). These 'surround' the reference time, indicating that the event is viewed by the speaker as 'ongoing' relative to the time referred to in the utterance:

(13) Progressive aspect: $B < R < F$

Tense and aspect can 'cut across' one another within the tense-aspect system. In other words, they can be combined to produce a large number of different permutations. Some of these are shown in example (14), together with the relevant SER 'timeline' diagrams:

(14) a. James Bond has outwitted the villain (now)
         ←———E——R = S——→   [present perfect]

     b. James Bond had outwitted the villain
         ←———E——R——S——→   [past perfect]

     c. James Bond will have outwitted the villain (by teatime)
         ←———S——E——R——→   [future perfect]

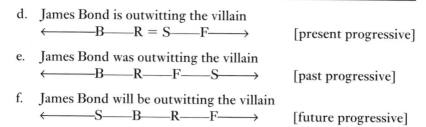

d. James Bond is outwitting the villain
  ←——————B——R = S——F——→   [present progressive]

e. James Bond was outwitting the villain
  ←——————B——R——F——S——→   [past progressive]

f. James Bond will be outwitting the villain
  ←——————S——B——R——F——→   [future progressive]

The aspect of each example can be identified according to whether the 'verb string' contains *have* (perfect) or *be* (progressive). Observe that these auxiliaries also require the verb that follows them to assume a particular form. The perfect auxiliary *have* requires the next verb to be in its **past participle** form. This term from traditional grammar is rather misleading since it implies that the past participle is restricted to past tense contexts. As examples (14a) and (14c) illustrate, this is not the case. It can also be difficult to identify the past participle because it often looks just like the past tense form (for example, *outwitted*), but certain verbs have distinct past tense/past participle forms (for example, *I wrote* [past tense] vs. *I have written* [past participle]). The progressive auxiliary *be* requires the verb that follows it to occur in the **progressive participle** form, which ends in *-ing*. These verb forms are called participles because they form a subpart of a tense-aspect configuration, and crucially they cannot 'stand alone' without an auxiliary verb (for example, *I written; *I singing).

The tense of each example can be identified by the form of the auxiliary verb. If this verb is present, past or future (marked by *will*), the whole clause has that tense property. For example, (14a) is in the present tense because the auxiliary *have* is in the (third person singular) present tense form *has*. Although the event is viewed as completed, it is viewed from the perspective of the moment of speaking; this is why present perfect configurations can be modified by the temporal expression *now*. Example (14b) is in the past tense because the auxiliary *have* is in its past tense form: *had*.

11.6.2 The tense-aspect system in Mental Spaces Theory

According to Mental Spaces Theory, the tense-aspect system participates in **discourse management**. Before we can look in detail at the Mental Spaces Theory analysis of tense-aspect systems, we need to establish some additional new terms: **viewpoint**, **focus** and **event**. These terms relate to the status of mental spaces in discourse. While the base represents the starting point for a particular stage in the discourse to which the discourse can return, the viewpoint is the space from which the discourse is currently being viewed and from which other spaces are currently being built. The focus is the space where new content is being added, and the event represents the time associated with the event being

Figure 11.15 *Jane is twenty*

described. While the focus and event spaces often coincide, as we will see, they can sometimes diverge. As discourse progresses, the status of mental spaces as base, viewpoint, focus or event can shift and overlap. In order to illustrate these ideas, consider the following text, in which the verb strings are underlined:

(15) Jane is twenty. She has lived in France. In 2000 she lived in Paris. She currently lives in Marseilles. Next year she will move to Lyons. The following year she will move to Italy. By this time, she will have lived in France for five years.

We will construct a Mental Spaces Theory representation of this text beginning with the base (B). The base space is also the initial viewpoint (V) and the focus (F), as we add new information to the base, namely that Jane is twenty. Time reference is now (E), as signalled by the present tense 'is'. This is illustrated in Figure 11.15, which represents the first space constructed by this text (space 1). In this section, we simplify the mental spaces diagrams by missing out the dialogue boxes, since our objective here is not to illustrate the establishment of elements, links, properties or relations, but to work out how the sentences in the discourse set up mental spaces that shift the status of previously constructed spaces with respect to base, viewpoint, focus and event.

The second sentence, *She has lived in France*, keeps the base in focus, as it adds new information of current relevance. This is signalled by the use of the present perfect *has lived*. The present tense auxiliary form *has* signals that we are building structure in space 1 which thus remains the focus space. However, the structure being built relates to an event that is complete (or past) relative to space 1, signalled by the past participle *lived*. This is set up as space 2. In this way, perfect aspect signals that focus and event diverge. Put another way, the present perfect *has lived* signals that knowledge of a completed event has current relevance. Because the focus space, 'now' (space 1), is also the perspective from which we are viewing the completed event, the focus space (space 1) is also the viewpoint. This is illustrated by Figure 11.16.

The third sentence, *In 2000 she lived in Paris*, contains the space builder *in 2000*. This sets up a new space, which is set in the past with respect to the viewpoint space which remains in the base (space 1). This new space (space 3) is therefore the event space. Because we have past tense marking, the focus shifts to the new space. This is illustrated in Figure 11.17.

MEANING CONSTRUCTION AND MENTAL SPACES

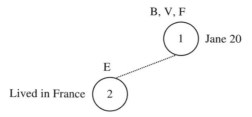

Figure 11.16 *She has lived in France*

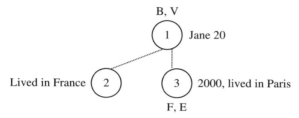

Figure 11.17 *In 2000 she lived in Paris*

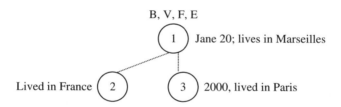

Figure 11.18 *She currently lives in Marseilles*

The fourth sentence, *She currently lives in Marseilles*, is marked for present tense. This returns the focus to the base space (space 1). The base also remains the viewpoint, because this is now the perspective from which the lattice is being viewed. Because the time reference relates to this space, this is also the event space. This is illustrated in Figure 11.18.

The fifth sentence, *Next year she will move to Lyons*, is marked for future tense. Together with the future tense, the space builder *next year* sets up a new space which is the current focus space (space 4). The event described in this space is future relative to the viewpoint, which remains in the base (space 1). This is illustrated in Figure 11.19.

In the penultimate sentence, *The following year she will move to Italy*, the space builder *the following year* sets up a new space which is the current focus space

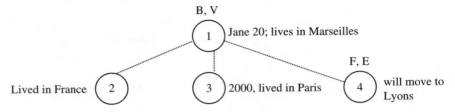

Figure 11.19 *Next year she will move to Lyons*

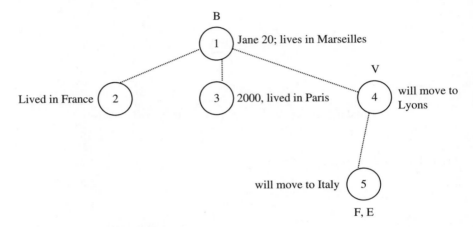

Figure 11.20 *The following year she will move to Italy*

containing the information that Jane will move to Italy (space 5). The future tense signals that the event is future relative to the base (space 1). However, the space builder *the following year* also shows that the new event space (space 5) is also future relative to space 4, from which the current space under construction is viewed. Hence, the viewpoint shifts from the base to space 4. This is illustrated in Figure 11.20.

In the final sentence, *By this time, she will have lived in France for five years*, the use of the future perfect auxiliary *will have* signals that the space in focus is the future space, space 5. However, the structure being built relates to a completed event, signalled by the past participle form *lived*. The future perfect *will have lived* therefore establishes an event space (space 6) that relates to a completed event: an event that is past with respect to the focus space. Thus the time of the event space diverges from the time of the focus space with respect to which it is relevant. This means that the focus remains in space 5 where structure is being added. The viewpoint remains in space 4 because it is from the perspective of her time in France that this sentence is viewed. At this point in the discourse, as Figure 11.21 illustrates, the base, viewpoint, focus and event all relate to distinct spaces.

MEANING CONSTRUCTION AND MENTAL SPACES

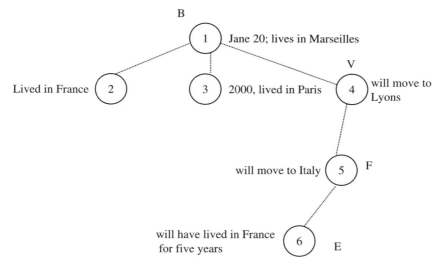

Figure 11.21 *By this time, she will have lived in France for five years*

The use of the future tense in this final sentence shows that the current space is still connected to the base space to which the discourse could return. For instance, if the discourse continued with the sentence *But at present Jane is happy in Marseilles*, this would return viewpoint, focus and event to the base.

As this discussion reveals, the tense-aspect system 'manages' the perspective from which an utterance is made. In particular, we have seen that while temporal adverbials like *in 2000* set up new spaces, it is the tense-aspect system that signals the perspective from which a particular space is viewed. Before completing this discussion of the tense-aspect system, we briefly mention progressive aspect. As noted earlier, this is signalled in English by the progressive auxiliary *be* and the progressive participle, ending in *-ing* (e.g. *Lily is writing a letter*, which illustrates the present progressive). As with perfect aspect, progressive aspect signals that event and focus spaces diverge. While the perfect signals that a completed event has current relevance in the focus space, progressive aspect signals that the focus space occurs during the event space. In other words, the focus space for the sentence *Lily is writing a letter* contains a schematic event that receives its complete temporal profile only in the event space. (For full details, see Cutrer (1994), a doctoral thesis that develops the Mental Spaces Theory account of the tense-aspect system.)

Table 11.1 summarises the functions of tense and aspect in terms of discourse management. In this table, X refers to a given mental space and the term 'simple' means that the relevant sentence that builds the space is not marked for aspect.

393

Table 11.1 The role of tense and aspect in discourse management

	Present (simple)	Past (simple)	Future (simple)	Perfect	Progressive
Focus	X	X	X	Not X	Not X
Viewpoint	X	X's parent	X's parent	X's parent or grandparent	X's parent or grandparent
Event	X equivalent to V	X before V	X after V	X is completed with respect to F	X contains F

11.6.3 Epistemic distance

In addition to its time reference function, tense can also signal epistemic distance. In other words, polysemy is not restricted to the open-class elements: tense, as part of the closed-class semantic system also exhibits polysemy. This means that the tense system has a range of distinct schematic meanings associated with it (Tyler and Evans 2001a). One illustration of this point relates to the use of tense in hypothetical constructions such as 'if A then B', which we briefly discuss in this section. Consider example (16).

(16) If the President agrees with the senator's funding request, then the senator has nothing to worry about

A and B refer to the two propositions that make up this complex sentence. In example (16), A stands for the antecedent: *the President agrees with the senator's funding request* and B stands for the consequent: *the senator has nothing to worry about*. According to Mental Spaces Theory, 'if A then B' constructions set up two successive spaces in addition to the base which is the reality space. The two successive spaces are the **foundation** space and the **expansion** space. The foundation space is a hypothetical space set up by the space builder *if*. The expansion space is set up by the space builder *then*. While the foundation space is hypothetical relative to the base, whatever holds in the expansion space is 'fact' relative to the foundation space, in the sense that it is entailed by the information in the foundation space (see Figure 11.22). In other words, if A (the foundation) holds, then B (the expansion) follows.

In order to uncover the role of 'if A then B' constructions in epistemic distance, consider the sentences in example (17).

(17) a. If I win the lottery, I will buy a Rolls-Royce.
b. If I won the lottery I would buy a Rolls-Royce.

MEANING CONSTRUCTION AND MENTAL SPACES

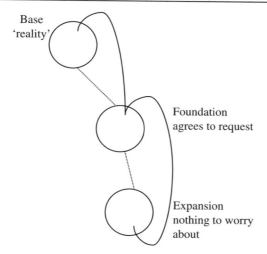

Figure 11.22 Foundation and expansion spaces

The first sentence expresses a neutral **epistemic stance** while the second expresses **epistemic distance**. Epistemic stance relates to the speaker's assessment of how likely a particular foundation-expansion sequence is relative to a particular reality base space. As we have seen, the term 'epistemic' relates to the speaker's knowledge or opinion concerning likelihood, possibility, certainty or doubt, and the terms 'epistemic stance' and 'epistemic distance' both rely on the speaker's metaphorical 'distance' from a particular state of affairs: the speaker's 'position' or judgement regarding the likelihood of a particular situation coming about. Notice that in sentence (17a), the *if* clause is in the present tense, although it refers to (hypothetical) future time. This example illustrates that the English present tense is not restricted to referring to present time. In (17a), the speaker is making no assessment in relation to epistemic distance; this sentence is purely hypothetical. In other words, the speaker takes a neutral or 'open' position with respect to the likelihood of winning the lottery. Observe that this sentence would be appropriate in a context in which the speaker regularly plays the lottery and therefore has a chance of winning.

The sentence in (17b) is also a hypothetical, but here the speaker is indicating epistemic distance by the use of the past tense in the *if* clause. This sentence might be uttered in a scenario in which the speaker doesn't actually play the lottery, or judges his or her chances of success as minimal or non-existent. This type of *if* . . . *then* . . . sentence, which refers to a non-existent situation, is called a **counterfactual**. Finally, compare the form of the modal verbs in the *then* clauses in these two examples. The form *will* in (17a) is traditionally described as the present tense form, while the form *would* in (17b) is described as the past tense form.

As the examples in (17) illustrate, the tense system can be used for more than signalling reference time. It can also be used to signal epistemic stance. The examples considered so far have not been marked for grammatical aspect: (17a) is in the 'simple present' and (17b) is in the 'simple past'. However, if we introduce perfect aspect into the *if* clause, the result is striking. Consider the following example:

(18) If I had won the lottery, I would have bought a Rolls-Royce.

This counterfactual example is in the past perfect form and is therefore marked for both past tense and perfect aspect. The result is increased epistemic distance. This example might be appropriate in a context where the speaker did in fact play the lottery but lost.

11.7 Summary

This chapter introduced **Mental Spaces Theory**, the cognitive semantics approach to meaning construction. This theory is associated most prominently with the influential work of Gilles Fauconnier. According to this view, meaning construction is a process that is fundamentally conceptual in nature. Sentences constitute **partial instructions** for the construction of highly complex and intricate conceptual **lattices** which are temporary, can be more or less detailed and are assembled as a result of ongoing discourse. These temporary domains, called **mental spaces**, are linked in various ways and contain elements that are also connected, allowing speakers to keep track of **chains of reference**. From this perspective, meaning is not a property of individual sentences nor of their interpretation relative to some objectively defined 'state of affairs' as in **formal semantics**. Instead, meaning arises from a dynamic process of meaning construction which we call **conceptualisation**. While our conceptualisations may or may not be about the 'real world', we keep track during ongoing discourse of elements, properties and relations in the complex mental space configurations assembled as we think and speak. From this perspective, sentences cannot be analysed in isolation from ongoing discourse, and **semantic** meaning, while qualitatively distinct, cannot be meaningfully separated from **pragmatic** meaning. From this perspective, meaning construction is a dynamic process, and is inseparable from context. Finally, because meaning construction is fundamentally conceptual in nature, we must also take account of the general cognitive processes and principles that contribute to this process. In particular, meaning construction relies on mechanisms of **conceptual projection** such as metaphors and metonymies and connectors. In this chapter, we saw how Mental Spaces Theory accounts for a diverse range of linguistic phenomena relating to meaning at the level of sentence and text, including referential ambiguities and the role of **tense** and **aspect** in **discourse management** and in **epistemic distance**.

Further reading

Foundational texts

- **Fauconnier (1994).** First published in English in 1985 based on a previously published French text, this is the foundational text that introduces the main tenets of Mental Spaces Theory. The 1994 edition provides a preface that traces some of the original motivations for the developments of the theory and provides an accessible introduction to some of the key ideas.
- **Fauconnier (1997).** This book is perhaps more accessible than *Mental Spaces*. Not only does it revise and extend the basic architecture, it also provides an overview of some of the key insights of the earlier work, and shows how the Mental Spaces framework has been extended giving rise to Blending Theory (discussed in the next chapter).

Applications of Mental Spaces Theory

- **Cutrer (1994).** In her doctoral thesis, Cutrer investigated how tense and aspect give rise to dynamic aspects of mental space construction.
- **Fauconnier and Sweetser (eds) (1996).** This volume contains a collection of articles by prominent cognitive semanticists who apply Mental Spaces Theory to a range of linguistic phenomena including grammar, metaphor, lexical polysemy, deixis and discourse.

Exercises

11.1 Assumptions of Mental Spaces Theory

What are the main assumptions of Mental Spaces Theory?

11.2 Space building

Provide an answer to each of the following questions, and illustrate with examples of your own:

(i) How are mental spaces set up?
(ii) How are they internally structured?
(iii) How are they related to each other?

11.3 Diagramming a mental space lattice

Provide a mental space configuration for the following text:

> *The witch is riding a unicorn. She thinks she's riding a horse and the horse has a blue mane.*

11.4 Referential ambiguity

Provide a mental spaces lattice for the following sentence. Based on the various connectors prompted for, explain how the referential ambiguity is accounted for.

> *I dreamed that I was Naomi Campbell and that I kissed me.*

11.5 Viewpoint, focus and event

Provide definitions of the terms viewpoint, focus and event, and illustrate with examples of your own.

11.6 Shift in viewpoint (advanced)

In view of your answers to exercise 11.5, provide a mental space configuration for the following text. In particular, provide an account of how tense signals a shift in the viewpoint, focus or event. (*Note:* In this example, *would* signals future perspective in the past.)

> *In 1995 John was living in London for the first time. In 1997 he would move to France. By this time he would have lived in London for two years.*

11.7 Foundation and expansion spaces

How are the following kinds of mental spaces different? Provide examples of your own to illustrate your answer.

(a) Base
(b) Foundation
(c) Expansion

11.8 Practice with foundation, expansion and possibility spaces

Once you have completed exercises 11.3 and 11.7, add to the mental space configuration you developed in exercise 11.3 the structure prompted for by the sentence below.

But she's flying through the air. If she were riding a horse, then she would not be flying through the air.

11.9 Hypotheticals versus counterfactuals

Does the mental space configuration constructed for exercise 11.7 involve a hypothetical or a counterfactual? What is the difference? How is this difference prompted for by language?

11.10 Foundation spaces again

Diagram a mental spaces lattice for the text given below. Explain how each sentence prompts for the addition of structure to the mental space lattice. Relative to which space is the foundation built? Explain your reasoning.

John has a pet cat. It's called Fred. Next year John will buy a dog. Maybe the cat will like the dog. If the cat doesn't like the dog, then John will have to keep them in separate parts of the house.

12

Conceptual blending

The subject of this chapter is the theory known either as **Conceptual Integration** or **Conceptual Blending Theory**. This approach, which we will call **Blending Theory**, derives from two traditions within cognitive semantics: Conceptual Metaphor Theory and Mental Spaces Theory, which we introduced in Chapters 9 and 11, respectively. In terms of its architecture and in terms of its central concerns, Blending Theory is most closely related to Mental Spaces Theory, and some cognitive semanticists explicitly refer to it as an extension of this approach. This is due to its central concern with **dynamic** aspects of meaning construction and its dependence upon mental spaces and mental space construction as part of its architecture. However, Blending Theory is a distinct theory that has been developed to account for phenomena that Mental Spaces Theory and Conceptual Metaphor Theory cannot adequately account for. Moreover, Blending Theory adds significant theoretical sophistication of its own. The crucial insight of Blending Theory is that meaning construction typically involves integration of structure that gives rise to more than the sum of its parts. Blending theorists argue that this process of **conceptual integration** or **blending** is a general and basic cognitive operation which is central to the way we think. For example, as we saw in Chapter 8, the category PET FISH is not simply the intersection of the categories PET and FISH (Fodor and Lepore 1996). Instead, the category PET FISH selectively integrates aspects of each of the source categories in order to produce a new category with its own distinct internal structure. This is achieved by conceptual blending.

One of the key claims of cognitive semantics, particularly as developed by conceptual metaphor theorists, is that human imagination plays a crucial role in cognitive processes and in what it is to be human. This theme is further developed by Gilles Fauconnier and Mark Turner, the pioneers of Blending Theory. Blending Theory was originally developed in order to account for linguistic structure and for the role of language in meaning construction, particularly 'creative' aspects of meaning construction like novel metaphors, counterfactuals and so on. However, recent research carried out by a large international community of academics with an interest in Blending Theory has given rise to the view that conceptual blending is central to human thought and imagination, and that evidence for this can be found not only in human language, but also in a wide range of other areas of human activity, such as art, religious thought and practice, and scientific endeavour, to name but a few. Blending Theory has been applied by researchers to phenomena from disciplines as diverse as literary studies, mathematics, music theory, religious studies, the study of the occult, linguistics, cognitive psychology, social psychology, anthropology, computer science and genetics. In their (2002) book, *The Way We Think*, Fauconnier and Turner argue that our ability to perform conceptual integration or blending may have been the key mechanism in facilitating the development of advanced human behaviours that rely on complex symbolic abilities. These behaviours include rituals, art, tool manufacture and use, and language.

12.1 The origins of Blending Theory

The origins of Blending Theory lie in the research programmes of Gilles Fauconnier and Mark Turner. While Fauconnier had developed Mental Spaces Theory in order to account for a number of traditional problems in meaning construction, as we saw in the previous chapter, Turner approached meaning construction from the perspective of his studies of metaphor in literary language. Fauconnier and Turner's research programmes converged on a range of linguistic phenomena that appeared to share striking similarities and that resisted straightforward explanation by either of the frameworks they had developed. Fauconnier and Turner both observed that in many cases meaning construction appears to derive from structure that is apparently unavailable in the linguistic or conceptual structure that functions as the input to the meaning construction process. Blending Theory emerged from their attempts to account for this observation.

We begin our overview of Blending Theory with an example of the kind of linguistic phenomenon that motivated the development of this approach. The following example is metaphorical in nature, and yet cannot be straightforwardly accounted for by Conceptual Metaphor Theory:

(1) That surgeon is a butcher.

Table 12.1 Mappings for SURGEON IS A BUTCHER

Source: BUTCHER	mappings	Target: SURGEON
BUTCHER	→	SURGEON
CLEAVER	→	SCALPEL
ANIMAL CARCASSES	→	HUMAN PATIENTS
DISMEMBERING	→	OPERATING

Within the conceptual metaphor tradition, examples like (1) have been explained on the basis of a mapping from a source domain onto a target so that the target is understood in terms of the metaphorically projected structure. Applying this explanation to the example in (1), the target domain SURGEON is understood in terms of the source domain BUTCHER. In the source domain we have a butcher, a cleaver and an animal's carcass that the butcher dismembers. In the target domain we have a surgeon, a scalpel and a live but unconscious patient on whom the surgeon operates. The mappings are given in Table 12.1.

The difficulty that this example poses for Conceptual Metaphor Theory is that the sentence in (1) actually implies a negative assessment (Grady, Oakley and Coulson 1999). Although butchery is a highly skilled profession, by conceptualising a surgeon as a butcher we are evaluating the surgeon as incompetent. This poses a difficulty for Conceptual Metaphor Theory because this negative assessment does not appear to derive from the source domain BUTCHER. While the butcher carries out work on dead animals, there is considerable expertise and skill involved, including detailed knowledge of the anatomy of particular animals, knowledge of different cuts of meat and so on. Given that butchery is recognised as a skilled profession, questions arise concerning the conceptual origin of the negative assessment arising from this example. Clearly, if metaphor rests on the mapping between pre-existing knowledge structures, the emergence of new meaning as a consequence of this mapping operation is not explained by Conceptual Metaphor Theory: how does the negative assessment of incompetence arise from conceptualising one highly skilled professional in terms of another?

This example points to powerful aspects of human cognition. Language and thought are not strictly compositional in the sense that they are **additive**. In other words, meaning construction cannot rely solely upon 'simple' conceptual projection processes like structuring one conceptual region in terms of another, as in the case of conceptual metaphors, or establishing connectors between counterparts in mental spaces. In example (1), the negative assessment is obvious and appears to be the driving force behind describing a surgeon as a butcher, yet this negative evaluation seems to be contained in neither of the input domains associated with the metaphor. Blending Theory accounts for the emergence of meanings like these by adopting the view that meaning construction involves **emergent structure**: meaning that is more than the sum of its component parts.

In this chapter, we present an overview of how Fauconnier and Turner draw together aspects of Conceptual Metaphor Theory and Mental Spaces Theory in order to account for these emergent aspects of meaning. We begin by mapping out the architecture of Blending Theory (section 12.2), and then look at how it is applied to both linguistic and non-linguistic examples of meaning construction (section 12.3). We then explore the cognitive basis of conceptual blending (section 12.4) and examine Fauconnier and Turner's claim that a small number of **integration networks** underlie the process of meaning construction (section 12.5). Finally, we look at the constraints on Blending Theory in terms of its theoretical machinery (section 12.6) and provide some explicit comparisons between Blending Theory and Conceptual Metaphor Theory (section 12.7).

12.2 Towards a theory of conceptual integration

In attempting to account for examples like the SURGEON AS BUTCHER metaphor, Fauconnier and Turner took aspects of the two frameworks they had developed and produced a theory of **integration networks**. An integration network is a mechanism for modelling how emergent meaning might come about. Fauconnier and Turner suggest that an integration network consists of **inputs** in which elements in each input are linked by mappings (see Figure 12.1). In this respect, Blending Theory draws upon Conceptual Metaphor Theory. Recall that Conceptual Metaphor Theory represents a two-domain model in which domains are linked by conventional mappings relating comparable elements.

From Mental Spaces Theory, Fauconnier and Turner took the idea that the conceptual units that populate an integration network should be Mental Spaces rather than domains of knowledge, as in Conceptual Metaphor Theory. As we have seen in previous chapters, the difference between the two is that domains of knowledge are relatively stable pre-existing knowledge structures, while mental spaces are temporary structures created during the on-line process of meaning construction. Therefore, the initial focus in Blending

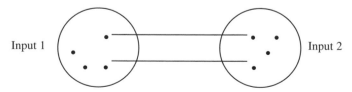

Figure 12.1 Mappings of elements across inputs

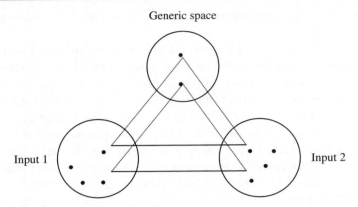

Figure 12.2 Addition of a generic space

Theory was to account for local and dynamic meaning construction, a focus that is inherited from Mental Spaces Theory.

Moreover, integration networks in Blending Theory are not simply two-space entities. Because these networks represent an attempt to account for the dynamic aspects of meaning construction, they are multiple-space entities, just like mental space lattices. One of the ways in which this model gives rise to complex networks is by linking two (or more) **input spaces** by means of a **generic space**. The generic space provides information that is abstract enough to be common to both (or all) the inputs. Indeed, Fauconnier and Turner hypothesise that integration networks are in part licensed by interlocutors identifying the structure common to both inputs that licenses integration. Elements in the generic space are mapped onto counterparts in each of the input spaces, which motivates the identification of cross-space counterparts in the input spaces. This is illustrated in Figure 12.2.

A further distinguishing feature of an integration network is that it consists of a fourth **blended space** or **blend**. This is the space that contains new or **emergent structure**: information that is not contained in either of the inputs. This is represented by the blended space in Figure 12.3. The blend takes elements from both inputs, as indicated by the broken lines, but goes further in providing additional structure that distinguishes the blend from either of its inputs. In other words, the blend **derives** structure that is contained in neither input. In Figure 12.3, this emergent structure or 'novel' meaning is represented by the elements in the blended space that are not connected to either of the inputs.

That surgeon is a butcher: the blending theory account

Having set out the basic architecture of the Blending Theory model, we outline an analysis of the SURGEON AS BUTCHER metaphor from a Blending Theory

CONCEPTUAL BLENDING

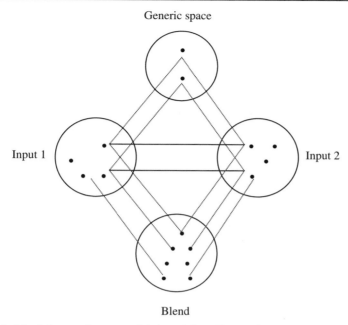

Figure 12.3 A basic integration network (adapted from Fauconnier and Turner 2002: 46)

perspective. As noted by Grady, Oakley and Coulson (1999), Blending Theory is able to account for the negative assessment associated with this utterance by allowing for emergent structure. This follows from the fact that, while a blend contains structure projected from both inputs, it also contains additional structure projected from neither. In the input space for BUTCHER, we have a highly skilled professional. However, in the blend, these skills are inappropriate for performing surgery on human patients. While surgeons attempt to save lives, butchers perform their work on dead animals. While the activity performed by butchers is dismembering, the activity performed by surgeons typically involves repair and reconstruction, and so on. The consequence of these contrasts is that in the blend a surgeon who is assessed as a butcher brings inappropriate skills and indeed goals to the task at hand and is therefore incompetent. This emergent meaning of incompetence represents the additional structure provided by the blend.

The emergent structure provided by the blend includes the structure copied from the input spaces, together with the emergent structure relating to a surgeon who performs an operation using the skills of butchery and is therefore incompetent. This individual does not exist in either of the input spaces. The structure in the blend is 'emergent' because it emerges from 'adding together' structure from the inputs to produce an entity unique to the blend. Furthermore, it is precisely by virtue of the mismatch between goal (healing)

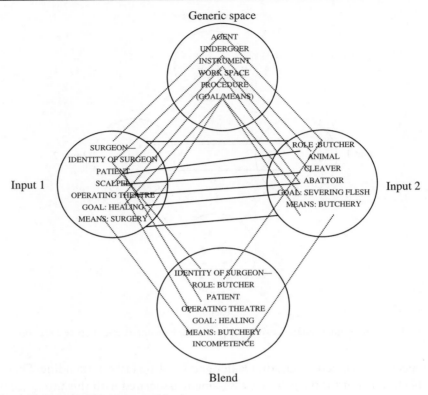

Figure 12.4 SURGEON as BUTCHER blend

and means (butchery), which exists only in the blend, that the inference of incompetence arises. This means that all the structure in the blend can be described as emergent, even though its 'ingredients' are provided by the input spaces. Finally, we address the role of the generic space in this integration network. As we noted earlier, the generic space contains highly schematic information which serves as a basis for establishing cross-space mappings between the two input spaces. In other words, the generic space facilitates the identification of counterparts in the input spaces by serving as a 'template' for shared structure. It is these counterparts that can then be projected to the blend. The integration network for this blend is illustrated in Figure 12.4.

While metaphors of this kind originally motivated Fauconnier and Turner's development of Blending Theory, this approach applies equally to non-metaphorical instances of meaning construction. Consider the counterfactual example (2), which we discussed in Chapter 5.

(2) In France, Bill Clinton wouldn't have been harmed by his relationship with Monica Lewinsky.

As with the SURGEON AS BUTCHER metaphor, this counterfactual prompts for a complex conceptualisation that is more than the sum of its parts. In particular, it involves the conceptual blending of counterparts in order to produce a blend in which Clinton is not politically harmed by his relationship with Lewinsky, an emergent meaning that does not exist in either of the inputs that give rise to it.

The integration network for this expression includes two inputs. One input space contains CLINTON, LEWINSKY and their RELATIONSHIP. This space is structured by the frame AMERICAN POLITICS. In this frame, there is a role for AMERICAN PRESIDENT, together with certain attributes associated with this role such as MORAL VIRTUE, a symbol of which is marital fidelity. In this space, marital infidelity causes political harm. In the second input space, which is structured by the frame FRENCH POLITICS, there is a role for FRENCH PRESIDENT. In this frame, it is an accepted part of French public life that the President sometimes has a MISTRESS. In this space, marital infidelity does not result in political harm. The two inputs are related by virtue of a generic space, which contains the generic roles COUNTRY, HEAD OF STATE, SEXUAL PARTNER and CITIZENS. The generic space establishes cross-space counterparts. The blended space contains BILL CLINTON and MONICA LEWINSKY, as well as the roles FRENCH PRESIDENT and MISTRESS OF FRENCH PRESIDENT, with which Clinton and Lewinsky are respectively associated. Crucially, the frame that structures the blend is FRENCH POLITICS rather than AMERICAN POLITICS. It follows that in the blend, Clinton is not politically harmed by his marital infidelity. However, because the inputs remain connected to the blend, structure in the blend can project back towards the inputs, giving rise to a **disanalogy** between the US and France. The integration network for this blend is represented in Figure 12.5.

The disanalogy between the United States and France is an important consequence of the counterfactual. The point of the utterance is to emphasise the difference between US and French attitudes, and perhaps moral values, with respect to the behaviour of their politicians in their personal lives. In the US, Clinton was censured for his attempts to keep his affair secret. In France, an affair would not have harmed him politically. The disanalogy is achieved by constructing a counterfactual through blending. An important advantage that Blending Theory has over Mental Spaces Theory, as we presented it in the previous chapter, is that we now have a mechanism that accounts for how structure is recruited and integrated in order to produce emergent structure: novel and highly creative scenarios like counterfactuals.

12.3 The nature of blending

As we saw in the previous section, metaphorical projection in the SURGEON AS BUTCHER metaphor is better accounted for by a conceptual integration network than by a two-domain mapping. This is because conceptual integration gives

COGNITIVE LINGUISTICS: AN INTRODUCTION

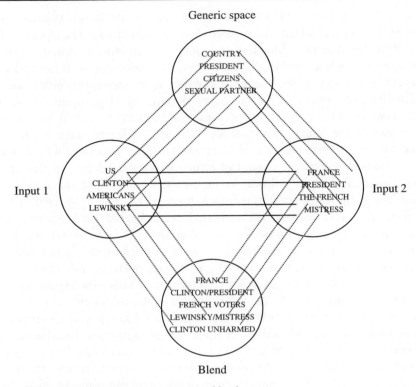

Figure 12.5 CLINTON as PRESIDENT OF FRANCE blend

rise to a blended space which provides a mechanism that accounts for the emergent structure not found in the input domains. We also saw that counterfactuals, like the CLINTON AS FRENCH PRESIDENT example, are accounted for by an integration network resulting in a blend. Since Fauconnier and Turner first advanced their theory in a seminal 1994 paper, a considerable amount of evidence for conceptual blending has been amassed from a range of non-linguistic phenomena. One of Fauconnier and Turner's central claims is that blending is a general and ubiquitous operation, central to human cognitive capabilities. In keeping with the Cognitive Commitment (Chapter 2), Fauconnier and Turner argue that conceptual blending is central not just to language, but to human thought in general. In this section we consider in more detail the elements involved in conceptual blending.

12.3.1 The elements of conceptual blending

We begin by sketching out the processes that give rise to conceptual blending and proceed in the next subsections to explore in detail how these processes apply in both linguistic and non-linguistic phenomena. We saw above that an integration network consists of at least four spaces: a generic space, two inputs and a blended space. We also saw that the generic space establishes **counterpart connectors** between input spaces, which are represented as the bold lines in integration network diagrams. These connections are established by **matching**, the conceptual operation responsible for identifying cross-space counterparts in the input spaces. Connectors between matched elements are then established, which, as we saw in the previous chapter, is a form of conceptual projection. Connectors can be established between matched elements on the basis of identity or role (as we saw in the CLINTON AS FRENCH PRESIDENT example), or based on metaphor (as we saw in the SURGEON AS BUTCHER example).

The input spaces give rise to **selective projection**. In other words, not all the structure from the inputs is projected to the blend, but only the matched information, which is required for purposes of local understanding. For example, in the CLINTON AS FRENCH PRESIDENT example, the fact that the role FRENCH PRESIDENT has a value in reality (currently Jacques Chirac) is not projected to the blend. Neither is the fact that Clinton speaks English rather than French, nor the fact that he is unlikely to have considered becoming president of France, nor the fact that he is ineligible, and so on. In other words, much of the structure in the inputs is irrelevant to, or even inconsistent with, the emergent meaning under construction. This type of information is therefore not projected into the blend. Selective projection is one reason why different language users, or even the same language user on different occasions, can produce different blends from the same inputs. In other words, the process of selective projection is not deterministic but flexible. However, projection, like the other aspects of blending, is subject to a set of **governing principles**. We return to this point later in the chapter (section 12.6).

In Blending Theory, there are three component processes that give rise to emergent structure: (1) composition; (2) completion; and (3) elaboration. The first involves the **composition** of elements from separate inputs. In the CLINTON AS FRENCH PRESIDENT example, composition brings together the value BILL CLINTON with the role FRENCH PRESIDENT in the blend, resulting in CLINTON AS FRENCH PRESIDENT. Similarly, the SURGEON AS BUTCHER blend composes the elements projected from the SURGEON input with those projected from the BUTCHER input. The second process, **completion**, involves schema induction. As we saw in the previous chapter, schema induction involves the unconscious and effortless recruitment of background frames. These complete the composition. For example, in the CLINTON AS FRENCH PRESIDENT example, the process of completion introduces the frames for FRENCH POLITICS and FRENCH MORAL ATTITUDES. Without the structure provided by these frames, we would lose the central inference emerging from the blend, which is that his

Table 12.2 Constitutive processes of Blending Theory

Matching, and counterpart connections
Construction of generic space
Blending
Selective projection

	Composition
Emergent meaning	Completion
	Elaboration

affair with Lewinsky would not harm Clinton in France. This process of schema induction is called 'completion' because structure is recruited to 'fill out' or complete the information projected from the inputs in order to derive the blend. Finally, **elaboration** is the on-line processing that produces the structure unique to the blend. This process is sometimes called **running the blend**.

A further consequence of conceptual blending is that any space in the integration network can, as a result of the blend, undergo **modification**. For example, because the inputs remain connected to the CLINTON AS FRENCH PRESIDENT blend, the structure that emerges in the blend is projected back to the input spaces. This is called **backward projection**, and is the process that gives rise to the disanalogy between the US and France. In other words, the inputs are modified by the blend: a powerful contrast is established between the nature of French and American moral attitudes governing the behaviour of politicians and this information may contribute to the encyclopaedic knowledge system of the addressee. In a related manner, although integration networks are typically set up in response to the needs of local meaning construction, blends can, if salient and useful, become conventionalised within a speech or cultural community. We will see an example of the conventionalisation of a blend later in the chapter (section 12.5).

The processes that we have discussed in this section represent the **constitutive processes** of Blending Theory and are summarised in Table 12.2. These processes together comprise conceptual integration and the conceptual blending that arises from integration. As we will see later in the chapter, these processes also serve to constrain conceptual blending in important ways (section 12.6).

12.3.2 Further linguistic examples

In this section, we consider some further examples of blending presented by Fauconnier and Turner, and look at how the processes described in the previous section might apply.

Boat race

Consider the example (3) from a news report in *Latitude 38*, a sailing magazine (Fauconnier and Turner 2002: 64).

(3) As we went to press, Rich Wilson and Bill Biewenga were barely maintaining a 4.5 day lead over the ghost of the clipper *Northern Light*.

This example relates to a 1993 news story in which a modern catamaran *Great American II*, sailed by Wilson and Biewenga, set out on a route from San Francisco to Boston. A record for this route had been set in 1853 by the clipper *Northern Light*, which had made the journey in 76 days and 8 hours. This record still held in 1993.

The utterance in (3) sets up an integration network in which there are two input spaces: one relating to the journey of the modern catamaran in 1993 and the other relating to the original journey undertaken by *Northern Light* in 1853. The generic space contains schematic information relating to BOATS and JOURNEYS, which motivates matching operations and thus cross-space connections between the two inputs. In the blend, we have two boats: CATAMARAN and NORTHERN LIGHT. Moreover, in the blend the two boats are engaged in a RACE, in which the CATAMARAN is barely maintaining a lead over NORTHERN LIGHT. As Fauconnier and Turner observe, no one is actually 'fooled' by the blend: we do not interpret the sentence to mean that there are actually two boats from two different periods in history engaged in a real side-by-side race. Despite this, we achieve valuable inferences as a result of setting up the conceptual blend. Indeed, it is only by virtue of blending that we can compare the progress of the catamaran against that of its 'rival' *Northern Light*, which set the original record over a century earlier. This blend is illustrated in Figure 12.6.

In achieving this blend, the first process to occur is selective projection from the inputs to the blend. Not all the information in the input spaces is projected. For example, information is not projected relating to weather conditions, whether the boats have cargo or not, the nature of the clipper's crew, what the crew ate for supper and so on. Instead, information is projected that is sufficient to accomplish the inference. For example, we only project the 1993 time frame. Secondly, the structure that is selectively projected into the blend is composed and completed. The schema induction that occurs at the completion stage adds the RACE frame to the blend and thus provides further structure: in a race there are two or more COMPETITORS and the first to complete the course is the WINNER. Next, upon running the blend, the additional structure emerges that has arisen as a result of composition and completion. In Figure 12.6, this emergent structure is appended to the blend in the box beneath the blended space. Once this has occurred, we can think of the two boats as competitors in a race

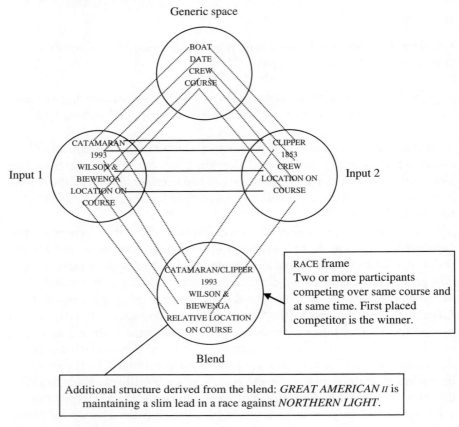

Figure 12.6 BOAT RACE blend

and compare their relative progress. Finally, as a result of backward projection the blend modifies the input spaces. For example, by 'living in the blend', the crew of the catamaran, their support team and others who are monitoring their progress can experience a range of emotions attendant upon participating in or watching a race, even though the 'race' is an imaginative feat.

XYZ constructions

In this section, we look at an example that shows how the conceptual blending approach can be applied to closed-class constructions. The XYZ construction is a grammatical construction specialised for prompting for conceptual integration. Consider the examples in (4) (Turner 1991: 199).

(4) a. Money is the root of all evil.
    b. Brevity is the soul of wit.

c. Politics is the art of the possible.
   d. Religion is the opiate of the masses.
   e. Language is the mirror of the mind.
   f. Vanity is the quicksand of beauty.
   g. Necessity is the mother of invention.
   h. Death is the mother of beauty.
   i. Children are the riches of poor men.

As Turner notes, these examples all share a form first noted by Aristotle in the *Poetics*. The form consists of three elements, which Turner labels X, Y and Z. These are all noun phrases, as illustrated in (5). Two of the elements, Y and Z, form a possessive construction (bracketed) connected by the preposition 'of'. The purpose of the construction is to propose a particular perspective according to which X should be viewed.

(5) Children are [the riches of poor men]
    [X]   [Y]   [Z]

In (5), for example, we are asked to view children as the riches of poor men, which results in a number of positive inferences relating to the 'value' of children. In addition to the elements X, Y and Z, the construction prompts for a fourth element, which Turner (1991) labels W. In order to understand children (X) in terms of riches (Y) we are prompted to construct a conceptual relation between children (X) and poor men (Z) and a parallel relation holding between riches (Y), and those who possess riches, namely rich men. This is the missing element (W), which is a necessary component to the interpretation of this construction: in the absence of a Y-W (RICHES-RICH MEN) relationship parallel to the X-Z (CHILDREN-POOR MEN) relationship, there is no basis for viewing children (X) and riches (Y) as counterparts. This idea is illustrated in (6).

(6) a. CHILDREN   ↔   POOR MEN
        [X]   [Z]
    b. RICHES   ←→   RICH MEN
        [Y]   [W]

Turner (1991) originally analysed XYZ constructions as metaphors. However, the development of Blending Theory offered a more revealing analysis. In the integration network for *children are the riches of poor men*, the two domains from Turner's original metaphor analysis are recast as input spaces. One input space contains the elements RICH MEN (W) and RICHES (Y), and the other input space contains the elements POOR MEN (Z) and CHILDREN (X). The generic space contains the schematic information MEN and POSSESSIONS. This generic structure

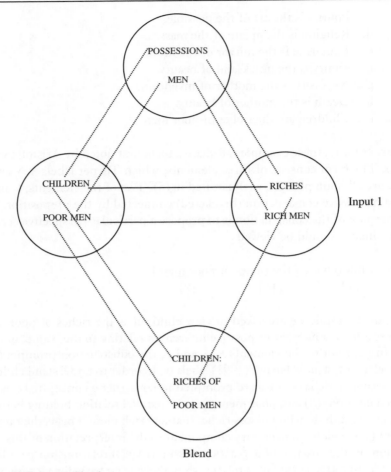

Figure 12.7 An XYZ blend

maps onto appropriate elements in both inputs and sets up cross-space connectors between counterparts in the input spaces, establishing cross-space commonalities and motivating integration within the blended space. In the blend, not only are certain elements from the inputs projected and integrated (the elements X, Y and Z), but their integration results in emergent structure that does not exist in either of the inputs: CHILDREN ARE THE RICHES OF POOR MEN. In neither of the inputs does there exist a conjunction between children of poor men and riches of rich men. This integration network is represented in Figure 12.7.

Formal blends

The XYZ blend is a **formal blend**. Formal blends involve projection of specific lexical forms to the blended space and rely, partly, upon formal (lexical or grammatical) structure for their meaning. In other words, part of the meaning of a given XYZ blend arises from the meaning conventionally associated with the XYZ **construction**. We will look in more detail at the meaning associated with grammatical constructions in Part III of the book.

A further example of formal blending is **compounding**, the process of blending two (or more) free morphemes to give rise to a new word. Recall from Chapter 4 that new words come into language on a remarkably regular basis. By providing an account of compounding, Blending Theory also offers an insight into this aspect of language change. The formal blend we consider here is the expression *landyacht*. According to Turner and Fauconnier (1995) this novel noun-noun compound relates to a large and expensive luxury car. It consists of two input spaces relating to the forms *land* and *yacht*, and the conventional range of meanings associated with these lexical items. However, projection to the blend is selective. Only a subset of the meanings associated with *land* and *yacht* are projected into the blend, together with the forms (the expressions *land* and *yacht*) themselves. In other words, Fauconnier and Turner suggest that linguistic forms as well as their associated lexical concepts can be projected into the blended space. When a lexical item is projected into the blend, this is known as **word projection**. As a result of composition, the forms as well as their projected meanings are integrated, giving rise to a new form *landyacht* with a distinct meaning: 'a large expensive luxury car'. Figure 12.8 illustrates the derivation of this compound, a process that could equally explain the PET FISH example that we discussed in Chapter 8 (Fodor and Lepore 1996).

12.3.3 Non-linguistic examples

The examples we have considered so far have illustrated how Blending Theory accounts for the on-line meaning construction arising from linguistic prompts, and have also illustrated how this approach can explain certain aspects of the meaning arising from formal linguistic units like grammatical constructions or compounds. However, although blending is a conceptual operation that can be invoked by language and that can also affect linguistic forms themselves, the blending operation itself, like the other cognitive processes that underlie cognitive semantics, is thought to be independent of language. In order to illustrate this, we consider some examples from the literature that illustrate conceptual blending at work in non-linguistic aspects of human thought and behaviour.

Computer desktop

When we interact with modern computers we do so via a computer 'desktop'. That is, we have icons on our computer screens that represent folders, files,

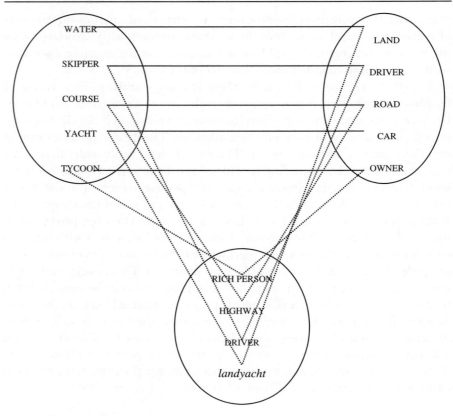

Figure 12.8 *Landyacht*

a wastepaper basket and so on. By selecting a particular icon from the computer 'desktop' we are able to tell the computer what we want it to do. The computer 'desktop' is a sophisticated blend which integrates structure from the domain of OFFICES and the WORKPLACE, including FILES, FOLDERS and WASTEPAPER BASKET. By providing an interface that translates the complex algorithmic operations that run the computer into simple commands, the blend allows us to understand and interact with our computer. However, the blend also features a range of novel characteristics that are unique to the blend. For instance, in 'real' offices we do not normally keep the wastepaper basket on our desktop. Moreover, as pointed out by Fauconnier and Turner (e.g. 2002), in the version of the desktop blend that applies to the Macintosh computer system, the 'trashcan', as well as facilitating file deletion, is also the means of ejecting the CD: in order to eject the disk, the user must drag the disk icon on the screen into the 'trashcan'. This directly contradicts knowledge from the domain of offices and workplaces where we are unlikely to place important disks in the bin in order to retrieve them.

Of course, the computer 'desktop' is facilitated by language in the sense that we rely upon linguistic expressions like *desktop*, *file* and *folder* to talk about our interaction with computers. Nevertheless, the blend is achieved by integrating conceptual structure from the domains of OFFICES and COMPUTER OPERATIONS, and relies upon iconic rather than linguistic representations, such as an image of a file or a folder, in order to prompt for these conceptual domains.

Talking animals

In many art forms, from oral and written literature from around the world to Disney cartoons, there are instances of talking animals. In his (1996) book *The Literary Mind*, in which he examines the conceptual basis of the parable story form, Turner observes that talking animals represent highly sophisticated conceptual blends. Consider, for instance, George Orwell's satirical parable *Animal Farm*. This novel describes an event in which farm animals lead a rebellion to overthrow the cruel farmer. In the novel, the animals talk, think, behave and feel in the same way as humans. In reality, we have no experience of talking animals. Although animals communicate in a number of sophisticated ways, we have no experience of animals manipulating a complex spoken symbolic system like human language for interactive communication (even parrots and mynah birds, which can mimic the sounds of human language, do not have conversations). Our ability to imagine talking animals is an example of **anthropomorphism**, where human characteristics are attributed to non-human entities, and is attested in human folklore all over the world. According to Turner, this fundamental aspect of human cognition arises from conceptual blending, where one of the input spaces is the HUMAN frame and the other is the frame relating to the non-human entity, here ANIMALS. In neither of the inputs do animals talk; this characteristic only emerges in the blend. This type of blend illustrates how Blending Theory can contribute to conventionalisation: it is not necessary for us to create a new blend each time we read about a fictional talking animal or watch one in a cartoon. Instead, we have a schematic blend for TALKING ANIMALS that is highly conventionalised in our culture and is continually reinforced and modified.

Rituals

Sweetser (2000) discusses the role of conceptual blending in human ritual. She argues that one purpose of ritual is to depict a particular scenario. If the ritual affects the scenario it represents, it is said to have a **performative function** or to exhibit **performativity**, an idea that derives from Austin's ([1962] 1975) influential work on speech acts. Sweetser argues that performativity is an important aspect of many rituals in the sense that the function of ritual is to bring about a desired state of affairs as a consequence of performing a physical or linguistic act. As an example of performative ritual, Sweetser discusses the Holy Communion service in the Christian Church. The consumption of the bread and the wine (which represent the body and the blood of Christ) represents a spiritual union between the human and the divine. In addition, Sweetser observes that 'it certainly must also be seen as intending to causally bring about this spiritual union via the consumption of the bread and the wine' (Sweetser 2000: 314). That is, the ritual of consuming bread and wine, through blending, is conceptualised as effecting union between the human and the divine. In one input space we have bread and wine and the ordinary act of consumption, in another we have the flesh and blood of Jesus Christ. In the blend, the bread and wine represent (or literally become) the flesh and blood of Christ, depending upon the denomination in question. In the blend, the act of consumption has a performative function, serving to bring about a union between the human worshipper and the sacred (Jesus Christ). This ritual is based on the events depicted in the New Testament relating to the Last Supper: a meal shared by Jesus and his disciples prior to his arrest. However, the Last Supper was itself a celebration of Passover, an ancient Jewish ritual in which the blood of a new-born lamb was ingested and the flesh eaten in order to commemorate the Angel of Death sparing Jewish newborn babies when the Jews were slaves in ancient Egypt. Thus the ritual of the Holy Communion is a complex blend, relying on historically earlier blends.

It is also the case that rituals often employ **material anchors** for the blend (Fauconnier and Turner 2002; Hutchins 1996). In other words, the material anchors embody and facilitate the blend. In the case of Holy Communion, the bread and the wine are material anchors, and our interaction with these both embodies and facilitates the blend (the union between the human and the divine). Similarly, the wedding ring in the Western marriage ritual is a material anchor. The ring both embodies the blend, representing an unbroken link and also has a performative function as part of a ritual: the act of placing the ring (which embodies an unbroken link) on the betrothed's finger serves, in part, to join two individuals in matrimony.

12.4 Vital relations and compressions

An important function of blending is the provision of **global insight**. In other words, a blend is an imaginative feat that allows us to 'grasp' an idea by viewing it in a new way. According to Fauconnier and Turner (2002), conceptual blending achieves this by reducing complexity to **human scale**: the scope of human experience. For example, imagine that you are attending a lecture on evolution

Table 12.3 Goals of blending

Overarching goal of blending
– Achieve human scale

Notable subgoals of blending
– Compress what is diffuse
– Obtain global insight
– Strengthen vital relations
– Come up with a story
– Go from many to one

and the professor says: 'The dinosaurs appeared at 10 pm, and were extinct by quarter past ten. Primates emerged at five minutes to midnight, Humans showed up on the stroke of twelve.' This represents an attempt to achieve human scale by blending the vast tracts of evolutionary time with the time period of a 24-hour day. This is achieved by 'compressing' diffuse structure (over 4.6 billion years of evolution) into a more compact, and thus less complex structure (a 24-hour day). This achieves human scale, because the 24-hour day is perhaps the most salient temporal unit for humans. This conceptual integration achieves global insight by facilitating the comprehension of evolutionary time, since we have no first-hand experience of the vast time scales involved. Indeed, Fauconnier and Turner argue that the primary objective of conceptual blending is to achieve human scale. This in turn relates to a number of subgoals (see Table 12.3).

By explaining blending in terms of these goals, Fauconnier and Turner subscribe to the view that blending provides humans with a way of 'making sense of' many disparate events and experiences. In this respect, the motivation for conceptual blending is not dissimilar from the explanation put forth in early Conceptual Metaphor Theory, which held that the human mind tends toward construal of the abstract in terms of the concrete, and that this tendency is an attempt to 'grasp' what is elusive in terms of what is familiar. In this section, we consider how blending achieves these goals by the **compression** of **vital relations**.

12.4.1 Vital relations

In the previous chapter, we saw that counterparts can be established between mental spaces, and that connectors are set up that link the counterparts. We described this process as a type of conceptual projection that involves mappings between spaces. In this chapter we have referred to the identification procedure as 'matching'. In Blending Theory, Fauconnier and Turner refer to the various types of connector as **vital relations**. A vital relation is a link that matches two

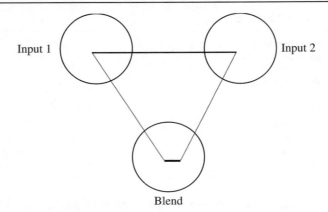

Figure 12.9 Compression of outer-space relation into inner-space relation in the blend (adapted from Fauconnier and Turner 2002: 94)

counterpart elements or properties. Fauconnier and Turner propose a small set of vital relations, which recur frequently in blending operations. From this perspective, what makes a connector a 'vital' relation is its ubiquity in conceptual blending.

Vital relations link counterparts in the input spaces and establish what Fauconnier and Turner call **outer-space relations**: relations in which two counterpart elements are in different input spaces. Vital relations can also give rise to **compressions** in the blend. In other words, the blend 'compresses the distance' or 'tightens the connection' that holds between the counterparts in the outer-space relation. This relation is compressed and represented as an **inner-space relation** in the blend: a counterpart relation inside a single mental space. As we saw earlier in relation to the example illustrating the blending of evolutionary time into the time-scale of a single day, the time-scale of evolution is **compressed** into the time-scale of a single day. This kind of compression, resulting in a reduced scale, is called **scaling**. The process of compression is illustrated in Figure 12.9. According to Fauconnier and Turner, it is by means of the mechanism of compression that blending achieves human scale, together with the various subgoals set out in Table 12.3. According to this perspective, conceptual blending represents an indispensable imaginative feat underlying human thought and reasoning.

12.4.2 A taxonomy of vital relations and their compressions

Fauconnier and Turner (2002) provide a taxonomy of vital relations together with a discussion of the ways in which they can be compressed. We consider some of these below.

Time

Because events are temporally situated, TIME can function as a vital relation that connects two (or more) events across input spaces. For example, in the BOAT RACE blend discussed above, the two input spaces relate to events from different time periods, 1853 and 1993. In the blend, this outer-space vital relation is compressed so that the two events are viewed as simultaneous. This is another example of scaling which reduces the 'distance' between individual events. TIME can also be compressed by **syncopation**. Syncopation reduces the number of events in a temporal 'string'. This is illustrated by example (7).

(7) My life has consisted of a few notable events: I was born, I fell in love in 1983 and was briefly happy, in 1990 I met my future husband. We got married a year later. As I look back the time since seems to have disappeared in housework and drudgery.

In this not altogether happy account, the narrator compresses time to reduce her life to 'a few notable events'. Compressions involving scaling and syncopation are also evident in non-linguistic phenomena. For example, a pictorial 'time-line' for evolutionary development can select just a few notable events in evolution, such as the emergence and extinction of the dinosaurs followed by the emergence of humans; this represents compression by syncopation.

Space

Also evident in the BOAT RACE blend is the scaling of the outer-space vital relation SPACE. In the two inputs, each of the boats occupies a unique spatial location. Indeed, the course followed by *Northern Light* may have been some miles distant from the course followed by *Great American II*. However, in the blend the outer-space relation is compressed so that the two boats are following the same course. As a result of the compression of SPACE, it is possible to talk about *Northern Light* 'catching up with' and even 'overtaking' *Great American II*. This is only possible if the two boats are following more or less the same spatial path.

Representation

Another kind of vital relation that can hold between input spaces is REPRESENTATION. While the vital relations discussed above relate counterparts of a similar kind (for instance, TIME relates two EVENTS), REPRESENTATION relates one entity or event with another entity or event that represents it, but may be of a different kind. For instance, imagine that a physics teacher is trying to explain the Solar System to a class of high-school children using coloured ping-pong balls to represent the Sun and the planets around the Sun:

(8) This yellow one, that's the Sun. This red one, that's Mars, it's the fourth planet from the Sun. Here's Earth, the blue one.

In the blend, the yellow ping-pong ball *is* the Sun. The outer-space relation has been compressed, and gives rise to the inner-space vital relation UNIQUENESS, which provides a way of understanding two spatially distinct entities as the same individual entity. This shows how an outer-space vital relation (in this case, REPRESENTATION) can give rise to a different inner-space vital relation in the blend (in this case, UNIQUENESS).

Change

The outer-space relation CHANGE can also be compressed into the inner-space relation UNIQUENESS. Consider the example of scaling in (9).

(9) The ugly duckling has become a beautiful swan.

In this example, CHANGE, which occurs over time, is compressed so that an ugly duckling and a beautiful swan are understood as the same individual.

Role-value

This is a vital relation that links roles with values. Compression of the ROLE-VALUE outer-space relation also results in UNIQUENESS in the blend. For example, consider the role QUEEN and the value ELIZABETH II. In the blend, compression results in UNIQUENESS so that the role and the value also result in a single entity which can be referred to as *Queen Elizabeth II*. Like the *landyacht* example, this is a formal blend that gives rise to a new expression as well as a new concept. Observe that once a series of such blends exists, for example KINGS OF ENGLAND, this series of individuals can be further compressed into an inner-space relation of UNIQUENESS, in which a series of individuals becomes conceptualised as a single unique individual. This is illustrated by example (10).

(10) After the Norman Conquest, the English King was French for centuries, until a quarrel with France. After that the King was English, and English once again became the language of Parliament.

In this example, compression into UNIQUENESS in the blend results in a single ENGLISH KING, who can be French at one point in time and English at another.

Analogy

ANALOGY is a vital relation established by ROLE–VALUE compression. Consider example (11).

(11) The city of Brighton is the closest thing the UK has to San Francisco.

In this example, there are two pre-existing blends in operation attached to two distinct integration networks. One blend contains the role CITY and the value BRIGHTON, and the other blend contains the role CITY and the value SAN FRANCISCO. Both blends are structured by the frame that relates to a cosmopolitan and liberal city by the sea. The compression of the role-value vital relations across these two blends from different integration networks establishes the ANALOGY between BRIGHTON and SAN FRANCISCO. Thus ANALOGY is an outer-space vital relation holding between the two blends from distinct integration networks. These blends themselves serve as the inputs for a third integration network. In the new blend analogy is compressed into IDENTITY. Brighton and San Francisco can be described as 'analogues' because they share identity in the blend.

Example (12) illustrates another way in which the outer-space relation ANALOGY can be compressed. Consider example (12) which relates to the destructive computer virus *My Doom*.

(12) My Doom is the latest in a series of large-scale computer viruses spread by opening an e-mail attachment.

The concept COMPUTER VIRUS is a conventional blend that emerges from the two input spaces DESTRUCTIVE COMPUTER PROGRAM and BIOLOGICAL VIRUS. The outer-space ANALOGY relation between DESTRUCTIVE COMPUTER PROGRAM and BIOLOGICAL VIRUS is compressed into a CATEGORY relation in the blend. The category relation is of the 'A is a B' type: DESTRUCTIVE COMPUTER PROGRAM is a VIRUS.

Disanalogy

The outer-space relation DISANALOGY can be compressed into the inner-space relation CHANGE. This can then be further compressed into UNIQUENESS in the blend. Example (13) illustrates this process.

(13) My tax bill gets bigger every year.

This example relates to a blend of a series of distinct and disanalogous (different) tax bills. As a result of the blend, the outer-space relation of DISANALOGY is compressed into CHANGE: in the blend the differences between the individual bills received each year are understood in terms of CHANGE as a result of the yearly increases. This inner-space relation can be further compressed into UNIQUENESS: in the blend there is a single tax bill that continues to change and increase. This shows how inner-space relations can also undergo compression ('reduction') into vital relations that further facilitate the process of achieving human scale.

Part–whole

Example (14) represents a part–whole metonymy uttered by someone who is looking at a photograph of a woman's face.

(14) That's Jane Smith.

This example represents a part–whole metonymy because the speaker is identifying the whole person simply by her face. By viewing the metonymy in terms of a blend, a clearer picture emerges of how the metonymy is working. Metonymies like this consist of two input spaces: JANE SMITH and her FACE. A PART–WHOLE vital relation establishes these elements as counterparts in two input spaces. In the blend, the PART–WHOLE relation is compressed into UNIQUENESS.

Cause–effect

The final vital relation we will examine is CAUSE–EFFECT. An example of this vital relation, provided by Fauconnier and Turner, is the distinction between a burning log in a fireplace and a pile of ash. These two elements are linked in an integration network by the outer-space CAUSE–EFFECT relation, which connects the burning log (the CAUSE) with the pile of ash (the EFFECT). The CAUSE–EFFECT relation is typically **bundled** with the vital relation TIME which undergoes scaling, and with CHANGE which is compressed into UNIQUENESS. For example, imagine that a speaker points to the ashes and utters the sentence in (15).

(15) That log took a long time to burn.

In this example, a blend has been constructed in which TIME has been scaled and the log and the ashes have been compressed into a single unique entity.

The CAUSE–EFFECT relation can also be compressed into the vital relation PROPERTY. For example, a consequence of wearing a coat is that the wearer is kept warm. However, when we describe a coat as 'warm', as in the expression *a warm coat*, we are compressing the CAUSE of wearing a coat with the EFFECT of being warm. In reality, the coat itself is not warm, but the vital relation is

CONCEPTUAL BLENDING

Table 12.4 Summary of vital relations and their compressions

Outer-space vital relation	Inner-space vital relation (compression)
TIME	SCALED TIME
	SYNCOPATED TIME
SPACE	SCALED SPACE
	SYNCOPATED SPACE
REPRESENTATION	UNIQUENESS
CHANGE	UNIQUENESS
ROLE–VALE	UNIQUENESS
ANALOGY	IDENTITY
	CATEGORY
DISANALOGY	CHANGE
	UNIQUENESS
PART-WHOLE	UNIQUENESS
CAUSE-EFFECT (bundled with TIME and CHANGE)	SCALED TIME
	UNIQUENESS
CAUSE-EFFECT	PROPERTY

compressed into PROPERTY of the coat in the blend. Table 12.4 provides a summary of the vital relations and their compressions discussed in this section, which represent only a subset of the vital relations proposed by Fauconnier and Turner (2002).

12.4.3 Disintegration and decompression

In the previous section, we saw that integration in the blend is a result of compression and observed that compressions provide human scale. We also saw that an important subgoal of this operation is to provide global insight. In this section, we briefly explore how compressions of outer-space relations achieve global insight as a consequence of the blend remaining connected to the rest of the integration network, including the input spaces.

Recall our discussion of the counterfactual CLINTON AS FRENCH PRESIDENT example. An important inference resulting from this blend is the DISANALOGY between the inputs. The ROLE-VALUE vital relation holding between CLINTON (value) and FRENCH PRESIDENT (role) in the input spaces is compressed into UNIQUENESS in the blend (where CLINTON is FRENCH PRESIDENT). At the same time, the process of **disintegration** can 'unpack' the blend which results in the backward projection of blended elements to the input spaces (section 12.3.1). Backward projection, or disintegration, results from the process of **decompression**, in which elements in the blend are separated. Observe that, although ANALOGY between France and the US motivates the blend (in the input spaces, both countries have a president for head of state, and both American and French presidents have famously had mistresses), the decompression of the blended elements gives rise to DISANALOGY. Indeed, while similarities can be exploited to create a blend, the same blend can be 'unpacked' to reveal dissimilarities. This follows from the fact that the elements projected back to the inputs have been 'affected' by blending. For example, the politically unharmed CLINTON as FRENCH PRESIDENT is projected back to the input space, in which he experiences political harm. This gives rise to an outer-space relation of DISANALOGY between the US space and the FRANCE space. In this way, the integration network provides global insight as a result of the implications that the blend has for the input spaces that gave rise to it in the first place.

12.5 A taxonomy of integration networks

One of the insights developed by Fauconnier and Turner (1998a, 2002) is the idea that there are a number of different kinds of integration network. Although Fauconnier and Turner propose a continuum that relates integration networks of various kinds, there are four points along the continuum that stand out. We briefly survey these four distinct types of integration network below.

12.5.1 Simplex networks

The simplest kind of integration network involves two inputs, one that contains a frame with roles and another that contains values. This is a **simplex network**. What makes this an integration network is that it gives rise to a blend containing structure that is in neither of the inputs. Consider example (16).

(16) John is the son of Mary.

This utterance prompts for an integration network in which there is one input containing a FAMILY frame with roles for MOTHER and SON. The second input contains the values JOHN and MARY. The integration network compresses the ROLE-VALUE outer-space relations into UNIQUENESS in the blend, so that JOHN is the SON and MARY the MOTHER, and so that JOHN IS MARY'S SON. The motivation for the cross-space connections is the generic space which contains the elements FEMALE and MALE. These elements identify potential counterparts in the inputs. To reiterate, only one of the inputs (input 1) contains a frame. The simplex network therefore represents an instance of basic **framing** (see Figure 12.10).

12.5.2 Mirror networks

According to Fauconnier and Turner, the defining feature of a **mirror network** is that all the spaces in the network share a common frame, including the blend.

CONCEPTUAL BLENDING

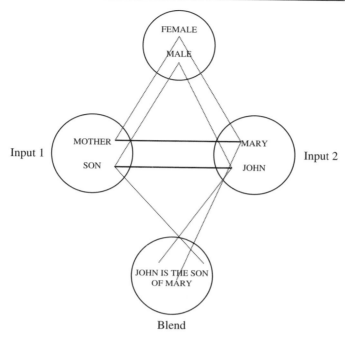

Figure 12.10 A simplex integration network

One example of a mirror network that we have already discussed in detail is the BOAT RACE blend (recall example (3) and Figure 12.6). Each of the spaces in this example contain the frame in which a boat follows a course, including the blend, which has the additional schema relating to a RACING frame.

12.5.3 Single-scope networks

While in the simplex network only one of the inputs is structured by a frame, and in the mirror network all the spaces share a common frame, in the **single-scope network** both inputs contain frames, but each is distinct. Furthermore, only one of the input frames structures the blend. Consider example (17).

(17) Microsoft has finally delivered the knock-out punch to its rival Netscape.

This sentence prompts for an integration network in which there are two inputs. In one input there are two business rivals, MICROSOFT and NETSCAPE, and Microsoft takes Netscape's market share. In the other input there are two BOXERS, and the first boxer knocks out the second. In the blend, MICROSOFT and NETSCAPE are BOXERS, and MICROSOFT KNOCKS OUT NETSCAPE

427

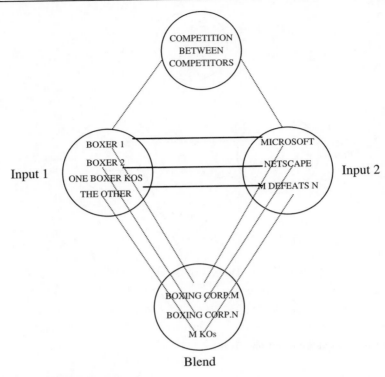

Figure 12.11 Single-scope network

(see Figure 12.11). What distinguishes this type of network is that only one frame (here, the BOXING frame rather than BUSINESS frame) serves to organise the blend. In other words, the framing input provides the frame, including the roles for BOXERS, while the focus input provides the relevant elements: the values MICROSOFT and NETSCAPE.

An important function of single-scope networks is to employ pre-existing compressions in the **framing input** (input 1 in Figure 12.11) to organise diffuse structure from the **focus input** (input 2 in Figure 12.11). The framing input is itself a blend that contains a number of pre-existing inner-space relations. These include compressions over TIME, SPACE and IDENTITY (different individuals perform as boxers, either as a hobby or as a career, and through shared identity give rise to the role BOXER), among others, which are then compressed into a BOXING frame. This pre-existing blend functions as the framing input for the single-scope network in Figure 12.11, where input 1 contains a tightly compressed inner-space relation that includes just two participants, a single boxing space, a limited period of time (for example, ten three-minute rounds), and a specific kind of activity. This inner-space relation, when projected to the blend, provides structure onto which a range of diffuse activities

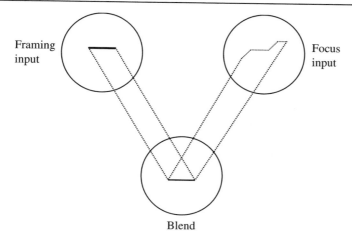

Figure 12.12 Structuring of focus input by inner-space projection from framing input (adapted from Fauconnier and Turner 2002: 130)

in the focus input can be projected: the input relating to BUSINESS RIVALRY between MICROSOFT and NETSCAPE. The blend compresses the diffuse nature of business rivalry as a result of the properties of the framing input. This function of single-scope networks in particular relates directly to one of the main subgoals of blending presented in Table 12.3: to compress what is diffuse. Figure 12.12 illustrates this subgoal.

Single-scope networks form the prototype for certain types of conceptual metaphor, such as compound metaphors and metaphors motivated by perceptual resemblance. In other words, the source-target mapping in a metaphor is part of an integration network that results in a blend. From this perspective, many conceptual metaphors may be more insightfully characterised as blends. However, it does not follow that all metaphors are blends. While compound metaphors like BUSINESS IS BOXING, or the more general mapping BUSINESS IS PHYSICAL COMBAT may be blends, it is less obvious that primary metaphors are blends. We return to this point below.

12.5.4 Double-scope networks

We turn finally to **double-scope networks**, in which both inputs also contain distinct frames but the blend is organised by structure taken from each frame, hence the term 'double-scope' as opposed to 'single-scope'. One consequence of this is that the blend can sometimes include structure from inputs that is incompatible and therefore **clashes**. It is this aspect of double-scope networks that makes them particularly important, because integration networks of this kind are highly innovative and can lead to novel inferences.

An example of a double-scope blend that we have already encountered, which does not involve clashes, is the COMPUTER DESKTOP blend. Fauconnier and Turner (2002) describe this blend in the following way:

> The Computer Desktop interface is a double-scope network. The two principle inputs have different organizing frames: the frames of office work with folders, files, trashcans, on the one hand, and the frame of traditional computer commands, on the other. The frame in the blend draws from the frame of office work – throwing trash away, opening files – as well as from the frame of traditional computer commands–'find', 'replace', 'save', 'print'. Part of the imaginative achievement here is finding frames that, however different, can both contribute to the blended activity in ways that are compatible. 'Throwing things in the trash' and 'printing' do not clash, although they do not belong in the same frame. (Fauconnier and Turner 2002: 131)

We can compare this example with a double-scope blend in which the two organising frames do clash. Consider example (18).

(18) You're digging your own grave.

This idiomatic expression relates to a situation in which someone is doing something foolish that will result in unwitting failure of some kind. For instance, a businessman, who is considering taking out a loan that stretches his business excessively, might be warned by his accountant that the business risks collapse. At this point, the accountant might say:

(19) You're digging your own financial grave.

This double-scope blend has two inputs: one in which the BUSINESSMAN takes out a LOAN his company can ill afford and another relating to GRAVE DIGGING. In the blend, the loan proves to be excessive and the company fails: the BUSINESSMAN and his BUSINESS end up in a FINANCIAL GRAVE. In this example, the inputs clash in a number of ways. For example, they clash in terms of **causality**. While in the BUSINESS input, the excessive loan is causally related to failure, in the GRAVE DIGGING input, digging a grave does not cause death; typically it is a response to death. Despite this, in the blend, digging the grave causes DEATH-AS-BUSINESS FAILURE. This is an imaginative feat that blends inputs from clashing frames. The reason the blend is successful, despite the clash, is that it integrates structure in a way that achieves human scale. Because the accountant's utterance gives rise to the DEATH-AS-BUSINESS FAILURE interpretation, the businessman is able to understand that the loan is excessive and will

CONCEPTUAL BLENDING

Table 12.5 Integration networks (based on Fauconnier and Turner 2002)

Network	Inputs	Blend
Simplex	Only one input contains a frame	Blend is structured by this frame
Mirror	Both inputs contain the same frame	Blend is structured by the same frame as inputs
Single-scope	Both inputs contain distinct frames	Blend is only structured by one of the input frames
Double-scope	Both inputs contain distinct frames	Blend is structured by aspects of both input frames

cause the business to fail. Hence the causal structure of the blend (the idea that digging the grave causes the failure) can be projected back to the first input space in order to modify it. In the BUSINESS input, the businessman can decide to decline the loan and thus save his business. In this way, the blend provides global insight, and thereby provides a forum for the construction and development of scenarios that can be used for reasoning about aspects of the world. According to Fauconnier and Turner, this enables us to predict outcomes, draw inferences and apply these insights back in the input spaces before the events constructed in the blend come about. For this reason, Fauconnier and Turner argue that blending, and double-scope blending in particular, is an indispensable tool for human thought. Table 12.5 summarises the properties of the four types of blend we have discussed in this section.

12.6 Multiple blending

While we have for the most part assumed that integration networks consist of four spaces (generic space, two input spaces and the blend), it is common, and indeed the norm, for blends to function as inputs for further blending and reblending. We illustrate this point in this brief section with a discussion of Fauconnier and Turner's (2002) example of the GRIM REAPER blend.

The Grim Reaper

This is a highly conventional cultural blend, in which DEATH is personified as the GRIM REAPER. This blend derives from an integration network consisting of three inputs, one of which is itself a blend consisting of two prior inputs. The Grim Reaper, as depicted in iconography since medieval times, is represented as a hooded skeleton holding a scythe.

Consider the three inputs to the GRIM REAPER blend. These relate to three AGENTS: (1) a REAPER, who uses a scythe to cut down plants;

(2) a KILLER, who murders a victim; and (3) DEATH, which brings about the death of an individual. Observe that the third AGENT is non-human: DEATH is an abstract AGENT. In other words, DEATH-AS-AGENT is itself a metaphoric blend, in which DEATH and AGENCY (human animacy and volition) have been blended, giving rise to the personification of death. In the GRIM REAPER blend, the AGENT is DEATH and this agent causes death by KILLING. The manner of killing is REAPING (the use of the scythe). The reaper is GRIM because death is the outcome of his reaping. This complex blend is illustrated in Figure 12.13.

Observe that the physical appearance of the Grim Reaper metonymically represents each of the three main inputs to the blend. The skeleton stands for DEATH, which is the outcome; the hood that hides the reaper's face represents the concealment that often characterises KILLERS; and the scythe stands for the manner of killing, deriving from the REAPER input. Finally, the Grim Reaper emerges from the blend rather than from any of the input spaces.

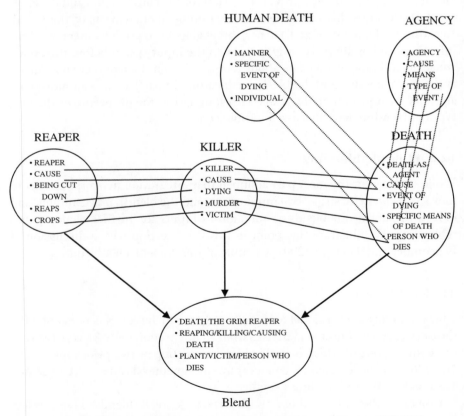

Figure 12.13 Death the Grim Reaper (adapted from Fauconnier and Turner 2002: 292)

12.7 Constraining Blending Theory

Of course, an important question that arises from Blending Theory concerns how this model of meaning construction is constrained. In particular, how is selective projection constrained so that we end up with the 'right' structure being projected to the blend? This is reminiscent of a similar question that arose in relation to Conceptual Metaphor Theory in Chapter 9 (for example, if THEORIES are BUILDINGS, why do they not have French windows?). In order to address this issue, Fauconnier and Turner (2002) propose a number of **governing principles**, also known as **optimality principles** (Fauconnier and Turner 1998a). We present these below (see Table 12.6), and briefly comment on just two of them in order to explain how selective projection is constrained.

Table 12.6 Governing principles of Blending Theory (Fauconnier and Turner 2002)

Governing principle	Definition
The topology principle	'Other things being equal, set up the blend and the inputs so that useful topology in the inputs and their outer-space relations is reflected by inner-space relations in the blend.' (F&T 2002: 327)
The pattern completion principle	'Other things being equal, complete elements in the blend by using existing integrated patterns as additional inputs. Other things being equal, use a completing frame that has relations that can be compressed versions of the important outer-space vital relations between the inputs.' (F&T 2002: 328)
The integration principle	'Achieve an integrated blend.' (F&T 2002: 328)
The maximisation of vital relations principle	'Other things being equal, maximize vital relations in the network. In particular, maximize the vital relations in the blended space and reflect them in outer-space vital relations.' (F&T 2002: 330)
The web principle	'Other things being equal, manipulating the blend as a unit must maintain the web of appropriate connections to the input space easily and without additional surveillance of composition.' (F&T 2002: 331)
The unpacking principle	'Other things being equal, the blend all by itself should prompt for the reconstruction of the entire network.' (F&T 2002: 332)
The relevance principle	'Other things being equal, an element in the blend should have relevance, including relevance for establishing links to other spaces and for running the blend. Conversely, an outer-space relation between the inputs that is important for the purposes of the network should have a corresponding compression in the blend.' (F&T 2002: 333)

These principles can be described as 'optimality' principles because blending is not a deterministic process. Instead, integration networks are established in order to achieve the goals we described in section 12.4. Thus, depending on the precise structure available in a given integration network and the purpose of integration, there may be competing demands on the selective projection of structure to the blend. For example, consider a scenario in which a child picks up a replica sword in a military museum. In response to the expression of alarm on the face of the parent the curator remarks, 'Don't worry, the sword is safe,' to which the parent rejoins, 'Not from him it isn't.' In this exchange, the curator intended that the sword would not cause the child harm. In this intended interpretation, the structure being projected relates to the potential harm that swords can cause, especially when handled by the inexperienced. However, the parent rejects this blend and proposes a new one in which it is the sword, rather than the child, that is at risk from potential harm. This blend arises because the parent projects his personal knowledge of the child, and the child's ability to inflict damage on anything they come into contact with. This example illustrates how it is possible to obtain different blends from the same, or very similar, input spaces by virtue of differential selective projection.

We briefly discuss two of the principles in Table 12.6 in order to give a sense of how projections from the inputs spaces to the blend are selected. In essence, these governing principles optimise with respect to each other in order to achieve the goals of blending that we summarised in Table 12.3. For instance, the topology principle ensures that **topology** (the relational structure between and within the input spaces) is preserved in the blended space. The default means of achieving this preservation of topology is by projecting relational structure as it occurs in the outer-space relation. For example, in the BOAT RACE blend, the distance travelled between San Francisco and Boston for both *Northern Light* and *Great American II* is preserved and projected unchanged to the blend. The preservation of this topology highlights the differences between inputs that we seek to understand via blending, such as the different spatial locations at a given temporal point in the BOAT RACE blend.

While the topology principle maintains the existing relational structure of the input spaces, this principle is at odds with the maximisation of vital relations principle. This principle serves, in part, to reduce outer-space vital relations to an undifferentiated single structure in the blend. This is the goal of compression. However, to fulfil the goals of blending, these two principles have to work in tandem, optimising the relative tensions they jointly give rise to in order to facilitate an optimal blend which best achieves the goals of blending. In this way, the governing principles work together to constrain, rather than to govern (in the sense of determining), what is projected to the blend by selective projection.

12.8 Comparing Blending Theory with Conceptual Metaphor Theory

When Blending Theory was first formulated, its proponents argued that it represented an alternative framework to Conceptual Metaphor Theory. However, there are good reasons to think that Blending Theory and Conceptual Metaphor Theory are complementary rather than competing theories, as explicitly argued by Grady, Oakley and Coulson (1999). In this section, we compare and contrast Blending Theory and Conceptual Metaphor Theory, and argue that as well as providing complementary perspectives, each theory addresses certain phenomena not accounted for by the other theory.

12.8.1 Contrasts

There are a number of ways in which Blending Theory is distinct from Conceptual Metaphor theory. We begin by addressing these contrasts between the two theories.

Not all blends are metaphorical

First of all, it is important to emphasise that not all blends are metaphorical. As we saw earlier in our taxonomy of integration networks (section 12.5), the prototypical metaphorical network is the single-scope integration network. The hallmark of metaphor and of single-scope blends is **frame-projection asymmetry**: while both inputs contain distinct frames, it is only the frame from one of these inputs (the 'source' in conceptual metaphor terms, the 'frame input' in blending terms) that is projected to the blend. Although single-scope networks are the prototypical kind for structuring metaphor, we have seen that other kinds of network may also produce metaphorical blends as in the case of the double-scope example: *You're digging your own financial grave*.

Blending does not involve unidirectional mappings

Unlike Conceptual Metaphor Theory, Blending Theory involves selective projection of structure from inputs to the blended space rather than unidirectional cross-domain mappings. In addition, structure from the blend can be projected back to the input spaces. Thus the two theories employ different architecture in order to model similar phenomena.

Spaces versus domains

Conceptual metaphors feature mappings (and domains) stored in long-term memory. These mappings hold between domains which are highly stable knowledge structures. In contrast, Conceptual Blending Theory makes use of mental spaces. As we saw in the previous chapter, mental spaces are dynamic and temporary conceptual 'packets' constructed 'on-line' during discourse. Despite this, blends can become conventionalised (for example, the GRIM REAPER blend), in which case the blend becomes established as a relatively stable knowledge structure in the conceptual system.

The many-space model

In their first Blending Theory paper, Fauconnier and Turner (1994) referred to conceptual integration or blending as the **many-space model**. This points to an obvious difference between Blending Theory and Conceptual Metaphor Theory: while Conceptual Metaphor Theory is a two-domain model, Blending Theory employs a minimum of four spaces.

Dynamic versus conventional

One consequence of the foregoing comparisons is that while Blending Theory emphasises the dynamic and mutable aspects of blending and its role in meaning construction, Conceptual Metaphor Theory emphasises the idea that there is a 'metaphor system' in which conceptual metaphors interact in order to provide relatively stable structure and organisation to the human conceptual system. This reflects the different emphases of the two traditions: metaphor theorists have been concerned with mapping the conventional patterns entrenched in conceptual structure, while blending theorists have been more concerned with investigating the contribution of conceptual integration to ongoing meaning construction. As we have seen, this does not entail that blending cannot give rise to conventionalised representations.

Difference in methodological emphasis

As a consequence of the previous contrast, while conceptual metaphor theorists have sought generalisations across a broad range of metaphoric expressions, conceptual blending theorists, while developing general principles based on specific examples, typically focus on the nature and particulars of those specific examples. This is because Blending Theory places emphasis upon a process of meaning construction rather than a system of knowledge.

Emergent structure

A particularly important difference between the two theories is that, while Blending Theory provides an account of emergent structure, Conceptual

Metaphor Theory does not. This follows from the fact that Conceptual Metaphor Theory relies upon a two-domain model. We discuss this issue in more detail below.

12.8.2 When is a metaphor not a blend?

We have seen that a constitutive process in conceptual blending involves matching, which identifies counterparts across input spaces. One of the motivations for matching is the presence of a generic space. Although a large subset of conceptual metaphors are blends, with counterparts established in the 'source' and 'target', as pointed out by Grady *et al.* (1999) there is a small but important subset of highly conventionalised conceptual metaphors that are not blends. These are the **primary metaphors** we discussed in Chapter 9. Recall that primary metaphors are based on a correspondence between concepts rather than entire domains (although the primary source and primary target concepts are in different domains). In addition, primary metaphors are established on the basis of close and highly salient correlations in experience which give rise to a pre-conceptual correlation rather than a matching operation at the conceptual level. However, while primary metaphors are not themselves blends, they can function as inputs to blending, as we will see in the discussion of the SHIP OF STATE metaphoric blend in the next section. In this way, an important achievement of Conceptual Metaphor Theory is to identify metaphoric mappings that are directly grounded in experience. Mappings of this kind, which are thought to be among the most foundational aspects of conceptual structure, are not blends, and are not therefore addressed by Blending Theory. In this respect, Conceptual Metaphor Theory retains an important role in the analysis of figurative thought and language.

12.8.3 What Blending Theory adds to Conceptual Metaphor Theory

There are two important contributions that Blending Theory makes to our understanding of conceptual metaphor. The first contribution is its account of emergent structure. As we saw earlier, one of the original motivations for Blending Theory was the failure of Metaphor Theory to account for emergent structure. In our discussion of the example *That surgeon is a butcher*, we saw that there were emergent inferences that could not be accounted for by a two-domain model. The second contribution that Blending Theory makes to our understanding of conceptual metaphor is its account of the derivation of compound metaphors. We saw in Chapter 9 that compound metaphors result from the integration or **unification** of more primitive primary metaphors. What Blending Theory provides is a means of understanding how this process of unification occurs, and how it results in a compound metaphorical blend.

In order to illustrate this process, we discuss the *ship of state* metaphorical blend.

Ship of state

Our discussion of this blend is based on proposals by Grady, Oakley and Coulson (1999). Consider the following attested examples provided by Grady *et al.* (1999: 108–9):

(20) With Trent Lott as Senate Majority Leader, and Gingrich at the helm in the House, the list to the Right could destabilize the entire Ship of State.

(21) Without the consent of our fellow citizens, we lose our moral authority to steer the ship of state.

(22) The [Sri Lankan] ship of state needs to radically alter course; weather the stormy seas and enter safe harbour.

The mappings for the NATION IS A SHIP metaphor are summarised in Table 12.7.

As we saw in Chapter 9, compound metaphors like THEORIES ARE BUILDINGS are derived from two more primitive primary metaphors. This also applies to the NATION IS A SHIP metaphor, which is derived from at least those indicated in Table 12.8. What the blending perspective offers is a way of seeing how the NATION IS A SHIP metaphor is derived. Each of the primary metaphors listed in Table 12.8 represents an input to the metaphoric blend. In addition, there is a SHIP input containing a SHIP, a CREW, a COURSE, SAILING MISHAPS and so on. In the blend, the SHIP input provides a single frame that structures the blend. Hence, in the metaphor ACTION IS SELF-PROPELLED MOTION, the nature of the self-propelled motion relates not just to any kind of entity that can undergo

Table 12.7 Mappings for NATION IS A SHIP

Source: NATION	mappings	Target: SHIP
NATIONAL POLICIES/ACTIONS	→	SHIP'S COURSE
DETERMINING NATIONAL POLICIES/ACTIONS	→	STEERING SHIP
NATIONAL SUCCESS/IMPROVEMENT	→	FORWARD MOTION OF SHIP
NATIONAL FAILURES/PROBLEMS	→	SAILING MISHAPS (e.g. FOUNDERING)
DIFFICULTIES HINDERING NATIONAL SUCCESS	→	OBSTACLES ENCOUNTERED (e.g. ROCKS)
CIRCUMSTANCES AFFECTING THE NATION (e.g. ECONOMIC OR POLITICAL)	→	SEA CONDITIONS

Table 12.8 Primary metaphors that serve as inputs to *ship of state* blend

ACTION IS SELF-PROPELLED MOTION
COURSES OF ACTION ARE PATHS
TIME IS MOTION
A SOCIAL RELATIONSHIP IS PHYSICAL PROXIMITY
CIRCUMSTANCES ARE WEATHER
STATES ARE LOCATIONS

self-propelled motion but is restricted to the kind of motion that characterises ships. The paths in the blend deriving from COURSES OF ACTION ARE PATHS are also restricted to the kind of path that characterises ships (a path across the sea rather than the land). In addition, the kind of physical proximity that is possible in the blend, due to the metaphor A SOCIAL RELATIONSHIP IS PHYSICAL PROXIMITY, is understood in terms of the possible configurations of physical proximity resulting from location on a ship, and so on.

A further important consequence of treating compound metaphors as blends is that we arrive at a means of understanding **metaphor mapping gaps**, first discussed in Chapter 9. Projection to the blended space is selective: while ships are steered and we also conventionally conceptualise nations as being *steered*, there are many aspects of a ship that are not projected to the highly conventional blend. For example, ships can have a mast or a crow's nest, yet we do not conventionally speak of a nation's mast or crow's nest. As we saw in section 12.7, selective projection arises from interaction between the optimality principles of Blending Theory.

12.9 Summary

In this chapter we have presented an overview of Blending Theory. This approach derives from Mental Spaces Theory and Conceptual Metaphor Theory, but differs from both in that it explicitly accounts for **emergent structure**: the idea that meaning construction often results in meaning that is 'more than the sum of its parts'. Blending is distinguished by an architecture that includes a **generic space**, two or more **input spaces** and a **blended space**. Counterparts between input spaces are connected by virtue of a **matching** operation, compressed and selectively projected to the blended space. Emergent meaning is derived via three constitutive processes called **composition, completion** and **elaboration**. While Blending Theory arose from concerns with linguistic structure and the role of language in meaning construction, conceptual blending is argued to be a fundamental cognitive operation that is central to general properties of human thought and imagination. Recent research suggests that blending may be fundamental to a wide range of non-linguistic human behaviour, including folklore and ritual among others. We concluded the chapter with a critical evaluation of the relative achievements of both Blending Theory and Conceptual Metaphor Theory and suggested that, while Blending Theory accounts for much of what was originally thought to fall within the remit of Conceptual Metaphor Theory, the latter nevertheless retains an important role in cognitive semantics in identifying primary metaphoric mappings that are directly grounded in experience.

Further reading

There is a vast literature relating to Blending Theory. The key text is Fauconnier and Turner's 2002 book, *The Way We Think*. The first blending paper, Fauconnier and Turner (1994), which is an unpublished technical report, is available on-line at: http://cogsci.ucsd.edu/cogsci/publications/9401.pdf. The blending website provides a sense of the diversity of the phenomena to which Blending Theory has been applied: http://markturner.org/blending.html.

The theoretical development of blending theory

- **Coulson (2000)**. Coulson is one of the leading scholars in Blending Theory. Her book addresses the role of blending in frame-shifting and on-line meaning construction.
- **Coulson and Oakley (2000)**. This special edition of the journal *Cognitive Linguistics* is devoted to articles on Blending Theory.
- **Fauconnier (1999)**. An important statement by Fauconnier on how Blending Theory embodies the assumptions and the methodology that characterise cognitive semantics.
- **Fauconnier and Turner (1998a)**. This paper examines some of the central principles, and is the definitive article-length treatment of Blending Theory.
- **Fauconnier and Turner (2000)**. This paper examines how blending achieves compressions over vital relations, and thereby achieves one of its important subgoals: the provision of global insight.
- **Fauconnier and Turner (2002)**. Chapter 16 of this book provides a far more detailed account of constraints on Blending Theory than we were able to present in this chapter.
- **Turner (2001)**. This book is a study of how Blending Theory can be applied to research in the social sciences.

Blending in grammar

- **Fauconnier and Turner (1996)**
- **Mandelbilt (2000)**
- **Turner and Fauconnier (1995)**

These articles apply Blending Theory to aspects of grammar like compounds and grammatical constructions.

Metaphor, metonymy and blending

- **Fauconnier and Turner (1999)**
- **Grady (2005)**
- **Grady, Oakley and Coulson (1999)**
- **Turner and Fauconnier (2000)**

The paper by Grady, Oakley and Coulson brings together Grady, a leading researcher in metaphor, and Oakley and Coulson, leading researchers in Blending Theory. This paper compares and contrasts Conceptual Metaphor Theory and Conceptual Blending theory, concluding that the two approaches treat related but complementary phenomena.

Blending and polysemy

- **Fauconnier and Turner (2003)**. This paper argues that blending is an importnat mechanism in the development of lexical polysemy.

Blending and literary theory

- **Oakley (1998)**
- **Turner (1996)**

Blending theory has provided literary theory with a new framework; the book by Turner has been highly influential in this field.

Exercises

12.1 Constitutive processes

What are the constitutive processes of Conceptual Blending Theory?

12.2 Practice with Blending Theory

Jokes, like other forms of meaning construction, crucially rely on Blending Theory. Consider the following joke. Provide an integration network in order to account for the joke. Taking into account the constitutive processes of Blending Theory, explain how the integration network you diagram gives rise to the humorous effect prompted for by the punchline.

> Q. What do you get if you cross a kangaroo with an elephant?
> A. Holes all over Australia.

Now make a list of or collect other jokes. Show how you can account for their humorous effects by applying Blending Theory. Can you use ideas from Blending Theory to provide a 'taxonomy' of different kinds of joke, based, for instance, on differences in form (e.g. 'Knock, knock' jokes, Q and A jokes, etc.), different sorts of punchlines, humorous effects and so on?

12.3 Vital relations and compressions

For each of the following expressions, identify the outer-space vital relations and the compressions that they give rise to in the blend.

(a) President Kim Jong-Il (of North Korea)
(b) 'Drive in the fast lane of the motorway!'
(c) Children provided with a solid primary education in mathematics today are tomorrow's techno whiz kids.
(d) That child is bigger every time I see him!
(e) 'I used to be Jane' (uttered by Peter, who used to be departmental secrtetary; Jane is currently the departmental secretary).
(f) 'The pronghorn runs as fast as it does because it is being chased by ghosts – the ghosts of predators past . . . As researchers begin to look, such ghosts appear to be even more in evidence, with studies of other species showing that even when predators have been gone for hundreds of thousands of years, their prey may not have forgotten them' (cited in Fauconnier and Turner 2000: 299).

12.4 Diagramming integration networks and backward projection

In 2003, David Blunkett, then British Home Secretary (secretary of state for national security and crime), unveiled controversial plans to launch national ID cards in order to tackle a potential rise of illegal immigration into the United Kingdom as well as terrorism and welfare abuse. The United Kingdom is one of the few European countries not to have some form of identity card. Despite misgivings, the British Prime Minister Tony Blair and his Labour government agreed that the Home Office could establish a feasibility study and draw up plans for implementation. David Blunkett hailed this as a major victory and compared his success to the failure of Barbara Castle, a former Labour Home Secretary, who in 1969 was unsuccessful in convincing the then Labour government to introduce new curbs on trade union powers. The breakdown of relations between the Labour government and the unions in 1970s Britain was widely held to have opened the door for Margaret Thatcher's Conservative Party to win the 1979 general election and for subsequently keeping the Labour Party out of office for eighteen years.

Consider the following quote from David Blunkett made in 2003. At the time of utterance a general election was less than two years away, in which Tony Blair would be seeking his third successive term in office as Prime Minister. Provide a diagram of the integration network that accounts for this utterance, and explain how it illustrates the constitutive processes of blending:

(a) 'We avoided me becoming the Barbara Castle of 2003'

From the perspective of David Blunkett, what inferences are we intended to draw for the present time, based on this blend? Describe how this is achieved in terms of backward projection to the political situation with respect to which this integration network is anchored?

12.5 Taxonomy of integration networks

For each of the following examples, state what kind of integration network is involved, and explain why.

(a) The integration network you devised in response to question 2.
(b) James is Mary's uncle.
(c) Jacques Chirac is President of France.
(d) Bill Clinton was acclaimed by many as the Pelé of politics. (This utterance, adapted from a newspaper report on the former US president, argued that Clinton was the most gifted US politician of his generation.)
(e) If the 1970 Brazilian World Cup winning team had played the Brazilian World Cup winners of 2002, the team of 1970 would have won.

12.6 Benefit tourist and welfare shopping

Consider expressions like *benefit tourist* and *welfare shopping*. Expressions like these emerged in the British media in 2003 and 2004. Their use by the right-wing press in particular relates to economic migrants, mainly from relatively poor former Eastern-bloc countries. Following the expansion of the European Union in late 2004, migrants who move to the United Kingdom may have the right to claim support from the welfare and social security system. The view expressed in the conservative press is that these migrants are *benefit tourists* who are *enjoying* their *welfare shopping*.

(i) *Benefit tourist* and *welfare shopping* are instances of lexical creation via compounding. Both these forms are also both instances of blending. What kinds of blends are they?
(ii) Provide separate integration networks for the expressions *benefit tourist* and *welfare shopping*. How many input spaces are required by each network? How do the networks you have diagrammed illustrate the constitutive processes of blending?
(iii) Discuss how backward projection from these blends might affect the way people react to and behave towards economic migrants from poorer countries. What does your discussion illustrate about the role of conceptual blending?

12.7 Metaphorical blends

(i) List the criteria that make a blend metaphorical.
(ii) Provide a blending analysis for the following two examples. What is the emergent structure associated with each? Are the blends you describe metaphorical according to your criteria?
   (a) 'One employer described an employee who phoned in apparently sick as *stealing* the company's time' (Lakoff and Johnson 1999).
   (b) 'Lies run sprints, but the truth runs marathons. The truth will win this marathon in court' (Michael Jackson in 2003 after his arrest on charges of child molestation).

13

Cognitive semantics in context

In this chapter, which concludes Part II of the book, we compare and contrast cognitive semantics with two other modern approaches to linguistic meaning: **truth-conditional** (or **formal**) **semantics** and **Relevance Theory**. As noted at various points in this book, cognitive semantics emerged and developed as a reaction against formal semantics. For this reason, we look in more detail at the truth-conditional approach to sentence meaning in this chapter and present some explicit points of comparison between the formal and cognitive approaches (section 13.1). We also provide an introduction to Relevance Theory, a modern approach that attempts to account for the pragmatic aspects of linguistic communication within a broader cognitive framework (section 13.2). Although this model explicitly adopts the formal view of language by assuming a modular theory of mind as well as a truth-conditional model of semantic meaning, it rejects some of the received distinctions assumed within formal approaches to linguistic meaning, such as a clear-cut division of labour between semantics and pragmatics. In this, Relevance Theory represents a formally oriented model that is in certain respects consonant with cognitive semantics. By drawing some explicit comparisons between cognitive semantics and these two models, we set the cognitive linguistics enterprise within a broader theoretical context. However, because cognitive semantics represents a collection of distinct theories, some of which examine quite distinct phenomena, this comparison will be limited to the areas that truth-conditional semantics and Relevance Theory are concerned with: while truth-conditional semantics is primarily concerned with meaning construction (or sentence meaning), Relevance Theory addresses word meaning, sentence meaning, pragmatic meanings and figurative language such as metaphor and irony.

13.1 Truth-conditional semantics

In this section we briefly present some of the ideas developed within the discipline called 'philosophy of language' that go under the name of **truth-conditional semantics**. As we saw briefly in Chapters 5 and 11, these ideas relate to meaning, truth and reality, and how meaning can be represented according to a formal metalanguage developed from logic. These ideas came to be highly influential in formal linguistics following the pioneering work of philosopher and logician Richard Montague in the 1960s and early 1970s. Montague argued that many of the ideas from the philosophy of language could be systematically applied to natural language. The tradition that grew up in linguistics following Montague's theory came to be known as truth-conditional or formal semantics.

13.1.1 Meaning, truth and reality

The philosophical interest in the relationship between meaning, truth and reality has a long and venerable tradition dating back to the ideas of the ancient Greek philosophers over 2,000 years ago. Since Aristotle, philosophers who have attempted to understand the concept of truth have equated this notion with reality as a guarantor of truth. This approach is called the **correspondence theory** and holds that a **truth bearer** (for example, a natural language sentence) is true if it corresponds to a state of affairs holding in the world. From this perspective, truth is a property of sentences that correspond to a reality they describe. The twentieth-century philosopher Alfred Tarski was influential in arguing that meaning could be equated with truth defined in terms of its correspondence with the world: if a sentence is true by virtue of its correspondence with some state of affairs, then this truth condition constitutes its meaning. Consider the following excerpt from Tarski's classic paper first published in 1944:

> *Semantics* is a discipline which ... *deals with certain relations between expressions of a language and the objects* (or 'states of affairs') *'referred to' by those expressions.* (Tarski [1944] 2004: 119; original emphasis)

From this perspective, linguistic meaning is truth defined in terms of correspondence to reality. Meaning can therefore be defined in terms of the conditions that hold for a sentence to be true.

13.1.2 Object language versus metalanguage

Tarski argued that truth can only be defined for those languages whose semantic structure has been exactly defined and that it is not possible to define the semantic structure of a language that is self-defining. For example, in a natural language, words are defined using other words in the language: if we 'define' *bachelor* as 'an unmarried adult male', we are using other words from the same language to define the word. According to Tarski, this fails to provide an objective definition, because it relies on words from the same language to understand other words. Tarski describes languages that are self-defining as **closed** because they fail to provide an objective definition of a particular term or expression. Therefore he argues that in order to establish the meaning of a sentence from a given natural language, we need to be able to translate the sentence from that **object language** into a **metalanguage**, a language that can be precisely and objectively defined. Tarski argues that **predicate calculus**, which was pioneered by the philosopher Gottlob Frege in his work on logic, provides a logic-based metalanguage for capturing the 'invariant' (semantic or context-independent) aspects of meaning. According to this view, predicate calculus, or a similar 'logical' language, provides a means of capturing meaning in a way that is objective, precisely stated, free from ambiguity and universal in the sense that it can be applied to any natural language.

13.1.3 The inconsistency of natural language

It is important to note that Tarski was concerned with the study of semantics (meaning in general) rather than specifically linguistic semantics. While Tarski thought that the truth-conditions for formal languages like logic could be precisely specified, he argued that the meaning of natural languages could not be precisely specified in terms of truth conditions. Tarski expresses this view in the following way:

> *The problem of the definition of truth obtains a precise meaning and can be solved in a rigorous way only for those languages whose structure has been exactly specified.* For other languages – thus, for all natural 'spoken' languages – the meaning of the problem is more or less vague, and its solution can have only an approximate character. (Tarski [1944] 2004: 121; original emphasis)

A particularly clear illustration of the way in which natural language resists precise definition in terms of truth conditions emerged from J. L. Austin's work on **speech acts**. This theory was developed in Austin's 1955 lectures, which were published posthumously in 1962. Austin observed that only certain types of sentence relate to 'states of affairs in the world'. This sentence type, which Austin called **constative**, is illustrated in examples (1) to (4).

(1) It is raining. [constatives]
(2) My cat is black and white.
(3) Tony Blair is Prime Minister.
(4) She doesn't feel very well today.

Compare examples (1)–(4) with what Austin called **performative** sentences, illustrated in examples (5)–(11).

(5) I bet you £10 it will rain tomorrow. [performatives]
(6) I hereby name this ship the HMS *Sussex*.
(7) I declare war on the citizens of Mars.
(8) I apologise.
(9) I dub thee Sir Walter.
(10) I hereby pronounce you man and wife.

Only sentences of the kind in (1) to (4) can be said to have truth conditions because they can be verified against the corresponding state of affairs that they describe. In contrast, it makes little sense to think of the sentences in (5) to (11) as 'describing' states of affairs because these sentences are performing verbal acts rather than describing situations. Observe that performatives license the adverb *hereby*, and are restricted to the first person present tense. If these sentences are changed to the third person and/or to the past tense, they become descriptions of states of affairs rather than performatives (11a), and do not license *hereby* (11b). Furthermore, only certain verbs function as performatives (11c).

(11) a. He sentenced you to ten years of hard labour yesterday.
b. He hereby sentenced you to ten years of hard labour yesterday.
c. I hereby love you.

As these examples illustrate, only a subset of sentence types can be understood in terms of their correspondence with 'states of affairs' or situations that they describe. Furthermore, this observation is not limited to the distinction between the types of examples illustrated here. For example, interrogative sentences like *Do you want a cup of tea?* and imperative sentences like *Shut the door!* cannot be described as 'true' or 'false' with respect to a given state of affairs in the world.

13.1.4 Sentences and propositions

Before exploring how truth-conditional semantics was developed into the basis of a formal approach to linguistic meaning, we first need to introduce the important distinction between **sentence** and **proposition**. A sentence is a linguistic object, a well-formed grammatical string of words that can be described according to its grammatical properties. The meaning 'carried' by a sentence is a proposition. Crucially, there is no one-to-one correspondence between sentence and proposition because the same sentence can carry different propositions (e.g. *I love you* expresses a different proposition depending on who *I* and *you* refer to), and the same proposition can be expressed by different sentences. This is illustrated by example (12), in which both the active sentence (12a) and the passive sentence (12b) describe the same state of affairs and thus represent the same proposition. This means that these two sentences have the same truth conditions.

(12) a. Shakespeare wrote *Romeo and Juliet*.
b. *Romeo and Juliet* was written by Shakespeare.

In truth-conditional semantics, it is the meaningful proposition that is the truth-bearer. In other words, truth conditions relate to the proposition expressed by a sentence rather than directly to the sentence itself.

13.1.5 Truth-conditional semantics and the generative enterprise

Despite reservations expressed by philosophers of language like Tarski and 'natural language philosophers' like Austin, the philosopher and logician Richard Montague (e.g. 1970, 1973) argued that natural language semantics could be modelled in terms of truth conditions. According to this perspective, a crucial aspect of natural language semantics relates to logical properties and relations so that natural language can be 'translated' into the metalanguage of predicate calculus, exposing its meaning to rigorous scrutiny and definition. In this section, we present an overview of this tradition.

Montague's ideas have appealed to formal linguists because of the precision offered by the application of truth-conditional semantics to natural language. In particular, this approach has appealed to scholars who have sought to integrate the field of linguistic semantics with the generative grammar model developed by Chomsky. As we have seen in earlier chapters, language is viewed as a modular system in the tradition pioneered by Chomsky (see Figure 13.1). Within this model, each module represents an encapsulated system of linguistic knowledge that contains principles operating over primitives of a specific kind. For example, while the syntax module operates over grammatical categories like noun, verb, tense and so on, the phonology module operates over speech sounds representing bundles of articulatory features. Many semanticists influenced by the generative enterprise sought to develop an approach to natural language semantics that could provide a semantic representation for the grammatical representation generated by the syntax module: the sentence.

COGNITIVE LINGUISTICS: AN INTRODUCTION

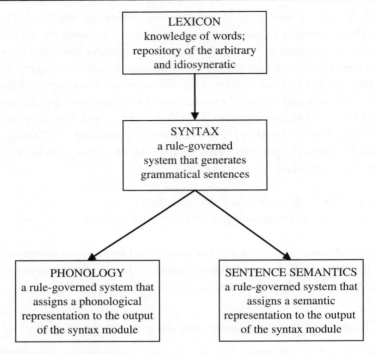

Figure 13.1 The generative model

13.1.6 Compositionality of meaning

Formal semanticists adopt the **Principle of Compositionality**. This principle states that the meaning of a complex expression is determined by the meanings of its parts, affected only by the grammatical structure in which these parts coexist. The fact that grammatical structure plays a role in linguistic meaning is illustrated by examples (13a) and (13b). These examples contain the same words, but express different propositions precisely because those parts are arranged differently within the syntactic configuration.

(13) a. Joe gave Sally a lift.
b. Sally gave Joe a lift.

The fact that syntax can affect the semantic interpretation of a sentence explains why, in the generative model, there is a semantic component that assigns a semantic representation to the output of the syntax module. While the lexicon accounts for a speaker's knowledge of word meaning, this model also requires a module that accounts for the meaning of a complex expression in which those words have been combined into a particular grammatical structure.

13.1.7 Translating natural language into a metalanguage

Predicate calculus, the logical metalanguage into which formal semanticists translate natural languages like English, contains a range of expressions. These expressions represent the meaning expressed by units of language like nouns, verbs and adjectives by means of **terms**. There are two kinds of terms: **individual constants** and **predicates**. Constants are expressions that relate to specific entities (like *James Bond* or *the spy*) and are represented by lower-case letters of the alphabet like *a, b, c* and so on. Predicates are expressions that represent processes (expressed by verbs like *eat*), properties (expressed by adjectives like *funny*), roles (expressed by nouns like *a top British spy*) and relations (expressed by prepositions like *under*). One-place predicates like *funny, die* or *a top British spy* only require a single participant to complete their meaning (e.g. *James Bond is funny; James Bond died; James Bond is a top British spy*), while two-place predicates like *appreciate* or *under* require two participants (e.g. *James Bond appreciates Miss Moneypenny; James Bond is under the desk*). Predicates are represented by upper-case letters of the alphabet, like A, B, C and so on. When constants and predicates are combined, this results in a **formula**. For example, the sentence in (14a) can be expressed by the formula in (14b), where upper-case S represents the predicate *sings* and lower-case f represents the constant *Fred*. By convention, the predicate occurs first in the predicate calculus formula, so the 'translation' does not reflect the word order of English.

(14) a. Fred sings.
     b. S(f)

Example (15) illustrates a formula in which a two-place predicate combines with two constants. The relative order of the constants is important, because this reflects the difference in meaning contributed by the syntactic structure: like the natural language sentence in (15a), the formula in (15b) says that Jane loves Tom, not that Tom loves Jane.

(15) a. Jane loves Tom.
     b. L(j, t)

In sentences like *Jane loves Tom* and *Tom loves Jane*, which consist of two or more conjoined clauses and thus express two or more propositions, the clauses are connected by natural language **connectives** like *and, or, but* and so on. In sentences like *Jane does not love Tom* or *Jane loves Tom but not Bill*, the negation word *not* is an **operator**, an expression that takes **scope** over some part of the sentence and affects its meaning. Natural language expressions like *all, every*

Table 13.1 Connectives and operators in predicate calculus

Connective	Syntax	English
∧	x ∧ y	X and y
∨	x ∨ y	X and/or y
$\vee_e$	x $\vee_e$ y	X or y but not both
→	x → y	If x, then y
≡	x ≡ y	X if and only if y

Operator	Syntax	English
¬	¬ x	not x
∀	∀ x	every/all x
∃	∃ x	some x

and *some* are also operators. These are **quantifiers** and take scope over some part of the sentence by quantifying it (for example, the sentences *Every policeman witnessed some crimes* and *Some policemen witnessed every crime* each express a different proposition due to the positions of the quantifiers, despite the fact that they contain the same predicates and constants). Connectives and operators are represented by the logical symbols in Table 13.1, where the column 'syntax' shows how these symbols can be combined with other units.

Example (16) shows how the sentence in (16a) is translated into a predicate calculus formula (16b). The expression in (16c) shows how the predicate calculus can be 'read'. In this example, *x* represents a **variable**. This is an expression that, like a constant, relates to an entity or group of entities (hence the lower-case symbol); unlike a constant, a variable does not indicate a specific entity. The lower case letters x, y and z are reserved for variables.

(16) a. Every pupil sat an exam
    b. ∀x (P(x) → S(x, e))
    c. For every entity x, if x is a pupil, then x sat an exam

13.1.8 Semantic interpretation and matching

Of course, the translation of a sentence from object language to metalanguage does not in itself tell us anything about what the sentence means. To accomplish this, the symbols in the metalanguage must be assigned a semantic interpretation or **value**, at which point the formula, which represents the proposition expressed by the original natural language sentence, must be **matched** with the state of affairs it describes. The process of assigning values and matching the proposition to the state of affairs it describes can be divided into four steps.

Assigning values

The first step is to assign the symbols of predicate calculus a semantic interpretation. This idea was implicit in the previous section, where we assigned the symbols a semantic value. For example, predicates expressed by *eat* and *love* are represented by E, L and so on, and constants expressed by proper nouns like *Jane* and *Tom* are represented by j, t and so on. Because natural language connectives and operators are closed-class expressions, these correspond to fixed logical symbols. In contrast, predicates and constants can be expressed by upper- or lower-case letters of the alphabet, with the exception of x, y and z, which by convention are reserved for variables.

Establishing a model of the world

The second step is the establishment of some model of the world against which the symbols in the metalanguage can be matched. Within formal semantics, models are typically represented in terms of **set theory**. For example, in a model of the world in which all women love chocolate, the sentence *All women love chocolate* would be true. However, in a model in which only a subset of women love chocolate, a further subset love chips and an intersection of these two subsets love both, the sentence all women love chocolate would be false, whereas the sentences *Some women love chocolate, Some women love chips, Some women love chocolate and chips* and *Not all women love chocolate* would be true. It is because the symbols are matched with a model of the world that this type of approach is also known as **model-theoretic semantics**. This idea is illustrated by Figure 13.2.

Matching formula with model

The third step is a matching operation in which the symbols are matched with appropriate entities in the model. This is called **denotation**: expressions in the metalanguage denote or represent elements in the model, and the meaning of the sentence is equivalent to its **denotatum**, or the sum of what it corresponds to in the model. Matching of predicates establishes the **extension** of individuals over which the predicate holds, which is represented in terms of sets. For example, in the sentence *All women love chocolate*, the predicate *love* represents a relation between the set of all entities described as *women* and the set of all entities described as *chocolate*. Once this matching operation has taken place, then the truth value of the sentence can be calculated.

COGNITIVE LINGUISTICS: AN INTRODUCTION

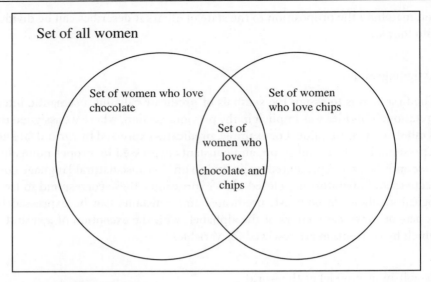

Figure 13.2 Set-theoretic model

Table 13.2 Steps for calculating truth conditions

Assigning values	This assigns a semantic value to the symbols in the formula. Upper case A, B, C correspond to predicates; lower case a, b, c to constants; and x, y, z to variables. Fixed symbols represent connectives and operators.
Establishing a model of the world	This set-theoretic model represents a 'state of affairs' against which the sentence is matched.
Matching formula with model	On the basis of correspondence theory, the denotatum of a sentence is its correspondence with the state of affairs represented by the model. The denotatum is composed of assignment of individual constant terms and the representation of predicates as a set (extension) of individuals over which the predicate holds.
Calculating truth values	Since meaning is defined in terms of truth and truth in terms of correspondence, the truth value is calculated on the basis of the correspondence between the sentence and the model.

Calculating truth values

The fourth step involves the calculation of truth values. If the formula matches the model, then the sentence is true. If it does not, then the sentence is false. These steps are summarised in Table 13.2. As this brief overview shows, in truth-conditional semantics the meaning of a sentence is equivalent to the conditions that hold for that sentence to be true, relative to a model of the world. Central to this approach is the correspondence theory of truth that

454

we considered earlier (section 13.1.1): meaning is defined in terms of the truth of a sentence, understood as conditions in the world (or a model of the world) to which the sentence corresponds.

We illustrate each of these steps with example (15), which is repeated here.

(15) a. Jane loves Tom.
b. L(j, t)

Once the sentence is translated into predicate calculus (15b), values are assigned to the logical symbols (e.g. j = Jane; t = Tom) and a model is established that identifies the entities corresponding to the linguistic expressions *Jane* and *Tom*. This model might represent the set of all people {Bill, Fred, Jane, Mary, Susan, Tom. . .}. Within this model is a domain or subset of entities who stand in the relation expressed by the predicate *love* (L). This is represented by (17), in which each ordered pair (inside angled brackets) stands in the relevant relation.

(17) L = {<Jane, Tom>, <Fred, Mary>, <Mary, Susan>}

Next, the formula is matched with the model so that constants and predicates are matched with entities and relations in the model. As (17) shows, this set contains an ordered pair, which means that Jane loves Tom. Finally, the truth condition of the proposition expressed by (15) is evaluated relative to this model. The rule for this evaluation process is shown in (18).

(18) $[L(j, t) = 1 \equiv [<j, t>] \in [L]]$

In this rule, the number '1' represents 'true' (as opposed to '0', which represents 'false'). This rule says '*Jane loves Tom* is true if and only if the ordered pair <Jane, Tom> is a member of the set L'. Since the set L contains the ordered pair <Jane, Tom> in the model, the sentence is true. Table 13.3 completes this brief overview of the truth-conditional approach to sentence meaning in formal semantics by summarising the properties that characterise this approach as it is conceived by generatively oriented semanticists.

13.1.9 Comparison with cognitive semantics

While the assumptions presented in Table 13.3 stand in direct opposition to those adopted within cognitive semantics, there are nevertheless some important similarities between the two approaches. Firstly, both approaches are concerned with explaining sentence meaning and with the nature of the relationships between the words in a sentence, as well as between the words

Table 13.3 Truth-conditional formal semantics

The nativist hypothesis is widely assumed.
The modularity hypothesis is widely assumed: linguistic knowledge emerges from an encapsulated cognitive system, and the language module itself has a modular structure.
Semantic (context-independent) knowledge is separable from pragmatic (context-dependent) and encyclopaedic (non-linguistic) knowledge.
A correspondence theory of truth is assumed, hence this approach is 'objectivist' in the sense that sentence meaning relies upon an objectively defined world or model of the world.
Sentence meaning can be modelled using a logical metalanguage.
The meaning of complex expressions is compositional. Figurative language is non-compositional and therefore exceptional.
In practice, this approach is focused upon the logical properties of a carefully selected set of declarative sentences.

and the grammatical structure in which they occur. Secondly, both formal semantics and cognitive semantics accept the existence of a real external world which bears upon the nature of linguistic meaning. For example, both theories distinguish between entities, properties, processes and relations. Thirdly, both approaches assume that humans have stable knowledge of the external world which is reflected in language, and attempt to model this knowledge. While the earliest truth-conditional models relied upon a direct link between language and external world (**referential** or **denotational** models), modern formal semantics attempts to model the system of human knowledge that mediates between linguistic symbols and external reality. Therefore, like cognitive semantics, formal semantics aims to construct a **representational** model.

Despite these important similarities, the differences remain significant. Beginning with fundamental assumptions, while formal semanticists assume an innate and modular system of specialised linguistic knowledge, cognitive semanticists reject this view in favour of a semantic system that provides 'prompts' to the rich conceptual system that it reflects. In adopting an objectivist approach to cognition, truth-conditional semanticists see human thought as 'disembodied' because linguistic meaning is conceived in terms of correspondence theory. In contrast, in adopting a broadly experientialist or empiricist approach to cognition, cognitive semanticists conceive meaning as the imaginative projection of bodily experience onto abstract cognitive models.

Turning to how each model views the nature of linguistic meaning, formal semanticists argue that one of the primary goals of a theory of linguistic meaning is to address the **informational significance** of language. From this perspective, language is used primarily to describe states of affairs in the 'world', which are thus central to the account of linguistic meaning, as we have seen. This idea is represented by Figure 13.3.

COGNITIVE SEMANTICS IN CONTEXT

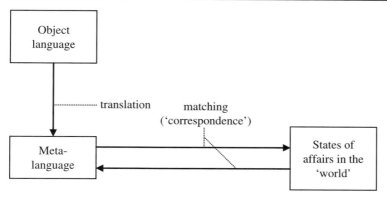

Figure 13.3 The construction of sentence-meaning in formal semantics

In Figure 13.3, the arrow from the object language to the metalanguage represents the translation process, which gives rise to a representation in the unambiguous and universally applicable language of predicate calculus. Meaning then derives from how well the values associated with the metalanguage correspond to a given state of affairs in the 'world', real or hypothetical.

In contrast, cognitive semanticists argue that the role of language is to prompt for conceptual representations (including simulations in the sense discussed in Chapter 7), so that meaning derives not from an objectively defined 'world' but from structured mental representations that reflect and model the world we experience as embodied human beings. According to the view in cognitive semantics, these mental representations are partly stable (stored) knowledge systems and partly dynamic (on-line) conceptualisations. It follows from this view that linguistic meaning resides not within a specialised system of linguistic knowledge but at the conceptual level itself. The cognitive view of the nature of linguistic meaning is represented by Figure 13.4.

Figure 13.4 represents the idea that two basic kinds of experience (sensory-perceptual experience of the external world and subjective experience from the introspective 'world') give rise to conceptual representations which can lead to simulations. Language prompts for these conceptual representations, serving as 'points of access' to relatively stable encyclopaedic knowledge (this is indicated by the arrow from 'language' to 'representation'). Conceptual representations are also subject to further processes of dynamic meaning construction. Meaning-construction can in turn have consequences for language, for example by giving rise to language change (this is indicated by the arrow from 'meaning-construction' to 'language'). For example, using the lexical item *mouse* to refer to a piece of computer hardware that 'resembles' a mouse is a consequence of single-scope blending; recall from the previous chapter that this involves the frame from one input space serving to organise the structure projected to the

457

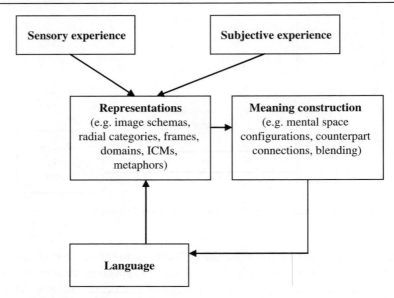

Figure 13.4 The nature of meaning construction in cognitive semantics

blended space. However, this blend has consequences for language: as a consequence of the perceived resemblance between a mouse and an item of computer hardware, the conceptual integration network that results affects conventional language use. Indeed, the conventional application of the lexical item *mouse* to the 'computer mouse' can be seen as testimony to the impact of blending on language. This illustrates the usage-based nature of the cognitive model, where language both gives rise to (= prompts for) conceptualisation (affecting our conceived 'reality') and in turn is modified and transformed by the resulting conceptual representations.

A further important difference relates to the nature of the relationship between semantics (context-independent meaning) and pragmatics (context-dependent meaning). As we have seen, cognitive semanticists adopt an encyclopaedic view of meaning together with a dynamic context-driven view of meaning construction, which entails that there is no principled distinction between semantic and pragmatic knowledge. In contrast, formal semanticists assume a sharp boundary between the two types of knowledge. According to this view, semantic knowledge is stable, conventionalised knowledge that is expressed by predictable form-meaning correspondences and is contained within the linguistic system. In contrast, pragmatic inferences cannot be predicted from linguistic form; pragmatic knowledge involves more generalised inferencing processes that do not relate specifically to language but operate over the output of the language system together with non-linguistic contextual factors. This is the issue that Relevance Theory addresses, to which we turn directly.

13.2 Relevance Theory

Relevance Theory was developed by psychologist Dan Sperber and linguist Deirdre Wilson, and develops key insights from the well-known theory of pragmatics proposed by Paul Grice (1975). We base our discussion here on the 1995 edition of their landmark book, *Relevance: Communication and Cognition*, which was originally published in 1986. Relevance Theory represents a modern approach to pragmatics that adopts an explicitly generative view of language, and aims to provide a mentalist account of communication that can be integrated with the generative model of language. Despite its generative orientation, in its emphasis on linguistic communication within the context of general cognition, Relevance Theory is consonant with cognitive semantics in a number of respects. For example, Sperber and Wilson reject the semantic decomposition account of word meaning that characterises the standard formal view, and argue in favour of the incorporation of encyclopaedic meaning within the lexical representation of words. In this section, we focus on the Relevance Theoretic account of meaning construction, or sentence meaning.

13.2.1 Ostensive communication

Relevance Theory is a theoretical approach to communication in general, which views **verbal communication** as one instance of **ostensive-inferential communication**. According to Sperber and Wilson, the defining characteristic of communication is that it involves revealing or making manifest a particular **communicative intention**. In other words, the communicator's intention is revealed by some kind of ostensive behaviour. For example, in response to the question *How are you getting home?* you can perform a manual gesture representing a car's steering wheel. This is a form of ostensive behaviour signalling a specific communicative intention, namely that the 'addressee' should infer that you will be driving home. Equally, if you are at a party that you wish to leave, you can raise your arm and tap your watch to indicate to your partner that it's time to go. In both cases, the act would fail as an instance of communication if it were not ostensive. For example, if you were sitting in the bathroom by yourself, the act of tapping your watch would fail to achieve ostensive-inferential communication.

13.2.2 Mutual cognitive environment

Of course, for speaker and hearer to communicate successfully, particularly where inference is concerned, they must rely upon shared information. For example, the person in our earlier example who indicates that s/he will be driving home relies upon the assumption that his or her 'addressee' knows that cars have steering wheels and can recognise the gesture that represents this.

Sperber and Wilson describe this shared knowledge upon which inferences depend as the 'mutual cognitive environment'. Consider the following excerpts from Sperber and Wilson.

> The cognitive environment of an individual is a set of facts that are manifest to him . . . A fact is manifest to an individual at a given time iff [if and only if] he is capable at that time of representing it mentally and accepting its representation as true or probably true . . . an individual's total cognitive environment is the set of all the facts he can perceive or infer . . . a function of his physical environment and his cognitive abilities . . . The total shared environment of two people is the intersection of their two total cognitive environments, i.e. the set of all facts that are manifest to them both. (Sperber and Wilson 1995: 39–41)

As these excerpts make explicit, inference depends upon the speaker's knowledge, and the knowledge s/he can assume on the part of the hearer.

13.2.3 Relevance

According to Sperber and Wilson, human cognition is driven by **relevance** in the sense that information (whether sensory-perceptual or linguistic) is selectively processed on the basis of the search for **contextual effects**: information that will affect our existing knowledge in some useful way or will allow us to construct an inference. For example, imagine driving down the road in your car with the radio on. In this context, you are bombarded with sensory-perceptual stimuli including visual stimuli as well as linguistic and non-linguistic sounds. Suppose that you have been worried about your car lately. In this context, you might 'tune out' the linguistic sounds coming from the radio and focus your attention on the sounds coming from under the bonnet. Depending on whether these sounds are out of the ordinary or not, this information will interact with what you already know about your car and allow you to draw some conclusions. In this context, the car's sounds are more relevant than the radio's sounds. Now imagine that you are late for work and concerned about the time. You transfer your attention to the linguistic sounds coming from the radio and listen for the newsreader to announce the time. In this context, the radio's sounds are more relevant than the car's sounds. As this simple example illustrates, the human mind constantly searches for relevant information. This idea is captured by the '**Cognitive Principle of Relevance**', which states that 'Human cognition tends to be geared to the maximisation of relevance' (Sperber and Wilson 1995: 158).

Sperber and Wilson argue that ostensive-inferential communication is driven by the presumption of relevance. In other words, a hearer will assume that any act of (linguistic or non-linguistic) ostensive-inferential communication is

COGNITIVE SEMANTICS IN CONTEXT

relevant, and moreover will search for the **optimally relevant** interpretation. It is this assumption that allows us to deduce or **infer** the communicative intention signalled by an act of ostensive communication. This idea is captured by the '**Communicative Principle of Relevance**', which states that 'Every act of ostensive communication communicates a presumption of its own optimal relevance' (Sperber and Wilson 1995: 260). 'Optimal relevance' is defined in the following way:

> Presumption of optimal relevance (Sperber and Wilson 1995: 158)
> 1. The set of assumptions **I** which the communicator intends to make manifest to the addressee is relevant enough to make it worth the addressee's while to process the ostensive stimulus.
> 2. The ostensive stimulus is the most relevant one the communicator could have used to communicate **I**.

Consider example (19) from Sperber and Wilson (1995: 189). Imagine that this utterance is made in a jeweller's shop in response to an enquiry from a customer about how long they might expect to wait for the watch to be repaired.

(19) It will take some time to repair your watch.

It is obvious that a watch repair must take 'some time' (as opposed to no time), so the customer assumes that the communicative intention behind the utterance cannot be to convey this uninformative and therefore irrelevant interpretation. Sperber and Wilson argue that our presumption of relevance in everyday communication guides us to a more appropriate interpretation of the utterance. If the customer knows that it usually takes about a week to get a watch repaired, then the most relevant reason for mentioning the time it will take is probably because the repair will take significantly longer than a week.

13.2.4 Explicature and implicature

Sperber and Wilson follow the formal view in distinguishing between what they call **explicature** and **implicature**. The term 'explicature' describes an assumption that is explicitly communicated. In relating to explicit or context-independent meaning, this term roughly corresponds to the traditional idea of semantic meaning. The term 'implicature', which is adopted from Grice (1975), relates to implicit or inferential (context-dependent) meaning, and corresponds to the traditional view of pragmatic meaning. Sperber and Wilson also follow the standard formal view in assuming that semantic 'decoding' takes place prior to the calculation of pragmatic inferences. However, they depart from the standard formal view in arguing that meaning construction relies to considerable extent upon inference, even in the 'decoding' of explicatures. This idea is illustrated by example (20) from Sperber and Wilson (1995: 186).

(20) The child left the straw in the glass.

This sentence is straightforwardly interpreted to mean that a child left a 'drinking tube' in a glass drinking vessel. This meaning is the explicature expressed by the sentence. However, as Sperber and Wilson observe, even this straightforward sentence requires some inferential work, because the expression *straw* is lexically ambiguous: it could mean the child left a 'cereal stalk' in the glass. To derive the more likely or accessible 'drinking tube' interpretation, the hearer has to access encyclopaedic information relating to children and the typical scenarios involving a 'straw' and a 'glass'. The availability of the most salient interpretation might also depend on contextual information, such as whether the child in question was in a kitchen or a farmyard. As this example illustrates, many explicatures will rely upon inference on the part of the hearer in order to retrieve the intended meaning. Indeed, all explicatures containing referential expressions like *that man* or *him* rely upon inference for **reference assignment**: matching a referring expression with the 'right' entity. Sperber and Wilson's model therefore departs from the standard formal model in emphasising the role of inference in deriving explicit meaning. The exchange in example (21) illustrates how an implicature is derived (Sperber and Wilson 1995: 194).

(21) a. Peter: Would you drive a Mercedes?
     b. Mary: I wouldn't drive ANY expensive car.

In this exchange, Mary fails to answer Peter's question directly (because Peter's utterance is a 'yes-no question' a straightforward 'yes' or 'no' would provide a direct answer). The presumption of relevance allows Peter to assume that Mary has answered the question in the most relevant way possible and to infer her intended meaning. Mary's utterance interacts with Peter's encyclopaedic knowledge and gives rise to the fact that a Mercedes is an expensive car. This fact interacts with Mary's assertion that she wouldn't drive ANY expensive car, and by a process of logical deduction gives rise to the explicature that Mary wouldn't drive a Mercedes. Mary's utterance counts as the optimally relevant way of answering Peter's question because it is maximally informative. Her utterance gives rise to a greater number of contextual effects than a direct 'no' response, because Peter now knows not only that Mary wouldn't drive a Mercedes, but also that she wouldn't drive a BMW, a Bentley, a Jaguar and so on. From this perspective, the extra effort or processing 'cost' involved in the retrieval of the implicature(s) is rewarded by the 'benefit' of a greater number of contextual effects.

13.2.5 Metaphor

Finally, we briefly consider Sperber and Wilson's account of figurative language, focusing on their discussion of metaphor. Sperber and Wilson argue that relevance and inference are also central to the interpretation of figurative language. Consider example (22) from Sperber and Wilson (1995: 236).

(22) This room is a pigsty.

According to the Relevance Theory account, the hearer is licensed to assume that the speaker is aiming for optimal relevance in uttering (22). Because the utterance is literally false (the room is not literally a pigsty), the literal interpretation is uninformative and therefore irrelevant. The hearer therefore assumes that the speaker intends some other interpretation and draws upon encyclopaedic knowledge and contextual knowledge in order to construct an inference. Encyclopaedic knowledge gives rise to the fact that a pigsty is associated with filth and untidiness. The resemblance between the encyclopaedic representation of a pigsty and the condition of the room (contextual information) allows the hearer to infer that the speaker intends to convey that the room is filthy and untidy. As Sperber and Wilson point out, the use of this metaphor carries additional contextual effects that could not be conveyed by the utterance *This room is filthy and dirty*. By comparing the room to a pigsty, the speaker provides a much richer representation of the condition of the room which might give rise to further implicatures (e.g. the filth and untidiness goes 'beyond the norm' for a room inhabited by humans rather than animals, the room smells bad, and so on). In this way, metaphor also rewards the hearer's extra processing cost with a richer set of contextual effects than a literal utterance: 'the wider the range of potential implicatures and the greater the hearer's responsibility for constructing them, the more poetic the effect, the more creative the metaphor' (Sperber and Wilson 1995: 236). Table 13.4 summarises the main assumptions of Relevance Theory.

13.2.6 Comparison with cognitive semantics

In many respects, the Relevance Theory view of meaning construction is similar to the view taken in cognitive approaches, including Mental Spaces Theory and Blending Theory. Both Relevance Theory and cognitive semantics are concerned with describing the mental processes involved in meaning construction. Like cognitive semantics, Relevance Theory focuses upon developing a psychologically plausible account of communication, and in emphasising inference, encyclopaedic knowledge and contextual knowledge, it relates to the processes that mental spaces and blending theorists refer to as projection, mapping, schema induction and integration. Furthermore, both Relevance Theory and cognitive

Table 13.4 Relevance Theory

Primarily concerned with accounting for ostensive-inferential communication; language is just one form of this.
Shared knowledge is the 'mutual cognitive environment'.
Cognition is driven by the search for relevance (Cognitive Principle of Relevance); relevance yields contextual effects.
Acts of ostensive communication (including utterances) presume their own optimal relevance.
Optimal relevance means that the information is worth retrieving and that the hearer has chosen the most relevant means of communicating.
While explicature and implicature roughly correspond to semantic and pragmatic meaning, respectively, both rely upon inference, which is relevance-driven.
Metaphors (and other forms of figurative language) are interpreted according to the same principles as literal utterances; they are relevance-driven in nature and provide a richer set of inferences than literal utterances.

semantics emphasise the idea that meaning construction is in large measure due to these mental processes rather than a simple matter of composing a sentence's meaning from its parts. Indeed, Sperber and Wilson explicitly reject what they call the 'code model' as a descriptively adequate account of communication. Furthermore, Sperber and Wilson claim that explicature, as well as implicature, require extensive inferencing (in processes such as disambiguation and reference assignment). In this respect, and in relying upon contextual and encyclopaedic information in these processes, Sperber and Wilson's view is consonant with the claim made by cognitive semanticists that words represent 'prompts' for meaning construction, and with the idea that a strict dividing line between semantics and pragmatics cannot be straightforwardly upheld. Finally, Sperber and Wilson argue that metaphor and other types of figurative language are unexceptional in the sense that they exploit the same cognitive processes by maximising relevance. In this respect, although the details of the Relevance Theoretic account of metaphor focus more on communication than on cognition, the integration of figurative and literal language is also consonant with the cognitive account.

Despite these areas of agreement, there are some fundamental differences between the two approaches. Most importantly, Relevance Theory assumes as its background a generative model of language; this model assumes the nativist hypothesis and the modularity hypothesis. In addition, Relevance Theory assumes a logical truth-conditional account of certain aspects of linguistic meaning. As a theory of communication, Relevance Theory provides an account of linguistic meaning with an emphasis on pragmatics, and sets out to account for the on-line process of meaning construction in more detail than it accounts for the stable knowledge systems that comprise knowledge of language or competence in the Chomskyan sense. In this respect, Relevance Theory accepts the distinction between linguistic knowledge and non-linguistic knowledge, and focuses on how the two interact to give rise to interpretation in communicative contexts. This relatively broad focus explains why certain aspects of the model resonate with cognitive approaches, despite starting assumptions that stand in direct opposition to the cognitive view. A further difference relates to the fact that Relevance Theory places the emphasis on communication (the speaker's intentions and the hearer's assumptions in deriving inferences), while cognitive semantics emphasises the nature of the conceptual system and conceptual processes. For example, while Relevance Theory emphasises the communicative aspects of metaphor, conceptual metaphor theorists emphasise the structural dimensions of metaphor within the conceptual system. Finally, each approach focuses on a largely distinct range of phenomena. Relevance Theory, although it develops a new perspective, is nevertheless concerned with accounting for the phenomena that have traditionally been of concern within approaches to linguistic meaning, such as ambiguity, the nature of the relationships between word meaning and sentence meaning, between explicit and implicit meaning, and between literal and figurative language. In contrast, cognitive semantics addresses a wider range of phenomena, and is concerned not only with addressing long-standing concerns within approaches to linguistic meaning, but also with phenomena revealed by other related disciplines that cast light upon the nature of the conceptual system.

13.3 Summary

In this chapter we compared and contrasted cognitive semantics with two other modern approaches to linguistic meaning: **formal (truth-conditional) semantics** and **Relevance Theory**. As we observed, while the assumptions of truth-conditional semantics stand in direct opposition to the assumptions of cognitive semantics, certain claims made within Relevance Theory are more consonant with the cognitive approach. Truth-conditional semantics takes an **objectivist** approach to meaning, and is concerned with modelling sentences in terms of their **correspondence** to the 'world'. This is achieved by first translating natural language sentences into a logical **metalanguage**, and then by establishing how the logical form derived corresponds to a particular model of reality, represented in terms of **set theory**. Formal semanticists have been primarily concerned with **sentence meaning**. Relevance Theory, in contrast, is a theory of **communication**. The main architects of the theory, Sperber and Wilson, emphasise the role of **ostensive-inferential communication, relevance** and **inference**. They argue that both explicit and implicit meaning construction relies upon contextual and encyclopaedic knowledge in giving rise to inferences, and that metaphor relies upon the same communicative goals as literal language. Despite these similarities, Relevance Theory assumes a generative model of language and therefore accepts the distinction between linguistic and non-linguistic knowledge. In these respects, Relevance Theory is formally oriented and rests upon guiding assumptions that stand in direct opposition to those of cognitive semantics.

Further reading

Readings in formal semantics

- **Bach (1989).** This is one of the most accessible book-length introductions to formal semantics.
- **Cann (1993).** This textbook is a challenging introduction for the novice, but is to be commended for attempting to introduce Montague's approach to natural language semantics without presupposing a particular theory of grammar.
- **Chierchia and McConnell-Ginet (2000).** A relatively accessible introduction to formal semantics.
- **Heim and Kratzer (1998).** This textbook explicitly attempts to relate formal semantics with grammatical phenomena from the perspective of Generative Grammar.
- **Portner (2005).** Another very accessible introduction to formal semantics.
- **Saeed (2003).** Saeed's excellent general introduction to semantics includes a chapter-length introduction to formal (truth-conditional) semantics. This is the most accessible chapter-length introduction around. Saeed also provides an overview of Jackendoff's Conceptual Semantics theory of linguistic meaning, which we briefly mentioned in Chapters 3 and 5. The reader is strongly encouraged to investigate Jackendoff's theory in order to gain insights into a non-truth-conditional formal model of linguistic meaning.

Relevance Theory

- **Carston (2002).** An extended application of Relevance Theory to a range of linguistic phenomena.
- **Sperber and Wilson (1995).** The seminal text by the architects of Relevance Theory, this book provides a remarkably accessible introduction.

Exercises

13.1 What's 'cognitive' about cognitive semantics?

In view of the discussion in Part II of the book, can you provide a rationale for the use of the term *cognitive* in cognitive semantics? In what respects can the formal or generatively oriented models we have discussed at various points in Part II of the book, as well as in this chapter, also be described as 'cognitive'?

13.2 Comparison between approaches

Make an annotated table of the points of similarity and contrast between the approaches compared in this chapter.

13.3 Propositions versus construals

One of the key distinctions between formal and cognitive approaches relates to their different views about grammatical structure. As we saw in Chapter 6, cognitive approaches view grammatical structure as independently meaningful while formal approaches do not. An important idea that we will discuss in detail in Part III relates to the notion of *construal*: the idea that different grammatical forms, like different words, give rise to distinct construals or 'ways of seeing'. Consider the following examples.

(a) John kicked the ball.
(b) The ball was kicked by John.

From the perspective of truth-conditional semantics, these sentences both encode the same proposition and therefore express the same 'meaning'. From what you have learned in this part of the book, (i) say what the difference in meaning is, and (ii) explain how it is encoded linguistically. How might these differences be accounted for within the formal approach? Comment on what these examples reveal in terms of differing assumptions between cognitive semantics and formal semantics.

13.4 Metaphor

Consider the following sentence.

> *John is a block of ice.*

Provide analyses of this example from the perspective of both Conceptual Metaphor Theory and Relevance Theory. In order to do so, you will need to be explicit about the context you are assuming. What do your analyses reveal about the similarities and differences between these two approaches?

Part III: Cognitive approaches to grammar

Introduction

This part of the book is entitled 'Cognitive approaches to grammar' rather than just 'Cognitive grammar' because Cognitive Grammar is the name of a specific cognitive theory of grammar (developed by Ronald Langacker), which we investigate alongside other cognitive approaches in this part of the book. Like cognitive semantics, cognitive approaches to grammar represent a collection of approaches united by theoretical assumptions rather than a single unified theory. As we saw in Part II, cognitive semantics is more an approach to the study of conceptual structure and organisation than an approach to modelling linguistic structure and organisation, although it necessarily maps out approaches to linguistic meaning because linguistic meaning is embedded within the broader conceptual system. In contrast, cognitive approaches to grammar focus directly upon the linguistic system. Moreover, because the symbolic thesis, which is central to all cognitive approaches to grammar, entails that sound, meaning and grammar are inextricably linked, the statements that comprise the theories addressed in this part of the book apply, in principle, to all these aspects of language.

In Chapter 14, *What is a cognitive approach to grammar?*, we address the two guiding principles of a cognitive approach to grammar: the symbolic thesis and the usage-based thesis. We introduce the idea that a speaker's knowledge of language is represented as a structured inventory of conventional symbolic units that subsumes both open-class and closed-class symbolic units. These represent qualitatively distinct endpoints on a lexicon–grammar continuum between specific (content) meaning and schematic (grammatical) meaning. This inventory is structured in part by schema-instance relations, which group specific instances together under a schematic representation of their shared properties. This chapter also introduces key grammatical terms and discusses the idea that the term 'grammar' refers not only to the structure of words and sentences but also to the 'mental grammar' or system of knowledge of language in the mind of the speaker, as well as to a theory of that system of knowledge.

In Chapter 15, *The conceptual basis of grammar*, we outline both Leonard Talmy's approach to language structure and Ronald Langacker's theory of Cognitive Grammar as they relate to the conceptual basis of grammar. Talmy proposes a 'Conceptual Structuring System' that consists of schematic systems relating to the structuring of perceptual (e.g. attentional) and kinaesthetic (force-dynamic) experience. We look at how these systems are reflected in the closed-class subsystem, for example in grammatical number and in the count-mass noun distinction. In our introductory sketch of Cognitive Grammar, we explore the cognitive processes that Langacker argues underpin the division of linguistic expressions into two major categories: nominal predications (THINGS) and relational predications (PROCESSES and STATES). We also explore two of the most important theoretical constructs in Langacker's theory: profile-base organisation and trajector (TR)-landmark (LM) organisation.

In Chapter 16, *Cognitive Grammar: word classes*, we look at how this approach divides linguistic expressions into two major categories: nominal predications and relational predications. The former accounts for nouns, which are schematically characterised as THING. Relational predications divide into two subcategories: temporal relations and atemporal relations. The former accounts for verbs, which are schematically characterised as PROCESS. Atemporal relations account for a number of word classes, including adjectives, adverbs, adpositions and non-finite verb forms, which can be schematically characterised as STATES. We also look at Langacker's account of determiners and quantifiers, which are characterised in terms of their grounding function.

In Chapter 17, *Cognitive Grammar: constructions*, we explore the structure of words, phrases and sentences. Cognitive Grammar defines a construction as any expression with complex symbolic structure, and approaches constituency and head-dependent relations from the perspective of valence, based on conceptual autonomy and conceptual dependence. This model of constituency accounts not only for phrase structure, but also for word structure. We also explore the Cognitive Grammar model of clause structure, and see how complements and modifiers are distinguished and how transitivity, grammatical functions and case receive a semantic account based on the action chain model. Finally, we look at passive constructions, which are analysed in terms of marked coding, which effects a figure-ground reversal.

In Chapter 18, *Cognitive grammar: tense, aspect, mood and voice*, we examine the Cognitive Grammar analysis of the English verb string, and see how the properties of lexical verbs, auxiliary verbs and tense morphemes are held to contribute to the meaning of the clause. The verb string is analysed in terms of a grounding predication – either a tense morpheme or a modal verb – and a clausal head, which can include a perfect construction, a progressive construction and a passive construction, as well as the content verb. In Cognitive Grammar, auxiliaries *have* and *be* are semantically related to non-auxiliary functions of the same verbs, and the past participle is also related to adjectival categories that share the same morphology. Tense and mood receive a unified semantic characterisation in terms of the epistemic model, and the polysemy of modals is accounted for in force-dynamic terms. Perfective and imperfective aspect share the same conceptual basis as count and mass nouns, and the passive voice, which effects a figure-ground reversal, is related to the semantic properties of the passive participle.

In Chapter 19, *Motivating a construction grammar*, we look beyond Cognitive Grammar to explore how a constructional account of grammar can be motivated. We compare a constructional account with the 'words and rules' account assumed in most generative models of language, and establish that a constructional account rests upon a single unified representation that links together syntactic, semantic, pragmatic and phonological information, rather than viewing these as the output of distinct components of the grammar. We look at how a constructional account is motivated by the properties of idiomatic expressions, which motivates the claim that grammatical constructions can be meaningful in part independently from the content words that instantiate them. We sketch out the influential Construction Grammar model (Kay and Fillmore 1999) which, although strictly a generative model, has been extremely influential in cognitive approaches to grammar.

In Chapter 20, *The architecture of construction grammars*, we explore how a constructional approach to grammar can be extended to deal with regular as well as idiomatic clausal grammatical patterns. We explore Goldberg's (1995) constructional approach and see that she defines a construction as any form-meaning pairing whose properties cannot be predicted by its subparts, a definition that includes simplex words. Like Langacker, Goldberg adopts the usage-based thesis, and assumes that knowledge of language consists of a structured inventory. Goldberg argues that certain clausal constructions have (schematic) meaning independent of the lexical items that instantiate them. Finally, we briefly compare two other constructional accounts: Radical Construction Grammar and Embodied Construction Grammar.

In Chapter 21, *Grammaticalisation*, we shift our focus from a synchronic to a diachronic perspective and focus on a type of language change known as grammaticalisation: a process that involves changes in the function or meaning of a linguistic unit, which evolves from content to grammatical or from grammatical to more grammatical. These changes may result in layering or polysemy at certain stages in the grammaticalisation process, and are often accompanied by correlated changes in the phonological and morphological form of the unit. We explore three cognitively oriented theories of grammaticalisation: metaphorical extension approaches, which hold that metaphor underlies the development of a new expression for a grammatical concept; Invited Inferencing Theory, which holds that the conventionalisation of pragmatic inference gives rise to new coded forms; and the subjectification approach, which takes a conceptual rather than contextual approach to grammaticalisation.

Finally, in Chapter 22, *Cognitive approaches to grammar in context*, we present some explicit comparisons between cognitive, generative and functional-typological approaches to grammar. We set out the assumptions, aims and methodology of each approach, and compare the cognitive and generative approaches in detail by revisiting some core grammatical phenomena, which have been explored from a cognitive perspective throughout Part III of the book, and comparing the cognitive and generative analyses of these phenomena. We conclude that while there are clearly significant points of divergence, there is a good deal of shared ground between cognitive and generative theories in terms of what they attempt to account for, as well as some similarities between their analyses. Although the starting assumptions of these approaches differ rather dramatically, cognitive and generative theories are united in the objective of modelling the representation of knowledge of language in the mind of the speaker.

14

What is a cognitive approach to grammar?

As we have observed elsewhere in this book, cognitive linguistics is a collection of approaches rather than a single unified framework. This is particularly evident in the cognitive approaches to the study of grammar. As we saw for cognitive semantics, cognitive linguists who study grammar typically have a diverse set of foci and interests. Some cognitive linguists are primarily concerned with mapping out the cognitive mechanisms and principles that might account for the properties of grammar, as Ronald Langacker does in his highly detailed theory **Cognitive Grammar**, and as Leonard Talmy does in developing his '**Conceptual Structuring System Model**'. Others are primarily concerned with characterising and delineating the linguistic units or constructions that populate a grammar; theories of this kind are called **construction grammars**. There are (at least) four distinct varieties of construction grammar, which we comment on later in this chapter. Finally, cognitive linguists who focus on grammatical change set out to explain the process of **grammaticalisation**, whereby open-class elements gradually transform into closed-class elements. Each of these paths of investigation are united by certain shared assumptions, which we set out in this chapter. We begin by identifying the guiding principles that underpin a cognitive approach to grammar (section 14.1) and present a brief overview of the distinct cognitive approaches to grammar that we will explore throughout Part III of the book (section 14.2). We then present an introduction to grammatical terminology (section 14.3). The purpose of this section is to provide an introduction to terms that are widely used in linguistics (not just in cognitive linguistics), which will be relied upon in the remainder of Part III. Finally, we examine some of the key characteristics, claims and assumptions that define cognitive approaches to grammar in general (section 14.4). This section provides a general overview of the **cognitive model of grammar**, and introduces the ideas that will be explored in detail throughout Part III of the book.

14.1 Guiding assumptions

In this section, we consider the two central guiding assumptions of a cognitive approach to grammar: the symbolic thesis and the usage-based thesis. We also sketch the architecture of the **cognitive model of grammar**. By 'cognitive model' we mean an approach to the study of language structure and organisation that assumes (1) the broad commitments of cognitive linguistics described in Chapter 2; (2) a cognitive semantics, particularly the assumptions described in Chapter 5; and (3) the guiding principles described below. Thus, when we use the term 'cognitive model of grammar', we do not have in mind a specific theory, but rather a model that generalises over the specific theories we discuss in this part of the book by drawing out what these theories share in common.

14.1.1 The symbolic thesis

The first guiding assumption is the symbolic thesis, which holds that the fundamental unit of grammar is a form-meaning pairing or **symbolic unit** (called a 'symbolic assembly' in Langacker's Cognitive Grammar framework or a 'construction' in construction grammar approaches). In Langacker's terms, the symbolic unit has two poles: a semantic pole (its meaning) and a phonological pole (its sound). The idea that language has an essentially symbolic function and that the fundamental unit of grammar is the symbolic unit has its roots in Saussure's theory of language. The Swiss linguist Ferdinand de Saussure (1857–1913) is often described as the 'father of modern linguistics'. Central to his theory was the view that language is a symbolic system in which the linguistic expression (**sign**) consists of a mapping between a concept (**signified**) and an acoustic signal (**signifier**), where both signified and signifier are psychological entities. While there are important differences between the Saussurean model and the cognitive model, the cognitive model adopts the idea of the Saussurean symbol. In the cognitive model, the semantic pole corresponds to the 'signified' and the phonological pole to the 'signifier'. These are both 'psychological entities' in the sense that they belong within the mental grammar (system of linguistic knowledge) in the mind of the speaker, which Langacker (1987: 57) describes as a **structured inventory of conventional linguistic units**. To illustrate, recall Figure 1.1 from Chapter 1 which is repeated here as Figure 14.1.

As we observed in Chapter 1, the visual image of the cat in the lower half of the figure represents the concept CAT, which is the semantic pole of a symbolic unit. The phonological pole of this symbolic unit is the speaker's knowledge of the string of speech sounds that correspond to the concept CAT, represented by

Figure 14.1 A symbolic unit

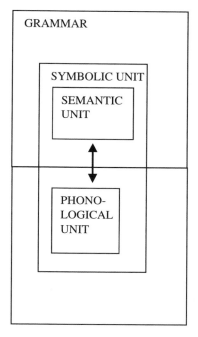

Figure 14.2 The symbolic unit (adapted from Langacker 1987: 77)

the International Phonetic Alphabet (IPA) symbols [kæt]. The symbolic unit is represented in Figure 14.2.

Of course, symbolic units can be expressed in different ways. In spoken language, the form is phonological: a string of speech sounds. However, language relies not only upon speech sounds but also upon written symbols, or manual gestures in the case of sign language. It follows that the idea of a symbolic unit does not relate solely to spoken language. The 'phonological' pole, in Langacker's terms, might therefore be realised in different ways, depending on the medium of communication.

The adoption of the symbolic thesis has an important consequence for a model of grammar. Because the basic unit is the symbolic unit, meaning achieves central status in the cognitive model. In other words, if the basic grammatical unit is a symbolic unit, then form cannot be studied independently of meaning. This means that the study of grammar, from a cognitive perspective, is the study of the full range of units that make up a language, from the lexical to the grammatical. For example, cognitive linguists argue that the grammatical form of a sentence is paired with its own (**schematic**) meaning in the same way that words like *cat* represent pairings of form and (**content**) meaning. Compare examples (1) and (2).

(1) Lily tickled George.   [active]

(2) George was tickled by Lily.   [passive]

In the English passive construction illustrated in (2), the entity that undergoes the action, which linguists call the PATIENT, is placed in subject position (before the verb). The sentence is also marked with a passive **verb string**, here *was tickled*. We can represent the generalised form of the passive construction as in (3).

(3) PATIENT 'passive verb string' *by* AGENT

According to cognitive linguists, this passive construction has its own schematic meaning that is independent of the specific words that 'fill' the construction. This meaning focuses attention on the PATIENT (e.g. what happened to George) rather than the AGENT (e.g. what Lily did). The idea that grammatical units are inherently meaningful is an important theme in cognitive approaches to grammar and gives rise to the idea of a **lexicon–grammar continuum**, in which content words like *cat* and grammatical constructions like the passive both count as symbolic units but differ in terms of the quality of the meaning associated with them. We return to this idea in more detail below (section 14.4), and it remains an important theme throughout Part III of the book.

14.1.2 The usage-based thesis

The second fundamental assumption of the cognitive approach to grammar is the usage-based thesis. As we saw in Chapter 4, the usage-based thesis holds that the mental grammar of the speaker (his or her knowledge of language) is formed by the abstraction of symbolic units from situated instances of language use. An important consequence of adopting the usage-based thesis is that there is no principled distinction between knowledge of language and use of language (competence and performance in generative terms), since knowledge emerges from use. From this perspective, knowledge of language is knowledge of how language is used.

14.1.3 The architecture of the model

The basic architecture of the cognitive model of grammar is represented in Figure 14.3. This diagram captures the idea that the act of deploying a symbolic unit in any given usage event involves both **semantic space** (meaning) and **phonological space** (form). In this diagram, the 'grammar' box represents the conventionalised knowledge of language in the mind of the speaker, and the 'usage' box represents the usage event or utterance. In intuitive terms, a usage event consists of speech sounds and their corresponding interpretations, hence the two boxes labelled 'conceptualisation' and 'vocalisation'. The horizontal arrows represent **coding links** or correspondences between the conventionalised units of knowledge in the mind of the speaker and the (vocal or conceptual) systems they interact with in instances of situated language use. In other words, the semantic pole of a linguistic expression corresponds to a concept, and the phonological pole of a linguistic expression corresponds to the string of sounds that realises it. The vertical arrows represent symbolic links which unite sound and meaning, or knowledge of sound and meaning. It is important to emphasise that, while knowledge of conventionalised units is represented in a separate box from usage events, this does not imply the distinction between competence and performance that is assumed in the generative approach.

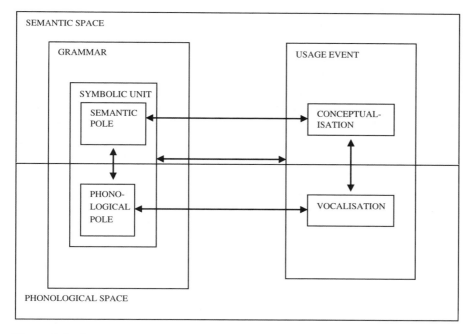

Figure 14.3 The cognitive model of grammar (adapted from Langacker 1987: 77)

According to the generative model, competence determines performance (which may also be affected by other factors). In the cognitive model, usage gives rise to knowledge, which in turn underlies usage. This is indicated by the double-headed horizontal arrows in Figure 14.3.

14.2 Distinct cognitive approaches to grammar

Having outlined the central assumptions of a cognitive approach to grammar, we now introduce some of the specific theories that represent this approach. We identify four main types of theoretical approach here, which we explore in detail in Part III of the book. These are listed below, followed by a brief overview of each type of approach.

1. The 'Conceptual Structuring System Model'
2. Cognitive Grammar
3. Constructional approaches to grammar
4. Cognitive theories of grammaticalisation

14.2.1 The 'Conceptual Structuring System Model'

This model, which has been developed by Leonard Talmy, assumes the symbolic thesis and, like other cognitive approaches to grammar, views grammatical units as inherently meaningful. However, this model is distinguished by its emphasis on the qualitative distinction between grammatical (closed-class) and lexical (open-class) elements. Indeed, Talmy argues that these two forms of linguistic expression represent two distinct conceptual subsystems, which encode qualitatively distinct aspects of the human conceptual system. These are the **lexical subsystem** and the **grammatical subsystem**. The 'conceptual structuring system' is another name for the grammatical subsystem. As we first saw in Chapter 1, while closed-class elements encode schematic or structural meaning, open-class elements encode meanings that are far richer in terms of content. We will explore the idea that grammatical meaning is schematic later in this chapter and in more detail in the next. Because Talmy assumes the bifurcation of the conceptual system into two distinct subsystems, this cognitive model of grammar focuses more on the closed-class system than it does on the open-class system. We will look in detail at Talmy's approach in Chapter 15.

14.2.2 Cognitive Grammar

Cognitive Grammar is the theoretical framework developed by Ronald Langacker. This is arguably the most detailed theory of grammar to have been developed within cognitive linguistics and to date has been the most influential.

Langacker's approach attempts to model the cognitive mechanisms and principles that motivate and license the formation and use of symbolic units of varying degrees of complexity. Like Talmy, Langacker argues that grammatical or closed-class units are inherently meaningful. Unlike Talmy, he does not assume that open-class and closed-class units represent distinct conceptual subsystems. Instead, as we saw earlier, Langacker argues that both types of unit belong within a single 'structured inventory of conventionalised linguistic units' which represents knowledge of language in the mind of the speaker. It follows that Langacker's model of grammar has a rather broader focus than Talmy's model. We will focus on Langacker's approach in detail in Chapters 15–18.

14.2.3 Constructional approaches to grammar

There are four main varieties of constructional approach to grammar. The first is the theory called **Construction Grammar** that was developed by Charles Fillmore, Paul Kay and their colleagues. While this theory is broadly generative in orientation, it set the scene for the development of cognitive approaches that adopted the central thesis of Fillmore and Kay's approach, namely that grammar can be modelled in terms of constructions rather than 'words and rules'. In part, Construction Grammar is motivated by the fact that certain complex grammatical constructions (e.g. idioms like *kick the bucket* or *throw in the towel*) have meaning that cannot be predicted on the basis of their sub-parts and might therefore be 'stored whole' rather than 'built from scratch'. We look in detail at Construction Grammar in Chapter 19, and in Chapter 20 we introduce three constructional approaches that are set firmly within the cognitive framework: (1) a model that we call **Goldberg's Construction Grammar**, developed by Adele Goldberg; (2) **Radical Construction Grammar**, developed by William Croft; and (3) **Embodied Construction Grammar**, a recent approach developed by Benjamin Bergen and Nancy Chang. It is worth pointing out that Cognitive Grammar could be also be classified as a constructional approach to grammar because Langacker also adopts a constructional view of certain types of grammatical unit. However, as we will see in later chapters, Langacker defines the construction in a different way from these models. Cognitive Grammar and constructional approaches to grammar share another feature in common: both are **inventory-based** approaches to the study of grammar. In other words, both types of approach view the grammar as an inventory of symbolic units rather than a system of rules or principles. This amounts to the claim that the language system does not work predominantly by 'building' structure (as in generative models of grammar) but by 'storing' it. We will return to this issue later in the chapter (section 14.4). Despite these important similarities, we have classified Langacker's model separately from constructional approaches because Cognitive Grammar places a

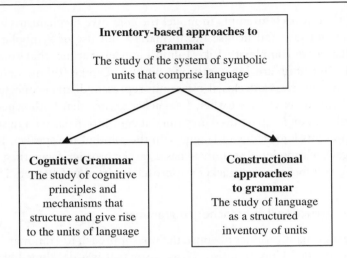

Figure 14.4 Inventory-based approaches to grammar

greater emphasis on the cognitive mechanisms and principles that underlie the grammar. Figure 14.4 summarises the main similarities and differences between Cognitive Grammar and constructional approaches to grammar.

14.2.4 Cognitive approaches to grammaticalisation

The final group of theories that we investigate in this part of the book are cognitive approaches to **grammaticalisation** (also called grammaticisation): the process of language change whereby grammatical or closed-class elements evolve gradually from the open-class system. Because it relates to language change, the process of grammaticalisation falls within the domain of historical linguistics. Grammaticalisation is also of interest to typologists, because patterns of language change can inform their explanations of current patterns in language. A subset of these historical linguists and typologists have developed models that are informed by cognitive linguistics, which attempt to explain the grammaticalisation process. In addition, Langacker has also made some proposals relating to the cognitive mechanisms that might give rise to the grammaticalisation process. There is a considerable literature in this area; we restrict ourselves to three representative types of approach: (1) **metaphorical extension approaches** (such as the model developed by Bernd Heine and his colleagues); (2) **Invited Inferencing Theory** (developed by Elizabeth Closs Traugott and Richard Dasher); and (3) the **subjectification model** developed by Ronald Langacker. Grammaticalisation is the topic of Chapter 21.

The four types of cognitive approach that we investigate throughout Part III of the book are summarised in Figure 14.5. (The parentheses around Fillmore

WHAT IS A COGNITIVE APPROACH TO GRAMMAR?

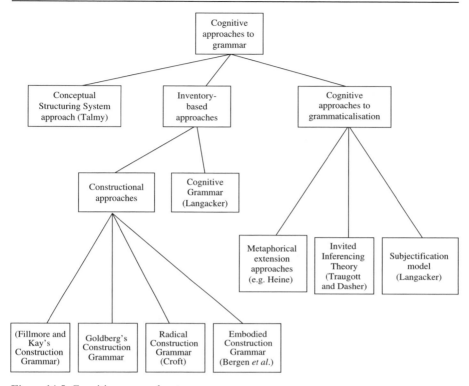

Figure 14.5 Cognitive approaches to grammar

and Kay's Construction Grammar indicate that this is not a fully 'cognitive' approach in the sense that we define it: while it subscribes to the symbolic thesis, it does not subscribe to the usage-based thesis.) As this diagram shows, the range of approaches that can be grouped together as 'cognitive' is considerable. We should emphasise that this diagram represents the way that we have grouped the approaches for the purposes of presentation in this book; while we have attempted to categorise these approaches on the basis of common themes or objectives, different taxonomies of cognitive approaches to grammar are certainly conceivable.

14.3 Grammatical terminology

All linguists, regardless of theoretical or descriptive orientation, rely upon a set of terms that enable them to describe and discuss the parts of language. In this section, we will introduce and define some key terms in the study of grammar. As we will see in the remainder of the book, not all these terms have equal status in different theories of language, but they nevertheless provide a core vocabulary that enable linguists of different theoretical orientations to communicate with one another and to understand grammatical descriptions of unfamiliar languages. We restrict ourselves here to the fundamentals and new grammatical terms will be elaborated as they are introduced in subsequent chapters.

14.3.1 Grammar

We begin with the term 'grammar', which we have taken largely for granted so far. This term has a number of different meanings. A grammar can be a written volume, such as a descriptive reference grammar prepared by a linguist for consultation by other linguists, or a teaching grammar prepared for language students. The term 'grammar' also refers to the discipline that focuses on morphology (word structure) and syntax (sentence structure), whether from the perspective of language learning (for example, French grammar, Latin grammar), from the perspective of language description, or from the perspective of general linguistics, where 'grammar' has the status of a subdiscipline alongside phonetics, phonology, semantics and so on. Indeed, an introductory 'grammar' course in a linguistics programme will usually focus solely upon word structure and sentence structure. If the approach taken is purely descriptive, this is known as 'descriptive grammar'. It is fair to point out, however, that even a 'purely descriptive' approach rests upon certain theoretical assumptions, even if these are not made explicit. The term 'grammar' is also used to refer to a theory of language such as Langacker's Cognitive Grammar or Chomsky's Generative Grammar. Finally, the term can also be used to refer to the psychological system that represents a speaker's knowledge of language. In these last two senses, the term is not (necessarily) restricted to word structure and sentence structure, but is applied to human language in general, and thus encompasses phonology and linguistic meaning as well as morphology and syntax.

14.3.2 Units of grammar

When grammarians break complex strings of language down into parts, they do so only as far as the smallest unit of meaning: the **morpheme**. Of course, individual speech sounds are smaller than most morphemes, but most individual speech sounds do not function as morphemes and therefore do not carry meaning. While 'grammar' in the broader sense might encompass a model of phonology, this area has its own complex set of terms that we do not explore here. The diagram in Figure 14.6 illustrates the grammatical units of varying sizes for which linguists have developed a set of terms. Some of these grammatical units should already be familiar from earlier chapters in the book. The sentence is represented as the largest grammatical unit because larger pieces of discourse consist of sentences joined together in a variety of ways.

Figure 14.6 Grammatical units

As we have seen, the morpheme is the smallest unit of language that can carry meaning. Some words, like *house*, consist of a single morpheme, while others, like *house-s* or *employ-ment* consist of more than one morpheme. The study of morphology, then, is the study of word structure. Morphemes that can stand alone, like *house*, are **free** morphemes, whereas those that need to attach to something, like plural *-s*, are **bound**. The simplest possible form of a content morpheme is called a **root**; this may be free, like *house*, or bound, like *pseudo-*. Bound morphemes like *-ment* or *-s* which do not have content meaning are called **affixes**. There are two types of affix: the **derivational** affix and the **inflectional** affix.

The derivational affix creates new words, often belonging to a different word class (we return to word classes below). In English, affixes that change word class are **suffixes**, which means that they attach to the end of words. For example, the verb *employ* plus the suffix *-ment* becomes a noun *employment*. The noun *nation* plus the suffix *-al* becomes the adjective *national*. Suffixes can be stacked; consider the noun *nation-al-is-ation*, for example. English also has some **prefixes** that do not affect word class, but do affect the meaning of the word (for example, *de-nationalise*, or *un-do*). These also fall within the category of derivational affixes.

The inflectional affix, which is also a suffix in English, does not change the category of the word, nor does it affect the content meaning. Instead, it marks a subclass of that word. Another way of saying this is that it marks a different grammatical form of the same lexical item. Some English inflectional morphemes are illustrated in Table 14.1. Some of the grammatical terms in the left-hand column will make more sense by the end of this section.

Of course, this brief discussion of morphology rests upon the assumption that we have a clear notion of what it means to describe something as a **word**. However, there are a number of different ways of defining this term (Trask 2004). We are used to thinking of a word in terms of an **orthographic word**: something that is written as a single unit. However, this does not necessarily tell us anything about spoken language, which is of primary interest to linguists. Orthographic systems are man-made and vary enormously, sometimes revealing little about the structure of the language they represent. A **phonological word** is a unit of pronunciation, defined according to the phonological rules of that language. In English, a phonological word usually contains one main stress. In rapid speech, some parts of an utterance are 'glued together' into single phonological words, which do not correspond to our idea of where

Table 14.1 English inflectional morphemes

plural -s	book*s*
possessive 's	Lily'*s* book
third person singular present -s	Lily read*s* well
progressive -ing	She is work*ing*
past tense -ed	She work*ed*
past participle -ed/-en	She has stud*ied*/brok*en*

the word boundaries lie from a meaning or grammar perspective. Trask (2004) provides the following example, where the bracketed units in (4b) correspond to phonological words:

(4) a. The rest of the books will have to go there.
   b. [The rest] [of the books'll] [have to] [go] [there].

As this example shows, the boundaries laid down by the system of pronunciation do not always correspond with the boundaries laid down by meaning or grammar. While the phonological word reveals much about the phonological structure of a language, it is less useful in the study of grammar.

A third definition of 'word' is **lexical item**, a term that we have relied upon throughout earlier parts of the book. This term means a unit of our mental 'dictionary' (or encyclopaedia), and this is the sense in which linguists use the term. A lexical item has a more or less identifiable meaning (like *cat*) or function (like *this*). However, recalling the discussion of inflectional morphology above, each lexical item may have a number of **grammatical word forms**. Nouns like *cat*, for example, have both a singular and a plural form (*cat–cats*), and verbs like *go* have a whole list of forms (*go, goes, went, going, gone*). The list for *be* is even longer (*be, am, are, is, was, were, being, been*). Adjectives like *big* also have a number of forms (*big, bigger, biggest*). We can think of each lexical item, then, as a bundle of forms, although some lexical items, like *my*, have only one form in English.

14.3.3 Word classes

Having arrived at a definition of 'word', we briefly introduce the notion of **word classes** or **parts of speech**. The idea that words can be straightforwardly grouped into classes is not uncontroversial, and some of these categories have a different status in different theories. In traditional descriptive grammar, where the word classes were inherited from Latin grammar via the traditional grammarians of the eighteenth and nineteenth centuries, English is usually described as having eight word classes: noun, pronoun, adjective, verb, adverb, preposition, conjunction and interjection. However, a new set of word classes has gradually emerged within modern descriptive linguistics which aims to present a more objective view of word classes from a cross-linguistic perspective. According to the **distributional approach** to word classes, words are grouped with certain classes mainly on the basis of their morphological and distributional behaviour: words of the same class will generally take the same sort of derivational and inflectional affixes (morphological behaviour), and will generally occupy the same positions or 'slots' in a sentence relative to members of other word classes (distributional behaviour). We illustrate here with English examples.

Nouns

Nouns often refer to entities, including people, and abstractions (like *war* and *peace*). Nouns typically take the inflectional plural affix *-s* (*cats, dogs, houses*) but there are exceptions (*mans, *peaces). Nouns also typically take the possessive affix *-'s* (*man's best friend*), and in terms of distribution, follow determiners like *your* and adjectives like *funny* (*your funny face*). Nouns can be divided into two main subclasses: common nouns and proper nouns. Proper nouns are names of people or places like *Lily* or *London*. Common nouns do not pick out particular individuals by name, but refer to classes. These are the 'ordinary' nouns like *cat, house* and *water*, and this subclass is the one that we are most concerned with in this book because common nouns represent one of the major linguistic categories. Common nouns can be divided into count nouns and mass nouns. Count nouns can be counted (*one book, two books*) and have to be preceded by a determiner like *the* when singular (compare *The book is on the table* with **Book is on the table*). In the plural, however, count nouns can occur without a determiner (*Books are expensive*). Mass nouns cannot be counted or pluralised (*two sands) and can occur with or without determiners. This classification of nouns is summarised in Figure 14.7.

Verbs

Verbs typically denote actions, processes or events, and take inflectional affixes including the third person singular (*he/she/it*) present tense *-s*, the past tense affix *-ed* and the progressive participle affix *-ing*. These are illustrated in example (5).

(5) a. She hopes
    b. She hoped
    c. She's hoping

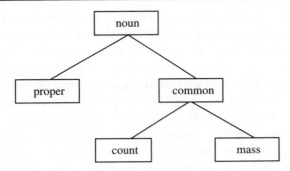

Figure 14.7 Noun categories

These verb forms reflect a number of properties relating to **agreement, tense** and **aspect** to which we return below. Verbs can often take derivational affixes like noun-forming *-er* (*employ–employer*) or adjective-forming *-able* (*employ– employable*). In terms of distribution, the English verb follows the subject.

Adjectives

Adjectives typically denote attributes or states, and some can inflect for grade (*tall, taller, tallest*). Adjectives can often be identified by the presence of a derivational affix like *-ful* (*careful*), *-y* (*funny*), or *-ish* (*selfish*). In terms of distribution, English adjectives occur in their **attributive** function preceding the noun or in **predicative** function following copular verbs like *be* or *become*:

(6) a. I love her funny face. [attributive]
    b. Her face was funny. [predicative]

The difference between the attributive and the predicative function of adjectives relates to how 'vital' the adjective is to the well-formedness of the grammatical unit. In (6a), we can remove the adjective and we still have a well-formed (although less informative) grammatical unit: *I love her face*. If we remove the adjective in (6b), we are left with an incomplete grammatical unit: *Her face was...*

Adverbs

Adverbs are words like *suddenly, repeatedly, hopefully* and *soon*. These typically express information relating to time, manner, place and frequency, and have a modifying function within the sentence (providing information, for example, about how, where or when something happened). Some are recognisable by the adverb-forming derivational affix *-ly*, and a few inflect for grade (*soon, sooner, soonest*), but on the whole these are difficult to identify by morphology or distribution because they have the widest distribution of all the English word classes. A further complication with this category is that members of other word classes can also perform the same function as adverbs. This is called an **adverbial** function, which means that something behaves in the same way as an adverb, providing modifying information about place, manner, time and so on, regardless of word class. For example, the expression *after supper* performs an adverbial function in the sentence *George arrived after supper*, but is not an adverb; it is a preposition phrase, consisting of a preposition and a noun phrase. The term 'adverbial' refers to a type of grammatical function (section 14.3.5).

The word classes introduced so far represent content words or open-class words. As we have seen, open-class words have a readily identifiable meaning and belong to classes that are large and constantly changing as new words are introduced and old words are lost. While open-class words provide the content meaning in utterances, there are several equally important word classes that contain grammatical words or closed-class words. These have a less readily identifiable meaning (often they are described as 'function words') and belong to classes that are small and more resistant to change. With the exception of some determiners (see below), none of these word classes has any inflectional or derivational properties in English, but they do show some predictable distributional patterns. The discussion of these categories in English rests upon some new terms like 'phrase' and 'clause' which will be discussed later in the section.

Prepositions

Prepositions are words like *on*, *with*, *under* and *beyond*, which combine with a noun phrase to form a preposition phrase (*on the table, with my best friend*). These are called prepositions because they precede the noun phrase. In some languages, they follow the noun phrase and are called postpositions. The general term for both prepositions and postpositions is 'adposition'.

Determiners

Determiners are words like *the, my* and *some*, which combine with a noun to form a noun phrase (*the garden, my cats, some flowers*). Apart from the determiners *this* and *that* which inflect for number (*these, those*), determiners have no other inflectional or derivational properties in English. It is important to remember that determiners are followed by nouns because some words can be both determiners (*I love these flowers*) and pronouns (*I love these*).

Pronouns

Pronouns are sometimes described as a subclass of nouns because they show the same pattern of distribution. In other words, pronouns substitute for nouns (hence the term 'pronoun'). However, pronouns can be viewed as a separate category from nouns because they belong to a closed class and because they provide what cognitive linguists call **schematic meaning** rather than **content meaning**. For example, you could probably draw a picture of *my favourite teacup* without having seen it, but you would be unable to draw a picture of *it* without having seen it. In isolation from context, *it* means 'a single inanimate object'. Of course, in reality we are never called upon to interpret *it* out of context, but this illustrates the difference between content meaning and schematic meaning. There are several different kinds of pronouns. To mention a few examples, personal pronouns are words like *you, me* and *her*; possessive pronouns are words like *mine* and *hers* (not to be confused with possessive determiners *my* and *her*), and demonstrative pronouns are words like *this/these* and *that/those*.

Auxiliary verbs

Finally, we mention the closed-class category of auxiliary verbs. In English, this category includes the modal auxiliaries (for example, *can, must* and *will*) which introduce **mood** into the sentence, and the primary auxiliaries (*have* and *be*) which introduce **aspect** and passive **voice**. We return to tense, aspect, mood and voice in more detail in Chapter 18, limiting the present discussion to the grammatical properties of the auxiliary verbs. The modal auxiliaries share few characteristics with 'ordinary' (lexical) verbs in English. They do not inflect for progressive aspect, for example (*musting) nor do they have a third person singular *–s* form (*she musts). They are called auxiliary verbs because they belong inside the **verb string** (this is bracketed in (7a)), because they must be followed by a verb phrase (VP), and because they can function as **operators**. This means that they can invert with the subject (*she*) to form a question:

(7) a. Lily [will sing] the blues.
    b. Will Lily sing the blues?

The primary auxiliaries *have* and *be* look more like 'ordinary' verbs. They inflect for tense, for example. As we saw briefly in Chapter 11, the auxiliary *have* introduces **perfect aspect** into the sentence, which means that the event is viewed as completed, and has to be followed by a perfect (traditionally called 'past') participle (*sung*):

(8) Lily has sung the blues all her life.

The auxiliary *be* can introduce **progressive** or **continuous aspect** into the sentence, which means that the event is viewed as ongoing. In this case, *be* has to be followed by a progressive (traditionally called 'present') participle (*singing*):

(9) Lily is singing the blues.

The auxiliary *be* can also introduce **passive voice**. As we saw earlier, this means that the person or thing that undergoes the event depicted by the verb appears in subject position (before the verb). Example (10a) shows an active sentence and (10b) its passive counterpart. Observe that the passive auxiliary *was* is followed by the same participle form as the perfect auxiliary *have* (e.g. *sung*). As we observed earlier, this is traditionally referred to as the 'past' participle:

(10) a. Elvis sang that song.
     b. That song was sung by Elvis.

Like the modal verbs, the primary auxiliaries can also function as operators (11). As example (11d) shows, the verb *do* also has an auxiliary function in English. This verb does not introduce its own aspect or voice into the clause. Instead, it occurs when the speaker wants to emphasise the truth of a statement (*Lily does like shellfish*), or when the sentence requires a verb that can function as an operator but lacks another modal or auxiliary to perform this function. For this reason, the auxiliary *do* is sometimes called a 'dummy' auxiliary.

(11) a. Has Lily sung the blues all her life?
     b. Is Lily singing the blues?
     c. Was that song sung by Elvis?
     d. Does Lily sing the blues?

The verbs *have*, *be* and *do* are not always auxiliaries. They can also be lexical verbs. If *have*, *be* or *do* is the only verb in the sentence, it is a lexical verb. This is illustrated by (12a). In some dialects of English, lexical *have* can invert with the subject to form a question (12b). If *have*, *be* or *do* is followed by another verb phrase, it is an auxiliary verb; the fact that these verbs can occur both as lexical and auxiliary verbs explains why it is possible to find a sequence of two instances of the 'same' verb in a single clause. This is illustrated by examples (12c)–(12e).

(12) a. I have two cats and a goldfish.
     b. Have you a pen I could borrow?
     c. Lily has had a headache.

d. George is being silly.
   e. Lily does do the washing up every morning.

When the verb *be* is a lexical verb, it is called the **copula**, which means 'linking verb'. It links the subject of the sentence (*Lily*) to the phrase that provides some information about the subject:

(13) a. Lily is [my best friend].
     b. Lily is [fond of fish and chips].
     c. Lily is [in the cellar].

Lexical *be*, the copula, can also function as an operator:

(14) a. Is Lily your best friend?
     b. Is Lily fond of fish and chips?
     c. Is Lily in the cellar?

As this discussion illustrates, the behaviour of the primary auxiliaries and their lexical counterparts is not entirely distinct. Another way of saying this is that lexical *have* and *be* are not prototypical lexical verbs.

There are several other closed-class categories that we will not discuss here, mainly including 'linking' categories that join sentences, like coordinating conjunctions (*and, but*), subordinating conjunctions (*although, because*), discourse connectives (*however, therefore*) and complementisers (for example, *that* in *she hoped that they would be married in the snow*). We will also have little to say about interjections, words like *yuk!* or *wow!* that form independent utterances and do not participate in grammatical structure.

14.3.4 Syntax

The term 'syntax' relates to the structure of phrases and sentences, the larger grammatical units. A **phrase** is a group of words that belong together as a group. Inside each phrase, there is one 'central' word or **head** which carries the main meaning of the phrase and which determines what other kinds of words the phrase can or must contain. These other words are traditionally called **dependents** and are divided into **complements** (a phrase required by the head to 'complete' it) and **modifiers** (an 'optional' phrase with a modifying function). **Constituency** is the term used to describe the grouping of words within phrases and the grouping of phrases within sentences. Phrases can be identified by constituency tests. There are various kinds of constituency test, but we will limit ourselves to three examples here: **substitution, coordination** and 'movement'. Example (15) illustrates the substitution test, where the bracketed constituents in (15a) are identified as phrasal units (NPs) because they can be substituted as a coherent unit by pronouns (15b):

(15) a. [That friend of George's with the glasses] pinched [Lily's bike].
b. [He] pinched [it].

The phrasal constituent *that friend of George's with the glasses* is identified as a noun phrase (NP) because it is headed by the noun *friend*. The same applies to the NP *Lily's bike*, which is headed by the noun *bike*.

Example (16) illustrates the **coordination** test, where a string of words is identified as a phrase by the fact that it can be coordinated with another phrase of the same category. For example, two NPs are coordinated in (16a), and two VPs are coordinated in (16b).

(16) a. He pinched [$_{NP}$ Lily's bike] and [$_{NP}$ her tent].
b. He [$_{VP}$ pinched Lily's bike] and [$_{VP}$ trashed her tent].

Example (17) illustrates the **'movement'** test. The idea behind the term 'movement' is that a phrase can occur in a 'special' position in order to become more prominent in the sentence. In English, the **cleft construction** is a productive means of achieving this kind of discourse prominence. The cleft construction is shown in schematic form in (17). Example (18a) shows an 'ordinary' (non-cleft) construction, and examples (18b)–(18e) show how different phrasal constituents can be 'clefted'.

(17) *It be* [CLEFTED PHRASE] *who/that* [REMAINDER OF CLAUSE]

(18) a. George gave food poisoning to his guests on Tuesday.
b. It was [$_{NP}$George] (who/that) gave food poisoning to his guests on Tuesday.
c. It was [$_{NP}$ food poisoning] (that) George gave to his guests on Tuesday.
d. It was [$_{NP}$ his guests] (that) George gave food poisoning to on Tuesday.
e. It was [$_{PP}$ on Tuesday] (that) George gave food poisoning to his guests.

The idea of constituency, which has been influential in linguistics at least since Bloomfield (1933), is open to different interpretations. In generative approaches, phrasal constituents are thought of as units of grammar that are 'built' on the basis of grammatical rules or principles. In contrast, the cognitive model rejects this idea and assumes that phrases and sentences are 'stored whole' as generalised patterns emerging from repeated experience of usage events. Despite this important theoretical difference, which is central to Part III of this book, cognitive linguists nevertheless recognise the existence of phrases within sentences and share this common vocabulary with linguists of other theoretical persuasions.

Another important term, which we have taken for granted so far, is **sentence**. This overlaps with the term **clause**. Linguists define the clause as a string of words containing a **subject** and a **predicate**. In the grammatical sense, the predicate corresponds to the verb phrase (everything apart from the subject). In example (19), *Lily* is the subject, and *loves George to distraction* is the predicate. The term 'subject' (like 'object', 'predicate' and 'adverbial') refers to a **grammatical function** (section 14.3.5).

(19) Lily loves George to distraction.

Strictly speaking, a clause consists of a single subject and a single predicate, while a sentence may be more complex. A **simple sentence**, like the ones we have seen so far, consists of a single clause; in this case, the terms 'clause' and 'sentence' are equivalent. A **complex sentence**, however, may consist of more than one clause. There are various kinds of relations that hold between the clauses in a complex sentence which we will not address here, but two examples of complex sentences are provided in (20), where clauses are bracketed.

(20) a. [Lily loves George] but [he is rather arrogant].
　　 b. Her friends said [he was no good].

Despite the distinction between the terms 'clause' and 'sentence', these are often used interchangeably by linguists.

14.3.5 Grammatical functions

Subject and object are types of grammatical function. In other words, these terms describe what phrases *do* in a sentence rather than describing what phrases *are* in terms of their **category** (NP, VP and so on). This is a useful distinction, because phrases of different categories can perform the same grammatical function, and phrases of the same category can perform different grammatical functions. For example, NP can function either as subject or object:

(21) [$_{\text{NP-SUBJECT}}$ George] wrote [$_{\text{NP-OBJECT}}$ several different love letters].

Table 14.2 Structural criteria for English subject

(Canonical) subject position in English is clause-initial
Subject inverts with auxiliary/modal verbs to form questions
Subject agrees with the verb in person and number
Subject pronoun shows subject (or nominative) case

While the category of a word or a phrase can usually be identified without context, the grammatical function of an expression can only be identified in the context of a particular sentence. This is because the same expression could be a subject in one sentence and an object in another. Compare (21) with (22):

(22) [$_{\text{NP-SUBJECT}}$ Several different love letters] arrived in the post.

Grammatical functions can be reflected in the word order of a language or by means of a **case** system (section 14.3.6). Many languages employ a combination of both word order and case.

Subject

The English subject, which is typically an NP but can also be a clause or a PP, can be characterised in terms of a number of morphological and distributional criteria which are summarised in Table 14.2.

We have already seen several examples of the clause-initial position of the English subject. It is worth observing, however, that a subject can be preceded by a topic (23a) or by an adverbial (23b), so that the subject is not always the very first element in the clause.

(23) a. [$_{\text{TOPIC}}$ That friend of George's], [$_{\text{SUBJECT}}$ she] talks rubbish.
     b. [$_{\text{ADVERBIAL}}$ Strangely], [$_{\text{SUBJECT}}$ George had an idea].

We have also seen examples of the inversion of subject with auxiliary verbs (section 14.3.3). We return below to case and agreement (section 14.3.6).

Predicate

The term 'predicate' refers to the main part of the sentence excluding the verb. Usually, this means the VP, or the verb plus its object(s). The idea that the sentence can be partitioned in this way is widespread in linguistics and reflects the idea that the verb phrase encapsulates the essence of the event that the sentence expresses while the subject is less crucial to defining the nature of the event. Compare the following examples:

(24) a. George ate cakes.
b. Lily ate cakes.
c. George ate bananas.

In (24a), the predicate *ate cakes* describes a cake-eating event that happens to involve George. If we change the subject (24b), the sentence still describes a cake-eating event, whereas if we change the object (24c), the sentence describes a different kind of event. It is also striking that idioms occur within the predicate of a sentence:

(25) a. George [threw in the towel].
b. Lily [threw in the towel].
c. George [threw in the flannel].

Observe that the idiomatic interpretation (meaning 'give up') is available in (25a) and (25b) regardless of the subject, but if the object is changed from *the towel* to *the flannel* the idiomatic interpretation is lost (25c).

Object

This grammatical function divides into two subtypes: direct object and indirect object. **Monotransitive** verbs like *eat*, *love* and *see* take a single object, which is the direct object. This is bracketed in the examples in (26).

(26) a. George eats [cake].
b. Lily loves [him].
c. Lily saw [George].

In contrast, **ditransitive** verbs like *give* require two objects. Consider example (27).

(27) George gave [Lily] [a box of chocolates].

In this example, *Lily* is the indirect object and *a box of chocolates* is the direct object. This type of construction is called a **double-object construction**. An alternative construction in English reverses the order of the two objects. When this happens, the indirect object (*Lily*) is expressed by a preposition phrase (*to Lily*).

(28) George gave [a box of chocolates] [to Lily].

Table 14.3 Structural criteria for English object

Object position in English is after the verb
Object can move to clause-initial position to become the grammatical subject of a passive sentence
Object pronoun shows object (or accusative) case
Indirect object precedes direct object, unless the indirect object is expressed as PP

Objects are typically NPs but can also be clauses. The English object can be characterised in terms of a number of structural criteria which are summarised in Table 14.3.

Examples (27) and (28) above illustrate the final property in Table 14.3. The second property is illustrated by example (29), which shows that either the direct object *a box of chocolates* or the indirect object *Lily* can become the subject of a passive sentence. We return to case below (section 14.3.6).

(29) a. A box of chocolates was given to Lily by George.
     b. Lily was given a box of chocolates by George.

Predicative complement

The predicative complement is a complement of the verb that is co-referential with or describes either the subject or the object, as in (30a) and (30b) but not (30c):

(30) a. George is [a heart-breaker].   subject complement
     b. Lily called George [a heart-breaker]. object complement
     b. Lily loves [a heart-breaker].   direct object

Unlike objects, predicative complements cannot move to clause-initial position to form a passive sentence. In example (31), *been* is the past participle of the copula *be* and *was* is the past tense form of the passive auxiliary *be*. The result is ungrammatical:

(31) *A heartbreaker was been (by George).

Adverbial

Finally, as we saw earlier, it is important to distinguish the term 'adverb' from the term 'adverbial'. While 'adverb' refers to a word class (for example, *suddenly, soon, fortunately*), 'adverbial' refers to a grammatical function that can be performed by various categories in addition to the adverb, as illustrated by the examples in (32).

(32) a. George [_ADVERB_ distractedly] wrote the letters.
b. George wrote the letters [_PP_ in the back garden].
c. [_CLAUSE_ Humming a happy tune], George wrote the letters.

For reasons that we will not pursue here, the expression *humming a happy tune* in (32c) is described as an embedded adverbial clause, even though it lacks a subject.

As these examples illustrate, adverbials are the 'optional' parts of sentence that modify the clause at some level and can be added or deleted without making the sentence ungrammatical. Adverbials typically express information about when, where or how something happened. It is difficult to pin down a set of structural criteria that characterise adverbials because they display considerable flexibility in terms of position. However, unlike the other grammatical functions, adverbials can be stacked (that is, can occur recursively):

(33) [_CLAUSE_ Humming a happy tune], George [_ADVERB_ distractedly] wrote the letters [_PP_ in the back garden].

14.3.6 Agreement and case

We saw earlier that the criteria for identifying subjects rest in part upon the notion of agreement. The term 'agreement' (known as **concord** in traditional grammar) describes the morphological marking of a grammatical unit to signal a particular grammatical relationship with another unit. Agreement involves grammatical features like person, number and gender and may interact with case. We will illustrate these grammatical features here with the personal pronouns, since they are the only nominal category in English to show a reasonably full range of distinct morphological forms. **Person** is the grammatical feature that distinguishes speaker (first person), hearer (second person) and third party (third person). Compare *I, you* and *she*. This feature participates in subject–verb agreement in English, but only in the present tense and only in the singular third person form. Consider the examples in (34).

(34) a. I love George.
b. You love George.
c. She loves George.
d. We love George.
e. They love George.

As these examples illustrate, it is only when the subject is a third person singular noun phrase (e.g. *he, she* or *Lily*) that the verb form changes. Person is a **deictic** category. As we saw in Chapter 7, deictic categories rely upon **context** in order to be fully interpreted. The aspects of context that are particularly relevant to deixis are space and time, and the speaker's location in space and time is central to how the deictic system works. For example, the use of open-class deictic expressions like the verbs *bring* and *take* or *come* and *go* are interpreted relative to the positions of speaker and hearer. *Bring* and *come* encode motion towards the speaker or hearer, while *take* and *go* encode motion away from the speaker or hearer's position at the moment of speaking. The adverbs *here* and *there* encode proximity to or distance from the speaker respectively, and the adverbs *now* and *then* are interpreted relative to the moment of speaking. The grammatical feature **person** is a deictic category because the meaning of personal pronouns shifts continually during conversational exchange, and you have to know who is speaking to know who these expressions refer to. Recall example (35), which we first saw in Chapter 7 (Levinson 1983: 55). Imagine you are on a desert island and you find this message in a bottle washed up on the beach.

(35) Meet me here a week from now with a stick about this big.

This example illustrates the dependence of deictic expressions on contextual information. Without knowing the person who wrote the message, where the note was written or the time at which it was written, you cannot fully interpret *me*, *here*, *a week from now*, or *a stick about this big*. The other major grammatical category that is deictic in nature is tense, which is interpreted relative to the moment of speaking.

Returning to agreement, **number** is the grammatical feature that distinguishes singular from plural. Compare *I* and *we*, which are both first person pronouns. Some languages have a more complex system; for example, Arabic distinguishes singular, dual and plural (three or more). **Gender** is the grammatical feature that distinguishes noun classes (commonly, 'masculine' and 'feminine'). Grammatical gender does not necessarily correlate with the biological sex of the referent. Strictly speaking, English does not have grammatical gender because common nouns are not subdivided into gender categories. Despite this, the pronouns *he/him/his* and *she/her/hers* are described as 'masculine' or 'feminine'. The fact that English lacks a system of grammatical gender explains why there is no gender agreement in English between nouns and other elements in the noun phrase. Compare the English and French phrases in examples (36) and (37). While the determiner and the adjective remain the same for *boy* and *girl* in English, these categories show distinct gender forms in French, a language with grammatical gender. In other words, the determiner and the adjective, which are dependents of the noun, **agree** with the noun in French.

(36) a. the little boy
      b. the little girl

Table 14.4 English personal pronouns

Person/number	Nominative	Accusative
1s	I	me
2s	you	you
3s	he/she/it	him/her/it
1PL	we	us
2PL	you	you
3PL	they	them

(37) a. la   petite   fille/table
       the.F   little.F   girl.F/table.F
       'the little girl/table'
    b. le   petit   garçon/chien
       the.M   little.M   boy.M/dog.M
       'the little boy/dog'

Finally, **case** is the grammatical feature that 'flags' the grammatical function of a word or phrase within a sentence (among other grammatical properties). For present purposes, we limit the discussion of case to subject case (**nominative**) and object case (**accusative**). Consider examples (38) and (39).

(38) a. Lily kissed George.
     b. The rocket scientist kissed the estate agent.
     c. She kissed him.

(39) a. George kissed Lily.
     b. The estate agent kissed the rocket scientist.
     c. He kissed her.

As these examples show, proper nouns and common nouns in English do not inflect for case: whether these occur as subject or object, their morphological form remains unchanged. In contrast, (most of) the English personal pronouns do show distinct case forms. The feminine singular form is *she* in subject position (nominative) and *her* in object position (accusative). The masculine singular form is *he* in subject position (nominative) and *him* in object position (accusative). Table 14.4 illustrates how these grammatical features interact within the English personal pronoun system.

14.4 Characteristics of the cognitive approach to grammar

In this section, we introduce some of the characteristics that identify cognitive theories of grammar. The ultimate objective of a cognitive theory of grammar is to model speaker knowledge of language in ways that are consistent with the two key commitments underlying the cognitive linguistics enterprise (Chapter 2). Recall that these are (1) the '**Generalisation Commitment**': a commitment to the characterisation of general principles that account for all aspects of human language; and (2) the '**Cognitive Commitment**': a commitment to establishing general principles for language that are consonant with what is known about the mind and brain from other disciplines. The cognitive model of grammar therefore represents an attempt to model speaker knowledge in ways that are compatible with these two commitments. From this perspective, language emerges from general cognitive mechanisms and processes. The ideas in this section have been most explicitly developed by Langacker and by Talmy, but we set them out here as representative assumptions that guide cognitive approaches to grammar in general.

14.4.1 Grammatical knowledge: a structured inventory of symbolic units

As we noted earlier, a central claim in some cognitive approaches to grammar is that knowledge of language (the mental grammar) is represented in the mind of the speaker as an inventory of symbolic units (Langacker 1987: 73). It is only once an expression has been used sufficiently frequently and has become **entrenched** (acquiring the status of a habit or a **cognitive routine**) that it becomes a **unit**. From this perspective, a unit is a symbolic entity that is not built compositionally by the language system but is stored and accessed as a whole. Furthermore, the symbolic units represented in the speaker's grammar are **conventional**. The **conventionality** of a linguistic unit relates to the idea that linguistic expressions become part of the grammar of a language by virtue of being shared among members of a speech community. Thus conventionality is a matter of degree: an expression like *cat* is more conventional (shared by more members of the English-speaking community) than an expression like *infarct*, which is shared only by a subset of English speakers with specialist knowledge relating to the domain of medicine (this expression refers to a portion of tissue that has died due to sudden loss of blood supply). The role of entrenchment and conventionality in this model of grammar emerge from the usage-based thesis.

Symbolic units can be **simplex** or **complex** in terms of their symbolic structure. For example, a simplex symbolic unit like a morpheme may have a complex semantic or phonological structure, but is simplex in terms of symbolic structure if it does not contain smaller symbolic units as subparts. The word *cat* and the plural marker *-s* are examples of simplex symbolic units. Complex units vary according to the level of complexity, from words (for example, *cats*) and phrases (for example, *Lily's black cat*) to whole sentences (for example, *George kicked the cat*).

The contents of this inventory are not stored in a random way. The inventory is **structured**, and this structure lies in the relationships that hold between the units. For example, some units form subparts of other units which in turn form subparts of other units (for example, morphemes make up words and words make up phrases which in turn make up sentences). This set of interlinking and overlapping relationships is conceived as a **network**. There are three kinds of relation that hold between members of the network: (1) **symbolisation** (the symbolic links between semantic pole and phonological pole that we described earlier); (2) **categorisation** (for example, the link between the expressions *rose* and *flower*, given that ROSE is a member of the category FLOWER; and (3) **integration** (the relation between parts of a complex symbolic structure like *flower-s*).

As a constraint on the model – in other words, a statement that places limits on how the model operates – Langacker (1987: 53–4) posits the **content requirement**. This requirement holds that the only units permissible within the grammar of a language ('grammar' in the sense of 'model') are (1) phonological, semantic and symbolic units; (2) the relations that hold between them (described above); and (3) **schemas** that represent these units. This requirement excludes abstract rules from the model. Instead, knowledge of linguistic patterns is conceived in terms of schemas. We return to this idea below (section 14.4.3).

14.4.2 Features of the closed-class subsystem

As we have seen, Talmy (2000) posits the bifurcation of linguistic knowledge into the open-class subsystem and the closed-class subsystem, also known as the grammatical subsystem. Closed-class elements may be **overt** or **implicit**. Overt elements can be bound (for example, inflectional morphemes) or free (for example, English determiners or prepositions). Implicit elements have no phonetic realisation but represent speaker knowledge of grammatical categories like noun and verb, subcategories (for example, count noun and mass noun) and grammatical functions (also known as 'grammatical relations') like subject and object. According to Talmy, the closed-class subsystem is semantically restricted and has a **structuring function**, while the open-class system is semantically unrestricted and has the function of providing conceptual content. To illustrate the restricted nature of the closed-class system, Talmy observes that while many languages have nominal inflections that indicate NUMBER, no language has nominal inflections that indicate COLOUR. For example, many languages have a grammatical affix like plural *-s* in English, but no language has a grammatical affix designating, say, REDNESS. Furthermore, the grammatical system reflects a restricted range of concepts within the relevant domain. For example, the grammatical NUMBER system can reflect concepts like SINGULAR, PLURAL or PAUCAL (meaning 'a few') but not concepts like MILLIONS or TWENTY-SEVEN.

Talmy accounts for such restrictions by means of the observation that grammatical categories display **topological** rather than **Euclidean** properties. This means that the meaning encoded by closed-class elements remains constant despite contextual differences relating to size, shape and so on. For example, the demonstrative determiner *that* in the expressions *that flower in your hair* and *that country* encodes DISTANCE FROM THE SPEAKER regardless of the expanse of that distance. Equally, the modal *will* in the sentences *I will fall!* and *The human race will become extinct* encodes FUTURE TIME regardless of the 'distance' of that future time. As these examples illustrate, the function of the closed-class system is to provide a 'pared-down' or highly abstract conceptual structure. This structure provides a 'scaffold' or a 'skeleton' over which elements from the open-class system are laid in order to provide rich and specific conceptual content. Consider example (40) which is similar to one we explored in Chapter 1.

(40) **These** cowboy**s are** ruin**ing my** flowerbed**s**.

In this example, the closed-class elements are in bold type and the open-class elements are in ordinary type. If we remove the content words, we end up with something like *these somethings are somethinging my somethings*. Although the meaning provided by the closed-class elements is rather **schematic**, it does provide the information that 'more than one entity close to the speaker is presently in the process of doing something to more than one entity belonging to the speaker'. This is actually quite a lot of information. If we exchange the content words for different ones, we can end up with a description of an entirely different situation but the schematic meaning provided by the closed-class elements remains the same:

(41) **These** angel**s are** paint**ing my** fingernail**s**.

As this example illustrates, the grammatical elements encode far less **specific** information than the content elements, and function to organise or structure the scene encoded by the utterance. This kind of information remains constant regardless of the content words.

As Talmy points out, however, there is not always a clear-cut distinction between open- and closed-class elements with respect to the kinds of concepts they encode. For example, while closed-class elements (auxiliary verbs like *will* or inflectional morphemes like *-ed*) encode past or future time in relation to the verb system, open-class elements (like the adjective *imminent*) encode these concepts in relation to the noun system. This point is illustrated by example (42).

(42) a. He will depart.
   b. his imminent departure

Talmy observes that while no inventory of concepts expressible by open-class forms can ever be specified (because there is no limit to human experience, knowledge and understanding), there is a restricted inventory of concepts expressible by closed-class forms. Each individual language has access to this inventory, but it does not follow that any given language will exploit all the available possibilities. Talmy (2000: 38) does not identify a single principle that accounts for the concepts that belong within the closed-class set but admits the 'strong possibility' that it may be partly innate.

14.4.3 Schemas and instances

A defining property of the cognitive model is that the characterisation of linguistic units as symbolic units is not restricted to the content system but also applies to the grammatical system. In other words, grammatical units are also seen as form-meaning pairings. As we have seen, while the meaning associated with open-class units is specific (rich in conceptual content), the meaning associated with closed-class units is schematic. From this perspective, there is no need to posit a sharp boundary in the grammar between open-class and closed-class units. Instead, specificity versus schematicity of meaning can be viewed as the poles of a **continuum**, according to which both open-class and closed-class expressions are meaningful, each making a distinct and necessary contribution to the **cognitive representation** prompted by the utterance. According to Langacker, the inventory of symbolic units is organised by **schema-instance relations**. A **schema** is a symbolic unit that emerges from a process of abstraction over more specific symbolic units called **instances**. In other words, schemas form in the mental grammar when patterns of similarity are abstracted from utterances, giving rise to a more schematic representation or symbolic unit. The relationship between a schema and the instances from which it emerges is the schema-instance relation. This relationship is hierarchical in nature.

Consider common nouns like *cats, dogs, books, flowers* and so on. Each of these expressions is a highly entrenched symbolic unit. For example, the symbolic unit *cats* might be represented by the formula in (43):

(43) [[[CAT]/[kæt]]-[[PL]/[s]]]

The representations in SMALL CAPITALS indicate the semantic poles and those in the International Phonetic Alphabet (IPA) font represent the phonological poles. The slash indicates the symbolic link between semantic and phonological poles, and the hyphen indicates the linking of symbolic units to form a complex

WHAT IS A COGNITIVE APPROACH TO GRAMMAR?

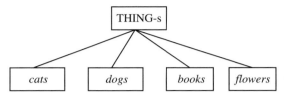

Figure 14.8 Schema–instance relations

structure. Given that there are many cases of regular plural nouns in the linguistic inventory, this regular pattern is captured by a schematic symbolic unit which contains only schematic information about the construction. This schema for plural nouns is represented in (44).

(44) [[[THING]/[...]]-[[PL]/[s]]]

In this schematic representation, the semantic pole THING indicates a noun but its corresponding phonological unit is left blank to indicate that this construction represents nouns in general. Each fully specified unit corresponding to this schema (for example, the expressions *cats, dogs, books, flowers*) represents an instance of the schema. The hierarchical relationship between a schema and its instances is captured in Figure 14.8.

It is important to point out here that the schema–instance relation is not restricted to symbolic units. For Langacker, the schema is any superordinate (more general) element in a taxonomy and the instance is any subordinate (more specific) element. In other words, the schema–instance relation represents a type of categorisation relation. In terms of phonological units, for example, the phoneme is the schema and its allophones are instances. In terms of semantic units, the concept FLOWER is schematic in relation to the instances ROSE, LILY and GERBERA. An instance is said to **elaborate** its schema, which means that it provides more specific meaning. For example, MAMMAL is more specific than ANIMAL, and in turn MONKEY is more specific than MAMMAL.

14.4.4 Sanctioning and grammaticality

Of course, any model of grammar must account for how speakers know what counts as a **well-formed** or **grammatical** utterance in his or her language. In the cognitive approach, well-formedness is accounted for on the basis of conventionality. Recall that the grammar is conceptualised not as an abstract system of rules, but as an inventory of symbolic units. Moreover, these symbolic units are derived from language use. The cognitive model captures generalisations and defines well-formedness on the basis of a categorisation process. For example, if the structure of an utterance produced by a speaker can be categorised as an instance of an existing schema, it is well-formed. Langacker uses the term **sanction** to refer to this categorisation process. For example, **coding** is the process whereby a speaker searches for a linguistic expression in order to express a concept. If the form the speaker arrives at matches forms existing in his or her inventory, this represents a case of sanction and thus well-formedness. The ability of language users to create novel forms according to the patterns of their language is accounted for by extrapolation from an existing pattern in the inventory, and this is when structure-building comes into the picture. Langacker (1987: 72) provides the example of a child describing a pie as *apricoty*. Although this is a novel form in the sense that it is not conventionalised, it clearly corresponds to a productive pattern in the inventory: many adjectives contain the derivational suffix *-y* (e.g. *fruity, funny, stinky*). Because well-formedness is conceived in terms of conventionality and conventionality is a matter of degree, it follows that well-formedness is also a matter of degree.

For example, Langacker demonstrates that acceptability of passive constructions depends on a number of factors, which give rise to **graded grammaticality judgements**. Consider the following examples of passive constructions. A question mark before the sentence indicates that the sentence is not perfectly well-formed but is acceptable. Two question marks indicate somewhat less acceptability. This convention is used in a system with asterisks which, as we have seen, indicate complete ungrammaticality.

(45) a. This view was enjoyed by Lily and George.
 b. ?A view was enjoyed by Lily and George
 c. ??Views were enjoyed by Lily and George

The examples in (45) become progressively less acceptable as the subject of the sentence moves from being definite or **individuated** to becoming less definite or individuated. In (46), the examples become progressively less acceptable the less the verb relates to a prototypical physical action.

(46) a. George was tickled by Lily.
 b. ?George was wanted by Lily
 c. ??George was resembled by his brother

This brief overview suffices to map out the general architecture of the cognitive model. We return to explore each of these issues in more detail in subsequent chapters.

14.5 Summary

In this chapter we have set out the characteristics of a **cognitive approach to grammar**. A cognitive approach adopts two fundamental assumptions:

the **symbolic thesis** and the **usage-based thesis**. The resulting model assumes that a speaker's knowledge of language or **mental grammar** emerges from his or her experience of situated usage events. We identified two main types of cognitive model: **inventory-based** approaches and the **'Grammatical Subsystem Approach'** developed by Talmy. The inventory-based approaches include **Cognitive Grammar** and **constructional approaches**, and are concerned with accounting for the entire inventory of symbolic units. In addition to these two types of model, we mentioned a number of cognitive approaches to **grammaticalisation** which are informed in various ways by cognitive linguistic theory. We also introduced some essential **grammatical terms** that we rely upon throughout Part III of the book. Finally, we set out some of the defining characteristics of a cognitive approach to grammar. We saw that a cognitive model represents knowledge of language in the mind of the speaker as a **structured inventory** of **conventional symbolic units**. Within this structured inventory, there is a qualitative distinction between **open-class** and **closed-class** symbolic units, a distinction that has also been expressed in terms of a distinction between **lexical** and **grammatical subsystems**. The inventory is structured by **schema-instance relations** in which more schematic symbolic units or **schemas** are abstracted from experience of more specific symbolic units or **instances**. The cognitive model we sketch here is not a specific theory, but is based on points of similarity across a number of cognitive approaches, each of which we explore in more detail in subsequent chapters.

Further reading

Introductory texts

- **Croft and Cruse (2004).** This textbook has useful chapters on construction grammars and the usage-based model.
- **Lee (2001).** This textbook provides a very basic introduction to cognitive linguistics. Some chapters relate to grammatical issues including constructions, nouns and verbs, and it also has a chapter on language change.
- **Taylor (2002).** This detailed and highly accessible textbook provides a comprehensive overview of Langacker's theory of Cognitive Grammar.

Foundational texts

The following are among the foundational book-length texts and articles that set out a cognitive approach to grammar. For purposes of accessibility, we have grouped this list by theory.

The 'Conceptual Structuring System Model'

- **Talmy (2000).** Volume I sets out Talmy's model in detail.

Cognitive Grammar

- **Langacker (1987).** This is the best and most comprehensive overview of the architecture of Langacker's theory of Cognitive Grammar.
- **Langacker (1991).** This volume applies the theoretical model developed in the first (1987) volume to a range of linguistic phenomena from English and other languages.
- **Langacker ([1991] 2002).** First published in 1991, this short volume is a collection of some of Langacker's most important papers and describes the architecture of Cognitive Grammar. Chapter 1 provides a particularly useful introduction to some of the key ideas that underpin the theory.
- **Langacker (1999b).** This volume is a collection of later papers that chart more recent developments in Cognitive Grammar.

Fillmore et al.'s Construction Grammar

- **Fillmore, Kay and O'Connor (1988).** This highly influential paper sets out the arguments for taking a constructional approach to grammar.

Embodied Construction Grammar

- **Bergen and Chang (2005).** This paper provides a detailed illustration of the architecture of Embodied Construction Grammar.

Goldberg's Construction Grammar

- **Goldberg (1992).** This article provides an illustration of Goldberg's approach based on a case study of the ditranstive construction.
- **Goldberg (1995).** This extremely accessible introduction to Goldberg's theory is perhaps the most influential and widely-read book on construction grammar to date.
- **Goldberg (1997).** This is a short and highly accessible encyclopaedia article which defines constructional approaches to grammar in general.

Cognitive approaches to grammaticalisation

- **Heine (1997).** This highly accessible book represents a powerful case for the experiential and conceptual basis of grammaticalisation.

- **Hopper and Traugott (2003).** An accessible introduction to modern grammaticalisation theory by two leading researchers in the field.
- **Traugott and Dasher (2002).** This book presents a theory of semantic change in grammaticalisation. The authors argue for a usage-based perspective: the Invited Inferencing Theory of semantic change.

Radical Construction Grammar

- **Croft (2002).** In this challenging but thought-provoking book, Croft presents a theory of construction grammar informed by his work as a linguistic typologist. We briefly discuss this theory in Chapter 20.

For beginners in grammatical terminology

- **Börjars and Burridge (2001).** An excellent introduction to descriptive grammar. While resting on broadly generative assumptions, this book remains largely theory-neutral. Highly recommended for students with little knowledge of grammatical terms.
- **Tallerman (1998).** An accessible descriptively-oriented introductory text with a fair amount of cross-linguistic data.
- **Trask (1993), Trask (1997), Trask (2000a).** Readers new to linguistic terminology will find these dictionaries invaluable.

Exercises

14.1 Defining a cognitive approach to grammar

Make a list of the key assumptions and characteristics of the cognitive approach to grammar. Now write a definition of this approach in no more than twenty words.

14.2 Morphemes and words

In the following sentences, identify the morphemes (bound and free), based on our discussion in section 14.3.2. State which of these units belong within the open-class subsystem and which belong within the closed-class subsystem.

(a) The rocket scientists have eaten the canapés.
(b) That old friend of his might stay the night.
(c) An estate agent fell asleep under the table.

14.3 Word classes

Label each word in the following sentence according to word class.

> That sparkly top Lily is wearing has suddenly given George a terrible headache.

14.4 Phrases

For each of the following sentences, (i) identify phrases; and (ii) state which word you think represents the head of each phrase. How did you reach your conclusions? What problems did you encounter?

(a) Lily is a rocket scientist.
(b) Lily is besotted with George.
(c) George bought Lily a packet of crisps.
(d) Hoping for a diamond ring, Lily was disappointed.

14.5 Grammatical functions

For the same set of examples (in exercise 14.4), identify the grammatical function of each of the phrases you identified and comment on case and agreement. (*Note*: you will need to substitute pronouns for proper nouns and common nouns to reveal case properties.)

14.6 The meaning of grammar

Consider the following sentence. Based on Talmy's distinction between the lexical and grammatical subsystems, (i) identify as many closed-class and open-class units as you can and divide them into two lists; (ii) for each unit, provide a semantic representation of its meaning.

> The estate agent has hidden Lily's slippers under the bed.

Do your findings support Talmy's claim regarding the semantic distinction between the lexical and grammatical subsystems?

14.7 Schema-instance relations

Recall from section 14.4.3 the idea that schema-instance relations hold between symbolic units. Based on Figure 14.8, make a diagram of schema-instance relations for the following two sets of symbolic units:

(a) man, boy, woman, girl, human
(b) human, adult, child, woman, man, girl, boy

In what ways are the two diagrams similar? How do they differ? What might this indicate about the way in which schemas and instances are related in the grammar?

15

The conceptual basis of grammar

In this chapter we consider the conceptual basis of grammar. The sense in which we use the term 'grammar' here refers to the closed-class or **grammatical subsystem**: grammatical words and morphemes, and grammatical categories and functions. To claim that grammar has a conceptual basis is to claim that grammar is meaningful. As we observed in the previous chapter, one way of defining 'grammar' is on the basis of the qualitative distinction in meaning between open-class and closed-class elements. In this chapter, therefore, we are primarily concerned with the semantics of the closed-class elements. The reason for this emphasis is that, in recognition of the distinction between closed and open classes, linguists have traditionally defined the closed-class elements of language in terms of structure, function and distribution rather than in semantic terms. In contrast, the cognitive model assumes the grammatical subsystem can be semantically characterised along the same lines as the open-class subsystem. This view entails a continuum between open- and closed-classes within the inventory that represents knowledge of language in the mind of the speaker, rather than two sharply distinct knowledge systems. Of course, to claim that closed-class elements are meaningful is not to claim that they are conventionally associated with rich meaning in the way that open-class elements are. Recall the distinction that was introduced in the previous chapter between **content meaning** and **schematic meaning** (which is also known as **structural meaning**). In this chapter, we begin to explore the kind of meaning that cognitive linguists associate with closed-class elements.

We begin the chapter by briefly setting this area of investigation within the context of the broader cognitive linguistics enterprise, as developed in earlier parts of the book (section 15.1). We then proceed to examine the conceptual basis of closed-class elements exploring the theoretical frameworks proposed by Leonard Talmy and Ronald Langacker. Both these researchers have been centrally concerned with the conceptual basis of grammar and with providing a description of how closed-class elements are meaningful. We begin by revisiting the '**Conceptual Structuring System Model**' proposed by Talmy (2000), which we introduced in Chapter 6 (section 15.2). We explore Talmy's thesis that the closed-class linguistic system reflects conceptual structure which, as we saw in Part II, reflects embodied experience. Talmy explores this thesis by examining how SPACE and TIME are configured by the grammatical subsystem, and by looking at how grammar encodes perspective, attention and force-dynamics. According to Talmy, the central function of the closed-class system is to encode these aspects of embodied experience, a view that entails that grammar has a conceptual basis and is therefore meaningful. We then turn to Langacker's theory of **Cognitive Grammar**, which complements Talmy's model in a number of ways (section 15.3). For example, Langacker argues that lexical classes like nouns and verbs reflect conceptually instantiated categories (which he calls THING and PROCESS) that derive ultimately from embodied experience. Finally, we re-examine the related issues of **categorisation** and **polysemy** from the perspective of the grammatical subsystem (section 15.4). We explore how closed-class elements reflect categorisation as a fundamental property of human cognition and how, like the open-classes categories, the closed-classes categories also display polysemy. Much of the discussion in this chapter will be familiar from Part II of this book, but we address these issues here with specific reference to how conceptual structure and organisation is encoded by the grammatical subsystem.

15.1 The grammatical subsystem: encoding semantic structure

According to the cognitive model that we sketched in the last chapter, knowledge of language is represented in the mind of the speaker in terms of a structured inventory. This inventory is structured in terms of a network of links between symbolic units. These symbolic units may be either lexical (open-class) or grammatical (closed-class) elements. This structured inventory represents **semantic structure**. Recall from Chapter 6 that cognitive linguists view semantic structure as constituting the conventional form that **conceptual structure** takes for expression in language. In Part II of the book, we began to map out semantic structure by exploring word meaning and sentence meaning. In that part of the book, our emphasis was on the open-class subsystem. In Part III of the book, we complement our exploration of the open-class subsystem by examining the closed-class or 'grammatical' subsystem.

In modern linguistics, the widespread view is that grammar is not independently meaningful. As we will see in detail in this chapter, cognitive linguists argue that grammar is independently meaningful because, like the open-class system, it has a conceptual basis. From this perspective, grammar derives from and reflects embodied experience in a similar way to open-class expressions; the difference between the open and closed classes relates only to the degree of semantic specificity or schematicity that a linguistic unit encodes. In this chapter, we explore the cognitive foundations of grammar posited by Talmy and Langacker and see how these models of the grammatical subsystem represent attempts to understand how grammar encodes concepts relating to TIME, SPACE and FORCE-DYNAMICS, and how it reflects cognitive phenomena like attention and perspective. Cognitive linguists argue that fundamental aspects of embodied experience have left an indelible imprint on the grammatical subsystem which provides (in Talmy's terms) the 'scaffolding' across which the open-class subsystem can drape its more specific content.

15.2 Talmy's 'Conceptual Structuring System Model'

We saw in the previous chapter that one of the claims that defines the cognitive model is that there is no **principled distinction** between the lexical (open-class) and grammatical (closed-class) subsystems. This was described in terms of a lexicon-grammar continuum. Nevertheless, Leonard Talmy has argued persuasively for a **qualitative distinction** between the lexical and grammatical subsystems, which is not incompatible with positing a lexicon-grammar continuum. As we also saw in Chapter 14, each of these subsystems provides a different kind of meaning: rich meaning versus schematic meaning. From this perspective, the schematic meaning provided by the grammatical subsystem forms a 'scaffold' that structures the rich content meaning provided by the lexical subsystem. According to the cognitive perspective, there is no need to posit grammatical 'rules' because the schematic meaning encoded by closed-class elements entails constrains upon how the units of grammar can be combined within complex constructions. Thus these two kinds of meaning encode and externalise distinct but equally important aspects of a particular scene as it is represented in conceptual structure by what Talmy calls the **cognitive representation** (CR). This is illustrated in Figure 15.1.

Talmy's research has primarily been concerned with examining the nature and the range of schematic or structural meaning encoded by the grammatical subsystem. We call Talmy's model the **'Conceptual Structuring System Model'** because he argues that the schematic structure encoded by closed-class elements can be divided into a series of different 'systems'. In this section, we explore each of these systems and see how they are claimed to account for some of the grammatical properties of language. This section complements the introduction to Talmy's model that we presented in Chapter 6, where we considered his claims from the perspective of cognitive semantics.

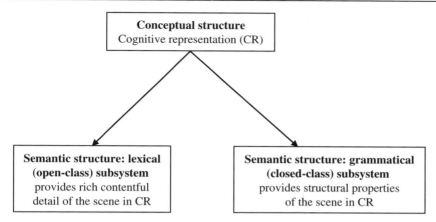

Figure 15.1 The bifurcation in semantic structure

15.2.1 The configuration of SPACE and TIME

We begin this section by revisiting the **domains** of SPACE and TIME. These conceptual domains have been recurring themes throughout the earlier parts of the book (see Chapters 3, 6 and 7 for example), and will continue to be prominent throughout Part III. Talmy views these as the primary basic domains.

The neutral term **quantity** is used by Talmy to refer to the content of these conceptual domains. The quantity that exists in the domain of SPACE is **matter**, which may be either **continuous** or **discrete**. We return to these terms directly, but for the time being we can think of 'continuous' matter as having no inherent 'segmentation' in its composition; this type of matter is **mass**, illustrated by AIR. **Discrete** matter, on the other hand, does have inherent 'segmentation', and this type of matter characterises **objects** which can be divided into parts, like the entity BIKE. The quantity that exists in the domain of TIME is **action**, which can also be continuous or discrete. Continuous action, like (to) SLEEP, is called **activity**. Discrete action, like (to) BREATHE, is described as an **act**. The difference between these two types of action is that it is not possible to describe the sub-parts of sleeping (unless you are a sleep specialist), while breathing is characterised by a series of distinct subparts (inhaling and exhaling). This partition of the domains of SPACE and TIME is summarised in Table 15.1.

The difference between the domains of TIME and SPACE is that while TIME has the property of **progression**, SPACE is **static**. 'Progression' means that the quantity within this domain is made up of a sequence of distinct representations because it changes from one instance to the next. By way of illustration, imagine photographing someone engaged in an activity like stroking a cat. Each of the photographs you take will be different from the previous one, and together they portray the activity. In contrast, change is not an inherent

Table 15.1 Matter and action (based on Talmy 2000: 42)

Domain	Continuous	Discrete
SPACE (matter):	mass	objects
TIME (action):	activity	acts

Table 15.2 Linguistic expressions relating to matter and action

Domain	Continuous	Discrete
SPACE (matter):	mass: *(the) air*	objects: *(a/the) cat(s)*
TIME (action):	activity: *(to) sleep*	acts: *(to) breathe*

property of objects, although of course objects can be involved in processes of change.

According to Talmy, these two conceptual domains are reflected in the way the grammatical subsystem encodes and externalises patterns of thought (the CR). In other words, the distinction between the domains of SPACE and TIME is reflected in grammatical structure. In the most general terms, verbs or verb phrases prototypically encode entities from the domain of TIME (activity and acts), while nouns or noun phrases prototypically encode entities from the domain of SPACE (masses and objects). This is illustrated by the examples in Table 15.2.

15.2.2 Conceptual alternativity

The membership of concepts within the domains of SPACE and TIME is not fixed, however. This is because TIME and SPACE are what Talmy describes as **homologous categories**, which means that they appear to share certain structural principles. As we have already seen, one of these relates to quantity: both SPACE and TIME can be conceived in terms of quantity. For example, in response to following question *How far is London from Brighton?* one could legitimately answer either *Fifty miles* (SPACE) or *About an hour* (TIME).

Talmy calls the ability to conceptualise a member of one domain in terms of another **conceptual alternativity**. Conceptual alternativity is reflected in the closed-class subsystem by grammatical categories. Conceptual alternativity is facilitated by a number of **conceptual conversion operations**. For example, **reification** is the name of the conversion operation that converts our conceptualisation of TIME (or action) into SPACE (or matter): an act can be converted into an object, or an activity into a mass. When a temporal concept is reified, it is expressed by a **nominal expression** (a noun phrase). Compare the examples in (1) and (2).

	An act	*reified as an object*	*(discrete)*
(1)	John washed her.	John gave her a wash.	

	Activity	*reified as a mass*	*(continuous)*
(2)	John helped her.	John gave her some help.	

In example (1), *washed* is a verb and encodes an act, while *a wash* is a noun phrase and encodes an act conceptualised as an object. In example (2), *helped* is a verb and encodes an activity, while *some help* is a noun phrase and encodes an activity conceptualised as a mass. When an act is construed as an object, it can be described in terms consistent with the properties of objects. For example, physical objects can be transferred: *to call (on the phone)* becomes *he gave me a call*. Physical objects can also be quantified: *to slap* becomes *he gave her two slaps*. As Talmy observes, however, there are constraints upon this process of reification. For example, a reified act or activity cannot be expressed in the same way that prototypical physical objects can. Example (3) illustrates that the reified act *a call* is incompatible with verbs that are prototypically physical.

(3) *George pushed/threw/thrust/slid Lily a call

The converse operation, which converts matter to action, is called **actionalisation**. When concepts relating to matter are actionalised, they are expressed by verb phrases. This operation is illustrated by the following examples adapted from Talmy (2000: 45).

	An object	*actionalised as an act*	*(discrete)*
(4)	Lily removed the pit from the olive.	Lily pitted the olive.	

	A mass	*actionalised as an activity*	*(continuous)*
(5)	George has a nosebleed.	George is bleeding from the nose.	

15.2.3 Schematic systems

As we first saw in Chapter 6, Talmy argues that the grammatical subsystem (or conceptual structuring system) is divided up into a number of **schematic systems**. By positing these systems, Talmy provides a way of modelling the different kinds of structural or schematic meanings associated with closed-class elements. In essence, Talmy's thesis is that closed-class elements encode different kinds of schematic meaning which cluster together within a single system of schematic meaning. Talmy elaborates four distinct schematic systems, although, as we noted in Chapter 6, there may well be others. The division of the conceptual structuring system into these four schematic systems is represented in Figure 15.2.

15.2.4 The 'Configurational Structure System'

The Configurational Structure System imposes structure upon the contents of the domains of SPACE and TIME. Closed-class elements perform an important role in encoding this configurational structure. We have already begun to see how this system works in our discussion of 'continuous' versus 'discrete' quantities of SPACE and TIME. Talmy proposes six further **schematic categories** within the Configurational Structure System. These are represented in Figure 15.3. These categories structure the **scenes** encoded by language and the **participants** that interact within these scenes. In the remainder of this section, we

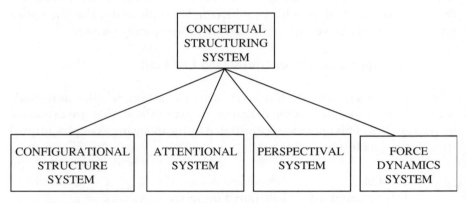

Figure 15.2 Four schematic systems within the conceptual structuring system

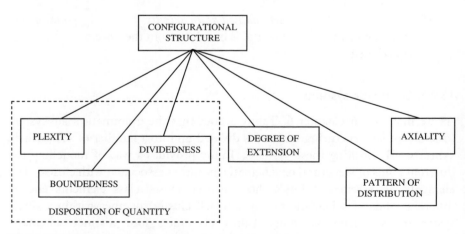

Figure 15.3 Schematic categories of configurational structure

examine each of the six schematic categories in turn, briefly illustrating the nature of the schematic meaning emerging from each category and establishing what kinds of closed-class elements encode this schematic meaning.

Plexity: number and aspect

Plexity relates to whether a quantity of TIME or SPACE consists of one (**uniplex**) or more than one (**multiplex**) equivalent elements. When related to SPACE (or matter), this is the basis of the grammatical category number. For instance, the singular count noun *slipper* represents **uniplex structure**, while the plural count noun *slippers* represents **multiplex structure**. Mass nouns like *champagne* also have multiplex structure. When related to the domain of TIME (or action), plexity forms part of the basis for the distinction between **semelfactive** versus **iterative** lexical aspect. **Lexical aspect** relates to the internal 'structure' of an event and is linguistically encoded in a number of ways. Consider example (6).

(6) a. George coughed. [semelfactive]
    b. George coughed for ten minutes. [iterative]

The verb *cough* encodes a **punctual** event which means that it is over almost as soon as it has begun. In the absence of any context that tells us that this event was drawn out over a period of time, we interpret the event as semelfactive, which means that it happened only once. This is the interpretation in (6a). When a punctual event is drawn out over a period of time, as in (6b), it becomes iterative. This means that it happens repeatedly. Clearly, semelfactive aspect has uniplex structure, while iterative aspect has multiplex structure. Observe that this type of aspect is built into the meaning of the verb itself rather than being grammatically marked. This is what distinguishes lexical aspect from grammatical aspect such as perfect or imperfect (progressive) aspect, which was introduced in the previous chapter. However, lexical aspect interacts in important ways with grammatical structure, as we will see at various points throughout this section.

Boundedness: count, mass and aspect

The term **boundedness** relates to whether a quantity is understood as having inherent boundaries (**bounded**) or not (**unbounded**). In the domain of SPACE, this is the basis of the count/mass noun distinction. For example, count nouns like *slipper* and *canapé* have bounded structure, in that each designates an entity with inherent 'edges' which can thus be individuated and counted. On the other hand, mass nouns like *champagne* and *oxygen* do not have inherent 'edges' and therefore cannot be individuated and counted. In the domain of TIME, boundedness is the basis of the distinction between **perfect** and **imperfect** grammatical aspect. Consider example (7).

(7) a. George has left the party.
  b. George is leaving the party.

Example (7a) is grammatically marked for perfect aspect by the presence of the perfect auxiliary *have* followed by the past participle *left*. As we saw in Chapter 14, perfect aspect encodes an event that is completed and can thus be thought of as bounded. Example (7b), on the other hand, is grammatically marked for imperfect (progressive) aspect by the progressive auxiliary *be* followed by the progressive participle *leaving*. Imperfect aspect encodes an event that is 'ongoing' and can thus be thought of as unbounded. It is important to point out that verbs can also be inherently bounded or unbounded in terms of their lexical aspect, which is traditionally described as **telicity**. **Telic** verbs like *win* entail what we can think of as an inherent 'endpoint' or 'goal', while **atelic** verbs like *sleep* do not. For our purposes, telicity can be thought of as boundedness and **atelicity** as unboundedness. Compare the following examples:

(8) a. Lily won the race in four minutes.
  b. *Lily slept in four minutes.

As Talmy points out, verbs that are inherently bounded are compatible with adverbial expressions like *in four minutes*, which denote a bounded period of time. This is illustrated by (8a). In contrast, verbs that are inherently unbounded are not compatible with this type of adverbial expression, as in (8b).

As with the conversion operations that mediate between the domains of SPACE and TIME, Talmy points out that it is possible to convert unbounded quantity (for example, *water* or *sleep*) into a bounded portion (for example *some water* or *some sleep*). This process is called **excerpting** and underlies expressions like *two champagnes* or *three sands and two cements*. Here, we rely upon the division of mass into bounded portions like glasses of champagne and sacks of sand and cement. The converse operation is called **debounding**. For example, the count noun *shrub* designates a bounded quantity while the mass noun *shrubbery* construes this as unbounded.

Dividedness

Dividedness relates to the internal segmentation of a quantity and underlies the distinction we introduced earlier between discrete and continuous matter: if matter can be broken down into distinct parts, it is **discrete**. If it cannot, it is **continuous**. It is important to emphasise that the properties 'unbounded' and 'continuous' are not the same, although they can correlate. For example, the mass noun *oxygen* is both continuous and unbounded. In contrast, the mass nouns *timber* and *furniture* are unbounded but have internally discrete structure. This property is not reflected in closed-class elements, unlike boundedness. As we have seen, though, Talmy relies upon this parameter for the broad division of the domains of SPACE and TIME into two subcategories.

Disposition of quantity: the role of closed-class elements

So far, we have seen that quantities of SPACE and TIME can be described in terms of plexity, boundedness and dividedness. Talmy describes the intersection between these three schematic categories in terms of **disposition of quantity**, as shown by the dotted box in Figure 15.3. We can think of disposition of quantity, then, as a 'bundle' of attributes that characterises certain conceptual categories and is reflected in the grammatical subsystem. For example, the mass noun *furniture* is matter, multiplex, unbounded and discrete, while the mass noun *water* is matter, multiplex, unbounded and continuous. Disposition of quantity is illustrated by Table 15.3. In this table, the two central columns represent the intersection of the three categories: plexity, dividedness and boundedness. The cell labelled A represents quantity that is [multiplex, discrete, unbounded]; cell B represents quantity that is [multiplex, discrete, bounded]; cell C represents quantity that is [uniplex, discrete, bounded]; cell 1 represents quantity that is [multiplex, continuous, unbounded]; and cell 2 represents quantity that is [multiplex, continuous, bounded]. Because a uniplex quantity consists of a single element, it is inherently bounded and discrete, which explains why cell 3 is labelled 'not applicable' and why there is no fourth row in the table illustrating the intersection of plexity with unboundedness.

Table 15.4 provides examples of linguistic expressions that reflect the 'bundles' of schematic properties represented in Table 15.3. The first example in each cell relates to matter (SPACE) and the second example to action (TIME).

Closed-class elements play a key role in the conversion of quantity from one state to another. Examples (9)–(12) illustrate some of the possibilities. In examples (9) and (10) it is the presence of the (plural/mass indefinite) determiner *some* that construes unbounded matter as bounded matter. In example (11) it is the plural noun suffix-*s* that construes uniplex matter as multiplex matter. In example (12), it is grammatical aspect, introduced by the progressive auxiliary *be* and the participial suffix-*ing*, that construes uniplex bounded action as multiplex unbounded action.

Table 15.3 Illustrating the schematic category DISPOSITION OF QUANTITY (adapted from Talmy 2000: 59)

	Discrete	Continuous	
Multiplex	A	1	Unbounded
Multiplex	B	2	Bounded
Uniplex	C	3 N/A	Bounded

Table 15.4 Illustration of lexical items that relate to DISPOSITION OF QUANTITY

	Discrete	Continuous	
Multiplex	*furniture* *(to) breathe*	*water* *(to) sleep*	**Unbounded**
Multiplex	*(a) family* *(to) molt*	*(a) sea* *(to) empty*	**Bounded**
Uniplex	*(a) slipper* *(to) sigh*	N/A	**Bounded**

(9) [multiplex, discrete, unbounded] → [multiplex, discrete, bounded]

furniture some furniture

(10) [multiplex, continuous, unbounded] → [multiplex, continuous, bounded]

water some water

(11) [uniplex, discrete, bounded]   →   [multiplex, discrete, bounded]
     slipper   slippers

(12) [uniplex, discrete, bounded]   →   [multiplex, discrete, unbounded]
     (to) sigh   be sighing

Degree of extension: aspect and preposition phrases

Degree of extension relates to how far quantities of SPACE or TIME 'stretch' over distance. This category interacts with boundedness, but introduces a more detailed structure that we can think of in terms of points on a continuum between bounded and unbounded. For example, SPACE or TIME can be either a **point** (*speck, die*), a **bounded extent** (*ladder, wash up*) or an **unbounded extent** (*river, sleep*). Focusing on the domain of TIME, example (13) illustrates that each of these degrees of extension (encoded by the verb) is compatible with different types of adverbial expressions.

(13) a. George's grandmother died [at four o'clock]. [point]
     b. Lily washed up [in ten minutes].   [bounded extent]
     c. Lily slept [for an hour].   [unbounded extent]
     d. *Lily slept [in an hour]
     e. *George's grandmother died [for an hour]

The differences between these verbs, as they relate to degree of extension, is once more a matter of lexical aspect. The verb *die* encodes a punctual event; as we saw earlier, this means that it is over almost as soon as it has begun. In contrast, *wash up* and *sleep* are **durative** events, which means that they extend over time. However, while *wash up* is telic (has an inherent endpoint), *sleep* is atelic. Observe that the adverbial expressions in (13) are preposition phrases, headed by closed-class elements like *at, in, for* and so on. Although these preposition phrases also contain noun phrases that encode the 'stretch' of time (*four o'clock, ten minutes, an hour*), it is the preposition that determines the compatibility of the adverbial expression as a whole with the meaning encoded by the verb, as illustrated by examples (13d) and (13e). However, these adverbial expressions can sometimes modify the degree of extension encoded by a verb. In example (14a), the verb *die* is construed in terms of a bounded extent (it took her an hour to die), and in (14b) it is construed in terms of an unbounded extent.

(14) a. George's grandmother died in an hour. [bounded extent]
     b. George's grandmother has been   [unbounded extent]
        dying for years.

Pattern of distribution: aspect and preposition phrases

Pattern of distribution relates to how matter is distributed through SPACE or how action is distributed through TIME. We illustrate this category by focusing on action through TIME, encoded by verbs. Table 15.5 provides examples of the various patterns of distribution.

These patterns can be explained as follows. While dying represents a change of state from which its participant cannot emerge, falling represents a change of state from which its participant can emerge (if you fall you can get up again, but getting up again is not a necessary part of falling). If a light flashes, it goes from dark to light and back to dark again, which represents a cyclical change of state. Repeating the cycle is not an intrinsic part of flashing (because a light can flash only once), while it is an intrinsic part of breathing. In contrast to all of these, which involve some internal change, sleep represents a steady or unchanging state. Like degree of extension, this category largely determines aspect, and is reflected in the compatibility or incompatibility of certain verbs with certain grammatical constructions. For example, these parameters explain why the examples in (15) are not well-formed.

(15) a. *George's grandmother kept dying
     b. *Lily fell out of bed for an hour

Because *die* is one-way non-resettable (at least under normal circumstances), you can only do it once. This is incompatible with the *keep* V-*ing* construction, which is restricted to events that either can be repeated (*Lily kept falling out of bed, Lily kept breathing*) or involve a steady state (*Lily kept sleeping*). Like *die, fall* is also one-way in the sense that it is unidirectional rather than cyclic, but unlike *die* it is resettable, so it is possible to do it more than once. However, because it is one-way rather than cyclic, it involves stopping and starting (you are not still falling while you are getting up again), so it cannot be done continually for an extended period of time (*for an hour*), unless it happens repeatedly.

As much of the discussion in this section has illustrated, aspect is a complex category, and reveals much about the interaction between grammar and meaning. We return to explore aspect in more detail in Chapter 18.

Table 15.5 Patterns of distribution

Pattern of distribution	Example
One-way non-resettable	*(to) die*
One way resettable	*(to) fall*
Full-cycle	*(to) flash*
Multiplex	*(to) breathe*
Steady-state	*(to) sleep*

THE CONCEPTUAL BASIS OF GRAMMAR

Axiality: degree modifiers

The final schematic category of configurational structure is **axiality**. This relates to the way a quantity of SPACE or TIME is structured according to a directed axis. For example, Talmy argues that the adjectives *well* and *sick* are points on an axis relating to HEALTH. On the axis, *well* is the endpoint, whereas *sick* is the remainder of the axis. This explains the different distribution of the closed-class **degree modifiers** like *almost* and *slightly* in relation to these adjectives. For example, while it is possible to be *slightly sick* or *almost well*, it is not possible to be *slightly well or *almost sick. This follows from the axiality model because it is not possible to be 'slightly' at an endpoint, nor 'almost' on the journey towards that endpoint. Axiality is illustrated in Figure 15.4.

In sum, we have seen that the 'Configurational Structure System' partitions quantities from the domains of SPACE and TIME according to the internal structural properties of those quantities, as well as in terms of how they are distributed within SPACE and TIME. We saw that quantity in SPACE is prototypically encoded by nouns, while quantity in TIME is prototypically encoded by verbs. We also saw that the processes of reification and actualisation can convert quantity from TIME to SPACE and from SPACE to TIME, respectively, and that grammatical categories (word classes) also play a role in this process. We saw that plexity (uniplex and multiplex structure) is reflected by the grammatical category number (singular and plural), and that boundedness is reflected by the distinction between count nouns and mass nouns, as well as playing a role in lexical aspect (telicity). In our discussion of disposition of quantity, we saw that closed class elements play a role in altering the way that quantities are construed; for example, determiners can impose boundedness on unbounded matter, the plural morpheme can convert uniplex matter to multiplex matter, and progressive aspect construes bounded action as unbounded action. In our discussion of both degree of extension and pattern of distribution, we saw that temporal expressions headed by prepositions play a key role in structuring lexical aspect. Finally, we saw an example of how the closed-class category degree modifiers reflect properties relating to axiality. The discussion in this section therefore begins to establish how the grammatical subsystem provides schematic or structural meaning to the linguistic expression of the cognitive representation.

Figure 15.4 Axiality (adapted from Talmy 2000: 65)

Table 15.6 Factors in the 'Attentional System'

Strength of attention	
Pattern of attention	focus of attention
	window of attention
	level of attention
Mapping of attention onto parts of a scene	

15.2.5 The 'Attentional System'

As we have seen, the 'Configurational Structure System' structures matter and action in SPACE and TIME. The 'Attentional System' governs the distribution of attention over matter and action (scenes and their participants) and is governed by three main factors. The first factor is **strength**, which relates to the relative prominence of referents: whether they are either backgrounded or foregrounded. The second factor is **pattern**, which concerns how patterns of attention are organised. For example, a **focus of attention** pattern gives rise to **figure-ground** organisation. Other patterns are **window of attention** and **level of attention**. The third factor is **mapping**, which governs the way in which parts of an attention pattern are mapped onto parts of the scene described. Table 15.6 summarises the three factors that govern the 'Attentional System'.

As we saw in Chapter 3, the figure-ground asymmetry, an attentional phenomenon, is fundamental to the nature of human perception. According to the cognitive model, attention is also fundamental to grammatical organisation. It is important to emphasise that the factors strength, pattern and mapping of attention should not be viewed as distinct types of attention. Instead, these are factors that interact to focus attention: prominence gives rise to patterns of attention which are then mapped onto scenes. We illustrate the interaction of these parameters here with examples of the three types of attention pattern: focus, window and level of attention.

Focus of attention pattern

Example (16) involves a COMMERCIAL EVENT frame (this was discussed in Chapter 7). In (16a) *the shop assistant*, corresponding to the SELLER role, is foregrounded. In other words, it is the **figure**. The BUYER and GOODS roles are backgrounded and together make up the **ground**. In (16b), *George*, which corresponds to the BUYER role, is the figure, and the SELLER and GOODS roles make up the ground.

(16) a. The shop assistant sold the champagne to George.
b. George bought the champagne from the shop assistant.

This example illustrates a focus of attention pattern. In terms of strength of attention, the foregrounding results from the mapping of attention onto a particular entity in the scene. The grammatical system encodes this in two ways: firstly, by the selection of one of several verbs relating to the event frame (*buy* versus *sell*, for example); and secondly by the associated word order. The prominence of the clause-initial position illustrates the phenomenon called grammatical **iconicity**, where some aspect of conceptual representation is 'mirrored' by grammatical structure. In this case, conceptual prominence is mirrored by grammatical prominence. The choice over which participant in the event is placed in this position is linked in part to the choice of verb and in part to the type of grammatical construction selected (e.g. active versus passive, or cleft versus unmarked declarative).

Windowing pattern

The windowing pattern involves the explicit mention of some part or parts of a scene (**windowing**), while other parts may be omitted (**gapping**). The windowing pattern differs from the figure-ground pattern because the figure-ground pattern concerns the organisation of aspects of the conceptual representation that are present in the linguistic representation. Like the figure-ground pattern, however, the windowing pattern represents a strategy for foregrounding (strength) and involves mapping. As we saw in Chapter 6, for example, a path of motion consists of a beginning, a middle and an end. In example (17), the whole path of motion is windowed whereas in the examples in (18) only the initial, medial or final portion of the path is windowed, respectively:

(17) The champagne cork shot out of the bottle, through the air, and into Lily's eye.

(18) a. The champagne cork shot out of the bottle. [initial]
b. The champagne cork shot through the air. [medial]
c. The champagne cork shot into Lily's eye. [final]

According to Talmy, the windowing pattern also accounts for grammatical behaviour, such as the division of the **complement** category into obligatory and optional complements. For example, one of the verbs relating to the COMMERCIAL EVENT frame, *spend*, only requires the MONEY role as an obligatory complement (together with BUYER). This is illustrated in (19a). The GOODS role can be realised as an optional complement (19b). However, the SELLER role is 'blocked' as a complement if this verb is selected (19c), because each choice of verb windows certain participants in the event frame.

(19) a. George spent £100.
   b. George spent £100 on that champagne.
   c. *George spent £100 on that champagne to the shop assistant.

Level of attention pattern

The examples in (20) illustrate two different level of attention patterns. As these examples show, this idea relates to whether the focus of attention is upon the group of friends as a whole, also known as a **Gestalt** representation (20a), or upon the internal structure or **componentiality** of the group (20b). This difference is encoded by grammatical construction.

(20) a. the group of friends
   b. the friends in the group

15.2.6 The 'Perspectival System'

We have seen that the 'Configurational Structure System' structures participants and scenes in space and time, and the 'Attentional System' governs the distribution of attention over those referents. The 'Perspectival System' establishes a **viewpoint** from which participants and scenes are viewed. This system relates to the conceptual 'perspective point' from which we view an entity or a scene and involves the four schematic categories: **location**, **distance**, **mode** and **direction**. These can be encoded by closed-class elements.

Perspectival location

This category relates to the location that a perspective point occupies relative to a given utterance. The linguistic system of **deixis**, for example, works by signalling perspective relative to the speaker's location, and deictic expressions are then interpreted with respect to that point of reference. As we saw in Chapter 14, the grammatical **person** system is an example of a deictic category, an idea that we explore in more detail later in the chapter (section 15.3.2).

Perspectival distance

In some languages, open- or closed-class expressions can signal 'proximal', 'medial' or 'distal' distance of a referent relative to speaker or hearer. This phenomenon therefore also relates to deixis. This is illustrated by the following examples from Hausa, a West African language belonging to the Chadic branch of the Afroasiatic family (Buba 2000). In this language, demonstrative determiners, pronouns and adverbs show a four-way deictic distinction, where distance interacts with location. The examples in (21) illustrate the behaviour of the pre-nominal demonstrative determiners:

(21) a. speaker proximal
   wannàn yārò   'this boy [near me]'
 b. addressee proximal
   wànnan yārò   'that boy [near you]'
 c. speaker/addressee medial
   wancàn yārò   'that boy [over there]'
 d. speaker/addressee distal
   wàncan yārò   'that boy [way over there]'

In these examples, the grave accent represents a low tone vowel whereas a vowel unmarked for tone is high. A macron indicates a long vowel whereas a vowel unmarked for length is short. As these examples demonstrate, Hausa is a tone language, where the relative pitch of the vowels can give rise to differences in meaning, both in terms of content and in terms of marking grammatical differences.

Perspectival mode

This schematic category relates to whether a perspective point is in motion or not. This interacts with perspectival distance, where 'distal' tends to correlate with 'stationary' and 'proximal' with 'moving'. If the perspective point is stationary, it is in **synoptic mode**. If the perspective point is moving, it is in **sequential mode**. Talmy argues that this category is also relevant to aspect. Perfect aspect encodes a perspective that is distal and stationary, because the event depicted is viewed as a completed whole. Progressive aspect, on the other hand, encodes an event that is proximal and 'moving', because the event is viewed as immediate and 'ongoing'. This is illustrated by the examples in (22).

(22) a. [synoptic]
   Lily had seen some houses through the window of the ambulance.
 b. [sequential]
   Lily kept seeing houses through the window of the ambulance.

Example (22a) invokes the perspective of a fixed vantage point. In contrast, example (22b) invokes a motion perspective, as a result of which the houses are seen one or some at a time.

Perspectival direction

The final schematic category relating to perspective point is perspectival direction. This category also interacts closely with attention and concerns the direction in which an event is viewed relative to a given perspective point. The direction can be **prospective** or **retrospective**. Consider the examples in (23).

(23) a. George finished the champagne   [prospective]
         before he went home.
     b. Before he went home, George   [retrospective]
         finished the champagne.

Observe that that it is not the order of the events themselves that distinguishes the two examples; in both cases, George first finishes the champagne and then goes home. The difference relates to the direction from which the two events are viewed, which is illustrated in Figures 15.5 and 15.6.

In Figure 15.5 the event-sequence is viewed from the perspective of the first event, event A. This is called a prospective direction because the perspective point is located at the temporally earlier event, from which the speaker looks 'forward' to the later event. In Figure 15.6 the event-sequence

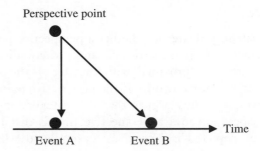

Figure 15.5 Prospective direction (adapted from Talmy 2000: 74)

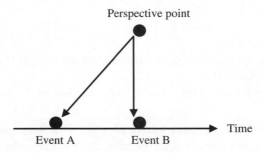

Figure 15.6 Retrospective direction (adapted from Talmy 2000: 75)

THE CONCEPTUAL BASIS OF GRAMMAR

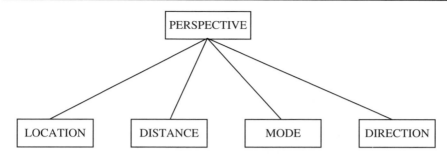

Figure 15.7 Schematic categories of the 'Perspectival System'

is viewed from the perspective of the second event, event B (going home). This is called a retrospective direction because the perspective point is located at the temporally later event (going home) and the viewing direction is 'backwards', towards the earlier event. Observe that perspectival direction rests upon the temporal sequence model of time that we discussed in Chapter 3. Figure 15.7 summarises the four schematic categories of the 'Perspectival System'.

15.2.7 The 'Force-Dynamics System'

The fourth schematic system proposed by Talmy is the 'Force-Dynamics System'. This system relates to our experience of how physical entities interact with respect to force, including the exertion and resistance of force, the blockage of force and the removal of such blockage. The 'Force-Dynamics System' encodes the 'naive physics' of our conceptual system (our intuitive rather than scientific understanding of force dynamics), and has implications not only for the expression of relationships between physical entities, but also for abstract concepts such as permission and obligation (**modal** categories).

The 'Force-Dynamics System' assumes two entities that exert force. The **agonist** is the entity that receives focal attention and the **antagonist** is the entity that opposes the agonist, either overcoming the force of the agonist or failing to overcome it. The force intrinsic to the agonist is either 'towards action' or 'towards rest', and the force intrinsic to the antagonist is the opposite. We illustrate this system with a set of examples that encode physical entities. The subscripts AGO and ANT represent 'agonist' and 'antagonist', respectively:

(24) a. [the glass]$_{AGO}$ kept rolling because of [the breeze]$_{ANT}$
b. [Lily]$_{AGO}$ kept standing despite [the gale]$_{ANT}$
c. [the glass]$_{AGO}$ kept rolling despite [the mud]$_{ANT}$
d. [the glass]$_{AGO}$ stayed lying on the slope because of [the grass]$_{ANT}$

In (24a), the force tendency of the agonist *the glass* is towards rest, but this is overcome by the greater force of the antagonist *the breeze*, which is towards motion and thus stands in a causal relationship with the agonist. In (24b), the force tendency of the agonist *Lily* is also towards rest, and in this case the agonist's force is greater. In (24c), the force tendency of the agonist, *the glass*, is towards motion, and the agonist's force is greater than the opposing force of the antagonist, *the mud*. Finally, in (24d), the force tendency of the agonist, *the glass*, is also towards motion, but this time the opposing force of the antagonist, *the grass*, is greater and prevents the motion. Observe that the force-dynamics of the interaction are expressed here by closed-class elements: the conjunctions *because of* or *despite*. While *because of* encodes the greater force of the antagonist, which overcomes the force of the agonist and thus entails causality, *despite* encodes the greater force of the agonist.

Talmy represents force dynamics with diagrams like Figure 15.8. The circle represents the agonist and the concave shape represents the antagonist. The symbol • represents the tendency towards rest, and the symbol > represents the tendency towards action. Finally, the symbol + represents the stronger of the two forces. This diagram represents the force-dynamics pattern in example (24a), where the inherent tendency of the agonist is towards rest but the greater force of the antagonist causes motion.

According to Talmy, the 'Force-Dynamics System' also underlies the behaviour of another major closed-class category: the **modal auxiliaries**. For example, *can* (in the **capacity** sense) encodes a tendency towards action (for example, *Lily can run a mile in four minutes*). In contrast, *must* encodes a tendency towards rest that is overcome by the force of the antagonist (for example, *You must pay your income tax*). In this example, the **deontic** reading encodes legal or social obligation and this obligation represents the antagonist.

In conclusion, we have seen how the four schematic systems proposed by Talmy are reflected in the grammatical subsystem of language. While the first three schematic systems ('Configurational Structure', 'Perspective' and 'Attention') relate most prominently to visual perception, the 'Force-Dynamics System' relates most prominently to kinaesthetic (motor) perception. In this respect, Talmy's theory reflects the embodied cognition thesis explored in earlier parts

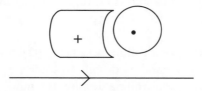

Figure 15.8 Force-dynamics encoded in sentences like (24a) (adapted from Talmy 2000: 415)

THE CONCEPTUAL BASIS OF GRAMMAR

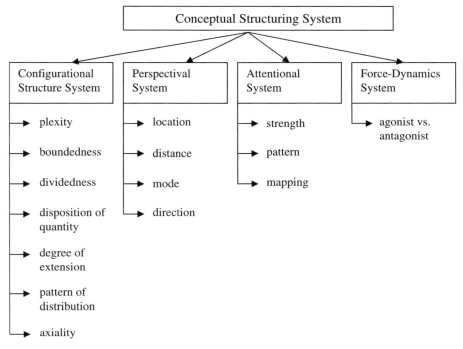

Figure 15.9 An overview of the conceptual structuring system

of in the book. The four schematic systems that comprise the 'Conceptual Structuring System', as presented here, are summarised in Figure 15.9.

15.3 Langacker's theory of Cognitive Grammar

In this section, we begin to explore Langacker's model of Cognitive Grammar, and provide a sketch of the theory with particular emphasis on its conceptual underpinnings. We return to fill in the details of this model in the next three chapters. Langacker's model complements Talmy's in that both of these researchers have been centrally concerned with the conceptual and embodied basis of language. In particular, both have been concerned with how the grammatical subsystem encodes concepts relating to domains like SPACE, TIME and FORCE-DYNAMICS, and how it encodes cognitive phenomena like attention and perspective. In this section we address some of the grammatical phenomena identified by Langacker from the perspective of their conceptual basis.

15.3.1 The conceptual basis of word classes

Like Talmy, Langacker argues that word classes have a conceptual basis. In other words, the linguistic categories noun, verb, adjective and so on are not 'purely grammatical' categories with 'purely formal' properties (such as the affixes they take or their patterns of distribution within phrases and sentences). Instead, Langacker argues that these categories have a conceptual basis and can therefore be semantically characterised. According to Langacker's model, linguistic expressions divide into two broad categories: **nominal predications** and **relational predications**. This distinction relates to the nature of the schematic meaning encoded by nouns and noun phrases (nominals) on the one hand, and by other lexical classes like verbs adjectives, prepositions and so on (relations) on the other. The term 'predication' relates to meaning and refers to the semantic pole of a symbolic unit. Nominal predications are **conceptually autonomous**, which means that they relate to conceptually independent entities like BED or SLIPPER: the expressions *bed* or *slipper* invoke concepts that are independently meaningful. In contrast, relational predications are **conceptually dependent**, which means that they rely on other units to complete their meaning, which are relational in nature. For example, in a sentence like *George hid the slipper under the bed*, the verb *hid* relates the conceptually autonomous entities GEORGE, SLIPPER and BED, establishing a relationship involving 'hiding' between them. Similarly, *under* establishes a spatial relation between SLIPPER and BED.

Nominal predications

Langacker argues that physical objects are the prototypical referents for the **noun** category, but as with any category, there are central and prototypical members. Langacker therefore proposes a highly schematic characterisation of the noun class: a noun encodes a region in some domain, and a count noun encodes a bounded region in some domain (recall our discussion of domains in Chapter 7). A region is defined as a 'set of **interconnected** entities' (Langacker 2002: 67). Sometimes the entities that comprise the region are **homogeneous** at least as far as the boundary (for example, *bleep, pond*), and sometimes they are **individuated** (for example, *bicycle, cat, piano, constellation*). This notion of homogeneity versus individuation is reminiscent of Talmy's parameter of dividedness, where quantities of SPACE and TIME are either continuous or discrete. As we saw earlier, a region is bounded if there is some inherent limit to the set of entities that constitute it. For example, a CONSTELLATION is bounded because it is a bounded region in a 'bigger picture' of SKY. A **mass noun** encodes an **unbounded region** in some domain. The concepts encoded by mass nouns can also differ in terms of how homogeneous or individuated the entities are that compose them (compare *water* and *furniture*, for example). Because count nouns are bounded they are **replicable**, which is why they can be counted; this property does not hold for mass nouns. As this brief sketch illustrates, Langacker relies upon a similar core of conceptual properties as Talmy in his characterisation of the noun category.

Relational predications

While nominal predications describe entities, relational predications describe relations between entities. Langacker divides the category of relational predications into two subcategories: **temporal** and **atemporal** relations. Temporal relations are **processes** and are encoded by verbs. The category of atemporal relations is a more disparate category and contains prepositions, adjectives, adverbs and non-finite verb forms (infinitives and participles). The domain of TIME underlies the distinction between temporal and atemporal relations. In describing the role of time in this distinction, Langacker distinguishes **conceived time** from **processing time**. We can think of processing time as 'real time', in the sense that any cognitive process requires processing time. In this sense, processing time is a **medium** of conceptualisation. On the other hand, conceived time refers to the cognitive representation of TIME, where time is an **object** of conceptualisation (see Evans 2004a, 2004b for a discussion of studies that take this approach). Within conceived time, Langacker distinguishes the processes of **summary scanning** and **sequential scanning**, where 'scanning' relates to how the aspects of a scene are perceived, visually or otherwise, and give rise to a conceptual representation. In summary scanning, aspects of a scene are scanned cumulatively and are simultaneously present in the conceptual representation. This gives rise to a **Gestalt representation** of time as a unified whole and characterises **static** scenes. In sequential scanning, the aspects of a scene are scanned in a sequential fashion, so that the aspects of the scene are not simultaneously present at any stage of the scanning. This gives rise to a conceptualisation of time as a dynamic **process** and characterises **events**.

Langacker likens the distinction between summary and sequential scanning to the difference between looking at a photograph (summary scanning) and watching a film (sequential scanning). While all aspects of a scene are simultaneously present in a photograph which presents a static scene, a film involves a sequence of scenes, each different from the next. Figure 15.10 summarises Langacker's model of word classes.

15.3.2 Attention

Having introduced Langacker's basic assumptions relating to the conceptual basis of word classes and how they relate to domains, including SPACE and TIME, we next consider how attention underpins language. Consider Langacker's definition of attention:

> Attention is intrinsically associated with the intensity or energy level of cognitive processes, which translates experientially into greater prominence or salience. Out of the many ongoing cognitive processes

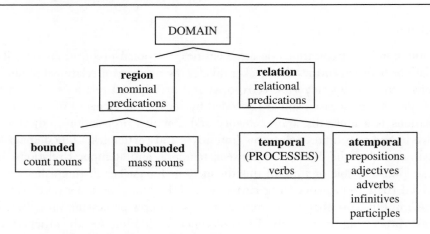

Figure 15.10 Langacker's model of word classes

that constitute the rich diversity of mental experience at a given time, some are of augmented intensity and stand out from the rest as the focus of attention. (Langacker 1987: 115)

Like Talmy, Langacker argues that grammar encodes schematic aspects of embodied experience and that attention, as a perceptual phenomenon, is one aspect of this.

Linguistic expressions relate to conceived situations or 'scenes'. As we have seen, attention is differentially focused on a particular aspect of a given scene. In Langacker's terms, this is achieved in language by a range of **focal adjustments** which 'adjust the focus' on a particular aspect of any given scene by using different linguistic expressions or different grammatical constructions to describe that scene. The visual metaphor that the expression 'focal adjustment' rests upon emphasises the fact that visual perception is central to how we focus attention upon aspects of experience. By choosing a particular focal adjustment and thus linguistically 'organising' a scene in a specific way, the speaker imposes a unique **construal** upon that scene. Construal can be thought of as the way a speaker chooses to 'package' and 'present' a conceptual representation, which in turn has consequences for the conceptual representation that the utterance evokes in the mind of the hearer. For example, as we have already seen, the active construction focuses attention upon the AGENT of an action (e.g. *George hid Lily's slippers*), while the passive construction focuses attention upon the PATIENT (e.g. *Lily's slippers were hidden by George*). Each of these constructions conventionally encodes a distinct construal.

Langacker distinguishes three parameters along which focal adjustments can vary: (1) **selection**; (2) **perspective**; and (3) **abstraction**. Together, these parameters provide different ways of focusing attention upon and thus construing

THE CONCEPTUAL BASIS OF GRAMMAR

a scene. In broad terms, these parameters are roughly equivalent to the first three of Talmy's schematic systems (the 'Configurational Structure System', the 'Attentional System' and the 'Perspectival System'). The interaction of these three parameters is illustrated in Figure 15.11 and addressed in more detail below.

Selection: profiling

Selection determines which aspects of a scene are attended to and relates to the notion of a conceptual domain. Recall from Chapter 7 that a conceptual domain is a body of related knowledge within the conceptual system. Langacker proposes a number of basic domains (those tied directly to preconceptual embodied experience), which are presented in Table 15.7.

One aspect of construal is the selection of a particular domain. This is illustrated by the following examples. In each example, the expression *close* selects

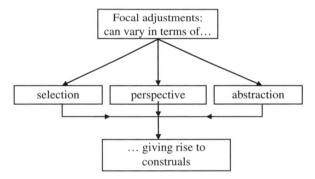

Figure 15.11 The relationship between focal adjustments and construal

Table 15.7 Basic domains proposed by Langacker (1987)

Basic domain	Pre-conceptual basis
SPACE	Visual system; motion and position (proprioceptive) sensors in skin, muscles and joints; vestibular system (located in the auditory canal; detects motion and balance)
COLOUR	Visual system
PITCH	Auditory system
TEMPERATURE	Tactile (touch) system
PRESSURE	Pressure sensors in the skin, muscles and joints
PAIN	Detection of tissue damage by nerves under the skin
ODOUR	Olfactory (smell) system
TIME	Temporal awareness
EMOTION	Affective (emotion) system

537

a different domain and therefore contributes to a very different construal in each sentence

(25) a. George's flat is quite close to Clapham Common. [SPACE]
b. It's close to Lily's birthday. [TIME]
c. Those roses are close to the colour she wants for her wedding dress. [COLOUR]
d. Lily and her cat are very close. [EMOTION]

Even within a single domain, an expression like *close* can give rise to distinct construals. For example, an expression can select for differences of **scale**. Langacker (1987: 118) illustrates this idea with the examples in (26), which relate to the domain of SPACE.

(26) a. The two galaxies are very close to one another.
b. San Jose is close to Berkeley.
c. The sulphur and oxygen atoms are quite close to one another in this type of molecule.

The expression *close* selects for different scales in each of these examples: the distance between the two elements in each example ranges from the distance between galaxies to the distance between the subparts of a single molecule.

A second aspect of selection, and one that is fundamental to Langacker's approach, relates to **profiling**. Earlier, we described profiling informally as 'conceptually highlighting' some aspect of a domain. As we saw in Chapter 7, profiling involves selecting some aspect of a **base**, which is a conceptual entity necessary for understanding what a word means. According to this perspective, words have **profile-base organisation**. For example, the expression *elbow* profiles a substructure within the larger structure ARM, which is its base. This idea is illustrated by Figure 15.12. As we saw earlier, Langacker calls the semantic pole of a symbolic unit its 'predication'. Because the predication necessarily includes both the profile and the base, the base represents the full **scope of predication** associated with an expression.

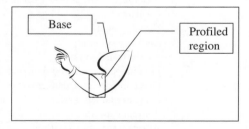

Figure 15.12 Profile-base organisation for *elbow*

The examples of selection we have discussed so far (selection of domain and profiling) relate to open-class elements. However, profiling is also reflected in the closed-class system. For example, active and passive constructions can give rise to different profiling possibilities. Consider the examples in (27).

(27) a. George opened the champagne.
     b. The champagne was opened.

The act of opening a bottle of champagne requires both an AGENT (the person opening the bottle) and a PATIENT (the bottle). These participants are both part of the scope of predication of a sentence describing this scene. While the example in (27a) profiles the full scope of predication, the example in (27b) selects the PATIENT for profiling while the AGENT remains part of the base. This is made possible by the passive construction, which allows the AGENT to remain unexpressed. This difference in terms of profiling is illustrated by figures 15.13 and 15.14 which represent examples (27a) and (27b), respectively. In these diagrams, the circles represent entities (AGENT and PATIENT) and the arrows represent energy transfer from AGENT to PATIENT. The fact that the AGENT is unshaded in Figure 15.14 represents the fact that that the AGENT is not profiled but is nevertheless present as part of the base.

Selection, particularly as it relates to profiling, is part of the process of **coding**. As we have seen, when a speaker wants to express a conceptual representation in language, he or she has choices over which linguistic expressions and constructions are used to 'package' the conceptual representation. Coding is the process of 'activating' these linguistic units. As Langacker (1991: 294) explains, the process of coding is closely interwoven with construal, because decisions about how a situation is construed have consequences for the linguistic expressions selected to code the conceptualisation. Consider the following examples, all of which might be appropriate descriptions of the same event. These are more complex than the examples in (27): in addition to an AGENT (*George*) and a PATIENT (*the TV*) they also involve an INSTRUMENT (*a shoe*), the entity used by the AGENT to carry out the action.

Figure 15.13 *George opened the champagne*

Figure 15.14 *The champagne was opened*

(28) a. George threw a shoe at the TV and smashed it.
b. George threw a shoe.
c. George smashed the TV.
d. The shoe smashed the TV.
e. The TV smashed.

These examples reflect different focal adjustments in terms of profiling and entail differences in terms of how much information the speaker intends to convey. The scope of predication (or base) is the 'background' against which the speaker construes the scene. Example (28a) profiles the entire scope of predication, as does (28c), although in less detail (this difference in detail relates to the focal adjustment **abstraction**, which we discuss below). Examples (28b) and (28e) have a narrower scope of predication, encompassing only the beginning of the event (28b) or the end of the event (28e). In other words, example (28b) only expresses information about George throwing a shoe; this sentence does not entail any consequences for the TV which is therefore not part of the scope of predication in this example. Equally, (28e) only tells us that the TV smashed but does not entail an AGENT or an INSTRUMENT (it may have fallen over). In contrast, (28d) does entail an AGENT as part of the scope of predication because a shoe is not an animate entity capable of smashing a TV without an AGENT. The scope of predication in turn has consequences for which participants are profiled. In (28a), AGENT, INSTRUMENT and PATIENT are all profiled. In (28b), only AGENT and INSTRUMENT are profiled. In (28c), AGENT and PATIENT are profiled, although the INSTRUMENT is 'understood' because we know that George must have used some INSTRUMENT to smash the TV, even if it was only his fist. This means that the instrument is part of the scope of predication in this example. Equally, in (28d), INSTRUMENT and PATIENT are profiled but an AGENT is understood as part of the base or scope of predication. Finally, in (28e), only PATIENT is profiled.

As these examples illustrate, the scope of predication or base of a given expression is determined by encyclopaedic knowledge. Compare the following examples (Langacker 1991: 332–5):

(29) a. An explosion woke me up.
b. A crowbar opened the window.

The conceptual representation or interpretation evoked by example (29a) does not necessarily entail an AGENT as part of its base, whereas the interpretation evoked by (29b) does. While the scope of predication in (29a) only includes the participants profiled by *an explosion* and *me*, the scope of predication of (29b) includes an unprofiled AGENT in addition to the two participants profiled by *a crowbar* and *the window*. This follows from the semantics of the expressions *an explosion* (which may occur without an external AGENT) and *a crowbar* (which cannot participate in an event without an external AGENT). In the same way, the unprofiled AGENT in (28d) arises from the semantics of *a shoe*.

Perspective: trajector-landmark organisation and deixis

The second parameter of focal adjustment is perspective. The perspective from which a scene is viewed has consequences for the relative prominence of its participants. Langacker argues that the grammatical functions subject and object are reflections of perspective and thus have a conceptual basis. He suggests that the distinction between subject and object relates to the prototype of an **action chain**, a cognitive model involving an active 'energy source' (AGENT) that transfers energy to an 'energy sink' (PATIENT). Langacker calls the semantic pole of the expression that fulfils the subject function the **trajector (TR)**, which reflects the observation that the prototypical subject is dynamic. The semantic pole of the expression that fulfils the object function is called the **landmark (LM)**. This reflects the observation that the prototypical object is stationary or inert. The terms 'trajector' and 'landmark' are familiar from earlier parts of the book and are used in a range of related ways in cognitive linguistics. As Langacker points out, TR-LM (or subject-object) organisation in linguistic expressions is an instance of the more general perceptual and attentional phenomenon of figure-ground organisation, a recurring theme throughout this book.

Langacker defines TR-LM organisation in terms of a conceptual asymmetry between participants in a profiled relationship: while the TR signifies the focal or most prominent participant, the LM represents the secondary participant. In an English sentence, the TR (subject) comes first and the LM (object) comes second. The familiar case of an active and passive pair of sentences illustrates this point. Consider example (30).

(30) a. George ate all the caviar.   [active]
     b. All the caviar was eaten by George. [passive]

In example (30a) the focal participant (TR) is *George* who is the AGENT of the action, and the secondary participant (LM) is *the caviar* which is the PATIENT. In (30b) the situation is reversed and the PATIENT is now the focal participant (TR). In a passive sentence, the AGENT is the secondary participant (LM), but it is not the object because passivised verbs do not take objects. Instead, the *by*-phrase that contains the object behaves more like a modifier and can be deleted without making the sentence ungrammatical. This difference between the active construction and the passive construction is represented by Figures 15.15 and 15.16.

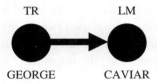

Figure 15.15 *George ate all the caviar*

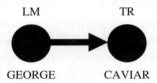

Figure 15.16 *All the caviar was eaten by George*

The distinction between these two sentences relates to a shift in perspective, which is effected by changing the relative prominence attached to the participants in the profiled relationship. While both participants are profiled, GEORGE is marked as TR in Figure 15.15, while the CAVIAR is marked as TR in Figure 15.16. The direction of the arrow remains the same in both diagrams because George is still the 'energy source', irrespective of whether he is the primary or secondary participant.

Although the term 'trajector' is derived from 'trajectory' (a path of motion), it is worth emphasising that this term is applied to all salient participants, regardless of whether the verb involves motion or not. For example, Langacker (2002: 9) illustrates the trajector-landmark asymmetry with the verb *resemble*. Consider example (31).

(31) a. [Lily's mum]$_{TR}$ resembles [Botticelli's angel]$_{LM}$.
 b. [Botticelli's angel]$_{TR}$ resembles [Lily's mum]$_{LM}$.

Although these two sentences are 'truth-conditionally equivalent' (the nature of the meaning of *resemble* is that it entails a mutual relationship: X resembles Y and vice versa), Langacker observes that they are not semantically equivalent. Example (31a) tells us something about *Lily's mum* (she resembles *Botticelli's angel*). Example (31b) tells us something about *Botticelli's angel* (it resembles *Lily's mum*). The verb *resemble* is a stative verb, which means that it describes an unchanging scene. Despite this, the TR-LM asymmetry is still evident.

Perspective also underpins the personal pronoun system. Recall from Chapter 14 that the grammatical feature **person** distinguishes speaker, hearer and third party. However, person is a **deictic** category because SPEAKER, HEARER and THIRD PARTY are not fixed properties of any given individual but shift continually during conversation. Consider the following short conversational exchange.

(32) George: I love caviar!
 Lily: I hate it!

In this short conversation, an individual referred to as *I* both loves and hates caviar. However, there is no contradiction in *I* both loving and hating caviar because the participants in the conversation know that the first person singular pronoun *I* refers to a different individual in each of the utterances. If George says *I*, it means GEORGE. If Lily says *I*, it means LILY. Speakers have no difficulty in 'keeping track' of who *I* or *you* refer to at any given point in a conversation. According to Langacker, it is our ability to adopt various viewpoints during a conversational exchange that underlies the ease with which we manipulate the person system: when George says *I*, Lily knows it means GEORGE the speaker and not LILY the hearer, because she momentarily adopts George's perspective as speaker.

Langacker argues that the parameter of perspective also gives rise to focal adjustments as a result of the distinction between **subjective construal** and **objective construal**, which relates to the asymmetry between perceiver and perceived. In order to illustrate this distinction, Langacker uses the example of a pair of glasses. If the wearer of the glasses takes them off, holds them in front of his or her face and looks at them, the glasses become the object of perception (perceived). In contrast, if the wearer has the glasses on and is using them to see some other object, the attention focused on the glasses themselves becomes far weaker to the extent that they become a subpart of the perceiver. In the same way, when an individual's attention is fully focused on some external entity, subjective construal (awareness of self) is backgrounded and objective construal is salient. When an individual's attention is fully focused on him or herself, subjective construal is foregrounded and objective construal is backgrounded. In reality, objective construal and subjective construal can be seen as extreme poles on a continuum, where the usual case is that an individual's attention is partly focused on objective construal and partly focused on subjective construal and one is more salient than the other. For example, objective construal is likely to be more salient than subjective construal when an individual is absorbed in watching a film or reading a gripping novel. However, subjective construal is likely to become more salient when an individual's attention is focused on riding a bike or threading a needle.

In order to see how this distinction between objective and subjective construal is related to perspective, and in turn how it is reflected in the grammatical system, we first introduce the term **ground**. In Langacker's model, this term describes any speech event, and includes the participants, the time of speaking and the immediate physical context. Deictic expressions make specific reference to ground, and Langacker divides them into two broad categories: those that place the ground 'offstage' or in the background, and those that focus attention upon the ground, placing it 'onstage'. For example, temporal deictics like *tomorrow* and *next week* place the ground offstage because they profile a point in time relative to the time of speaking, but the time of speaking which makes up part of the ground ('now') is backgrounded or implicit. In contrast, deictic expressions like *now* (temporal), *here* (spatial) and *you* (person deixis) place the ground onstage because they focus explicit attention upon aspects of the ground: time, place and participant(s). The greater the attention upon the ground, the greater the objectivity of construal. Speaker and hearer are usually subjectively construed or 'off stage', and only become objectively construed or 'on stage' when linguistically profiled by expressions like *I* or *you*. For example, if George utters the first person pronoun *I*, he places himself in the foreground as an object of perception. In this way, the speaker is **objectified**. According to Langacker, then, the difference between explicit mention of the ground (objective construal) and implicit dependence upon the ground (subjective construal) is a difference of perspective. As we will see in Chapter 21, this aspect of perspective forms the basis of Langacker's theory of grammaticalisation.

Abstraction: profiling

This focal adjustment operation relates to how specific or detailed the description of a scene is. This also has consequences for the type of construction selected. Recall our earlier examples in (28), two of which are repeated here as (33).

(33) a. George threw a shoe at the TV and smashed it.
 b. George smashed the TV.

The example in (33b) is more abstract (less detailed) than the example in (33a). As we saw earlier, both of these examples share the same scope of predication, which involves an AGENT, a PATIENT and an INSTRUMENT. However, the more abstract description only profiles the AGENT and the PATIENT and leaves the INSTRUMENT as an unprofiled part of the base. In this way, abstraction, which relates to the level of attention paid to a scene, is paralleled by the kinds of linguistic constructions available to us in terms of level of detail.

15.3.3 Force-dynamics

Recall from section 15.2.7 that the term 'force-dynamics' relates to our experience of motion energy. This experience gives rise to the knowledge that while

THE CONCEPTUAL BASIS OF GRAMMAR

Figure 15.17 The prototypical action chain model (adapted from Langacker 2002: 211)

some entities have an inherent capacity for energy, other entities only receive energy from external entities. In Langacker's model, the 'prototypical action' is characterised in terms of the transfer of energy from AGENT to PATIENT resulting in a change of state of the PATIENT. As we noted above, this is called the action chain model, and is illustrated in Figure 15.17, where A represents AGENT and P represents PATIENT.

We have already seen examples in the previous section of how this action chain model is manifested linguistically in our discussion of profiling and the TR-LM asymmetry. Langacker argues that the unmarked status of the active transitive sentence with third-person participants arises from the fact that it represents the prototypical action from a canonical viewpoint perspective. As we have seen, Langacker also argues that the grammatical notions of subject and object have their basis in this prototypical action chain model. The prototypical subject (TR) is the volitional 'energy source', and the prototypical object (LM) is the passive 'energy sink'. We also saw that the TR-LM reversal effected by the passive construction could be modelled in terms of a shift of attention from AGENT to PATIENT within the action chain model. We return to these issues in greater detail in Chapters 17 and 18, observing for the time being that grammatical features like voice and person, together with the grammatical functions subject and object, receive an experientially based semantic account within Langacker's theory of Cognitive Grammar.

15.4 Categorisation and polysemy in grammar: the network conception

Having outlined the conceptual bases of grammar according to Talmy and Langacker, in this final section we revisit the related issues of categorisation and polysemy from a grammatical perspective. Recall from Chapter 8 the idea that conceptual categories display prototype structure, and from Chapter 10 Lakoff's proposal that words, like concepts, are represented in the mind as radial categories that typically exhibit polysemy. As we saw, this approach was held to account not only for open-class words like nouns but also for closed-class words like prepositions.

Langacker develops a **network model** that represents the structure of categories. In this model, members of a category are viewed as nodes in a complex network. This can be seen as analogous to Lakoff's radial category model. In Langacker's model, the links between nodes in a network arise from

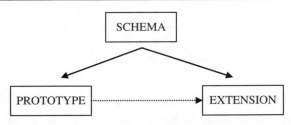

Figure 15.18 Network model (adapted from Langacker 2002: 271)

a number of different kinds of **categorising relationships** that hold between the symbolic units stored in the grammatical inventory. One categorising relationship is **extension from a prototype**, represented as [A] ⇢ [B], where A is the prototype and B shares some but not all attributes of A and is thus categorised as an instance of that category. A second type of categorising relationship is the relationship between **schema and instance**, represented as [A] → [B]. Recall from Chapter 14 that entrenched units that share a structural pattern give rise to a schematic representation of that structure, reflecting the usage-based nature of the model. The schema structures those related units as a category within the network, and novel expressions can be compared against such categories. According to Langacker, the network can grow 'upwards' via schematisation, 'outwards' via extension and 'downwards' as more detailed instances are added. Figure 15.18 captures the basis of the network model.

According to Langacker, the network model characterises not only polysemous open-class elements, but also underlies other kinds of linguistic categories, including those relating to sound as well as meaning and grammar. This means that morphemes, word classes and grammatical constructions are also envisaged as nodes in a network. It follows that while some nodes (like morphemes) are structurally simplex, other nodes (like the phonological poles of symbolic units or phrase- or sentence-level grammatical constructions) themselves have complex internal structure. For example, the English past tense morpheme is represented by the (partial) model in Figure 15.19.

At the semantic pole, PROCESS represents the verb and PAST represents the past tense morpheme. The next level in the network represents the various phonological instantiations of this schema, which are themselves schematic representations of the next level in the hierarchy, where specific instances of each schema are represented.

Polysemy and prototype structure in grammatical categories

If grammatical categories like closed-class words or bound grammatical morphemes are represented as discrete nodes in a complex network, and if

THE CONCEPTUAL BASIS OF GRAMMAR

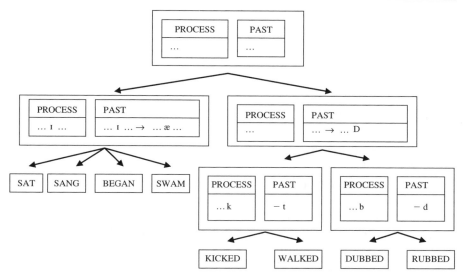

Figure 15.19 (Partial) network model of the English past tense morpheme (adapted from Langacker 2002: 283)

such categories are by nature organised with respect to a prototype, we might expect closed-class categories to display polysemy in the same way that open-class categories do. Langacker argues that this is the case. For example, as we saw in Chapter 14, the **modal auxiliaries** represent a closed-class and show a fixed and predictable participation in the grammatical behaviour of the verb string. In terms of their semantic contribution to the clause, however, the modal auxiliaries present a striking case of polysemy. Compare the examples in (34), which illustrate the polysemous nature of the modal verb *can*.

(34) a. Lily can cook.
 b. You can leave now.
 c. You can tidy up that mess before you leave.

In example (34a), the modal has a **capacity** reading: it conveys the speaker's judgement concerning Lily's capabilities. This is a kind of **epistemic** modality, which means it relates to the speaker's knowledge. The interpretations in the other two examples are quite different: these reflect **deontic** modality, which relates to obligation and permission. While (34a) encodes permission, (34b) encodes obligation.

Langacker's (1991: 185) discussion of **gender** morphology in Spanish illustrates prototype structure in grammatical categories. Spanish has two

547

nominal gender suffixes, *-o* and *-a*. According to Langacker, the prototypical values of these morphemes are MALE (*-o*) and FEMALE (*-a*), as the examples in (35) illustrate.

(35) a. *hermano*   'brother'   b. *hermana*   'sister'
     c. *oso*   'male bear'   d. *osa*   'female bear'
     e. *muchacho* 'boy'   f. *muchacha* 'girl'

However, these affixes also attach to nouns that have no inherent sex, such as the inanimate objects *mesa* 'table' and *cerro* 'hill'. Langacker argues that these affixes are still meaningful, but have a highly schematic meaning in the latter type of example where they mean THING. Recall that this corresponds to Langacker's representation of the noun category. This schematic semantic representation is therefore consistent with the behaviour of these gender morphemes as noun-forming affixes. At the 'centre' of this category are the semantically rich instances which encode MALE/FEMALE and ANIMACY as well as having the status of entity or THING. At the periphery of the category are the schematic instances which share the semantic characterisation THING but lack the richer semantic characterisation.

15.5 Summary

We began this chapter by describing cognitive approaches to grammar as attempts to 'complete the picture' of semantic structure suggested by cognitive semantics. We then outlined Talmy's theory concerning the conceptual basis of grammar. This discussion focused on Talmy's proposals concerning the four **schematic systems** that comprise the **conceptual structuring system**, and explored how each of these is reflected in the grammatical subsystem. We saw that the three 'perceptual' systems – the **'Configurational Structure System'**, the **'Attentional System'** and the **'Perspectival System'** – are argued to have consequences for the grammatical subsystem. For example, the configurational parameters of **plexity, boundedness** and **dividedness** are argued to underlie the system of grammatical **number** as well as the **count-mass** noun distinction; the 'Attentional System' underpins the relative **prominence** of referents in a construction, encoded by means of the **figure-ground** pattern and the **windowing of attention** pattern; and the 'Perspectival System' underpins the behaviour and interpretation of **deictic expressions**. We also introduced Talmy's 'Force-Dynamics System', which is argued to arise from **kinaesthetic** experience and involves the transfer or resistance of energy between **agonist** and **antagonist**. Talmy's approach thus addresses the **embodied** nature of conceptualisation as reflected in the grammatical subsystem. For example, Talmy argues that the 'Force-Dynamics

System' forms the basis of a characterisation of the **modal auxiliaries**. In our introductory sketch of Langacker's theory of Cognitive Grammar, we saw how this model complements Talmy's approach in a number of ways, and how lexical classes are held to have a conceptual basis relating to basic domains like TIME and SPACE. We saw that Langacker argues for division of linguistic expressions into two major categories: **nominal predications** and **relational predications**. We also saw how **attention** is encoded in language by virtue of **focal adjustments**, which have consequences for the **construal** and **coding** of a conceptual representation in language. We focused in particular on the three parameters according to which focal adjustments can vary: **selection, perspective** and **construal**. Two of the most important theoretical constructs in Langacker's theory are the notion of **profile-base** organisation, which relates to the parameter of selection, and **trajector (TR)-landmark (LM)** organisation, which relates to the parameter of perspective. We then saw how Langacker addresses the issue of force-dynamics by proposing the **prototypical action chain model**. Finally we re-examined the related issues of **categorisation** and **polysemy** as they relate to grammar and outlined Langacker's **network model**.

Further reading

Introductory texts

- **Radden and Dirven (2005).** Chapter 2 of this very accessible book looks at nominal predications, relational predications and grounding. Chapter 3 explores construal, scanning, perspective and the figure-ground asymmetry. This chapter also presents inference and mental spaces as two 'mental operations' alongside that of construal. Chapter 8 focuses on aspect, although it uses different terms from the ones used in this chapter.
- **Taylor (2002).** Part 2 of Taylor's excellent book (Chapters 7–13) concentrates on 'basic concepts'. Particularly relevant to the discussion in this chapter are Chapter 10 (which focuses on profile, base and domain) and Chapter 11 (which focuses on nominal and relational predications). Part 4 of Taylor's book (Chapters 18–21) discusses nouns, verbs and clauses, and Chapter 23 addresses networks.

Talmy's 'Conceptual Structuring System Model'

- **Talmy (2000).** Volume I, Chapter 1 outlines the fundamental schematic systems and provides examples of how these are reflected in the grammatical system. Chapter 4 focuses in more detail on windowing of attention, and Chapter 5 on figure and ground organisation. Chapter 7 discusses force-dynamics.

Langacker's Cognitive Grammar

- **Langacker (1987).** Chapter 3 outlines the cognitive basis of Langacker's model, focusing on scanning, attention and perspective. Chapter 4 outlines Langacker's assumptions concerning a model of mind, and discusses domains, configurational structure, the encyclopaedic nature of meaning, the role of linguistic expressions as points of access and abstract motion. Chapters 5–7 provide more detail on linguistic categories and relations, to which we return in more detail in subsequent chapters. Chapter 6 includes a discussion of the trajector-landmark asymmetry, and Chapter 7 explains profiling. The first section of Chapter 9 introduces action chains. Chapter 10 addresses categorisation and the network model, and Chapter 11 explores how the network model ensures well-formed expressions by means of the sanction process.
- **Langacker (1991).** Chapter 7 explains the relationship between coding and construal, and explores profiling in relation to the clause in some detail.
- **Langacker (1999b).** Chapter 2 provides an overview of and evidence for some of the theoretical constructs discussed in this chapter, including profiling and scope of predication. Chapter 7 addresses perspective and construal.
- **Langacker ([1991] 2002).** Chapter 1 provides a sketch of the model and introduces the profile-base relation. Chapter 3 maps out the foundational character of the domains of space and time in the characterisation of word classes. Chapter 10 outlines the network conception, and Chapter 12 explores subjectification and grounding as processes driven by perspective point.

Exercises

15.1 Comparing models of grammar

(i) What are the main claims of Talmy's model of grammar?
(ii) What are the main claims of Langacker's model?
(iii) What are the main points of similarity and difference between the two models?

15.2 The conceptual basis of grammar

What does it mean to say that grammar has a conceptual basis? How does this view differ from the traditional or formal view, as far as you are aware?

15.3 The 'Configurational Structure System'

Recall Talmy's classification of quantities of SPACE and TIME in terms of disposition of quantity, which relates to dividedness, plexity and boundedness. On the basis of Table 15.4, provide four additional examples of linguistic expressions that illustrate each of the ten possible combinations. Explain how you reached your conclusions.

15.4 The 'Force-Dynamics System'

Recall our discussion of force-dynamics in relation to example (24) in the text. On the basis of Figure 15.8, provide force dynamics diagrams for examples (24b)–(24d).

15.5 Nominal and relational predications

In Cognitive Grammar, nominal predications profile conceptually autonomous regions of a domain (or domains), while relational predications profile interconnections between those regions. Relational predications can be further subdivided into temporal and atemporal relations. In the light of these distinctions, categorise each of the words in the following examples according to whether you think they are nominal or relational (temporal or atemporal) predications. Keep a note of any difficulties you encounter; these should be addressed in the next chapter.

(a) Lily is making George a big surprise.
(b) Lily knows the date of George's birthday.
(c) Lily is fond of knitting.

15.6 Profile-base and TR-LM organisation

Invoking Langacker's notions of profile and base and TR-LM organisation, consider the following sentences which all relate to aspects of the same scene. How do these examples give rise to different construals?

(a) George punctured Lily's valentine balloon.
(b) Lily's valentine balloon was punctured by George.
(c) Lily's valentine balloon got punctured.

Now consider the following example:

(d) Lily's valentine balloon is punctured.

What is the difference between (d) and (c)? How might you account for this in terms of profile and base organisation?

16

Cognitive Grammar: word classes

So far in Part III of this book we have sketched the basic assumptions of a cognitive approach to grammar. We have also explored the conceptual basis of both Talmy's and Langacker's models. In this and the next two chapters we explore Langacker's theory of Cognitive Grammar in more detail. As we saw in Chapter 14, **symbolic units** can be minimal or simplex, as in the case of **morphemes**, or complex to varying degrees, as in the case of morphologically complex **words, phrases** or **sentences**. Any unit having complex symbolic structure – as opposed to complex semantic or phonological structure – is called a **construction** in Langacker's theory. In other words, Langacker does not refer to simplex symbolic units as constructions (although some cognitive linguists do, as we will see in Chapters 19 and 20). Figure 16.1 represents a taxonomy of symbolic units according to Langacker.

In this chapter we will begin our detailed survey of Cognitive Grammar by looking at Langacker's model of word classes, which we touched on briefly in the previous chapter. We will look at the properties of grammatical constructions in the next chapter. Of course, as Figure 16.1 shows, there is some overlap between words and constructions, given that complex words count as constructions in Langacker's model. It follows, therefore, that there will be some overlap between this chapter and the next in that both will have something to say about words. In this chapter, we will concentrate on describing word classes and on exploring Langacker's semantic account of these linguistic categories. This will involve some discussion of the morphemes that help to identify certain word classes, but we defer a detailed discussion of word structure for the next chapter.

This chapter is divided into five main sections in which we examine in more detail Langacker's schematic characterisation of the **open classes** noun, verb, adjective and adverb, as well as the **closed classes** adposition and determiner.

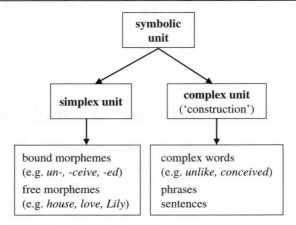

Figure 16.1 Symbolic units in Cognitive Grammar

We begin by setting Langacker's approach to word classes in a broader context in terms of **categorisation** (section 16.1). We then proceed to explore in more detail Langacker's characterisation of nouns in terms of **nominal predications**, an approach that was briefly introduced in Chapter 15 (section 16.2). This is followed by a discussion of the differences between nominal and **relational predications** (section 16.3). As we will see, the distinction between **profile** and **base** is central to this account, an idea that was also introduced in Chapters 7 and 15. Within the category of relational predications, we explore the distinction between and **temporal** and **atemporal relations**, where we will see that the distinction between the cognitive processes of **summary scanning** and **sequential scanning** is fundamental to Langacker's account (section 16.4). Finally, we look at determiners and quantifiers and explore Langacker's accounts for these closed-class expressions in terms of **grounding** (section 16.5). As we will see, Langacker exploits independently motivated cognitive phenomena, particularly those related to attention, in order to develop an account of word classes that emerges from a generalised model of human cognition.

16.1 Word classes: linguistic categorisation

Recall that the scope of predication of a linguistic expression is its base, and its profile is what the expression designates from within that base. We have also seen that symbolisation is the link between the phonological and semantic poles of a linguistic unit, while coding is the link between a linguistic unit and a speech event. As we saw in Chapter 14, the cognitive model views lexicon and grammar in terms of a continuum of symbolic units within the inventory rather than in terms of separable subsystems of language. Indeed, Langacker was an early pioneer in developing this view. At the open-class end of the continuum, units have rich and specific content meaning, and at the closed-class end of the continuum, units have schematic meaning. Despite broad acceptance of the distinction between open- and closed-class expressions, Langacker (1987: 18–19) cautions against viewing these as discrete categories. He argues that just as conceptual categories relating to content words have fuzzy boundaries, so do grammatical categories. This entails that certain linguistic expressions may fall at the periphery – or near the middle of the continuum. For example, the expression *thing*, which has been called a 'conceptual shell' (Schmid 2000), is an open-class word, but lacks the semantic specificity of a prototypical open-class word like *cat*. Langacker also points out that while the closed classes are resistant to change, they are not immune to it. In other words, the closed classes are not entirely closed. He provides the example of the Southern US expression *y'all* (second person plural), which has entered the 'closed' class of personal pronouns, an expression that has the counterpart *yous* in certain dialects of British English.

In contrast to the **distributional approach** to the characterisation of word classes (see Chapters 14 and 22), Langacker adopts the position that semantic characterisations of the major word classes are possible. Furthermore, Langacker supports the cognitive semantics model of categorisation, arguing that the formal view of category membership in terms of necessary and sufficient conditions should be abandoned in favour of a **prototype** model (see Chapter 8). Langacker argues that grammatical categories, like conceptual categories, display prototype effects and that a semantic characterisation of the category prototypes is therefore uncontroversial. In other words, it is only problematic to define nouns in terms of THINGS (matter) and verbs in terms of PROCESSES (action) if we assume that these rather specific semantic properties should hold for all members of the category, an idea that follows from a necessary and sufficient conditions model of categorisation. It is for this reason that a semantic characterisation of word classes is traditionally disfavoured in comparison to a structural characterisation based on morphological features and syntactic distribution.

However, the idea that prototypical nouns and verbs might have a semantic characterisation is not at the heart of Langacker's proposal. The crux of his proposal is rather that all nouns and verbs have a 'schematic semantic characterization' (Langacker 2002: 60), and furthermore that these characterisations are universal. To illustrate the idea that word classes can be described in terms of schematic meaning, consider the following examples:

(1) a. George loves poodles.
    b. George's love for poodles is rather worrying.

(2) a. Lily destroyed the letters.
    b. Her destruction of the letters was regrettable.

Although the verb *love* in (1a) and the noun *love* in (1b) might be difficult to distinguish in terms of content meaning, Langacker argues that they do encode different meanings because they encode different **construals** of the scene. The same argument applies to the verb *destroy* in (2a) and the noun *destruction* in (2b). As we saw in Chapter 15, construal is central to the choices that speakers make about how a scene is linguistically 'packaged', and this in turn explains the availability of related yet distinct constructions. For example, the nominal expressions in (1b) and (2b) involve the process of reification, which construes what Langacker calls a PROCESS (action) in terms of what he calls a THING (matter). As we will see in this chapter, construal is central to Langacker's theory of word classes.

16.2 Nominal predications: nouns

The challenge for a semantic account of the noun class is to provide a characterisation of a category that includes a very wide range of concept types. Consider the underlined nouns in the following examples.

(3) a. Lily sent a <u>letter</u> to her <u>lover</u>.
 b. Her <u>car</u> was making a funny <u>noise</u>.
 c. Lily tried to teach George the Arabic <u>alphabet</u>.
 d. The only good thing about George was his <u>height</u>.
 e. The <u>explosion</u> in her engine made her late for work.
 f. Lily's <u>love</u> for George began on a <u>Tuesday</u>.

While some nouns (like *letter* and *car*) are objects, others (like *lover*) encode a relation between two people or things. The noun *noise* expresses a physical sensation, while a noun like *alphabet* refers to a group of interconnected yet discrete entities. The noun *height* expresses a scalar concept, while the noun *explosion* describes an event. The noun *love* encodes an emotion, while the noun *Tuesday* refers to a point in time. As this small set of examples illustrates, the content meanings of members of the noun class is extremely disparate, and it is unlikely that a semantic account of the noun class that rests upon content meaning is an achievable goal. However, Langacker argues that a semantic account is not impossible. Recall that Langacker views meaning in terms of a continuum ranging from the highly specific to the highly schematic. If we move along the scale towards schematicity, it appears that a schematic semantic characterisation of the noun class is possible. Langacker states his schematic characterisation of the noun class as follows (Langacker 2002: 63):

(4) a. A 'noun' designates a 'region' in some domain.
 b. A 'count noun' designates a 'bounded region' in some domain.

In our discussion in the previous chapter we very briefly exemplified this claim with respect to basic domains like TIME and SPACE. For example, count nouns that designate a region in the domain of TIME include *moment* and *period*, and count nouns that designate a region in the domain of SPACE include *line, triangle* and *circle*. However, some nouns evoke a combination of domains. For example, *flash* profiles a region in the domains TIME, COLOUR and VISION. As Langacker observes, *flash* is bounded in TIME but not in VISION. In other words, a flash must be very brief in terms of time, but can expand to take up our whole visual field, so bounding need only apply in one of the domains evoked by the expression. Langacker also points out that count nouns like *second, hour, week, month* and *year* do not evoke the basic domain of TIME directly, but evoke abstract domains that humans have constructed in order to 'measure' time. We might refer to these domains as CLOCK (in the case of *seconds, minutes* and *hours*), or CALENDAR (in the case of *days, weeks, months* and *years*), although the two are not necessarily distinct.

16.2.1 Bounding

Langacker (2002: 65–9) raises a number of important points in relation to the notion of bounding. Firstly, bounding must be defined **within** rather than **by** the scope of predication or domain evoked by the expression. He illustrates this point with the visual example *I see NP*, which limits the scope of predication of the NP to whatever is contained within the speaker's VISUAL FIELD. Langacker's examples concern a scene in which the speaker is standing in front of a wall upon which a large red spot is painted against a white background. If the speaker is standing far enough from the wall to see both the red spot and the white background, the speaker will describe what he or she sees in the following way:

(5) I see a red spot.

Observe that *a red spot* is a noun phrase (NP) with the count noun *spot* as its head. This is consistent with the fact that the red area is bounded within the field of vision (because the speaker can see the 'edges' of the red spot) which is equivalent to the scope of predication of the NP. Now imagine that the speaker approaches the wall and stands so close to it that the red spot fills the visual field. The speaker may now describe what he or she sees as follows:

(6) I see red.

In this example, *red* is a mass noun, which is consistent with the fact that the red area is unbounded within the visual field of the speaker which is equivalent to the scope of predication of the NP. Crucially, in both scenarios, the red spot is bounded *by* the visual field in the sense that it is not experienced outside the visual field. However, this is not sufficient for the speaker to construe the red spot as bounded in both scenarios. Indeed, if this were the case, any noun relating to the domain of VISION would have to be bounded and therefore a count noun, which is clearly not the case. As this example illustrates, bounding must occur within the relevant domain.

Langacker's second point concerning bounding is that it is does not necessarily entail sharp boundaries. In other words, while some count nouns like *January* or *tummy button* designate regions with sharp and clearly defined boundaries, others, like *season* or *tummy*, designate regions with fuzzy boundaries. This is consistent with the nature of categories in general, as we saw in Chapter 8.

Langacker's third point is that bounding is often a function of construal rather than of objective reality. In other words, whether a region of a given domain is bounded or not sometimes depends on how we construe it rather than upon its inherent properties. Consider the examples in (7).

(7) a. Lily rubbed frantically at the spot on her mum's best rug.
    b. Lily and George met at their favourite spot on the downs.

In example (7a), *spot* designates an area on the rug that has inherent boundaries perceived within the visual field. We know that if we spill something on a rug, the stain has 'edges'. On the other hand, *spot* in (7b) does not designate an area with inherent or readily perceivable boundaries. Instead, bounding is imposed upon the area by construal.

Finally, Langacker observes that the term **region** must be defined as 'a set of interconnected entities' in order to account for count nouns like *team, group, family* and so on. This is because it is less straightforward to think of these as regions in the sense of having a clear 'shape' or in the sense that they occupy a distinct area of space from other categories. For example, you can think of your mum, dad, brother and sister as making up your FAMILY even when each of them is on a different continent. This illustrates the importance of **interconnection** to the notion of 'region'. Given the discussion so far, Langacker (2002: 69) revises his schematic characterisation of the noun category that we saw in (4) as follows:

(8) a. A 'count noun' designates a region that is bounded within the scope of predication in its primary domain.
    b. A 'mass noun' designates a region that is NOT specifically bounded within the scope of predication in its primary domain.

The modifier 'specifically' in (8b) relates to the fact that the nominal expression itself does not specify bounding, regardless of whether the mass evoked by the expression is bounded in reality. Consider the following examples:

(9) a. This scientist studies sand.
 b. Lily bought sand yesterday for her building project.

As these examples demonstrate, the mass noun *sand* can be used to refer to a mass that has no boundaries imposed on it by the context, as in (9a). In this case, the result is **generic construal**, which means that the noun designates 'sand in general'. However, mass nouns are often used to designate a mass that does have boundaries imposed on it by the context, as in (9b). In this context, Lily did not buy 'sand in general' but a specific amount, probably measured in sacks. The contrast between these two examples illustrates that the regions designated by mass nouns can be externally bounded. However, this does not affect their conception as unbounded masses.

As we saw in Chapter 15, however, unbounded masses can be construed as bounded masses, which is reflected in expressions like *one sand and two cements*. Langacker argues that bounding is only one parameter that distinguishes the regions designated by count and mass nouns. Other parameters include **homogeneity** versus **heterogeneity**, **expansibility** and **contractibility**, and **replicability**.

16.2.2 Homogeneity versus heterogeneity

As we saw in Chapter 15, homogeneity relates to whether a region consists of entities that are all alike (like *oxygen* or *water*, which have the property of homogeneity), or entities that are dissimilar (like *bicycle*, which consists of heterogeneous subparts including *wheel, frame, handlebars* and so on, and thus has the property of heterogeneity). While the entities that constitute the regions designated by mass nouns are typically homogeneous, the entities that constitute the regions designated by count nouns are typically heterogeneous. Of course, there are exceptions to this generalisation. While *bicycle* is a good example of a count noun that has heterogeneous structure, *pond* is not. Equally, while *water* is a good example of a mass noun that has homogeneous structure, *furniture* is not. Since Langacker's characterisation does not rest upon necessary and sufficient conditions, however, such exceptions are unproblematic, and indeed might be expected on the assumption that linguistic categories reflect the prototype structure of the conceptual categories that they evoke.

16.2.3 Expansibility and contractibility versus replicability

Expansibility and contractibility are properties of the regions designated by mass nouns. For example, *sand* can designate an entire desert or a single grain of sand, and *water* can designate a whole sea or a single drop of water. It follows that any subpart of the region designated by a mass noun is still an instance of that category: a grain of sand is still SAND. It is clear, then, that the property of expansibility and contractibility is interwoven with homogeneity and the absence of bounding. The same is not true for typical count nouns. If we contract a BICYCLE to its smallest subpart, we might get a cog or a spring or a screw: this is not still a *bicycle*. If we expand BICYCLE, we don't get more BICYCLE, because a bicycle has inherent boundaries. Instead, we get more bicycles. In other words, an increase to the region designated by *bicycle* results in what Langacker calls replicability. This is interwoven with bounding, as we saw in Chapter 15, and is reflected in the linguistic system by the fact that count nouns can be counted and pluralised, and can co-occur with the indefinite article.

16.2.4 Abstractions

It is important to point out that Langacker's schematic characterisation of nouns in terms of bounded or unbounded regions does not necessarily mean that nouns refer to physical objects. As we saw in Chapter 7, many domains do not relate to physical entities but to abstractions like LOVE, HOPE and HAPPINESS. Langacker does not have a fully developed theory of **abstract nouns**, but he does observe (1987: 207) that the fact that the count/mass distinction holds for abstract nouns suggests that these might also be characterised in terms of bounded/unbounded regions. For example, *hope* can be pluralised and can take the singular indefinite article (e.g. *She hasn't got a hope; her hopes and dreams*), while *happiness* cannot (**a happiness;* *happinesses).

Nouns like *hope* are called **deverbal nominalisations**, which means that they are nouns derived from verbs. These are argued to have a PROCESS (action rather than matter) as their base, and encode an 'episode' bounded in time by a beginning and a finish. Langacker (1987: 208) compares the count noun *jump* with the mass noun *jumping*. The count noun profiles a single episode of the process that makes up its base, while the mass noun, because it is unbounded in time, gives rise to a **generic** reading (*jumping is silly*).

Figure 16.2 summarises the conceptual properties that distinguish the regions designated by mass nouns and count nouns.

As we have noted previously (see Chapters 4 and 14), entrenched patterns of use give rise to **schemas** in Langacker's theory. The noun class schema is represented in (10). Langacker uses the term THING to represent the schematic conceptual content of the noun schema at the semantic pole, and because this is a maximally general schema, the content of the phonological pole is unspecified.

(10) [[THING]/[. . .]]

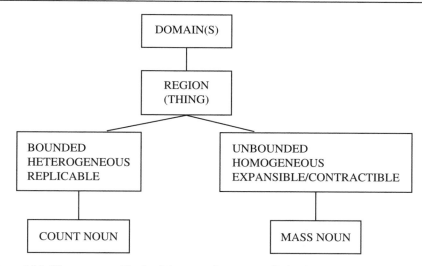

Figure 16.2 The conceptual basis of the count/mass noun distinction

16.3 Nominal versus relational predications

Turning to relational predications, recall our discussion of examples (1) and (2) above, which related to the extent to which a semantic characterisation of word classes is possible given the existence of pairs like *love* (V) and *love* (N) or *destroy* and *destruction*. The semantic similarity of pairs like these forms part of the argument by formal linguists against the possibility of a semantic characterisation of word classes. According to Langacker, however, the difference between *destroy* and *destruction* does not lie in their specific or content meaning. In this respect, Langacker's view is consistent with the formal view. Instead, Langacker argues that the difference lies in how each member of the pair construes and profiles that content meaning. Langacker summarises the difference between nominal and relational predications as follows:

> A nominal predication presupposes the interconnections among a set of conceived entities, and profiles the region thus established. On the other hand, a relational predication presupposes a set of entities, and profiles the interconnections among these entities. (Langacker 2002: 74–5)

Langacker illustrates this distinction by comparing the noun *group* with the adverb *together*. These expressions share the same conceptual content, which is represented in Figure 16.3(a). The circles represent the entities and the lines the interconnections. The noun *group* profiles the entities and the whole that they comprise (the region occupied). This is indicated by the bold type in

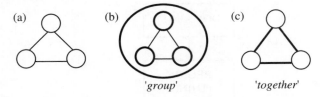

Figure 16.3 Nominal versus relational predication (adapted from Langacker 2002: 75)

Table 16.1 Trajector-landmark combinations in relational predications

Trajector (TR)	Landmark (LM)	Examples
THING	THING	*on, (to) love*
PROCESS	THING	*fast*
PROCESS	PROCESS	*before*
THING	PROCESS	*want, think*

Figure 16.3(b). In contrast, the adverb *together* profiles the interconnections between the entities and is thus a relational predication (Figure 16.3(c)).

It follows from this characterisation of nominal versus relational predications that while nominal predications designate a region, relational profiles designate an interconnection, which typically involves two or more entities. According to Langacker, there is always an **asymmetry** between the interconnected entities, and this asymmetry relates to **prominence**, which in turn relates to the TR-LM organisation. This idea was discussed in the Chapter 15. Langacker describes relations as **conceptually dependent** because they profile interconnections which cannot be conceived independently of the entities they connect. Furthermore, relational predications bring with them the schematic representation of the entities that they interconnect, which display a TR-LM asymmetry. Langacker (1987: 219) states that there are four possible patterns in terms of TR-LM combinations, which are summarised in Table 16.1.

We illustrate the four possibilities in example (11).

(11) a. the letter <u>on</u> the table
 b. Lily writes <u>fast</u>.
 c. Lily went out <u>before</u> George came in.
 d. Lily <u>thinks</u> George is a darling.

With the exception of (11b), these examples are self-explanatory. According to Langacker, the adverb *fast* in (11b) profiles a relation between a PROCESS (*writes*) and a THING, where the region on a scale of speed profiled by *fast* is construed as a THING and is implicit in the relational predication itself. We revisit the idea that adjectives and adverbs are characterised by implicit TRs below (section 16.4.2). As we saw in Chapter 15, Langacker divides relational predications into two subcategories: temporal relations and atemporal relations. We now look at each of these in more detail.

16.4 Temporal versus atemporal relations

In the previous chapter we briefly introduced Langacker's distinction between summary scanning and sequential scanning. Scanning is viewed as a type of cognitive processing that occurs in two distinct modes. In the summary scanning mode, the stimulus is scanned **cumulatively** which gives rise to a **static** cognitive representation. In Langacker's terms, atemporal relations (encoded by adpositions, adjectives, adverbs, infinitives and participles) fall into this category. It is important to emphasise that 'atemporal' does not mean that the linguistic expression is prohibited from making reference to the domain of TIME. On the contrary, many linguistic expressions that Langacker characterises as atemporal relations evoke the domain of TIME. Instead, the term 'atemporal' can be thought of as equivalent to **static in time**. Consider the examples in (12).

(12) a. George is late.
     b. George is annoying her now.

In (12a) the adjective *late* evokes the domain of TIME, as does the adverb *now* in (12b). However, neither of these expressions evokes a PROCESS. Instead, they construe time in terms of a property (12a) or a point (12b), both of which are static.

In the sequential scanning mode, on the other hand, the stimulus is scanned **sequentially**. Crucially, no two subparts of the resulting cognitive representation are the same, which gives rise to the cognitive representation with the status of a **process**. This scanning mode is evoked by temporal relations, which Langacker therefore calls PROCESSES. This is how Langacker characterises finite verb forms. As we have seen, language users (or conceptualisers) are not at the mercy of objective reality in terms of how they describe scenes using language. While objective reality exists, speakers have choices over how they decide to portray this reality in linguistic terms. In other words, summary versus sequential scanning is a matter of construal, although some situations lend themselves more readily to one type of construal than the other. Consider the examples in (13).

(13) a. Lily destroyed the letters secretively.
     b. Her destruction of the letters was secretive.

Example (13a) construes the scene as a PROCESS, and thus employs sequential scanning. In this example, *destroyed* is conceived as a dynamic PROCESS that is carried out in a certain manner, expressed by the adverb *secretively*. In contrast, (13b) construes the scene as a STATE, and thus employs summary scanning. Here, *destruction* is conceived as a THING that has the property expressed by the predicative adjective *secretive*.

16.4.1 Temporal relations: verbs

Langacker characterises finite verb forms (PROCESSES) in the following terms:

> A processual predication involves a continuous series of states . . ., each of which profiles a relation; it distributes these states through a continuous span . . . of conceived time; and it employs sequential scanning for accessing this complex structure. A process contrasts with the corresponding atemporal relation by having a 'temporal profile', defined as the span of conceived time through which the profiled relationship is scanned sequentially. (Langacker 2002: 81)

Although Langacker presents no direct psychological evidence that verbs are processed differently from other parts of speech, he does offer several motivations for this characterisation. Firstly, it captures the fact that verbs typically (although not always) express dynamic events. This in turn explains the 'temporal' nature of verbs, and explains why verbs are directly marked for time by means of the **tense system**, an issue to which we return in Chapter 18. Furthermore, Langacker argues that the schematic characterisation of verbs is in keeping with the objectives of the theory in the sense that 'conceptual content is less important than how this content is construed and accessed' (Langacker 2002: 81). In other words, recall that Langacker's objective is not to provide a specific semantic characterisation for prototypical nouns and verbs (although we may observe that this falls out from his analysis). Rather, his aim is to provide a schematic characterisation for all members of the word classes, which relies crucially upon independently established cognitive operations and conceptual representations. Langacker also argues that this theory achieves **descriptive adequacy** by distinguishing different kinds of relational predications and by enabling generalisations to be made about the behaviour of certain grammatical categories. A model of language achieves descriptive adequacy if it accurately models the tacit knowledge that underlies speaker intuitions about what is possible in language.

Simple and complex temporal relations

Temporal relations (processes) can be divided into two subcategories: **simple temporal relations** and **complex temporal relations**. Consider the examples in (14).

(14) a. Lily loves chocolate [simplex]
b. Lily is eating the chocolate [complex]

Both examples involve temporal relations (PROCESSES) because they construe scenes that hold over a given span of time (in both cases, the span of time includes the time of speaking, hence the present tense). The difference between the two examples is that while (14a) designates a **stative** PROCESS, (14b) designates a **dynamic** PROCESS. The terms 'stative' and 'dynamic' refer to types of lexical aspect. 'Stative' means that the situation remains constant throughout the time span, while 'dynamic' means that the situation involves some change over time. In (14a), for example, *love* designates a PROCESS that involves a stable and constant relation between the TR *she* and the LM *chocolate*. In (14b), on the other hand, *eat* designates a PROCESS that involves inherent change in the relation between the TR *she* and the LM *the chocolate*. The PROCESS of eating involves initial, medial and final stages, and at each of these stages the relation between the TR and the LM is different. Processes that involve no internal change are therefore described as 'simple', while processes that involve internal change are described as complex. As this discussion indicates, the TR–LM organisation that is evident in the structure of clauses emerges from the schematic TR–LM organisation that is part of the meaning of a verb, given that a verb expresses a relation (see Table 16.1).

We will have much more to say about the properties of verbs in the next chapter, given their central status in the clause. We return there to a fuller characterisation of tense and aspect, as well as looking in more detail at the nature of the relations that hold between TR and LM in temporal relations. For the time being, the class schema for verbs is represented in (15).

(15) [[PROCESS]/[. . .]]

16.4.2 Atemporal relations

Unlike the nominal and temporal categories, each of which characterises a single word class, the atemporal relation subsumes a range of word classes. These classes have two properties in common. Firstly, they profile a RELATION rather than a THING and are thus distinct from nouns. Secondly, as we saw earlier, the relation they profile is atemporal in the sense that it is cumulatively scanned and gives rise to a cognitive representation that is **static** in time. In this respect, atemporal relations are distinct from finite verb forms. However, in the same way that the sequentially scanned temporal relation can be simple or complex, the cumulatively scanned atemporal relation can also be simple or complex.

Simple and complex atemporal relations

A **simple atemporal relation** designates a STATE. Some examples are given in (16).

(16) a. That rocket scientist is <u>beautiful</u>.
     b. That <u>beautiful</u> rocket scientist.
     c. She writes <u>beautifully</u>.
     d. The letters <u>in</u> the sink.

The predicative adjective in (16a) describes a STATE, as does the attributive adjective in (16b). The difference between these two examples is that (16a) is a clause, where the adjective collaborates with the copular verb in forming the predicate of the clause. (We discuss this type of construction in Chapter 17.) In contrast, (16b) is a noun phrase that profiles a THING (*scientist*), and the adjective **modifies** the head noun. We return to heads and modifiers in the next chapter, but for the time being we can describe the attributive adjective as having a noun as its TR. In contrast, while the adverb in (16c) also describes a STATE, it modifies a verb, or takes a PROCESS as its TR.

A **complex atemporal relation** encodes a complex static scene. Compare the examples in (16) with those in (17).

(17) a. the sand all <u>over</u> the floor
     b. the last contestant <u>to reach the finishing line</u>

Observe that the preposition *over* in (17a) involves a **multiplex TR**. It follows that the relation encoded by this preposition is complex, because it profiles all the points in space at which the TR *the sand* and the LM *the floor* are related. In this example, the atemporal relation is still cumulatively scanned but gives rise to a more complex cognitive representation which consists of a 'bundle' of properties. A second example of a complex atemporal relation is the *to*-infinitive in the noun phrase in (17b). The base of this infinitival subordinate clause is a PROCESS, but due to summary scanning this expression is relational and atemporal and can therefore take on a modifying role, rather like an adjective. In other words, like an adjective, this infinitival subordinate clause has a noun (*contestant*) as its TR. We return to non-finite verb forms below.

Adjectives and adverbs

At this juncture, we might pause to consider how adjectives and adverbs are considered to be relations, given that they only seem to interact with a single participant. In other words, in examples (16a) to (16c) the adjective and the adverb only describe the state of a single entity or act: the rocket scientist or the act of writing. Given Langacker's claim that relational predications always have a prominent participant (the TR), we treat the rocket scientist or the act of writing as the TR in these examples. According to Langacker, the LM is **implicit** in the relational predications themselves. For example, we might paraphrase (16a) in terms of the scientist being 'in a state of beauty'. In this sense, the scientist is the TR and the LM is 'the state of beauty', which is part of the relational predication itself. While this is the typical case for adjectives, compare example (16a) with example (18).

(18)   That rocket scientist is fond of chips.

In this example, the predicative adjective *fond* participates in profiling a relation between two entities: *that rocket scientist* and *chips*. We use the expression 'participates' here because we have yet to establish what the copular verb contributes to the clause, a point to which we return in Chapter 17. Adjectives like these are sometimes described as 'transitive adjectives' because, like transitive verbs, they can take a complement. Other examples include *proud* and *envious*. It is also worth emphasising that an atemporal relation that profiles a STATE may well have a PROCESS as its base. For example, in the sentence: *That cup is broken*, the adjective *broken* profiles the end state in a PROCESS. This explains why past participle forms can often function as adjectives.

Adpositions

Example (16d) provides us with a more prototypical case of a relational predication. In this example, the preposition *in* profiles a spatial relation between the TR (*the letters*) and the LM (*the sink*). Furthermore, this is a simple atemporal relation because it describes a STATE.

Participles

Recall from our discussion of example (17b) that Langacker analyses non-finite verb forms as atemporal relations. As we saw in Chapter 14, participles are verb forms like *written* and *eating* that cannot occur as the main verb in a sentence, but require an auxiliary verb. This property of participles is illustrated by examples (19)–(21).

(19)   a.   Lily has written many love letters.
       b.   *Lily written many love letters

(20)   a.   That love letter was written by Lily.
       b.   *That love letter written by Lily

(21) a. Lily is writing a love letter.
    b. *Lily writing a love letter

In example (19a), *has* is the perfect aspect auxiliary and is followed by the participle *written*. In example (20a), *was* is the passive voice auxiliary, and is also followed by the participle *written*. In example (21a), *is* is the progressive aspect auxiliary and is followed by the participle *writing*. As the (b) examples show, the participles cannot occur as the main verb in a sentence without the relevant auxiliary. Participles are described as non-finite verb forms because they are not marked for tense. In each of the (a) examples in (19) to (21), it is the auxiliary verb that is marked for tense. In (19a), *has* is in its present tense form. Observe that if we change the auxiliary to the past tense form, the participle stays the same, which explains why it is described as non-finite:

(22) Lily had written many love letters.

Equally, in (20a), the passive auxiliary is in its past tense form, and in (21a) the progressive auxiliary is in its present tense form. The fact that participles are non-finite means that they can only occur without an auxiliary verb in subordinate clauses, where they often perform a modifying function. Compare example (20b) with example (23).

(23) The love letter written by Lily burst into flames.

In this example, the passive participle *written* heads an adverbial 'subordinate clause' which modifies *love letter*. The fact that it is a modifier explains why it can be removed from the sentence without affecting its well-formedness (*The love letter burst into flames*). The main verb in this sentence is *burst*, which is a finite (past tense) verb form. The subordinate clause describes a property of *the love letter* but profiles a STATE rather than a PROCESS. However, like the adjective *broken*, the passive participle *written* in (23) has a PROCESS as its base and profiles the end STATE in that PROCESS. In Langacker's model, participles are derived from PROCESSES by the affixation of the relevant morphology (*-ing*, *-en* and so on), and this has the effect of 'suspending the sequential scanning of the verb stem' (Langacker 2002: 82). This changes a PROCESS into an ATEMPORAL RELATION. We will have more to say about auxiliaries and participles in Chapter 18, where we return to a fuller discussion of tense, aspect and mood.

Infinitives

Infinitives occur in two forms. The *to*-infinitive is illustrated in example (24a). This is restricted to occurring in embedded clauses in English. The 'bare infinitive', which is the same as the *to*-infinitive minus the *to*, is illustrated in (24b) and (24c). As these examples show, the bare infinitive occurs after modal verbs (24b) and in imperative clauses (24c).

(24) a. She wants George to write her more love letters.
  b. He can write beautiful love letters.
  c. Write her one more love letter, George!

The infinitive is so called because it is another non-finite verb form: like participles, it is not marked for tense. If past tense forms are substituted for the infinitives in (24), the results are not well-formed.

(25) a. *She wants George to wrote her more love letters
  b. *He can wrote beautiful love letters
  c. *Wrote her one more love letter, George!

Often, it is difficult to recognise the bare infinitive in English because it takes the same form as most present tense forms. Observe, though, that it does not show agreement with the subject:

(26) *He can writes beautiful love letters

Langacker extends the same analysis to infinitives as to participles, viewing both types of non-finite verb form as atemporal relations. While the English bare infinitive is restricted to occurring with modal auxiliaries and in imperative clauses, the *to*-infinitive patterns in a similar way to participles, occurring in subordinate clauses.

As this section illustrates, while the notion of atemporal relations enables a characterisation of adjectives, adverbs, adpositions and non-finite verb forms, we should be cautions about viewing these word classes as discrete and mutually exclusive categories in Langacker's model. As atemporal relations, these word classes are characterised as members of one broad category whose properties may overlap. For example, we have seen that adjectives, adverbs and prepositions can all profile STATES, and we have also seen that expressions headed by different word classes can modify nouns. The examples in (27) illustrate the latter point.

(27) a. those <u>lovely</u> shoes
  b. those shoes <u>in Lily's wardrobe</u>
  c. those shoes <u>bought in haste</u>

In (27a), the noun *shoes* is modified by the adjective (phrase) *lovely*. In (27b), the same noun is modified by the preposition phrase *in Lily's wardrobe*. In (27c), *shoes* is modified by an adverbial subordinate clause headed by the passive participle *bought*.

Our final challenge in this section is to establish the schematic representation of the atemporal relation. This is shown in (28), where we represent the atemporal relation as STATE.

(28) [[STATE]/[. . .]]

16.4.3 Class schemas

Summarising Langacker's model of word classes so far, we have seen that a noun designates a THING and a verb designates a TEMPORAL RELATION (a PROCESS). We look at verbs in more detail in the next chapter. Nouns and verbs therefore comprise two of Langacker's major word classes. The third major class contains ATEMPORAL RELATIONS. An adjective designates an ATEMPORAL RELATION and has a THING as its TR, while an adverb designates an ATEMPORAL RELATION and has a RELATION as its TR. The relation can either be a PROCESS or an ATEMPORAL RELATION, since adverbs can also modify adjectives (for example, *incredibly funny*). The two subclasses adjective and adverb are 'special' in the sense that their LM is implicit in the relation itself. An adposition designates an ATEMPORAL RELATION that has its LM elaborated by the nominal predication that either precedes it (in the case of postpositions) or follows it (in the case of prepositions). Non-finite verb forms designate ATEMPORAL RELATIONS that have either a THING or a PROCESS as their LM, since these expressions can modify either nouns, verbs or clauses.

We have also established three basic class schemas, which are represented in (10), (15) and (28). These are represented by the diagrams in Figure 16.4, which summarise the schematic conceptual content of each of the three major categories.

In Figure 16.4(a), the circle represents the THING that a nominal predication designates. In Figure 16.4(b), the ATEMPORAL RELATION is represented as a line connecting TR and LM, which are part of the schematic representation of an ATEMPORAL RELATION. For example, if the ATEMPORAL RELATION is a preposition, the TR and the LM are the two nouns related by the preposition. In Figure 16.4(c), the TEMPORAL RELATION or PROCESS is also represented as a relation connecting TR and LM (we'll see more about how verbs do this in the next chapter). The crucial difference between the ATEMPORAL RELATION (STATE) and the TEMPORAL RELATION (PROCESS) is that the latter is specified as having a temporal profile. In other words, the PROCESS is sequentially scanned

COGNITIVE GRAMMAR: WORD CLASSES

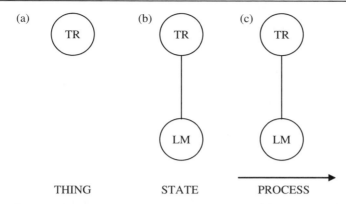

Figure 16.4 Conceptual representation of the three major word classes (adapted from Langacker 1987: 220)

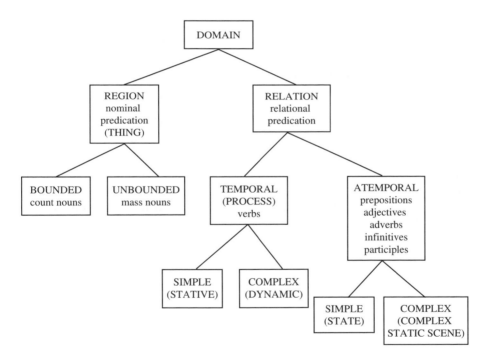

Figure 16.5 Langacker's model of word classes

through time; this is represented by the arrow in the diagram. We conclude this section with the diagram in Figure 16.5, a simplified version of which was originally introduced in Chapter 15. This diagram summarises Langacker's model of word classes.

16.5 Nominal grounding predications

The earlier parts of this chapter have focused mainly on the major categories or open word classes, although we have also seen how Langacker's model accounts for adpositions, which represent a closed class. It might be argued that adpositions represent a 'special' type of closed class. The closed classes are typically characterised not only by their relative resistance to change, but also by their lack of independent semantic content. As we saw in Chapter 14, for example, while it is relatively straightforward to draw a picture of *cat* or *happy* or *jump*, most people would struggle to draw a picture of *if* or *the*. Of course, this 'draw a picture' test vastly oversimplifies the conceptual content of linguistic expressions, but it serves to illustrate the distinction traditionally drawn between the open and closed classes. Against this background, prepositions present a striking case. While undeniably a closed class, many prepositions do have readily accessible semantic content. It would not be particularly difficult to draw a picture of *above, behind* or *under* (see Tyler and Evans 2003, for instance, where the **proto-scenes** relating to these prepositions are diagrammed), although cases like *of* present more of a challenge. From this point of view, it is unsurprising that prepositions fit rather well into Langacker's model of word classes, where word classes are characterised in terms of schematic meaning.

The question that naturally arises at this juncture is how Langacker accounts for some of the other much less 'contentful' closed word classes. In particular, the question is whether these categories can be integrated into the existing categories or whether they receive a 'special' account. Given that the cognitive model views grammatical elements as part of the same continuum as the open class elements, we might expect an integrated account, and this is the basis of Langacker's approach. Indeed, we have already seen that adpositions receive the same characterisation as open-class words like adjectives and adverbs in Langacker's model. In this section, we explore Langacker's account of determiners and quantifiers.

16.5.1 Determiners and quanitifers

As we saw in Chapter 14, determiners are words like *a, the* and *my* which form noun phrases with nouns. Some nouns require determiners in order to form noun phrases. This is the case for singular count nouns, as illustrated by example (29).

(29) a. Lily bought a book yesterday.
     b. *Lily bought book yesterday

On the other hand, plural count nouns (30) and mass nouns (31) can form noun phrases by themselves, which means that they occur optionally with determiners.

(30) a. Lily bought slippers yesterday.
　　 b. Lily bought those slippers yesterday.

(31) a. George bought champagne yesterday.
　　 b. George bought that champagne yesterday.

There are a number of categories of determiners which we briefly describe here.

Articles

The words *a* and *the* are described as **articles**, which can be **definite** (*the*) or **indefinite** (*a, some*). Definite articles are used when the noun phrase refers to something familiar to the speaker and hearer, perhaps because it has already been a topic of conversation or because it is salient in terms of shared knowledge. Indefinite articles are used when the speaker wants to introduce something new or unfamiliar into the conversation. While the English definite article does not inflect for number (for example, *the book, the books*), the indefinite article has two distinct forms (for example, *a book, some books*). When the word *some* is an indefinite article, it is unstressed. Its stressed counterpart is a **quantifier** (see below). As example (32) illustrates, a noun phrase headed by a plural count noun does not have to rely on the presence of a determiner to give rise to an indefinite noun phrase.

(32) a. Lily bought some wedding magazines yesterday.
　　 b. Lily bought wedding magazines yesterday.

Furthermore, noun phrases can be **generic** with either definite articles (33a), with indefinite articles (33b) or with no article at all (33c). Observe that this is the one situation in which (some) singular count nouns are licensed to occur without a determiner (33d). A generic noun phrase is interpreted as referring to a whole class of entities in general rather than a **specific** instance of that category.

(33) a. The cat makes an ideal childhood pet.
　　 b. A cat makes an ideal childhood pet.
　　 c. Cats make ideal childhood pets.
　　 d. Man is a complex species.

Demonstrative determiners

Demonstrative determiners are deictic expressions like *this* and *that*, both of which inflect for number:

(34) a. Give me back that letter!
 b. These letters are mine.

Possessive determiners

These are expressions like *my* and *your*. In English, these inflect for number and person features of the possessor rather than the possessed.

(35) a. My cat is sick.
 b. Your cat has fleas.
 c. Their cat is vicious.

Interrogative determiners

These are expressions like *which, whose* and *what*. These are used for asking questions, as illustrated by the examples in (36).

(36) a. Which rocket scientist goes out with the estate agent?
 b. Whose champagne is that?

Observe that the determiner *whose* in (36b) is possessive as well as interrogative.

Quantifiers

Finally, nouns can also be preceded by quantifiers. These are expressions like *any, no,* (stressed) *some, enough, every* and *each*, which quantify the noun in terms of number or amount. Some linguists treat quantifiers as a subtype of determiner because many quantifiers cannot co-occur with other determiners. The formal approach to identifying a determiner is to see whether it can co-occur with the articles, which are viewed as the prototypical determiners. The English noun phrase only permits a single determiner, so any element that meets the distributional criterion of forming a noun phrase together with a singular count noun and cannot co-occur with an article is classified as a determiner. This approach is illustrated by the following examples:

(37) a. *the my cat
 b. *the this cat
 c. *the which cat

Clearly, some quantifiers meet these criteria, as illustrated by the examples in (38).

(38) a. *those every cats
b. *the some cats

However, other quantifiers can co-occur with (precede) determiners, as shown by the examples in (39).

(39) a. <u>all</u> my friends
b. <u>both</u> my children

For some linguists, who favour a strictly distributional approach to word classes, these are described as **pre-determiners**. However, many linguists prefer to treat quantifiers as a separate class of their own on the basis of their semantic properties.

16.5.2 Grounding

We have already had a glimpse of Langacker's account of determiners in Chapter 15 where these were briefly discussed in relation to the notion of grounding. Recall that each speech event involves a ground, which consists of place and time of speaking, the participants in the speech event and the shared knowledge between them. As we saw, grounding is the process whereby linguistic expressions are linked to the ground, and determiners are one example of a grammatical element that serves this function. According to Langacker, determiners ground nominal expressions by profiling an instance of a category (*a cat*) and by indicating information such as whether participants are already familiar with the referent (*the cat*) or whether the referent is present in the immediate physical context (*that cat*). This explains why many of the determiner subcategories have **deictic** properties, particularly the demonstrative and possessive determiners, which encode spatial deixis and person deixis respectively. Like determiners, quantifiers also perform a grounding function by profiling the number or amount of the entity out of a larger mass. Expressions that perform a grounding function are called **grounding predications**, but these are not viewed as a distinct word class. Instead, grounding predications are seen as schematic categories for the class that they interact with. For example, Langacker (2002: 322) argues that 'the grounding predication of a nominal profiles a thing and is thus itself a schematic nominal'. In other words, the determiner or quantifier is represented not as a distinct category, but as a highly schematic noun phrase, inextricably linked to the category of nominal predications. This characterisation is consistent with the fact that the same determiner and quantifier forms can often function as pronouns, a common pattern cross-linguistically. This is illustrated by the examples in (40) and (41).

(40) a. Lily loves these slippers.
b. Lily loves these.

(41) a. Lily wants some champagne.
b. Lily wants some.

Langacker argues that the base of a grounding predication is a **grounding relation**, which is revealed by the fact that these expressions can be paraphrased in terms of atemporal relations which also reveal the schematic meaning associated with these closed-class elements. This idea is illustrated by the examples in (42). Observe that these paraphrases reveal that the base of a grounding predication like *my* is a relation between the nominal (X) and the speaker (*me*).

(42) a. the X   'X known to us'
b. this X   'X near me'
c. that X   'X far from me'
d. my X   'X belonging to me'

Langacker (2002) argues that the difference between the determiner on the one hand and the atemporal relations that paraphrase it on the other is an issue of construal rather than conceptual content. While the atemporal relation makes **explicit** the ground, which makes the ground a matter of **objective construal**, the profile of the determiner is restricted to the grounded entity. In the latter case, then, the ground is **implicit** and a matter of **subjective construal**. Furthermore, although the base of a grounding predication is a relation, the grounding predication itself profiles a schematic **grounded entity**. When the grounding predication combines with a noun, the noun elaborates the grounded entity and contributes its content meaning to the NP. The schematic representation of a nominal grounding predication is shown in (43).

(43) [[[GROUND]/[. . .]] / [[THING]/[. . .]]]

This schematic representation differs from the ones we have seen so far in its complexity. This is because it represents a schematic phrase rather than a schematic word. Of course, the question that arises at this point is how we can account in more detail for the nature of the relationships between words and phrases. This question is addressed in the next chapter.

16.6 Summary

In this chapter we explored the Cognitive Grammar model of **word classes**. We began the chapter by recalling Langacker's claim that open-class and closed-class expressions belong at different points on a **continuum** of symbolic units within the inventory that represents speaker knowledge of language. We saw that Langacker advocates a semantic characterisation of both open-class and closed-class expressions, where the former are characterised by content or **specific** meaning and the latter by **schematic** meaning. In this approach, linguistic expressions are divided into two major categories: **nominal predications** and **relational predications**. The former accounts for nouns, which are schematically characterised as THINGS. While nominal predications profile a region in some domain and can be described as **conceptually autonomous**, relational predications profile relations between those entities upon which they are **conceptually dependent**. Relational predications therefore have a schematic TR and LM as part of their representation. Relational predications divide into two subcategories: **temporal relations** and **atemporal relations**. The former accounts for finite verb forms which are schematically characterised as PROCESSES. Atemporal relations can be schematically characterised in terms of STATES and account for a number of word classes including adjectives, adverbs, adpositions and non-finite verb forms. Finally, we saw how Langacker's model can be extended to account for determiners and quantifiers, which are characterised in terms of their **grounding** function but which do not constitute an independent category. Instead, these are viewed as schematic nominals or noun phrases.

Further reading

Introductory texts

- **Radden and Dirven (2005).** Chapter 2 introduces nominal and relational profiles and presents a useful preliminary discussion of sentence structure, to which we return in Chapter 18. Chapter 4 elaborates the conceptual basis of nominal predications and discusses the count noun/mass noun distinction as well as the reification of abstractions. Chapter 5 provides an in-depth discussion of the role of determiners in grounding, and Chapter 6 focuses on quantification. Chapter 7 investigates the role of atemporal relations as nominal modifiers. Chapter 8 presents a detailed discussion of the aspectual distinctions between situation types which subsumes our discussion of temporal relations.
- **Lee (2001).** Chapter 8 provides an accessible introduction to the cognitive view of the count noun/mass noun distinction.

- **Taylor (2002).** A number of chapters of this excellent textbook provide more in-depth discussion of the material covered in this chapter. Chapter 9 presents a range of approaches to word classes and Chapter 11 maps out Langacker's theory of nominal and relational predications. Chapter 17 presents a taxonomy of symbolic units and Chapters 18–19 explore nominal predications in more detail.

Langacker's Cognitive Grammar

- **Langacker (1987).** Chapter 5 of this book focuses on nominal predications, and Chapter 6 on atemporal relations. Chapter 7 explores complex atemporal relations and processes.
- **Langacker (1991).** Part I of this book (Chapters 1–4) concentrates on nominal predications in greater descriptive detail than Volume I (Langacker 1987). Quantification is discussed in Chapters 2 and 3.
- **Langacker ([1991] 2002).** Chapter 3 of this book presents an overview of Langacker's theory of word classes and defines nominal and relational predications.

Exercises

16.1 Nominal predications

Consider the nouns listed below.

(a) group   (f) ice
(b) dot   (g) equipment
(c) wine   (h) grass
(d) swarm   (i) computer
(e) archipelago   (j) stone

On the basis of our discussion in section 16.2 (see Figure 16.2), classify each noun in terms of the following distinctions:

(i) bounded or unbounded;
(ii) heterogeneous or heterogeneous;
(iii) expansible/contractible or replicable.

On the basis of your classification, state whether each noun is a count noun or a mass noun. What kind of linguistic evidence can be used to support your conclusions? Finally, comment on any difficulties you encountered.

16.2 Relational predications

Consider the underlined expressions in the following examples, all of which represent relational predications.

(a) Lily <u>thought</u> George was <u>genuine</u>.
(b) <u>Eating</u> her toast, she <u>read</u> his letters <u>avidly</u>.
(c) She <u>slept</u> <u>with</u> his letters <u>under</u> her pillow.
(d) Those <u>lovely</u> letters <u>resembled</u> poetry.
(e) She <u>sprinkled</u> his aftershave <u>over</u> her pillow.
(f) She <u>wanted</u> <u>to marry</u> him.
(g) She <u>walked</u> <u>through</u> the park <u>in</u> a dream.

On the basis of our discussion in section 16.3, categorise these relational predications in terms of the following distinctions:

(i) temporal or atemporal;
(ii) simple or complex.

Now consult Table 16.1. How would you classify each of these relational predications in terms of the properties of their TR and LM? Finally, comment on any difficulties you encounter.

16.3 Nominal versus relational predications

Based on your answers to the previous two questions and on other relevant ideas in the previous two chapters, compare and contrast the following symbolic units:

(a) THING
(b) PROCESS
(c) COMPLEX STATIC SCENE
(d) team
(e) (to) cross
(f) across

16.4 Atemporal relations: prepositions

Recall from section 16.4.2 the idea that some adjectives like *fond* require both their TR and LM to be explicit, while others like *silly* do not. Now consider the underlined prepositions in the following examples.

(a) Lily tripped <u>over</u> her new slippers.
(b) George took her <u>to</u> hospital.
(c) She was <u>under</u> the anaesthetic for hours.
(d) George drank six pints <u>with</u> Billy.

For each example, identify the TR and LM, and state whether the preposition can or cannot occur without an explicit LM. Now collect a further four examples of each type (those that can occur without an explicit LM and those that cannot). Can you identify any patterns that explain your findings?

16.5 Grounding predications

Consider the following compound nouns. In these expressions, one of the nouns heads the compound, and the other is a modifier.

(a) rocket scientist
(b) estate agent

When these expressions occur with a grounding predication, only the head noun is grounded (recall our discussion of heads of phrases in Chapter 14) and may show agreement with the grounding predication. Investigate the agreement relations between grounding predications and each of these nouns, then (i) state which noun is the head and explain how patterns of agreement between grounding predication and noun reveal the head; and (ii) explain why you think the modifying noun is ungrounded.

17

Cognitive Grammar: constructions

In this chapter, we set out the Cognitive Grammar account of **grammatical constructions** and find out how the relationships between the component parts of complex words, phrases and clauses are accounted for. We begin by focusing on the nature of grammatical constructions at the phrase level (section 17.1) and at the word level (section 17.2). We explore the nature of the units that comprise grammatical constructions and the nature of the relationships between them. In traditional terms, this relates to the relationships between **heads** and **dependents**. As we will see, in Cognitive Grammar the head of a construction is described as the **profile determinant**, and the relations between the components of a construction are described in terms of conceptual **autonomy** and **dependence**, both of which are accounted for in semantic terms. The nature of **agreement** is also discussed. We will then see how autonomy and dependence give rise to clauses, and how **valence, transitivity, grammatical functions** and **case** are accounted for in Cognitive Grammar (section 17.3). In this section, we also look at **marked coding**, where the properties of **passive constructions** are analysed from a Cognitive Grammar perspective. In some respects, we take a rather traditional approach here, beginning with a discussion of the smaller grammatical units (words and phrases) before proceeding to discuss the more complex grammatical units (clauses). As we will see, there are sound reasons for approaching the Cognitive Grammar model in this way, because the properties of complex grammatical constructions are viewed in Cognitive Grammar as emerging from the properties of their components. However, in accounting for **constituency** (the internal structure of complex constructions), Cognitive Grammar emphasises not 'structure building' but the semantic relationships between the component parts of a complex structure.

17.1 Phrase structure

Recall from the previous chapter (Figure 16.1) that symbolic units are divided into simplex units and complex units in Cognitive Grammar. In this theory, it is only complex symbolic units that are called 'constructions'. In this section, we begin to explore these constructions. We approach the idea of a construction in Cognitive Grammar by looking at how words combine to make phrases, and find out how Langacker accounts for the relationships within the phrase that are traditionally described in terms of heads and dependents and in terms of **valence**. We will then come back to words in the next section and investigate how the Cognitive Grammar account of phrase structure can be extended to **morphological structure** or word-level constructions. While it may seem counter-intuitive to go from word classes (in the previous chapter) to phrases and then back to words again, our rationale for approaching constructions in Cognitive Grammar in this way is that it is often easier to think about relationships between words than to think about relationships between subparts of words. We therefore establish the Cognitive Grammar approach to phrases first and then apply the same line of reasoning to words.

In Cognitive Grammar a complex composite symbolic structure is a construction, which could be a complex word, a phrase or a clause. It follows that **constituency** – the combination of smaller subparts into larger, more complex units – is the result of the combination of symbolic structures. As Langacker (2002: 293) observes, 'in this regard, the only difference between morphology and syntax resides in whether the composite phonological structure . . . is smaller or larger than a word.' Most theories of grammar explicitly attempt to account for constituency, because for many theorists constituency represents a fundamental structural property of language. In Cognitive Grammar, constituency receives a semantic account in terms of TR-LM organisation.

For example, a phrase like *pink fish* brings together two semantic poles: *pink* designates (in other words, profiles) a subpart of the COLOUR SPECTRUM, and brings with it as part of its structure a schematic TR. This schematic TR is specified only as PHYSICAL OBJECT, which is a schematic instance of THING. In other words, part of the meaning of *pink*, which is an instance of the lexical class adjective, is that it relates to some entity, a TR, which is pink. While the TR is not specified, we know that pink is relational in this way (it has to be a property of something), which is part of what it means for *pink* to be an adjective. *Fish* designates a specific type of PHYSICAL OBJECT among its other far richer semantic specifications. The association of these two semantic poles within the phrase maps the semantically specific *fish* onto the schematic semantic TR of *pink*. At the phonological pole, the association of the two simplex symbolic units entails that they are pronounced sequentially, one after the other.

17.1.1 Valence

Let's now look in more detail at what happens inside phrase-level grammatical constructions. Grammatical constructions are **composite structures** consisting of **component structures** between which **valence relations** hold. The term 'valence' (or 'valency') usually refers to the number of participants a verb requires in order to complete its meaning. For example, a verb like *die* only involves a single participant, (for example, *He died*) whereas a verb like *love* involves two (for example, *Lily loves George*). More generally the term 'valence' can also be used to encompass all instances of what is traditionally described as the **head-dependent relation**, and this is the sense in which Langacker uses the term. These ideas are illustrated by Figure 17.1 which shows the structure of the PP *under the bed*. This diagram shows that the composite structure (PP) *under the bed* is comprised of the component structures *under*, *the* and *bed*, which are related by valence or the head-dependent relation (we explain the latter point in more detail below).

It is useful to revisit the traditional terms 'head' and 'dependent' before looking at how Langacker accounts for these phenomena in Cognitive Grammar. As we saw in Chapter 14, the 'head' of a phrase is a word-level constituent (a single word) that determines the categorical status of the phrase (for example, a noun heads a noun phrase). In addition, the head determines the core meaning of the phrase, and **selects** its dependents (the elements it co-occurs with inside the phrase). Consider the following example:

(1) a girl at the bus stop knitting a scarf

This is a noun phrase that contains three nouns: *girl*, the compound noun *bus stop*, and *scarf*. However, only one of these heads the phrase. The head of the phrase can be uncovered by our intuitions about what this phrase describes. It describes a kind of girl, not a kind of scarf or a kind of bus stop. These nouns

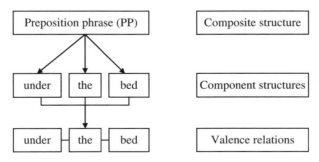

Figure 17.1 Composite and component structures

are parts of the dependents of the head. In addition, the head of a noun phrase is revealed by **subject–verb agreement**:

(2) The girl knitting scarves is over there.

Observe that it is the singular noun *girl* rather than the plural noun *scarves* that agrees with the verb, which is in the singular third person form *is* rather than the plural *are*. It follows that we can reduce the noun phrase to its head (plus determiner in the case of a single count noun) and preserve its basic import: *a girl*.

In traditional terms, dependents divide into two main categories: complements and modifiers. **Complements** are phrase-level units that 'complete' the head both in semantic and structural terms. For example, a preposition is often incomplete without the noun phrase that follows it, in which case the noun phrase is the complement of the preposition. **Modifiers**, on the other hand, are 'optional' phrase-level units that provide additional information of a more incidental kind. In example (1), *at the bus stop* and *knitting scarves* modify the noun *girl*. For any theory of grammar, then, it is necessary to model these phrase-internal relationships. It is worth observing that determiners and quantifiers can also be classified as a third kind of dependent within the noun phrase which generative linguists describe as a **specifier**, a dependent that specifies the applicability of the head (for example, to one specific referent or to a whole class of referents). Recall that we addressed the Cognitive Grammar account of determiners and quanitifiers in the previous chapter, where we saw that these are analysed in terms of **grounding predications** rather than dependents. For this reason, we will have little to say about the role of determiners and quantifiers within noun phrases in this chapter beyond their role in agreement (section 17.2.6).

In Langacker's model, there are four main factors that determine valence: (1) correspondence; (2) profile determinacy; (3) conceptual autonomy versus conceptual dependence; and (4) constituency. We address each of these below.

17.1.2 Correspondence

The term 'correspondence' relates to the idea that the component structures within a composite structure or construction share some common aspects of their structure. This sharing or correspondence arises from the ways in which the TR–LM organisation of the component structures interacts. For example, consider the preposition phrase *under the bed* from Figure 17.1. While the NP *the bed* is a nominal predication, the preposition *under* is a relational predication, which means that it only becomes fully meaningful when it relates two entities which are represented as part of its meaning in terms of a schematic TR and LM. There is a correspondence between the LM of *under* and the profile of *the bed*. The LM of *under* is a schematic representation of some THING in SPACE. The profile of *the sofa* 'fills in' or **elaborates** this schematic LM and the preposition phrase as a whole inherits its specificity or content meaning from the noun phrase. In the same way, a noun elaborates the schematic TR of an attributive adjective, as we saw earlier in our discussion of the NP *pink fish*.

17.1.3 Profile determinacy

Profile determinacy relates to which of the component structures determines the profile of the composite structure as a whole. Consider once more the preposition phrase *under the bed*. This construction contains *under*, which profiles a RELATION, and *the bed*, which profiles a THING, but the phrase as a whole *under the bed* profiles a RELATION rather than a THING in the sense that it describes a property of some entity in terms of its location in space. The composite structure has this meaning because the preposition is the profile determinant of the construction. Profile determinacy relates to the traditional grammatical term 'head', which determines the core meaning as well as the grammatical category of the phrase that it heads. In Cognitive Grammar terms, the profile determinant (the Cognitive Grammar counterpart of 'head') is the element that determines the profile of the entire phrase that it participates in. The term 'profile' of course relates to meaning, but since word classes receive a schematic semantic characterisation within Cognitive Grammar, the term 'profile determinant' also subsumes category or word class.

Now consider what happens if our PP *under the bed* occurs as a component structure of a yet more complex construction like *that slipper under the bed* (3). When the preposition phrase *under the bed* modifies a noun (like *slipper*), the profile determinant of the whole construction is the noun *slipper*, which means that the construction as a whole is a noun phrase, a construction that profiles a THING. At this point, it is useful to introduce labelled brackets which show the subparts of the construction.

(3) [$_{NP - THING}$ that slipper [$_{PP - RELATION}$ under [$_{NP - THING}$ the bed]]]

As this example shows, phrasal constructions have a 'layered' structure, where each 'layer' has a different profile determinant.

17.1.4 Conceptual autonomy versus conceptual dependence

In Cognitive Grammar, conceptual autonomy versus conceptual dependence explains the asymmetry that is traditionally described in terms of heads and dependents. Of course, these terms are already familiar from the previous chapter, where we saw that nominal predications are defined in terms of conceptual autonomy, whereas relational predications are defined in terms of conceptual dependence. This idea is extended to account for the relationships between the subparts of a construction. Langacker defines dependence in the following terms:

> One structure, D, is dependent on the other, A, to the extent that A constitutes an elaboration of a salient substructure within D. (Langacker 1987: 300)

This means that the component structure that provides the elaboration is conceptually autonomous (like *fish* in *pink fish*, or *the bed* in *under the bed*), while the structure that is elaborated is dependent, because it requires elaboration in order to become fully meaningful (like *pink* in *pink fish*, or *under* in *under the bed*). Langacker calls the schematic aspect of a component structure that is elaborated in a valence relation the **elaboration site**. As we have seen, there are two main types of dependent: complements and modifiers.

Complements

In Cognitive Grammar, a complement is a 'component structure that elaborates the head' (Langacker 2002: 297). In other words, when the dependent component is the profile determinant and the profile determinant is elaborated and thus dependent on the structure that elaborates it, we have what is traditionally described as a **head-complement structure**. For example, in a preposition phrase like *under the bed* the preposition *under* (the profile determinant) is dependent and its complement is the autonomous noun phrase *the bed* which elaborates its LM. In this conception of the head-complement relation, the complement is conceptually autonomous and the head or profile determinant is conceptually dependent because it relies upon the complement to elaborate its LM. In one sense, this is consistent with the traditional term 'complement', which means a constituent that 'fills out' or 'completes' the meaning of a head within a phrase.

Modifiers

In Cognitive Grammar, a modifier is a 'component structure that is elaborated by the head' (Langacker 2002: 297). In this case, the autonomous component is the profile determinant because the head does not require the modifier to complete its meaning, either because it is conceptually autonomous (the prototypical nominal predication) or because it is a relational predication (e.g. a verb) that already has its meaning completed by a complement. In contrast, the modifier is a relational predication that requires the head to elaborate some aspect of its schematic structure. This type of relationship gives rise to what is traditionally described as a **head-modifier structure**. For example, in the NP *that slipper under the bed*, the profile determinant *slipper* is autonomous and *under the bed* is dependent (having a schematic TR that requires elaboration). The head *slipper* elaborates the schematic TR of *under the bed*. This means that *under the bed* is a modifier rather than a complement. Table 17.1 summarises the Cognitive Grammar model of heads and dependents in terms of conceptual autonomy and conceptual dependence.

It is important to point out how Langacker's view of dependence differs from the traditional view. The traditional view is that complements and modifiers depend upon the head rather than the other way around. The term 'dependent' has its roots in a selection-based theory which is favoured by formal models. For example, the presence of a preposition entails the presence of a noun phrase, so the preposition is said to 'select' or 'subcategorise' for the noun phrase. Equally, the presence of a transitive verb entails the presence of an object which is a type of complement, so the verb is said to select or subcategorise for that phrase. This information is stored in the lexicon in 'selection frames'. An example of how the selection frame for *in* might look is given in (4). Of course, the lexical entry for this item would also contain further detail relating to meaning and pronunciation but these aspects do not concern us here.

(4) *in*   P   [_ NP]

This lexical entry says that *in* is a member of the category preposition and occurs in a syntactic context where it is followed by a noun phrase (the underscore represents the position of the preposition itself within the resulting structure: it precedes the noun phrase). In the selection model of head-dependent relations, both complements and modifiers are dependent upon the head for their presence in the structure: complements are selected and modifiers are added optionally to provide additional information about the head. In Cognitive Grammar, the head is dependent upon the complement to elaborate its schematic LM, but the modifier is dependent upon the head to elaborate some schematic aspect of its structure.

Table 17.1 Head-dependent relations in Cognitive Grammar

Complement	A conceptually autonomous component structure that **elaborates** the profile determinant, which is conceptually dependent. This gives rise to a head-complement structure.
Modifier	A conceptually dependent component structure that is **elaborated by** the profile dependent, which is conceptually autonomous. This gives rise to a head-modifier structure.

17.1.5 Constituency

The final factor that contributes to the Cognitive Grammar account of valence is constituency. This relates to the construction of progressively more complex composite structures. For example, consider again the NP *the slipper under the bed*. In traditional terms, the head of this noun phrase is *slipper* and the preposition phrase *under the bed* is a modifier. Within the preposition phrase *under the bed*, the preposition *under* is the head and *the bed* is its complement. In Cognitive Grammar terms, the NP's profile determinant *slipper* is autonomous while its modifier *under the bed* is dependent because it relies on *slipper* to elaborate its schematic TR. In contrast, the profile determinant of the PP *under the bed* is dependent because it requires the autonomous unit *the bed* to elaborate its schematic LM. These **constituents** or component structures together give rise to the composite grammatical construction *the slipper under the bed*, which is a nominal predication. In Cognitive Grammar, constituency is a feature of all constructions from complex words to phrases to clauses.

17.1.6 The prototypical grammatical construction

Of course, the four factors we have just discussed (correspondence, profile determinacy, autonomy versus dependence and constituency) are not independent properties of constructions. In particular, correspondence is closely linked to the properties of autonomy and dependence, which in turn give rise to constituency. These factors should therefore be viewed as interrelated aspects of what it means to define a grammatical relationship in terms of valence. Furthermore, Langacker argues that the four factors discussed here are not of equal importance to valence relations. He argues that correspondence is a central factor because it participates in every kind of valence relation. In contrast, not all composite grammatical structures have a readily identifiable profile determinant (consider the compound noun *puppy dog*, for example), and it is equally difficult to establish an autonomous and a dependent component in such constructions. For this reason, Langacker suggests that it is only meaningful to refer to a profile determinant in cases where there is a clear asymmetry between component structures: because *puppy* and *dog* each profile a THING and together profile a THING, it may not be useful to identify one or the other as profile determinant. Langacker (1987: 185) describes this type of valence relation in terms of **apposition**, which means that both component parts of a construction designate the same entity (see Taylor 2002: 235–8). Langacker also argues that constituency is not fundamental to valence because a given complex construction might be arrived at via various routes. In other words, the 'order' in which constructions are 'built' is not important in this model because of its usage-based nature. As we have seen, in Cognitive Grammar many complex constructions are stored as units, which in turn give rise to schemas. These schemas do not contain step-by-step 'instructions' for the composition of novel instances (for example, 'build head-complement structures before adding modifiers'), because the instances give rise to the schema and not vice versa.

According to Langacker, the prototypical grammatical construction involves two component structures, an idea that reflects the assumption that the combination of component structures into composite structures is binary in nature. In other words, regardless of the 'order' in which constructions are composed, their internal constituency tends to reflect 'layers' that can be described in terms of binary relations. For example, in the NP *that slipper under the bed*, the PP 'layer' *under the bed* involves a relation between P *under* and NP *the bed*, while the larger NP 'layer' *that slipper under the bed* involves a relation between N *slipper* and PP *under the bed*. In the prototypical grammatical construction, one of the component structures is a RELATION and the other a THING. The RELATION is dependent and is the profile determinant. The THING is autonomous and serves to elaborate the schematic aspect of the dependent unit's structure. This prototype represents the head-complement structure, which corresponds to the preposition phrase (PP) structure and, as we will see below, it also corresponds to the structure built around a verb and its arguments which gives rise to the clause (section 17.2). Observe that the head-modifier structure departs from the prototype despite the fact that it is a frequently attested structure. The head-modifier structure departs from the prototype because it involves a RELATION (the modifier) that is not the profile determinant. Langacker (1987: 326) accords the head-modifier relation the status of a 'secondary prototype'. The valence relation apposition (which we discussed above in relation to the expression *puppy dog*) represents a more extreme departure from the prototype, since it relates two autonomous THINGS, does not contain a RELATION and lacks a profile determinant.

17.2 Word structure

In this section, we will see how the notions of autonomy, dependence and elaboration can be applied to complex words. We concentrate the discussion in this section on bound grammatical morphemes. As we saw in Chapter 14, these are generally divided into two subcategories: (1) **derivational morphemes**, which are typically category changing (for example, *employ-ment*, which derives a noun from a verb); and (2) **inflectional morphemes**, which mark a grammatical subclass of the category (for example, *slipper-s*, which marks the plural subclass of the noun).

17.2.1 Phonological autonomy and dependence

In keeping with cognitive assumptions (recall our discussion of the Generalisation Commitment in Chapter 2), Langacker (2002: 291) does not assume a 'sharp dichotomy' between inflectional and derivational grammatical morphemes. In Cognitive Grammar, grammatical morphemes are maximally specific at the phonological pole, which means that they are specified for phonetic content. This is because the schema for PLURAL NOUN (or THING), for example, will have a relatively stable phonological realisation (regular plural nouns in English end in some allophone of -s), while the schema for NOUN (or THING) does not have a predictable phonological form. In Cognitive Grammar, affixes and non-segmental morphemes (for example, tones) are described as **phonologically dependent**. A **root** is the smallest phonologically autonomous unit within a composite structure, and a **stem** is defined as 'an autonomous phonological structure at any level within a word' (Langacker 1987: 345). For example, in the word *character-istic-ally*, *character* is the root as well as the stem to which *-istic* is affixed, and *characteristic* in turn forms the stem to which *-ally* is affixed. It is worth pointing out that Langacker's defininitions of the terms 'root' and 'stem' differ significantly from how these terms are traditionally defined (see, for example, Trask 1993, and compare the term 'base').

17.2.2 Semantic autonomy and dependence

At the semantic pole, grammatical morphemes have only schematic meaning. For example, the derivational morpheme *-er* has the information AGENTIVE NOUN (or THING) at the semantic pole, and the inflectional morpheme *-en* as in *broken* has CHANGE OF STATE PROCESS as its base and profiles the end result STATE. Because (most) derivational morphemes determine the category of the composite structure, they are the profile determinants. This means that category-changing derivational morphemes are themselves schematic instances of the word class that they derive. For example, *-er* (as in *driver*) is a schematic THING, *-ise* (as in *stigmatise*) is a schematic PROCESS, *-y* (as in *sticky*) is a schematic STATE, and so on. Langacker also extends this analysis to inflectional morphemes. For example, it is the plural morpheme *-s* that lends its profile PLURAL NOUN (or THING) to the composite grammatical construction; the inflectional morpheme is therefore the profile determinant and is itself a schematic NOUN or THING. Of course, the English derivational prefixes (e.g. *un-* in *unlikely*) are not category-changing, which suggests that they do not qualify as profile determinants. Taylor (2002: 274) suggests that these might best be analysed as modifiers, a point to which we return below (section 17.2.4).

Table 17.2 Properties of prototypical stems and affixes

Stem	Affix
Phonologically autonomous	Phonologically dependent
Has greater phonological 'weight'	Has less phonological 'weight'
Semantically autonomous	Semantically dependent
Semantically specific	Semantically schematic
Open-class elements	Closed-class elements

17.2.3 Prototypical stems and affixes

Langacker observes that autonomy and dependence tend to mirror one another at the semantic and phonological poles. In other words, if a unit is phonologically dependent it is likely to be semantically dependent as well, and if it is phonologically autonomous, it is also likely to be semantically autonomous. This means that it is often possible to describe whole symbolic units as autonomous or dependent. On the basis of the observations made thus far in relation to morphological structure, the properties of prototypical stems and affixes are summarised in Table 17.2.

Despite these patterns, there are exceptions to the generalisation that both poles of a symbolic structure will have the same status with respect to autonomy or dependence. Langacker provides the example of unstressed clitic pronouns attached to prepositions (for example, when *with her* is pronounced *with'er*, with main stress on the preposition). The clitic pronoun is phonologically dependent but remains semantically autonomous regardless of whether it has clitic status because it elaborates the preposition's schematic LM.

17.2.4 Composite structure

Langacker applies the same theory of valence to complex words as he does to phrases. We have already seen that stems and affixes are described in terms of autonomy and dependence, and that category-changing or inflectional grammatical morphemes can be analysed as profile determinants. Constituency or composite structure within words is thus accounted for in terms of correspondences between (semantically) autonomous and dependent units, and these correspondences are accounted for in terms of elaboration.

For example, the agentive nominal suffix *-er* is a semantically (and phonologically) dependent profile determinant which has a schematic PROCESS as its TR. This schematic PROCESS is elaborated by the stem (e.g. *teach*), which is semantically (and phonologically) autonomous. In several respects, the relationship between the dependent head and the autonomous stem is rather like the relationship between the dependent head and the autonomous complement within phrases, where the dependent head requires additional structure to complete its meaning. In contrast, the derviational prefix *un-* in *unlikely* or *unlovable*, while both semantically and phonologically dependent, is not a profile determinant. While this prefix also requires a semantically autonomous stem to elaborate its schematic TR (a schematic STATE), it does not head the resulting construction. In this respect, the prefix behaves more like a modifier within a phrase: it is a dependent unit that adds additional information to an autonomous head. As this discussion illustrates, word structure mirrors phrase structure in a number of important respects, an idea that is also influential in formal morphological theory (see Spencer 1991).

17.2.5 Constructional schemas

We are now in a position to address the schematic representation of constructions containing the grammatical morphemes we have discussed in this section. Two examples of constructional schemas that capture the properties of complex words are represented in (5). While (5a) represents the schema for agentive nouns (illustrating derivational morphology), (5b) represents the schema for plural nouns (illustrating inflectional morphology).

(5) a. Agentive noun (e.g. *driver, lover, singer*)
 $[_{THING} [PROCESS/. . .] [ER/-er]]$
 b. Plural noun (e.g. *slippers, scientists, canapés*)
 $[_{THING} [THING/. . .] [PLURAL/-s]]$

As we saw in Chapter 14, the information on the left of the slash represents the semantic pole and the information on the right represents the phonological pole. Each component unit is contained inside square brackets. For example, the constructional schema in (5a) says that the construction, which as a whole represents a schematic THING, consists of two component parts. The first component part is some member of the verb class, which is schematically represented at the semantic pole as PROCESS but which has no phonological specification because this is a generalised class schema. The second component part is a unit that also has a schematic representation at the semantic pole (ER represents the semantics of AGENTIVE NOUN) but has a specific representation at the phonological pole.

It is worth emphasising that in Cognitive Grammar constructional schemas do not belong in their own separate 'box' within the inventory of linguistic knowledge. In this theory, there is no principled distinction between the schemas that capture generalised patterns of structure and the specific instances that give rise to those schemas. The only difference lies in the extent to which the representation is semantically specific. Therefore the schemas belong within

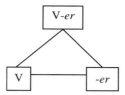

Figure 17.2 The agentive nominal schema (adapted from Langacker 1987: 446)

the same complex network that contains the instances. For example the schema for agentive nouns (5a) is connected within the network to specific instances, as well as to other schemas to which it is related (for example, the schema for plural nouns in (5b), given constructions like *lovers*), and ultimately to the noun class schema.

Sanctioning

In Cognitive Grammar, novel constructions are sanctioned (licensed) by schemas. Langacker argues that the schema is what enables the speaker to recognise a pattern and hence judge the well-formedness of a novel construction by analogy. For example, consider the following units which are recognised as pairs due to links within the network: *search – searcher; lecture – lecturer; examine – examiner; complain – complainer*. As we have seen, these and others like them give rise to the schema represented in Figure 17.2 which enables the speaker to derive a novel construction like *striver*. Of course, the idea that novel instances are sanctioned by schemas is not restricted to word-level constructions but can also be applied to phrases and clauses.

17.2.6 Grammatical morphemes and agreement

Langacker (2002: 301–2) suggests that one of the reasons why closed-class or grammatical elements are traditionally placed in a separate category from open-class or content elements is that grammatical elements often encode information that 'overlaps' with information already present elsewhere in the construction and are therefore viewed as 'semantically empty' because they do not contribute independent meaning to the construction. This observation relates to the issue of agreement. Recall from Chapter 14 that agreement relates features like person, number and gender. For example, if a noun is already marked as plural by the plural morpheme *-s* (for example, *slipper-s*), the presence of a plural demonstrative determiner that 'agrees' with the plural noun (for example, *those cats*) duplicates the same information. Equally, if the subject of a clause is marked for third person singular by the pronoun *he, she* or *it*, the presence of the third person singular suffix on the present tense verb form (for example, *she love-s*) also duplicates the same information.

Since agreement morphemes are (inflectional) grammatical morphemes, it follows from our discussion in this section that agreement morphemes are represented in Cognitive Grammar as independent symbolic units, which have independent but schematic meaning. Langacker (2002: 308) represents the agreement construction schema as follows:

(6) Agreement construction schema
[[[A/a][X/x]] [[B/b][X´/x´]]]

The elements [A/a] and [B/b] represent the words that carry the agreement morphemes, and [X/x] and [X´/x´] represent the agreement morphemes themselves. Recall that the information on the left of the slash represents the semantic pole and the information on the right the phonological pole. The substructures of this highly schematic schema can be instantiated by members of any word class. For example, in the plural noun phrase *those slippers*, [A/a] is instantiated by *those* and [B/b] is instantiated by *slippers*. [X/x] represents the plural feature of *those*. Observe that this is not a readily 'detachable' morpheme, because this word shows **fusional** morphology (this means that each of its features is not represented by a separate morpheme). [X´/x´] represents the plural morphology on *slippers*. The construction *those slippers*, an instance of the constructional schema in (6), is shown in (7):

(7) Instance of the agreement construction schema: *those slippers*
[$_{THING}$[$_{THING}$ [GROUND/ðəʊz] [PLURAL/Ø]] [$_{THING}$ [SLIPPER/slɪpə] [PLURAL/-s]]]

The semantic pole of the determiner is represented as GROUND because the determiner is a grounding predication. Its plural morpheme is represented as Ø because the plural determiner shows fusional morphology.

17.3 Clauses

Having set out how Cognitive Grammar accounts for the structure of words and phrases, we turn our attention in this section to the structure of clauses and sentences. We begin with a discussion of how valence operates at the clause level and compare clauses headed by prototypical content verbs with those headed by the copular verb *be*. We also briefly consider embedded clauses in the structure of complex sentences. We will then look at Langacker's account of transitivity, grammatical functions and case, and turn finally to the Cognitive Grammar analysis of passive constructions.

17.3.1 Valence at the clause level

As we have seen, valence is described in Cognitive Grammar in terms of correspondences between the component structures that make up a grammatical construction. These correspondences are accounted for in terms of autonomy, dependence and elaboration. As an illustration of how these ideas can be applied to the clause, consider the verb *see*. This verb expresses a temporal relation or PROCESS and has a schematic TR and LM as part of its representation. The schematic TR and LM are the elaboration sites. In a clause like (8), the NP *George* elaborates the schematic TR of *see* and the NP *his childhood sweetheart* elaborates the schematic LM.

(8) George saw his childhood sweetheart.

Because the verb relies on the two NPs to elaborate its schematic TR and LM, the verb is conceptually dependent and the two NPs are conceptually autonomous. As this example illustrates, the Cognitive Grammar account of the constituency of the clause rests upon the same assumptions as the account of word structure and phrase structure. The only difference is that the component parts of these grammatical constructions become increasingly complex as we move from word, via phrase, to clause. As we saw in the last section, the pattern of autonomy and dependence illustrated by example (8) represents the prototypical grammatical construction where a dependent RELATION (here, the verb) relies on an autonomous THING to elaborate some schematic aspect of its structure. While *see* is a 'typical' transitive verb and has two elaboration sites, intransitive verbs like *die* have only one elaboration site which corresponds to the TR, and ditransitive verbs like *give* have three elaboration sites. We return to each of these clause types in more detail below. As we will see in the next section, in Cognitive Grammar a **subject** is the unit that corresponds to the TR of the verb and the **object** is the unit that corresponds to its LM.

Recall that Langacker describes valency relations as binary, an idea that captures the 'layered' structure of complex constructions (section 17.1.6). This idea also accounts for the 'layered' structure of the clause. The dependent verb and the autonomous object combine to form a complex unit (VP), of which the verb is the profile determinant. In other words, the VP profiles a PROCESS. The object elaborates the schematic LM of the verb. This is an instance of the head-complement relation. However, the resulting PROCESS (VP) remains a dependent unit, since the verb still has a schematic TR that requires elaboration. The VP then combines with the subject which elaborates its TR. This also represents the prototypical valence relation between a dependent RELATION and an autonomous THING, despite the fact that it is not strictly speaking a head-complement relation since the dependent relation is itself a complex

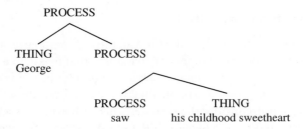

Figure 17.3 PROCESS as profile determinant of the clause

grammatical construction. In Cognitive Grammar terms, however, the valence relation between V and object NP is the same as the valence relation between VP and subject NP: in both cases, the profile determinant is a PROCESS, and in both cases that PROCESS requires the NP to elaborate some aspect of its schematic structure. The resulting construction still profiles a PROCESS, which means that the verb is the profile determinant or head of the clause as a whole, an idea that is central to many theories of grammar. The structure of the prototypical transitive clause is illustrated in Figure 17.3.

Of course, this account suggests that the construction of a clause is a 'step-by-step' process. However, recall that the usage-based model assumes that even clause-level constructions are stored as wholes if they represent well-entrenched units, and thus give rise to schemas that mirror their structure. From this perspective, viewing a construction in terms of the 'order' in which it is 'built' is rather meaningless. However, Langacker (1987: 319) suggests that the natural intonation pattern of a basic transitive clause supports the traditional partition of the clause into subject and predicate, or NP and VP, an idea that we discussed in Chapter 14.

Arguments versus modifiers

The participants that are required by the verb to complete its meaning are described as the **arguments** of the verb, which is the **predicate** or semantic core of the clause. (This sense of the term 'predicate' should not be confused with the traditional grammar sense of the term which refers to the verb phrase.) As we have seen, arguments of the verb are autonomous. Consider the examples in (9).

(9) a. Lily put the letters in the bin
 b. Lily saw George across a crowded room

In (9a), the verb *put* requires three participants to complete its meaning and therefore has three elaboration sites, a schematic TR (elaborated by *Lily*), a schematic LM (elaborated by *the letters*) and a schematic 'destination', also known as the **secondary landmark**. In Cognitive Grammar terms, the verb is conceptually dependent upon the arguments that elaborate its sites. Arguments are often given labels in terms of **semantic roles**. For example, the NP *Lily* that elaborates the schematic TR has the role AGENT, which describes an entity that acts with volition and intention. The NP *the letters* that elaborates the LM has the role MOVER, which is Langacker's term for an entity that undergoes a change of location. The relational unit (PP) *in the bin* elaborates the schematic destination or secondary landmark and has a LOCATION role. We discuss semantic roles in more detail in the next section, where we will see how they interact with the grammatical functions subject and object. While Langacker calls both the subject *Lily* and the object *the letters* in (9a) **nominal complements**, he calls the PP *in the bin* a **relational complement** because its profile determinant is a RELATION.

Example (9b) has a rather different structure from example (9a). This is because the PP *across a crowded room* does not elaborate any part of the verb's substructure. The verb *see* requires two arguments to elaborate its schematic TR, (*Lily*), and its schematic LM, (*George*). This means that the PP is a clausal modifier or adverbial. In other words, the verb is autonomous in relation to this PP. It is the modifier that is conceptually dependent in this type of relationship. The modifier relies on the verb plus its arguments, the clause-level PROCESS, to elaborate its schematic PROCESS TR. This explains the optionality of such modifiers in contrast to the obligatory presence of subjects and objects. Clause-level modifiers are not always PPs. They can also be adverb phrases (*very sincerely*), NPs (*these days*) or other clauses (*humming a happy tune*). What these all have in common is that they have a schematic PROCESS as part of their meaning which is elaborated by the (main) clause.

Copular clauses

Recall from Chapter 14 that some clauses do not contain a prototypical content verb. These are clauses headed by the copular or 'linking' verb *be*, which takes a subject predicative complement. Consider the examples in (10).

(10) a. Lily is [$_{NP}$ a rocket scientist].
  b. Lily is [$_{AP}$ drunk].
  c. Lily is [$_{PP}$ under the table].

As we saw in Chapter 14, the bracketed constituent in each example is called the subject predicative complement, because it defines or describes the subject, occurs as the complement of *be* and is predicative in the sense that it contributes the semantic core of the clause. Some grammars limit the term 'predicative complement' to the NP (10a) and the AP (10b), describing the PP (10c) instead as an adverbial complement because it expresses the same kind of meaning as an adverbial (for example in describing the place of the subject).

Given that most grammatical theories, including Cognitive Grammar, view the lexical verb as the 'heart' of the clause, questions arise concerning the formation of copular clauses. In some formal approaches, the copula *be* is treated as a semantically empty verb that does not have independent argument structure (in other words, does not behave as a predicate). This semantically empty verb licenses the combination of subject and predicate (NP, AP or PP) by enabling the formation of a clause, which it does by contributing finiteness (in English, main clauses have to be finite). From this perspective, the copula 'mediates' between subject and predicate by licensing a well-formed finite clause structure, complete with agreement, tense and so on. This entails that the copular verb has a subject and a complement in structural terms, even though the constituents in these positions are not semantically related to the copula. Instead, the subject and the predicative complement are semantically related.

The Cognitive Grammar account is reminiscent of the formal account in the respect that the verb *be* is described as maximally schematic. According to this analysis, the verb *be* designates a schematic stative PROCESS. However, the verb *be* has its own semantic structure in Cognitive Grammar: like a contentful lexical transitive verb, it has a schematic TR and LM. These are elaborated by subject NP and by either nominal (10a) or relational predications (10b)–(10c), respectively. In other words, *a rocket scientist* is a nominal complement of the verb *be* in (10a), while *drunk* in (10b) and *under the table* in (10c) are relational complements of the verb *be*. Langacker points out that the correspondences between the subparts of the copular construction entail that the subject is not only subject of the verb *be*, but also subject of each component part of the VP and subject of the VP as a whole. In this way, the Cognitive Grammar analysis captures the same intuition as the formal analysis concerning the predicative nature of the subject complement and the semantic relationship between subject and complement.

It is important to point out that the status of the bracketed units in (10) is rather different from the status of these units in other types of constructions. For example, while it is not unusual for a NP like *a rocket scientist* to occur either as subject or object of a clause headed by a lexical verb, there is an important difference between its occurrence in that type of construction and its occurrence as a subject complement. This difference relates to **grounding**. Consider the following examples:

(11) a. I met a rocket scientist the other day.
     b. A rocket scientist appeared on the news.

In (11a), *a rocket scientist* is the direct object. In (11b), the same NP is the subject. In both these examples the NP is grounded. In other words, the indefinite article *a* grounds the noun and the resulting NP and designates a specific individual, even though this individual is not familiar to the hearer. In cases like these, the NP is a referring expression, which means that the speaker uses the expression to pick out a specific individual in the world. In contrast, the NP *a rocket scientist* in (10a) is not grounded. An indefinite predicate nominal, in other words an indefinite noun phrase that functions as subject complement, is not a referring expression. Instead, it describes a property of the subject and in this respect might more accurately be characterised as a RELATIONAL PREDICATION than a NOMINAL PREDICATION (see Langacker 1991: 65–6). This idea captures its predicational role in the clause but raises questions about the nature of the autonomy-dependence relation between the predicate nominal and the copula.

Observe that if the predicate nominal is marked as definite, something interesting happens to the clause:

(12) a. Lily is the rocket scientist.
  b. The rocket scientist is Lily.

The definite NP *the rocket scientist* in (12a) is now grounded, which means that it now functions as a referring expression and picks out a particular individual that is equivalent to *Lily*. This type of copular clause is described as **equative** rather than **predicative** and is characterised by **reversibility** (12b). As Taylor (2002: 361–2) observes, it is not clear why the ungrounded predicate nominal in examples like (10a) should have an article at all in light of this discussion. Indeed, languages including French, German and Spanish omit the article in constructions like this, and some English expressions also license the absence of the article:

(13) a. She was Queen of England.
  b. George will be chair today.
  c. Lily was crayon monitor at school.

Observe that what the constructions in (13) share in common is that the predicate nominal designates a unique role: the expectation is that there is only one Queen of England, chairperson, crayon monitor and so on. Despite these differences between predicate nominals on the one hand and subjects and objects on the other, the NPs in all these functions share the same property of conceptual autonomy in relation to the verb (with the possible exception of the indefinite predicate nominal), hence their status as complements in Cognitive Grammar.

In contrast, an attributive adjective (14a) and a predicative adjective (14b), while both profiling ATEMPORAL RELATIONS, do not have the same status with respect to autonomy and dependence. The attributive adjective *daft* in (14a) is dependent in relation to the autonomous noun and therefore a modifier. The predicative adjective *daft* in (14b) is autonomous in relation to the dependent copula verb and is therefore a complement:

(14) a. that daft estate agent
b. That estate agent is daft.

It is worth observing here that not all adjectives can occur in both attributive and predicative positions (15). Furthermore, some that can occur in both positions take on a distinct sense in each position (16).

(15) a. ?The unwell estate agent
b. The estate agent is unwell.

(16) a. my late grandfather
b. My grandfather was late.

Embedded clauses

As we saw briefly in Chapter 14, clauses can function as subparts of complex sentences. When this happens, the (subpart) clause is called an **embedded clause** or a **subordinate clause**. This is illustrated by the examples in (17) where the embedded clauses are bracketed.

(17) a. Lily thought [that George was a dreamboat].
b. [That George was a dreamboat] was a well known fact.
c. the idea [that George was a dreamboat]
d. the love letters [that she burnt]
e. George could make [Lily cry].
f. Lily wanted [George to be happy].
g. [Falling for men like George] was not unusual.
h. [Wiping her eyes], she got on her bike.
i. [Worshipped by women everywhere], George went about his daily business.

While these examples do not represent every type of embedded clause that is possible in English, they do provide a representative sample. Although these constructions look very complicated, they actually represent construction types that we have already seen. This is because clauses can perform the same kinds of grammatical functions that smaller grammatical constructions like nominal expressions can perform. In (17a), the embedded clause is the object, while in (17b) it is the subject. In (17c), the embedded clause is the complement of a noun because it completes its meaning, while in (17d) it is a nominal modifier. This is called a relative clause and is identifiable as such by the fact that we can substitute *which* for *that* (*the love letters which she burnt*), which is not possible in (17d) (**the idea which George was a dreamboat*). In examples (17a) to (17d), the embedded clauses are all finite. This means that, with the exception of the relative clause, they are all capable of functioning as independent clauses. This means that they are PROCESSES.

In examples (17e) to (17i), the embedded clauses are non-finite. In (17e), for example, the embedded verb is in its bare infinitive form (compare **George could make Lily cries*), while in (17f) the embedded verb is in its *to*-infinitive form. In both of these examples, the embedded clause is the object. In examples (17g) to (17i), the embedded clauses contain participles and lack subjects. While in (17g) the embedded clause is the subject, in examples (17h) and (17i) the embedded clause is an adverbial clause or modifier. Recall from Chapter 17 that Langacker treats infinitives and participles as ATEMPORAL RELATIONS. We will leave it for the reader to work out the autonomous and dependent units in these examples (see exercise 17.4).

17.3.2 Grammatical functions and transitivity

Recall the **prototypical action chain model** that we introduced in Chapter 15. According to this model, the prototypical TR is the 'energy source' and the prototypical LM is the 'energy sink'. Langacker (2002: 208) describes the prototypical action in terms of what he calls the 'billiard-ball model'. This idea relates to the fact that we experience motion, that motion is driven by energy, and that while some entities have an inherent capacity for energy, other entities only receive energy from external entities. The 'billiard-ball' metaphor expresses the idea of energy transfer from one entity to another. According to this model, energy is transferred from AGENT to PATIENT, and results in a change of state for the PATIENT. Figure 15.18 is repeated here as Figure 17.4.

According to Langacker, the unmarked active **transitive** clause with third-person participants represents the prototypical action from a canonical viewpoint or perspective. This means that a sentence like *George tickled Lily* represents the prototypical action.

Figure 17.4 The prototypical action model (adapted from Langacker 2002: 211)

Grammatical functions and the transitive clause

This characterisation of the transitive clause has implications for how the **grammatical functions** subject and object are viewed in Cognitive Grammar. Although we have taken the terms 'subject' and 'object' largely for granted in the discussion so far, grammatical functions are a rather controversial issue in the debate between cognitive and formal approaches. Briefly, in rather the same way that formal approaches reject a semantic characterisation of word classes, they also reject a semantic characterisation of subject and object. Instead, these core grammatical functions are described in terms of **distribution** (for example, in English, the subject precedes the main verb and the object follows it) and **morphology** (for example, the subject pronoun is marked for nominative case while the object pronoun is marked for accusative case). We introduced the structural criteria for English subjects and objects in Chapter 14 (see Tables 14.2 and 14.3).

In contrast, as we mentioned earlier in the chapter, the grammatical functions subject and object (like the word classes) receive a schematic semantic characterisation in Cognitive Grammar. In the prototypical action chain, the subject, which elaborates the schematic TR of the verb, is characterised as the volitional energy source. The object, which elaborates the schematic LM of the verb, is the passive energy sink. The transfer of energy between the participants in a scene is described in terms of the action chain model. As we saw in Chapter 15, different participants in this action chain can be profiled, which has consequences for how the clause is structured. Consider the examples in (18), which reflect different construals of the same scene.

(18) a. George ignited the love letters with a match.
b. A match ignited the love letters.
c. The love letters ignited.

The act of igniting love letters involves an AGENT (*George*), a PATIENT (*the love letters*) and an INSTRUMENT (*a match*). In example (18a), each component of this action chain is profiled. The energy is transferred from the AGENT, *George*, via the INSTRUMENT, *a match*, to the PATIENT, *the love letters*. In (18b), on the other hand, only the INSTRUMENT, *a match*, and the PATIENT, *the love letters*, are profiled. Despite this, the AGENT is understood as part of the base (or scope of predication) of (18b) because we know that matches generally lack the inherent energy required for independent action. In (18c), only the PATIENT is profiled, but nevertheless the AGENT and the INSTRUMENT are understood as part of the base. The action chain that underlies all these clauses can be represented as in Figure 17.5, where the circles represent each of the participants and the arrows represent the transfer of energy.

Figure 17.5 Prototypical action chain (adapted from Langacker 2002: 217)

Figure 17.6 Action chain for (18a) (adapted from Langacker 2002: 217)

Figure 17.7 Action chain for (18b) (adapted from Langacker 2002: 217)

Figure 17.8 Action chain for (18c) (adapted from Langacker 2002: 217)

As we have seen, the difference between the clauses in (18) concerns which elements of the action chain are profiled. Langacker represents profiling in these diagrams with bold type. For example, because all the participants in the action chain are profiled in (18a), all parts of the diagram are in bold (Figure 17.6). Examples (18b) and (18c) are represented by Figures 17.7 and 17.8 respectively.

As these diagrams show, the subject of the clause in each case is the participant that is closest to the energy source out of the participants profiled. In other words, when the active clause profiles both AGENT and INSTRUMENT as well as PATIENT, the AGENT (as energy source) will be subject. When the clause profiles only INSTRUMENT and PATIENT, the INSTRUMENT will be subject and so on. A number of researchers, including Fillmore (1968), have proposed a **thematic hierarchy**, which makes predictions concerning the likelihood of a given semantic role occurring as subject of a clause. The hierarchy can be understood as a prototypicality scale, with prototypical subjects on the left and less prototypical subjects on the right. An example of a simple thematic hierarchy is given in (19). This hierarchy predicts that if a language permits any given semantic role in subject position (e.g. INSTRUMENT), it will also allow every semantic role to the left in subject position (PATIENT, BENEFACTIVE and AGENT).

(19) AGENT > BENEFACTIVE > PATIENT > INSTRUMENT > LOCATION

Of course, not all researchers agree about how many semantic roles there are, nor about what labels they should be given. Another point of debate concerns whether semantic roles should be viewed as semantic primitives or as discrete and bounded categories. Dowty (1991) is among those researchers who have attempted to explain semantic roles in terms of a PROTOTYPE model rather than as a set of discrete categories. Dowty (1991: 572) proposes an AGENT proto-role and a PATIENT proto-role. Each proto-role is characterised not by a set of necessary and sufficient conditions, but by a set of properties that a prototypical AGENT or PATIENT will display, whereas less prototypical AGENTS or PATIENTS will display only a subset of these characteristics. The prototypical AGENT is characterised by volition, sentience and movement and by causing a change of state to be effected in relation to another participant. In contrast, the prototypical PATIENT is characterised by being stationary relative to another participant, by being causally affected by another participant and by undergoing a change of state that may be incremental in nature. As the discussion above illustrates, Langacker's model captures the predictions stated by Fillmore's (1968) thematic hierarchy in terms of the 'billiard-ball' or transfer of energy model.

Of course, the examples we have seen so far encode physical acts and therefore energy transfer. These examples therefore lend themselves well to illustrating the 'billiard-ball' model. As Langacker observes, however, not all clauses can be so easily characterised in terms of energy transfer. Consider the examples in (20).

(20) a. She saw George across a crowded room.
 b. She thought about his lovely blue eyes.
 c. She loved him.

In these clauses, the subject *she* is not an AGENT but an EXPERIENCER. This semantic role describes a conscious and sentient participant who participates in mental or emotional rather than physical activity. Verbs of perception and cognition therefore have EXPERIENCER subjects. We can describe the object (*George; his lovely blue eyes; him*) as the STIMULUS. There is no sense in which the subject of these clauses acts with volition or transfers energy in the direction of the object. Despite this, Langacker (2002: 221) suggests that these clauses display the same asymmetry found in clauses describing the prototypical action. While the asymmetry in an action chain arises from the direction of the energy flow, the asymmetry in the EXPERIENCER–STIMULUS relation arises from the fact that the EXPERIENCER is conscious and sentient and is thus responsible for establishing mental 'contact' with the STIMULUS by creating a cognitive representation of the experience.

However, some clauses do not encode this asymmetry in terms of 'directionality' (either in terms of energy or 'mental contact'). Consider the examples in (21).

(21) a. Her childhood sweetheart resembles George Clooney.
 b. George Clooney resembles her childhood sweetheart.

These clauses are stative. Furthermore, the participants in the relations that they profile are reversible. Recall that this property is also characteristic of the equative copular clauses that we discussed in the previous section. These properties suggest that there is no inherent asymmetry between the participants in these relations. Despite this, the fact that we can reverse the clauses shows that it is possible to construe either participant as TR, which means that the clauses still encode TR-LM asymmetry. In this case, the asymmetry arises from the construal (which participant the speaker chooses to focus attention upon) rather than from the semantics of the verb. As this discussion shows, while the prototypical subject has properties like inherent energy and volition, not all subjects have these properties. The property that all subjects do share in Langacker's model is that they construe a given participant as TR. In the same way, the prototypical direct object is the PATIENT or 'energy sink', but not all objects can be characterised in this way. For example, in the sentence *George's jokes amuse Lily enormously*, the object *Lily* is an EXPERIENCER rather than a PATIENT. The property that all objects share in Cognitive Grammar is that they construe a given participant as 'second-most prominent' (Langacker 2002: 225). In other words, the object designates the most prominent aspect of the **ground** which is the **primary landmark**.

Intransitive clauses

Given the prototypical action chain, the subject is 'upstream' in terms of energy flow and the object is 'downstream' from the subject. It follows that a clause can have a subject but no object, but not vice versa. According to Langacker, this is because an object is only meaningful in relation to a subject, while a subject, as TR, is independently meaningful and can thus participate in processes where there is no second participant. Langacker suggests that this explains the properties of intransitive clauses. Here, the subject does not interact with a second participant in some PROCESS. Instead, the subject interacts with itself, as in (22a), or interacts with the ground by undergoing a change of state that 'changes the world' as in (22b). This explains why intransitive verbs, like predicative adjectives, still profile a RELATION.

(22) a. Lily slept.
 b. Lily's oven exploded.

Ditransitive clauses

We turn finally to ditransitive clauses, which are also called **double-object constructions**. Recall that these contain two objects, a direct object and an indirect object, as illustrated by example (23).

(23) George gave Lily a kiss.

The question that arises in relation to double-object constructions concerns how the **indirect object** (*Lily*) might be semantically characterised, given the schematic semantic account of subjects and direct objects adopted in Cognitive Grammar. Langacker (1991: 326) argues that a thematic characterisation is most appropriate for indirect objects, since this function shows a greater thematic consistency than the functions subject and object. In other words, Langacker suggests that this grammatical function might be characterised in terms of its semantic role, which means that it is associated with a less schematic (or more specific) meaning than subjects and objects. For example, verbs of transfer like *give* or *send* have an indirect object with the semantic role RECIPIENT (24a), while verbs of perception typically have an indirect object with the role EXPERIENCER (24b):

(24) a. George sent Lily the wrong love letter.
     b. George showed Lily his bank statement.

We return to ditransitive clauses in Chapter 20, where we will see that this type of construction has been of particular interest to researchers who take a constructional approach to grammar (particularly Goldberg 1995).

17.3.3 Case

As we saw in Chapter 14, case is often described as the grammatical feature that 'flags' the grammatical function of a word or phrase within a clause. This view of case is most widespread in descriptions of nominative or accusative case, although some other types of case such as locative or instrumental case (both found in Basque) are more transparently semantic in nature. As we have seen, the grammatical functions subject and object are traditionally viewed as 'purely grammatical' notions that cannot be semantically characterised because a wide range of different semantic roles can occur in the subject and object positions. It follows from this view that the types of case that 'flag' these functions are also 'purely grammatical' features of language, which cannot be semantically characterised but are seen to arise from purely structural factors within the clause. Indeed, in the generative model, nominative and accusative case in English receive a purely configurational characterisation. Nominative case is licensed in the subject position of a finite clause and accusative case is licensed in the complement position of lexical verbs and prepositions.

Correlated and uncorrelated case systems

Langacker proposes that case, like the grammatical functions subject and object, can be semantically characterised. According to Langacker, there are two types of case system. A **correlated** system is based on the relative 'degrees of prominence' of each of the participants. For example, nominative (subject case) and accusative (object case) in English might be viewed in these terms, where nominative or subject case corresponds to the TR and object or accusative case to the LM. An **uncorrelated** case system is based on semantic role archetypes rather than grammatical functions. For example, Basque has instrumental case and locative case, which are examples of case marking that rest on semantic roles rather than grammatical functions. In reality, most languages represent some combination of the two systems.

Correlated case systems: a case study

We focus our discussion here on two examples of correlated cases systems, since these arguably represent a greater challenge to a semantic account of case than uncorrelated systems. Langacker proposes a cognitive account of the typological difference between **nominative/accusative** case systems and **ergative/absolutive** case systems in terms of how the case system marks the relative degrees of prominence of each of the participants in the clause. Although English has a nominative/accusative case system, this is only evident in the personal pronouns, as we have seen. We will therefore illustrate this discussion with a comparison of two languages that mark case on noun phrases headed by common nouns as well as pronouns. These two languages are German, which has a nominative/accusative case system, and Basque, which has an ergative-absolutive case system. In order to simplify the comparison, the subject of a transitive verb is labelled A (for AGENT). The object of a transitive verb is labelled O (for object). The subject of an intransitive verb is labelled S (for subject). Clearly, a case system only needs to distinguish A and O (the subject and object of a transitive clause), since S and A cannot co-occur (a clause cannot simultaneously be transitive and intransitive) and S and O do not co-occur (an intransitive clause does not have an object). If a language marks S and A in the same way, but marks O differently, this is a nominative/accusative case system. This is illustrated by the following German examples. Observe that German marks case on the NP by marking determiners and adjectives with case morphemes, rather than the head noun (examples (25) and (26) are both from Tallerman 1998: 154–5).

(25) a. Der   gross-e   Hund   knurrt
         the.NOM   big.NOM   dog   growled
         '[The big dog]$_S$ growled'
    b. Der   gross-e   Hund   biss   den   klein-en   Mann
         the.NOM   big.NOM   dog   bit   the.ACC   small.ACC   man
         '[The big dog]$_A$ bit [the small man]$_O$.'

Like English, the German case system has one type of case for subjects (nominative), regardless of whether the clause is transitive (25b) or intransitive (25a), and another type of case for objects (accusative). In contrast, if a language marks the intransitive subject S and the object O in the same way (absolutive), but marks the transitive subject A differently (ergative), this is an ergative/absolutive system. This is illustrated by the following Basque examples:

(26) a. Gixona-k   liburûa   erosi   dau
         man-ERG   book.ABS   buy   AUX.3s
         '[The man]$_A$ has bought [the book]$_O$.'
    b. Gixonâ   etorri   da
         man.ABS   come   AUX.3s
         '[The man]$_S$ has come'
    c. Gixonâ   ikusi   dot
         man.ABS   see   AUX.1s
         '[I]$_A$ have seen [the man]$_O$.'

As these examples show, Basque is an SOV language. Example (26c) begins with the object because the subject is not expressed in this clause. Like many languages with a rich inflectional system (notice that the AUX word is marked with the person and number of the subject), the subject can be left out of the main clause as long as it can be retrieved from the context. Languages that license implicit subjects in main clauses are often described as **pro-drop** languages.

According to Langacker (2002: 247), there are two important similarities between the two types of system. Firstly, both systems encode the relative prominence of participants by distinguishing subject and object where these co-occur: in the transitive clause. Secondly, both systems reflect the asymmetry that Langacker characterises in terms of the action chain. The difference between nominative/accusative languages and ergative/absolutive languages, according to Langacker, can be characterised in terms of the 'starting point' each case system reflects. A nominative/accusative system 'starts' with the energy source (subject), hence both transitive and intransitive subjects are marked in the same way (nominative) and a distinct case is only necessary if a second 'downstream' participant is involved. In contrast, an ergative/absolutive language 'starts' with the relationship between the verb and its 'closest' argument. In an intransitive clause, this is the subject, but in a transitive clause this is the object. Hence, an ergative/absolutive system marks object and intransitive subject with the same case (absolutive), and a distinct case (ergative) is only necessary if a further participant is involved moving 'outwards' from the core of the clause. In intuitive terms, then, the nominative/accusative system works 'from the top down', while an ergative/absolutive system works 'from the middle out'. Of course, this account relies upon the assumption that a verb is most closely associated with its object, an idea that is reflected in the traditional partition of the clause into subject and predicate and an idea that remains prominent in most current theories of grammar. For example, derivational or 'structure building' generative theories 'build' the verb-complement structure before 'adding' the subject. As we have seen, Langacker's (2002: 172, 296) account of compositionality in a prototypical transitive clause reflects a broad agreement with this partition of the clause, despite the non-derivational nature of the Cognitive Grammar model.

17.3.4 Marked coding: the passive construction

So far, we have been discussing **unmarked** clause types. The passive construction is one example of what Langacker calls **marked coding**. 'Markedness' in this sense refers to the extent to which a given construction can be described as 'typical' or 'representative' of the grammar of a language. It is a widely held view that the active transitive declarative clause represents the unmarked clause type. Indeed, typologists classify languages in terms of word order patterns by looking at the properties of this clause type. For example, English is described as an SVO language because the active transitive declarative clause has subject, verb and object in that order (despite the fact that marked constructions like clefts may reflect a different order). Furthermore, transformational models within generative grammar have always taken the active transitive declarative clause as the 'underlying' structure from which other clause types are derived (see Chapter 22).

Of course, the question that arises here concerns how we might define 'typical' or 'representative' grammatical constructions. Typologists define markedness according to a number of parameters, including **distributional potential** (Croft 2003). A construction with greater distributional potential is unmarked in comparison with a construction that has a more restricted distributional potential. For example, this definition of markedness can be applied to **voice**: while most verbs can occur in the **active voice**, a more restricted set of verbs can occur in the **passive voice**. Therefore, active voice is unmarked while passive voice is marked. An asymmetry in terms of frequency of use is predicted and statistical corpus studies often form the basis of typological approaches to markedness. Langacker (2002: 226) characterises an unmarked construction as 'the most natural construal of an event on the basis of its conceptual content'. For example, the active transitive clause views the energy source as the figure, or most prominent participant. Passive clauses, in contrast, represent an alternative or marked construal of a given event. This is motivated by discourse goals: the speaker intends to draw the hearer's **attention** to a given participant by making that participant **prominent**. Compare the examples in (27).

(27) a. George deceived Lily.
b. Lily was deceived by George.

In example (27a) the AGENT is prominent (TR): this clause construes the event from the perspective of what George did. In example (27b), the PATIENT is prominent (TR): this clause construes the event from the perspective of what happened to Lily. The passive clause represents an instance of **TR-LM reversal**, so that the PATIENT is construed as the TR and realised as the subject of the clause while the AGENT is demoted to background status and realised as a dependent modifier. We return to the details of the passive construction in the next chapter, where we investigate the Cognitive Grammar account of the verb string.

17.4 Summary

In this chapter, we have explored two main themes in Cognitive Grammar. Firstly we looked at grammatical constructions and saw that Cognitive Grammar approaches **constituency** and **head-dependent relations** from the perspective of **valence** and by relying upon the idea of conceptual **autonomy** and conceptual **dependence**. Crucially, autonomy versus dependence is independent from the status of a given component as **profile determinant**, so that the latter notion only partially overlaps with the traditional notion of head. We saw that this model of constituency is held to account not only for **phrase structure** but also for **word structure**. We also briefly outlined the Cognitive Grammar account of **agreement**. We then looked at how autonomy and dependence give rise to clause level constructions, as well as accounting for the distinction between **complements** and **modifiers** at the clause level. In this section we also saw how **transitivity, grammatical functions** and **case** are semantically characterised in Cognitive Grammar by means of the **action chain** model. Finally, we briefly addressed **passive constructions** and saw that these are analysed in Cognitive Grammar in terms of **marked coding** which effects a **TR-LM reversal**.

Further reading

Introductory texts

- **Radden and Dirven (2005).** A number of chapters in this textbook are relevant to the present chapter. In particular, Chapter 2 provides a basic introduction to clause structure from a cognitive perspective and Chapter 11 looks at marked coding.
- **Lee (2001).** Chapter 5 of this textbook focuses on grammatical constructions but only the early part of Lee's chapter is relevant to the present discussion. He focuses mainly on the Construction Grammar approach, to which we return in Chapters 19 and 20.
- **Taylor (2002).** Chapter 12 of this textbook explores valence, autonomy and dependence and constituency. This chapter contains particularly useful discussion of the differences between complements and modifiers. Chapter 12 discusses grammatical constructions, valence, heads and dependents. Chapters 14–16 focus on morphological structure and Chapter 21 focuses on clause structure. Chapter 17 is also extremely useful as it presents a typology of symbolic units according to the properties of content/schematicity, autonomy/dependency, valence and complexity.

Cognitive Grammar

- **Langacker (1987).** Chapter 8 of this volume discusses valence, compositionality, autonomy and dependence. Chapter 9 concentrates on morphological structure.
- **Langacker (1991).** Part II of this volume is dedicated to clause structure. Chapter 4 includes a discussion of nominal inflection and agreement. Chapter 7 discusses transitivity and outlines a schematic characterisation of the major grammatical functions subject, direct object and indirect object. Chapter 8 focuses on marked coding and Chapter 9 discusses case systems. Chapter 10 discusses embedded clauses, along with other complex sentence types, in much more detail than has been possible here.
- **Langacker ([1991] 2002).** Chapter 3 looks at aspect and Chapter 4 focuses on the passive construction. Chapter 6 sets out Langacker's theory of valence. Chapter 9 discusses transitivity, grammatical functions and case, and includes a detailed discussion of semantic roles. Chapter 11 discusses constructional schemas, constituency, the head–dependent relation, grammatical morphemes and agreement.

- **Talmy (2000)**. Although the present chapter focuses on Langacker's Cognitive Grammar framework, it is important to emphasise that Langacker's model shares many important assumptions in common with Talmy's 'Conceptual Structuring System Model', as we saw in Chapter 15. Chapter 5 of this volume, for example, focuses on the implications of figure and ground for clause structure.

Background reading: semantic roles

- Dixon (1991)
- Dowty (1991)
- Fillmore (1968)
- Jackendoff (1987)

These sources provide a range of perspectives on semantic roles and argument structure.

Exercises

17.1 Phrase structure

Consider the following phrases:

(a) post the letters in the morning
(b) fond of chips
(c) the demolition of Lily's flat

In each case, begin by bracketing phrases within phrases. Then, for each phrase that you have identified, (i) identify the profile determinant and label the phrase accordingly (e.g. NP_{THING} or $PP_{RELATION}$); (ii) work out which are the autonomous and dependent elements; and (iii) explain what elaborates what. Finally, in the light of your findings, comment on the status of the prepositions in these examples.

17.2 Word structure: stems and affixes

Consider the following complex words:

(a) tickled
(b) undress
(c) boorish
(d) transmit
(e) astronaut

Recall the properties of prototypical stems and affixes summarised in Table 17.2. In the light of these properties, (i) identify stems and affixes in these complex words; (ii) identify profile determinant; (iii) work out which are the autonomous and dependent units; and (iii) explain what elaborates what. Finally, how does Langacker's (1987: 359) observation that 'the stem/affix distinction may not always be clear-cut' reflect upon these examples?

17.3 Word structure: compound nouns

In the last exercise of the previous chapter, we looked at the two compound nouns *rocket scientist* and *estate agent*. One of the tasks was to identify the head. In this chapter, we saw that it is not possible to identify the head in some compounds (like *puppy dog*), which are described in terms of apposition. Consider the following examples of compound nouns:

(a) boyfriend
(b) bluebottle
(c) angel face
(d) fighter pilot
(e) hatchback

In each case, state whether the compound represents a case of apposition or whether it is possible to identify a head or profile determinant. If the latter, what type of valence relation holds between the two elements? Explain how you reached your conclusions.

17.4 Autonomy and dependence: embedded clauses

Recall our discussion in section 17.2.1 of the complex sentences in example (17) which are repeated here. Work out the autonomous and dependent units in these constructions and explain what elaborates what.

(a) Lily thought [that George was a dreamboat].
(b) [That George was a dreamboat] was a well known fact.
(c) the idea [that George was a dreamboat]
(d) the love letters [that she burnt]
(e) George could make [Lily cry].
(f) Lily wanted [George to be happy].
(g) [Falling for men like George] was not unusual.
(h) [Wiping her eyes], she got on her bike.
(i) [Worshipped by women everywhere], George went about his daily business.

17.5 Double object constructions

In this chapter we saw that Langacker characterises indirect objects in terms of the semantic role RECIPIENT or EXPERIENCER. Consider the following examples of double object constructions:

(a) George accidentally gave Lily a black eye.
(b) George promised Lily a packet of crisps.
(c) Lily knitted George a jumper.
(d) George gave Lily a mean look.
(e) Lily sang George a lullaby.
(f) George refused Lily a kiss.

For each of these examples, identify the indirect object and state whether the semantic role RECIPIENT or EXPERIENCER best describes the semantic properties of the indirect object. Do some indirect objects represent more prototypical RECEIPIENTS or EXPERIENCERS than others? Comment on what might explain this pattern.

17.6 Marked coding: the passive construction

In this chapter (section 17.3.4), we observed that only a subset of transitive verbs can be passivised. Consider the following examples:

(a) George was admired (by women everywhere).
(b) Lily was tickled mercilessly (by George).
(c) The truth about their relationship was known (by Billy).
(d) Adonis was resembled (by George).

Rank these examples in terms of their acceptability. What kind of pattern emerges from your findings? How might this be accounted for in Cognitive Grammar?

18

Cognitive Grammar: tense, aspect, mood and voice

We continue our investigation of the Cognitive Grammar account of the clause in this chapter by focusing on the **verb string**, a central feature of the English clause. In the last chapter, we set out the Cognitive Grammar account of grammatical constructions, and looked at how the relationships between the lexical verb and its dependents are captured in terms of autonomy and dependence. We now focus more closely on both the structural and semantic properties of the verb group within the clause. We begin with a short section that summarises the properties of English verb forms, which are central to the discussion in this chapter (section 18.1). We then explore what Langacker calls the **clausal head**, which is a string of verbs that can include a **perfect construction**, a **progressive construction** and a **passive construction**, as well as the **content verb** (section 18.2). As we see in this section, Cognitive Grammar views **auxiliaries** *have* and *be* as semantically related to non-auxiliary functions of the same verbs, as well as suggesting a unified analysis for the range of forms that show 'past' participle morphology. We then look at how **tense** and **mood** are analysed in terms of a **grounding predication**, and receive a semantic account in terms of the **epistemic model** (section 18.3). We also see that the **polysemy** of **modal verbs** can be accounted for in force dynamic terms. Finally, we look in detail at Langacker's account of **lexical aspect** in verbs, which Langacker describes in terms of two broad categories: **perfective PROCESSES** and **imperfective PROCESSES** (section 18.4). This aspectual distinction is accounted for in a similar way to the **count** and **mass** noun distinction: in terms of the nature of the component parts of the PROCESS and in terms of bounding.

18.1 English verbs: form and function

Recall from Chapter 14 that lexical verbs like *adore* are open-class elements, while auxiliary verbs belong to a closed class. We also saw in Chapter 14 that the English auxiliaries can be divided into two subcategories: modal auxiliaries and primary auxiliaries. Modal auxiliaries like *can* and *must* are responsible for introducing epistemic mood (relating to knowledge) or deontic mood (relating to obligation or permission) into the clause. The primary auxiliaries *have* and *be* introduce grammatical aspect and passive voice: while *have* introduces perfect aspect, *be* introduces either progressive aspect or passive voice. Each type of auxiliary requires the verb that follows it to occur in a certain form: the modal requires a bare infinitive (e.g. *must write*); the perfect auxiliary requires a 'past' or perfect participle (e.g. *have written*) and the passive auxiliary requires the same form (e.g. *be written*). The progressive auxiliary requires a 'present' or progressive participle (e.g. *be writing*). As we mentioned in Chapter 11, the traditional labels 'past' and 'present' participle are rather misleading because participles are not finite verb forms, which means they can occur in past, present or future contexts.

As these examples indicate, each English verb has a number of different grammatical forms. For example, the verb *write* has five morphologically distinct forms, which represents the typical case for English verbs: (*to*) *write, writes, wrote, writing, written*. The verb *go* also has five forms: (*to*) *go, goes, went, going, gone*. In contrast, the verb *put* only has three morphologically distinct forms: (*to*) *put, puts, putting*. The list for *be* is the longest with eight distinct forms: (*to*) *be, am, are, is, was, were, being, been*. Table 18.1 summarises the prop-

Table 18.1 English verb forms

Infinitive	(to) be	(to) write	(to) go	(to) sing	(to) put
1s present	am	write	go	sing	put
2s present	are	write	go	sing	put
3s present	is	writes	goes	sings	puts
1pl present	are	write	go	sing	put
2pl present	are	write	go	sing	put
3pl present	are	write	go	sing	put
1s past	was	wrote	went	sang	put
2s past	were	wrote	went	sang	put
3s past	was	wrote	went	sang	put
1pl past	were	wrote	went	sang	put
2pl past	were	wrote	went	sang	put
3pl past	were	wrote	went	sang	put
Progressive participle	being	writing	going	singing	putting
Past participle	been	written	gone	sung	put

erties of these forms. Each distinct form is marked once in bold type. As this table shows, a single verb form in English is typically compatible with subjects that reflect a wide range of different person and number features, exceptions being the third person singular present tense form, and the richer set of forms representing the verb *be*. This explains why English does not usually license implicit subjects in main clauses (because the person and number features of the subject are not usually marked on the verb). Recall that only past and present tense forms are finite (marked for tense), while infinitives and participles are non-finite.

18.2 The clausal head

According to Langacker (1991), the traditional partition of the verb string into auxiliary verb(s) on the one hand and lexical verb on the other does not correctly reflect the semantic division of labour within the verb string. While it is clear that auxiliary verbs have a number of properties that distinguish them from lexical or content verbs, Langacker proposes that the verb string should be partitioned into **grounding predication** and **clausal head**. The grounding predication is the part of the verb string that is responsible for finiteness. In English, this is either the first element in the verb string (a modal verb) or is attached to the first element in the verb string (a tense morpheme). The remainder of the verb string, including any other auxiliary verb(s) together with the content verb, makes up the clausal head. This is illustrated by example (1).

(1) George   [must]   [have been singing]   the blues
               GROUNDING PREDICATION   CLAUSAL HEAD

Recall that if the sentence contains a modal, as in example (1), none of the other verbs in the string are finite (marked for tense). In the absence of the modal, the first verb in the string is finite. In example (2a), the perfect auxiliary *have* is finite (present tense) while in example (2b) the progressive auxiliary *be* is finite (past tense). If the lexical or content verb is the only verb in the string, this verb is finite; in (2c) the lexical verb *sing* occurs in its past tense form.

(2) a. George has been singing the blues.
     b. George was singing the blues.
     c. George sang the blues.

As the examples in (2) demonstrate, it is not always possible to separate a tense morpheme from the verb. This is because some English verb forms mark grammatical distinctions by vowel changes (ablaut). Cases where the tense morpheme can be separated from the verb are the third person singular present tense form of most verbs (*walk-s*), and the past tense form of some verbs (*walk-ed*). In this section, we concentrate on Langacker's account of the clausal head. We return to the grounding predication in the next section.

Within the clausal head, the lexical or **content verb** provides the content meaning. The leftmost verb functions as the profile determinant for the entire clause; Langacker calls this verb the **grounded verb**, because it is under the direct control of the grounding predication. For example, as we saw in (2), the leftmost verb, lexical or auxiliary, is marked for tense in the absence of a modal verb. Langacker's (1991: 198) representation of the organisation of the complex clausal head in English is represented in (3a).

(3) a. [*have* [PERF$_4$ [*be$_1$* [*-ing* [*be$_2$* [PERF$_3$ [V]]]]]]]
    b. George must [have been being stalked].

In the representation in (3a), *have* is the perfect auxiliary and PERF represents the 'past' or perfect participle morphology (for example, *-ed* or *-en*). The subscripts represent the different senses of the perfect participle in perfect constructions (PERF$_4$) and passive constructions (PERF$_3$), which are elaborated below. Equally, the subscripts on the *be* auxiliaries indicate the different senses of this verb in progressive constructions (*be$_1$*) and passive constructions (*be$_2$*). The morpheme *-ing* represents the 'present' or progressive participle. Finally, V represents the content verb. Observe that the modal is not included in this representation of the verbal complex. Recall that this is because, if the clause contains a modal verb, it functions as the grounding predication. The example in (3b) illustrates a verb string that contains all the elements in this complex clausal head (bracketed). The verb *have* is the perfect auxiliary; the verb *been* is the perfect participle (PERF$_4$) form of the progressive auxiliary (*be$_1$*). The verb *being* is the progressive participle (*-ing*) form of the passive auxiliary (*be$_2$*). Finally, the verb *stalked* is the 'past' participle (PERF$_3$) form of the lexical verb (V), which is required by the passive auxiliary. As we explain below (section 18.2.1), this verb form, which we call the 'passive participle', is semantically related to the perfect participle in Langacker's analysis, hence its 'PERF' label.

As this example shows, the elements that make up the complex clausal head (3a) do not occur separately in the verb string, but are 'glued together' by morphology. Of course, not all clausal heads are as complex as the example in (3). Some clauses might just contain a single finite lexical verb. We look at each of the component parts of this complex clausal head in more detail below.

As Langacker (1991: 199) observes, a number of striking patterns emerge from the representation in (3a). To begin with, moving from right to left, the elements alternate between phonologically autonomous and phonologically dependent units. For example, the content verb is phonologically autonomous while the 'past' participle morphology is a dependent (affixal) form. The passive auxiliary is phonologically autonomous while the progressive participle morphology *-ing* is a dependent (affixal) form, and so on. Secondly, recall from Chapter 16 that participles, as non-finite verb forms, are classified as ATEMPORAL RELATIONS. This means that the phonologically dependent forms in (3a) (participial morphemes) have ATEMPORAL RELATIONS at their semantic poles. In contrast, Langacker argues that the phonologically autonomous forms (the auxiliary verbs *have* and *be*, together with the content verb) have TEMPORAL RELATIONS (PROCESSES) at their semantic poles. In other words, the Cognitive Grammar analysis of the primary auxiliaries *have* and *be* relates these semantically to their non-auxiliary (lexical) counterparts. Thirdly, at each 'level' in the increasingly complex clausal head construction, the leftmost element functions as the profile determinant, so that the constructions within this representation also alternate between PROCESS and ATEMPORAL RELATION, which in turn has consequences for how the complex construction can function within a larger construction. While a construction with the status of PROCESS can function as a clausal head, a construction with the status of ATEMPORAL RELATION can modify a noun. The possibilities are illustrated in Table 18.2.

Observe that the clausal examples all contain a modal verb. This makes it easier to show the clausal head independently from the grounding predication. As this pattern illustrates, the only example that is not well-formed concerns the penultimate combination, where $PERF_4$ is restricted to co-occurring with the perfect auxiliary *have*. As Langacker observes, this model of the verb string explains why auxiliaries and their participles have to occur in pairs within a clause: *have* + V-$PERF_4$ make a **perfect construction** (e.g. *have betrayed*); be_1 + V-*ing* make a **progressive construction** (e.g. *be betraying*); and be_2 + V-$PERF_3$ make a **passive construction** (e.g. *be betrayed*). The reason why these elements have to occur in pairs within the clause is because each construction headed by the participle (without its auxiliary) has the status of an ATEMPORAL

Table 18.2 Clausal head complex (based on Langacker 1991: 198–9)

Example	Construction	Relation type
George will **betray** Lily	[V]	PROCESS
a **betrayed** lover	[$PERF_3$ [V]]	ATEMPORAL RELATION
Lily should not **be betrayed**	[be_2 [$PERF_3$ [V]]]	PROCESS
any lover **being betrayed**	[-*ing* [be_2 [$PERF_3$ [V]]]]	ATEMPORAL RELATION
Lily must **be being betrayed**	[be_1 [-*ing* [be_2 [$PERF_3$ [V]]]]]	PROCESS
*a lover **been being betrayed**	[$PERF_4$ [be_1 [-*ing* [be_2 [$PERF_3$ [V]]]]]]	ATEMPORAL RELATION
Lily must **have been being betrayed**	[*have* [$PERF_4$ [be_1 [-*ing* [be_2 [$PERF_3$ [V]]]]]]]	PROCESS

RELATION. While this can occur as modifier, it cannot occur as a clausal head. Therefore the auxiliary verb is required to contribute its own profile to the construction, which then has the status of a PROCESS and can head a clause. Of course, as we will see in more detail below (section 18.3), the PROCESS profile contributed by the auxiliary still requires grounding (by a modal or a tense morpheme), which explains why a non-finite auxiliary cannot occur as the first element in a verb string (e.g. *George have betrayed Lily).

18.2.1 The passive construction: [be$_2$ [PERF$_3$ [V]]]

In Cognitive Grammar auxiliaries are not viewed as 'purely grammatical' elements but represent an extension from the other uses of that verb. This means that they receive a semantic characterisation, as does the participial morphology. According to Langacker, the morpheme PERF$_3$, which gives rise to what we will call the passive participle (the 'past' participle that occurs in passive constructions like *be betrayed*), imposes a **construal** upon the construction whereby a TR-LM reversal is effected. In order for this to be possible, the content verb must have both a TR and a LM, which means it must be a transitive (or ditransitive) verb. The morpheme PERF$_3$ (e.g. *-ed* in *betrayed*) is both phonologically and conceptually dependent and has a schematic PROCESS as its TR, which in turn has a schematic TR and LM. The autonomous content verb (e.g. *betray*) elaborates the schematic PROCESS of PERF$_3$. The resulting construction [PERF$_3$ [V]] (e.g. *betrayed*) is headed by PERF$_3$, and, as a participle, has the status of an ATEMPORAL RELATION that specifies the TR-LM reversal characteristic of a passive construction. According to Langacker, the passive participle morpheme PERF$_3$ belongs to a network of PERF morphemes that have related yet distinct meanings. We have already seen that PERF$_4$ is the perfect participle morpheme that occurs in perfect constructions, to which we return below. Langacker proposes that PERF$_1$ is the form found in stative adjectival constructions like (4). This form is related to intransitive verbs.

(4) Lily's heart is brok**en**.

Of course, the verb *break* can also be transitive (e.g. *George broke Lily's heart*), but the fact that we can say *Lily's heart broke* shows that it can also be intransitive. The fact that PERF$_1$ combines with intransitives means that, unlike the passive PERF$_3$, this form does not involve any TR-LM reversal, because intransitive verbs do not specify an independent LM. In contrast, the form that Langacker calls PERF$_2$ also participates in stative adjectival constructions but is related to transitive verbs and does effect a TR-LM reversal:

(5) George left Lily betray**ed** (by his cowardice).

It is because PERF₂ relates to transitive verbs that we interpret *Lily* as the LM of the transitive verb *betray* and *his cowardice* as the TR. As a result of the TR-LM reversal *Lily* occurs here elaborating the TR of the atemporal relation *betrayed* and *his cowardice* need not be present in the construction.

Turning to the other key component of the passive construction, the passive auxiliary *be*₂, this also occurs in a network of related yet distinct uses of the same verb, which profiles a schematic imperfective PROCESS. The basic from of *be*, which Langacker calls *be*₁, functions as the copula, which (as we saw in Chapter 17) construes a NOMINAL or simple ATEMPORAL RELATION as a PROCESS. According to Langacker *be*₁ also functions as the progressive auxiliary, which construes a complex ATEMPORAL RELATION as a PROCESS. When the passive auxiliary *be*₂ combines with the passive participle containing PERF₃, its role is also to construe an ATEMPORAL RELATION as a PROCESS.

18.2.2 The progressive construction: [be₁ [-ing [v]]]

Consider Langacker's description of this construction:

> ... the progressive construction always views a perfective process from an internal perspective and thereby renders it imperfective ... Viewing a process from an internal perspective is a matter of restricting its profile to a series of component states that does not include the initial and final states. (Langacker 1991: 209)

According to this view, the meaning of the progressive participle morpheme *-ing* lies in its restriction of the sequence of events that make up a PROCESS to just the 'middle' stages, thereby construing it as an ongoing event. Langacker suggests that *-ing* also renders the sequence of states identical and thereby 'suspends sequential scanning' which results in a complex atemporal relation. By itself, the resulting construction [*-ing* [v]] (e.g. *fretting*) can perform a modifying function (e.g. *fretting, she opened the envelope*), but cannot head a clause (**Lily fretting at the moment*). When *be*₁ combines with [*-ing* [v]], it imposes its PROCESS profile on the resulting construction [*be*₁ [*-ing* [v]]] (e.g. *be fretting*), which can then head a clause.

18.2.3 The perfect construction: [have [PERF₄ [v]]]

Finally, we turn to the perfect construction, in which PERF₄ represents the 'past' or perfect participial morphology (e.g. *-ed* in *fixed*), which gives rise to an ATEMPORAL RELATION. The perfect auxiliary *have* imposes its PROCESS profile upon this construction, giving rise to the perfect construction [*have* [PERF₄ [v]]] (e.g. *have fixed*), which can then function as clausal head. As we have seen, the perfect construction encodes an event as 'completed' with respect to a given reference point in time. Compare example (6a), which is in the present perfect, with (6b), which is in the simple past.

(6) a. The gas men have fixed Lily's heating.
    b. The gas men fixed Lily's heating.

The difference between these two examples is clear: while example (6a) might be used in a context where the gas men have just left, example (6b) locates the event at a more distant point in time. Langacker (1991: 212) therefore suggests that the function of the perfect is to encode **current relevance**. In considering how the perfect auxiliary *have* is linked to other senses of the same verb, Langacker compares it to examples like (7).

(7) We have a lot of windy weather.

This example illustrates the content meaning of lexical *have*. Prototypically, lexical *have* encodes a relationship of POSSESSION, but in example (7) it also evokes two other salient aspects of meaning. Firstly, example (7) makes reference to a **spatial reference point**, where the spatial location of the object (*a lot of windy weather*) is interpreted on the basis of the spatial location of the subject (*we*), which is salient because it makes reference to a human or a group of humans and thus serves as a spatial reference point. Secondly, this type of construction also encodes what Langacker calls **potential relevance**, in the sense that the construction does not necessarily describe any current relation between the subject and the object (we could utter (7) on a calm sunny day, for example) but describes a relation that is a potential (if not an actual) aspect of the subject's experience.

On the basis of a comparison of the perfect auxiliary *have* with lexical *have* in constructions like (7), Langacker identifies the semantics of auxiliary *have* as evoking (1) a temporal (rather than spatial) reference point, and (2) current (rather than potential) relevance. These semantic properties of auxiliary *have* are related to the spatial reference point and the potential relevance that lexical *have* evokes in (7). Of course, the perfect construction does not always occur in the present tense, as it does in (7). This is illustrated by example (8), which is in the past perfect.

(8) The gas men had fixed Lily's heating.

It is important to emphasise that 'current relevance' is not restricted to the present tense (the time of speaking). Instead, this term is understood relative to the temporal reference point that is evoked by the construction or provided by the context. In other words, whether the temporal reference point is in the past, present or future, the completed event is construed as 'currently relevant' relative to that temporal reference point (recall our discussion of tense and aspect from the perspective of Mental Spaces Theory in Chapter 11). In order to illustrate this point, observe that the perfect construction is often used in contexts where the relevance of the completed event to some immediately preceding or following event is emphasised. This is illustrated by the contrast between (9a) and (9b).

(9) a. The gas men had just fixed Lily's heating when the sun came out.
    b. The gas men fixed Lily's heating. Then the sun came out.

Turning to the 'division of labour' between the component parts of the perfect construction [*have* [PERF$_4$]], we can observe that perfect participle morpheme PERF$_4$, like the other instances of PERF, imposes its profile as an ATEMPORAL RELATION on the content verb. Unlike the passive PERF$_3$, it does not impose a TR–LM reversal. Instead, Langacker characterises the meaning of PERF$_4$ as **temporal anteriority**. It is important not to confuse this with past tense, which also makes reference to past time: as we have seen, the perfect construction can occur in the present tense. Regardless of its tense properties, the perfect construction construes an event as completed and furthermore emphasises the event's completion. In (9a), for example, it is the end stage of the fixing event that is salient in relation to the temporal reference point (when the sun came out). In this respect, Langacker argues that PERF$_4$ shares a further aspect of its meaning with the other instances of PERF, which is that all four variants emphasise the terminal stage of an event.

Although we have discussed the passive, progressive and perfect constructions individually in this section, it is important to emphasise that these are not separate or unrelated constructions in the sense that they all form part of a network of verb-string constructions that may display greater or lesser complexity. As we saw in Table 18.2 and example (3), for example, it is possible for all three constructions to co-occur in one complex construction, where their properties are closely interwoven.

As the discussion in this section illustrates, while auxiliary verbs are recognised in Cognitive Grammar as having distinct properties from content verbs, they are not sharply distinguished from content verbs. Both auxiliary and content verbs are represented as PROCESSES, but differ with respect to their relative schematicity or specificity at the semantic pole. According to Langacker, the prototypical English auxiliary verb is *do*, which is maximally schematic. Unlike the other auxiliary verbs discussed in this section, *do* does not contribute any meaning to the construction beyond imposing its profile as a schematic

PROCESS. This explains the distribution of this verb, which, as we saw in Chapter 14, occurs only when the sentence lacks another auxiliary or modal verb licensed to participate in question or negation constructions, or when the speaker wants to emphasise the truth of a statement.

18.3 The grounding predication: mood and tense

As we saw in the last section, Langacker divides the verb string into grounding predication and clausal head. Example (1) is repeated here as (10).

(10) George   [must]   [have been singing]   the blues
              GROUNDING PREDICATION   CLAUSAL HEAD

In the last section, we simplified this division by relying mainly on examples that contained modal verbs. Because the English modal is phonologically autonomous, examples like these provide a straightforward illustration of the different roles played by the grounding predication and the clausal head. Of course, not all sentences contain modal verbs, in which case the grounding predication is a tense morpheme, which is phonologically dependent and attaches to the first verb that makes up the clausal head. In many constructions, then, the distinction between grounding predication and clausal head is blurred by the fact that they may be morphologically bound together.

We first discussed the idea of a grounding predication in Chapter 16, where we saw that this idea forms the basis of the Cognitive Grammar account of the relationship between nouns and determiners. Nouns and verbs are widely recognised as the two universal linguistic categories, hence their central status in Cognitive Grammar. Furthermore, verbs head clauses, which represent a universal construction type. According to Langacker, what noun phrases and finite clauses share is the fact that they are **grounded**. As we have seen, each speech event involves a **ground**, which consists of place and time of speaking, the participants in the speech event and so on: grounding is the process whereby linguistic expressions are linked to the ground. Determiners ground nominal expressions by profiling an instance of a category (*a rocket scientist*), and by indicating information such as whether participants are already familiar with the referent (*the rocket scientist*), or whether the referent is present in the immediate physical context (*this/that rocket scientist*). This view explains why many determiners (in particular, demonstrative and possessive determiners) have **deictic** properties, which means that they rely upon aspects of the ground in order to be fully interpreted. As we saw in Chapter 16, **grounding predications** do not make up a distinct word class but are represented as schematic categories for the class that they interact with.

In the same way that nouns are grounded by determiners, finite clauses are grounded by **tense** and by **modals** which link the PROCESS designated by the clause to the specific usage event. As a deictic category, tense situates the PROCESS relative to the time of speaking, while the modal verbs establish the 'reality' status of the designated PROCESS from the speaker's perspective. In the same way that a nominal grounding predication is represented in the network as a schematic nominal or THING, the clausal grounding predication is represented in the network as a schematic verb or PROCESS.

18.3.1 Mood

Mood (or modality) is usually divided into two broad categories: epistemic and deontic. As we saw in Chapter 11, epistemic modality is a type of grammatical marking that encodes the speaker's judgement relating to his or her knowledge about the possibility, likelihood or certainty of the proposition expressed by the sentence. This is what the English modal verbs *will, can* and *might* express in the sentences in (11).

(11) a. George will be here soon. [epistemic modality]
 b. Lily can cook a mean risotto.
 c. She might open a bottle of champagne.

Deontic modality expresses the speaker's judgement relating to obligation (moral or social), permission or prohibition. This is what the English modal verbs *must* and *should* express in the examples in (12).

(12) a. Lily must get away from all this. [deontic modality]
 b. George should try to be on time.

However, the English modal verbs cannot be divided neatly into two categories according to which type of modality they express because their interpretation can be rather fluid and depends on the context in which they occur. For example, *must* expresses epistemic modality in example (13a) and *can* expresses deontic modality in example (13b).

(13) a. George must have been held up. [epistemic]
 b. You can go now. [deontic]

As we saw in Chapter 14, the modal auxiliaries do not inflect in the usual way for tense or aspect (**musted, *musting*), nor do they have a third person singular *-s* form (**she musts*). Modals also lack an infinitive form (**to must*), and must occur as the first verb in a verb string (**I am musting . . .*), followed by the bare infinitive form of the next verb in the string (*she must went). With the exception of *must*, the English modals occur in pairs (*can – could; may – might; shall – should; will – would*). These are traditionally described as present and past tense forms on the basis of 'sequence of tense' patterns. For example, a past tense verb in a main clause tends to require a past tense verb in a complement clause; compare (14a) with (14b). As examples (14c) and (14d) show, the modals sometimes pattern in a similar way.

(14) a. I thought Lily loved shellfish.
 b. *I thought Lily loves shellfish.
 c. I thought Lily could cook a risotto.
 d. *I thought Lily can cook a risotto

However, it is worth observing that what might be called 'past' modal forms are not restricted to past tense contexts, which means that the traditional classification of modals into 'past' and 'present' forms is not a matter of consensus. For example, consider uses like *I'd like to help* or *I could do it if you'd let me*. Despite these difficulties in pinning down the tense properties of the modal verbs, they are usually referred to finite verb forms because they pattern together with tensed verb forms in licensing a main clause verb string. As we will see, this 'licensing' is conceived in terms of **grounding** in Cognitive Grammar.

18.3.2 Tense

As we saw in some detail in Chapter 11, tense refers to the **grammatical marking** of time relative to the time of speaking. In other words, a language is only described as having tense if it has a distinct **morphological verb form** that indicates past/present/future time. English is usually described as having two tenses: past and present (non-past). While past tense describes an event that took place prior to the point of speaking (15), the present tense is not restricted to describing an event that is concurrent with the moment of speaking (16).

(15) a. Lily called George yesterday.
 b. George pretended he was out.

(16) a. George is under the bed.
 b. You'll never guess what happened to Lily yesterday. She walks into George's flat and this woman comes out of the bedroom . . .
 c. George leaves tomorrow.
 d. Lily eats passion fruit.

While (16a) describes an event that is taking place at the time of speaking, (16b) illustrates the **historical** use of the simple present, where it can be used to narrate a sequence of events that took place in past time. The simple present form in (16c) describes an event that is located in future time. Finally, the simple present in (16d) is not interpreted as meaning that Lily is eating passion fruit right now, but that she **habitually** eats it. Notice that the mass noun contributes to this interpretation. If the noun had an indefinite article (*Lily eats a passion fruit*), the sentence would instead have the flavour of a 'stage direction', where it describes a specific eating event, but observe that sentences like this are quite unnatural in ordinary spoken English when referring to present time. Instead, we use the present progressive (for example, *Lily is eating a passion fruit*). The fact that the simple present in English can be used to refer to past, present and future time, as well as encoding habitual events (a type of aspect), means that some linguists prefer the label **non-past**.

Future tense expresses reference to future time, but English has no future tense, since it lacks a verb form inflected for future. Instead, English has a range of different ways of referring to future time, some of which are illustrated in (17).

(17) a. Lily will leave tomorrow.
 b. Lily is leaving tomorrow.
 c. Lily leaves tomorrow.
 d. Lily is going to leave tomorrow.

18.3.3 The epistemic model

In order to provide a semantic account of tense and the modals, Langacker invokes an **idealised cognitive model** or ICM (see Chapter 8), which he calls the **epistemic model**. Recall that the term epistemic relates to knowledge systems. The epistemic model is illustrated in Figure 18.1.

In this model, the large circle represents immediate reality, which we can think of as 'here and now'. This represents the **ground** in which the speech event occurs. The small shaded circle represents the language user. Of course, 'reality' is used here in the sense pertaining to the knowledge represented in the conceptual system of the individual rather than to an objective external reality. The horizontal line running through the centre of the diagram represents TIME, which Langacker (1991: 242) describes as 'the axis along which reality evolves'. The dotted line represents TIME 'until now' and the continuous line represents TIME 'after now'. Although in reality 'now' is momentary, speech events tend not to be momentary, so that 'now' as construed for linguistic purposes may be a significant period of time. This is represented by the portions of the time line inside the large circle. Of course, Cognitive Grammar

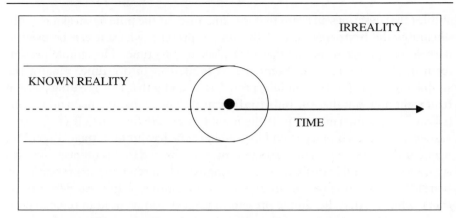

Figure 18.1 The epistemic model (adapted from Langacker 1991: 242)

is not the first or only model of language to represent tense in terms of a time line. Indeed, this approach is reminiscent of the well known model developed by Reichenbach (1947) which we discussed in Chapter 11.

It is already clear how this model accounts for tense, at least in its core uses. While present tense refers to time inside the ground, past tense refers to the portion of time within known reality but outside the ground, and future tense refers to time in irreality, beyond the ground. This model represents a version of an ego-based model for TIME that we saw in Chapter 3. In Cognitive Grammar, the epistemic model also accounts for the modal verbs. While the absence of a modal verb indicates that the speaker construes the event as part of known reality (for example, *Lily is a rocket scientist*), the presence of a modal verb indicates that the speaker construes the event as part of irreality (for example, *Lily might be a rocket scientist; Lily will be a rocket scientist*).

Given the basis of this model in TIME and SPACE, reality and irreality also vary along parameters of distance, in terms of whether they are immediate (close) or non-immediate (distant) relative to the ground. This type of variation is called **epistemic distance** (which was also discussed in relation to Mental Spaces Theory in Chapter 11). For example, in the case of the modals, Langacker argues that the pairs that we observed earlier (for example, *can* and *could*) represent immediate versus distant irreality, respectively. Immediate irreality is close to known reality, and characterises verbs like *must, will* and *can* which encode a strong degree of obligation, likelihood or possibility. These contrast with their distal counterparts like *would* and *could*, which encode a much weaker sense of possibility, a much stronger sense of doubt, and so on. In the same way, Langacker analyses the past tense morpheme *-ed* as a **distal morpheme**, since it evokes a portion of time that belongs to known reality, but is distant relative to the 'here and now' of the ground. In contrast, the simple present prototypically refers to the 'here and now' of the ground, and its prototypical morpheme is treated as an **unmarked** form, or **zero morpheme** ø. This model therefore predicts that there will be four basic types of grounding predication, which are listed in (18).

(18) a.   IMMEDIATE REALITY   e.g. ø, -s
     b.   NON-IMMEDIATE REALITY   e.g. -ed
     c.   IMMEDIATE IRREALITY   e.g. can
     d.   NON-IMMEDIATE IRREALITY   e.g. could

While (18a) and (18b) relate to present and past tense, respectively, (18c) and (18d) relate to modality. Beyond these parameters of variation, tense and modality essentially belong to the same category in Cognitive Grammar: they are both types of **grounding predication**. As we saw in Chapter 16, the nominal grounding predication specifies an instance of a category and is itself a schematic nominal or THING. In the same way, a clausal grounding predication is a schematic PROCESS, and specifies an instance of the PROCESS category. Compare the two examples in (19).

(19) a.   Lily was a rocket scientist.
     b.   Lily to be a rocket scientist

Example (19a) is a finite clause and is therefore grounded. This means that the location of the event described in the clause is established relative to the ground in terms of (ir)reality. As a consequence, the clause is realised as a PROCESS and can stand alone as a communicative speech event. Example (19b), in contrast, is a non-finite clause. Because it is not grounded, which means that its reality status has not been established, it cannot stand alone as a communicative speech event. Clauses like this can only occur as embedded clauses, where the main clause is grounded (20). In this way, Langacker accounts for the fact that main clauses have to be finite.

(20) George never wanted Lily to be a rocket scientist.

Explaining the grammatical behaviour of the modals

Cognitive Grammar exploits this epistemic account of tense and mood to explain the 'special' characteristics of the modal verb that were outlined above. Firstly, the fact that the modal does not inflect to form a participle or an infinitive is consistent with its role as a grounding predication: participles and infinitives are ATEMPORAL RELATIONS, while the modal is a schematic PROCESS. Secondly, this analysis also explains the fact that the modal does not participate in subject–verb agreement (*Lily musts succeed*). This is because the third person present tense morpheme *-s* is itself a grounding predication with an opposing reality value, so the two are not expected to co-occur. Finally, the fact that the modal has to be followed by the bare infinitive form of the next verb in the string is accounted for on the basis that a grounding predication and its grounded element must match in terms of category. In other words, given that the modal represents a schematic PROCESS, its grounded element must also be a PROCESS. Of course, this claim cannot be maintained if the verb form that follows the modal is described as a 'bare infinitive', given that the infinitive represents an ATEMPORAL RELATION. In Langacker's model, the verb form that follows the modal is described as a **simple verb**, which counts as a PROCESS. In other words, it encodes a temporal relation, but is uninflected because the modal performs the grounding function.

Potential and projected reality

In the context of the epistemic distance model, the modals are characterised in terms of **potential reality** and **projected reality**. The distinction between these explains the difference between the future time epistemic modals *will* and *may*. The modal *will* encodes projected reality (in IMMEDIATE IRREALITY), and therefore gives rise to the future time interpretation. In contrast, *may* encodes only potential reality (although still in IMMEDIATE IRREALITY), hence a weaker epistemic reading. Along with Talmy (1985) and Sweetser (1990), Langacker adopts a **force-dynamics** model to capture this distinction between projected and potential reality. If the event is construed as having sufficient 'momentum' that the speaker can be confident that it will reach the predicted reality status, this is projected reality. In contrast, an event that is construed as having weaker momentum has only potential reality status. The distal counterparts of these modals are analysed along the same lines, but involve a temporal reference point more distant from the ground. As we saw in Chapter 11, essentially similar considerations motivate the Mental Spaces approach to counterfactuals.

The polysemy of the modal verbs is also explained in force-dynamics terms. The distinction between the deontic and epistemic readings, which is often not a clear-cut distinction, relates to whether the source of the momentum is **salient**. If the source of the momentum is salient, this gives rise to deontic interpretations (involving obligation, permission and so on). This is illustrated by examples (21a) and (22a), where the source of the momentum or force is understood as the speaker or some other authority. If the source of the momentum is not salient, as in (21b) and (22b), this gives rise to the epistemic reading. The fact that modals are frequently ambiguous with respect to epistemic versus deontic interpretations illustrates that these are not discrete categories. For example, the sentence *George must be kind* is open to either a deontic or an epistemic interpretation.

(21) a. You may kiss my hand.   [deontic (permission)]
     b. Lily may be too sad to dance. [epistemic]

(22) a. You must kiss my hand.   [deontic (obligation)]
     b. Lily must be too sad to dance. [epistemic]

18.4 Situation aspect

In this section, we will look at the type of aspect that is inherent in certain content words rather than the type of aspect that is imposed on a construction by auxiliary verbs. As we have seen, aspect that is imposed upon a clause by auxiliary verbs (perfect and progressive, in English) is called **grammatical aspect**, while aspect that is inherent in the semantics of a content word is called **lexical aspect**. Verbs are not the only linguistic category to have lexical aspect, but they have received the most attention in the literature, given their central role in the clause. This is usually described in terms of **situation aspect**.

18.4.1 Situation types

A particularly influential account of lexical aspect is Vendler's (1967) account of **situation types**, where verbs, and hence the clauses that they head, are classified into four major categories in terms of a set of aspectual features. This model is represented in Table 18.3.

These terms are familiar from our discussion in Chapter 15. Briefly, a verb is stative if it describes an event that remains constant through time and, crucially, does not involve internal change or action. A prototypical stative verb is *resemble*. In contrast, a dynamic verb involves internal change (for example, *grow*), or action (for example, *eat*). The distinction between durative and punctual aspect relates to whether the event described by the verb is over almost as soon as it has begun, in which case it is punctual (for example, *flash*), or extends over time, in which case it is durative (for example, *resemble, love, grow*). The distinction between telicity and atelicity relates to whether the event described by the verb has an inherent endpoint or goal as part of its meaning, in which case it is telic (for example, *die*). In contrast, stative verbs like *love* express atelic

Table 18.3 Situation types (Vendler 1967)

Situation type	Stative/dynamic	Durative/punctual	Telic/atelic
state	stative	durative	atelic
activity	dynamic	durative	atelic
achievement	dynamic	punctual	telic
accomplishment	dynamic	durative	telic

events. Of course, it is rather misleading to suggest that verbs in isolation determine the situation type of the clause. This is because other parts of the clause, particularly objects and temporal adverbials, also participate in determining the aspectual properties of the clause as a whole.

These aspectual features together give rise to the taxonomy represented by Table 18.3. The examples in (23) illustrate each of the situation types.

(23) a. Lily knows her times tables.   [state]
     b. Lily's eyes sparkled.   [activity]
     c. George arrived at midnight.   [achievement]
     d. George walked home in twenty minutes. [accomplishment]

Example (23a) is stative because knowing something does not involve internal change; it is durative because it extends across time, and it is atelic because we do not expect the situation to reach some inherent endpoint. Example (23b) is dynamic because sparkling involves inherent change. This event is also durative and atelic; of course, activities can come to an end (Lily's eyes can stop sparkling), but this endpoint is not an intrinsic part of the meaning of *sparkle*. Example (23c) is dynamic because arriving involves action, and it is punctual because the act of arriving somewhere is achieved in the moment of arriving, hence its inherent endpoint or telicity. This explains why the event cannot be drawn out across time (e.g. **George arrived for hours*). Finally, example (23d) is also dynamic and telic, involving action towards an inherent endpoint or goal, but it is durative because it is extended across time. In the remainder of section 18.4, we will explore the Cognitive Grammar account of these situation types.

18.4.2 Perfective and imperfective PROCESSES

According to Langacker, the basic aspectual distinction is between **perfective** and **imperfective**, and the semantic basis of this aspectual distinction can be described in terms of scanning. As we have seen, Langacker (2002: 86) defines a PROCESS as 'a series of profiled relations . . . distributed through conceived time and scanned sequentially.' This definition as it stands makes no reference to aspectual distinctions, so it applies equally to both perfective and imperfective PROCESSES. However, an imperfective PROCESS is characterised by the fact that each relation that makes up the cognitive representation is the same as the next, which means that the situation described remains **constant** through time. In contrast, a perfective PROCESS is characterised by a sequence of relations where each is different from the last, which means that the situation described involves **change** through time.

Langacker (2002: 86) describes verbs like *jump*, *kick* and *arrive* as 'canonical' or prototypical perfectives, and verbs like *resemble*, *have* and *know* as prototypical imperfectives. Langacker relies on well-established grammatical tests for distinguishing between the two. Prototypical imperfectives like *resemble* can occur in the simple present (24a) but not in the progressive (24b).

(24) a. Lily resembles her mother.
     b. *Lily is resembling her mother

In contrast, while prototypical perfectives like *build* can occur in the progressive (25a), they are unnatural in the simple present (25b), unless this gives rise to the habitual or 'narrative' senses of the simple present.

(25) a. George is building a canoe.
     b. ?George builds a canoe

However, there are not always clear-cut distinctions between perfective and imperfective categories. As we mentioned above, context can alter the construal of aspect. For example, while perfectives are often odd in the simple present, an appropriate context can license this usage and give rise to a habitual interpretation, which construes the situation as imperfective. Compare the following conversational exchanges:

(26) George: What are you doing?
     Lily: ?I eat an orange
(27) George: How come *you* never catch a cold?
     Lily: I eat an orange every morning

In example (27), the context of Lily's utterance, together with her use of the expression *every morning*, gives rise to a habitual interpretation. Despite this broad division between perfective and imperfective PROCESSES, some verbs can occur quite naturally in both the simple present and the progressive, illustrating that they can be interpreted as either imperfective or perfective, respectively. This is illustrated by example (28).

(28) a. Lily loves *Gone with the Wind*.   [imperfective]
     b. Lily is loving *Gone with the Wind*. [perfective]

Example (28a) describes a situation that remains constant over time: Lily has loved the book (or the film) for some time, and this is not expected to end. In contrast, (28b) describes an ongoing experience: Lily is enjoying reading the book or watching the film at the moment, and at some point this activity will come to an end.

18.4.3 Aspect and the count/mass distinction

Langacker proposes that the perfective/imperfective distinction can be modelled in the same terms as the count/mass distinction. In other words, the aspectual distinction relates to the nature of the component parts of the PROCESS, and to the presence or absence of **bounding**. Of course, aspect relates to bounding in TIME rather than bounding in SPACE. Langacker summarises this idea in the following way:

> The component states of a process (each profiling a relation) are analogous to the component entities constituting the region profiled by a noun. For a process, time is the primary domain with respect to which the presence vs. absence of bounding is determined.(Langacker 2002: 87)

The diagrams in Figure 18.2 represent Langacker's model of aspect. The box represents the scope of predication. A perfective event (Figure 18.2(a)) is bounded within this scope and involves internal change which is represented by a squiggly line. In contrast, an imperfective event (Figure 18.2(b)) is unbounded and does not involve internal change, remaining constant both within and beyond the scope of predication. This is represented by a straight line. The arrow represents the passage of time.

Perfective

The perfective PROCESS is likened to a count noun in that both are bounded and in that both are replicable. For count nouns, replicability gives rise to pluralisation. For perfective processes, replicability can give rise to iterative aspect. This is illustrated by (29a). Example (29b) shows that the imperfective PROCESS is incompatible with an iterative interpretation.

(29) a. Lily read the letter over and over again.
     b. ?Lily knew the truth over and over again

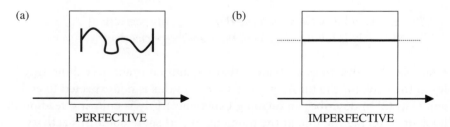

Figure 18.2 Perfective and imperfective aspect (adapted from Langacker 2002: 88)

The incompatibility of a perfective PROCESS with the simple present tense is explained by Langacker's (2002: 89) definition of tense: in the case of present tense, a 'full instantiation of the profiled process occurs and precisely coincides with the time of speaking.' In the case of past tense, a 'full instantiation of the profiled process occurs prior to the time of speaking.' As we have seen, a perfective PROCESS is bounded, which means that a full instantiation includes the beginning and end points of the PROCESS. This explains why perfective PROCESSES are typically incompatible with the simple present which encodes an event coextensive with the moment of speaking: it is not usually possible for all the distinct subparts of a perfective PROCESS to coincide with the moment of speaking. Furthermore, because perfective PROCESSES involve internal change and therefore do not consist of identical subparts, a single 'moment' in the PROCESS cannot serve as a representation of the PROCESS as a whole. **Punctual** events represent an exception to this generalisation: verbs like *flash*, *sneeze* or *blink* encode bounded events that are over almost as soon as they have begun, which explains why they can be modified by temporal expressions that pinpoint a moment in time (e.g. *Lily sneezed at midnight*). **Performative** verbs like *promise* or *declare* also represent an exception to this generalisation: while perfective and therefore bounded, the act of promising or declaring is instantaneous (punctual) and can therefore coincide with the moment of speaking. This explains why performatives are licensed in the simple present. As Taylor (2002: 401) observes, bounded processes that are not punctual can be described as **extended**: these are compatible with temporal expressions that express a bounded period of time (e.g. *George built a canoe in two weeks*).

We saw above that there are other contexts in which the simple present is licensed for perfectives, but it is striking that these contexts require a 'special' interpretation to license the use of the simple present: as we saw in example (16), the simple present can be used to refer to the imminent future or the past, and can also give rise to a habitual interpretation. In Cognitive Grammar, these 'special' interpretations are a matter of construal. Langacker argues that the imminent future use of the simple present situates the whole bounded event at some point in the future, preserving its bounded nature, but that the present tense emphasises the planned status of the future event, which remains constant through time. He further argues that a habitual reading construes a PROCESS as constant through time and thus imperfective, while the historical present construes a past (bounded) event as though it were happening in the present.

Recalling Vendler's situation type taxonomy (Table 18.3), it is clear that the perfective PROCESS is necessarily telic, because bounded events entail an endpoint, and necessarily dynamic, because perfective PROCESSES involve internal change. While some perfective processes (e.g. *sneeze*) are punctual, others are extended or 'durative' (e.g. *build*). This means that Langacker's perfective aspect corresponds to **achievement** (punctual) and **accomplishment** (durative) in Vendler's system.

Imperfective

The imperfective PROCESS is likened to a mass noun, because in the same way that the component parts of a mass noun are homogeneous, the component states of a prototypical imperfective PROCESS are identical. Furthermore, in the same way that a mass noun is expansible or contractible, any given subpart of an imperfective PROCESS is still an instance of that PROCESS. This explains why a prototypical imperfective PROCESS is compatible with the simple present, because a subpart of the PROCESS that is coextensive with the moment of speaking can serve as a representation of the PROCESS as a whole. As we saw in Chapter 15, this follows directly from the property of homogeneity. This is illustrated by (30a). Unlike the perfective, the prototypical imperfective PROCESS is incompatible with the progressive, because the function of the progressive is to construe an event as imperfective. It is therefore redundant to mark an imperfective process as progressive (although see Taylor 2002: 404 for further discussion of this point). This is illustrated by example (29b).

(30) a. Lily knows her times tables.
 b. ?Lily is knowing her times tables

As we saw earlier (section 18.2), the inflectional *-ing* morpheme derives an ATEMPORAL RELATION from a PROCESS. This explains why progressive participles of imperfective PROCESSES are licensed in adverbial clauses (31a), despite the fact that an imperfective PROCESS cannot occur in the progressive (31b). This is because the progressive auxiliary *be* imposes a PROCESS reading on the ATEMPORAL RELATION.

(31) a. Having a broken nose, George was not supermodel material.
 b. *George was having a broken nose

Recalling Vendler's situation type aspect system once more (Table 18.3), it is clear that Langacker's imperfective PROCESS is atelic, because unbounded processes do not specify an inherent endpoint. The imperfective PROCESS is also necessarily durative, since it is in the nature of an unbounded PROCESS that it endures across time. As we have seen, the prototypical imperfective PROCESS is the stative PROCESS, which involves no internal change (e.g. *resemble, know, have*). This corresponds to Vendler's **state**. Taylor (2002: 402) also suggests that **activities** can be classified as a type of imperfective PROCESS. Although these do involve internal change (e.g. *Lily's eyes sparkled*) and are therefore

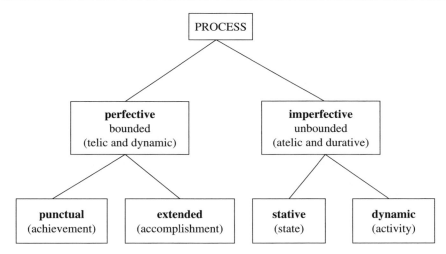

Figure 18.3 Perfective and imperfective situation types

dynamic, activities are durative and atelic, hence unbounded. Unlike states, activities are compatible with the progressive (e.g. *Lily's eyes were sparkling*). Figure 18.3 summarises the interaction of Langacker's aspectual system with the four situation types identified by Vendler.

18.5 Summary

The main theme of this chapter was **grounding**. We began the chapter with a short discussion of English verb forms and then looked in detail at the **verb string**, a central feature of the structure of the English clause. We saw that the verb string is analysed in terms of a **grounding predication** – either a tense morpheme or a modal verb – and a **clausal head**, which can include a **perfect construction**, a **progressive construction** and a **passive construction**, as well as the **content verb**. By looking at each of these constructions in turn, we saw that Langacker's model treats **auxiliaries** *have* and *be* as semantically related to non-auxiliary functions of the same verbs. In Cognitive Grammar, the 'past' (perfect or passive) participle is also semantically related to adjectives that share the same morphology. We then looked at the Cognitive Grammar account of **tense, aspect** and **mood**, and saw that **tense** and **mood** receive a unified semantic characterisation in terms of the **epistemic model**, and that the **polysemy** of modals can be accounted for in force-dynamics terms. Finally, we looked at the Cognitive Grammar account of the aspectual properties of **situation types**. These are accounted for in terms of a broad distinction between **perfective** and **imperfective** aspect, which, like **count** and **mass** nouns, can be characterised in terms of **homogeneity** versus **heterogeneity** and in terms of **bounding**.

Further reading

Introductory texts

- **Radden and Dirven (2005)**. Chapters 8 and 9 of this textbook focus on tense, aspect and mood. Chapter 11 discusses marked coding.
- **Lee (2001)**. Chapter 9 of this textbook provides a short and accessible introduction to Langacker's approach to perfective and imperfective aspect in verbs.
- **Taylor (2002)**. Chapter 20 of this textbook focuses on tense, aspect and mood. Taylor's discussion of situation aspect from a Cognitive Grammar perspective is particularly useful.

Cognitive Grammar

- **Langacker (1987)**. Chapter 7 of this volume focuses on temporal relations, and includes a discussion of perfective and imperfective processes.
- **Langacker (1991)**. Part II of this volume is dedicated to clause structure. Chapters 5 and 6 focus on the contribution of auxiliary verbs to the clause, and discuss voice, tense, aspect and mood. Chapter 8 addresses marked coding.
- **Langacker ([1991] 2002)**. Chapter 3 sets out Langacker's model of aspect and Chapter 4 focuses on the passive construction. Chapter 12 briefly discusses tense as a grounding predication.
- **Talmy (2000)**. Although this chapter has once more focused on Langacker's Cognitive Grammar framework, Langacker's model shares many important assumptions in common with Talmy's Conceptual Structuring System Model, as we saw in Chapter 15. Chapter 7 of this volume elaborates Talmy's force-dynamics approach to the modal verbs, which is adopted by Langacker.

Exercises

18.1 English verb forms

Recall from section 18.1 (Table 18.1) that the form of an English verb rarely reveals much about its person and number features, nor even its tense in some cases. In these sentences, describe tense, person and number features of the verb forms and explain how you reached your conclusions. What role does the clausal context play in determining the interpretation of these features?

(a) Lily let George's friend borrow her bike.
(b) George's parents let him eat too may sweets when he was a boy.
(c) George lets Lily polish his shoes on Sundays.

18.2 Perf

In Cognitive Grammar, as we saw in section 18.2, the morphology PERF is analysed in terms of a network of related forms ($PERF_1$ to $PERF_4$). Consider the unit *broken* in each of the following examples. In each case, state which variant of PERF plays a role in the example and explain how you reached your conclusions. What properties do these variants share, and how do they differ?

(a) George has broken her heart.
(b) Her heart was broken by George.
(c) How can she mend her broken heart?
(d) Her heart was broken for years.

18.3 Be

As we also saw in section 18.2, Langacker's analysis unites the lexical and auxiliary functions of the verb *be*. According to this analysis, there are two variants of *be*. In each of the following examples, state which variant(s) of *be* you have identified and explain how you reached your conclusions. Why do you think Langacker proposes two variants of *be* instead of a single verb?

(a) Lily was exhausted.
(b) George was being silly.
(c) Lily felt she was being persecuted.

18.4 The non-present present

Langacker suggests that the 'historical present' and the use of the simple present to refer to the immediate future may both be related to a shift in **perspective**. Explain how an analysis along these lines might work, basing your discussion on the epistemic model (Figure 18.1). Illustrate your discussion with examples of your own.

18.5 Deontic and epistemic mood

For each of the following examples, state whether the modal verb gives rise to an epistemic or a deontic reading. Are any of these examples ambiguous?

Explain what role context (linguistic or otherwise) plays in the interpretation. How might your findings be explained in force-dynamics terms?

(a) Lily said George could call her at the office.
(b) Lily said George could cook.
(c) George must be the luckiest man alive.
(d) George should be more careful.
(e) George should be home by now.
(f) George may have a bath later.

18.6 Perfective and imperfective processes

In section 18.4, we saw that the Cognitive Grammar account of situation types rests on a broad distinction between perfective and imperfective processes, which in turn is related to the count/mass noun distinction, particularly in relation to bounding. We also observed that verbs cannot always be classified in terms of one particular situation type aspect, because other parts of the clause contribute to its aspectual properties. For each of the following examples, state which situation type you have identified and explain how you reached your conclusions. You may find it helpful to consult Figure 18.3.

(a) George winked (at beautiful women) all night long.
(b) Lily discovered the truth about George.
(c) George is very handsome.
(d) Lily's heart sank.

Now explain why the following examples are not well-formed.

(a) *George winked in an hour
(b) *Lily discovered the truth about George for an hour
(c) *George is being very handsome
(d) *Lily's heart sank in an hour

19

Motivating a construction grammar

So far in Part III of this book we have sketched out the characteristics of a cognitive approach to grammar (Chapter 14) and have investigated the main claims made by cognitive linguists relating to the conceptual basis of grammar (Chapter 15). We have also explored in some detail Cognitive Grammar, the influential theory developed by Langacker (Chapters 16–18). As we have seen, the **construction** has a central place in Cognitive Grammar, in the sense that any symbolically complex unit is 'stored whole' in the **structured inventory** that represents a speaker's knowledge of language. In this chapter, we set about explaining how a constructional account can be **motivated**, something we have taken largely for granted up to this point. We will begin by comparing a constructional account with the **'words and rules'** account assumed in most generative models of language (section 19.1). We then look in some detail at **idiomatic expressions**, linguistic units that display **idiosyncratic** as well as **regular** properties and cannot therefore be fully accounted for by a model of language that focuses on accounting for what is 'regular' (section 19.2). We explore two idiomatic grammatical constructions in detail: the *let alone* construction (Fillmore, Kay and O'Connor 1988) and the *what's X doing Y* construction (Kay and Fillmore 1999). As we will see, in addition to displaying some regular grammatical properties, these constructions have grammatical, semantic and pragmatic properties that are not fully predictable from their subparts. This discussion sets the scene for the development of the idea that grammatical constructions can be **meaningful**, in part independently of the content words that realise specific instances of the construction. Having explored the empirical motivation for a constructional approach to grammar, we sketch out the theory of **Construction Grammar** proposed by Kay and Fillmore (1999), and compare and contrast this approach with both generative and cognitive approaches to language (section 19.3). Finally, we consider Construction Grammar in the light of the **'Generalisation Commitment'** (section 19.4).

19.1 Constructions versus 'words and rules'

In their influential 1988 paper, Fillmore, Kay and O'Connor challenge the 'words and rules' approach assumed by the standard generative model. According to this model, the properties of language can be accounted for by a system of 'words and rules', where the words are the individual lexical items in the speaker's lexicon, and these words are subject to rules of different types within the language system. Phonological rules govern the assembly of complex strings of sounds. Syntactic rules govern the assembly of words into grammatical structures such as phrases and sentences, while semantic rules assign a semantic interpretation to the clause according to the **principle of compositionality**. As we saw in Part II of the book, this principle holds that the meaning of a sentence arises from the meanings of the words it contains, together with the way in which these words are syntactically arranged. This gives rise to propositional meaning, a 'purely semantic' meaning that is independent of context. In addition to syntactic and semantic rules, speakers also have knowledge of pragmatic principles that map propositional meaning onto context and guide the hearer in drawing the relevant inferences. Crucially, as we saw in Part I of the book, this approach is **modular** in the sense that syntax, semantics (and phonology) are encapsulated subsystems that only communicate with one another via linking rules. This type of model can be represented by the diagram in Figure 19.1.

Observe that there is no 'pragmatics box' in this model. As we saw in Part II of the book, this is because the standard generative model views pragmatic

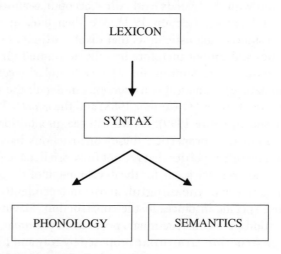

Figure 19.1 A modular view of the language system

knowledge as peripheral to linguistic knowledge 'proper' in the sense that pragmatic knowledge involves the interface between language and other systems of knowledge and information processing. This model of speaker knowledge only accounts for what is **regular** in language, and leaves aside **idiomatic** units, which, according to (Fillmore *et al.* 1988: 504), have the status of an 'appendix to the grammar'. In other words, in the standard generative model, the only the complex units that are 'stored whole' are those whose properties cannot be predicted on the basis of the regular rules of the grammar. As we saw in Chapter 1, idiomatic expressions like 'kick the bucket' fall into this category.

According to Fillmore *et al.*, this appendix is not only very large, but also has the potential to reveal much about how language works. For this reason, as we will see in the next two sections, they propose a model of language that accounts for idiomatic constructions not as an exception to the norm, but as a central feature of human language. Furthermore, Fillmore *et al.* propose that the same theoretical machinery should be held to account for both regular and idiomatic grammatical units.

19.2 Exploring idiomatic expressions

In their 1988 paper, Fillmore, Kay and O'Connor argue in favour of a model in which, like the lexical item, the complex grammatical construction (the phrase or the clause), has semantic and pragmatic properties directly associated with it. In this section, we explore Fillmore *et al.*'s typology of idiomatic expressions and look in detail at two complex constructions that provide the empirical basis of the constructional approach to grammar.

19.2.1 Typology of idiomatic expressions

Idiomatic expressions are those that a speaker cannot 'work out' simply by knowing the grammar and the vocabulary of a language. This is why idiomatic expressions are described as 'non-compositional'. Instead, a speaker has to 'learn them whole', rather like individual lexical items. Fillmore *et al.* develop a typology of idiomatic expressions based on four main parameters: (1) decoding and encoding idioms; (2) grammatical versus extragrammatical idioms; (3) substantive versus formal idioms; and (4) idioms with and without pragmatic point.

Decoding and encoding idioms

Decoding idioms like *kick the bucket* have to be decoded or 'learnt whole' in the sense that the meaning of the expression cannot be worked out on first hearing. In contrast, encoding idioms like *wide awake* may be understood on the first hearing: the adjective *wide* functions as a degree modifier, and it is possible to work out that this expression means 'completely awake'. However, the speaker would not be able to predict the conventionality of the expression. In other words, there is nothing in the 'rules' of English that enables a speaker to predict the existence of this expression as opposed to, say, *narrow awake*, *narrow asleep* or *wide alert*. Encoding idioms also include expressions that are perfectly regular but just happen to represent the conventional way of saying something. For example, the expression *driving licence* is an encoding idiom in the sense that it represents the conventional way of describing a document that could be (but is not) called a *driving permit* or a *driving document* (Taylor 2002: 547). Since encoding idioms are expressions that the speaker cannot predict the conventionality of, it follows that decoding idioms are also encoding idioms.

Grammatical versus extragrammatical idioms

Grammatical idioms are expressions that obey the usual rules of grammar. For example, in the grammatical idiom *spill the beans*, a verb takes a noun phrase complement. In contrast, extragrammatical idioms like *all of a sudden* do not obey the usual rules of grammar. In this expression, the quantifier *all* is followed by a preposition phrase, where we would expect to find a noun phrase. Furthermore, an adjective, *sudden*, occurs after a determiner where we might expect to find a noun.

Substantive versus formal idioms

The third distinction is between substantive and formal idioms. Substantive idioms, like most of those we have seen so far, are 'lexically filled', which means that they have fixed lexical items as part of their composition. For example, *kick the mop* does not have the same communicative impact as *kick the bucket*, and *spill the beans* does not have the same communicative impact as *spill the champagne*. Both *kick the bucket* and *spill the beans* are substantive idioms because most or all of the substantive or content expressions involved are intrinsic to the idiom. In contrast, formal idioms provide syntactic 'frames' into which different lexical items can be 'inserted'. An example of a formal idiom is the *let alone* construction. As the following examples illustrate, the frame provided by this construction can be filled with all sorts of lexical items. In other words, this type of idiom is **productive**.

(1) a. George doesn't understand maths, let alone rocket science.
　　b. George can't wash up, let alone cook.
　　c. I wouldn't describe George as mildly amusing, let alone hilarious.

Table 19.1 Distinctions in idiom types

Idiom type	Meaning	Example
Decoding	Neither meaning nor conventionality can be predicted	*kick the bucket*
Encoding	Meaning may be predicted, but not conventionality	*wide awake*
Grammatical	Obeys the rules of grammar	*spill the beans*
Extragrammatical	Does not obey the rules of grammar	*all of a sudden*
Substantive	Lexically filled	*spill the beans*
Formal	Lexically open	the *let alone* construction
Pragmatic point	Specific pragmatic function	*How do you do?*
No pragmatic point	Pragmatically neutral	*by and large*

Idioms with and without pragmatic point

Some idiomatic expressions have a very clear pragmatic function, such as greeting (*How do you do?*) or expressing a particular attitude (*What's your car doing in my parking space?*). In contrast, other idiomatic expressions appear to be pragmatically neutral, in the sense that they can be used in any pragmatic context. Expressions like *by and large* and *on the whole* fall into this category. Table 19.1 summarises these four distinctions. As this table shows, a single idiom can be classified according to each of these four parameters. For example, the expression *by and large* is a decoding idiom that is extragrammatical (a preposition is coordinated with an adjective), substantive and pragmatically neutral.

In addition to setting out the distinctions summarised in Table 19.1, Fillmore *et al.* provide a typology of idiomatic expressions. An adapted version of this typology is represented in Figure 19.2.

According to this typology, idioms can consist of either familiar or unfamiliar linguistic expressions (familiar in the sense that they occur in non-idiomatic expressions). These expressions can be arranged in either familiar (regular) or unfamiliar (irregular) grammatical patterns. Two of the four resulting possibilities (those relating to familiar components) can then be further subdivided into formal (lexically open) or substantive (lexically filled) idioms, which may or may not have specific pragmatic point.

Familiar pieces familiarly arranged

In this case, lexical items that are commonly used outside the idiom are arranged in a way that reflects the regular grammatical patterns of the language. It follows that such expressions will have a literal as well as an idiomatic meaning (e.g. *kick the bucket, spill the beans, throw in the towel, take a running*

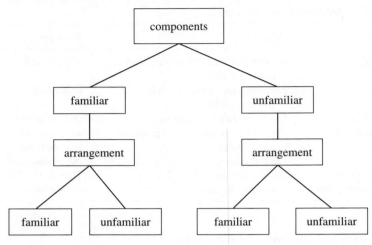

Figure 19.2 Typology of idioms

jump). What makes expressions like this idiomatic is that one meaning of the expression cannot be predicted from the principle of compositionality. As we have seen, expressions like *kick the bucket* are substantive idioms. An example of a formal idiom that illustrates this type is the *Is the X a Y?* construction, exemplified by the expression *Is the Pope a Catholic?* This construction has regular syntax (e.g. *Is Lily a rocket scientist?*), which is filled by regular expressions (*Pope, Catholic*), yet gives rise to an interpretation that emphasises the overwhelming certainty that a particular state of affairs will come to pass. This construction is typically used in response to a question. Consider the short conversational exchange in (2).

(2) Lily: Will Liverpool beat Tranmere Rovers in the FA cup?
    George: Is the Pope a Catholic?

From George's response, Lily infers that the answer to her question is a definite yes.

Familiar pieces unfamiliarly arranged

In idioms of this kind, familiar words are arranged in ways that do not conform to the regular grammatical patterns in the language. As we have seen, the substantive idiom *all of a sudden* consists of lexical items that are widely used in English, but which are arranged in a way that is unique to this idiom (compare **all of a fortunate*). Another substantive example in this category is the expression *by and large* (versus **by and small*).

Unfamiliar pieces familiarly arranged

In this category, we might place expressions that show regular syntax but that contain expressions that do not occur outside the idiom. Examples (Taylor 2002: 550) include *take umbrage at* (compare *take offence at* or *take exception to*), *in cahoots with* (compare *in collusion/collaboration/trouble with*), *by dint of* (compare *by virtue/necessity of*) and *wend one's way* (compare *make/trudge/climb one's way*). The expressions *umbrage*, *cahoots*, *dint* and *wend* are not found outside these idioms, yet their syntax is not restricted to these idioms. By definition, members of this category are substantive idioms, because a formal or lexically unfilled idiom is productive as a result of being filled by familiar expressions. However, it is important to point out that Fillmore *et al.* (1988: 506) do not include this category in their typology, since they argue that unfamiliar pieces are 'by definition' unfamiliarly arranged 'because, if the pieces are themselves unfamiliar or unique, there can be no standard principles for arranging them in larger patterns.' This suggests that expressions like *umbrage*, *cahoots*, *dint* and *wend* are not recognised as members of any word class and therefore cannot participate in regular syntax. Observe, however, that *umbrage* shows recognisable noun-forming morphology (compare *plumage*, *acreage* or *wattage*), and *cahoots* might plausibly be a plural noun. In addition, each of these examples can be assigned to a word class by comparing their distribution with other familiar expressions in the same context. Indeed, *wend* can occur in the past tense (*He wended his way home*), suggesting that it fills a verb slot in the construction. There might therefore be reasonable grounds for including this category in the typology of idioms.

Unfamiliar pieces unfamiliarly arranged

Idioms of this kind consist of expressions not found outside the idiom arranged in syntactic patterns that are also not found outside the idiom. According to our typology, this category is also by definition restricted to substantive idioms, for the same reason as the previous category. However, Fillmore *et al.* (1988: 506–7) place the formal idiom *the X-er the Y-er* in this category, which is illustrated by expressions like *the more the merrier* and *the fewer the better*. Although the 'slots' in this construction can be filled with familiar expressions, Fillmore *et al.* suggest that in addition to its irregular syntax, the instances of *the* in this construction are not in fact definite determiners but descendants of the Old English instrumental demonstrative *ðy*. Because Fillmore *et al.* reject the idea that unfamiliar pieces can be familiarly arranged, they also place substantive examples like *kith and kin* in this category, which are similar to *take umbrage with*, *in cahoots with* and so on.

As the discussion in this section suggests, the category 'idiom' (in the broad sense of any expression whose meaning cannot be predicted from the principle of compositionality) subsumes a wide range expressions, not all of which are straightforwardly classified. In fact, Taylor (2002: 550) casts doubt on the idea that linguistic expressions can even be categorised according to whether they exhibit fully compositional meaning or not: 'Strict compositionality is rarely, if ever, encountered. Most expressions (I am tempted to say: *all* expressions), when interpreted in the context in which they are uttered, are non-compositional to some degree.' Furthermore, as Taylor also points out, if we were to include idioms of encoding within the taxonomy, the number of expressions that would be described as idiomatic (the conventional way of describing something) would increase dramatically to include a far wider range than those captured by the taxonomy set out in this section. It follows that the success of any attempt to 'organise' idioms into categories depends to a large degree on the definition of 'idiom' that it rests upon. Despite this note of caution, a relatively stable empirical generalisation to emerge from this discussion is the distinction between substantive and formal idioms; it is the latter category that represents the focus of Fillmore *et al.*'s (1988) study.

19.2.2 Case study I: the *let alone* construction

Fillmore *et al.* are particularly interested in accounting for formal idioms because, while it is at least plausible that speakers might learn substantive idioms item by item rather like learning individual words, it is not plausible that a speaker learns each instance of a formal idiom item by item. In principle, the number of instances of formal idiom constructions is infinitely large. Despite this, such constructions often have a clearly identifiable pragmatic force. For this reason, formal idioms pose a particularly interesting challenge to the 'words and rules' model of grammar: they are productive and therefore rule-based (systematic), yet often defy the 'usual' rules of grammar. Fillmore *et al.* therefore took as their case study the idiomatic *let alone* construction.

According to Fillmore *et al.*, the *let alone* construction can be described in terms of its structural, semantic and pragmatic properties, some of which are regular and some of which are idiosyncratic. The *let alone* construction displays regular syntactic properties, and is characterised by the presence of the coordinating conjunction *let alone*, which coordinates two prosodically prominent (stressed) expressions. This construction is illustrated by example (1a), which is repeated here as (3). In this example, the expressions in boldtype, *maths* and *rocket science* (labelled as A and B respectively), are prosodically prominent and are coordinated by *let alone*.

(3) George doesn't understand **maths**, let alone **rocket science**
                                    [A]   [B]

In semantic terms, the construction has the idiosyncratic property that the coordinated expressions are interpreted as contrasted points on a scale, where the second conjunct (*rocket science*) has greater emphatic force than the first (*maths*). In the context of knowing that Lily, the famous rocket scientist besotted with George, often tells George about her work, we might ask whether George in fact understands rocket science. The utterance in (3), as a result of the *let alone*, conveys they information that because George doesn't understand maths, he is even less likely to understand rocket science. This rests upon the assumption that 'understanding maths' is a prerequisite for 'understanding rocket science'.

Closely related to this property of the construction is the fact that *let alone* can be described as a **negative polarity item**. This means that it can only occur in negative contexts, whether this is determined by a morphosyntactic negation, as it is in example (3), or by a lexical item like *doubt*, which brings with it a negative interpretation. This is illustrated by example (4).

(4) I doubt George can ride a bike, let alone drive a car

The *let alone* construction has pragmatic point. Not only does the construction reject a particular proposition (for example, that George understands rocket science or can drive a car), but it does so by providing additional relevant information. The relevant information relates to the first conjunct (A) and establishes an **implicational scale** between the expressions conjoined by *let alone*. If George doesn't understand maths (A) this implies that he doesn't understand rocket science (B). The pragmatic impact of this construction is that by first rejecting a weaker proposition, the proposition that our attention is focused upon (e.g. whether George understands rocket science) is more forcefully rejected than it would otherwise have been. These idiosyncratic properties of the *let alone* construction are in fact shared among a 'family' of similar constructions. Some examples are provided in (5).

(5) a. George can't make a slice of toast, never mind cook a lobster.
    b. Lily doesn't approve of canned tomatoes, much less pot noodles.

In light of their findings concerning the *let alone* construction, Fillmore *et al.* argue against the 'words and rules' view (which they call the 'atomistic' view) of grammatical operations, where lexical items are assembled by phrase structure rules into complex units that are then assigned compositional meaning and only subsequently subjected to pragmatic processing. In other words, they argue against a modular view of the language system. Instead, Fillmore *et al.* (1988: 534) argue that speakers have, as part of their linguistic knowledge or competence, 'clusters of information including, simultaneously, morphosyntactic patterns, semantic interpretation principles to which these are dedicated, and, in many cases, specific pragmatic functions in whose service they exist.' In other words, speakers have access to **constructions**.

At this point, we should pause to consider the various senses of the term 'construction'. In traditional grammar, this term refers to a clause type, such as the 'passive construction' or the 'cleft construction'. These labels apply to the sentence as a whole, which can be classified as construction X or construction Y on the basis of certain morphosyntactic or semantic properties. For example, the passive construction (6a) is identified by the fact that the subject is interpreted as the PATIENT, while the (optional) *by*-phrase expresses the AGENT. In addition, it is identified by the presence of the passive auxiliary *be* and the past participle form of the content verb. This information can be schematically represented as in (6b). In a similar way, the (subject) cleft construction (7a) can be captured by the schematic representation in (7b).

(6) a. Lily was betrayed by George.
  b. $NP_{PATIENT}\ be_{AUX\text{-}PASS}\ V_{P.PART}\ (by\ NP_{AGENT})$

(7) a. It was George who betrayed Lily.
  b. $It\ be_{COPULA}\ NP_{FOCUS}\ who/that\ VP$

In the Chomskyan generative model, these constructions have the status of 'taxonomic epiphenomena' (Chomsky 1991: 417). In other words, the model of grammar does not need to contain whole constructions because these can be predicted on the basis of the words and rules that the grammar contains. This means that most generative linguists use the term 'construction' as a shorthand for describing certain types of syntactic structures that have certain identifiable properties (for example, 'the passive construction' or 'the *wh*-construction'), but these constructions are not themselves primitives in the model. Instead, they are the output of the 'words and rules' model and as such are not of central importance. Instead, the emphasis in this model is upon characterising the rules that give rise to the constructions.

Against this background, it is clear that Fillmore *et al.*'s proposal reflects a very different view of how language should be modelled. Instead of a model in which syntactic, semantic, phonological and pragmatic knowledge is represented in encapsulated subsystems, the constructional model proposes that all this information is represented in a **single unified representation**, which is the construction. In the next section, we will look in detail at a representation of an idiomatic construction, and we will discuss in more detail what it means to develop a constructional model of language and in what sense this type of approach can be held to account for both regular and idiomatic properties of language. Indeed, the constructional model proposed by Charles Fillmore, Paul Kay and their colleagues, grounded in their work on idioms, provided the empirical basis for the symbolic thesis which, as we saw in Chapter 14, is central to a cognitive approach to grammar.

However, it is important to emphasise that Fillmore *et al.*'s discussion of the *let alone* construction is situated within a broadly generative paradigm rather than a cognitive linguistics paradigm. For example, part of their paper is concerned with the rules that might underlie the *let alone* construction, and this theoretical context also explains the separation of semantic and pragmatic meaning in their discussion of the construction. Nevertheless, their proposal that speaker knowledge is not 'compartmentalised' but 'clustered' around individual constructions represented an important shift in terms of how speaker knowledge could be modelled and set the scene for the emergence of constructional models of grammar.

19.2.3 Case study II: the *what's X doing Y* construction

In a later paper, Kay and Fillmore (1999) map out the details of the new framework that they call Construction Grammar. Although, as we have noted, their approach remains situated in a broadly generative formal framework, their model has more in common with a non-transformational generative theory like Head-driven Phrase Structure Grammar (HPSG) than it does with transformational generative models such as Principles and Parameters Theory or the Minimalist Program (see Chapter 22 for some comparison of these theories). We begin by looking at the idiomatic construction that Kay and Fillmore choose to illustrate their theoretical framework and we then sketch out the details of the framework itself.

The idiomatic construction that Kay and Fillmore discuss in their 1999 paper is called the *what's X doing Y* construction, which they abbreviate to the 'WXDY construction'. This construction is illustrated by the examples in (8).

(8) a. What's [$_X$ George] doing [$_Y$ kissing that woman]?
 b. What are [$_X$ these dishes] doing [$_Y$ in the sink]?
 c. What was [$_X$ Lily] doing [$_Y$ with my nightie on]?
 d. What's [$_X$ George] doing [$_Y$ with those silver candlesticks]?
 e. What was [$_X$ Lily] doing [$_Y$ without a solicitor]?
 f. What is [$_X$ Lily] doing [$_Y$ covered in spaghetti]?
 g. What is [$_X$ Lily] doing [$_Y$ naked]?

As these examples illustrate, the construction lends itself to a wide range of specific examples. The Y part of the construction is particularly flexible, and can be headed by various categories including participial verb forms (*kissing;*

covered), prepositions (*in; with; without*) or adjectives (*naked*). We explore the properties of the construction in more detail below.

Kay and Fillmore motivate the existence of this idiomatic construction with a discussion of the familiar 'fly in the soup joke' (Kay and Fillmore 1999: 4):

(9) Diner: Waiter, what's this fly doing in my soup?
    Waiter: Madam, I believe that's the backstroke

As we discussed in Chapter 1, this joke turns on the fact that there are two possible interpretations of the diner's question. One is that it is a straightforward information question, while the other is that it is an expression of what Kay and Fillmore call the **incongruity** of the situation described. The latter reading identifies the WXDY construction. Each interpretation can be paraphrased differently, as shown by the following examples (Kay and Fillmore 1999: 4):

(10) a. How come there's a fly in my soup?
     b. What's this fly in my soup doing?

The paraphrase in (10a) identifies the WXDY construction, which is what the diner in (9) intended. In contrast, the paraphrase in (10b) identifies the straightforward information question interpretation. This is the interpretation that the waiter chooses to respond to and it is this 'mismatch' between what the diner intended and how the waiter responds that gives rise to the joke.

Like the *let alone* construction, the WXDY construction is a productive formal idiom that has identifiable structural and pragmatic properties. As we have seen, what is 'special' about the WXDY construction in pragmatic terms is the incongruity judgement it gives rise to. In structural terms, the WXDY construction is characterised by certain idiosyncratic grammatical properties. We will examine a few of these here. To begin with, Kay and Fillmore demonstrate that in order to achieve the incongruity reading, the construction must contain the verb *do*. While (11a) is ambiguous between the straightforward information question interpretation and the incongruity interpretation, the latter interpretation is not available for examples (11b) and (11c), despite the fact that these are (rather unnatural but grammatical) paraphrases of (11a) (examples adapted from Kay and Fillmore 1999: 5):

(11) a. What was she doing under the bed?
     b. What activity was she engaged in under the bed?
     c. What act was she performing under the bed?

Secondly, the WXDY construction requires the verb *do* to appear in the progressive participle form, as illustrated by example (12). Observe that if the verb *do* occurs in the simple past, for example (12b), the sentence becomes ill-formed.

(12) a. What was Lily doing eating fish and chips?
     b. *What did Lily do eating fish and chips?

Thirdly, the construction does not allow either *be* (13a) or *do* (13b) to be negated, unlike an ordinary information question (e.g. *What isn't Lily doing at work?*). Observe, though, that negation of the Y part of the construction is possible (13c). This example gives rise to the interpretation that Lily is expected to be eating fish and chips.

(13) a. *What isn't Lily doing eating fish and chips?
     b. *What is Lily not doing eating fish and chips?
     c. What's Lily doing not eating fish and chips?

19.3 Construction Grammar

In this section, we sketch out the architecture of Kay and Fillmore's theory of Construction Grammar. It is not our objective to provide a detailed account of this model here as this would take us into a discussion of formal models that is beyond the scope of this book. Instead, we will try to give a sense of how this model departs from the 'words and rules' approach that characterises most generative approaches to grammar and how it thus sets the scene for the emergence of usage-based constructional models in the cognitive linguistics paradigm, which we turn to in detail in the next chapter. Kay and Fillmore (1999: 7) state that the Construction Grammar approach has 'the ability to demonstrate the smooth interaction of relatively idiomatic constructions, like WXDY, with the more familiar constructions in licensing the sentences of the language.' In other words, the Construction Grammar commitment is to provide an integrated account of both the regular and the idiomatic properties of language.

19.3.1 The Construction Grammar model

Kay and Fillmore's Construction Grammar model is **monostratal**. This means that it contains only one level of syntactic representation rather than a sequence of structures linked by transformations, a feature that characterises transformational generative models like Principles and Parameters Theory. Furthermore, the representations in Construction Grammar contain not only syntactic information but also semantic information relating to argument structure as well as pragmatic information.

Kay and Fillmore's Construction Grammar contains a number of generalised constructions that underlie more specific constructions like the WXDY construction. Because this is a non-derivational monostratal model, it does not have any phrase structure rules that assemble words into phrases and sentences. Instead, it has constructions that represent syntactic patterns. For example, the model has a head-complement construction that represents the structural relationship between a lexical head (for example, a verb) and its complement(s) (for example, the object(s)). This construction captures the basic structural relationship that holds across different categories (for example, VP, AP, PP, NP). The model also has a subject-predicate construction, which captures relationships between, for example, subject NP and predicate VP. In addition to various construction types, the model also contains a number of principles that ensure, for example, that categorial features of a lexical head will be shared with the constituent headed by that phrase (e.g. a verb heads a verb phrase), or that constituents local to a head, with the appropriate features, can be recognised as complements. We do not concern ourselves with the details of these principles here, beyond pointing out that readers familiar with HPSG will notice a number of striking similarities between these two models.

The various constructions that make up Kay and Fillmore's Construction Grammar model are linked together via an **inheritance** relation. This means that more specific constructions inherit the properties of more general constructions. For example, the VP construction inherits all the information in the head-complement construction, but adds further information concerning the category of the head and the fact that the VP requires a subject in order to complete the valence requirements of the head. As we will see, the WXDY construction also inherits the properties of several more generalised constructions. To illustrate these properties, we will look in detail at the Construction Grammar representation of the WXDY construction, which is represented in Figure 19.3.

Although this diagram appears rather complex, we provide a 'translation' below, based on example (14).

(14) What is Lily doing under my bed?

According to Kay and Fillmore, this construction is headed by the verb *be* (the form *is* in our example), and the category ('cat') of the construction as a whole is therefore V. Like HPSG, and indeed like Langacker's Cognitive Grammar, Kay and Fillmore's Construction Grammar approach views the verb as the head of the sentence. This is the information that appears in the top set of brackets in Figure 19.3, marked 'syn'. This is an abbreviation of 'syntax' and labels the construction as a whole in terms of its categorial status. Kay and Fillmore's claim that the verb *be* heads the construction rests upon their view

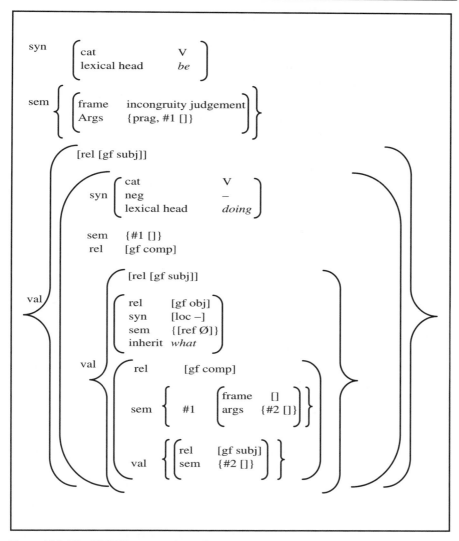

Figure 19.3 The WXDY construction (after Kay and Fillmore 1999: 20)

that this is not a progressive auxiliary but a copula, a discussion that we do not pursue here.

The next set of brackets, labelled 'sem', provides information about the semantic and pragmatic properties of the construction. The information 'frame: incongruity judgement' provides information about the pragmatics of the construction. The term 'frame' refers to the scene described by the sentence over which the pragmatic value 'incongruity judgement' is held by a 'judge'. This 'judge' is labelled 'prag', which means that the identity of this judge has to be pragmatically resolved. In other words, the identity of the 'judge' may or may not be the speaker, depending on the context in which the construction is uttered. In our example, the 'judge' is likely to be the speaker, but if this example were a case of reported speech (e.g. . . . *and then she said 'What is Lily doing under my bed?'*) the judge would be the person referred to by the speaker as *she*. The incongruity judgement is held by this 'judge' with respect to a situation labelled #1. In our example, we can paraphrase the situation over which the incongruity judgement is held as 'Lily being under my bed'.

The largest set of curly brackets, labelled 'val' (valence), provides information about the structure of the construction. The first part, [rel [gf subj]], identifies a unit with the relation (rel) of grammatical function (gf) subject (subj). This is the X in the WXDY construction. In our example, the X corresponds to *Lily*, the subject of the verb *be*.

The largest set of square brackets, below the information about the subject, corresponds to the rest of the construction, headed by *doing*. In other words, Kay and Fillmore argue that *doing Y* forms a constituent. In our example, the string *doing under my bed* is the relevant part of the construction. Within these square brackets, the top brackets labelled 'syn' tell us that this part of the construction is headed by *doing*, which has the category V. Observe that this element is marked as having a negative value for negation ('neg –'). This is how Kay and Fillmore capture the fact that the WXDY construction does not license negation of the verb *doing*, as we saw in example (11b). Indeed, we might ask why they *be* is not marked in the same way, given example (11a). The next set of brackets marked 'sem' tells us that the semantics of this part of the construction correspond to the situation over which the incongruity judgement holds. In other words, it is the *Y* part of *doing Y* that fills in the information that the 'judge' holds to be incongruous. The next set of brackets marked 'rel' tell us that this part of the construction, *doing Y*, is the complement of the verb *be*. Again, this rests on Kay and Fillmore's views concerning the constituent structure of the construction which we do not pursue here.

The second largest set of curly brackets, labelled 'val', tells us about the structure of *doing Y*. One important aspect of the construction that this model needs to account for is how the main clause subject comes to be understood as the subject of the Y predicate. In other words, in our example, *Lily* is understood not only as the subject of the verb *be*, but also as the subject of *under my bed*. This means that both instances of [gf subj] have to be linked in the construction. In Kay and Fillmore's model, this linking is done by an independent construction that they call the **coinstantiation construction** (Kay and Fillmore 1999: 23). We do not concern ourselves here with the details of this.

A further point of interest relating to this part of the construction concerns the element *what*, which is not yet accounted for. Observe that the construction starts by telling us about the head (*be*), the subject (*Lily*) and the complement (*doing* Y), but has not so far told us anything about *what*, the first element in the construction. This is because *what* is part of a ***wh*-dependency** relation. The label '*wh*' is shorthand for interrogative words like *what, who, where* and so on. The term 'dependency' relates to the idea that two positions in a structure are related. Consider example (15):

(15) Q: What is Lily doing?
    A: Lily is doing some dusting.

Although the question word *what* occurs in clause-initial position in the question, it is nevertheless interpreted as the object of *doing*, which is illustrated by the fact that the answer to the question, *some dusting*, occurs in the object position (after the verb). The question of how to account for dependencies like this is a recurring theme in models of grammar. Transformational generative models like Principles and Parameters Theory account for *wh*-dependencies by means of a **movement analysis**: the *wh*-phrase is moved by transformation from object position to clause-initial position. Non-transformational theories, which are monostratal, favour a different explanation. Like HPSG, Kay and Fillmore's Construction Grammar model accounts for the *wh*-dependency by means of a **filler-gap analysis**. This means that the construction simultaneously represents the *wh*-phrase (filler) in its clause-initial position and the position in which the *wh*-phrase is interpreted (gap), and links the two together.

Kay and Fillmore's account of the *wh*-dependency rests on what they call the **left isolation** construction. The term 'left isolation' expresses the fact that (at least in languages like English) the *wh*-phrase occurs in the leftmost position in the clause, and is 'isolated' in the sense that it is separate from local constituents. The left isolation construction has two 'daughters', which means that it consists of two main constituents. (The term 'daughter' is inherited from syntactic theories that rely upon tree diagrams.) The left daughter is the *wh*-phrase, which is linked or unified with one of the arguments that is required by a predicate within the right daughter. In other words, the *wh*-phrase satisfies one of the valence requirements of a non-local predicate. In the WXDY construction, *what* is interpreted as the object of *doing*.

Let's look again at Figure 19.3 in order to see how this information is represented. Observe that the set of square brackets that tells us about the object of *doing* marks this part of the construction as [loc−]. This means that the unit that satisfies the object requirement of *doing* is non-local. The information 'inherit *what*' tells us that the non-local constituent *what* is to be unified with this position in the structure, thus fulfilling the valence requirement of *doing*. This account of the filler-gap dependency can account not only for part of the syntax of the WXDY construction, but also for *wh*-interrogatives in general. This illustrates an important aspect of Kay and Fillmore's theory, which is that 'regular' and 'idiomatic' constructions should in large part be accounted for by the same theoretical machinery.

However, the expression *what* that occurs in the WXDY construction, despite sharing syntactic properties with the expression *what* that occurs in 'ordinary' questions, does not share the referential features of the 'ordinary' question word *what*. In other words, observe that *what* in the 'ordinary' question in example (15Q) picks out some entity or event (the thing that Lily is doing) whose identity or value is questioned. This licenses an answer along the lines given in (15A), which fills in the required information about that entity or event. In contrast, given the pragmatic import of the WXDY construction, which identifies the function of the construction as the expression of an incongruity judgement, the *what* in this construction does not have **referential value**. In other words, it does not pick out an entity or event in the same way that the 'ordinary' question word *what* does, because it does not require an answer like 'dusting' or 'hiding from George'. This explains why an answer like the waiter's in example (9) is not licensed. This 'special' feature of *what* is marked in the WXDY construction by the information {[ref Ø]}, which tells us that *what* does not have referential value. Observe, however, that the position of *what* is not marked in the WXDY construction. This is because the left isolation construction is an independent construction from which the WXDY construction inherits certain properties. This illustrates how certain generalised properties of idiomatic constructions are accounted for in Kay and Fillmore's model.

The next set of square brackets provides information about the Y part of the construction. In our example, this part corresponds to *under my bed*. The representation tells us that Y is a complement of *doing* in Kay and Fillmore's analysis. The representation also links the semantics of Y to the situation over which the incongruity judgement holds, which is tagged as #1 throughout the representation of this construction, as we have seen. The information about the semantics of the frame or situation is left blank in this diagram because this is a generalised representation of the WXDY construction rather than a representation of a specific example. In other words, one of the features that identifies the WXDY as a formal idiom is the fact that it provides a syntactic 'template' into which a potentially infinite set of specific lexical items can be inserted. The construction tells us that this unspecified Y constituent will contain at least one argument that corresponds to the subject requirement of the Y predicate. This is why the information about the argument of Y and its subject are linked by the tag #2.

In sum, the WXDY construction has a number of **'regular'** syntactic features, which it inherits from other less specific constructions. Firstly, the WXDY construction contains head-complement structures and subject-predicate structures, which means that it inherits the properties of these two fundamental constructions that underlie 'regular' as well as idiomatic constructions.

Furthermore, the properties of these basic constructions are inherited in turn by the specific categorial instantiations of these generalised constructions, such as VP, PP and so on. In these respects, the idiomatic construction shares much in common with all 'regular' constructions. Secondly, the WXDY construction inherits the syntactic properties of the left isolation construction, which is involved in 'regular' interrogatives as well as in this idiomatic construction. Thirdly, the WXDY construction inherits the properties of the coinstantiation construction, in order to link a single NP to the subject valence requirement of two distinct predicates.

In addition to its 'regular' properties, the WXDY construction also has a number of features that identify it as **idiomatic**. As we have seen, these features not only involve the morphosyntax of the construction (in terms of the form of the verb *doing*, or the restriction on negation, for example), but also crucially involve the meaning of the construction, which brings with it a striking and idiosyncratic interpretation that cannot be straightforwardly predicted from the parts that make up the construction.

19.3.2 Construction Grammar: a 'broadly generative' model

It is important to reiterate the fact that Kay and Fillmore's Construction Grammar model is a **formal** model. In other words, it requires the statement of exhaustive, precise and unambiguous theoretical machinery that is intended to be sufficient in accounting for the properties of human language. As we have observed, and as Kay and Fillmore themselves acknowledge, their model is reminiscent of other broadly generative formal models, particularly models like HPSG that assume a non-transformational monostratal syntax. What 'broadly generative' theories have in common is that they assume Universal Grammar as a working hypothesis, and attempt to build a model that represents this knowledge. In other words, these are not usage-based theories.

The differences between non-transformational generative models on the one hand and the transformational generative model on the other are obvious. While the transformational model captures phenomena like the *wh*-dependency by means of two syntactic representations linked by a transformation or movement operation, the monostratal generative models assume a single syntactic representation and build into that representation features or 'tags' that capture the same linguistic phenomena. We have seen how Kay and Fillmore's Construction Grammar accounts for the *wh*-dependency, for example, by means of the left-isolation construction.

The differences between Kay and Fillmore's Construction Grammar and another monostratal model like HPSG, however, may be less obvious but are no less important. The HPSG model is, like the transformational generative model, a 'words and rules' model. In other words, it assumes a lexicon in which items are tagged with a complex and detailed set of features (including category, valence, number and so on) and a set of rules that assemble those lexical items into a syntactic structure. For example, HPSG assumes a head-complement rule and a subject-predicate rule, which are 'structure building rules' in the same sense as the phrase structure rules that operate within the transformational model. Although meaning and grammar are arguably more closely integrated in HPSG than in the transformational generative model, HPSG can still be described as a modular theory, particularly given that it assumes the autonomy of syntax.

In contrast, as we have seen in this chapter with respect to Construction Grammar, and as we saw in our discussion of Cognitive Grammar in Chapters 16–18, a constructional model does not assume 'words and rules' but instead assumes 'ready-made' grammatical constructions, some of which are highly detailed and some of which are highly generalised. A further important difference between the HPSG model and Kay and Fillmore's Construction Grammar is that the latter assumes that non-compositional meaning (such as the incongruity judgement associated with the WXDY construction) is directly linked to the grammatical construction itself. Furthermore, this meaning is linked to the construction as a whole rather than being derived from some subpart of the construction. This is important because it shows that the constructional model is not modular. In other words, constructions contain information about syntax, morphology, semantics and pragmatics (and, in principle, phonology) within a single integrated representation.

19.3.3 Comparing Construction Grammar with Cognitive Grammar

In this section, we compare Kay and Fillmore's Construction Grammar with Langacker's Cognitive Grammar. It will already be clear that the two theories share a number of important assumptions. Firstly, both approaches agree that idiomatic expressions should have central rather than peripheral status in a model of grammar. Secondly, both approaches agree that the most explanatory model of language is one that assumes constructions. In other words, both approaches favour a unified representation that links together syntactic, semantic, pragmatic (and phonological) information rather than representing these as properties of distinct components of the grammar, as in a 'words and rules' generative model. This means that both approaches subscribe to the symbolic thesis as construed by cognitive linguists. Of course, all theories of language adopt some version of the symbolic thesis in the sense that words are widely recognised as form-meaning pairings. As we have seen throughout Part III of this book, the cognitive model extends this idea to complex constructions, and furthermore accords the symbolic unit a central status by rejecting syntactic rules. From this perspective, we can describe Cognitive Grammar as a type of construction grammar. Thirdly, as we saw in Chapter 14, an important similarity between Cognitive Grammar and construction grammars is that they take an **inventory** approach to the psychological representation of grammar. This type of approach assumes that the language system does not work predominantly by **building** structure, but by **storing** it in a complex network of interlinked constructions.

Despite these important points of agreement, however, there is an important difference between the two approaches. As we have seen, the Construction Grammar model developed by Kay and Fillmore rests upon broadly **generative** assumptions and therefore assumes Universal Grammar as a working hypothesis. In other words, Construction Grammar sets out to develop a set of statements, albeit stated in terms of constructions, which underlie competence or knowledge of language in the Chomskyan sense. In contrast, Cognitive Grammar is a usage-based theory, a feature that unites the cognitively oriented constructional approaches that we discuss in the next chapter. As we saw in Chapter 4, and throughout Part III of the book, usage-based models of language reject the Universal Grammar hypothesis, and argue instead that knowledge of language emerges from language use. Finally, as we have seen in previous chapters, Langacker's emphasis is on mapping out the cognitive principles and mechanisms that give rise to the units of language and to the relationships that hold between these units. In contrast, the Construction Grammar approach developed by Kay and Fillmore focuses directly upon the formal properties of the constructions that make up the structured inventory assumed by both approaches.

19.4 The 'Generalisation Commitment'

In this final section, we comment briefly on the Construction Grammar approach in the light of the Generalisation Commitment. Recall from Chapter 2 that cognitive linguists are committed to characterising general principles that are responsible for all aspects of human language. We saw in Chapter 14 that cognitive approaches to grammar adopt this commitment. This means that a cognitive approach aims to characterise not only 'general' or 'regular' properties of language but also 'irregular' or 'idiosyncratic' properties. In according idiomatic expressions a central place in a model of grammar, the Construction Grammar approach goes some way towards meeting the Generalisation Commitment, despite the fact that Construction Grammar is strictly characterised as a formal rather than a cognitive approach. In several respects, then, Kay and Fillmore's Construction Grammar has been extremely influential in cognitive approaches to grammar, despite the fact that it is not a usage-based model and cannot therefore be classified as a cognitive approach to grammar. In the next chapter, we explore several approaches to grammar that build upon the insights developed in Construction Grammar within a usage-based and thus fully cognitive model.

19.5 Summary

In this chapter, we explored the empirical **motivation** for a constructional model of grammar. We compared a constructional account with the **'words and rules'** account assumed in most generative models of language, and established that a constructional account rests upon a single unified representation that links together syntactic, semantic, pragmatic (and, in principle, phonological) information, rather than viewing these as the output of distinct components of the grammar. We then turned our attention to **idiomatic expressions**, linguistic units that display **idiosyncratic** as well as **regular** properties and cannot therefore be fully accounted for by a model of language that focuses on accounting for what is 'regular'. We explored in detail two idiomatic grammatical constructions, the *let alone* **construction** (Fillmore, Kay and O'Connor 1988) and the *what's X doing Y* **construction** (Kay and Fillmore 1999). In addition to displaying some regular grammatical properties, these constructions have grammatical, semantic and pragmatic properties that are not fully predictable from their subparts. This finding motivates the claim that grammatical constructions can be **meaningful** in part independently from the content words that make up instances of the construction. This claim is central to the cognitive approaches to construction grammar explored in the next chapter. Having explored the empirical motivation for a constructional approach to grammar, we sketched out the **Construction Grammar** model proposed by Kay and Fillmore (1999), and compared and contrasted this approach with both generative and cognitive approaches to language. Finally, we considered the Construction Grammar model in the light of the **'Generalisation Commitment'**. We established a number of important similarities and differences between these models, and observed that Construction Grammar, although strictly a generative model, has been extremely influential in cognitive approaches to grammar. The constructional models that have emerged in response to the claims made by Construction Grammar are the topic of the next chapter.

Further reading

Introductory texts

- **Croft and Cruse (2004)**. Chapter 9 of this book, entitled 'From idioms to construction grammar', describes a range of idiom types and discusses the challenges posed by such expressions to a modular theory of language. Chapter 10 presents an overview of a range of

constructional accounts, including the Construction Grammar model of Fillmore *et al.* (1988) and Kay and Fillmore (1999), and Langacker's Cognitive Grammar model. This chapter also discusses a number of other constructional theories, to which we return in the next chapter.
- **Taylor (2002)**. Chapter 27 provides a detailed description of idioms, formulas and fixed expressions, and discusses the status of these types of expressions in formal linguistics. Chapter 28 discusses the status of constructions in Cognitive Grammar and includes some discussion of the literature reviewed in this chapter.

Construction Grammar

- **Fillmore (1985b); Fillmore (1988)**. These two papers map out Fillmore's early ideas about Construction Grammar.
- **Fillmore, Kay and O'Connor (1988); Kay and Fillmore (1999)**. These two papers, which provide the basis of the discussion in the present chapter, represent the seminal primary sources for Construction Grammar.
- **Östman and Fried (2005a)**. This paper provides a useful introduction to Construction Grammar, and the volume in which it appears includes papers on a range of constructional approaches (see Chapter 20).

Background reading on HPSG

- **Borsley (1996)**. This book provides an accessible textbook treatment of HPSG and is probably the best place to start for readers unfamiliar with the model.
- **Borsley (1999)**. This book presents a balanced and neutral comparative introduction to HPSG and Principles and Parameters Theory and compares the two approaches side by side, examining how each theory builds syntactic structure as well how each theory accounts for phenomena like binding (referential dependencies), passive constructions and *wh*-dependencies.
- **Pollard and Sag (1994)**. This book represents the major primary source for HPSG.

Exercises

19.1 The main assumptions of Kay and Fillmore's approach

Based on the discussion in this chapter, what were the main motivations for claiming that grammar has a constructional basis? Summarise the main claims of Kay and Fillmore's approach in the form of an annotated table.

19.2 Idioms

In your own words, provide definitions of the following terms, and provide four examples of your own to illustrate each type of idiom:

  (i) encoding and decoding idioms
  (ii) grammatical versus extragrammatical idioms
  (iii) substantive versus formal idioms
  (iv) idioms with versus idioms without pragmatic point

Now classify the following idioms based on the definitions you have devised:

  (a) break a leg
  (b) fast asleep/sound asleep
  (c) bite the dust
  (d) full of yourself
  (e) pull a fast one
  (f) top of the morning
  (g) under the weather

19.3 Let alone

In the text, *let alone* was described as a negative polarity item. However, it is possible to find the *let alone* construction in contexts like the following:

  (a) A: Was Lily surprised that George remembered her birthday?
      B: Lily was flabbergasted, let alone surprised!

  (b) A: I doubt George has enough champagne for all the guests
      B: He's got enough for a small army, let alone the guests!

Can the characterisation of *let alone* as a negative polarity item be maintained, given these examples?

19.4 Let alone again

Now consider the order in which the prosodically focused elements are conjoined in examples like those given in Exercise 19.3. It seems that when *let alone* occurs in a construction with an overt negative expression, it organises the scale it sets up in a different way from when it occurs without an overt negative expression. Explain how the scale is organised differently. Can you account for this observation? Finally, consider the extent to which the following two expressions pattern in the same way when they occur in the related construction types that we illustrated in example (5) in the text:

(a) never mind
(b) much less

19.5 Comparing Construction Grammar with Cognitive Grammar

Compare and contrast Kay and Fillmore's approach as presented in this chapter with Langacker's theory of Cognitive Grammar. In what ways are the approaches, assumptions and claims of these two models similar? In what ways do they diverge? Summarise your comparison in the form of an annotated table.

20

The architecture of construction grammars

In the last chapter, we explored motivations for a constructional approach to grammar. We established that a constructional account rests upon a single unified representation that links together all aspects of the meaning and form of an utterance, rather than viewing these as the output of distinct components of the grammar. In that chapter, we concentrated on the model of Construction Grammar developed by Paul Kay and Charles Fillmore, a broadly generative model that claims that grammatical constructions can be meaningful, in part, independently of the words that 'fill' them. As we will see in this chapter, this claim has been central to the constructional approaches developed within cognitive approaches to grammar. We will concentrate our discussion here mainly on the framework developed by Adele Goldberg, particularly in her 1995 book, Constructions (section 20.1). As we will see, Goldberg's approach focuses on the **argument structure** of sentence-level constructions such as the English **ditransitive construction** (for example, *Lily knitted George a jumper*) and the English **resultative construction** (for example, *Lily drank herself stupid*). Although most instances of these constructions are not idiomatic in the sense that they do conform to the 'regular' patterns of language, Goldberg argues that these constructions contain meaning that cannot be attributed to the lexical items that fill them. In this way, the constructional approach is extended to account for regular instances as well as idiomatic instances. As we will see, however, Goldberg's model departs from Kay and Fillmore's Construction Grammar in that it is fully usage-based. Having discussed Goldberg's approach in some detail, we will then briefly compare two other cognitively oriented constructional approaches: **Radical Construction Grammar**, developed by William Croft (section 20.2) and the most recent approach known as **Embodied**

Construction Grammar, developed by Benjamin Bergen and Nancy Chang (section 20.3). Finally, we will draw some explicit comparisons between the various constructional approaches to grammar that we have explored in Part III of this book (section 20.4).

20.1 Goldberg's construction grammar

The contribution of Fillmore *et al.* and Kay and Fillmore in developing Construction Grammar was to establish the symbolic thesis from first principles. These researchers observed that the 'words and rules' approach to grammar, while accounting for much that is regular in language, had failed to account for the irregular, which represents a significant subset of language. They then set out to explain the irregular first, on the assumption that once principles have been developed that account for the irregular, then the same principles should be able to explain the regular. As we saw, their approach centred upon the construction. In this way, these researchers motivated the extension of the symbolic thesis from words to complex grammatical constructions on the basis of the generalisation commitment: a commitment to a common set of principles that accounts for all the units that comprise a language, including sound, meaning, lexicon and grammar.

The next stage in developing the constructional perspective is to apply this approach to what is regular in the grammar. Perhaps the most important development in this area has been Adele Goldberg's work, most notably her landmark 1995 book. Influenced both by the work of Kay and Fillmore and by the early work of George Lakoff on constructions (in particular his 1987 case study of *there* constructions), Goldberg developed a construction grammar that sought to extend the constructional approach from 'irregular' idiomatic constructions to 'regular' constructions. In order to do this, Goldberg focused on **verb argument constructions**. In other words, she examined 'ordinary' sentences, like transitives and ditransitives, and built a construction grammar on the patterns she found there.

20.1.1 Assumptions

The central thesis of Goldberg's theory is that sentence-level constructions 'themselves carry meaning, independently of the words in the sentence' (Goldberg 1995: 1). According to this view, constructions are themselves theoretical primitives rather than 'taxonomic epiphenomena' (Chomsky 1991: 417), as we saw in the last chapter. Although Goldberg does not deny that word-level units contribute a great deal to the meaning and structure of sentences (section 20.1.2), she argues that a purely 'bottom-up' or lexically driven model of grammar fails to provide the whole picture.

As Goldberg observes, the issue of **argument structure alternations** has received a considerable amount of attention in twentieth-century linguistics. We will look in more detail at argument structure alternations in the next section, but for the time being consider the examples in (1) and (2).

(1) a. George brought Lily some breakfast.
    b. George brought some breakfast to Lily.

(2) a. *George brought the table some breakfast
    b. George brought some breakfast to the table.

As these examples illustrate, the ditransitive verb *bring* can occur in two different construction types. Examples like (1a) and (2a) are called **double object constructions** (or **dative shift** constructions) because the verb is followed by two nominal objects. In examples (1b) and (2b), which we will call the **prepositional construction** (Goldberg 1995: 8), the indirect object (*Lily* or *the table*) is instead represented by a preposition phrase (PP). The point of interest here relates to the fact that while the prepositional construction allows the recipient to be either animate (1b) or inanimate (2b), the double object construction requires that it be animate (compare (1a) with (2a)). The issue that arises from this observation is how these differences are best captured in the model of the grammar. Goldberg argues that the most explanatory account associates these semantic restrictions directly with the grammatical construction itself, rather than stating the information in the lexical entries of individual verbs.

Before proceeding with the discussion of Goldberg's theory, it is important to point out that her definition of a construction differs somewhat from the definition assumed by Langacker in his theory of Cognitive Grammar. Recall that Langacker defines a construction as any unit with a complex symbolic structure (a complex word, a phrase consisting of more than a single free morpheme or a sentence). Compare Goldberg's definition:

> C is a CONSTRUCTION iff C is a form-meaning pair <F_i, S_i> such that some aspect of F_i or some aspect of S_i is not strictly predictable from C's component parts or from other previously established constructions. (Goldberg 1995: 4)

In this definition, F stands for 'form' and S stands for 'semantics', so that <F, S> represents a symbolic unit. The subscripts represent the symbolic link between form and meaning. Crucially, this definition of construction hinges on the issue of **predictability**, which in turn is related to compositionality, but in a different way from Langacker's definition. If any aspect of either the form or the meaning of a unit cannot be shown to be predictable from the properties of its component parts, then it has the status of a construction in Goldberg's model. It follows that both bound morphemes (like plural-*s*) and free morphemes (simplex words like *cat*) are constructions in Goldberg's theory, while they do not have construction status in Langacker's theory. For Goldberg, neither the form nor the meaning of a morpheme is predictable from its component parts, since it lacks compositional structure. It also follows from Goldberg's definition of a construction that a complex word, phrase or sentence (which are all constructions in Langacker's theory), will only count as a construction in Goldberg's model if some aspect of its form or meaning cannot be predicted from its subparts.

Given that the central status of constructions blurs the boundaries between lexicon and syntax, Goldberg, like other cognitive linguists, assumes the **lexicon-grammar continuum**. Because Goldberg makes no distinction between simplex and complex symbolic units (since either kind may count as a construction) she refers to the lexicon-grammar continuum as the **constructicon** (the repository of constructions). Goldberg (1995: 5) also assumes that knowledge of language is represented as a 'highly structured lattice of interrelated information'. This view is consonant with Langacker's description of knowledge in terms of a structured inventory. Furthermore, Goldberg (1995: 5) assumes that 'knowledge of language is knowledge'. In other words, in keeping with the **Cognitive Commitment**, she rejects the idea that knowledge of language is separate and distinct in nature from other kinds of knowledge and experience. Instead, like other cognitive linguists, Goldberg argues that the properties of language directly reflect human experience, conceptual organisation and **construal**. Finally, as we have already mentioned, Goldberg's theory in part rests upon the theory of Construction Grammar that we explored in the last chapter. As we saw there, Construction Grammar is a monostratal generative model. While Goldberg's model can also be described as monostratal in the sense that it does not involve transformations, it cannot be described as a generative model because it assumes the **usage-based** thesis. In these core respects, then, Goldberg's construction grammar is a cognitive approach to grammar.

20.1.2 Advantages of a constructional approach to verb argument structure

Goldberg argues that there are a number of advantages to adopting a constructional approach to verb argument structure.

Avoids implausible verb senses

Firstly, the constructional approach avoids the necessity of positing several distinct senses for one verb (which is necessary in a lexically driven model), in order to account for all the constructions it can appear in; some of these might be implausible senses. Consider the examples in (3).

(3) a. Lily sneezed.
b. Lily sneezed the birthday cards off the mantelpiece.

The verb *sneeze* is a prototypical intransitive verb (3a). That is, it normally occurs with a single argument: the subject (*Lily*). Despite this fact, *sneeze* can occur in a syntactic construction like (3b), which can be represented as X CAUSES Y TO MOVE Z BY SNEEZING: [$_X$ Lily] causes [$_Y$ the birthday cards] to move [$_Z$ off the mantelpiece] by sneezing. As Goldberg points out, if we assume that this 'cause to move by sneezing' sense is a property of the verb itself, then we might expect to find a language (or languages) somewhere in the world with a lexical item specialised for this meaning, yet the existence of a verb sense of this kind is not attested.

Avoids circularity

Secondly, Goldberg argues that a constructional account has the advantage of avoiding circularity. If we assume that verbs are 'in charge' of everything that happens in a sentence – for example, how many participants are required and in what order – we are forced to posit as many senses for a verb as there are constructions in which that verb can occur:

> It is claimed that *kick* has an *n*-argument sense on the basis of the fact that *kick* occurs with *n* complements; it is simultaneously argued that *kick* occurs with *n* complements because it has an *n*-argument sense. This is where the circularity arises. (Goldberg 1995: 11)

Goldberg argues that if the properties of the constructions in which a verb can occur are seen as the properties of the construction itself rather than properties determined by the verb, this problem is avoided.

Semantic parsimony

The third advantage that Goldberg claims for a constructional approach is that it enables semantic parsimony. In other words, if the range of constructions in which a verb can occur – as well as the subtle differences in meaning associated with different possibilities – can be accounted for directly in relation to the construction itself rather than by positing long lists of senses for individual verbs, the resulting explanation is more **economical**. For example, because the verb *kick* can appear in the eight different verb argument constructions illustrated in (4), a lexically driven approach would be forced to posit eight different senses or lexical entries for this verb.

(4) a. George kicked the car.
   b. George kicked the bin over.
   c. George kicked Lily's slippers under the sofa.
   d. George kicked at the car.
   e. George kicked his foot against the wall.
   f. George kicked Lily her slippers.
   g. George's mum's horse kicks.
   h. George kicked his way out of the dentist's surgery.

In contrast, a constructional approach places the burden of explanation on the syntactic construction itself rather than on the verb.

Compositionality

The fourth advantage claimed by Goldberg is that a constructional account preserves compositionality, albeit in a weakened form. In other words, while all linguists would agree that words contribute to the meaning of sentences, there is considerable disagreement about what and how much they contribute. As we have seen, in a lexically driven approach, words (particularly verbs) are assumed to contribute not only their content meaning, but also their 'requirements' concerning the syntactic structure of the sentence. In a constructional approach, Goldberg argues, the problems inherent in a lexically driven approach can be avoided while preserving the point of agreement: words do contribute meaning to sentences, but not *all* the meaning. Put another way, sentence-level constructions have their own conventional schematic meaning independent of the verbs and other lexical items that are embedded in them. These sentence-level constructions represent symbolic units in their own right, much like the formal idioms discussed in the previous chapter, which can be lexically filled in a number of ways. In the next section, we will set out in more detail how this set of claims is substantiated in Goldberg's theory.

20.1.3 The relationship between verbs and constructions

Goldberg (1995: 24) explores the nature of the relationship between verbs and constructions by posing three questions which we discuss here in turn. We then consider a number of other issues relating to the relationship between verbs and the sentence-level (verb argument) constructions that they fill.

What is the nature of verb meaning?

Goldberg argues in favour of a **Frame Semantics** view of verb meaning (e.g. Fillmore 1977, 1982). As we saw in Chapter 7, this account of word meaning holds that the rich and detailed meaning of individual words is understood against the background of a particular conceptual frame (or domain, in Langacker's terms). Goldberg argues that an account like this is necessary, among other reasons, for explaining the distribution of adverbial expressions. Consider the examples in (5).

(5) a. Lily staggered into the kitchen slowly.
    b. ??Lily bounded into the kitchen slowly

Goldberg argues that a frame provides the basis of our understanding of the nature and manner of the motion involved, which explains why *slowly* can be felicitously applied to *stagger* but not to *bound*.

What is the nature of constructional meaning?

Within the speaker's knowledge of language or constructicon, Goldberg argues that constructions form a network. Within this network, constructions have related and sometimes overlapping meanings. This means that constructions are not individually represented with unique fixed meanings, but that they interact with other constructions in a rather fluid network of relationships (section 20.1.4). This view predicts that constructions, just like words, will exhibit **polysemy**. Consider the examples in (6).

(6) a. Lily gave George a kiss.
    b. Lily knitted George a jumper.
    c. Lily owes George a fiver.

Observe that all the examples in (6) are instances of the ditransitive construction. While example (6a) implies SUCCESSFUL TRANSFER of *a kiss* to *George*, example (6b) only implies intended transfer (it's possible that Lily will suffer a crisis of confidence and George will never see the jumper). In example (6c), it is also unclear whether George will ever receive the money, or indeed whether Lily even intends to repay it. According to Goldberg, SUCCESSFUL TRANSFER (6a) represents the central or **prototypical** sense of the ditransitive construction, while the other examples share aspects of the prototypical sense (TRANSFER) while departing from it in other respects (the TRANSFER may only be intended or potential). These examples also effectively illustrate the contribution of both the construction and the verb itself to the overall meaning of the sentence. While the construction determines what the possible meanings are (TRANSFER, successful or otherwise), the verb determines which of these possible meanings is realised. According to Goldberg, the central or prototypical sense associated with a construction is salient because it represents a basic aspect of human experience. She captures this view by positing the **scene encoding hypothesis**:

> *Scene encoding hypothesis*: Constructions which correspond to basic sentence types encode as their central senses event types that are basic to human experience. (Goldberg 1995: 39)

According to this view, a basic 'scene' of experience involves TRANSFER of an enitity from one person to another. This is a scene that we participate in and witness scores of times every day, which therefore represents a basic and fundamental aspect of human experience.

When can a given verb occur in a given construction?

In explaining what governs the interaction of particular verbs with particular constructions, Goldberg argues that while verbs are associated with **participant roles**, constructions have **argument roles**. In other words, the frame semantics of a given verb means that it is associated with frame-specific participants. For example, the verb *buy* might be associated with the participant roles BUYER, SELLER and GOODS, while the verb *sing* might be associated with the participant roles SINGER and SONG. As these examples illustrate, participant roles are associated with rather specific meanings that are related to their underlying frame or domain of experience. Furthermore, Goldberg adopts Langacker's (1987) view that a particular verb **profiles** particular participants within the frame or conceptual domain that underlies the meaning of that verb. Recall from Chapter 7 the distinction in profiling between *buy* and *sell*, for example. Goldberg discusses a similar example, which we saw in Chapter 5, comparing the verbs *rob* and *steal*. Consider the examples in (7) and (8).

(7) a. George robbed Lily (of hope).
    b. *George robbed hope (from Lily)

(8) a. George stole hope (from Lily).
    b. *George stole Lily (of hope)

While *rob* obligatorily profiles THIEF (*George*) and TARGET (*Lily*), *steal* obligatorily profiles THIEF (*George*) and (metaphorical) GOODS (*hope*). While either verb may optionally represent the third participant as a peripheral prepositional phrase (7a; 8a), the sentences become ungrammatical if this optional participant is represented as the direct object (7b; 8b). Goldberg (1995: 45) represents the profiling properties of the two verbs as follows.

(9) a. *rob* <THIEF TARGET GOODS>
    b. *steal* <THIEF TARGET **GOODS**>

The relatedness of the two verbs is captured by the fact that each is associated with the same set of participant roles by virtue of being associated with the same (or similar) frame or conceptual domain. The difference between the two verbs is captured in terms of their profiling properties, represented in bold type.

Argument roles

In contrast to the relative specificity of participant roles, the argument roles that are associated with sentence-level constructions in Goldberg's model are of a more general semantic kind, and are familiar from a range of approaches to sentence structure that assume **semantic roles**. We have already encountered semantic roles in Part III of the book (recall our discussion of grammatical functions in Chapter 17, for example). As we have seen, this type of approach rests upon the semantic partition of the clause into **predicate** and **arguments**. Recall that this sense of the term 'predicate' is different from the traditional grammar sense, in which the predicate is everything in a clause apart from the subject (that is, the verb and any objects or modifiers it may have). In the semantic roles sense, the predicate is usually a word-level unit that can be thought of as the semantic 'head' of the sentence. This word expresses the action, event, property or relation that the clause describes. Prototypically, the predicate of a clause is the lexical or content verb, which explains the central status of the verb in many approaches to explaining the relationship between grammar and meaning. As we saw in Chapter 17, the predicate can be a predicative adjective, a predicate nominal or a preposition in sentences with a copular verb.

Depending on the semantics of the predicate, it will take a certain number of arguments which are the participants or **entities** that the predicate requires in order to complete its meaning: a verb like *die* only involves a single participant, while a verb like *love* involves two and a verb like *put* involves three. The number and type of arguments that a predicate requires is traditionally referred to in terms of valence, as we saw in Chapter 17; **argument structure** is an alternative term for valence. Parts of the sentence that are not required by the predicate, but that provide 'incidental' or circumstantial information (typically, expressions of place, manner, time and so forth), fall outside the argument structure of that predicate, which explains why expressions like this are optional.

The semantic roles approach goes beyond the number of arguments required by a predicate and also looks at the types of arguments required in terms of their semantic properties. For example, the verb *die* requires a participant capable of living in the first place, while the verb *love* requires at least one of its participants to be a conscious and sentient being. On the other hand, it is difficult to say that you love someone or something 'on purpose', while purpose and intention are certainly involved if you slap someone. In order to try and capture these semantic restrictions, various proposals have been put forth concerning the **semantic roles** played by these arguments or participants, some of which are familiar from the discussion in previous chapters. Another name for semantic roles is **thematic roles**. Some examples are given in (10).

(10) Semantic roles
    a. AGENT   volitional initiator of action
    b. PATIENT   undergoes effect of action; change of state
    c. THEME   moved by action or whose location is described
    d. EXPERIENCER sentient and aware of action/state but not in control
    e. BENEFICIARY for whose 'benefit' action is performed
    f. INSTRUMENT means by which action is performed
    g. LOCATION   place in which event takes place
    h. GOAL   entity towards which something moves
    i. SOURCE   entity from which something moves

Example (11) illustrates a prototypical AGENT and PATIENT.

(11) [George] ate   [the caviar].
     AGENT   PATIENT

The idea of semantic roles has been very influential in modern linguistics, and both formal and cognitive models rely upon this notion in terms of addressing the nature of the relationship between grammar and meaning. In transformational generative approaches like Principles and Parameters Theory, semantic roles are listed in the lexicon as part of the lexical entry of a predicate. In the cognitive model, of course, this partition between lexicon and grammar is not admitted. As we saw in Chapter 17, semantic roles play a crucial role in Langacker's Cognitive Grammar account of the grammatical functions subject and object via their participation in the prototypical action chain model, where AGENT is conceived in terms of 'energy source' and PATIENT in terms of 'energy sink'. This model underlies unmarked active declarative sentences as well as explaining the properties of passive constructions on the basis of marked coding or TR-LM reversal. In this respect, Langacker's model is rather similar to Goldberg's, in that AGENT and PATIENT are not linked directly to individual verbs but to some underlying representation that structures the clause. However, while Langacker focuses on the cognitive model that underlies the clause, Goldberg focuses on the grammatical construction itself that arises from this cognitive model.

  In the remainder of the chapter, we will see how semantic roles also play a crucial role in constructional approaches to grammar. As we have seen, it is

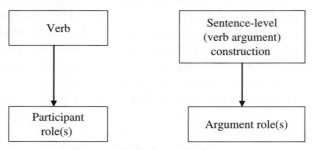

Figure 20.1 Participant roles and argument roles

important to emphasise the difference between the 'standard' view of argument structure and Goldberg's view. In most frameworks that assume semantic roles, these are associated directly with a particular lexical item, usually the verb. In Goldberg's model, semantic roles or argument roles are associated instead with the sentence-level construction. Thus, while a verb is conventionally associated with its own participant roles, a sentence-level construction has its own independent argument roles. This idea is represented by Figure 20.1.

Constructional profiling

While each verb determines which of its participant roles is lexically profiled or conceptually highlighted, sentence-level constructions also profile their argument roles. However, the **constructional profiling** of argument roles is more flexible. Goldberg suggests that only the argument roles that are linked to a grammatical function (subject, direct object or indirect object) are constructionally profiled. As we saw in the case of examples (7a) and (8a), other argument roles may optionally be present in the sentence but represented as prepositional phrases, sometimes called **oblique objects**. In Goldberg's sense of the term, these are not constructionally profiled: 'Every argument role linked to a direct grammatical relation (SUBJ, OBJ or OBJ$_2$) is constructionally profiled' (Goldberg 1995: 48). This reveals the distinction between **lexical profiling** and **constructional profiling** in Goldberg's model. Lexical profiling relates to the aspect of an expression's meaning that is made explicit by some expression (recall our discussion of profile and base in Chapter 15). In other words, in the sentence *George bought some champagne*, the expressions *George* and *some champagne* lexically profile (express in language) two participant roles relating to the semantic frame of the verb *buy* (BUYER and GOODS, respectively). Constructional profiling in Goldberg's model relates to the realisation of argument roles in terms of core grammatical relations. This means that other arguments may be explicit (lexically profiled) yet not constructionally profiled.

Fusion

Having set out the semantic and structural properties that the individual verb and the grammatical construction each bring to the sentence, questions naturally arise concerning how the two are integrated or **fused**, in Goldberg's terms. Goldberg posits two principles that govern the association of a verb's participant roles with a construction's argument roles: (1) the **Semantic Coherence Principle**; and (2) the **Correspondence Principle**. These are reproduced below:

> *The Semantic Coherence Principle*: Only roles which are semantically compatible can be fused. Two roles r_1 and r_2 are semantically compatible if either r_1 can be construed as an instance of r_2, or r_2 can be construed as an instance of r_1 . . . Whether a role can be construed as an instance of another role is determined by general categorization principles.
>
> *The Correspondence Principle*: Each participant role that is lexically profiled and expressed must be fused with a profiled argument role of the construction. If a verb has three profiled participant roles, then one of them may be fused with a nonprofiled argument role of a construction. (Goldberg 1995: 50)

The Semantic Coherence Principle works by matching a participant role with an argument role and seeing if the two overlap sufficiently for one to be construed as an instance of the other. For example, general categorisation principles enable us to determine that the THIEF participant role of the verb *steal* overlaps sufficiently with the argument role AGENT, because both share semantic properties such as ANIMACY, INTENTION, CAUSATION and so on.

The Correspondence Principle states that profiled argument roles are obligatorily matched with profiled participant roles, but builds some flexibility into the system by allowing that one of the participant roles may or may not be constructionally profiled in the case of a verb with three participant roles. Equally, a ditransitive construction can contribute a third role to a two–participant verb. These ideas are illustrated by Figure 20.2, which represents the CAUSE-RECEIVE ditransitive construction.

In this representation of the construction, 'Sem' represents the semantic structure of the construction in terms of argument roles, and 'Syn' represents the syntactic structure of the construction in terms of how the grammatical functions subject and object(s) realise the argument roles. 'PRED' represents the potential for any given verb to be mapped onto the construction, and the empty angled brackets represent the potential for that verb's participant roles to be fused onto the argument roles of the construction. The dotted line represents

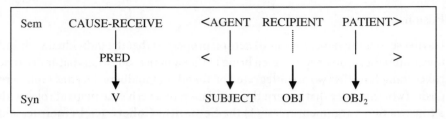

Figure 20.2 Ditransitive construction (adapted from Goldberg 1995: 50)

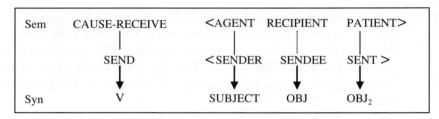

Figure 20.3 Ditransitive + *send* (adapted from Goldberg 1995: 51)

the argument role that may or may not be constructionally profiled in the case of a three-participant verb, or the argument role that can be contributed by the construction in the event that this third participant is not part of the verb's independent specification. This means that two-participant or three-participant verbs can be inserted into the construction (because the construction obligatorily profiles AGENT and PATIENT, strict one-participant verbs are not compatible with this construction). Consider the examples in (12) which illustrate how this works.

(12) a. George sent Lily the tickets.
  b. George sent the tickets (to Lily).
  c. (*)George sent Lily
  d. George wrote Lily a letter.
  e. George sang Lily a song.

In example (12a), the three participant roles of the verb *send* (SENDER, SENDEE and SENT) are mapped onto the three argument roles of the ditransitive construction (AGENT, RECIPIENT and PATIENT, respectively). In this case, all three profiled participant roles are constructionally profiled. In (12b), on the other hand, only the SENDER and SENT participant roles are mapped onto argument roles; the SENDEE role is optionally represented as a PP, which means that it is not constructionally profiled because it is not represented as a direct object nor as an indirect object. These possibilities are represented in Figure 20.3. Observe that the construction also rules out (12c), on the ungrammatical interpretation that Lily is the RECIPIENT (George sent Lily something). Because AGENT and PATIENT roles are obligatorily profiled, if one of these fails to be realised, the result is ungrammatical. Observe that (12c) is grammatical on the interpretation that Lily is the PATIENT (George sent Lily somewhere).

While we might describe *send* as a prototypical three-participant verb, it is not obvious that the verbs *write* and *sing* would also be described in this way. For example, both can occur in an intransitive frame (for example, *George writes; George sings* vs. **George sends*), as well as in a monotransitive frame (for example, *George wrote a novel; George sang the blues*). As examples (12d) and (12e) illustrate, however, these verbs are licensed to occur with an 'extra' argument (the RECIPIENT) by virtue of their occurrence in the ditransitive construction. As these examples show, the construction contains the flexibility, while the verb determines which of the possibilities provided by the construction are realised. Furthermore, while the verb *send* permits both possibilities presented by the construction – in other words instances of the construction in which the recipient either is (12a) or is not profiled (12b) – a verb like *hand* permits only the first option.

(13) a. Lily handed George a napkin.
     b. *Lily handed a napkin

The difference between the two verbs can be captured in terms of which participant roles they obligatorily lexically profile, as we saw above. The square brackets around the SENDEE participant role in the representation of *send* (14a) illustrates that this participant role is **optionally lexically profiled**, while all three of its participant roles are **obligatorily lexically profiled** by the verb *hand* (14b).

(14) a. *send*   <SENDER [SENDEE] SENT>
     b. *hand*   <HANDER HANDEE HANDED>

Of course, as well as explaining how the participant roles of particular verbs are mapped onto the argument roles of particular constructions, Goldberg's model must also explain how the 'right' verbs are matched with the 'right' constructions in the first place. In other words, the model must explain how examples like (15) are ruled out:

(15) *George saddened Lily the letter
     'George gave Lily a letter, which made her sad.'

An example like (15) might result if we were licensed to map the verb *sadden* onto the ditransitive construction, merging the three-participant semantics of *sadden*

(X CAUSES Y TO BE SAD BY SOME MEANS Z) onto the three-role semantics of the ditransitive (X CAUSES Y TO RECEIVE Z). As Goldberg points out, it is necessary to restrict the linking of certain constructions to certain classes of verbs by explaining which aspects of a verb's meaning license the linking. Although Goldberg does not state a specific principle that governs this licensing, she suggests that certain aspects of verb meaning are salient in this licensing process. For example, if a verb's meaning denotes a subtype of the **event type** represented by the semantics of the construction, this will license the linking. For example, *give*, *hand* and *send* are all subtypes of the CAUSE-RECEIVE event. Alternatively, a verb's meaning might denote the **means** by which the event designated by the construction is brought about. This is illustrated in (16):

(16) a. George threw Lily the can of tomatoes.
     b. George rolled Lily the can of tomatoes.
     c. George slid Lily the can of tomatoes.

In each of these examples, the mapping of the verbs *throw*, *roll* and *slide* onto the ditransitive construction is licensed because the verb expresses the means by which George caused Lily to receive the tomatoes (by throwing, rolling or sliding them). This approach therefore goes some way towards explaining why strict one-participant verbs like *die* are not mapped onto the ditransitive construction: it is difficult to think of a strict one-participant verb that encodes the semantics of the TRANSFER event type, because this event type by definition requires the profiling of at least two participants, if not three.

20.1.4 Relationships between constructions

Having set out what kinds of factors govern the relationships between constructions and individual verbs, Goldberg's next task is to explain what governs the relationships between constructions themselves. As we saw earlier in the chapter, Goldberg, like Langacker, assumes that constructions interact within a **structured network** of relations rather than comprising an unordered set. In Goldberg's model, relationships between constructions are captured in terms of **motivation** and **inheritance**. Consider the following principle:

> *The Principle of Maximized Motivation:* If construction A is related to construction B syntactically, then the system of construction A is *motivated* to the degree that it is related to construction B semantically . . . Such motivation is maximized. (Goldberg 1995: 67)

The term 'motivation' reflects the degree to which the properties of a given construction are **predictable**. In other words, given the premise that grammatical

THE ARCHITECTURE OF CONSTRUCTION GRAMMARS

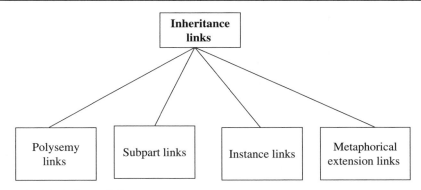

Figure 20.4 Inheritance links

constructions are meaningful, it follows that constructions that share grammatical properties will to some extent also share semantic properties. The Principle of Maximised Motivation is a psychological principle. In order to explain how language observes this principle, Goldberg posits **inheritance links** within the network of constructions that comprise knowledge of language: 'construction A motivates construction B iff B inherits from A' (Goldberg 1995: 72). There are four different kinds of inheritance links, which are shown in Figure 20.4. We will examine each of these in turn.

Polysemy links

Goldberg observes that a given sentence-level construction (a syntactic pattern conventionally associated with a meaning, and thus a symbolic unit) can be associated with a range of related senses. For example, the ditransitive construction is associated with a range of senses that all share the semantics of TRANSFER, but which also differ in systematic ways. Consider the following examples (based on Goldberg 1995: 75).

(17) a. X CAUSES Y TO RECEIVE Z (central sense)
       Lily gave George a helicopter.
   b. CONDITIONS OF SATISFACTION IMPLY X CAUSES Y TO RECEIVE Z
       Lily promised George a yacht.
   c. X ENABLES Y TO RECEIVE Z
       Lily allowed George a 50cc scooter.
   d. X CAUSES Y NOT TO RECEIVE Z
       Lily refused George a motorbike.
   e. X INTENDS TO CAUSE Y TO RECEIVE Z
       Lily built George a skateboard.
   f. X ACTS TO CAUSE Y TO RECEIVE Z AT SOME FUTURE POINT IN TIME
       Lily commissioned George a jet.

As we saw earlier in the chapter, although examples like these share in common the salient feature of the central sense, they differ in terms of whether the TRANSFER is actual or intended, permitted or prohibited, and so on. Just like lexical polysemy, **constructional polysemy** can be modelled in terms of a semantic network (recall Chapter 10).

Subpart links

If one construction is a **proper subpart** of another construction but exists independently, the two constructions are related by a subpart link. Consider the following example.

(18) a. Lily flew George to the conference.
b. George flew.

Example (18a) is an instance of the **caused motion construction**. Example (18b) is an instance of the **intransitive motion construction**. While (18a) lexically profiles the argument roles CAUSE (*Lily*), THEME (*George*) and GOAL (*the conference*), (18b) lexically profiles only the THEME (*George*). In this sense, the construction illustrated in (18b) is a proper subpart of the construction in (18a). Thus the relationship between the two constructions is captured by a subpart inheritance link.

Instance links

An instance link exists where one construction is a **special case** of a related construction. This type of link explains substantive idioms of the kind that we saw in Chapter 19. Recall that substantive idioms are lexically filled, which means that the idiomatic interpretation is only available if one of a restricted set of expressions is present within the construction. Compare the following examples (based on Goldberg 1995: 79).

(19) a. Lily drank herself silly.
b. George drove Lily mad/loopy/round the bend/up the wall.
c. *George drove Lily sad/cross/ecstatic/bored
d. George rowed Lily round the bend.

The example in (19a) is an instance of the **resultative construction**. When the verb *drive* occurs in this construction, one of a particular set of expressions must fill the result 'slot', as in (19b), and the construction takes on an idiomatic reading that can be paraphrased as 'make somebody crazy'. If the 'wrong' expressions are chosen to fill the construction, the result is ungrammatical

(19c) or fails to be recognised as the resultative construction (19d) and therefore loses its idiomatic interpretation. The idiomatic construction in (19b), then, is a special case or instance of the 'ordinary' resultative construction illustrated by example (19a).

Metaphorical extension links

Goldberg argues that some constructions are metaphorical extensions of other constructions, and that metaphorical extension therefore gives rise to a further type of inheritance link. For example, she argues that the resultative construction in (20a) is a metaphorical extension of the caused motion construction in (20b).

(20) a. George tickled her senseless.
b. George threw her onto the sofa.

The similarity between these two construction types revolves around the interpretation of the result phrase (the adjective phrase (AP) *senseless* in example (20a)) as a type of metaphorical GOAL, parallel to the actual GOAL expressed by the PP in the caused motion construction (*onto the sofa*, in example (20b)). In other words, the resultative construction encodes a metaphorical movement towards a GOAL or a metaphorical change of location. As Goldberg observes, this parallel is further supported by the fact that resultatives do not permit GOAL PP phrases. This can be accounted for by the fact that the result phrase already expresses the (metaphorical) GOAL, so the expression of an additional GOAL is redundant. This is illustrated by the unacceptability of example (21).

(21) *George tickled her senseless off the sofa

Despite this metaphorical inheritance link, Goldberg argues that it is important to recognise the caused motion construction and the resultative construction as distinct, albeit linked, constructions. This is because each construction places different restrictions on what verbs can occur in the construction. For example, while the resultative construction licenses *make* (22a), the caused motion construction does not (22b).

(22) a. George made her happy.
b. *George made her onto the sofa

In contrast, the caused motion construction licenses *move* (23a), while the resultative construction does not (23b).

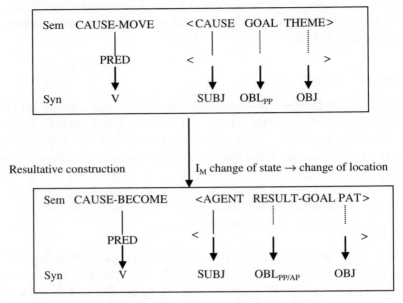

Figure 20.5 Metaphorical inheritance link (adapted from Goldberg 1995: 88)

(23) a. George moved her across the dance floor.
     b. *George moved her happy

The metaphorical inheritance link (I_M) between the two constructions is shown in Figure 20.5.

In sum, constructions can be related in a number of ways within a complex network of inheritance links, and any given construction might be linked to a number of other constructions or families of constructions via a number of different types of links. Although the set of links must be learnt for each 'family' of constructions, any frequently occurring links will license novel instances of the construction. This is reminiscent of Langacker's notion of entrenchment and emphasises the usage-based nature of Goldberg's model.

20.1.5 Case studies

In this section, we will look more closely at three constructions that have been studied in detail by Goldberg (1995). As we will see, Goldberg develops a strict methodology in her model of construction grammar which can be summarised in terms of the following five stages: (1) Goldberg first motivates the existence of these constructions by demonstrating that each has certain semantic and/or syntactic properties that cannot be predicted on the basis of the lexical items that fill the construction; (2) she then posits the central sense of the construction; (3) she posits the syntactic frame that identifies it; (4) she establishes the mapping between the argument roles of the construction and the participant roles of the lexical verb that fills the construction; and finally (5) she explores inheritance links within the construction, focusing mainly on polysemy and metaphor.

The English ditransitive construction (X CAUSES Y TO RECEIVE Z)

The ditransitive construction, which is sometimes called the double object construction, is associated with the syntactic frame [SUBJ [V OBJ OBJ$_2$]] (*e.g. George gave Lily flowers*), where both objects are noun phrases (NPs). The ditransitive construction is not associated with the syntactic frame [NP [V NP PP]] (e.g. *George gave flowers to Lily*), which identifies the distinct **prepositional construction**. These two constructions are distinct (although related in the network by shared aspects of form and meaning) because any difference in form or meaning signifies a distinct construction in Goldberg's model.

Goldberg lists a number of properties that are specific to the ditransitive construction, which cannot be predicted either from the lexical items that fill the construction or from other constructions in the language. Recall that this issue of predictability is important, because the presence of unique or unpredictable semantic or syntactic properties is what identifies a construction in Goldberg's theory. The properties of the ditransitive construction are summarised in Table 20.1.

We examine each of these properties in turn. Beginning with the TRANSFER semantics of the construction, compare examples (24a) and (24b).

(24) a. Lily knitted George a jumper.
 b. Lily knitted a jumper.

It is clear that while (24a) has the semantics of TRANSFER (that is, Lily knitted the jumper with the intention of giving it to George), this interpretation is missing from example (24b). If the semantics of TRANSFER were directly associated with the verb *knit*, we would expect both the ditransitive sentence in

Table 20.1 Properties of the English ditransitive construction (Goldberg 1995)

The English ditransitive construction: X CAUSES Y TO RECEIVE Z

Contributes TRANSFER semantics that cannot be attributed to the lexical verb
The GOAL argument must be animate (RECIPIENT rather than PATIENT)
Two non-predicative NPs are licensed in post-verbal position
The construction links RECIPIENT role with OBJ function
The SUBJ role must be filled with a volitional AGENT who intends TRANSFER

(24a) and the monotransitive sentence in (24b) to share this aspect of meaning. The fact that they do not suggests that this aspect of the meaning of (24a) is contributed by the construction. The same point is illustrated by examples (12d) and (12e), which are headed by the verbs *write* and *sing*, neither of which has any inherent semantics of TRANSFER.

The ungrammaticality of example (25a) illustrates the second property of the ditransitive construction: the GOAL argument must be animate. Observe that if the alternative prepositional construction is chosen, this restriction does not hold (25b).

(25) a. *Lily knitted George's winter wardrobe a jumper
    b. Lily knitted a jumper for George's winter wardrobe.

It is clear from examples like (24a) that the ditransitive construction licenses two non-predicative NPs in post-verbal position. Goldberg's claim is that the ditransitive construction is unique in having this property. Compare (26a) with (26b), for example.

(26) a. Lily cooked George a lobster.
    b. *George put the lobster the plate

The verb *put* is superficially similar to the verbs that are licensed to occur in the ditransitive construction in that it is a verb with three participant roles. Unlike these verbs, however, *put* is not licensed to occur in the ditransitive construction, but can only occur followed by NP + PP:

(27) George put the lobster on the plate.

Despite superficial similarities, (27) is an instance of the distinct caused motion construction, which is characterised by distinct semantics and syntax from the ditransitive construction. This last point is also related to the fourth characteristic of the ditransitive construction: the construction is also unique in linking the RECIPIENT role with the OBJ function. As example (27) illustrates, the object of *put* cannot be interpreted as RECIPIENT, but as PATIENT (the lobster does not receive anything in example (27) but directly undergoes the action of being put somewhere).

Goldberg's final claim, that the SUBJ role must be filled with a volitional AGENT who intends TRANSFER, is apparently contradicted by examples like those in (28).

(28) a. Lily gave George food poisoning.
    b. The lobster gave him food poisoning.
    c. The sound of the ambulance gave him some relief.

In these examples, the subject is either animate but (probably) lacking intention (28a), or inanimate and thus incapable of acting with intention (28b, 28c). Goldberg argues that these examples are motivated by metaphorical extension from the ditransitive construction. The conceptual metaphor in question is CAUSAL EVENTS ARE PHYSICAL TRANSFERS. In other words, in this metaphor, the event that causes the outcome (food poisoning or George's relief) is construed as an entity that transfers that outcome.

The English caused-motion construction (X CAUSES Y TO MOVE Z)

This construction has the syntax [SUBJ [V OBJ OBL]], where OBL (short for 'oblique') denotes a directional PP. The construction has the semantics X CAUSES Y TO MOVE Z, where Z designates a path of motion expressed by the directional PP. The construction is illustrated by the following examples.

(29) a. Lily sneezed the birthday cards off the mantelpiece.
b. Lily coaxed George into the kitchen.
c. George led Lily up the garden path.

The properties of this construction are summarised in Table 20.2.

The examples in (30) illustrate the fact that the CAUSED MOTION semantics cannot be attributed to the lexical verb. As (30b) shows, the verb *laugh* does not independently give rise to a CAUSED MOTION interpretation. It is only when this verb occurs in the caused motion construction (30a) that this interpretation is licensed.

(30) a. Lily laughed him out of the bedroom.
b. Lily laughed.

Of course, given that this construction is also characterised by a prepositional phrase, a question also arises concerning whether it is the preposition itself that licenses the CAUSED MOTION semantics. Although it is fair to say that the prepositions that occur in this type of construction are typically **directional** in the sense that they encode motion along a path (for example, *across, towards, into*), it is also possible to find prepositions occurring in this construction that are not independently directional, but **locational**. Consider example (31), in which

Table 20.2 Properties of the English caused-motion construction (Goldberg 1995)

The English caused-motion construction: X CAUSES Y TO MOVE Z
Contributes CAUSED MOTION semantics that cannot be attributed to the lexical verb
Contributes CAUSED MOTION semantics that cannot be attributed to the preposition
The CAUSER argument cannot be an INSTRUMENT

the preposition *under* is independently locational in the sense that it describes a static location rather than a path of motion.

(31) George sat under the table.

However, the same preposition takes on a directional interpretation when it occurs in the caused motion construction, as illustrated by example (32). The sentence in (32) can be interpreted to mean that due to Lily's coaxing, George ended up under the table. Goldberg argues that the caused motion construction **coerces** the essentially locative preposition into a directional interpretation.

(32) Lily coaxed George under the table.

Finally, the examples in (33) illustrate that while the CAUSER argument can be either an AGENT (33a) or a natural force (33b), it cannot be an INSTRUMENT (33c). The English caused motion construction is represented in Figure 20.6.

(33) a. George tickled Lily off the sofa (with a feather duster).
     b. The wind blew Lily's hair into her eyes.
     c. *The feather duster tickled Lily off the sofa

Like the ditransitive construction (recall example (17)), Goldberg argues that the caused motion construction has a number of related senses. These are illustrated in example (34).

(34) a. X CAUSES Y TO MOVE Z (central sense)
        [$_X$ Lily] persuaded [$_Y$ him] [$_Z$ into the aeroplane].
     b. CONDITIONS OF SATISFACTION ENTAIL X CAUSES Y TO MOVE Z
        [$_X$ Lily] ordered [$_Y$ him] [$_Z$ into the bedroom].
     c. X ENABLES Y TO MOVE Z
        [$_X$ Lily] allowed [$_Y$ him] [$_Z$ into her boudoir].
     d. X PREVENTS Y FROM MOVING Z
        [$_X$ Lily] barricaded [$_Y$ him] [$_Z$ into the kitchen].

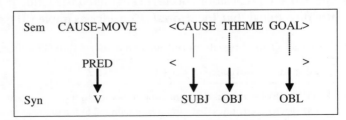

Figure 20.6 The English caused-motion construction (adapted from Goldberg 1995: 78)

e. X HELPS Y TO MOVE Z
[$_X$ Lily] gently guided [$_Y$ him] [$_Z$ towards the washing up].

Senses (34b) to (34e) represent **polysemy inheritance links** to the central sense (34a).

Finally, it is worth observing that Goldberg (1992: 69) analyses the prepositional construction that we saw in example (1b) as related to the caused motion construction. Observe that these constructions show the same syntax. Goldberg motivates the link between the two constructions on the basis of a metaphorical link between CAUSED MOTION and POSSESSION, a feature of the prepositional construction. Because there is also semantic overlap between POSSESSION and TRANSFER, this explains why the classes of verbs associated with the ditransitive and the prepositional (caused motion) construction overlap.

The English resultative construction

Recall from our earlier discussion that Goldberg argues that the English resultative construction is a metaphorical extension of the caused motion construction. However, she views the resultative construction as a distinct construction because it licenses different verbs from the caused motion construction (recall example (20)). The English resultative construction is illustrated by the examples in (35).

(35) a. They laughed themselves silly.
b. George drank himself into oblivion.
c. George tickled Lily senseless.

Goldberg argues that the existence of the resultative construction is characterised by a number of unique properties which are summarised in Table 20.3. As this table suggests, X corresponds to the AGENT (subject) NP, Y to the PATIENT (object) NP and Z to the RESULT argument, which may be realised either by an adjective phrase (AP) like *silly* (35a) or by a preposition phrase (PP) like *into oblivion* (35b).

Table 20.3 Properties of the English resultative construction (Goldberg 1995)

The English resultative construction: X CAUSES Y TO BECOME Z
Subject argument has to be an animate AGENT
Object argument has to be PATIENT (undergoes change of state)
Verb has to encode direct causation
Resultative adjective has to designate the endpoint of a scale (binary adjectives)
Resultative adjective cannot be deverbal

The ungrammaticality of the examples in (36) illustrates that the subject role has to be mapped onto an AGENT. Example (36a) is ungrammatical because the verb *become* is specified for a PATIENT subject, and is therefore incompatible with the resultative construction. Example (36b) is ungrammatical because the subject is not animate.

(36) a. *Lily became herself happy
　　 b. *The feather duster tickled Lily senseless

The ungrammaticality of example (37a) illustrates that the object argument must be a PATIENT, that is an entity capable of undergoing the change of state denoted by the RESULT argument.

(37) a. *Lily cooked the sausages dead
　　 b. Lily cooked the sausages to death.

Example (37a) is not acceptable because the sausages are already 'dead' and therefore fail to undergo a change of state resulting in death. In contrast, (37b) is acceptable because *to death* can mean 'to excess', and sausages are capable of being cooked to excess. As these examples demonstrate, the status of an expression as PATIENT depends not only upon its inherent meaning, but also upon the meaning of the other expressions in the construction.

The claim that the verb has to encode direct causation is supported by the interpretation of resultatives, where the result state is understood to be immediately effected as a consequence of the action of the verb. We cannot interpret (38), for example, to mean that George shot the seagull and it died a week later from its injuries. We can only interpret the sentence to mean that George shot the seagull and it died instantly.

(38) George shot the seagull dead.

Goldberg's claim that the resultative adjective has to designate the endpoint of a scale is related to the fact that 'binary' adjectives are more frequently attested in the resultative construction than 'gradable' adjectives. This is illustrated by example (38), in which *dead* is a binary adjective (*dead* is equivalent to 'not alive', and it is not possible to be somewhere in between dead and alive, nor to be *rather dead* or *slightly dead*). Goldberg argues that when gradable adjectives like *silly* or *stupid* are licensed in the construction, they are interpreted in the same way, as an endpoint or resultant state. Compare (39a) and (39b). In (39a), *stupid* designates a state of extreme drunkenness. The unacceptability of (39b) relates to the fact that *tipsy* does not designate an extreme state.

(39) a. Lily drank herself stupid.
b. *Lily drank herself tipsy

Finally, the claim that the resultative adjective cannot be deverbal is supported by examples like (40), which demonstrate that adjectives derived from participial forms of verbs are not licensed in the resultative construction, although Goldberg does not offer any explanation for this grammatical restriction imposed by this construction. The resultative construction is represented in Figure 20.7.

(40) *George talked/tickled her bored/excited/thrilled/captivated

Once again, the dotted lines indicate the potential of the construction to add arguments that are not specified independently by the verb. In the case of a two-participant verb like *tickle*, for example, the construction adds the RESULT-GOAL argument. In the case of a one-participant verb like *laugh* (35a), the construction adds both the PATIENT argument and the RESULT-GOAL argument.

Unlike the other two constructions we have seen in this section, the resultative construction does not display polysemy. It is not possible, for example, to derive an interpretation whereby the realisation of the result state depends upon satisfaction conditions (41a) or permission (41b).

(41) a. *Lily promised George senseless
b. *Lily allowed George senseless

According to Goldberg (1995: 84), the absence of polysemy in this construction is predicted by the analysis that the resultative construction itself is a metaphorical extension of the caused motion construction. In particular, Goldberg argues that the result phrase of the resultative construction is a metaphorical GOAL. This follows from the independently motivated metaphors CHANGE IS MOTION and STATES ARE LOCATIONS (recall Chapter 9). According to this perspective, the resultative construction encodes metaphorical caused

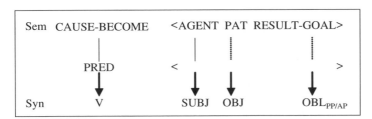

Figure 20.7 The English resultative construction (adapted from Goldberg 1995: 189)

motion resulting in a change of state (expressed in terms of a metaphorical GOAL); this means that the resultative is related to the caused motion construction, which encodes literal caused motion resulting in a literal change in location. Because the resultative is metaphorically extended from the central sense of the caused motion construction, it is predicted that it will fail to exhibit the range of polysemy exhibited by that construction.

In summary, Goldberg adopts aspects of the approach to Construction Grammar developed by Fillmore *et al.* (1988) and Kay and Fillmore (1999), in claiming that certain aspects of the meaning of a sentence, as well as certain restrictions upon its structure, arise directly from the properties of the skeletal grammatical construction rather than from the properties of the lexical verb. In addition, the verb contributes its own rich and specific (frame semantic) meaning, as well as bringing with it participant roles. It is in the interaction between the properties of the verb and the properties of the construction that both semantic and syntactic properties of these classes of sentences receive an explanation. Furthermore, Goldberg adopts the Construction Grammar notion of inheritance in accounting for generalisations across constructions and relationships between constructions. Goldberg develops this idea into a taxonomy of inheritance links that enable certain shared properties between constructions to be explained in terms of polysemy and conceptual metaphor, as well as in terms of more straightforward and predictable similarities (subpart and instance links).

However, Goldberg's construction grammar model departs from Kay and Fillmore's Construction Grammar in assuming a usage-based model. As we have seen throughout Part III of the book (also recall Chapter 4), this is one of the central distinctions between a cognitive approach to grammar and a generative approach. In the next two sections, we will explore two other cognitively oriented constructional approaches to grammar, and we draw some further comparisons between constructional approaches in section 20.4.

20.2 Radical Construction Grammar

The Radical Construction Grammar (RCG) model is developed by Croft (2001), and sets out to explore the implications of linguistic typology for syntactic theory. As we saw in Chapter 3, linguistic typology is the subdiscipline of linguistics that examines the structural properties of language from a cross-linguistic perspective and describes patterns of similarity as well as observing points of diversity. Although typological studies can in principle be theory-neutral, relying on large-scale comparisons and statistical findings, explanations for the patterns observed are usually couched in **functional** terms. As we saw in Chapter 3, **functional typology** is in certain respects rather compatible with the approach adopted by cognitive linguists, and it is this link that Croft seeks to exploit in developing a model of language that marries typological insights with a meaning-based model of language structure. In this section, we present a brief overview of Croft's model.

20.2.1 Taxonomy of constructions

RCG is in many respects compatible with Langacker's Cognitive Grammar. For example, Croft assumes the lexicon-grammar continuum, the continuum between specific and schematic meaning and the representation of the mental grammar in terms of a structured inventory. Croft also adopts the usage-based approach and the idea of entrenchment. However, in Croft's model, everything from a morpheme to a sentence is a construction. Croft's definition of construction is therefore different both from Langacker's definition and from Goldberg's definition. Table 20.4 represents Croft's taxonomy of constructions. We return below to the reason why the penultimate line of the table is shaded.

20.2.2 Emphasis on diversity

Croft argues that instead of taking grammatical **universals** across the world's languages as a starting point and building a model of language that assumes a universal grammar (the formal approach), we should instead take grammatical **diversity** as a starting point and build a model that accounts adequately for patterns of typological variation. Croft argues that a constructional approach is best placed to provide this type of model, since a constructional approach enables the articulation of the **arbitrary** and the unique, in contrast to most formal approaches which place the emphasis on **generalisation**.

20.2.3 Five key features of RCG

Croft (2001: 362–3) states that RCG can be summed up in five key points, which we briefly summarise in this section.

Table 20.4 RCG taxonomy of constructions (adapted from Croft 2001: 17)

Construction type	Traditional name	Example
Complex and (mostly) schematic	Syntax	[NP *be*-TENSE VERB-*en by* NP]
Complex and (mostly) specific	Idiom	[*pull*-TENSE NP's *leg*]
Complex but bound	Morphology	[NOUN-*s*], [VERB-TENSE]
Atomic and schematic	Word classes	[NOUN], [VERB]
Atomic and specific	Lexical items	[*the*], [*jumper*]

Primitive units

Firstly, Croft assumes that the construction is the only **primitive** unit in the grammar, and may therefore be either simplex or complex in terms of structure and either specific or schematic in terms of meaning. However, only overt (which is to say fully substantive) constructions, such as independent words, can be recognised as **atomic** in Croft's model. This means that grammatical categories (for example, word classes like noun and verb, or grammatical functions like subject and object) have no independent status, but are defined in relation to the constructions within which they occur. This explains why the relevant line in Table 20.4 is shaded: in RCG, word classes do not exist as primitive categories. This does not mean that words do not exist, but that words cannot be categorised into word classes that have any independent reality. Instead, words are just part of individual constructions. In this respect, the RCG model is diametrically opposed to the 'words and rules' model, where the words are the primitives and the constructions are epiphenomenal. In the RCG model, constructions are the primitives, and word classes, as they emerge from constructions, are epiphenomenal. From this perspective, it is to be expected that the types of word classes that we observe from one language to another might be significantly different, and because no universal word classes are posited, this cross-linguistic variation is not only unproblematic but predicted. Croft therefore argues against the traditional **distributional** approach to word classes (Chapter 14), which holds that they can be identified by morphological and syntactic properties. In support of this position, Croft (2001: 29) points out that some languages lack some of the relevant features that define the distributional approach (the lack of inflectional morphology in Vietnamese, for example), and that other languages might have the relevant features but reveal such different patterns of distribution that it is difficult to arrive at meaningful distributional criteria. Croft therefore argues against universal primitives, and also argues against the independent existence of word classes within any given language. Instead, Croft argues in favour of language-specific constructions, and in favour of construction-specific **elements** (grammatical subparts) and **components** (semantic subparts).

Syntactic relations and constituent structure

Secondly, the only syntactic relations admitted in the RCG model are the part-whole relations that hold between the construction as a whole and the syntactic elements that fill it. In other words, the model does not recognise **grammatical relations** (grammatical functions) like subject and object as having any independent reality outside of individual constructions. Instead, to the extent that grammatical functions emerge from constructions, these also have the status of construction-specific epiphenomena. In this model, constituency is conceived in terms of **grouping**, where grammatical units are identified in terms of contiguity and prosodic unity, and heads receive a semantic characterisation as **primary information bearing units** or PIBUs (Croft 2001: 258). Croft adopts Langacker's account of relationships between heads and dependents in terms of semantic valence and in terms of **instantiation** (Fillmore and Kay 1993), which is a property of constructions that links semantic components to their syntactic counterparts or elements.

Symbolic relations

Thirdly, the form and the meaning of a construction are linked in RCG by **symbolic** relations, in Langacker's sense of the term. In other words, each construction as a **whole** is a form-meaning pairing in the same way that each lexical item is a form-meaning pairing in the conventional view of the lexicon. As we have seen, this is a defining feature of constructional approaches (including Langacker's Cognitive Grammar approach).

Functional prototypes

Croft's fourth point relates to how RCG describes typological generalisation and variation. In Croft's model, both are characterised in terms of **categorisation** and in terms of how **function** is linguistically encoded. In other words, cross-linguistic similarities and differences are described in terms of **functional typological prototypes**: while referring expressions relate to OBJECTS, attributive expressions relate to PROPERTIES and predicative constructions relate to ACTIONS (Croft 2001: 87). Of course, OBJECTS, PROPERTIES and ACTIONS are semantic or conceptual categories, and these prototypes underlie the parts of speech in the world's languages. However, RCG does not specify the boundaries of these categories, which may vary from one language to another (Croft 2001: 103).

Explaining linguistic universals

Finally, RCG explains linguistic universals (linguistic generalisations, in Croft's terms), not by assuming of a set of universal grammatical primitives, but by assuming a **universal conceptual space**. In this respect, the RCG approach, which inherits much from functional typology (Croft 2003), reflects one of the core assumptions of cognitive approaches to grammar: cross-linguistic patterns of grammatical structure, such that they exist, are motivated by meaning, which in turn emerges from conceptual structure.

As we mentioned in Chapter 3, many typologists adopt some version of a **semantic map model**. A semantic map is a language-specific typological pattern which rests upon a universal conceptual space or system of knowledge. Croft defines conceptual space as follows:

> Conceptual space represents a universal structure of conceptual knowledge for communication in human beings. (Croft 2001: 105)

> The categories defined by constructions in human languages may vary from one language to the next, but they are mapped onto a common conceptual space, which represents a common cognitive heritage, indeed the geography of the human mind. (Croft 2003: 139)

To take a concrete example, recall our discussion of case-marking systems from Chapter 17, where the subject of a transitive verb is labelled A (for AGENT), the object of a transitive verb is labelled O (for object) and the subject of an intransitive verb is labelled S (for subject). We saw that a case system need only distinguish A and O (the subject and object of a transitive sentence), since S and A cannot co-occur (a sentence cannot simultaneously be transitive and intransitive), and S and O do not co-occur (an intransitive sentence does not have an object). The conceptual space that represents these participants is represented in Figure 20.8. This diagram represents the universal conceptual space that underlies language-specific patterns for marking these participants morphologically.

We saw in Chapter 17 that if a language marks S and A in the same way but marks O differently, this is a nominative/accusative case system (for example, German). In contrast, if a language marks the intransitive subject S and the object O in the same way (absolutive) but marks the transitive subject A differently (ergative), this is an ergative/absolutive system (for example, Basque). The semantic maps for these two systems are represented in Figure 20.9.

Although this overview of RCG is necessarily brief, it should be clear why Croft describes his model as **radical**. This approach questions basic

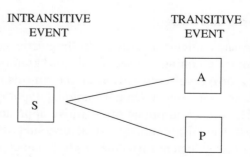

Figure 20.8 Conceptual space for transitive/intransitive participant roles (adapted from Croft 2003: 145)

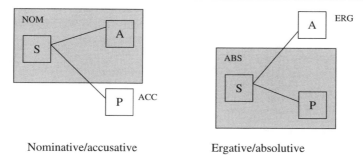

Figure 20.9 Semantic maps for nominative/accusative and ergative/absolutive (adapted from Croft 2003: 145)

assumptions that have defined theoretical and descriptive linguistics throughout the history of the discipline, such as the existence of word classes and grammatical functions. From this perspective, what many linguists think of as the building blocks of language (its grammatical units) are epiphenomenal. In the place of these cross-linguistic universals, the RCG model emphasises the universality of the conceptual system and explains typological patterns on this basis.

20.3 Embodied Construction Grammar

Embodied Construction Grammar (ECG) is a recent theory of construction grammar developed by Benjamin Bergen and Nancy Chang, together with various collaborators. In this section, we present a very brief overview of this model based on Bergen and Chang (2005). This approach assumes that all linguistic units are constructions, including morphemes, words, phrases and sentences.

20.3.1 Emphasis on language processing

In this model, the emphasis is on **language processing**, particularly language **comprehension** or **understanding**. In other words, while the approaches we have discussed thus far place the emphasis on modelling linguistic knowledge rather than on on-line processing, the ECG model takes it for granted that constructions form the basis of linguistic knowledge, and focuses on exploring how constructions are processed in on-line or **dynamic** language comprehension. Moreover, ECG is centrally concerned with describing how the constructions of a given language relate to embodied knowledge in the process of language understanding. Therefore much of the research to date in ECG has been focused on developing a formal 'language' to describe the constructions of a language like English; this formal language also needs to be able to describe the embodied concepts that these constructions give rise to in dynamic language comprehension.

20.3.2 Analysis and simulation

ECG claims that when a hearer hears an utterance, he or she has two distinct tasks to perform. The first is **analysis** (parsing), which involves the hearer mapping the stimulus (the utterance) onto the structured inventory of constructions in his or her grammar and recognising which constructions are instantiated by the utterance. The second task is **simulation**, which involves the activation of conceptual representations that underlie the interpretation of the utterance and the 're-enactment' of these conceptual representations (recall our discussion of Barsalou's research on perceptual symbol systems in Chapter 7). It is this process of simulation, together with contextual factors, that gives rise to the hearer's response. According to ECG, the conceptual representations that are accessed and simulated during language understanding are **embodied schemas** like the SOURCE-PATH-GOAL schema that we saw in Chapter 7. In other words, it is embodied experience that gives rise to these conceptual representations, and during language processing constructions are specified to prompt for these conceptual representations that arise from embodied experience. This explains why the approach is called 'Embodied Construction Grammar'. To take a concrete example, consider how a hearer might process the following utterance.

(42) Lily passed me a dead frog.

In terms of the analysis stage, each of the phonetic forms maps onto a construction (form-meaning pairing) in the hearer's inventory of constructions, at morpheme, word, phrase and construction level. The hearer recognises the ditransitive construction, which brings with it the semantics of TRANSFER, as we saw in our discussion of Goldberg's theory of Construction Grammar. The mapping of participant roles onto argument roles in the construction contributes to the interpretation of the utterance, and the context of the utterance enables the referent of the expression *me* to be identified (as the speaker). Recall from Chapter 11 that Mental Spaces Theory provides a cognitive account of how this process of reference assignment takes place.

At the simulation stage, Bergen and Chang argue that the interpretation of a ditransitive utterance like this activates three embodied schemas: FORCE APPLICATION, CAUSE-EFFECT and RECEIVE. Each of these is associated with schematic events and schematic roles such as ENERGY SOURCE and ENERGY SINK (Langacker 1987), and it is the mapping of constructions onto these schematic events and roles that gives rise to the simulation process. For example, in (42) the construction instantiated by *Lily* is ENERGY SOURCE, and the construction instantiated by *me* is ENERGY SINK. This simulation process gives rise to an ordered set of **inferences**, some of which are represented in (43), where SMALL CAPS indicate participants and event schemas (based on Bergen and Chang (2005):

(43) a. SPEAKER does not have FROG
 b. LILY exerts force via PASS
 c. FROG in hand of LILY
 d. LILY moves FROG towards SPEAKER
 e. FROG not in hand of LILY
 f. LILY causes SPEAKER to receive FROG
 g. SPEAKER has received FROG

Although these inferences seem rather obvious in terms of deconstructing the meaning of the utterance, it is nevertheless important for a model of language processing to explain how such inferences arise in utterance comprehension. According to the ECG model, it is the hearer's own embodied experience, which results in conceptual representations of that experience in terms of embodied schemas, that gives rise to these inferences via a simulation process. In this way, the hearer mentally re-enacts the event designated by the utterance.

Although we do not go into further detail on the ECG approach here, this brief overview provides a sense of how a constructional approach can be extended to account not only for knowledge of language but also for the dynamic processing of language, while taking seriously the role of embodied knowledge and the notion of mental simulations as the outcome of language comprehension.

20.4 Comparing constructional approaches to grammar

Constructional approaches to grammar (among which we include Langacker's Cognitive Grammar) share two key features in common, by definition. Firstly, despite differences in how 'construction' is defined, these approaches recognise grammatical constructions as **symbolic units**. From this perspective, linguistic knowledge consists of constructions 'stored whole' rather than 'built' (as they are in the 'words and rules' model). The second shared feature is the assumption of a structured inventory (Langacker 1987). All constructional approaches reject the idea that knowledge of language consists of an unordered set of constructions. Instead, these approaches make some statement concerning the nature of relationships between constructions within a complex network of links; these links rest not only upon shared structure, but also upon shared meaning (such as polysemy links or metaphorical extension links).

With respect to the differences between constructional approaches, we saw that each approach takes a different position on how the notion of construction is to be defined. Langacker views any unit with complex symbolic structure as a construction, regardless of whether its structure or meaning can be predicted from the properties of its subparts. In defining the term 'construction' in this way, Langacker's conception of a construction comes closest to the traditional sense of the term. Goldberg defines the construction as any symbolic unit, complex or otherwise, that is to some degree arbitrary (unpredictable) in terms of meaning or structure. In this respect, Goldberg's model is closer to the Construction Grammar model that we discussed in the previous chapter than it is to Cognitive Grammar. In Radical Construction Grammar, every linguistic unit is a construction, regardless of complexity or arbitrariness. Indeed, everything in RCG is arbitrary, if we take this model to its logical conclusion, since everything is construction-specific. In Embodied Construction Grammar, a similar view of constructions is taken, although the emphasis in this model is on language processing and on the nature of the embodied knowledge with which the language system interacts rather than on the nature of the language system itself.

A further point of contrast concerns whether or not constructional approaches can be described as usage-based. While the constructional approaches discussed in this chapter, in addition to Cognitive Grammar, can all be described as usage-based approaches, the Kay and Fillmore theory of Construction Grammar, as we saw in the previous chapter, is not a usage-based model. This represents a fundamental division between Kay and Fillmore's theory on the one hand, and what can be classed together as cognitively oriented constructional approaches on the other. As we saw in Chapter 4, the usage-based thesis is central to cognitive approaches, because the usage-based thesis goes hand in hand with the cognitive model's rejection of the Universal Grammar hypothesis, which in turn is related to the rejection of the idea that linguistic knowledge is a specialised or encapsulated knowledge system.

Finally, a parameter of comparison between constructional approaches that is discussed by Croft and Cruse (2004) relates to the issue of whether these theories can be described as **reductionist** or **non-reductionist**. The difference between a reductionist model and a non-reductionist model relates to the directionality of the relationship between part and whole (for example, the relationship between grammatical units like subject, object and verb (parts) and the construction in which they occur (whole)). In a non-reductionist model, the whole (construction) is the primitive unit, rather like a Gestalt, and the parts emerge as properties of that whole. In a reductionist model, the parts are the primitives and the whole is constructed from the parts. Croft and Cruse describe RCG as non-reductionist, because the whole construction is the primitive and the parts are defined in relation to that whole. In contrast, Kay and Fillmore's Construction Grammar is reductionist, because although it recognises constructions, it still views these as composed of smaller atomic units. Recall from Chapter 19 that in Construction Grammar complex constructions in part inherit their properties from basic constructions such as the head-complement construction and the subject-predicate construction, for example. Croft and Cruse describe Goldberg's participant roles as non-reductionist, because she assumes frame semantics. In other words, the frame is the primitive unit and the participant roles emerge from that frame. In contrast, her analysis of syntactic roles and relations is described as reductionist by Croft and Cruse, since she relies upon atomic primitives such as subject, object and verb in describing the syntactic properties of each construction, without positing an independent account of the origins of these primitives. In contrast, Cognitive Grammar views grammatical units such as subject and object as emerging from TR-LM organisation, which in turn derives from figure-ground organisation at the conceptual level. It follows that Cognitive Grammar is a non-reductionist model.

20.5 Summary

In this chapter, we have explored how a constructional approach to grammar can be extended to deal with regular sentence-level grammatical patterns. We looked in detail at Goldberg's (1995) constructional approach, and saw that she defines a construction as any form–meaning pairing whose properties cannot be **predicted** by its subparts, if it has subparts (recall that Goldberg's definition of construction subsumes symbolically simplex units). This entails a different definition of construction from that assumed in Cognitive Grammar, where a construction is any unit with a complex symbolic structure. We saw that Goldberg assumes, like Langacker, that knowledge of language consists of a **structured inventory** or **constructicon**, in which constructions are related to one another by **inheritance links**. In addition, Goldberg's model is **usage based**, unlike Kay and Fillmore's Construction Grammar. In looking in detail at Goldberg's case studies, we saw that her model focuses on sentence-level constructions such as the **ditransitive construction**, the **caused-motion construction** and the **resultative construction**, all of which are argued to have (schematic) meaning independent of the lexical items that instantiate them. Goldberg argues that verbs are associated with rich **Frame Semantics**, which gives rise to **participant roles** that are mapped onto the **argument roles** provided by the construction. Restrictions on sentence-level constructions can be accounted for in terms of **profiling**, either by the verb (lexical profiling) or by the construction itself (constructional profiling). In some cases, constructions can add roles, supporting Goldberg's hypothesis that argument structure alternation is governed to a large extent by sentence-level constructions rather than purely by the semantic properties of individual verbs.

Having set out Goldberg's constructional approach in some detail, we then provided a brief overview of Croft's (2001) Radical Construction Grammar and Bergen and Chang's (2005) Embodied Construction Grammar. We saw that the different focus of each approach (typological variation versus language processing, respectively) gives rise to a constructional model with different emphases. Finally, we compared the cognitive constructional approaches discussed in this chapter with one another and with Cognitive Grammar and Kay and Fillmore's Construction Grammar, and found that while these approaches often differ in terms of how a construction is defined as well as in terms of what the model sets out to account for, the main properties that define the cognitive constructional approach are (1) its adoption of the **symbolic thesis**; (2) its view that knowledge of language is represented as a **structured inventory**; and (3) and its adoption of the **usage-based** thesis.

Further reading

Introductory texts

- **Croft and Cruse (2004)**. Chapter 10 includes a discussion of Goldberg's approach and how it has been influenced by Lakoff's (1987) analysis of the English *There*-construction (see below). This chapter also provides concise points of comparison between different constructional models.
- **Goldberg (1997)**. This short encyclopaedia article provides an accessible overview of the constructional approach to grammar.
  Lee (2001). Chapter 5 of this textbook introduces Construction Grammar, focusing mainly on Goldberg's model.
- **Taylor (2002)**. Chapter 28 includes some discussion of the difference between the Cognitive Grammar view of constructions and Goldberg's view.

Background reading: semantic roles

- **Dixon (1991)**
- **Dowty (1991)**
- **Fillmore (1968)**
- **Jackendoff (1987)**

These sources provide a range of perspectives on semantic roles and argument structure.

Goldberg's constructional approach to grammar

- **Goldberg (1992); Goldberg (1995).** These represent the primary sources for the material discussed in section 20.1 of this chapter. Goldberg's (1995) monograph is very readable and includes a thorough comparison of her model with formal models, transformational and non-transformational, to which we have been unable to do justice here.
- **Lakoff (1987).** This monograph contains an in-depth discussion of the English *There*-construction (Case Study 3), for which Lakoff proposes a constructional analysis. Goldberg acknowledges the influence of Lakoff's analysis upon her theory; Lakoff's analysis therefore represents essential background reading for anyone interested in exploring constructional approaches in more depth.

Radical Construction Grammar

- **Croft (2001).** This monograph sets out Croft's RCG model in detail and makes for stimulating reading. Croft's close attention to cross-linguistic data is particularly appealing for readers bored with looking at English examples.

Embodied Construction Grammar

- **Bergen, Chang and Narayan (2004); Bergen and Chang (2005).** These two papers set out the Embodied Construction Grammar framework in more detail than has been possible here. The interested reader is referred in particular to Bergen and Chang (2005), which presents a detailed analysis of the stages involved in the comprehension of a ditransitive sentence of the type discussed in this chapter.

Collection of articles on constructional approaches to grammar

- **Östman and Fried (eds) (2005b).** This volume includes papers on a range of constructional approaches.

Exercises

20.1 Constructions

Consider the following expressions. In each case, state whether the expression is a construction according to (i) Langacker, and (ii) Goldberg. Explain how you reached your conclusions. Finally, summarise the differences between the two definitions of construction.

   (a) champagne
   (b) canapés
   (c) under the bed
   (d) George tickled Lily
   (e) She blew him a kiss

20.2 Mapping verbs onto constructions

Recall our discussion of the ditransitive (X CAUSES Y TO RECEIVE Z) construction in section 20.1.3. As we saw, Goldberg claims that certain aspects of verb meaning are salient in licensing the linking of verbs to constructions. Consider the verbs listed below. In each case, state whether the verb can be mapped onto the ditransitive construction and provide an example to support your conclusion. Do any semantic patterns emerge from your findings?

   (a) type
   (b) cook
   (c) swim
   (d) drive
   (e) tickle
   (f) donate

20.3 Identifying constructions

The following sentences are all examples of the sentence-level (verb argument) constructions studied by Goldberg. For each sentence, (i) identify which type of construction it belongs to and explain how you reached your conclusions; (ii) state whether it counts as an instance of the prototypical or central meaning associated with the construction in question and explain how you reached your conclusions; and (iii) if you think the example is not an instance of the central sense, explain how it is motivated by or related to the central sense.

   (a) Lily's best friend refused George an invitation.
   (b) Lily beckoned George into the bedroom.
   (c) George moved her to tears.
   (d) The French windows lent Lily's kitchen an air of sophistication.
   (e) He gave her a headache.
   (f) She drove him mad.
   (g) She kissed him unconscious.

20.4 The conative construction

Consider the following examples, which illustrate the conative construction.

(a) George tore at the wrapping paper.
(b) Lily hacked at the loaf of bread.
(c) George kicked at the cat.
(d) Lily lashed out at him.

On the basis of these examples, work out the syntax and semantics associated with this construction. Based on your model, how would you now account for example (e)? Does this relate to the central sense, a peripheral sense or a completely different construction? Explain how you reached your conclusions.

(e) She shouted at him.

Finally, consider example (f):

(f) She threw a glance at the other driver.

Is this example related to the example in (e), or does it relate to a different construction? Explain your reasoning.

20.5 The *way* construction

The following examples illustrate what Goldberg (1995) calls the *way* construction. On the basis of these examples, work out the syntax and semantics of this construction. Goldberg suggests that there are two main senses associated with this construction, connected by a polysemy inheritance link. How might these two senses be described?

(a) Lily fought her way to the top.
(b) George flirted his way into Lily's life.
(c) George whistled his way home.
(d) Lily drank her way through the party.
(e) Lily clattered her way down the stairs.

20.6 Comparing construction grammars

In the form of an annotated table, present a comparison of the key similarities and differences between the following constructional approaches to grammar: (i) Cognitive Grammar; (ii) Construction Grammar; (iii) Goldberg's approach; (iv) Radical Construction Grammar.

20.7 Lexical and constructional polysemy

Based on the discussion in this chapter, how does the notion of polysemy relate to sentence-level constructions? Provide a definition of polysemy, and compare and contrast lexical polysemy with constructional polysemy. Are these distinct or related phenomena, given the cognitive view of the mental representation of linguistic knowledge?

21

Grammaticalisation

So far in this book, we have taken a mainly **synchronic** approach to language. In other words, we have concentrated our discussion upon languages as they are now, in the early years of the twenty-first century. In particular, we have focused on Modern English. As we saw in Chapter 4, the process of language change is continuous. Historical linguists take a **diachronic** view of language, describing patterns of change and attempting to account for those changes. The findings of historical linguistics have implications for most areas of modern linguistics, because language change affects phonology, semantics and grammar, and can therefore inform synchronic theories about these core areas of language. In addition, as we saw in Chapter 4, the causes of language change can often be attributed to socio-linguistic forces, which entails a close link between historical linguistics and socio-linguistics. There is also a close interrelationship between historical linguistics and linguistic typology (see Chapter 3), since it is largely by looking at patterns of language change and discovering the directions that such changes follow that typologists can form a view on the directions that typological patterns are likely to follow.

Some types of language change move at a more rapid pace than others. For example, the lexicon of a language changes more rapidly than its phonology or its grammar, with new words coming into the language, old words falling out of use and existing words taking on different meanings. The sound patterns of a language change more rapidly than its grammar (compare the modern 'Received Pronunciation' accent of British English with its 1950s counterpart, for example). Finally, the slowest type of language change is grammatical change. For example, as we will see in section 21.3, while the English verb *must* was a full content verb in Old English, as attested in the Old English corpus (for example, in the epic poem *Beowulf*, written sometime in the eighth century), in Modern English it functions as a modal auxiliary. These two points in the history of this symbolic unit are separated by over a thousand years. As Heine *et al.* (1991: 244) observe, the time span involved in grammatical change depends on the kinds of grammatical elements involved.

The type of language change we focus upon in this chapter is **grammaticalisation**. (Some linguists prefer the term 'grammaticisation'.) This is the process whereby lexical or content words acquire grammatical function or existing grammatical units acquire further grammatical functions. Grammaticalisation has received a great deal of attention within cognitive linguistics. This is because grammaticalisation is characterised by interwoven changes in the form and meaning of a given construction and can therefore be seen as a process that is essentially grounded in meaning. Furthermore, cognitive linguists argue that semantic change in grammaticalisation is grounded in usage events, and is therefore itself a usage-based phenomenon. There are a number of different cognitive theories of grammaticalisation, each of which focuses on the semantic basis of the process. After providing an overview of the nature of grammaticalisation (section 21.1), we discuss three of these theories below: metaphorical extension approaches (section 21.2); Invited Inferencing Theory (section 21.3); and Langacker's subjectification approach (section 21.4). Finally, we present a brief comparison of the three approaches (section 21.5).

21.1 The nature of grammaticalisation

In this section, which owes much to Croft (2003), we will provide a descriptive overview of the grammaticalisation process. Although the term 'grammaticalisation' suggests a type of grammatical change, grammaticalisation in fact involves **correlated** changes in sound, meaning and grammar. In other words, the process of grammaticalisation affects the phonology, morphosyntax and meaning or function of a given symbolic unit. Grammaticalisation can therefore be described as a kind of language change that involves form-meaning reanalysis. Grammaticalisation is essentially the process whereby contentful or lexical constructions (including words) develop grammatical functions, or already grammaticalised constructions evolve further grammatical functions. Grammaticalisation, like many kinds of language change, is **unidirectional** and **cyclic** (Croft 2003: 253). It is described as 'unidirectional' because the direction of this type of change is from the lexical to the grammatical (from the open class to the closed class), and not vice versa. The cyclic nature of grammaticalisation is evident in the fact that linguistic units enter a language as open-class lexical items, evolve into closed-class items via the process of grammaticalisation, eventually leave the language via a process of loss and are replaced by new open-class items. For example, a common process involves the evolution of a lexical verb meaning 'want' or 'intend' into a modal auxiliary, then into a bound inflectional (e.g. future) morpheme that may eventually be lost as its function is taken over by a new open-class item.

Another example is provided by Heine *et al.* (1991) from Yoruba, a Nigerian language that belongs to the Kwa branch of the Niger-Congo language family. The Yoruba verb *kpé* 'say' evolved into a complementiser (1a), was then replaced by another verb *wí* 'say' that was also grammaticalised into a complementiser and compounded with *kpé* (1b), and then this form was lost as a new 'say' verb *ní* emerged (1c). Examples from Lord (1976) cited in Heine *et al.* (1991: 246–7).

(1) a. ó   sɔ   kpé   adé   lɔ
       he   say   say/that   Ade   go
       'He said that Ade went'
    b. ó   sɔ   wí-kpé   adé   lɔ
       he   say   say-say/that   Ade   go
       'He said that Ade went'
    c. ó   ní   adé   lɔ
       he   say   Ade   go
       'He said that Ade went'

One reason for the cyclic nature of grammaticalisation is the phenomenon of **renewal**. For example, the English degree modifiers or 'intensifiers' (e.g. *very* in *Lily's knowledge of rocket science can be very intimidating*) are particularly prone to renewal. As Hopper and Traugott (2003) observe, at different points over the last 200 years the following degree modifiers have been particularly fashionable: *frightfully, terribly, incredibly, really, pretty, truly*. Renewal is motivated by the tension that holds between informativeness and routinisation. Routinisation relates to frequency of use and thus predictability: a form becomes highly predictable in linguistic contexts in which it occurs frequently. Because grammaticalisation ensures a more limited distribution of a grammaticalised form, grammaticalised elements tend to become highly predictable. However, predictability entails a reduction in the informational significance of a particular form. This is attested by the phenomenon of **phonological attrition**, which is the endpoint of morphological fusion and coalescence (discussed below). This process, which eventually results in the complete loss of phonological form, is well attested in the languages of the world (see Hopper and Traugott 2003). Renewal reflects a natural shift towards new forms in order to increase informativeness, by avoiding forms that, as a result of routinisation, have reduced informational significance. This process manifests itself in innovation in language use and contributes to the cyclical nature of the grammaticalisation process.

Grammaticalisation is effected through a shift in the meaning associated with the linguistic unit element from the specific to the schematic. According

Table 21.1 Common grammaticalisation patterns (adapted from Croft 2003: 254)

content verb > auxiliary > tense-aspect-mood affix
verb > adposition
noun > adposition
adposition > case affix
adposition > subordinator
emphatic personal pronoun > clitic pronoun > agreement affix
cleft sentence marker > focus marker
noun > classifier
verb > classifier
demonstrative > article > gender/noun class marker

demonstrative or article > complementiser or relativiser
numeral 'one' > indefinite article
numerals 'two' or 'three' > dual/paucal/plural affix
collective noun > plural affix
demonstrative > copula
positional verb > copula

to the cognitive perspective, the grammaticalised unit takes on meaning associated with the usage event, and is thus a fundamentally usage-based change. The most frequent patterns of grammaticalisation are listed in Table 21.1.

In the next two subsections, we consider in more detail a number of characteristics associated with the grammaticalisation process.

21.1.1 Form change

As Table 21.1 illustrates, a common pattern in grammaticalisation is one in which free morphemes become bound or fused together. In other words, a grammaticalised unit undergoes a tighter integration of morphophonological form. This is known as **coalescence**, which is a process whereby two words become one. For example, Modern English derivational affixes *-hood*, *-dom* and *-ly* evolved from nouns meaning 'condition', 'state, realm' and 'body, likeness', respectively. Consider the following examples from Hopper and Traugott (1993: 41):

(2) a. cild–had 'condition of a child' > childhood
    b. freo–dom 'realm of freedom' > freedom
    c. man–lic 'body of a man, likeness of a man' > manly

The process of coalescence is accompanied by **reduction** or **loss**: a process whereby a morpheme or sound segment is either shortened or lost altogether. This process is illustrated by the English *(be) going to* construction, which has undergone syllabic reduction (from trisyllabic to disyllabic) and has also undergone coalescence, resulting in a fused form *gonna*. Observe that the form associated with the FUTURE meaning has undergone reduction while the form associated with the ALLATIVE (motion) meaning has not. This is illustrated by the acceptability of *gonna* with a FUTURE meaning (3a), but not with an ALLATIVE meaning (3b) (compare *I'm going to the fish and chip shop*).

(3) a. I'm gonna be a rocket scientist when I grow up. [FUTURE]
   b. *I'm gonna the fish and chip shop [ALLATIVE]

Moving from morphophonological form to morphosyntactic form, grammaticalised units display **rigidification** of morpheme/word order (Croft 2003: 257). For example, consider the position of the French clitic pronoun *l'* (a grammaticalised unit) with the position of its full NP counterpart *le livre* in example (4).

(4) a. Je   l'ai   lu
       1s   3s-AUX.1s   read.P.PART
       'I've read it.'

   b. J'ai   lu   le   livre
      1s-AUX.1s   read.P.PART   DEF.MS   book
      'I've read the book.'

As French evolved from its ancestor Latin, the relatively free Latin word order (which tended towards a default SOV pattern in transitive clauses) became rigidified along two parameters in French: an SOV pattern became fixed in the case of (clitic) object pronouns (4a), while an SVO pattern became fixed in the case of free nominals (4b).

In Croft's (2003: 259) terms, grammaticalised units also undergo **paradigmaticisation**, whereby they move from membership of an open class to membership of a closed class, and **obligatorification**, whereby an optional element in a construction becomes obligatory. The latter process is illustrated by the French negation particle *pas*. This open-class word means 'footstep', and was originally introduced into the French negative construction *ne* V as an emphatic object of verbs of movement (5a). Over time, this element was reanalysed as an optional negation particle in negated verb of movement constructions: *ne* V-movement (*pas*). The negation particle was then extended to occur optionally in all negated verb constructions: *ne* V (*pas*), and then obligatorily: *ne* V *pas* (5b). Finally, in spoken French, the element *pas* retained its obligatory status (in the absence of another negative morpheme like *rien* 'anything' or *jamais* '(n)ever'), while the earlier negation particle *ne* became optional, giving rise to the construction (*ne*) V *pas*. In some current spoken varieties, *ne* has now been lost altogether, resulting in the construction V *pas* (5c). This path of change is schematically represented in (5d). The example in (5) is based on Hopper and Traugott (1993: 58).

(5) a. Il   ne   va   (pas)
       3MS   NEG   go   step
       'He doesn't go (a step).'

   b. Il   ne   sait   pas
       3MS   NEG   know   NEG
       'He doesn't know.'

   c. Il   sait   pas
       3MS   know   NEG
       'He doesn't know.'

   d. *ne* V-movement (*pas*) > *ne* V (*pas*) > *ne* V *pas* > (*ne*) V *pas* > V *pas*

The disappearance of the negation particle *ne* in varieties of modern spoken French illustrates the final stage in the life-cycle of a grammatical morpheme: **grammatical loss**.

21.1.2 Meaning change

A key characteristic of grammaticalisation, which accompanies and indeed can be said to give rise to form changes, is change in the meaning or function associated with a linguistic form. While some grammaticalisation scholars argue that this semantic change is the result of meaning loss, called semantic **bleaching** or **attenuation** (weakening), others argue that it is more accurate to describe the semantic change that characterises grammaticalisation, particularly in the early stages of grammaticalisation, as an instance of **polysemy**. Croft (2003: 262) describes this polysemy as 'a chain of related meanings or uses', and illustrates this point by means of the English word *that*, which has four functions. This coexistence of related meanings which emerged at historically different periods is sometimes called **layering** in grammaticalisation theory. The four functions of *that* are illustrated by the following examples.

(6) a. Pass me that.   [deictic demonstrative]
    b. George snored and she hated that.   [anaphoric demonstrative]
    c. Lily said that she'd had enough.   [complementiser]
    d. George was the man that she loved.   [relativiser]

As Croft observes, there may not be a single meaning that underlies all four functions of *that*. Nevertheless, we might plausibly argue that the demonstrative function has been extended from the domain of physical space (6a) to the domain of linguistic organisation (6b). In other words, the anaphoric demonstrative 'points to' another element in the discourse (the fact that George snores) rather than an entity in the physical context. Similarly, the function of *that* as a complementiser is to 'point to' or introduce what is coming next (6c), while the relativiser *that* in (6d) 'points to' some characteristic of the noun *man* which the relativiser *that* introduces. From this perspective, the four uses of *that* are plausibly related to and motivated by **deixis**. Cognitive linguists therefore argue that the semantic change that characterises grammaticalisation involves not necessarily 'semantic bleaching', but the shift from lexical or content meaning to grammatical or schematic meaning, which at certain stages in the grammaticalisation process gives rise to a set of overlapping form–meaning pairings along the continuum between content units and grammatical units.

Finally, just as grammaticalisation involves phonological and morphological loss, it can also involve semantic or functional loss. To illustrate this point, we can return to the French negation construction that we saw in example (5). Here, the emphatic meaning of *pas* is lost as it becomes a fully grammaticalised negation particle and the negation function of *ne* is lost as it is superseded by *pas*. This explains why it eventually slips out of certain varieties of the language altogether.

The study of grammaticalisation has a rich history, dating back at least as far as the eighteenth century (Heine *et al.* 1991: 5). This area of language change has received most attention from philologists (historical linguists with a particular interest in establishing language families) and from typologists, and has therefore been approached more from a functional than a formal perspective. More recently, a number of cognitively oriented theories of grammaticalisation have emerged. We limit ourselves in this section to introducing and discussing three cognitively oriented theories of grammaticalisation. As we will see, these theories differ in a number of ways, but what they share is the view that grammaticalisation is essentially grounded in meaning and the view that grammaticalisation is a usage-based phenomenon. These three approaches are represented in Figure 21.1.

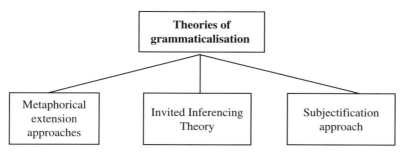

Figure 21.1 Cognitive models of the grammaticalisation process

21.2 Metaphorical extension approaches

Of the three of approaches to grammaticalisation addressed here, the metaphorical extension approach is probably the most widely adopted and is therefore associated with a considerable number of researchers. We have chosen to illustrate this approach by focusing on a representative study by Heine, Claudi and Hünnemeyer (1991). The further reading section lists a number of other studies that can be broadly grouped under the heading of metaphorical extension.

The evolution of grammatical concepts

In their 1991 book, Heine, Claudi and Hünnemeyer argue that grammaticalisation is one of a number of strategies that speakers rely upon in developing new expressions for concepts, together with other strategies such as coining new words, borrowing words from other languages and so on. Of course, the emergence of new grammatical forms is more gradual than the emergence of new lexical forms. Grammaticalisation results in the development of expressions for **grammatical concepts**. According to Heine *et al.* (1991: 28), grammatical concepts share a number of properties: they are relatively abstract in nature, lack semantic autonomy, contribute to the structure rather than content of the cognitive representation encoded in language, belong to closed classes and tend to lack morphological autonomy. In these respects, Heine *et al.*'s characterisation of grammatical concepts is reminiscent of the cognitive view of grammatical concepts, which has been a recurring theme throughout Part III of this book.

Metaphorical extension

According to Heine *et al.*, grammaticalisation essentially arises from human **creativity** or problem-solving. In developing a new expression for a grammatical concept, speakers 'conceptualize abstract domains of cognition in terms of concrete domains' (Heine *et al.* 1991: 31). In particular, these researchers adopt the view that this process involves **metaphorical extension** emerging from the mapping of **image schematic** concepts from source to target domain (Heine *et al.* 1991: 46). Given the gradual nature of grammaticalisation, Heine *et al.*'s approach involves the **reconstruction** of **dead** or **frozen** metaphors, in the sense that the synchronic grammatical forms are often no longer transparently recognisable as metaphors but are argued to have originated from the same cognitive metaphorical mapping processes as 'living' metaphors.

Heine *et al.* argue that basic **source concepts** have a strong tendency to be **concrete objects, processes** or **locations**, and to involve frequently used expressions such as body part terms, verbs expressing physical states or processes such as *sit, lie* or *go*, and verbs expressing core human activities such as *make, do, have* or *say* (Heine *et al.* 1991: 32–5). They therefore suggest that **egocentricity** (or embodiment) is a central feature uniting source concepts. In these respects, they argue that grammaticalisation emerges from human construal, and thus they take an explicitly **experientialist** stance. Heine *et al.* (1991: 48) propose a metaphorical source domain **hierarchy**, which is represented in (7). According to this hierarchy of basic concepts, any (less abstract) concept on the hierarchy can be used to metaphorically structure any other (more abstract) conceptual domain to its right. In this way, the hierarchy captures the **unidirectionality** that is characteristic of grammaticalisation.

(7) PERSON > OBJECT > ACTIVITY > SPACE > TIME > QUALITY

As Heine *et al.* observe, some basic source concepts are difficult to place in this hierarchy. For example, they argue that POSSESSION might be located somewhere to the right of SPACE. An alternative representation of this hierarchy is shown in Figure 21.2.

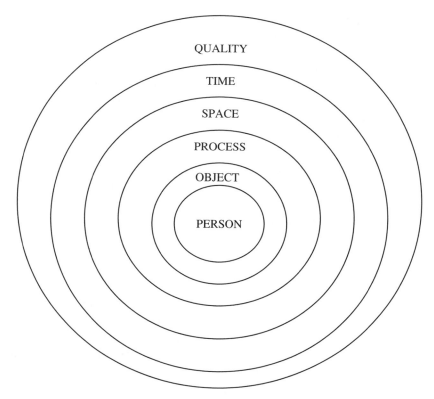

Figure 21.2 Source domain hierarchy (Heine *et al.* 1991: 55)

The grammaticalisation continuum

In accounting for the transition from source concept to target concept in grammaticalisation, Heine *et al.* observe that the continuum between less and more grammatical meaning might be considered a potential problem for their metaphor extension account. In other words, because Heine *et al.*'s account assumes that the underlying motivation for grammaticalisation is metaphoric extension from a more concrete source domain to a more abstract target domain, then examples that fall somewhere between source and target domains might be seen as counterevidence for the metaphorical extension account. For example, the conceptual metaphor TIME IS SPACE motivates the grammaticalisation of the *(be) going to* construction, which evolves from its ALLATIVE meaning towards its more abstract and hence more grammaticalised FUTURE meaning. Consider the examples in (8) (adapted from Heine *et al.*, 1991: 70).

(8) a. Lily is going to town.
    b. George: Are you going to the library?
        Lily: No, I'm going to eat.
    c. George is going to do his very best to make Lily happy.
    d. It is going to rain.

As Heine *et al.* observe, while *be going to* in (8a) has an ALLATIVE meaning and *be going to* in (8d) reflects a purely FUTURE meaning, the examples in (8b) and (8c) are intermediate between these two senses. For example, Lily's use of *be going to* in (8b) encodes what Heine *et al.* call an INTENTION meaning, with a secondary sense of PREDICTION; they also suggest that there is a 'relic' of the spatial (ALLATIVE) meaning in examples like this. This contrasts with (8c) which encodes INTENTION and PREDICTION, but no spatial (ALLATIVE) sense is apparent in this example. Examples like (8b) and (8c) are potentially problematic for a metaphor account because they illustrate that grammaticalisation involves a continuum of meanings rather than a clear-cut semantic shift from one domain (SPACE) to another (TIME).

The role of discourse context

Heine *et al.* argue that the metaphorical extension approach can account for this continuum between more and less grammaticalised meanings by taking into account the role of discourse context. While conceptual metaphors like TIME IS SPACE structure the directionality associated with grammaticalisation, the process of grammaticalisation itself is effected by discourse-related processes including **context-induced** re-interpretations which arise as a result of **metonymy**. Heine *et al.* use the term 'metonymy' in a similar way to

Barcelona (e.g. 2003c), whose account we presented in Chapter 9 in this book. Some cognitive linguists refer to the context-induced metonymy that gives rise to language change in terms of **experiential correlation** (see Tyler and Evans 2001a, 2003). For example, in an exchange in which George meets Lily by chance and asks her where she's going, Lily might reply 'I'm going to town' In this utterance, Lily refers to the act of moving in the direction of town. At the same time, this act is due to her intention to move in the direction of town. This example illustrates a close correlation between the experience of moving towards a particular goal and the intention to reach that goal. Experiential correlations of this kind can be described as metonymic in the sense that the motion event described as *be going to* 'stands for' the closely related intention. From this perspective, the semantic shift from an ALLATIVE interpretation to an INTENTION interpretation is metonymic, induced by a context-based interpretation. Further shifts of this kind may eventually result in a FUTURE interpretation, because intentions are future-oriented.

The microstructure and macrostructure of grammaticalisation

According to Heine *et al.*, more local-level discourse context processes (referred to as the **microstructure** of grammaticalisation) manage the process of semantic change resulting in grammaticalisation. However, the microstructure is guided by the conceptual metaphor, which is part of the **macrostructure**. This account of context-induced reinterpretation and metonymy emphasises the **usage-based** nature of this model: it is discourse that effects the grammaticalisation process, because forms take on new meanings as a result of speakers' communicative goals. In emphasising the relationship between pragmatic and cognitive factors in grammaticalisation, Heine *et al.* present a perspective that is in many ways consonant with the model of grammaticalisation developed by Elizabeth Closs Traugott and her collaborators (section 21.3). For example, Heine *et al.*, invoke Traugott's notion of **pragmatic strengthening**, the conventionalisation of situated inference or implicatures that results in new meanings. Consider the examples in (9), which are from Heine *et al.* (1991: 77).

(9) a. Bill died sooner than John. [temporal sense]
　　b. I'd sooner die than marry you. [preference sense]

According to Traugott and König (1991), the preference sense of *sooner* evolves from the temporal sense as a result of conversational implicature that is driven by pragmatic strengthening. Over time, the new preference sense becomes conventionalised, and may coexist alongside the original sense so that the form *sooner* becomes polysemous. In adopting the view that grammaticalisation is

Table 21.2 Macrostructure and microstructure in grammaticalisation (Heine *et al.* 1991: 103)

Macrostructure	Microstructure
Conceptual domains	Context
Similarity; analogy	Conversational implicatures
Transfer between conceptual domains	Context-induced reinterpretation
Metaphor	Metonymy

discourse-driven in this way, Heine *et al.* also point out that their model is compatible with Hopper's (1987) idea of **emergent grammar**. According to Hopper, the grammar of a language is not most insightfully conceived as a fixed or stable system that precedes discourse, but as a system that is in a constant state of flux, and 'comes out of discourse and is shaped by discourse as much as it shapes discourse in an ongoing process' (Hopper 1987: 142). Once more, this emphasises the usage-based nature of this group of grammaticalisation theories.

In sum, Heine *et al.*, argue that metaphor and context-induced reinterpretation involving metonymy are inextricably linked in the process of grammaticalisation. However, they suggest that the two are 'complementary' in the sense that one is likely to figure more prominently in any given case of grammaticalisation than the other:

> The more prominent the role of context-induced reinterpretation, the less relevant the effect of metaphor . . . the more remote the sense along any of the channels of conceptualization described . . ., the more plausible an analysis in terms of metaphor is. (Heine *et al.* 1991: 96)

Table 21.2 summarises the macrostructure and the microstructure of grammaticalisation according to Heine *et al.* While the macrostructure relates to cognitive domains (conceptual structure) and involves linking processes between domains that emerge from conceptual similarities, the microstructure relates to the pragmatic domain (discourse context).

21.2.1 Case Study: OBJECT-TO-SPACE

Having presented an overview of the framework developed by Heine *et al.*, we now consider some evidence that these researchers discuss which illustrates the nature of the grammaticalisation process. Recall that the first historical stage in the grammaticalisation cycle is the stage when a lexical item takes on a new grammatical sense, and recall also the source domain hierarchy in (7). Heine *et al.* argue that the OBJECT-TO-SPACE metaphor represents this early stage in the grammaticalisation process, and this is evident in languages where **body-part terms** have evolved into **locative adpositions**. While there is a strong tendency for these body-part terms to relate to the human body (the **anthropomorphic model**), body-part terms in some languages are also related to the animal body (the **zoomorphic model**).

Heine *et al.* conducted a study based on 125 African languages, representing the four major language families of Africa (Afroasiatic, Congo-Kordofanian, Khoisan and Nilo-Saharan). Their findings were striking. Among other prominent patterns, it emerged that in eighty of these languages, the adposition BEHIND had evolved from the body-part term for BACK. In fifty-eight of these languages, the adposition INSIDE had evolved from the body-part term for STOMACH. In forty-seven of these languages, the adposition IN FRONT OF had evolved from the body-part term for FACE. Finally, in forty of these languages, the adposition ON had evolved from the body-part term for HEAD. Consider the following examples from Swahili (Guthrie 1967–71, cited in Heine *et al.* 1991: 139). The left column represents source morphemes reconstructed for Proto-Bantu; the asterisk here represents proto-forms rather than ungrammaticality. The right column shows current Swahili adpositions.

(10) a.   *-bééde 'breast, udder, milk'   mbele 'front, before'
     b.   *-numá 'back, rear'   nyuma 'behind'
     c.   *-da 'intestines, abdomen'   ndani 'inside'

In some languages, the same modern form is polysemous between a body-part term and a spatial adposition. Consider the following examples from Hausa (Afroasiatic – Chadic; Jaggar 2001: 675–6). The bound morpheme *-n* is a genitive linker.

(11) a.   cikī   'stomach'   ciki-n   'in'
     b.   bāyā   'back'   bāya-n   'behind'
     c.   jìkī   'body'   jìki-n   'against (the side of)'
     d.   kâi   'head'   kân   'on'

21.2.2 Case study: SPACE-TO-POSSESSION

The next stage in the grammaticalisation process involves an already grammaticalised form acquiring further grammatical senses or functions. Moving further along the source domain hierarchy in (7), the evolution of possession markers from spatial terms (SPACE-TO-POSSESSION) represents this stage of grammaticalisation. Heine (1997) also argues that, in the case of POSSESSION, grammaticalisation cannot be fully characterised in terms of the evolution of a single morpheme or word, but involves the whole possessive **construction**.

This is because the syntax of possessive constructions often shows properties that are distinct from canonical syntactic patterns within the language. Heine argues that this is because possessive constructions are structured in terms of **event schemas** (these are similar to Goldberg's verb-argument constructions, which are motivated by the scene encoding hypothesis, as we discussed in Chapter 20). The structure of the relevant schema is reflected in the syntax of the construction. Consider the following examples (Heine 1997: 92–5).

(12) a. Estonian (Lehiste 1969: 325)
        Isal   on   raamat   [location schema]
        father.ADDESSIVE 3s.be   book.NOM
        'Father has a book' (lit: 'the book is at father')
  b. Russian (Lyons 1967: 394)
        U   menja   kniga   [location schema]
        at   me   book
        'I have a book' (lit: 'a book is at me')
  c. Mupun (Afroasiatic – Chadic; Frajzyngier 1993: 264)
        War   kə   siwol   [companion schema]
        3F   with   money
        'She has money' (lit: 'she is with money')
  d. French
        Le   livre   est   à   moi   [goal schema]
        the   book   is   to   me
        'The book is mine' (lit: 'the book is to me')

Heine (1997) classifies these examples in terms of various event schemas. For example, he describes (12a) and (12b) in terms of the **location schema**, (12c) in terms of the **companion schema** and (12d) in terms of the **goal schema**. What these examples all share in common, however, is that they rely upon a grammatical unit that relates to SPACE in order to express POSSESSION. While example (12a) relies upon an adessive case morpheme (expressing adjacency) to express POSSESSION, (12b) relies upon a locative preposition. Both examples express POSSESSION in terms of location in SPACE. Example (12c) relies upon an associative preposition and expresses POSSESSION in terms of proximity or contiguity in SPACE. Finally, example (12d) relies upon a preposition that encodes motion towards a goal in order to express POSSESSION.

In summary, Heine *et al.* (1991) develop a theory of grammaticalisation that relies predominantly upon the idea of metaphorical extension along a continuum from more concrete to more abstract domains. The unidirectionality of grammaticalisation is explained in terms of this metaphorical extension, which provides the macrostructure of grammaticalisation. According to this model, discourse goals giving rise to context-induced reinterpretation are also inextricably linked with grammaticalisation and provide the microstructure of the process.

21.3 Invited Inferencing Theory

In this section, we discuss a theory of semantic change in grammaticalisation proposed by Elizabeth Closs Traugott and Richard Dasher, focusing on the presentation in Traugott and Dasher (2002). This theory is called the Invited Inferencing Theory of Semantic Change because its main claim is that the form-meaning reanalysis that characterises grammaticalisation arises as a result of situated language use. In other words, semantic change is usage-based. Traugott and Dasher argue that pragmatic meaning or **inferences** that arise in specific contexts come to be reanalysed as part of the conventional meaning associated with a given construction. Inferences of this kind are **invited** in the sense that they are suggested by the context.

From invited inference to coded meaning

According to Invited Inferencing Theory, semantic change occurs when invited inferences become **generalised**. The distinction between an invited inference and a generalised invited inference is that a generalised invited inference is not simply constructed on line, but is preferred without yet being conventionalised. Inferences that subsequently become conventionalised are called **coded meanings**. According to this model, semantic change follows the path indicated in Figure 21.3.

The difference between a generalised invited inference and a coded meaning is that while a generalised invited inference can be **cancelled**, a coded meaning cannot. The ability to be cancelled is a property of inferences that can be eradicated by subsequent context. Consider example (13).

(13) After the trip to Paris, Lily felt very tired.

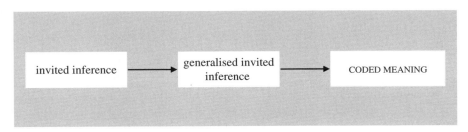

Figure 21.3 From invited inference to coded meaning

An inference associated with the temporal expression *after* is that Lily felt tired as a result of the trip. In other words, there is an inference of **causality**: the trip to Paris made Lily tired. However, causality is not a coded meaning associated with *after*, because it can be cancelled. In example (14), further information is given relating to the cause of Lily's tiredness.

(14) After the trip to Paris Lily felt very tired. It turned out she had been unwell for some time.

In this example, the second sentence cancels the inference of causality associated with *after* by providing explicit information concerning the cause of Lily's tiredness.

Subjectification

Traugott and Dasher (2002) argue that the range of semantic changes apparent in grammaticalisation are most insightfully conceived in terms of shifts from more objective to more subjective meaning. This process is called **subjectification** (not to be confused with Langacker's approach to grammaticalisation, which we call the Subjectification Approach; this is discussed in the next section). In Traugott and Dasher's sense of the term, subjectification involves a shift from a construction encoding some speaker-external event to a construction encoding the speaker's perspective in terms of location in space and time, or in terms of the speaker's attitude to what is being said. This is called the **grounding** of the speaker's perspective, which is thereby **lexicalised** (becoming part of the coded meaning). For example, while the ALLATIVE meaning of *be going to* represents a concrete and objective event, the FUTURE meaning grounds the assertion with respect to the speaker's subjective perspective. The FUTURE sense of the construction encodes the speaker's 'location' in TIME relative to the event described in the utterance. Consider example (3a) again, which is repeated here as (15).

(15) I'm gonna be a rocket scientist when I grow up.

In this example, *gonna* indexes the speaker's present location in TIME, marking the assertion as future-oriented from the speaker's perspective. In this way, the grammaticalisation of *be going to* from ALLATIVE to FUTURE involves a shift from a more objective meaning to a more subjective meaning. Like other examples of subjective meaning, this involves deixis. As we have seen, **deictic** expressions encode information that is grounded in the speaker's perspective. For example, **spatial deixis** grounds an entity relative to speaker location, as in expressions like *here* and *there*, whose reference can only be fully understood relative to the speaker's location. Similarly, **temporal deixis** concerns the subjective grounding of speaker 'location' in TIME, as reflected in the use of tense and temporal adverbials such as *yesterday* and *tomorrow*, as well as in the future sense of the *be going to* construction. These expressions can only be fully interpreted if we know 'where' in time the speaker is located. As we have seen, **person deixis** governs the use of personal pronouns like *I* versus *you*, which are also grounded in speaker perspective. Another class of expressions that are subjective in this sense are the **modal verbs**, which encode information relating to possibility, necessity and obligation (among others), and thus encode these aspects of the speaker's perspective.

Intersubjectification

A subsequent grammaticalisation process is **intersubjectification**. This relates to a shift from objective meaning to a meaning that grammatically encodes the relationship between speaker and hearer. For instance, Traugott and Dasher (2002) discuss **social deixis** in relation to the Japanese verb *ageru*, 'give'. They note that until recently *ageru* was an **honorific** verb, which means that it had to be used by a speaker (giver) who was of an inferior social status to the (hearer) recipient. In other words, part of the meaning of the verb was to signal the recognition of differential social status. More recently, this verb has begun to be used to express politeness, regardless of the relative social status of the giver and recipient. In other words, a shift has occurred in which the expression has acquired a different intersubjective meaning, evolving from an honorific expression to a marker of politeness.

Other examples of intersubjective meaning include pronoun forms in languages like French, which has 'polite' and 'familiar' variants of the second person singular pronoun (*vous* and *tu* respectively). The choice of pronoun is grounded in intersubjective perspective and encodes the social relationship that holds between interlocutors. Explicit markers of the speaker's attention to the hearer, including politeness markers like *please* and *thank you* and honorific titles like *Doctor* and *Sir* also express intersubjective meaning. Figure 21.4 summarises the evolution of subjectivity in the semantic change that underlies grammaticalisation.

As this diagram shows, objectivity and intersubjectivity represent the extreme poles of the continuum. The more objective an expression, the more likely it is to be unmarked for modality (speaker attitude) and the least likely it is to be dependent on inference for full interpretation. It follows that Grice's (1975) maxim of **quantity** predominates in this type of expression: the hearer assumes that the speaker has given as much information as is required, and is licensed to infer that what is not said is not the case. In contrast, as an expression moves along the continuum from objectivity to subjectivity, the more likely it is to be marked for speaker perspective, including modality, spatial and

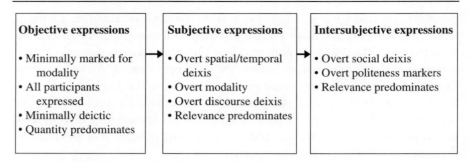

Figure 21.4 The evolution of subjectivity in grammaticalisation (based on Traugott and Dasher 2002: 22–3)

temporal deixis, and discourse deixis. The latter relates to expressions that link back explicitly to portions of preceding discourse (recall our discussion of example (6b)), or link pieces of discourse by means of connectives such as *so, if* or *because*. Furthermore, the more subjective the expression, the more dependent it is upon inference. Traugott and Dasher (2002) argue that the Grice's (1975) maxim of **relevance** therefore predominates in this type of expression: the hearer assumes that more is meant than is said. At the most subjective point on the continuum, intersubjectivity, expressions are characterised by overt social deixis (for example, honorifics) and overt politeness markers (for example, hedges like *I suppose* and expressions like *please* and *thank you*). Relevance also predominates in this type of expression.

As we mentioned earlier, Traugott and Dasher's use of the terms 'subjectification' and 'subjectivity' differs from Langacker's use of the same terms in Cognitive Grammar. Recall from Chapters 15 and 16 that, in Cognitive Grammar, subjectivity is related to **perspective** or **vantage point**, and is a property of 'off-stage' or implicit concepts, while objectivity is a property of 'on-stage' or explicit concepts. While subjectification relates to speaker perspective in both approaches, for Langacker subjectivity correlates with the absence of overt expression while for Traugott and Dasher subjectivity correlates with the presence of an overt expression that signals subjectivity. Furthermore, as Traugott and Dasher (2002: 98) point out, Langacker's model focuses upon the conceptual representation of event structures and how they are construed by the speaker. In contrast, the invited inference model focuses upon discourse, and therefore subjectivity is seen as contextually determined rather than as an inherent property of constructions.

The status of metaphor in Invited Inferencing Theory

Traugott and Dasher (2002) observe that metaphor has sometimes been considered the predominant force behind the semantic change that underlies grammaticalisation. As we have already seen, because metaphor 'was conceptualized as involving one domain of experience in terms of another and operating "between domains" . . . changes motivated by it were conceptualized as primarily discontinuous and abrupt' (Traugott and Dasher 2002: 77). As we noted earlier, the linguistic evidence does not support this view, and Heine *et al.*'s metaphorical extension account has to allow the language user a significant role in grammaticalisation in order to overcome this potential problem. For Heine *et al.*, metaphor represents a macrostructure in grammaticalisation: it is within the conceptual frame established by a conceptual metaphor that grammaticalisation occurs. From this perspective, a conceptual metaphor provides the underlying schema that facilitates context-induced semantic change.

In contrast, Traugott and Dasher argue that many, perhaps most, of the regular semantic changes involved in grammaticalisation do not involve metaphor. Instead, semantic change arises from the usage-based processes we described above, in which invited inferences become generalised before becoming conventionalised as coded meaning. From this perspective, the changes involved are smaller-scale, mediated by context and language use. These changes are therefore metonymic in the sense that one concept 'stands for' another closely related concept rather than one concept being understood in terms of another as a result of a metaphorical mapping from one domain to another (recall our discussion of the examples in (8)). Traugott and Dasher's model is summarised in Figure 21.5.

While studies that focus on metaphor complement the Invited Inferencing approach, Traugott and Dasher argue that the predominance of metaphor-based explanations in theories of grammaticalisation results from a tendency to focus on the beginning and endpoints of the process of change (the bottom of the model in Figure 21.5), without fully investigating the pragmatic processes that drive the process of change (the top of the model in Figure 21.5). In this respect, Traugott and Dasher tend towards the view that metaphor is epiphenomenal in the context of grammaticalisation: a 'side effect' of the grammaticalisation process rather than an underlying cause.

21.3.1 Case study: the evolution of *must*

Having set out the path of change predicted by Invited Inferencing Theory, we now consider a case study of grammaticalisation discussed by Traugott and Dasher (2002). Traugott and Dasher observe that modal verbs (see Chapter 18) follow a unidirectional path of evolution. Firstly, content expressions evolve into deontic modals which express obligation, permission or prohibition (e.g. *George must learn to be on time*). The same expressions then evolve into epistemic modals which relate to knowledge and belief (e.g. *George must be home by now*). Evolution in the opposite direction (from content expression

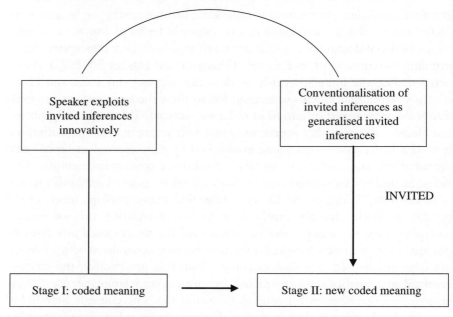

Figure 21.5 Invited inferencing model (adapted from Traugott and Dasher 2002: 38)

Table 21.3 The evolution of modal verbs (Traugott and Dasher 2002)

deontic > epistemic
narrow scope > wide scope
less subjective > more subjective

to epistemic to deontic) is not attested. Secondly, the path of evolution is from narrow scope (over some subpart of the proposition) to wide scope (over the whole proposition), not vice versa. This point is illustrated below. Finally, this path of grammaticalisation correlates with increased subjectivity. The properties that characterise the evolution of modal verbs are summarised in Table 21.3.

In order to illustrate this path of evolution, Traugott and Dasher present a case study of the English modal *must*. In Modern English, this modal verb has both deontic readings (16a) and epistemic readings (16b).

(16) a. You must stop talking.   [deontic modality]
     b. Lily must love him, I suppose.   [epistemic modality]

Traugott and Dasher describe the evolution of this verb from Old English to Modern English in terms of three stages:

STAGE I: *must₁*: ability (17); permission (18) (Old English)

(17) Ic hit þe þonne gehate pæt þu on Heorote *most* sorhleas swefan.
 I it you then promise that you in Heorot will.be.able anxiety.free sleep
 'I promise you that you will be able to sleep free from anxiety in Heorot.'
 (Eighth century, *Beowulf*; Traugott and Dasher 2002: 122 [Visser 1969])

(18) þonne rideð ælc hys weges mid ðan feo & hyt *motan* habban eall.
 then rides each his way with that money and it be.permitted have-INF all
 'Then each rides his own way with the money and can keep it all.'
 (c.880, *Orosius*; Traugott and Dasher 2002: 123 [Traugott 1989: 37])

STAGE II: *must₂*: obligation/deontic (Late Old English – Early Middle English)

(19) Ac ðanne hit is þin wille ðat ic ðe loc ofrin *mote*.
 but then it is thy will that I thee sacrifice offer must
 'But then it is Thy will that I must offer Thee a sacrifice.'
 (c.1200, *Vices and Virtues*; Traugott and Dasher 2002: 124 [Warner 1993])

STAGE III: *must₃*: epistemic (Middle English – Modern English)

(20) For yf that schrewednesse makith wrecches, than *mot* he nedes ben moost wrecchide that longest is a schrewe.
 'For if depravity makes men wretched, then he must necessarily be most wretched that is wicked longest.'
 (c.1380, Chaucer, *Boece*; Traugott and Dasher 2002: 129)

(21) There ys another matter and I *must* trowble you withal . . . hit ys my lord North . . . surely his expences cannott be lytle, albeyt his grefe *must* be more to have no countenance at all but his own estate.
 'There is another matter I must trouble you about . . . It is my Lord North . . . surely his expenses can't be small, although it must be an even greater grief to him that he has no standing other than his own estate.'
 (1586, Dudley; Traugott and Dasher 2002: 129)

As these examples demonstrate, *must* originated in Old English as a content verb meaning 'be able' and evolved into a deontic verb expressing permission and then obligation (19). Traugott and Dasher (2002) argue that the evolution from permission to obligation correlates with increased subjectivity. The earliest uses of the obligation (deontic) sense of *must* appear to have been **participant-external**. That is, the obligation (deontic) sense arose in contexts where permission was being granted to a third person referent. In such contexts, particularly when the person or entity granting the permission, such as a King or the Church, is in a position of authority, there is a context-induced implication of obligation. Once the deontic sense was established, more subjective **participant-internal** (first person) uses began to emerge, as illustrated by the example in (19). The shift from deontic to epistemic senses in (20)–(21) also follows a path from objective to subjective uses. According to Traugott and Dasher (2002: 132), therefore, there is no basis for a metaphorical account of the evolution of this modal verb since an invited inferencing analysis provides an explanatory account.

In summary, the main claim of Invited Inferencing Theory is that contextual (pragmatic) meaning is reanalysed as inherent (coded) meaning. While the role of metaphor is recognised, it is not seen as the central force behind the grammaticalisation process.

21.4 The subjectification approach

Langacker (1999b, 1999c) argues that **subjectification** is central to grammaticalisation. As we have already mentioned, Langacker uses the term 'subjectification' in a different way from Traugott and Dasher. In Langacker's model, subjectification relates to **perspective**. For example, speaker and hearer are usually subjectively construed or 'off-stage', and only become objectively construed or 'on-stage' when linguistically profiled, for example by the use of pronouns such as *I* or *you*. Langacker argues that subjective construal is **immanent** in (subsumed by) objective construal, because whether or not the conceptualiser is on-stage (objectified), his or her perspective in terms of participation in **scanning** is part of the conceptualisation process. This idea is illustrated by Figure 21.6.

In Figure 21.6, the circle marked C represents the conceptualiser who is **mentally scanning** the interaction between trajector (TR) and landmark

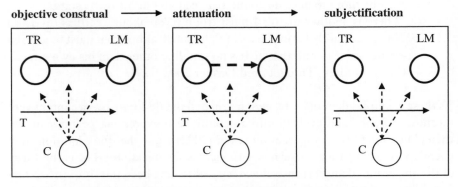

Figure 21.6 Subjectification, or the attenuation of objectivity (adapted from Langacker 1999b: 6)

(LM). This scanning process is represented by the arrows between C and TR and LM, and takes place across processing time, which is represented by the horizontal arrow marked T. The difference between the three diagrams in Figure 21.6 is the arrow that connects TR and LM, which represents the profiling of the relationship between TR and LM. In the case of objective construal, this arrow is unbroken. This represents the idea that the relationship between TR and LM is highly objectively salient. In the central diagram in Figure 21.6, this arrow is broken, which represents **attenuation** or weakening of the objective salience of the relationship between TR and LM. When subjectification occurs, the arrow representing the relationship between TR and LM is absent, which represents the idea that there is no longer any objective salience in the relationship between TR and LM. Although the two are still related, the relationship holds only within the conceptualiser's construal.

The examples in (22) provide some linguistic evidence for this rather abstract idea. Langacker (1999b) compares two different senses of the expression *across* in order to illustrate subjectification or the attenuation of objectivity.

(22) a. Lily ran across the street.
 b. There is an off-licence across the street.

In example (22a), the TR *Lily* is in motion, and the expression *across* encodes her path of movement which is therefore objectively salient. In contrast, TR *an off-licence* in example (22b) is a static entity, and the expression *across* only encodes its location. Although both examples involve the same perspective point for the conceptualiser, who mentally scans the path across the street, the objective salience of this path is weaker in (22b) because of the absence of a moving TR. Furthermore, while the entire path is profiled in (22a), only the endpoint of the path is profiled in (22b). The idea behind immanence is that subjective construal is 'there all along', but only comes to the fore when objective construal is attenuated or weakened.

Langacker claims that subjectification or the attenuation of objectivity gives rise to grammaticalised forms over a period of time, and that in the intervening stages on the gradual path to grammaticalisation, a number of layered senses or functions of a single form may coexist. Langacker argues that attenuation is evident in four main patterns of change, which are summarised in Table 21.4.

Table 21.4 Patterns of attenuation (Langacker 1999b)

Status	actual → potential; specific → generic
Focus	profiled → unprofiled
Domain	physical → experiential/social
Locus of activity	on-stage → off-stage

Langacker (1999b) provides a number of examples of how the attenuation process evolves grammaticalised forms. In the remainder of this subsection, we revisit the *be going to* construction from the perspective of Langacker's subjectification approach, and look at how this model accounts for the evolution of auxiliary verbs.

21.4.1 Case study: *be going to*

As we saw earlier, this construction is associated with an ALLATIVE (motion) sense and a FUTURE sense. Consider example (23), which is ambiguous between these two senses.

(23) George is going to buy some champagne

In this example, George may be walking across the street towards the off-licence (ALLATIVE sense) or sitting on Lily's sofa planning his drinks party (FUTURE sense). Langacker (1999b) argues that the FUTURE sense arises from subjectification, in that the conceptualiser mentally scans George's motion through TIME rather than SPACE, and this scanning becomes salient in the conceptualiser's construal because the motion along this path is not objectively salient (there is no physical motion). In a number of ways, the FUTURE sense of *be going to* shows the defining properties of attenuation that are set out in Table 21.4.

21.4.2 Case study: the evolution of auxiliaries from verbs of motion or posture

The Spanish auxiliary verb *estar* 'be' evolved from a content verb meaning 'stand'. Langacker (1999b: 309) provides the following examples, which show that this verb behaves like the English copula in that it can take subject (or adverbial) complements, for example AP (24a) or PP (24b). The verb *estar* also functions like the English progressive auxiliary, in that it can also take a present participle (24c).

(24) a. Está   enfermo
         be.3s   ill
         'He is ill'
     b. Está   en   la   cocina
         be.3s   in   the   kitchen
         'He is in the kitchen'
     c. Está   trabajando
         be.3s   working
         'He is working'

Recall from Chapter 16 that what distinguishes TEMPORAL RELATIONS (PROCESSES) from ATEMPORAL RELATIONS in Cognitive Grammar is **sequential scanning**. We also saw in Chapters 17 and 18 that, in the Cognitive Grammar analysis, the role of the verb *be* is to impose PROCESS status upon otherwise ATEMPORAL RELATIONS like adjectives, prepositions and participles. Langacker (1999b) argues that the path of change from a verb of posture to a *be* verb involves attenuation of objectivity resulting in loss of subject control and consequent subjectification. The path of evolution proposed by Langacker is schematically represented in (25) and the English examples in (26) provide an illustration of this claim.

(25) [*stand* + PARTICIPLE] → [*stand'* + PARTICIPLE] → [*be* + PARTICIPLE]

(26) a. Lily stood there gazing into his eyes.
 b. The bomb stood ticking on the mantelpiece.
 c. The bomb was ticking.

In (26a), the situation designated by *stand* is salient and the event designated by the adverbial subordinate clause *gazing into his eyes*, headed by the participle *gazing*, is less salient hence its status as a modifier. In (26b), which contains an attenuated instance of *stand* that Langacker represents as *stand'*, the situation designated by *stand* is still salient, but its objectivity is attenuated because its TR is a static and inanimate entity. It is in this example that the notion of loss of subject control becomes clear: the extent to which the objective construal of the construction is attenuated is closely linked to the properties of the subject (or TR) in terms of animacy, potential for motion and so on. The further attenuation of *stand* results in a sense that is also devoid of orientation in SPACE, and it is at this point that Langacker suggests the verb of posture evolves into a *be* verb which has lost its original content meaning but retains its PROCESS (verbal) essence which designates sequential scanning. At this stage, the *be* verb and the participle merge in terms of expressing a single event or situation (26c).

In summary, although we have only been able to provide a brief sketch of how Langacker's notion of subjectification (or attenuation of objectivity) may give rise to grammaticalisation, a number of points of contrast emerge in relation to the other approaches to grammaticalisation we have discussed in this chapter. To begin with, while both the metaphorical extension approach and Invited Inferencing Theory place the burden of explanation on metaphor and pragmatic inferencing respectively, Langacker's explanation has little to say about either of these factors, but focuses the account entirely on how the conceptual system might give rise to grammaticalisation as a consequence of perspective and construal. Secondly, Langacker's account – most obviously his account of the evolution of *be* – can be described as a version of the **semantic bleaching** account that is largely rejected by other cognitively oriented grammaticalisation researchers. Indeed, Langacker ([1991] 2002: 324) explicitly equates semantic attenuation and semantic bleaching.

21.5 Comparison of the three approaches: *be going to*

We conclude our discussion of the three theoretical approaches to grammaticalisation presented in this chapter with a brief comparison of how each approach accounts for the *be going to* construction. As we saw in section 21.2, the metaphorical extension approach analyses the shift from ALLATIVE to FUTURE in terms of metaphorical extension from the more concrete domain of SPACE to the more abstract domain of TIME. Because the *be going to* construction exhibits polysemy, which is potentially problematic for a metaphorical extension account, the analysis developed by Heine *et al.* (1991) also takes into account the role of discourse context, which gives rise to context-induced reinterpretations based on metonymy or experiential correlation, for example between motion and intention: recall our discussion of example (8).

In contrast, we saw in section 21.3 that Invited Inferencing Theory (Traugott and Dasher 2002) rejects the metaphorical extension account, and analyse the *be going to* construction in terms of a shift from a construction encoding a speaker-external event towards a construction encoding speaker perspective relative to TIME and SPACE. According to this theory, the ALLATIVE sense encodes a concrete and objective event, while the FUTURE sense relates to the speaker's location in TIME and is therefore more subjective: recall example (15).

Finally, we saw in section 21.4 that the subjectification approach developed by Langacker (1999b) analyses the evolution of the ALLATIVE into the FUTURE sense in terms of the nature of the conceptual processes that underlie each interpretation. While the ALLATIVE sense involves the conceptualiser scanning actual physical motion through space, objective construal is salient and subjective construal remains backgrounded. In contrast, the FUTURE sense lacks physical motion and therefore objective construal is attenuated, which enables subjective construal to become salient.

In many ways, the fully usage-based character of the theory proposed by Traugott and Dasher, which views metaphor as epiphenomenal, is in keeping with some of the most recent trends within cognitive linguistics, which focus increasingly upon 'bottom-up' or usage-based explanations of 'dynamic' aspects of language use, rather than upon 'top-down' or structural explanations that are characteristic of conceptual metaphor theory. As we will briefly see in Chapter 23, the changing status of conceptual metaphor in grammaticalisation theory has parallels in the recent movement within cognitive semantics to question the status of conceptual metaphor as an explanatory construct. This is evident in research on conceptual blending (e.g. Turner and Fauconnier 1995, and indeed to some extent Grady *et al.* 1999); in research on cognitive lexical semantics (e.g. Evans 2004a); and in research on conceptual projection (Zinken *et al.* forthcoming). Moreover, research outside cognitive linguistics (e.g. Stern 2000) is increasingly critical of the absence of a serious account of context and use in conceptual metaphor theory. As we saw in our discussion of Heine *et al.*'s 'metaphor' account of grammaticalisation, a descriptively adequate account of grammaticalisation cannot ignore the context of language use, which, at least in part, contributes to the process of grammaticalisation.

While not strictly an account of grammaticalisation *per se*, which is a historical and usage-based phenomenon, Langacker's account represents a serious attempt to model the kinds of mental processes that result in the form-meaning reanalysis characteristic of grammaticalisation. It follows that Langacker's account complements (rather than competes with) the usage-based accounts proposed by Heine *et al.* and by Traugott and Dasher.

21.6 Summary

The discussion in this chapter took a **diachronic** perspective on language and focused on the type of language change known as **grammaticalisation**. We began by presenting a descriptive overview of grammaticalisation and saw that this **unidirectional** and **cyclic** process involves changes in the function or **meaning** of a linguistic unit, which evolves from content to grammatical or from grammatical to more grammatical. These changes may result in **layering** or **polysemy** at certain stages in the grammaticalisation process. Such changes are often accompanied by **correlated** changes in the phonological and morphological **form** of the unit in terms of **reduction** and **loss of autonomy**. Having set out the features of grammaticalisation, we explored three main cognitively oriented theories of grammaticalisation. We saw that the main claim of **metaphorical extension theory**, which until recently was probably the most widely adopted cognitive theory, is that human **creativity** gives rise to the process of **metaphorical extension** in developing a new expression for a grammatical concept. Metaphorical extension involves the mapping of **image schematic** concepts from concrete **source** domain to abstract **target** domain. According to this theory, **egocentricity** and embodiment are central to defining source concepts, and this theory therefore takes an explicitly **experientialist** stance. In contrast, the main claim of Invited Inferencing Theory is that the **generalisation** and ultimately the **conventionalisation** of **pragmatic inference** gives rise to new **coded forms**, a process that it characterised by increasing **subjectification** (in a pragmatic sense). While this theory admits the contribution of metaphor to grammaticalisation, its proponents argue that the metaphorical extension approach only presents a partial explanation of the phenomenon, since it is communicative goals that ultimately give rise to grammaticalisation. Finally, we saw that Langacker's subjectification approach takes a conceptual rather than contextual approach, and holds that the subjectification (in a conceptual sense) that characterises grammaticalisation is **immanent** in a conceptualiser's construal of a scene encoded in language and is revealed by the **attenuation** of objective salience.

Further reading

Introductory texts and background reading

- **Croft (2000).** This book sets out Croft's theory of language change, which we discussed in Chapter 4. Chapter 6 of this book addresses grammaticalisation, where Croft argues in favour of a pragmatic inference account. This usage-based perspective on language change is compatible with Croft's (2001) Radical Construction Grammar framework which we discussed in Chapter 20.
- **Croft (2003).** Chapter 8 of this excellent textbook provides a descriptive overview of grammaticalisation, and briefly discusses explanations for the functional process that underlies it, including Heine's theory and Traugott's theory.
- **Heine and Kuteva (2002).** This book provides a theory-neutral reference source for grammaticalisation data. The authors have collected examples from over 500 languages, which illustrate over 400 grammaticalisation processes. Each entry states the source and target of the grammaticalisation process (for example, OBLIGATION > FUTURE), and lists a number of representative examples, together with sources of the data.
- **Hopper and Traugott ([1993] 2003).** This extremely accessible textbook, now in its second edition, is written by two leading researchers in the field of grammaticalisation and provides an introductory overview of the field. The textbook takes a theory-neutral approach and includes chapters on the history of research into grammaticalisation, the mechanisms of language change, pragmatic inferencing and unidirectionality. The authors explore grammaticalisation processes both internal to and across the clause, and present examples from over eighty languages from around the world.
- **Lee (2001).** Chapter 7 of this textbook provides a brief overview of language change, and includes a discussion of the evolution of the English modals *can* and *may* with a good range of examples from Old English and Middle English.

- **Trask (1996)**. This textbook introduces historical linguistics from a general linguistics perspective.
- **Trask (2000b)**. A reference guide to key terms in historical and comparative linguistics, including information on language families and notes for further reading.

Metaphorical extension approaches

- **Bybee, Perkins and Pagliuca (1994)**. Like Heine *et al.* (1991), Bybee *et al.* argue that metaphorical extension and the conventionalisation of implicature are central to grammaticalisation. This study focuses on the evolution of grammatical markers of tense, aspect and mood in a large-scale sample of over ninety languages, and argues for an explicitly usage-based account of the findings.
- **Heine (1997)**. This book presents an introduction to Heine's metaphorical extension theory of grammaticalisation. This book is conceived as a teaching text so there is less emphasis on theoretical detail but considerable emphasis on cross-linguistic data representing grammatical evolution in areas such as spatial deixis, indefinite articles and possession.
- **Heine, Claudi and Hünnemeyer (1991)**. The discussion in section 21.2 of this chapter is largely based on this book. The first chapter of this book traces the development of grammaticalisation theory from the eighteenth to the late twentieth centuries. Chapter 2 addresses the metaphorical basis of grammaticalisation and Chapter 3 addresses the contribution of discourse goals. Chapter 4 sets out Heine *et al.*'s framework and Chapters 5 and 6 present case studies of grammaticalisation processes. Chapter 7 focuses on the experiential basis of metaphor in grammaticalisation and Chapter 8 focuses on the cyclic nature of grammaticalisation. The concluding chapter argues for a 'panchronic' view of grammaticalisation, according to which the phenomenon can only be fully explained by merging diachronic and synchronic perspectives.
- **Svorou (1994)**. Svorou's typological study focuses on grammatical morphemes that encode spatial concepts and relies upon a sample of twenty-six unrelated languages. Like Heine *et al.* (1991), Svorou argues that the semantic change that underlies grammaticalisation is essentially metaphorical in nature.
- **Sweetser (1988); Sweetser (1990)**. Like Heine *et al.* (1991), Sweetser also argues that the semantic change that underlies grammaticalisation involves metaphorical extension. In her (1990) book, Sweetser explores the basis of lexical polysemy, pragmatic ambiguity and semantic change, and argues that all these features of language can be explained

in terms of metaphor. For example, she proposes a metaphor-based account of the transition from deontic or 'root' modality, to epistemic and speech act uses of the modal verbs. Her 1988 paper focuses more closely on the connection between metaphorical extension and grammaticalisation.

Invited Inferencing Theory

- **Traugott and Dasher (2002)**. This book, on which section 21.3 of this chapter is based, sets out Traugott and Dasher's Invited Inferencing Theory of Semantic Change. Chapter 2 provides an overview of the literature on semantic change in grammaticalisation, Chapter 3 presents a discussion of the evolution of modal verbs. In Chapter 4, the authors extend the same approach to adverbial discourse markers such as *indeed, in fact* and *actually*, which, like the modals, are argued to evolve epistemic from non-epistemic senses. The evolution of performative verbs and social deictics represent the other major topics addressed in this volume.

Langacker's subjectification approach

- **Langacker (1991)**. Chapters 5 and 6 of this volume include discussions of the impact of subjectification on the evolution of auxiliary and modal verbs.
- **Langacker (1999b)**. In Chapter 10 of this book, on which the discussion in section 21.4 of this chapter is based, Langacker sets out his theory concerning the impact of subjectification (revealed by attenuation of objectivity) upon grammaticalisation. An earlier version of this paper is also published as Langacker (1999c).
- **Langacker ([1991] 2002)**. Chapter 12 of this book includes a discussion of the relationship between subjectification and attenuation that presents the same basic argumentation as Langacker (1999) but includes a discussion of some different examples, such as the evolution of verbs of possession into markers of perfect aspect.

Exercises

21.1 Cognitive accounts of grammaticalisation

Based on the discussion in this chapter and on your own reading, what are the main similarities and differences in the three approaches to grammaticalisation discussed here? Present your conclusions as an annotated table.

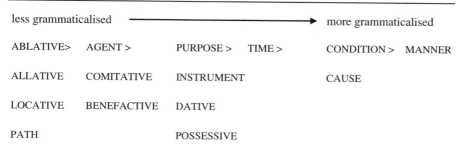

Figure 21.7 Chain of increasing grammaticalisation: case functions (adapted from Heine *et al.* 1991: 159)

21.2 Metaphorical extension theory: *for*

Consider the following examples (adapted from Heine *et al.* 1991: 152).

(a) Lily couldn't see the wood for the trees.
(b) Lily cooked supper for George.
(c) Lily set off for work.
(d) Lily worked hard for her wages.

In English, prepositions often perform the functions that are carried out by case markers in other languages. According to Heine *et al.*, the polysemy of *for* illustrates the evolution of more grammaticalised concepts from existing grammaticalised concepts which can be plotted on the chain of evolution of case functions represented in Figure 21.7. Recall that Heine *et al.* claim that grammaticalisation involves metaphorical extension from more concrete to more abstract concepts, so that more abstract senses are 'more grammaticalised'.

Firstly, work out which case functions in Figure 21.7 are illustrated by the examples above. You may need to consult a dictionary of grammatical terms. Next, work out the predictions of Heine *et al.*'s theory for the relative order in which each of the functions of *for* might have emerged. Once you have mapped out your hypothesis, you may consult an etymological dictionary in order to find out whether your hypothesis receives any support from the historical facts.

21.3 Metaphorical extension theory: *with*

Consider the following examples (adapted from Heine *et al.* 1991: 164). Develop the same kind of analysis for the preposition *with* in these examples as you did for the preposition *for* in the previous exercise.

(a) George opened the champagne with his teeth.
(b) George ate the oysters with gusto.
(c) Lily went to the restaurant with George.

21.4 Invited Inferencing Theory: speech act verbs

Consider the following examples from Traugott and Dasher (2002: 201–22).

(a) insist < Latin *in* 'in, on' + past participle of *sta-* 'stand'
(b) concur < Latin *con* 'with' + *curr-* 'run'
(c) concede < Latin *con* 'with' + *ced-* 'go away, withdraw'
(d) promise < Latin *pro* 'forward' + past participle of *mitt-* 'send'

Traugott and Dasher argue that the evolution of speech act verbs from verbs with spatial senses illustrates the process of subjectification. Explain how these examples might be analysed from the perspective of Invited Inferencing Theory. In particular, how do you think this type of grammaticalisation illustrates subjectification in Taugott and Dasher's sense of the term?

21.5 Comparing and contrasting metaphorical extension with Invited Inferencing

Prepositions often have temporal meanings conventionally associated with them. Try to provide plausible explanations of how the following temporal senses may have been derived from the earlier spatial senses, employing (i) the metaphorical extension account and (ii) the Invited Inferencing account. Comment on any problems that these examples present for either of these accounts.

(a) the TEMPORAL CONTAINER sense
    The election will be held in May.
(b) the AMOUNT OF TIME REQUIRED/TAKEN sense
    Lily ran the marathon in three hours.
(c) the AMOUNT OF TIME UNTIL OCCURRENCE sense
    I'll call you in twenty minutes.

Now check the OED (*Oxford English Dictionary*), or some other suitable source, in order to establish the historical accuracy of the paths of evolution you proposed.

21.6 Langacker's subjectification approach: *get*

Consider the following examples (some of which are adapted from Langacker 1999b: 312).

(a) Lily got two bags of chips.
(b) Lily got a pay rise.
(c) Lily got a black eye.
(d) Lily got herself promoted.
(e) George got fired.
(f) Lily's bike got stolen.

According to Langacker, the grammaticalisation of *get* from a full lexical verb in (a)–(c) to a function close to that of the passive auxiliary in (d)–(f) is a result of subjectification, particularly with respect to the attenuation of subject control. How do these examples illustrate Langacker's analysis? Has *get* been fully grammaticalised as a passive auxiliary? Provide additional examples to support your discussion and comment on any problems that you encounter.

21.7 Accounting for the future

In English, one way of referring to future time is by means of the modal verb *will*. This evolved from the Old English form *willan*, which was a full lexical verb meaning 'to want or desire'. Experiential accounts suggest that the future meaning arose because desire is inherently future-oriented. In other words, an invited inference of desiring something is that the act of obtaining the desired object lies in the future; therefore, the statement of a desire implicates future attainment. Languages employ a variety of constructions to grammaticalise futurity; some examples are provided below. In each, the first (functional) English gloss provides the English free translation equivalent, while the second (literal) gloss explains the original and literal sense of the construction. For each example, (i) try to provide a plausible experiential motivation for the shift to future meaning, and (ii) employ one of the theories discussed in this chapter to account for how this shift might have occurred.

(a) Danish   Jeg skal komme i morgen
    Functional: 'I'll come tomorrow'
    Literal: 'I must come tomorrow'

(b) Inuit   Atuarniarpara
    Functional: 'I'm going to read it'
    Literal: 'I try to read it'

(c) Modern Greek   Tha pao stin Athina
    Functional: 'I'll go to Athens'
    Literal: 'I want to go to Athens'

(d) Spanish Voy a comprarlo
  Functional: 'I'm going to buy it'
  Literal: 'I am going (somewhere) in order to buy it'

(e) Hungarian Jól fogunk szórakozni
  Functional: 'We'll enjoy ourselves'
  Literal: 'We catch a good time'

22

Cognitive approaches to grammar in context

Throughout Part III of this book, we have developed an insight into cognitive approaches to various topics in the study of grammar. In this chapter, we set the cognitive approach in a wider theoretical context. In particular, we compare the assumptions, aims and methodology of cognitive approaches with other influential approaches to grammar. Our discussion will focus on generative and functional-typological approaches, because the former represents the paradigm against which cognitive approaches were originally defined, and the latter shares certain core assumptions with aspects of cognitive approaches and indeed informs the cognitive approach in various ways. We begin our comparison by setting out the assumptions, aims and methodology adopted by each of the three broad traditions, and by comparing the models according to these parameters (section 22.1). We then present a comparative discussion of cognitive and generative approaches to core issues in the study of grammar, including word classes, constituency, grammatical functions and case, and tense, aspect, mood and voice (section 22.2). As we will see, cognitive and generative approaches have a great deal in common in terms of the phenomena they set out to explain, and a certain amount in common in terms of how they explain those phenomena. However, as we have emphasised throughout this book, there are important differences between these approaches in terms of their foundational assumptions. These in turn give rise to differences in descriptive terminology and in some of the phenomena that are investigated.

22.1 Theories of grammar: assumptions, objectives, methodology

In this section, we compare cognitive approaches to grammar with generative and functional-typological approaches. Of course, these are not the only theories of grammar that exist, and there are approaches that might not align themselves directly with any of these three traditions. It is also important to emphasise that each of these theoretical approaches can be subdivided into a number of approaches that might differ significantly. Indeed, this has become clear not only in our discussion of different cognitive approaches but in our investigation of the 'broadly generative' Construction Grammar approach developed by Kay and Fillmore, which departs in significant ways from the transformational approach developed by Chomsky, which we elaborate on below. A further complication is that these broad categories may overlap significantly. For example, the work of certain researchers could in principle be classified as partly cognitive and functional-typological in nature. We might characterise the work of William Croft and Bernd Heine in this way. We could also place Kay and Fillmore's Construction Grammar theory on the borderline between cognitive and generative approaches in a number of respects, as we saw in Chapters 19 and 20. As these comments suggest, approaches to grammar, like most categories, have fuzzy boundaries. Despite these difficulties in drawing sharp dividing lines, these three broad categories represent the prominent traditions in approaches to grammar. Furthermore, these three approaches are interrelated in important ways, as we will elaborate below.

The emphasis in this section is on the assumptions, aims and methodology adopted by each of these approaches. Figure 22.1 represents the three traditions and includes a representative but non-exhaustive list of theories and/or researchers associated with that tradition. The reader will observe that most of the comparative discussion in this book has focused on the differences between cognitive and generative ('formal') approaches while we have had little to say about the functional-typological approaches. This is because the cognitive approach to grammar originally grew out of a reaction against the generative approach and defined itself explicitly against that tradition. However, functional approaches to language, particularly the functional-typological approach, have also informed and influenced the cognitive approach in a number of important ways, although this influence is not always explicitly acknowledged in the literature. We therefore include in this section a brief discussion of this type of approach.

Any theory of grammar can be characterised along three parameters: assumptions, objectives and methodology. The **assumptions** of a theory reflect the philosophical orientation of that theory in terms of how it sees the nature of the relationship between language, thought and world. The **objectives** of a theory reflect what that theory seeks to establish, describe or explain. The **methodology** of a theory reflects the ways in which it sets about meeting those objectives.

COGNITIVE APPROACHES TO GRAMMAR IN CONTEXT

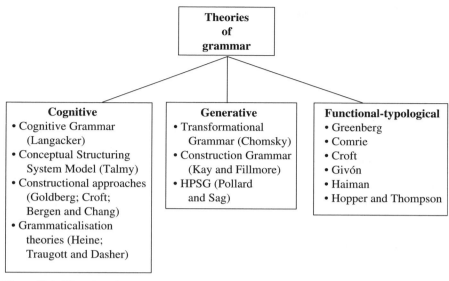

Figure 22.1 Theories of grammar

22.1.1 Cognitive approaches to grammar

Given our investigation of cognitive approaches to grammar throughout Part III of this book, we can summarise the key characteristics of this type of approach as shown in Table 22.1. We revisit some of these characteristics in our comparison of cognitive and generative approaches to grammar below.

22.1.2 Generative approaches to grammar

In this section, we present an overview of the characteristics of generative approaches to grammar. As we will see, the most prominent generative approach is that of Transformational Grammar developed by Chomsky, but there are also several other broadly generative approaches that are non-transformational, including Kay and Fillmore's Construction Grammar, Head-driven Phrase Structure Grammar (HPSG) and Lexical Functional Grammar (LFG).

Prior to the emergence of the Chomskyan model, the prominent approach in twentieth-century American linguistics was the **behaviourist** approach, which viewed linguistics as the study of observable linguistic behaviour. This approach is associated with the American structuralists, such as Leonard Bloomfield (1887–1949), whose work focused upon field linguistics and characterising directly observable linguistic phenomena such as phonological and grammatical form. Bloomfield's 1933 book *Language* is regarded by many linguists as a model of careful and precise linguistic description. However, this approach had little to say about unobservable phenomena such as meaning or

Table 22.1 Characteristics of a cognitive approach to grammar

Assumptions	Objectives	Methodology
• Empiricist view • Cognitive Commitment • Generalisation Commitment • Embodied cognition thesis • Symbolic thesis • Usage-based thesis: schemas reflect use • Grammar is a structured inventory • Lexicon-grammar continuum • Constructions have meaning: scaffolding metaphor • Redundancy is natural	• To demonstrate that grammar is meaningful • To account for both regular and irregular phenomena • To develop a model of language that reflects cognition	• Search for converging evidence • Take account of diachronic evidence • Examine both regular and irregular patterns • Avoid extreme formalism • Prohibit 'underlying' representations in accounting for grammatical phenomena

about the mental representation of language. As we saw in Chapter 4, the behaviourist psychologist B. F. Skinner (1904–90), in his (1957) book *Verbal Behaviour*, outlined the behaviourist theory of language acquisition, which held that children learnt language by imitation and that language has the status of stimulus-response behaviour, conditioned by positive reinforcement (rather like Pavlov's dog).

The generative framework has its origins in Chomsky's (1957) book *Syntactic Structures*, in which he proposed – contrary to the behaviourist theory of language prevalent at that time – that human beings are predisposed for language acquisition by virtue of a designated cognitive system that later came to be known as **Universal Grammar**. As we saw in Chapter 4, in his (1959) review of Skinner's book, Chomsky argued (among other things) that the behaviourist theory failed to explain how children produce utterances that they have never heard before, as well as utterances that contain errors that are not present in the language of their adult caregivers. Chomsky's theory was the first **mentalist** or cognitive theory of human language, in the sense that it attempted to explore the psychological representation of language and to integrate explanations of human language with theories of human mind and cognition. For this reason, Chomsky's early work is often described as one of the catalysts of the 'cognitive revolution', coinciding with the birth of cognitive science as a discipline in its own right, uniting through common goals and research questions disciplines such as philosophy, psychology, linguistics and artificial intelligence.

As we saw in Chapter 4, the generative model rests upon the hypothesis that there is a specialised and innate cognitive subsystem that represents unconscious knowledge of language, or **competence**. The idea that linguistic knowledge arises from 'drawing out what is innate in the mind' (Chomsky 1965: 51) is described by philosophers as the **rationalist** view, and contrasts with the **empiricist** view, which holds that linguistic knowledge is constructed on the basis of experience and is independent of any specialised cognitive system. Universal Grammar is the model of the **initial state** of the innate language faculty: in other words, the system of linguistic knowledge that all humans bring to the process of acquiring their first language. In developing this mentalist theory of language, Chomsky asserts that the only revealing object of linguistic study, given the objective of characterising competence, is the system of linguistic knowledge in the mind of the idealised individual speaker. This system of internalised linguistic knowledge is known as **I-language** (Chomsky 1986: 19–56). From this perspective, the externalised language of the speech community (E-language) is merely epiphenomenal, in the sense that it arises as the output of individual I-languages.

In the generative model, this innate language system is viewed as 'encapsulated' or **modular** and patterns of **selective impairment**, particularly when these illustrate **double dissociation**, are often seen as evidence for the encapsulation of such cognitive subsystems. Of course, the interpretations of such patterns are open to a range of interpretations (see the discussion of Tomasello's review of these issues in Chapter 4). In addition, the language module itself is viewed as a modular system. In other words, the linguistic subsystems such as syntax, semantics and phonology are seen as independent **submodules** within the language system. This view rests upon the premise that the principles and processes, and the primitives over which they operate, are different in kind from one area of language (for example, phonology) to another (for example, syntax). In addition, selective impairment within the language system itself is a frequent consequence of acquired left-hemisphere brain damage. For example, certain types of acquired **aphasia** (or language disorder) such as anomia (loss of content words) or agrammatism (loss of or damage to grammatical units and structures) appear to target different aspects of the language system. This type of selective language impairment is often interpreted as evidence for the plausibility of a model in which subtypes of linguistic knowledge are organised separately within the cognitive system as well as being localised separately within the physical brain. A simple model of the language module is shown in Figure 22.2. The levels of phonological form (PF) and logical form (LF) operate over the output of the syntax (sentence-level structures) with respect to phonological and semantic principles, respectively.

Within the generative model, as we saw in Chapter 4, the existence of a language module is held to account for the rapid **acquisition** of language by

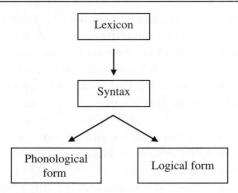

Figure 22.2 A modular view of the language system

human infants and for the existence of linguistic **universals**. The system is not open to conscious introspection, nor does it correlate with any single local function in the physical brain. Rather, it is one aspect of a complex model of 'mind', and can only be reconstructed on the basis of its output: human language itself. For this reason, **native speaker intuition** and judgement play a central role in this model. While speakers can rarely explain the rules that govern their native language, they can (often) rapidly judge what is possible in the language and what is not, thereby providing a body of data on the basis of which the linguist can attempt to model the system of knowledge that underlies those judgements. The generative model consists in part of a set of **principles** of language: statements that account for all possible (**grammatical**) linguistic structures, and which also rule out impossible (ungrammatical) structures within each of the submodules. This system of principles is described as 'generative' because it makes explicit the underlying knowledge that gives rise to the output.

There are a number of current generative theories of language. These theories tend to focus on the directly 'measurable' structural aspects of language such as morphology, syntax and phonology, although some approaches (notably Jackendoff's theory of Conceptual Semantics) attempt to integrate theories of linguistic meaning into a formal generative framework. While all generative theories assume Universal Grammar as a common working hypothesis, they differ in terms of how they model the system. For example, some theories of grammar such as Head-driven Phrase Structure Grammar (HPSG) and Lexical Functional Grammar (LFG) place the burden of explanation on information stored in the lexicon and assume only a single **monostratal** level of syntactic representation. Others, such as the Transformational Grammar model, place the burden of explanation on the syntax, and therefore assume a **multistratal** system where 'underlying' and 'surface' syntactic structures are linked by generalised derivational processes. As we saw in Chapter 19, theories that we might describe as 'broadly generative' can differ in significant ways.

Table 22.2 Characteristics of a generative approach to grammar

Assumptions	Objectives	Methodology
• Rationalist view • Universal Grammar • Modularity thesis • Autonomy of syntax thesis ('words and rules') • Computational system: rules build structure • Constructions are epiphenomenal: building-block metaphor • Economy prohibits redundancy • Competence determines performance	• To describe Universal Grammar • To account for grammaticality • To uncover and explain generalisations • To develop a formal model	• Native speaker intuition • Small-scale cross-linguistic comparison • Focus on 'core' phenomena • Often rely upon 'underlying' representations in accounting for grammatical phenomena

We might describe Transformational Grammar and Kay and Fillmore's Construction Grammar as extreme poles on a continuum of 'broadly generative' theories, given the substantial differences between them. Table 22.2 summarises the key characteristics of a generative ('formal') approach to grammar. We revisit some of these characteristics in the discussion that follows.

Transformational Grammar

The Transformational Grammar model was first proposed by Chomsky in the late 1950s, since when it has itself undergone a number of transformations, resulting at various historical stages in models known as Transformational Generative Grammar, Standard Theory, Extended Standard Theory, Revised Extended Standard Theory, Government and Binding Theory, Principles and Parameters Theory and, most recently, the Minimalist Program. The transformational model is not only the most prominent generative model but is also the model against which the cognitive approach to grammar defined itself in the early stages of its development. For this reason, the terms 'generative model' and 'formal model' have largely been equivalent to 'transformational model' for the purposes of our discussion in this book. As we have seen, however, the transformational model is not the only generative model, and generative models are not the only models of language that rely upon a significant amount of formalism.

Within the transformational framework, lexical items are stored in the lexicon together with information about their phonological, semantic and core syntactic properties (such as word class and valence requirements, for example). As a result of its interaction with generalised syntactic principles, this information gives rise to 'deep structures': syntactic structures in which the core requirements of the lexical items are satisfied in accordance with the syntactic principles. Deep structures typically correspond to unmarked active declarative sentences, the clause type that is traditionally viewed as the 'canonical' or 'basic' syntactic structure within any given language. 'Non-canonical' clause types such as passives and interrogatives – where these involve syntactic reordering – are then derived by means of syntactic 'movement' or 'transformation' and give rise to 'surface structures'. As a simple illustration of these ideas, consider the relationship between the declarative clause in (1a) and the interrogative clause in (1b):

(1) a. Lily has met another man.
    b. Has Lily met another man?

In the transformational model, the declarative structure in (1a) corresponds to the 'deep structure'. If the speaker intends to make a statement, no transformation is necessary and this deep structure is equivalent to the surface structure that the speaker actually produces. However, if the speaker intends to ask a question, the interrogative structure in (1b) is **derived** from the deep structure in (1a) by a syntactic transformation that raises the auxiliary verb *has* to a position in front of the subject *she*. This transformation is illustrated by the tree diagram in Figure 22.3, which shows how the auxiliary verb raises to a clause-initial position created by the transformation. We return to discuss the status of tree diagrams in the generative model in more detail below.

A version of the model in Figure 22.2 that incorporates a transformational syntax is shown in Figure 22.4. This corresponds to the model of

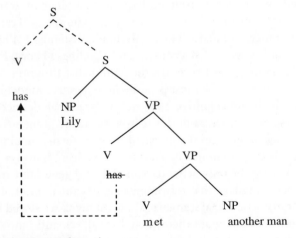

Figure 22.3 A syntactic transformation

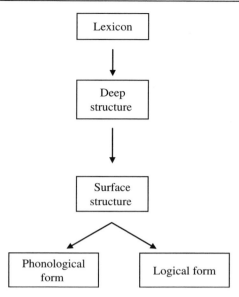

Figure 22.4 A transformational model

transformational syntax assumed most recently within Government and Binding Theory and within Principles and Parameters Theory.

Within the syntactic component, there are **phrase structure rules** that build syntactic structures. **X-bar syntax** is one approach to the statement of generalised phrase structure rules that was introduced into the transformational model during the 1970s, and a version of this approach remains in the current model. The X-bar model replaces category-specific phrase structure rules (separate sets of rules for building NPs, VPs and so on) with a small set of category neutral rules, where hierarchical (head-dependent) relationships are universal but linear precedence (word order) relations are subject to cross-linguistic variation. The existence of a small set of category-neutral rules within Universal Grammar is motivated on the basis of **economy of representation**: a small set of category-neutral rules eliminates **redundancy** and thus accounts for the efficiency of the language system both in terms of how it is acquired and in terms of how it underlies language use. A small set of category-neutral rules is also motivated on the basis of **learnability**: the fewer the rules, the more rapidly the child will fully acquire the grammatical system of his or her native language. The tree diagram in Figure 22.5 represents the structure that is built by X-bar rules, where X^0 is the head of a phrase and XP its phrasal level. An important constraint on this structure is that it is (maximally) binary branching. This constraint is also motivated on the basis of learnability: the fewer the structures the grammar can build, the more rapidly the child can fully acquire the system of his or her native language.

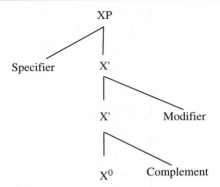

Figure 22.5 The X-bar structure

In the X-bar model, X is a variable that can be instantiated by any word class. For example, if X is a noun, XP is a noun phrase; if X is a verb, XP is a verb phrase, and so on. The structure in Figure 22.5 is used to model the relationships between heads and dependents. As we have seen, specifiers, complements and modifiers are types of dependent. In principle, the phrase is limited to a single specifier (e.g. the determiner in a noun phrase), head and complement, but may contain an unlimited number of modifiers. Of course, the existence of ditransitive verbs has proven a challenge to this highly constrained syntactic model. It is important to point out that certain parts of this structure are 'optional' in the sense that not every phrase will contain some, all or any dependents, and some phrases will contain more that one adjunct (or modifier). The minimal requirement for a phrase is the head.

An important development within the transformational framework was the extension of the X-bar structure from content phrases such as the noun phrase (NP) and the verb phrase (VP) to grammatical units such as the determiner phrase (DP) and the clause or 'tense phrase' (TP). This means that the same basic X-bar structure is used to model clauses as well as phrases; indeed, the extension of the X-bar model to a range of functional categories was one of the defining features of the Principles and Parameters framework. According to this model, the universal properties of human language are attributable to the shared principles of Universal Grammar, while cross-linguistic variation relates to 'parameter setting': the typological characteristics of each language arise from 'options' within a set of well defined parameters of variation.

Since the early 1990s, Chomsky has proposed some radical changes to the transformational model, which together constitute the basis for the ongoing research framework known as the **Minimalist Program**. Figure 22.6 represents the Minimalist model of the grammar.

An important difference between the Minimalist model and the model assumed within the Principles and Parameters framework concerns the

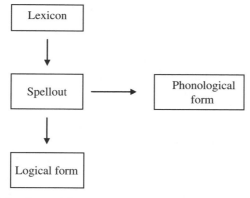

Figure 22.6 The Minimalist model

elimination of 'deep structure' and 'surface structure' as distinct levels of syntactic representation. Instead, a single syntactic component described as the 'computational system' derives syntactic structures from sets of lexical items (including both lexical and functional categories) and maps these structures onto two distinct 'interface' levels: the phonological level (PF), which interfaces with the articulatory-perceptual performance system (phonology), and the semantic level (LF), which interfaces with the conceptual-intentional performance system (meaning). In principle, the lexical items themselves, consisting of phonological, semantic and formal features, encode all the information required for the derivation, so that principles operating over the derivation remain maximally simple and general.

Indeed, according to Chomsky (2000b) there are only two basic operations occurring within the computational system: **Merge** and **Agree**. Merge is a basic structure-building operation that is driven largely by the lexical properties of the predicational item(s) within the set of lexical items. This operation assembles phrase markers (tree structures) from pairs of syntactic objects, beginning with the head-complement structure, then merging the resulting structure with its specifier, and finally combining the resulting phrase markers into larger structures. The second operation Agree matches the morphosyntactic features of two elements within the structure. This process involves features like category selection features, phi-features (person, number, gender), case, tense-aspect, and interrogative or *wh*-features. The 'matching' of these features has to take place within a local configuration, and it is this requirement that motivates syntactic transformations. Consider the examples in (2) by way of illustration.

(2) a. George asked whether Lily really loved him.
    b. George asked [what] Lily really wanted [___].

In example (2a), the embedded clause selected by the verb *ask* has an interrogative feature because of the semantics of *ask*. This explains why it takes a complementiser with a *wh*-feature (*whether* as opposed to *that*). The transformational model assumes that *wh*-expressions like *what* in (2b) have a *wh*-feature that needs to be locally 'matched' with a functional head (complementiser) in the left periphery of the clause. This explains why the *wh*-expression, which 'originates' as the object of *want*, raises to clause-initial position. Because the *wh*-feature only needs to be spelled out or made explicit by one unit in the clause (the fronted *wh*-expression), the *wh*-complementiser remains implicit in (2b), which explains why the sentence **George asked what whether Lily really wanted* is ungrammatical in English, although other *wh*-fronting languages allow both components to be spelled out. This is an example of parametric variation. This analysis is the transformational equivalent of the filler-gap analysis of *wh*-dependencies that we saw in our discussion of Construction Grammar in Chapter 19.

The transformational model is open to a range of interpretations. At one extreme, it can be interpreted as underlying a literal step-by-step process of (unconscious) sentence construction in the mind of the speaker, where syntactic trees are planted and pruned during the processing of each sentence uttered. There is little evidence to support this view: indeed, the transformational model is not intended as a model of language processing, but as a model of linguistic knowledge that interfaces with performance (production and comprehension) systems. At the other extreme, the syntactic transformation can be viewed as a metaphor that attempts to capture similarities between related constructions both within and between languages, and attempts to model the invisible and mysterious aspects of human cognition that underlie those similarities.

It follows that both the generative approach and the cognitive approach are 'cognitive' in the sense that they seek to model the psychological representation of language. However, the two frameworks approach this in radically different ways, as we have seen. While the formal model views language as an innate, encapsulated and computational system, the cognitive model views language as an emergent system, inextricably linked with general processes of communication and conceptualisation, with meaning at its core. We look in more detail at the key differences between the two models in the next section.

22.1.3 Cognitive versus generative models

As we have seen, the formal approach assumes the **modularity thesis** and, within this, the **autonomy of syntax thesis**. Cognitive approaches, on the other hand, assume that language is not an encapsulated system but a system embedded within and inextricable from generalised cognitive processes.

Lexicon–grammar continuum versus autonomous syntax

It follows from this view that, in the cognitive model, syntax is not autonomous. Instead, the syntax forms part of a **continuum** together with lexicon and morphology. This continuum consists of **symbolic units** of varying shapes and sizes. In the formal model, the syntactic component mediates between form and meaning, whereas in the cognitive model, grammatical structures are just another kind of meaningful symbolic unit, albeit of a schematic nature. Furthermore, in formal approaches, particularly Chomsky's approach, the symbols that operate within the syntax or 'Computational System' are meaningless. Indeed, many of the features that drive the Merge and Move operations in the Minimalist Program are described as 'uninterpretable' features, which have to be eliminated in the course of the derivation to avoid an ungrammatical output. Category selection features work in this way, for example: a verb 'arrives' in the syntax with selection features (e.g. the verb *kiss* selects a complement with the category N), and this selection feature is eliminated from the representation when the verb is merged with a complement bearing the appropriate feature. If this feature is not eliminated (in other words, if the verb fails to get its NP object), the result is ungrammatical. These features thus serve only to create a well-formed grammatical output, regardless of the semantics of that output, and the syntax operates blindly and automatically over these features. As we have seen, in cognitive approaches the symbolic units that comprise the grammatical system are meaningful and serve a structuring function.

Inventory versus derivational system

Despite the fact that cognitive approaches to grammar share with formal approaches the ultimate objective of modelling speaker knowledge, and despite the fact that both assume a 'dynamic' model (that is, not a static body of knowledge), the architecture of the two models differs considerably in nature. While the formal model posits a computational system that **generates** (builds or derives) well-formed grammatical structures without recourse to meaning, the cognitive model posits an **inventory** of symbolic units containing 'schematic templates'. These templates are formed as a consequence of regular use and are thus entrenched. When a speaker forms or interprets new structures, he or she does so not by applying a set of rules or principles, but by comparing the new structure with existing templates, and by taking into account the goals of the communicative exchange, the context and so on. In this sense, the cognitive model is a **problem-solving** model. While formal models capture generalisations and define well-formedness (or grammaticality) as the output of precisely stated rules and principles, the cognitive model captures generalisations and defines well-formedness (or conventionality) as the result of a **categorisation** process.

The status of constructions

As we have seen, the term **construction** is used rather differently in cognitive and formal approaches. In the cognitive approach, it refers to a symbolic unit, which may be as small as a word or as big as a clause, that is stored 'whole' within the inventory of symbolic units that represents the speaker's knowledge of language. In the cognitive approach, the construction is primitive, in the sense that it does not represent the output of any more fundamental linguistic unit or process. In formal approaches, the term 'construction' is usually applied only to clauses, and, in derivational theory, carries with it the sense that the structure has been 'built' by the application of grammatical structure-building rules and transformational rules. In this type of model, the construction is epiphenomenal, because it emerges as the output of more fundamental primitives and processes (the 'words and rules' model). As we saw in Chapter 21, however, Kay and Fillmore's Construction Grammar is a 'broadly generative' theory that takes a rather different view of constructions in the sense that constructions, although inheriting certain properties from other constructions, are 'stored whole' rather than built from syntactic rules. In this respect, the Construction Grammar view of constructions shares more in common with the cognitive model than with other generative models.

Schemas versus rules

A further point of contrast between the two theories concerns the distinction between **schemas** and **rules**, which follows from a number of points that we have already discussed. Consider the Cognitive Grammar schema for plural nouns in (3).

(3) [[[THING]/[. . .]]-[[PL]/[s]]]

The question that arises here concerns how the presence of a **schema** like (3) in the grammar is different from a derivational **rule**, since both aim to capture the same aspect of speaker knowledge. The difference lies in the directionality of the relationship between the schema or rule on the one hand, and the specific expressions that correspond to it on the other. In the generative model, the rule precedes and thus determines the specific expressions that instantiate it. In the cognitive model, the schema does not give rise to the instance but follows from it: the schema represents a pattern that emerges from entrenched units as a consequence of usage. Of course, novel uses represent an exception to this generalisation in the sense that they are sanctioned by existing schemas. In these ways, both models account for well-formedness.

Redundancy versus economy

In the cognitive model, generalisations result from recurring patterns of usage that enable the speaker to arrive at a 'higher-order' schema. This means that both schemas (the cognitive counterpart of rules) and instances of those schemas (lists of specific constructions) coexist in the grammar, and the schema is therefore an expression of the generalisation that emerges from patterns of usage. In contrast, generative linguists argue that that forms that can be derived from the application of a generalised rule need not be listed in the grammar. For example, if the rule N + *s* derives plural nouns, then specific instances like *philanderers*, *lovers* and *deceivers* need not be listed in the grammar in addition to their singular counterparts, because the singular nouns plus the generalised rule can straightforwardly derive the plural forms. This rule/list dichotomy is motivated on the basis of economy: it is argued that language must be a maximally economical system in order to be acquired and manipulated so rapidly, hence the model should avoid **redundancy**. Indeed, this **economy-driven** approach lies at the core of Chomsky's Minimalist approach.

Conventionality versus regularity

A related difference between the formal and cognitive approaches concerns the nature of the phenomena each model attempts to account for. Formal approaches to grammar have tended to focus on the statement of general rules that account for grammaticality or well-formedness in any given language, and in human language in general. For this reason, generative theories of grammar tend not to be concerned with 'conventional' or 'fixed' expressions, just as formal theories of meaning have not been concerned with 'non-compositional' or 'figurative' language. Since conventional or idiomatic expressions like *by and large* or *kick the bucket* clearly have complex syntactic structure, they are atypical lexical items. Since such expressions often fail to conform to general patterns of syntactic structure, they are not accounted for by this component of the grammar either. In the formal approach, such expressions are considered peripheral and uninteresting because they do not reveal general and productive patterns. Instead, the formal model focuses upon 'core' phenomena (word order, major clause types, case and agreement patterns and so on). This is because generalisation is a primary objective of this approach, which emerges as a consequence of its central research goal, which is to characterise Universal Grammar. In contrast, the cognitive approach views conventional and idiomatic expressions as a central part of what it means to know and to use a language. Indeed, these 'irregular' expressions are not viewed as unusual or problematic because the cognitive model does not assume a rule/list dichotomy. Instead, all expressions, 'regular' or 'irregular', form part of a speaker's inventory of linguistic knowledge and must be accounted for.

'Scaffolding' versus 'building blocks'

This point of contrast relates to the status of compositional structure within the model. As we have seen, the generative model assumes that rules give rise to constructions, which Langacker (1987) describes in terms of the **building-block metaphor**. In other words, formal models view linguistic elements as having a componential structure: elements from speech sounds to sentences are viewed as having a complex internal structure, which may consist of structural 'building blocks' like articulatory features, morphemes or grammatical categories, or which may consist of semantic 'building blocks' like semantic primitives or meaning components. In Langacker's view, while these 'building blocks' may serve a useful practical function as classificatory features, and may even have cognitive reality, they are epiphenomenal. In other words, they are a 'symptom' of the status of that linguistic expression within a complex network of meanings and forms, but are not themselves the foundations of either meaning or structure within linguistic expressions.

In contrast, as we have seen, the usage-based model holds that entrenched instances give rise to schemas. Despite this important difference, Langacker's model of grammatical constructions acknowledges that complex structures are recognised by speakers as having compositional structure. Indeed, it is the recognition of recurring structural patterns that enables speakers to create novel grammatical constructions. Langacker proposes an alternative to the building-block metaphor that encompasses both compositional and non-compositional units: the **scaffolding metaphor**. In Cognitive Grammar, component structures are described as **immanent** in the complex grammatical construction, regardless of whether the compositionality is recognised by the speaker. Langacker argues that entrenchment decreases the salience of compositionality. For example, we are less aware of the well-entrenched noun *computer* as a complex construction than we are of a less well-entrenched or novel instance like *striver*. The compositional structure of a grammatical construction may be essential to the initial creation or construction of that expression, but once the construction is entrenched and gains the status of a unit, this compositional scaffolding is no longer required. Despite this, the compositional structure remains immanent: we may still recognise the compositionality of well-entrenched units, but it does not follow that we 'build them from scratch' each time we use them. The fact that certain complex constructions do not conform to the prototypical patterns of compositionality does not present a problem in this model. For example, we might argue that the compound noun *bluebottle* or the idiomatic expression *have a butchers at* represent cases where the individual components are no longer recognised as making a contribution to the construction as a whole, and that these expressions have therefore been reanalysed as simplex units, at least at the semantic pole.

Constraints on the model

Cognitive and generative approaches also differ to a considerable extent in terms of the constraints placed upon the model. Because of its emphasis on economy and generalisation, the formal model places strict constraints upon grammatical constructions and processes. This is particularly evident in its emphasis on the relatedness of constructions. For example, the transformational model assumes that all clause types are constructed according to the same general principles and share a similar underlying structure. Furthermore, it is assumed that non-canonical clause types like interrogative clauses, passive clauses and cleft clauses are related to, and therefore derived from, more basic underlying clause structures. In order to preserve these assumptions, the transformational model admits 'invisible' and semantically empty elements. Invisible elements lack phonetic realisation but are thought to be present for semantic or structural reasons. Consider the examples in (4).

(4) a. George wanted [Lily to see the world].
    b. George wanted [___ to see the world].

Example (4a) contains an embedded clause, and the NP *Lily* is the subject of the embedded clause (she is doing the seeing). In the transformational model, example (4b) is also thought to include an embedded subject (interpreted as co-referential with *George*) that has no phonetic realisation. This invisible embedded subject is represented by the underscore. This assumption preserves the view that both examples share a parallel structure. Semantically 'empty' elements include so-called 'dummy' elements. For example, the 'dummy' subject *it* in *it surprised her that he turned up at all* has no referential content. We also saw in Chapter 14 that the auxiliary verb *do* is described as a 'dummy' auxiliary since it is conditioned by certain grammatical requirements but does not bring its own contribution to the clause in terms of aspect or voice. In sum, while the formal model places severe constraints on grammatical constructions and processes, it allows a proliferation of 'invisible' and 'dummy' elements in order to preserve generalisations.

In contrast, the cognitive model adopts the opposite position: 'invisible' or 'semantically empty' elements are not permitted, but constructions, related and unrelated, proliferate. For example, in Cognitive Grammar, the **Content Requirement** prohibits invisible or semantically empty elements, although symbolic units can be implicit (for example, class schemas). However, even implicit symbolic units are meaningful, albeit schematic. The Content Requirement also prohibits abstract 'underlying' structures. However, the cognitive model does not emphasise constraints upon grammatical constructions, which proliferate. Because the cognitive model views redundancy as natural and is less concerned with generalisation, it requires less theoretical machinery.

Emphasis on formalism

This brings us to our final point of contrast in this section. An important difference between formal and cognitive approaches, as the term 'formal approach' itself suggests, is a different degree of emphasis on formalism. Formalism in linguistics is the practice of adopting a metalanguage for the description of natural language phenomena, and often involves the manipulation of abstract symbols and rules. As we saw in Chapter 13, formal semantics adopts logic as a metalanguage for the description of linguistic meaning, and the tree diagrams, transformational rules and abstract features of Chomsky's approach to syntactic theory are also components of a formal metalanguage for describing the grammatical properties of human language. In formal theories, formalism has a status beyond description, however. It is also the basis of the model of speaker knowledge and must therefore work like a perfect 'machine' (efficient, economical and automatic) to generate the correct forms and interpretations. The formal approach therefore necessarily involves a level of abstraction. Although the adoption of an abstract metalanguage and a computational or algorithmic system of rules has certain advantages (it is precise, unambiguous and universally applicable), cognitive linguists (among others) argue that the level of abstraction adopted within the transformational model in fact obscures or misrepresents the reality of human language. Cognitive models of grammar therefore avoid the use of abstract symbols and rules on the whole, although we have seen that cognitive models do rely upon a fair number of complex diagrams aimed at representing the links between grammar and cognition.

22.1.4 Functional-typological approaches to grammar

As we mentioned earlier, functional approaches to language, particularly the functional-typological approach, have informed and influenced the cognitive approach in a number of important ways. In this section, we briefly summarise the characteristics of functional-typological approaches to grammar. This section owes much to Croft (2003). For sources that provide a more in-depth introduction to functional-typological approaches, we refer the reader to the further reading section at the end of the chapter.

A functional approach to language is any approach that places particular emphasis on the **communicative** and **social** functions of language, and attempts to explain the grammatical properties of language in terms of how it is used. In this respect, functional approaches tend to be less concerned with the psychological representation of language as a system of knowledge and more concerned with its **use**. Functional approaches therefore characterise grammatical phenomena in terms of discourse, pragmatic, sociolinguistic and cultural properties. One of the best-known functional approaches to grammar is Michael Halliday's **Systemic Functional Grammar**, which holds that language is organised to reflect **ideational** and **interpersonal** meaning (Halliday 1994: xiii). According to Halliday, ideational meaning reflects the speaker's attempt to understand his or her environment while interpersonal meaning reflects the speaker's objectives in terms of influencing other people within that environment. This approach is called 'systemic' because it conceives the grammar as a system of choices available to the speaker in achieving his or her goals. Unlike many other approaches to grammar, the systemic functional approach does not stop at the sentence but looks at both written and spoken **texts**, since it is only by analysing the interaction between speakers within these larger pieces of language that one can discover the communicative functions of language. In this respect, the functional approach lends itself to discourse analysis (see the critical approach to discourse analysis proposed by Norman Fairclough (e.g. 2001) for an influential application of Halliday's approach to discourse). Within Halliday's framework, a clause is analysed in terms of three 'strands of meaning' (Halliday 1994: 33): **message** (the information conveyed by the clause), **exchange** (the communicative transaction between speaker and hearer represented by the clause, for example offer versus request) and **representation** (the way in which the clause represents a construal of some aspect of human experience, for example saying, thinking or doing).

The functional-typological approach shares these concerns to the extent that it attempts to explain typological patterns in terms of language **use**. Croft (2003: 4–5) points out that while generative grammar emerged as a reaction against behaviourist psychology, linguistic typology emerged as a reaction against 'anthropological relativism': the idea that languages can vary in arbitrary and unconstrained ways. As we saw in Chapter 3, linguistic typologists have discovered that while languages can and do vary, cross-linguistic variation is constrained. From the perspective of linguistic typology, it is the constraints on variation that make up the universals of language, rather than a set of universal principles. This means that while generative linguists assume an innate

Universal Grammar as the basis of linguistic universals, functional typologists appeal to functional and cognitive explanations for these universals. For example, two of the major explanations posited by typologists to account for cross-linguistic patterns are **economy** and **iconicity**. Croft (2003: 116) argues that both relate to language use in the sense that they relate to language processing. For example, it is economical for a language to shorten frequently used forms (recall from Chapter 21 that grammaticalised forms are reduced or shortened). Iconicity refers to the way that language 'mirrors' experience. For example, the tendency for some languages to present old information before new information in an utterance represents iconicity between language and experience, because new experiences happen later than old ones (Croft 2003: 202). As we have seen, a number of typologists also adopt some version of a **semantic map model** in accounting for typological patterns (Croft 2003: 133). A semantic map is the language-specific typological pattern which rests upon a universal conceptual space (recall our discussion of the semantic map for case systems in Chapter 21). Finally, as we saw in Chapter 21, the term **emergent grammar** coined by the functional typologist and grammaticalisation scholar Paul Hopper (1987) sums up the cognitive and usage-based nature of the functional-typological approach:

> The notion of Emergent Grammar is meant to suggest that structure, or regularity, comes out of discourse and is shaped by discourse as much as it shapes discourse in an on-going process [. . . Grammar's] forms are not fixed templates, but are negotiable in face-to-face interaction in ways that reflect the individual speakers' past experience of these forms, and their assessment of the present context, including especially their interlocutors, whose experiences and assessments may be quite different. Moreover, the term Emergent Grammar points to a grammar which is not abstractly formulated and abstractly represented, but always anchored in the specific concrete form of an utterance. (Hopper 1987: 142, cited in Croft 2003: 289)

As Croft (2003: 5) points out, there is considerable agreement between generative and functional-typological approaches with respect to the existence of cross-linguistic universals, the search for what defines a 'possible human language' and the close attention to linguistic form. However, the functional-typological tradition departs from the generative tradition and is more closely aligned with cognitive approaches in its rejection of specialised innate linguistic knowledge (Universal Grammar), in its appeal to non-linguistic aspects of cognition to explain the properties of language, and in its emphasis on language function and use. Finally, as we saw in Chapter 21, large-scale samples of the kind compiled by typologists have also formed the basis of grammaticalisation

Table 22.3 Characteristics of a functional-typological approach to grammar

Assumptions	Objectives	Methodology
• Semantic map model • Cognitive economy • Iconicity • Variation in language is natural • Language is dynamic • Use shapes language • The properties of language can be explained on the basis of human cognition and language use	• To describe linguistic universals • To state generalisations in terms of implicational universals	• Large-scale cross-linguistic samples • Close attention to linguistic form

studies, which represent one of the areas in which the functional-typological approach has been particularly influential in the development of cognitive approaches to grammar. Table 22.3 summarises the key characteristics of the functional-typological approach.

22.2 Core issues in grammar: comparing cognitive and generative accounts

Having set out the characteristics of the major theoretical models in the previous section, we turn our attention in this section to core issues in the study of grammar which we have discussed throughout Part III of this book, and compare the ways in which these shared areas of interest are approached by different theories of grammar. Our discussion in this section focuses on a comparison between cognitive and generative approaches, since both types of approach share the same objective of modelling the representation of knowledge of language in the mind. This shared objective, together with the fact that the cognitive approaches to grammar originally emerged as a reaction against the generative model, means that both types of approach have often focused on explaining similar core grammatical phenomena.

22.2.1 Word classes

An important difference between formal and cognitive approaches relates to the characterisation of word classes. As we have seen, the cognitive approach favours a semantic characterisation. In Cognitive Grammar, for example, symbolic units vary in terms of specificity versus schematicity at the semantic pole. While content words are maximally specific, grammatical categories like NOUN are maximally schematic but both specific and schematic units belong within the same inventory. The major word classes receive a semantic characterisation: for example, the category NOUN is characterised at the semantic pole by the schema [THING] and the category VERB by the schema [PROCESS]. We also saw that closed classes like determiners and auxiliary verbs received a semantic account in terms of grounding predications.

In contrast, the formal approach argues against a semantic characterisation and defines word classes on the basis of morphological and distributional properties. As we saw in Chapter 14, this represents the traditional **distributional** approach to word classes. This type of approach is either explicitly adopted or taken for granted by most formal theories of language which reject a semantic characterisation of word classes on the basis that such an approach inevitably results in a description so vague as to be meaningless. In addition, the formal approach takes the position that a semantic characterisation cannot adequately distinguish word classes because members of two different categories can have the 'same' meaning. Consider the following examples:

(5) a. Women everywhere love George.
    b. Their love of George is legendary.

(6) a. He destroyed the photos.
    b. His destruction of the photos was heartbreaking.

According to the formal approach, the verbs in (5a) and (6a) are not semantically distinct from the nouns in (5b) and (6b), respectively. *Love* describes the same emotion in both (5a) and (5b), and *destroy* and *destruction* in (6a) and (6b) both describe the same kind of act. For this reason, a distributional approach is widely favoured because the structural characteristics of word classes are readily identifiable and, although not without exception, are also more or less predictable. Of course, most linguists would agree that there is some semantic basis to word classes. Speakers recognise that nouns typically describe things, verbs typically describe actions, adjectives typically describe properties and prepositions typically describe relations. According to the formal model, however, these rather 'vague' semantic characterisations are insufficient grounds upon which to base a model of language. Given that the aim of most modern theories of language is to describe a speaker's psychological representation of language, the structural features of word classes are generally thought to lend themselves more readily to a model of this psychological representation of language, particularly in a modular system where morphology and syntax operate independently of meaning. According to the position adopted in cognitive linguistics, however, these distributional properties are epiphenomenal.

22.2.2 Constituency: heads and dependents

As we saw in Chapter 17, heads are described as profile determinants in Cognitive Grammar and the head-dependent relation is characterised in terms of conceptual autonomy and dependence. This view is consonant with the formal view in a number of ways. In particular, both models hold that the head or profile determinant lends its features to the phrase that contains it. The difference, of course, is that in Cognitive Grammar the features of the head relate to its schematic meaning (e.g. PROCESS or THING), while in formal approaches the features of the head relate to its grammatical category (e.g. V or N), which, as we have seen, is defined in structural rather than semantic terms. A second similarity between Cognitive Grammar and the formal approach relates to the metaphor of 'dependency' relations. In both models, the relationships between the component parts of a phrase are modelled in terms of dependence. Again, the difference lies in the fact that the Cognitive Grammar view of dependence relates to conceptual dependence, whereas the formal view relates to categorial selection. Of course, formal models also posit semantic selection in order to ensure, for example, that a verb like *love* selects an animate subject, but this process often operates independently of the grammatical component. This means that in the formal model of syntax, a head is entirely 'autonomous' within its phrase, in the sense that it **selects** all its dependents, some obligatorily (e.g. complement) and others optionally (e.g. modifiers). The Cognitive Grammar view is rather different: a head can be conceptually dependent on its 'dependents' if they elaborate some aspect of its structure. As we saw in Chapter 17, the head-complement construction illustrates this prototypical dependence relation. A further important difference follows from points that we have already discussed: in the cognitive model, constituency emerges from the properties of constructions, which are primitive. In the formal model, constructions emerge from 'words and rules', which are primitive.

22.2.3 The status of tree diagrams

As we have seen, tree diagrams are used in transformational generative approaches such as Principles and Parameters Theory and the Minimalist Program, as well as in other non-transformational generative models such as Head-driven Phrase Structure Grammar (HPSG). Tree diagrams have a special status in generative theories. They are not just a convenient 'shorthand' for representing grammatical structure. Tree diagrams represent instantiations of the grammatical rules or principles that generative grammarians posit as the basis of a speaker's knowledge of language. Consider the example in Figure 22.7.

Tree diagrams like this represent a range of information. They represent the word class of each element within the phrase, and they also represent

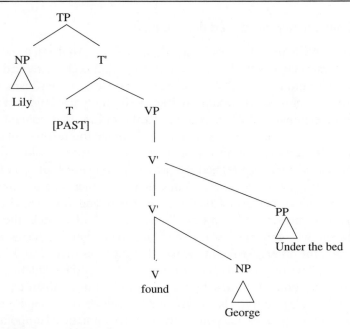

Figure 22.7 Sentence tree diagram

constituency, showing how the elements are grouped together into increasingly complex constructions. In addition, tree diagrams encode information about the kinds of relationships that hold between the subparts of the phrase. For example, in the X-bar model, complements are shown as sister to the head: both *found* and *George* are immediately dominated by the same node in the tree, which is the lowest V-bar level. In contrast, modifiers like *under the bed* are more remote from the head. Tree diagrams also represent linear order. The triangles in this diagram represent phrases whose internal structure is not 'unpacked'. Finally, this tree is labelled TP because tense is viewed as the head of the clause in the current transformational generative approach. We return to this point below (section 22.2.5).

Cognitive approaches to grammar reject tree diagrams as part of their theoretical model. For example, in Cognitive Grammar, the Content Requirement prohibits tree diagrams. Of course, nothing prevents the cognitive grammarian from sketching out tree diagrams as a convenient shorthand, but they are not admitted as a model of speaker knowledge. This is because the only kind of abstract representation that the Cognitive Grammar model permits (as a usage-based model) is the schema that emerges from entrenched patterns. However, Langacker asserts that his model accounts for the same information that is captured by tree diagrams. As we have seen, the Cognitive Grammar model captures word class by means of class schemas. Furthermore, constituency is viewed as 'just a matter of the step-by-step assembly [. . .] of progressively more elaborate symbolic structures' (Langacker 2002: 296). It is important to remember, however, that entrenched constructions are not assembled 'from scratch' each time they are used in a speech event. Instead, frequently occurring constructions are stored 'whole'. Finally, Cognitive Grammar captures linear order by viewing it as temporal order within phonological space. In other words, when two or more units are combined to make a larger construction, the composition of these units at the phonological pole specifies a linear order.

22.2.4 Grammatical functions and case

The terms 'subject' and 'object' are frequent expressions in any linguist's vocabulary. However, the status of these terms in cognitive models of language is rather different from their status in traditional and formal approaches. As we saw in Chapter 17, a subject is defined in Cognitive Grammar as a unit that corresponds to the TR of the verb, while the object is the unit that corresponds to its LM. While the prototypical subject is the 'energy source' and the prototypical object is the 'energy sink', these roles can be reversed in passive clauses to effect a marked pattern of attention, where the 'energy sink' occurs as the TR or focus of attention. This means that grammatical functions are defined in terms of their schematic meaning and in terms of how they provide a linguistic reflection of attention patterns. In contrast, as we saw in Chapter 14, the traditional (and formal) approach to grammatical functions is rather like the traditional (and formal) approach to word classes. Grammatical functions are defined according to morphological and distributional criteria: case, agreement, position in a sentence, or ability to undergo certain syntactic processes such as passivisation (which identifies an object) or subject-auxiliary inversion (which identifies a subject). According to the cognitive view, these grammatical features are only superficial 'symptoms' of the primitive semantic properties of the construction.

Furthermore, as we also saw in Chapter 14, case is traditionally described as the grammatical feature that 'flags' the grammatical function of a word or phrase within the sentence. Since the formal approach treats grammatical functions in terms of their structural features, it follows that case also receives a **configurational** account within the transformational generative model. This means that case is associated with position and with the locality of the case-bearing element to other 'influential' heads in the sentence. Consider the examples in (7).

(7) a. I expect [she eats shellfish].
    b. I expect [her to eat shellfish].

Observe that the case of the subject of the embedded clause (bracketed) appears to be conditioned by the tense properties of the embedded clause: if the embedded verb is finite (7a), the embedded subject occurs in the nominative form (*she*). If the embedded verb is non-finite (7b), the embedded subject occurs in the accusative form (*her*). Despite this, the embedded subject stands in the same semantic relationship with the embedded verb in both examples: *she* (or *her*) is the AGENT of the verb *eat*. In transformational generative approaches, this interdependence is captured by assuming that finite inflection assigns nominative case to the subject of the sentence, or that case features on the subject are licensed or 'checked' by virtue of its local relationship with finite inflection, which, in this model, is viewed as the head of the clause. Accusative case is assigned to objects (or licensed or 'checked') by virtue of the NP's local relationship with a verb or a preposition, which are therefore viewed as accusative case-assigners. Observe that in (8a) the accusative pronoun *him* is local to the verb, and in (8b) the accusative pronoun *him* is local to the preposition.

(8) a. Lily gave him the love letters.
    b. Lily gave the love letters to him.

Without going into details about how the locality of a given case-bearing NP is to its case-licensing head is described, the simplified tree diagram in Figure 22.8 represents this configurational account. The pattern illustrated by the examples in (7) is accounted for in terms of the inability of non-finite inflection to assign nominative case. This means that the closest available case-licensing head (the verb) case-marks the embedded subject instead, which explains why it surfaces with accusative case.

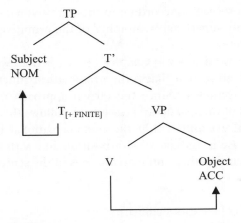

Figure 22.8 Case assignment

This very brief account omits a considerable amount of detail but provides a sense of one formal account of case, which is viewed in terms of a meaningless and automatic licensing process governed by position and locality to the appropriate case-licensing head. From this perspective, case features are just one of the formal mechanisms that ensure that licensed or grammatical structures are built. We elaborate this idea below in our discussion of the transformational account of passive constructions (section 22.2.6).

As this discussion illustrates, the formal approach to grammar is more concerned with accounting for grammatical details of constructions than accounting for their semantic properties. This follows from the autonomy of syntax. In contrast, the cognitive approach views grammatical features of constructions as 'symptoms' of their semantic properties. For example, the cognitive approach to the pattern illustrated by the examples in (7) would be to treat these as distinct constructions with distinct semantic properties (see Exercise 22.3).

22.2.5 The verb string: tense, aspect and mood

As we saw in Chapter 18, Cognitive Grammar analyses the verb string in terms of a grounding predication (either a tense morpheme or a modal verb) and a clausal head, which can include a perfect construction, a progressive construction and a passive construction, as well as the content verb. This partition of the verb string is illustrated by example (9).

(9) Lily   [must]   [have been eating]   shellfish
            [GROUNDING PREDICATION]   [CLAUSAL HEAD]

As Langacker (1991: 195) acknowledges, this partition of the verb string is also reflected in the transformational model, where the modal or finite auxiliary occupies the head of the TP or tense phrase (in other words, heads the sentence as a whole), and the remainder of the verbs occupy positions within an extended verb phrase. This is represented in Figure 22.9.

According to the transformational model, the fact that the finite verb precedes negation means that negation heads its own functional phrase in between TP and VP. This pattern is illustrated by the examples in (10).

(10) a. Lily has not been eating shellfish.
      b. Lily was not eating shellfish.
      c. *Lily ate not shellfish.
      d. Lily did not eat shellfish.

Example (10a) shows that when the clause lacks a modal verb, the next verb in the string (here, the perfect auxiliary *have*) becomes finite. According to the

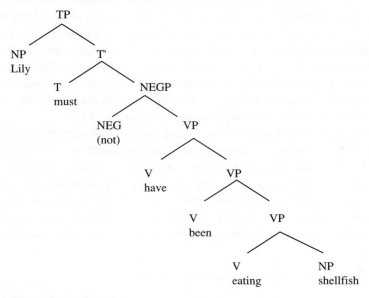

Figure 22.9 The clause: tense phrase

transformational model, this means that the relevant verb raises to the T position, an analysis that receives some support from the relative ordering of the finite verb and the negation element. Example (10b) makes the same point, but in this case it is the progressive auxiliary *be* that raises. Example (10c) shows that the lexical verb cannot raise; in this case, the 'dummy auxiliary' *do* is inserted into the clause in order to provide morphological 'support' for tense, which is a bound morpheme. This means that in a sentence like *Lily ate shellfish*, the tense morpheme must 'lower' to attach to the lexical verb. The same pattern is evident in the behaviour of lexical and content verbs with respect to (negated) subject–auxiliary inversion, as illustrated by example (11). Observe that when negation is cliticised to the verb (*-n't*), it raises with the verb to a position preceding the subject. When the negation element remains a free morpheme (*not*), it does not raise with the auxiliary verb.

(11) a. Must(n't) Lily (not) have been eating shellfish?
b. Has(n't) Lily (not) been eating shellfish?
c. Was(n't) Lily (not) eating shellfish?
d. *Ate(n't) Lily (not) shellfish?
e. Did(n't) Lily (not) eat shellfish?

The Cognitive Grammar account and the transformational account are similar in that both accord a special status to the first element in the verb string, and this special status relates to its tense properties. Both models provide an account based on the idea that that it is finiteness that licenses the clause to stand alone as an independent grammatical unit. However, the Cognitive Grammar account analyses tense/mood properties in terms of what they contribute to the semantics of the clause as a grounding predication, while the transformational account emphasises the grammatical behaviour of the finite verb. Secondly, the Cognitive Grammar account analyses the string of verbs containing the lexical or content verb as the clausal head, while the transformational account views tense features (roughly, the equivalent of Langacker's grounding predication) as the clausal head. It is worth emphasising that only the transformational model accords this status to tense features. Other formal models, including HPSG, view the lexical verb as the head of the clause.

22.2.6 The passive construction

As we have seen throughout this book, the passive construction has received a great deal of attention in the literature. This interest in the passive construction is not restricted to cognitive approaches. Indeed, in the transformational generative model, the passive construction is described in terms of what is probably one of the best-known syntactic transformations (the other being the '*wh*-movement' transformation that derives *wh*-questions in languages like English).

As we saw in Chapters 17 and 18, the cognitive account emphasises the meaningful aspects of the construction, in particular the fact that it effects a TR-LM (figure-ground) reversal. Furthermore, the Cognitive Grammar account holds that the passive participle is synchronically related to other functions of the same form (PERF). In contrast, the formal account holds that examples (12a) and (12b) have distinct structures and that the category of the verbs is different in each case, as is the category of *broken*.

(12) a. Her heart was broken for years.
 b. Her heart was broken by that cad.

According to this view, example (12a), as an active clause, is not derived by transformation but has the standard structure of the canonical English active declarative clause. The copular verb is treated as a lexical verb because it is the only verb in the clause, but it is viewed as a 'defective' lexical verb because it lacks semantic content and because it displays the same 'operator' features as auxiliary verbs. It can invert with the subject to form an interrogative clause (*Was her heart broken?*), and it can carry negation (*Her heart wasn't broken*). In this example, *broken* is a predicative adjective. Example (12b), on the other hand, is derived by transformation from an active clause. In this example, *was* is an auxiliary verb and *broken* is the past participle form of the lexical verb.

In the transformational model, the fact that an active clause like *George betrayed her* paraphrases a passive clause like *She was betrayed by George* (and vice versa) motivates the view that the passive clause is related to the active clause by derivation. This relatedness is captured by a lexical process that passivises the verb by altering its morphology as well as its argument structure: the passive verb only requires a single argument, which is interpreted as PATIENT but surfaces in subject position. To capture this pattern, the transformational analysis holds that the PATIENT argument originates as the complement of the passive verb, because PATIENT arguments are canonically linked to complement positions of transitive verbs. This accounts for the fact that the single obligatory argument in a passive clause is interpreted as a PATIENT despite the fact that it appears in the position that is typically associated with the AGENT. The PATIENT raises to subject position by transformation. As we mentioned earlier, this transformation relies upon case as a mechanism. The fact that the PATIENT of a passive verb is not licensed to occur in its canonical post-verbal position is interpreted as an indication that passivisation affects the case-marking qualities of a verb. This claim is supported by the ungrammaticality of example (13a), in which *it* is a 'dummy subject' that is inserted into the example to discount the possibility that the sentence is ungrammatical because it has an empty subject position. According to this analysis, the PATIENT NP raises to subject position in order to be case-licensed and surfaces with nominative case (13b).

(13) a. *(It) was betrayed her (by George).
     b. [She] was betrayed [___] (by George).

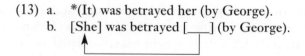

The *by*-phrase is viewed as 'underlying' subject in the sense that the subject of the active form of the verb is 'absorbed' by the passive morphology. This means that if the optional *by*-phrase occurs in a passive clause, it has the status of a modifier. Finally, the active and passive counterparts are seen as truth-conditionally synonymous (if one is true, the other is true, and vice versa). Any meaning-related distinction (that is, the fact that the speaker chooses to make one referent rather than another discourse prominent) falls within the domain of pragmatics and thus beyond the immediate 'responsibility' of the grammar. All the grammar has to do is to generate a well-formed output. Hence, the emphasis in the formal approach is upon accounting for the syntax of the construction rather than accounting for its discourse function.

According to Langacker, the intuition behind the transformational account is valid in the sense that *broken* in (12a) describes a STATE, whereas in (12b) it describes a PROCESS. Furthermore, Langacker's claim that the copula verb *be* designates a schematic imperfective PROCESS but contains no other information means that the Cognitive Grammar analysis is also rather similar to the formal account in this respect, as we saw in Chapter 17. A final similarity relates to the modifier status of the *by*-phrase in both accounts. Despite these similarities, the cognitive account differs from the formal account in emphasising the different semantic properties of the active and passive constructions: from this perspective, their grammatical properties are only superficial 'symptoms' of the cognitive representation that these constructions evoke.

22.3 Summary

In this chapter, we presented some explicit comparisons between cognitive, generative and functional-typological approaches to grammar. We began by setting out the assumptions, aims and methodology of the cognitive approach, and compared these with the generative approach and the functional-typological approach. We found that each approach can be distinguished by its starting assumptions, which in turn gives rise to different objectives and methodologies. We then revisited some core grammatical phenomena which have been explored from a cognitive perspective throughout Part III of the book, and compared the cognitive account with the generative view of these phenomena. As the discussion in this chapter illustrates, while there are substantial differences, there is also a good deal of shared ground between cognitive and generative theories in terms of what they attempt to account for. Moreover, there are sometimes rather striking similarities between the resulting analyses, at least superficially. Although the starting assumptions of these approaches differ rather dramatically, it is important to recognise that linguists of any theoretical persuasion are united in the objective of uncovering the mysteries of human language, and furthermore that the modern theoretical approaches discussed here are also united in the objective of modelling the representation of knowledge of language in the mind of the speaker.

Further reading

Introductory texts

- **Taylor (2002).** The first four chapters of this excellent book are directly relevant to the discussion in this chapter. Taylor provides some particularly useful historical perspective, as well as some explicit comparisons between formal and cognitive approaches.

Generative grammar: introductory texts

- **Adger (2003).** This textbook provides an introductory level interpretation of the Minimalist Program.

- **Carnie (2002).** A recent introduction to the transformational framework, this textbook focuses mainly on the Principles and Parameters approach and includes useful chapters on two non-transformational theories: Lexical Functional Grammar (see Bresnan 2001) and Head-driven Phrase Structure Grammar (see Borsley 1996).
- **Pinker (1994).** Written for the layperson, this popular book provides an accessible introduction to the key assumptions and arguments of the nativist hypothesis. For an alternative perspective, see Sampson (2005).
- **Radford (1997a, 1997b, 2004).** These texts introduce the Minimalist Program. The (1997a) book is the most accessible, but the (2004) book is the most up to date.

Generative Grammar: primary sources

- **Chomsky (1957, 1959, 1965, 1981, 1986, 1995, 2000a).** These sources provide some insight into the development of Chomsky's transformational model over almost fifty years. The (2000a) book provides a recent and accessible overview.

Background reading: modularity, specific impairment and aphasia

- **Bishop (1997).** Chapter 9 of this book, entitled 'Modularity and interaction in language development disorders', takes a critical view of the idea that specific language impairment entails modularity.
- **Bishop and Mogford (eds) (1993).** This book contains a collection of papers exploring child language acquisition under 'exceptional circumstances'. It includes papers on language acquisition by hearing-impaired and sight-impaired children, by hearing children of deaf parents, by children with Down's syndrome, Williams syndrome and autism, as well as looking at bilingualism and language development in twins.
- **Caplan (1992).** This book focuses on language processing (comprehension and production of language) and on acquired language disorders.
- **Fodor (1983, 2000).** The 2000 paper provides a short overview of Fodor's influential 1983 book. The Cummins and Cummins (2000) volume that contains Fodor's paper also includes other papers of interest. For example, Part II of the book is dedicated to papers on 'the mind as neural network', where you can read about models of information processing that inform the cognitive linguistics framework.

- **Jackendoff (1997).** This book sets out Jackendoff's views concering the organisation of linguistic knowledge.

Systemic Functional Grammar

- **Halliday (1994).** This introductory-level textbook sets out Halliday's framework.

Functional-typological approaches

- Anderson (1987)
- Comrie (1989)
- Croft (2001, 2003)
- Givón (1979, 1991)
- Haiman (1983)
- Haiman (ed.) (1985)
- Haspelmath (1997)
- Kemmer (1993)
- Stassen (1997)

These sources provide some insight into functional typological approaches, including those that assume some version of the semantic map model.

Exercises

22.1 The poverty of the stimulus

Chomsky (1959) put forth a number of arguments against B. F. Skinner's behaviourist theory of language acquisition. One argument rested upon what Chomsky has called the 'poverty of the stimulus': the idea that children are able to produce grammatical structures that they have not been exposed to sufficient data to acquire by imitation. How might the cognitive model account for Chomsky's observation?

22.2 Transformations

How do you think the transformational approach might analyse the derivation of the following sentence? What is the semantic motivation for the transformational analysis? Can you see any morphological evidence for a transformational approach? How might the intuitions behind this analysis be captured in a cognitive analysis?

Whom will Lily marry?

22.3 Exceptional case marking

In section 22.2.4, we presented the (transformational) generative analysis of the examples in (7). Examples like (7b) are described in this model as 'exceptional case marking' or ECM constructions, because an embedded subject surfaces with object (accusative) case. In our discussion of example (7), we suggested that a cognitive account might view these pairs as distinct constructions with distinct semantic properties. How do you think the semantic differences between the examples in (7) might be described? It may help to consider the contexts in which each example would be most natural. Now consider the following pairs of examples, which also illustrate the ECM phenomenon. In each pair, it is the second example that illustrates the ECM construction.

(a) Lily found that George/he was less than honest.
(b) Lily found George/him to be less than honest.

(c) Lily knew that George/he was an estate agent.
(d) Lily knew George/him to be an estate agent.

(e) Lily believed that George/he was her knight in shining armour.
(f) Lily believed George/him to be her knight in shining armour.

Can your account be extended to these examples? If not, why not?

22.4 'False dichotomies'

Table 22.4 lists a number of what Langacker (1987: 18) describes as the 'false dichotomies' assumed by formal approaches. In the light of our comparison between cognitive and generative approaches in this chapter, discuss how and why you think the generative approach might uphold these dichotomies, and how and why the cognitive approach rejects them.

Table 22.4 'False dichotomies' (based on Langacker 1987: 18)

Synchrony versus diachrony
Competence versus performance
Grammar versus lexicon
Semantics versus pragmatics
Grammatical versus ungrammatical
Derivational versus inflectional morphology

Part IV: Conclusion

23

Assessing the cognitive linguistics enterprise

In this final short chapter we examine in very general terms both the achievements of the cognitive linguistics enterprise and its remaining challenges. As we will see, this relatively new theoretical approach to language has given rise to an integrated and increasingly influential model that reasserts the empiricist view of language. However, the cognitive approach faces a number of challenges, including the extension of the model to a fully developed account of language beyond semantics and grammar, and the development of empirically falsifiable methodology.

23.1 Achievements

An integrated model of language and thought

In Chapter 2, we introduced the two central commitments of the cognitive linguistics enterprise: the 'Generalisation Commitment' and the 'Cognitive Commitment'. These commitments have given rise to an integrated approach to linguistic and conceptual organisation. Since the mid-1980s, when cognitive linguistics began to emerge as a distinct approach, the integrated model adopted by cognitive linguists has given rise to a collection of detailed investigations of a wide range of cognitive and linguistic phenomena. This has been particularly evident in cognitive semantics and cognitive approaches to grammar, the two areas we have focused upon in this book. Other areas, such as cognitive approaches to phonology, cognitive approaches to pragmatics and applications of cognitive linguistics to areas such as psycholinguistics and language teaching, while increasingly the focus of research in cognitive linguistics, remain at this point less well developed. Moreover, while many of the individual theories we have encountered in this book rely upon sophisticated argumentation, the integrated nature of the theories, a consequence of the two key commitments, entails that linguistic phenomena are discussed in terms that are relatively accessible to neighbouring disciplines.

In addition to its influence in linguistics and in cognitive science, cognitive linguistics has also been influential in the social sciences and in the humanities. This means that the enterprise has gained influence relatively quickly. Indeed, in the twenty-five years or so since the publication of *Metaphors We Live By*, the first well-known book to explicitly advocate a cognitive linguistics approach to language and the mind, the field has covered considerable ground.

Re-examination of the empiricist thesis

The rationalist view that underpins generative approaches to language has dominated the field of linguistics for over half a century. A notable achievement of the cognitive linguistics enterprise has been to refocus interest on the empiricist perspective, and thus to reopen channels of investigation into language and mind that take into account embodiment, experience and usage while remaining firmly committed to the mentalist approach.

Focus on conceptual phenomena

Cognitive linguistics has also contributed to extending the range of conceptual phenomena studied by cognitive scientists. For example, the idea of conceptual projection, which is addressed by the frameworks of Conceptual Metaphor Theory, Mental Spaces Theory and Conceptual Blending Theory, attempts to model the richness and complexity of the human imagination. Until relatively recently, it was assumed either that the human imagination was peripheral to cognition or that it could not be systematically studied. The cognitive linguistics enterprise has provided an approach for studying the imagination, and has shown that language reveals systematic processes at work in human imagination which cognitive linguists have argued are central to the way we think.

Integration of formalist and functionalist concerns

A further achievement of the cognitive linguistics enterprise has been to integrate formalist and functionalist concerns. While formalists are particularly concerned with developing descriptively adequate accounts of linguistic phenomena and with modelling the representation of knowledge of language in the mind, functionalists have been primarily concerned with exploring the social and communicative functions of situated language use. Cognitive linguistics, while functionalist in spirit, is concerned both with achieving descriptive adequacy and with modelling language as a cognitive phenomenon.

23.2 Remaining challenges

Competing cognitive theories

The fact that cognitive linguistics is a collection of distinct theories and approaches rather than a single theoretical perspective means that there is often overlapping terminology (compare Talmy's approach with Langacker's, for example). There are also differences of opinion over the appropriate division of labour between stored lexical knowledge versus dynamic processes of meaning construction (compare Lakoff's 1987 'full-specification' approach to word meaning with more recent approaches to cognitive lexical semantics, e.g. Tyler and Evans 2003). There is also a tendency towards distracting disputes about whether theoretical approaches address the same issues or whether they actually relate to distinct phenomena, whether theories compete with each other or complement one another, and so on. For example, in an early paper on Blending Theory, Turner and Fauconnier (1995) suggested that Blending Theory was competing with Conceptual Metaphor Theory in terms of the nature of the phenomena it was seeking to account for. However, more recently, Grady, Oakley and Coulson (1999) have argued explicitly for the view that these two theories complement one another. This development has, in part, come about as Blending Theory has matured and a clearer view has emerged of the phenomena it is best equipped to address. Of course, in this respect, cognitive linguistics is no different from any collection of approaches that share a common set of assumptions, and the theories we have presented in this book continue to evolve. Given the objectives of this book, we have not attempted to present contrary perspectives within the cognitive approach, but have primarily concentrated on demonstrating how the collection of approaches discussed here have evolved from common starting assumptions.

However, some theories within cognitive linguistics have been the subject of intense criticism and debate, involving scholars both within and outside the cognitive framework. For example, critics of Conceptual Metaphor Theory have expressed concern about the remoteness of the generalisations made by metaphor scholars from the linguistic data they rely upon (e.g. Evans 2004a; Stern 2000). Others have expressed concern that Conceptual Metaphor Theory underplays the role of context (e.g. Leezenberg 2001; Stern 2000), and that Conceptual Metaphor Theory fails to account for the role of conventionalisation in patterns of conceptual projection (Evans and Zinken 2005). A further concern is that as a result of the increasingly abstract level at which conceptual metaphors are stated, conceptual metaphor theory precludes the possibility of identifying meaningful cross-linguistic differences and fails to adequately take account of socio-cultural context (Evans 2004a; Leezenberg 2001). Others have observed that the view from discourse suggests that metaphor, rather than being an underlying conceptual structure in the sense of Lakoff and Johnson, is more insightfully conceived as a form of elaboration or evolution of particular ideas. This is achieved through ongoing discourse, resulting in conventional collocations that become stabilised in memory (for various representative views see Evans 2004a; Musolff 2004; Zinken, Hellsten and Nerlich forthcoming). From this perspective, conceptual projection may have a 'symbolic' rather than a 'pre-symbolic' basis. As the specific theories we have examined in this book mature and achieve a more solid empirical basis, they will continue to be exposed to scrutiny from other perspectives. Of course, this challenge is to be expected, because no theory evolves without constant reappraisal.

Lack of empirical rigour

A criticism that has been levelled against cognitive linguistics, particularly early on in the development of the enterprise, related to a perceived lack of empirical rigour. This criticism arose in response to some of the early foundational studies conducted under the banner of cognitive semantics. For example, while intuitively appealing, early research on lexical polysemy networks (see Brugman and Lakoff 1988) and early research on conceptual metaphors (Lakoff and Johnson 1980) was largely based on speaker intuition and interpretation. The studies on *over* by Brugman and Lakoff, for instance, were criticised for lacking a clear set of methodological decision principles (see Sandra 1998), particularly given semantic network analyses of the same lexical item often differed quite radically from one theorist to another (see Sandra and Rice 1995 for a review). In recent years, the empirical foundations of cognitive linguistics have become stronger. For example, research by Ray Gibbs (e.g. 1994) and Lera Boroditsky (e.g. 2000) has begun to provide an empirical basis for drawing conclusions about conceptual metaphor. Research by Seana Coulson (e.g. 2000) has begun to provide an empirical basis for assessing conceptual integration networks. Research by psycholinguists Sandra and Rice (1995) and Cuyckens *et al.* (1997), together with cognitively oriented corpus studies as illustrated by Gries and Stefanowitsch (2005) have begun to strengthen the empirical basis of cognitive approaches to lexical semantics, and research by Tyler and Evans (e.g. 2003), among others, has begun to provide a sound methodological basis for investigating lexical polysemy. With respect to cognitive approaches to grammar, William Croft's (e.g. 2002) proposals concerning the integration of typological methods with cognitive linguistic theory has strengthened the empirical basis of constructional accounts of grammar. More generally, the last few years have witnessed an increase in the influence of empirical methods from neighbouring disciplines upon cognitive linguistics, including brain-scanning techniques from experimental psychology. The increased concern with empirical methods is attested by Gonzales-Marquez *et al.* (forthcoming), a collection of papers emerging from a recent workshop entitled 'Empirical Methods in Cognitive Linguistics'.

Despite these advances, outstanding challenges remain. For example, Gibbs (2000: 349) observes that many psychologists complain that work in cognitive linguistics that attempts to infer 'aspects of conceptual knowledge from an analysis of systematic patterns of linguistic structure leads to theories that appear to have a post hoc quality'. In other words, psychologists have argued that cognitive linguistic theories are not predictive but assume without adequate evidence that the conceptual system has certain properties in order to account for the properties of language. For example, Blending Theory purports to be a theory about conceptual processes but is forced to posit underlying mental spaces and integration networks in order to account for linguistic expressions. In other words, it infers the conceptual structures that it attempts to demonstrate evidence for rather than seeking independent evidence for these conceptual structures (from psychology or psycholinguistics, for example). This means that the theory cannot be empirically falsified, since it does not make predictions about the properties of conceptual structure that can be empirically tested. Of course, philosophers of science have argued that falsifiability is a necessary property of any theory that seeks to achieve scientific rather that purely ideological status. It follows that if cognitive linguistic accounts of conceptual structure are to achieve a theoretical status beyond ideology, it will be necessary for them to develop the means by which they can be empirically tested.

There remain, of course, other kinds of challenges that are not unique to the cognitive linguistics enterprise. It is worth emphasising that none of the theoretical constructs described in this book can be taken as proven fact. Indeed, the detailed and precise claims made by cognitive linguists about conceptual organisation are largely based on the properties of language and are therefore, for the most part, inferential. Until we learn a good deal more about the human mind and brain, this remains a sobering caveat for any theory that attempts to model the cognitive representation of language.

Bridges between cognitive and generative theories

Finally, because cognitive linguistics originally emerged as a reaction against generative approaches, particularly the highly influential transformational model developed by Chomsky, argumentation by cognitive linguists has sometimes relied upon an over-simplified and outdated representation of the formal model. Newmeyer (1999) argues that the differences between the two approaches might not be as significant as they sometimes appear. For example, Newmeyer (1999: 4) observes that the generative approach is consonant with the Cognitive Commitment in the sense that it seeks to develop a model of language that is consistent with a 'neuropsychologically real overall theory of mind-brain'. He further observes that although cognitive linguists often suggest that the Chomskyan model of language entails model-theoretic semantics, Chomsky himself has rejected this idea, and Jackendoff has developed a theory of linguistic meaning that is in many ways consonant with cognitive approaches (in particular, it is 'non-objectivist'), despite explicitly defending the autonomy of syntax thesis. While the starting assumptions of each approach do stand in direct opposition, as we have seen in this book, both approaches stand to benefit from the recognition of the fact that they share many important concerns, not least in relation to the development of a theory of linguistic meaning that reflects human construal of external reality.

23.3 Summary

It remains for the cognitive approach to continue to develop a stronger empirical basis, and to develop detailed accounts of areas beyond grammar and semantics. Nevertheless, in a relatively short period of time, cognitive linguistics has achieved theoretical sophistication in a range of areas and significant influence in neighbouring disciplines. It now represents one of the most exciting and rapidly expanding theoretical enterprises in linguistics and cognitive science.

Appendix

List of Tables

Table 1.1 Properties of the lexical and grammatical subsystems
Table 3.1 Figure-ground segregation, as encoded in language (adapted from Talmy 2000: 183)
Table 3.2 Hydraulic system model (based on Gentner and Gentner 1982: 110)
Table 3.3 Moving crowd model (based on Gentner and Gentner 1982: 120)
Table 4.1 Past tense endings of selected verbs in 1982 (based on Bybee and Slobin 1982)
Table 4.2 Key ideas in the Generalised Theory of Selection (Croft 2000)
Table 4.3 Terms for Generalised Theory and linguistic equivalents (Croft 2000)
Table 4.4 Causal mechanisms involved in language stability and change (Croft 2000)
Table 4.5 Holophrases (Tomasello 1992; adapted from Tomasello 2003: 36–7)
Table 4.6 Examples of utterance schemas (based on Tomasello 2003: 66)
Table 4.7 Human pattern-finding skills (Tomasello 2003)
Table 4.8 Human intention reading abilities (Tomasello 2003)
Table 4.9 Phrase structures in English
Table 5.1 The guiding principles of cognitive semantics
Table 6.1 Some sensory-perceptual systems
Table 6.2 Shared characteristics of FORCE schemas
Table 6.3 A partial list of image schemas
Table 7.1 The dictionary view of key distinctions in the study and representation of meaning
Table 7.2 Four kinds of knowledge that relate to the centrality of encyclopaedic knowledge of word meaning
Table 7.3 The valence of the verbs relating to the COMMERCIAL EVENT frame (adapted from Fillmore and Atkins 1992: 79)
Table 7.4 Partial inventory of basic domains

Table 7.5 Attributes of basic domains
Table 7.6 Distinctions between basic domains and image schemas
Table 8.1 Semantic features or markers for the category CHAIR
Table 8.2 Problems for the classical theory of categorisation
Table 8.3 Example of a taxonomy used by Rosch *et al.* (1976) in basic-level category research
Table 8.4 Six of the taxonomies used by Rosch *et al.* (1976) as stimuli
Table 8.5 Examples of attribute lists (based on Rosch et al. 1976: appendix I)
Table 8.6 Motor movements for categories at three levels of inclusiveness (based on Rosch et al. 1976: appendix II)
Table 8.7 Comparison between levels of categorisation
Table 8.8 A selection of goodness-of-example ratings (based on Rosch 1975: appendix)
Table 8.9 Comparison of some attributes for ROBIN and OSTRICH
Table 8.10 Summary of some metonymic ICMs
Table 9.1 Mappings for LOVE IS A JOURNEY
Table 9.2 The Event Structure Metaphor
Table 9.3 Mappings for AN ABSTRACT ORGANISED ENTITY IS AN UPRIGHT PHYSICAL OBJECT
Table 9.4 Constraints on possible vehicles in metonymy (Kövecses and Radden 1998)
Table 10.1 Assumptions of cognitive lexical semantics
Table 10.2 Schemas proposed by Lakoff (1987) for *over* in addition to the central schema
Table 10.3 The main findings of the full-specification approach (Lakoff 1987)
Table 10.4 Distinct senses for *over* identified in Tyler and Evans (2003)
Table 10.5 Some senses of *fly*
Table 11.1 The role of tense and aspect in discourse management
Table 12.1 Mappings for SURGEON IS A BUTCHER
Table 12.2 Constitutive processes of Blending Theory
Table 12.3 Goals of blending
Table 12.4 Summary of vital relations and their compressions
Table 12.5 Integration networks (based on Fauconnier and Turner 2002)
Table 12.6 Governing principles of Blending Theory (Fauconnier and Turner 2002)
Table 12.7 Mappings for NATION IS A SHIP
Table 12.8 Primary metaphors that serve as inputs to *ship of state* blend
Table 13.1 Connectives and operators in predicate calculus
Table 13.2 Steps for calculating truth conditions
Table 13.3 Truth-conditional formal semantics
Table 13.4 Relevance Theory
Table 14.1 English inflectional morphemes
Table 14.2 Structural criteria for English subject
Table 14.3 Structural criteria for English object

Table 14.4 English personal pronouns
Table 15.1 Matter and action (based on Talmy 2000: 42)
Table 15.2 Linguistic expressions relating to matter and action
Table 15.3 Illustrating the schematic category DISPOSITION OF QUANTITY (adapted from Talmy 2000: 59)
Table 15.4 Illustration of lexical items that relate to DISPOSITION OF QUANTITY
Table 15.5 Patterns of distribution
Table 15.6 Factors in the 'Attentional System'
Table 15.7 Basic domains proposed by Langacker (1987)
Table 16.1 Trajector-landmark combinations in relational predications
Table 17.1 Head-dependent relations in Cognitive Grammar
Table 17.2 Properties of prototypical stems and affixes
Table 18.1 English verb forms
Table 18.2 Clausal head complex (based on Langacker 1991: 198–9)
Table 18.3 Situation types (Vendler 1967)
Table 19.1 Distinctions in idiom types
Table 20.1 Properties of the English ditransitive construction (Goldberg 1995)
Table 20.2 Properties of the English caused-motion construction (Goldberg 1995)
Table 20.3 Properties of the English resultative construction (Goldberg 1995)
Table 20.4 RCG taxonomy of constructions (adapted from Croft 2001: 17)
Table 21.1 Common grammaticalisation patterns (adapted from Croft 2003: 254)
Table 21.2 Macrostructure and microstructure in grammaticalisation (Heine et al. 1991: 103)
Table 21.3 The evolution of modal verbs (Traugott and Dasher 2002)
Table 21.4 Patterns of attenuation (Langacker 1999b)
Table 22.1 Characteristics of a cognitive approach to grammar
Table 22.2 Characteristics of a generative approach to grammar
Table 22.3 Characteristics of a functional-typological approach to grammar
Table 22.4 'False dichotomies' (based on Langacker 1987: 18)

List of Figures

Figure 1.1 A symbolic assembly of form and meaning
Figure 1.2 Levels of representation
Figure 1.3 Possible trajectories for *The cat jumped over the wall*
Figure 1.4 The interactive function
Figure 1.5 *The cat is on the chair*
Figure 1.6 Figure and gound
Figure 2.1 Some members of the category CUP
Figure 2.2 Profiling
Figure 2.3 The study of meaning and grammar in cognitive linguistics
Figure 3.1 Figure-ground segregation

Figure 3.2 Columns of dots
Figure 3.3 Rows of dots
Figure 3.4 Columns of shapes
Figure 3.5 A triangle and three black circles
Figure 3.6 Two rectangles
Figure 3.7 A black cross
Figure 3.8 Taxonomy of reference frames in the languages of the world (adapted from Talmy 2000: 213)
Figure 3.9 Simple cityscape scene
Figure 3.10 Ground-based reference
Figure 3.11 Field-based reference
Figure 3.12 Guidepost-based reference
Figure 3.13 Projector-based reference
Figure 3.14 Approximately three seconds of data from eight EEG electrodes
Figure 3.15 Taxonomy of cognitive models for time
Figure 3.16 The moving time model
Figure 3.17 The moving ego model
Figure 3.18 The temporal sequence model
Figure 3.19 The division of spatial scenes in English (adapted from Bowerman and Choi 2003: 393)
Figure 3.20 The division of spatial scenes in Korean (adapted from Bowerman and Choi 2003: 394)
Figure 3.21 The field-based spatial terms of Guugu Yimithirr (Haviland 1993)
Figure 3.22 The slope model (adapted from Shinohara 2000: 5)
Figure 3.23 Spatial primes (adapted from Boroditsky 2001)
Figure 4.1 An instantiation of a schema (adapted from Langacker 2000: 10)
Figure 4.2 Partial sanction by a schema (adapted from Langacker 2000: 10)
Figure 4.3 Schema-instance relations
Figure 4.4 Frequency effects and entrenchment of instances
Figure 4.5 Frequency effects and entrenchment of schemas
Figure 4.6 The structure of language change
Figure 4.7 The synchronic and diachronic dimensions of language change
Figure 4.8 The use of a linguistic symbol in a triadic relationship expressing a communicative intention (adapted from Tomasello 2003: 29)
Figure 4.9 The X-bar approach to phrase structure
Figure 5.1 Some members of the category CUP
Figure 6.1 From embodiment to linguistic meaning
Figure 6.2 CONTAINER image schema
Figure 6.3 Image schema for OUT1
Figure 6.4 Image-schema for OUT2
Figure 6.5 A bottle or a lightbulb? (Adapted from Vandeloise 1994)
Figure 6.6 The PATH image schema

APPENDIX

Figure 6.7   The COMPULSION image schema
Figure 6.8   The BLOCKAGE image schema
Figure 6.9   The COUNTERFORCE image schema
Figure 6.10   The DIVERSION image schema
Figure 6.11   The REMOVAL OF RESTRAINT image schema
Figure 6.12   The ENABLEMENT image schema
Figure 6.13   The ATTRACTION image schema
Figure 6.14   The bifurcation in the cognitive representation (CR)
Figure 6.15   The lexicon–grammar continuum
Figure 6.16   The key schematic systems within the conceptual structuring system
Figure 6.17   Degree of extension for matter (adapted from Talmy 2000: 61)
Figure 6.18   The path associated with an object falling out of a plane
Figure 7.1   Bucket or pail?
Figure 7.2   Identifying knowledge types which give rise to centrality
Figure 7.3   A partial frame for CAR (adapted from Barsalou 1992a: 30)
Figure 7.4   Partial COMMERCIAL EVENT frame
Figure 7.5   Location of the lexical concept KNUCKLE in a hierarchy of domain complexity
Figure 7.6   Scope for the concept HYPOTENUSE
Figure 7.7   Different profiles derived from the same base
Figure 7.8   Familial network in which UNCLE is profiled
Figure 7.9   Active zones for the sentences in (5)
Figure 8.1   The human categorisation system
Figure 8.2   Radial network for the category MOTHER
Figure 9.1   Primary metaphor
Figure 9.2   Compound metaphor
Figure 9.3   Comparison between metaphor and metonymy
Figure 10.1   A radiating lattice diagram ('semantic network') for modelling radial categories
Figure 10.2   The central schema for *over* (adapted from Lakoff 1987: 419)
Figure 10.3   The bird flew over the yard (Schema 1.X.NC) (adapted from Lakoff 1987: 421)
Figure 10.4   The plane flew over the hill (Schema 1.VX.NC) (adapted from Lakoff 1987: 421)
Figure 10.5   The bird flew over the wall (Schema 1.V.NC) (adapted from Lakoff 1987: 421)
Figure 10.6   John walked over the bridge (Schema 1.X.C) (adapted from Lakoff 1987: 422)
Figure 10.7   John walked over the hill (Schema 1.VX.C) (adapted from Lakoff 1987: 422)
Figure 10.8   Sam climbed over the wall (Schema 1.V.C) (adapted from Lakoff 1987: 422)

Figure 10.9 Instances of schema 1, the central image schema (adapted from Lakoff 1987: 423)
Figure 10.10 St Paul's Cathedral is over the bridge (Schema 1.X.C.E) (adapted from Lakoff 1987: 424)
Figure 10.11 John lives over the hill (Schema 1.VX.C.E) (adapted from Lakoff 1987: 423)
Figure 10.12 The distinction between polysemy and vagueness
Figure 10.13 Division of the vertical axis into subspaces by prepositions
Figure 10.14 The proto-scene for *over* (Tyler and Evans 2003)
Figure 10.15 The semantic network for *over* (based on Tyler and Evans 2003: 80)
Figure 11.1 *In that play, Othello is jealous*
Figure 11.2 *In the picture, a witch is riding a unicorn*
Figure 11.3 Schema induction
Figure 11.4 A lattice of mental spaces
Figure 11.5 Linking counterparts
Figure 11.6 Directionality of connectors
Figure 11.7 Hitchcock and the movie
Figure 11.8 Two connectors to one element
Figure 11.9 Roles and values
Figure 11.10 *Fido sees a tortoise*
Figure 11.11 *He chases it*
Figure 11.12 *He thinks that the tortoise is slow*
Figure 11.13 *But it is fast*
Figure 11.14 *Maybe the tortoise is really a cat*
Figure 11.15 *Jane is twenty*
Figure 11.16 *She has lived in France*
Figure 11.17 *In 2000 she lived in Paris*
Figure 11.18 *She currently lives in Marseilles*
Figure 11.19 *Next year she will move to Lyons*
Figure 11.20 *The following year she will move to Italy*
Figure 11.21 *By this time, she will have lived in France for five years*
Figure 11.22 Foundation and expansion spaces
Figure 12.1 Mappings of elements across inputs
Figure 12.2 Addition of a generic space
Figure 12.3 A basic integration network (adapted from Fauconnier and Turner 2002: 46)
Figure 12.4 SURGEON as BUTCHER blend
Figure 12.5 CLINTON as PRESIDENT OF FRANCE blend
Figure 12.6 BOAT RACE blend
Figure 12.7 An XYZ blend
Figure 12.8 *Landyacht*
Figure 12.9 Compression of outer-space relation into inner-space relation in the blend (adapted from Fauconnier and Turner 2002: 94)

Figure 12.10 A simplex integration network
Figure 12.11 Single-scope network
Figure 12.12 Structuring of focus input by inner-space projection from framing input (adapted from Fauconnier and Turner 2002: 130)
Figure 12.13 Death the Grim Reaper (adapted from Fauconnier and Turner 2002: 292)
Figure 13.1 The generative model
Figure 13.2 Set-theoretic model
Figure 13.3 The construction of sentence-meaning in formal semantics
Figure 13.4 The nature of meaning construction in cognitive semantics
Figure 14.1 A symbolic unit
Figure 14.2 The symbolic unit (adapted from Langacker 1987: 77)
Figure 14.3 The cognitive model of grammar (adapted from Langacker 1987: 77)
Figure 14.4 Inventory-based approaches to grammar
Figure 14.5 Cognitive approaches to grammar
Figure 14.6 Grammatical units
Figure 14.7 Noun categories
Figure 14.8 Schema-instance relations
Figure 15.1. The bifurcation in semantic structure
Figure 15.2 Four schematic systems within the conceptual structuring system
Figure 15.3 Schematic categories of configurational structure
Figure 15.4 Axiality (adapted from Talmy 2000: 65)
Figure 15.5 Prospective direction (adapted from Talmy 2000: 74)
Figure 15.6 Retrospective direction (adapted from Talmy 2000: 75)
Figure 15.7 Schematic categories of the 'Perspectival System'
Figure 15.8 Force-dynamics encoded in sentences like (24a) (adapted from Talmy 2000:415)
Figure 15.9 An overview of the conceptual structuring system
Figure 15.10 Langacker's model of word classes
Figure 15.11 The relationship between focal adjustments and construal
Figure 15.12 Profile-base organisation for *elbow*
Figure 15.13 *George opened the champagne*
Figure 15.14 *The champagne was opened*
Figure 15.15 *George ate all the caviar*
Figure 15.16 *All the caviar was eaten by George*
Figure 15.17 The prototypical action chain model (adapted from Langacker 2002: 211)
Figure 15.18 Network model (adapted from Langacker 2002: 271)
Figure 15.19 (Partial) network model of the English past tense morpheme (adapted from Langacker 2002: 283)
Figure 16.1 Symbolic units in Cognitive Grammar
Figure 16.2 The conceptual basis of the count/mass noun distinction

Figure 16.3 Nominal versus relational predication (adapted from Langacker 2002: 75)
Figure 16.4 Conceptual representation of the three major word classes (adapted from Langacker 1987: 220)
Figure 16.5 Langacker's model of word classes
Figure 17.1 Composite and component structures
Figure 17.2 The agentive nominal schema (adapted from Langacker 1987: 446)
Figure 17.3 PROCESS as profile determinant of the clause
Figure 17.4 The prototypical action model (adapted from Langacker 2002: 211)
Figure 17.5 Prototypical action chain (adapted from Langacker 2002: 217)
Figure 17.6 Action chain for (18a) (adapted from Langacker 2002: 217)
Figure 17.7 Action chain for (18b) (adapted from Langacker 2002: 217)
Figure 17.8 Action chain for (18c) (adapted from Langacker 2002: 217)
Figure 18.1 The epistemic model (adapted from Langacker 1991: 242)
Figure 18.2 Perfective and imperfective aspect (adapted from Langacker 2002: 88)
Figure 18.3 Perfective and imperfective situation types
Figure 19.1 A modular view of the language system
Figure 19.2 Typology of idioms
Figure 19.3 The WXDY construction (after Kay and Fillmore 1999: 20)
Figure 20.1 Participant roles and argument roles
Figure 20.2 Ditransitive Construction (adapted from Goldberg 1995: 50)
Figure 20.3 Ditransitive + *send* (adapted from Goldberg 1995: 51)
Figure 20.4 Inheritance links
Figure 20.5 Metaphorical inheritance link (adapted from Goldberg 1995: 88)
Figure 20.6 The English caused-motion construction (adapted from Goldberg 1995: 78)
Figure 20.7 The English resultative construction (adapted from Goldberg 1995: 189)
Figure 20.8 Conceptual space for transitive/intransitive participant roles (adapted from Croft 2003: 145)
Figure 20.9 Semantic maps for nominative/accusative and ergative/absolutive (adapted from Croft 2003: 145)
Figure 21.1 Cognitive models of the grammaticalisation process
Figure 21.2 Source domain hierarchy (after Heine *et al.* 1991: 55)
Figure 21.3 From invited inference to coded meaning
Figure 21.4 The evolution of subjectivity in grammaticalisation (based on Traugott and Dasher 2002: 22–3)
Figure 21.5 Invited inferencing model (adapted from Traugott and Dasher 2002: 38)
Figure 21.6 Subjectification, or the attenuation of objectivity (adapted from Langacker 1999b: 6)
Figure 21.7 Chain of increasing grammaticalisation: case functions (adapted from Heine et al. 1991: 159)
Figure 22.1 Theories of grammar
Figure 22.2 A modular view of the language system

APPENDIX

Figure 22.3 A syntactic transformation
Figure 22.4 A transformational model
Figure 22.5 The X-bar structure
Figure 22.6 The Minimalist model
Figure 22.7 Sentence tree diagram
Figure 22.8 Case assignment
Figure 22.9 The clause: tense phrase

References

Achard, Michel and Susanne Niemeier (eds) (2000) 'Special issue on language acquisition', *Cognitive Linguistics*, 11, 1–2.
Achard, Michel and Susanne Niemeier (2004) *Cognitive Linguistics, Second Language Acquisition, and Foreign Language Teaching*. Berlin: Mouton de Gruyter.
Adger, David (2003) *Core Syntax: A Minimalist Approach*. Oxford: Oxford University Press.
Aitchison, Jean (1996) *Words in the Mind*. Oxford: Blackwell.
Allwood, Jens (2003) 'Meaning potentials and context: some consequences for the analysis of variation in meaning', in H. Cuyckens, R. Dirven and J. Taylor (eds), *Cognitive Approaches to Lexical Semantics*. Berlin: Mouton de Gruyter, pp. 29–66.
Allwood, Jens and Peter Gärdenfors (eds) (1999) *Cognitive Semantics: Meaning and Cognition*. Amsterdam and Philadelphia: John Benjamins.
Alverson, Hoyt (1994) *Semantics and Experience: Universal Metaphors for Time in English, Mandarin, Hindi and Sesotho*. Baltimore, MD: Johns Hopkins University Press.
Anderson, Lloyd B. (1987) 'Adjectival morphology and semantic space', in B. Need, E. Schiller and A. Bosch (eds), *Papers from the Twenty-third Annual Regional Meeting of the Chicago Linguistic Society, Part One: The General Session*. Chicago: Chicago Linguistic Society, pp. 1–17.
Armstrong, Sharon Lee, Lila Gleitman and Henry Gleitman (1983) 'What some concepts might not be', *Cognition*, 13, 263–308.
Austin, J. L. ([1962] 1975) *How to Do Things with Words*, 2nd edn, eds J. O. Urmson, and S. Marina. Oxford: Oxford University Press.
Bach, Emmon (1989) *Informal Lectures on Formal Semantics*. New York: SUNY.
Barcelona, Antonio (2003a) *Metaphor and Metonymy at the Crossroads: A Cognitive Perspective*. Berlin: Mouton de Gruyter.

Barcelona, Antonio (2003b) 'Introduction. The cognitive theory of metaphor and metonymy', in A. Barcelona (ed.), *Metaphor and Metonymy at the Crossroads: A Cognitive Perspective*. Berlin: Mouton de Gruyter, pp. 1–30.

Barcelona, Antonio (2003c) 'On the plausibility of claiming a metonymic motivation for conceptual metaphor', in A. Barcelona (ed.), *Metaphor and Metonymy at the Crossroads: A Cognitive Perspective*. Berlin: Mouton de Gruyter, pp. 31–58.

Barlow, Michael and Suzanne Kemmer (2000) *Usage-Based Models of Language*. Stanford, CA: CSLI Publications.

Barsalou, Lawrence (1983) 'Ad-hoc categories', *Memory and Cognition*, 11: 211–27.

Barsalou, Lawrence (1992a) 'Frames, concepts and conceptual fields', in A. Lehrer and E. Kittay (eds), *Frames, Fields and Contrasts*. Hillsdale, NJ: Lawrence Erlbaum, pp. 21–74.

Barsalou, Lawrence (1992b) *Cognitive Psychology: An Overview for Cognitive Scientists*. Hillsdale, NJ: Lawrence Erlbaum.

Barsalou, Lawrence (1999) 'Perceptual symbol systems', *Behavioral and Brain Sciences*, 22, 577–609.

Barsalou, Lawrence (2003) 'Situated simulation in the human conceptual system', *Language and Cognitive Processes*, 5/6, 513–62.

Bartlett, F. C. (1932) *Remembering: A Study in Experimental and Social Psychology*. Cambridge: Cambridge University Press.

Bechtel, William and George Graham (eds) (1999) *A Companion to Cognitive Science*. Oxford: Blackwell.

Bergen, Benjamin K. and Nancy Chang (2005) 'Embodied construction grammar in simulation-based language understanding', in J.-O. Östman and M. Fried (eds), *Construction Grammars: Cognitive Grounding and Theoretical Extensions*. Amsterdam: John Benjamins, pp. 147–190.

Bergen, Benjamin K., Nancy Chang and Shweta Narayan (2004) 'Simulated action in an embodied construction grammar', in *Proceedings of the Twenty-Sixth Annual Conference of the Cognitive Science Society*. Mahwah, NJ: Lawrence Erlbaum, pp. 108–13.

Berlin, Brent, Dennis Breedlove and Peter Raven (1974) *Principles of Tzeltal Plant Classification*. New York: Academic Press.

Bishop, Dorothy (1997) *Uncommon Understanding: Development and Disorders of Language Comprehension in Children*. Hove: Psychology Press.

Bishop, Dorothy and Kay Mogford (eds) (1993) *Language Development in Exceptional Circumstances*. Hillsdale, NJ: Lawrence Erlbaum.

Bloom, Paul, Mary A. Peterson, Lynn Nadel and Merrill F. Garrett (eds) (1996) *Language and Space*. Cambridge, MA: MIT Press.

Bloomfield, Leonard (1933) *Language*. New York: Henry Holt.

Börjars, Kersti and Kate Burridge (2001) *Introducing English Grammar*. London: Arnold.

Boroditsky, Lera (2000) 'Metaphoric structuring: understanding time through spatial metaphors', *Cognition*, 75, 1, 1–28.

Boroditsky, Lera (2001) 'Does language shape thought? Mandarin and English speakers' conceptions of time', *Cognitive Psychology*, 43, 1–22.
Borsley, Robert (1996) *Modern Phrase Structure Grammar*. Oxford: Blackwell.
Borsley, Robert (1999) *Syntactic Theory: A Unified Approach*, 2nd edn. London: Arnold.
Bowerman, Melissa (1973) *Early Syntactic Development*. Cambridge: Cambridge University Press.
Bowerman, Melissa (1996a) 'Learning how to structure space for language: a crosslingusitic perspective', in P. Bloom, A. Peterson, L. Nadel and M. Garrett (eds), *Language and Space*. Cambridge, MA: MIT Press, pp. 385–436.
Bowerman, Melissa (1996b) 'The origins of children's spatial semantic categories: cognitive versus linguistic determinants', in J. Gumperz and S. Levinson (eds), *Rethinking Linguistic Relativity*. Cambridge: Cambridge University Press, pp. 145–76.
Bowerman, Melissa and Soonja Choi (2003) 'Space under construction: language-specific spatial categorization in first language acquisition', in D. Gentner and S. Goldin-Meadow (eds), *Language in Mind: Advances in the Study of Language and Thought*. Cambridge, MA: MIT Press, pp. 387–428.
Braine, Martin (1976) 'Children's first word combinations', *Monographs of the Society for Research in Child Development*, 41, 1.
Bresnan, Joan (2001) *Lexical-Functional Syntax*. Oxford: Blackwell.
Brown, Donald (1991) *Human Universals*. Boston, MA: McGraw-Hill.
Brugman, Claudia (1981) *The Story of 'over': Polysemy, Semantics and the Structure of the Lexicon*, MA thesis, University of California, Berkeley (published New York: Garland, 1988).
Brugman, Claudia (1990) 'What is the invariance hypothesis?', *Cognitive Linguistics*, 1, 257–66.
Brugman, Claudia and George Lakoff (1988) 'Cognitive topology and lexical networks', in S. Small, G. Cottrell and M. Tannenhaus (eds), *Lexical Ambiguity Resolution*. San Mateo, CA: Morgan Kaufman, pp. 477–507.
Buba, Malami (2000) 'The pragmatics of addressee-based Hausa demonstratives', *South African Journal of African Language*, 20, 3, 239–59.
Bybee, Joan and Dan Slobin (1982) 'Rules and schemas in the development and use of the English present tense', *Language*, 58, 265–89.
Bybee, Joan, Revere Perkins and William Pagliuca (1994) *The Evolution of Grammar: Tense, Aspect and Modality in the Languages of the World*. Chicago: Chicago University Press.
Cann, Ronnie (1993) *Formal Semantics*. Cambridge: Cambridge University Press.
Caplan, David (1992) *Language: Structure, Processing and Disorders*. Cambridge, MA: MIT Press.
Carnie, Andrew (2002) *Syntax: A Generative Introduction*. Oxford: Blackwell.
Carpenter, Malinda, Nameera Akhtar and Michael Tomasello (1998) 'Sixteen-month-old infants differentially imitate intentional and accidental actions', *Infant Behavior and Development*, 21, 315–30.

Carston, Robyn (2002) *Thought and Utterances: The Pragmatics of Explicit Communication*. Oxford: Blackwell.

Carston, Robyn, Sun Song Nam and Seji Uchida (1998) *Relevance Theory: Applications and Implications*. Amsterdam: John Benjamins.

Chierchia, Gennaro and Sally McConnell-Ginet (2000) *Meaning and Grammar*, 2nd edn. Cambridge, MA: MIT Press.

Chilton, Paul and George Lakoff (1995) 'Foreign policy by metaphor', in C. Schäffner and A. Wenden (eds), *Language and Peace*. Aldershot: Dartmouth, pp. 37–59.

Choi, Soonja and Melissa Bowerman (1991) 'Learning to express motion events in English and Korean: the influence of language-specific lexicalization patterns', *Cognition*, 41, 83–121.

Chomsky, Noam (1957) *Syntactic Structures*. The Hague: Mouton.

Chomsky, Noam (1959) 'Review of B. F. Skinner, *Verbal Behaviour* (New York: Appleton-Century Crofts, 1957)', *Language*, 35, 26–58.

Chomsky, Noam (1965) *Aspects of the Theory of Syntax*. Cambridge, MA: MIT Press.

Chomsky, Noam (1968) *Language and Mind*. New York: Harcourt Brace Jovanovich.

Chomsky, Noam (1970) 'Remarks on nominalization', in R. A. Jacobs and P. S. Rosenbaum (eds), *Readings in English Transformational Grammar*. Waltham, MA: Ginn, pp. 184–221.

Chomsky, Noam (1981) *Lectures on Government and Binding*. Dordrecht: Foris.

Chomsky, Noam (1986) *Knowledge of Language: Its Nature, Origin and Use*. New York: Praeger.

Chomsky, Noam (1991) 'Some notes on economy of derivation and representation', in Robert Freidin (ed.), *Principles and Parameters in Comparative Grammar*, Current Studies in Linguistics No. 20. Cambridge, MA: MIT Press, pp. 417–54.

Chomsky, Noam (1995) *The Minimalist Program*. Cambridge, MA: MIT Press.

Chomsky, Noam (2000a) *New Horizons in the Study of Language and Mind*. Cambridge: Cambridge University Press.

Chomsky, Noam (2000b) 'Minimalist inquiries', in R. Martin, D. Michaels and J. Uriagereka (eds), *Step by Step: Essays on Minimalist Syntax in Honor of Howard Lasnik*. Cambridge, MA: MIT Press, pp. 89–156.

Chomsky, Noam (2002) *On Nature and Language*, eds A. Belletti and Luigi Rizzi. Cambridge: Cambridge University Press.

Cienki, Alan (1998) 'STRAIGHT: an image schema and its metaphorical extensions', *Cognitive Linguistics*, 9, 2, 107–50.

Cienki, Alan (1999) 'Metaphoric gestures and some of their relations to verbal metaphoric expressions', in J. P. König (ed.), *Discourse and Cognition: Bridging the Gap*. Stanford, CA: CSLI Publications, pp. 189–204.

Clark, Andy (1997) *Being There: Putting Brain, Body, and World Together Again*. Cambridge, MA: MIT Press.

Clausner, Timothy C. and William Croft (1999) 'Domains and image schemas', *Cognitive Linguistics*, 10, 1, 1–31.

Comrie, Bernard (1989) *Language Universals and Linguistic Typology*, 2nd edn. Chicago: Chicago University Press.

Coulson, Seana (2000) *Semantic Leaps*. Cambridge: Cambridge University Press.

Coulson, Seana and Todd Oakley (eds) (2000) 'Special issue on conceptual blending', *Cognitive Linguistics*, 11, 3/4, 175–360.

Coventry, Kenny and Simon Garrod (2004) *Saying, Seeing and Acting: The Psychological Semantics of Spatial Prepositions*. Hove: Psychology Press.

Croft, William (1993) 'The role of domains in the interpretation of metaphors and metonymies', *Cognitive Linguistics*, 4, 335–70.

Croft, William (1998) 'Mental representations', *Cognitive Linguistics*, 9, 2, 151–74.

Croft, William (2000) *Explaining Language Change: An Evolutionary Approach*. London: Longman.

Croft, William (2001) *Radical Construction Grammar: Syntactic Theory in Typological Perspective*. Oxford: Oxford University Press.

Croft, William (2002) *Radical Construction Grammar: Syntactic Theory in Typological Perspective*. Oxford: Oxford University Press.

Croft, William (2003) *Typology and Universals*, 2nd edn. Cambridge: Cambridge University Press.

Croft, William and D. Alan Cruse (2004) *Cognitive Linguistics*. Cambridge: Cambridge University Press.

Cruse, D. Alan (1986) *Lexical Semantics*. Cambridge: Cambridge University Press.

Cruse, D. Alan (2000) *Meaning in Language*. Oxford: Oxford University Press.

Cruse, D. Alan (2002) 'Aspects of the micro-structure of word meanings', in Y. Ravin and C. Leacock (eds), *Polysemy: Theoretical and Computational Approaches*. Oxford: Oxford University Press, pp. 30–51.

Cummins, Robert and Denise Dellarosa Cummins (eds) (1999) *Minds, Brains and Computers: The Foundations of Cognitive Science*. Oxford: Blackwell.

Cutrer, L. Michelle (1994) *Time and Tense in Narrative and in Everyday Life*. Doctoral thesis, Department of Cognitive Science, University of California, San Diego; available as Technical Report No. 9501 from the Department of Cognitive Science (ordering details available online at: www.cogsci. ucsd.edu).

Cuyckens, Hubert and Britta Zawada (2001) *Polysemy in Cognitive Linguistics*. Amsterdam: John Benjamins.

Cuyckens, Hubert, René Dirven and John Taylor (2003) *Cognitive Approaches to Lexical Semantics*. Berlin: Mouton de Gruyter.

Cuyckens, Hubert, Dominiek Sandra and Sally Rice (1997) 'Towards an empirical lexical semantics', in B. Smieja and M. Tasch (eds), *Human Contact Through Language and Linguistics*. Frankfurt: Peter Lang, pp. 35–54.

Deane, Paul (forthcoming) 'Multimodal spatial representation: on the semantic unity of *over*', in B. Hampe (ed.), *From Perception to Meaning: Image Schemas in Cognitive Linguistics*. Berlin: Mouton de Gruyter.

Dewell, Robert (1994) 'Over again: image-schema transformations in semantic analysis', *Cognitive Linguistics*, 5, 4, 351–80.
Dirven, René and Ralph Pörings (2002) *Metaphor and Metonymy in Comparison and Contrast*. Berlin: Mouton de Gruyter.
Dirven, René and Marjolin Verspoor (2004) *Cognitive Exploration of Language and Linguistics*, 2nd edn. Amsterdam: John Benjamins.
Dixon, R. M. W. (1991) *A New Approach to English Grammar on Semantic Principles*. Oxford: Oxford University Press.
Dowty, David (1991) 'Thematic proto-roles and argument selection', *Language*, 67, 574–619.
Dunbar, George (2001) 'Towards a cognitive analysis of polysemy, ambiguity, and vagueness', *Cognitive Linguistics*, 12, 1, 1–14.
Evans, Vyvyan (2004a) *The Structure of Time: Language, Meaning and Temporal Cognition*. Amsterdam: John Benjamins.
Evans, Vyvyan (2004b) 'How we conceptualise time', *Essays in Arts and Sciences*, 33–2, 13–44.
Evans, Vyvyan (2005) 'The meaning of *time*: polysemy, the lexicon and conceptual structure', *Journal of Linguistics*, 41, 1, 33–75.
Evans, Vyvyan and Andrea Tyler (2004a) 'Rethinking English "prepositions of movement": the case of *to* and *through*', in H. Cuyckens, W. De Mulder and T. Mortelmans (eds), *Adpositions of Movement* (*Belgian Journal of Linguistics*, 18). Amsterdam: John Benjamins.
Evans, Vyvyan and Andrea Tyler (2004b) 'Spatial experience, lexical structure and motivation: the case of *in*', in G. Radden and K. Panther (eds), *Studies in Linguistic Motivation*. Berlin: Mouton de Gruyter, pp. 157–92.
Evans, Vyvyan and Jörg Zinken (2005) *Figurative Language in a Modern Theory of Meaning Construction: A Lexical Concepts and Cognitive Models Approach*. Paper presented at New Directions in Cognitive Linguistics, University of Sussex, October 2005.
Fairclough, Norman (2001) *Language and Power*, 2nd edn. Harlow: Longman.
Fauconnier, Gilles ([1985] 1994) *Mental Spaces*. Cambridge: Cambridge University Press.
Fauconnier, Gilles (1997) *Mappings in Thought and Language*. Cambridge: Cambridge University Press.
Fauconnier, Gilles (1999) 'Methods and generalizations', in T. Janssen and G. Redeker (eds), *Cognitive Linguistics: Foundations, Scope and Methodology*. Berlin: Mouton de Gruyter, pp. 95–128.
Fauconnier, Gilles and Eve Sweetser (1996) *Spaces, Worlds and Grammar*. Chicago: University of Chicago Press.
Fauconnier, Gilles and Mark Turner (1994) *Conceptual Projection and Middle Spaces*, Technical Report No. 9401, Department of Cognitive Science, University of California, San Diego (available online at: www.cogsci.ucsd.edu/research/files/technical/9401.pdf).

Fauconnier, Gilles and Mark Turner (1996) 'Blending as a central process of grammar', in A. Goldberg (ed.), *Conceptual Structure, Discourse, and Language*. Stanford, CA: CSLI Publications, pp. 113–30.

Fauconnier, Gilles and Mark Turner (1998a) 'Conceptual integration networks', *Cognitive Science*, 22, 2, 33–187.

Fauconnier, Gilles and Mark Turner (1998b) 'Principles of conceptual integration', in J.-P. Koenig (ed.), *Discourse and Cognition*. Stanford, CA: CSLI Publications, pp. 269–83.

Fauconnier, Gilles and Mark Turner (1999) 'Metonymy and conceptual integration', in K. Panther and G. Radden (eds), *Metonymy in Language and Thought*. Amsterdam: John Benjamins, pp. 77–90.

Fauconnier, Gilles and Mark Turner (2000) 'Compression and global insight', *Cognitive Linguistics*, 11, 3–4, 283–304.

Fauconnier, Gilles and Mark Turner (2002) *The Way We Think: Conceptual Blending and the Mind's Hidden Complexities*. New York: Basic Books.

Fauconnier, Gilles and Mark Turner (2003) 'Polysemy and conceptual blending', in B. Nerlich, V. Herman, Z. Todd and D. Clarke (eds), *Polysemy: Patterns of Meaning in Mind and Language*. Berlin: Mouton de Gruyter, pp. 79–94.

Fillmore, Charles (1968) 'The case for case', in E. Bach and R. Harms (eds), *Universals in Linguistic Theory*. New York: Holt, Reinhart & Winston, pp. 1–81.

Fillmore, Charles (1975) 'An alternative to checklist theories of meaning', *Proceedings of the First Annual Meeting of the Berkeley Linguistics Society*. Amsterdam: North Holland, pp. 123–31.

Fillmore, Charles (1977) 'Scenes-and-frames semantics', in A. Zampolli (ed.), *Linguistic Structures Processing*. Amsterdam: North Holland, pp. 55–82.

Fillmore, Charles (1982) 'Frame semantics', in Linguistic Society of Korea (ed.), *Linguistics in the Morning Calm*. Seoul: Hanshin Publishing, pp. 111–37.

Fillmore, Charles (1985a) 'Frames and the semantics of understanding', *Quaderni di Semantica*, 6, 222–54.

Fillmore, Charles (1985b) 'Syntactic intrusions and the notion of grammatical construction', *Proceedings of the Berkeley Linguistics Society*, 11, 73–86.

Fillmore, Charles (1988) 'The mechanisms of construction grammar', *Proceedings of the Berkeley Linguistics Society*, 14, 35–55.

Fillmore, Charles and Beryl T. Atkins (1992) 'Toward a frame-based lexicon: the semantics of RISK and its neighbors', in A. Lehrer and E. F. Kittay (eds), *Frames, Fields and Contrasts*. Hillsdale, NJ: Lawrence Erlbaum, pp. 75–102.

Fillmore, Charles and Beryl T. Atkins (2000) 'Describing polysemy: the case of *crawl*', in Y. Ravin and C. Leacock (eds), *Polysemy: Theoretical and Computational Approaches*. Oxford: Oxford University Press, pp. 91–110.

Fillmore, Charles and Paul Kay (1993) *Construction Grammar Coursebook, Chapters 1 thru 11 (Reading Materials for Ling. X20)*. University of California, Berkeley.

Fillmore, Charles, Paul Kay and Mary Katherine O'Connor (1988) 'Regularity and idiomaticity: the case of let alone', *Language*, 64, 3, 501–38.

Flaherty, Michael (1999) *A Watched Pot: How We Experience Time*. New York: New York University Press.

Fodor, Jerry A. (1975) *The Language of Thought*. Cambridge, MA: Harvard University Press.

Fodor, Jerry A. (1983) *The Modularity of Mind*. Cambridge, MA: MIT Press.

Fodor, Jerry A. (1998) *Concepts: Where Cognitive Science Went Wrong*. Oxford: Oxford University Press.

Fodor, Jerry A. (2000) 'Precis of *The Modularity of Mind*', in R. Cummins and D. Dellarosa Cummins (eds), *Minds, Brains and Computers: The Foundations of Cognitive Science*. Oxford: Blackwell, pp. 493–9.

Fodor, Jerry A. and Ernie Lepore (1996) 'The red herring and the pet fish: why concepts still can't be prototypes', *Cognition*, 58, 253–70.

Foley, William (1997) *Anthropological Linguistics: An Introduction*. Oxford: Blackwell.

Frajzyngier, Zygmunt (1993) *A Grammar of Mupun*. Berlin: Dietrich Reimer Verlag.

Freeman, Margaret (2003) 'Poetry and the scope of metaphor: toward a cognitive theory of literature', in A. Barcelona (ed.), *Metaphor and Metonymy at the Crossroads*. Berlin: Mouton de Gruyter, pp. 253–81.

Frege, Gottlob ([1892] 1975) 'On sense and reference', in D. Davidson and G. Harman (eds), *Semantics of Natural Language*. Encino, CA: Dickenson, pp. 116–28.

Fromkin, Victoria, Rodney Rodman and Nina Hyams (2002) *An Introduction to Language*, 7th edn. New York: Holt, Rinehart & Winston.

Gavins, Joanna and Gerard Steen (2003) *Cognitive Poetics in Practice*. London: Routledge.

Geeraerts, Dirk (1993) 'Vagueness's puzzles, polysemy's vagaries', *Cognitive Linguistics*, 4, 3, 223–72.

Geeraerts, Dirk (1994) *Diachronic Prototype Semantics*. Oxford: Oxford University Press.

Geeraerts, Dirk (1995) 'Cognitive linguistics', in J. Verschueren, J. Oestman and J. Blommaert (eds), *Handbook of Pragmatics: A Manual*. Amsterdam: John Benjamins, pp. 111–16.

Geeraerts, Dirk and Hubert Cuyckens (2005) *Oxford Handbook of Cognitive Linguistics*. Oxford: Oxford University Press.

Gentner, Dedre and Donald R. Gentner (1982) 'Flowing waters or teeming crowds: mental models of electricity', in D. Gentner and A. Stevens (eds), *Mental Models*. Hillsdale, NJ: Lawrence Erlbaum, pp. 99–129.

Gentner, Dedre and Susan Goldin-Meadow (2003) *Language in Mind: Advances in the Study of Language and Thought*. Cambridge, MA: MIT Press.

Gibbs, Raymond W. (1994) *The Poetics of Mind*. Cambridge: Cambridge University Press.

Gibbs, Raymond W. (2000) 'Making good psychology out of blending theory', *Cognitive Linguistics*, 11, 3/4, 347–58.

Gibbs, Raymond W. and Herbert Colston (1995) 'The cognitive psychological reality of image schemas and their transformations', *Cognitive Linguistics*, 6, 4, 347–78.

Gibbs, Raymond W. and Teenie Matlock (2001) 'Psycholinguistic perspectives on polysemy', in H. Cuyckens and B. Zawada (eds), *Polysemy in Cognitive Linguistics*. Amsterdam: John Benjamins, pp. 213–39.
Gibbs, Raymond W. and Gerard Steen (1999) *Metaphor in Cognitive Linguistics*. Amsterdam: John Benjamins.
Givón, Talmy (1979) *On Understanding Grammar*. New York: Academic Press.
Givón, Talmy (1991) *Syntax: A Functional-Typological Introduction*. Amsterdam: John Benjamins.
Goldberg, Adele (1992) 'The inherent semantics of argument structure: the case of the English ditransitive construction', *Cognitive Linguistics*, 3, 1, 37–74.
Goldberg, Adele (1995) *Constructions: A Construction Grammar Approach to Argument Structure*. Chicago: Chicago University Press.
Goldberg, Adele (ed.) (1996) *Conceptual Structure, Discourse, and Language*. Stanford, CA: CSLI Publications.
Goldberg, Adele (1997) 'Construction grammar', in E. K. Brown and J. E. Miller (eds), *Concise Encyclopedia of Syntactic Theories*. New York: Elsevier Science.
Gonzalez-Marquez, Monica, Irene Mittelberg, Seana Coulson and Michael J. Spivey (eds) (forthcoming) *Empirical Methods in Cognitive Linguistics*. Amsterdam: John Benjamins.
Goossens, Louis (1990) 'Metaptonymy: the interaction of metaphor and metonymy in expressions for linguistic action', *Cognitive Linguistics*, 1, 3, 323–40.
Grady, Joseph (n.d.) '*Foundations of Meaning*' (unpublished manuscript).
Grady, Joseph (1997a) *Foundations of Meaning: Primary Metaphors and Primary Scenes*. Doctoral thesis, Linguistics Dept, University of California, Berkeley (available from UMI Dissertation Services: www.il.proquest.com/ umi/dissertations/).
Grady, Joseph (1997b) 'THEORIES ARE BUILDINGS revisited', *Cognitive Linguistics*, 8, 4, 267–90.
Grady, Joseph (1998) 'The conduit metaphor revisited: reassessing metaphors for communication', in J. P. König (ed.), *Conceptual Structure, Discourse and Language II*. Stanford, CA: CSLI Publications, pp. 205–18.
Grady, Joseph (1999) 'A typology of motivation for conceptual metaphor: correlation vs. resemblance', in R. W. Gibbs and G. Steen (eds), *Metaphor in Cognitive Linguistics*. Amsterdam: John Benjamins, pp. 79–100.
Grady, Joseph (2005) 'Primary metaphors as inputs to conceptual integration', *Journal of Pragmatics*, 37, 1595–614.
Grady, Joseph, Todd Oakley and Seana Coulson (1999) 'Blending and metaphor', in R. W. Gibbs and G. Steen (eds), *Metaphor in Cognitive Linguistics*. Amsterdam: John Benjamins, pp. 101–124.
Grady, Joseph and Christopher Johnson (2000) 'Converging evidence for the notions of "subscene" and "primary scene"', *Proceedings of the 23rd Annual Meeting of the Berkeley Linguistics Society*. Berkeley, CA: Berkeley Linguistics Society, pp. 123–36 (reprinted in Dirven and Pörings 2002).

Grady, Joseph, Sarah Taub and Pamela Morgan (1996) 'Primitive and compound metaphors', in A. Goldberg (ed.), *Conceptual Structure, Discourse and Language*. Stanford, CA: CSLI Publications, pp. 177–87.

Green, David (ed.) (1996) *Cognitive Science: An Introduction*. Oxford: Blackwell.

Greenberg, Joseph (1990) *On Language: Selected Writings of Joseph Greenberg*, eds S. Kemmer and K. Denning. Stanford, CA: Stanford University Press.

Grice, H. Paul (1975) 'Logic and conversation', in P. Cole and J. L. Morgan (eds), *Syntax and Semantics 3: Speech Acts*. New York: Academic Press, pp. 41–58.

Gries, Stephan Th. (2005) 'Corpus-based methods and cognitive semantics: the many meanings of *to run*', in S. Gries and A. Stefanowitsch (eds), *Corpora in Cognitive Linguistics: The Syntax-Lexis Interface*. Berlin: Mouton de Gruyter.

Gries, Stefan Th. and Anatol Stefanowitsch (eds) (2005) *Corpora in Cognitive Linguistics. The Syntax-Lexis Interface*. Berlin: Mouton de Gruyter.

Gumperz, John and Stephen Levinson (1996) *Rethinking Linguistic Relativity*. Cambridge: Cambridge University Press.

Guthrie, Malcolm (1967–71) *Comparative Bantu: An Introduction to the Comparative Linguistics and Prehistory of the Bantu Languages*. Farnborough: Gregg International.

Haiman, John (1980) 'Dictionaries and encyclopedias', *Lingua*, 50, 329–57.

Haiman, John (1983) 'Iconic and economic motivation', *Language*, 59, 781–819.

Haiman, John (ed.) (1985) *Iconicity in Syntax*. Amsterdam: John Benjamins.

Halliday, M. A. K. (1994) *An Introduction to Functional Grammar*, 2nd edn. London: Arnold.

Hampe, Beate (ed.) (forthcoming) *From Perception to Meaning: Image Schemas in Cognitive Linguistics*. Berlin: Mouton de Gruyter.

Haser, Verena (2005) *Metaphor, Metonymy, and Experientialist Philosophy: Challenging Cognitive Semantics*. Berlin: Mouton de Gruyter.

Haspelmath, Martin (1997) *From Space to Time: Temporal Adverbials in the World's Languages*. Munich: Lincom Europa.

Haviland, John (1993) 'Anchoring, iconicity, and orientation in Guugu-Yimidhirr pointing gestures', *Journal of Linguistic Anthropology*, 3, 3–45.

Heider, Eleanor (1972) 'Universals in color naming and memory', *Journal of Experimental Psychology*, 93, 10–20.

Heim, Irene and Angelika Kratzer (1998) *Semantics in Generative Grammar*. Oxford: Blackwell.

Heine, Bernd (1997) *Cognitive Foundations of Grammar*. Oxford: Oxford University Press.

Heine, Bernd and Tania Kuteva (2002), *World Lexicon of Grammaticalization*. Cambridge: Cambridge University Press.

Heine, Bernd, Ulrike Claudi and Friederike Hünnemeyer (1991) *Grammaticalization: A Conceptual Framework*. Chicago: Chicago University Press.

Herskovits, Annette (1986) *Language and Spatial Cognition*. Cambridge: Cambridge University Press.

Hopper, Paul (1987) 'Emergent grammar', *Proceedings of the Berkeley Linguistics Society*, 13, 139–57.
Hopper, Paul and Elizabeth Closs Traugott (1993) *Grammaticalization*. Cambridge: Cambridge University Press.
Hopper, Paul and Elizabeth Closs Traugott (2003) *Grammaticalization*, 2nd edn. Cambridge: Cambridge University Press.
Hunt, Earl and Franca Agnoli (1991) 'The Whorfian hypothesis: a cognitive psychology perspective', *Psychological Review*, 98, 377–89.
Hutchins, Edwin (1996) *Cognition in the Wild*. Cambridge, MA: MIT Press.
Jackendoff, Ray (1983) *Semantics and Cognition*. Cambridge, MA: MIT Press.
Jackendoff, Ray (1987) 'The status of thematic relations in linguistic theory', *Linguistic Inquiry*, 18, 369–411.
Jackendoff, Ray (1990) *Semantics Structures*. Cambridge, MA: MIT Press.
Jackendoff, Ray (1992) *Language of the Mind*. Cambridge, MA: MIT Press.
Jackendoff, Ray (1997) *The Architecture of the Language Faculty*. Cambridge, MA: MIT Press.
Jackendoff, Ray (2002) *Foundations of Language: Brain, Meaning, Grammar, Evolution*. Oxford: Oxford University Press.
Jaeger, Jeri and John Ohala (1984) 'On the structure of phonetic categories', *Proceedings of the 10th Annual Meeting of the Berkeley Linguistics*. Berkeley, CA: Berkeley Linguistics Society, pp. 15–26.
Jaggar, Philip J. (2001) *Hausa*, London Oriental and African Language Library No. 7. Amsterdam and Philadelphia: John Benjamins.
Janssen, Theo and Gisela Redeker (1999) *Cognitive Linguistics: Foundations, Scope, and Methodology*. Berlin: Mouton de Gruyter.
Jesperson, Otto ([1909–49] 1961) *A Modern English Grammar on Historical Principles*. London: Allen & Unwin.
Johnson, Mark (1987) *The Body in the Mind: The Bodiliy Basis of Meaning, Imagination and Reason*. Chicago: Chicago University Press.
Johnson, Mark (1992) 'Philosophical implications of cognitive semantics'. *Cognitive Linguistics*, 3, 4, 345–66.
Johnson, Mark (1994) *Moral Imagination: Implications of Cognitive Science for Ethics*. Chicago: University of Chicago Press.
Katz, Jerrold J. (1972) *Semantic Theory*. New York: Harper & Row.
Katz, Jerrold J. and Paul M. Postal (1964) *An Integrated Theory of Linguistic Descriptions*. Cambridge, MA: MIT Press.
Kay, Paul and Charles Fillmore (1999) 'Grammatical constructions and linguistic generalizations: the *What's X doing Y* construction', *Language*, 75, 1–34.
Kay, Paul and Willett Kempton (1984) 'What is the Sapir-Whorf Hypothesis?', *American Anthropologist*, 86, 1, 65–79.
Keller, Rudi (1994) *On Language Change: The Invisible Hand in Language*. London: Routledge.

Kemmer, Suzanne (1993) *The Middle Voice*. Amsterdam: John Benjamins.

Komatsu, Lloyd (1992) 'Recent views of conceptual structure', *Psychological Bulletin*, 112, 3, 500–26.

Kövecses, Zoltán (2000) *Metaphor and Emotion*. Cambridge: Cambridge University Press.

Kövecses, Zoltán (2002) *Metaphor: A Practical Introduction*. Oxford: Oxford University Press.

Kövecses, Zoltan and Gunter Radden (1998) 'Metonymy: developing a cognitive linguistic view', *Cognitive Linguistics*, 9, 1, 37–77.

Kreitzer, Anatol (1997) 'Multiple levels of schematization: a study in the conceptualization of space', *Cognitive Linguistics*, 8, 4, 291–325.

Labov, William (1966) *The Social Stratification of English in New York City*. Washington, DC: Center for Applied Linguistics.

Labov, William (1994) *Principles of Linguistic Change. Volume 1: Internal Factors*. Oxford: Blackwell.

Lakoff, George (1987) *Women, Fire and Dangerous Things: What Categories Reveal About the Mind*. Chicago: University of Chicago Press.

Lakoff, George (1990) 'The invariance hypothesis: is abstract reason based on image-schemas?', *Cognitive Linguistics*, 1, 1, 39–74.

Lakoff, George (1991) *Metaphor and War: The Metaphor System Used to Justify War in the Gulf. An open letter to the Internet*. Available from the Center for the Cognitive Science of Metaphor Online: http://philosophy.uoregon.edu/metaphor/metaphor.htm.

Lakoff, George (1993) 'The contemporary theory of metaphor', in A. Ortony (ed.), *Metaphor and Thought*, 2nd edn. Cambridge: Cambridge University Press, pp. 202–51.

Lakoff, George (2002) *Moral Politics: How Liberals and Conservatives Think*, 2nd edn. Chicago: University of Chicago Press.

Lakoff, George and Mark Johnson (1980) *Metaphors We Live By*. Chicago: Chicago University Press.

Lakoff, George and Mark Johnson (1999) *Philosophy in the Flesh: The Embodied Mind and Its Challenge to Western Thought*. New York: Basic Books.

Lakoff, George and Mark Turner (1989) *More than Cool Reason: A Field Guide to Poetic Metaphor*. Chicago: University of Chicago Press.

Lakoff, George and Rafael Núñez (2000) *Where Mathematics Comes From: How the Embodied Mind Brings Mathematics into Being*. New York: Basic Books.

Langacker, Ronald (1987) *Foundations of Cognitive Grammar, Volume I*. Stanford, CA: Stanford University Press.

Langacker, Ronald (1991) *Foundations of Cognitive Grammar, Volume II*. Stanford, CA: Stanford University Press.

Langacker, Ronald (1993) 'Reference-point constructions', *Cognitive Linguistics*, 4, 1–38.

Langacker, Ronald (1999a) 'Assessing the cognitive linguistic enterprise', in T. Janssen and G. Redeker (eds), *Cognitive Linguistics: Foundations, Scope, and Methodology*. Berlin: Mouton de Gruyter, pp. 13–60.

Langacker, Ronald (1999b) *Grammar and Conceptualization*. Berlin: Mouton de Gruyter.

Langacker, Ronald (1999c) 'Losing control: grammaticization, subjectification and transparency', in A. Blank and P. Koch (eds), *Historical Semantics and Cognition*. Berlin: Mouton de Gruyter, pp. 147–175 (revised version, Chapter 10 in *Grammar and Conceptualization*).

Langacker, Ronald (2000) 'A dynamic usage-basked model', in M. Barlow and S. Kemmer (eds), *Usage-Based Models of Language*. Stanford, CA: CSLI Publications, pp. 1–64.

Langaker, Ronald ([1991] 2002) *Concept, Image, Symbol: The Cognitive Basis of Grammar*, 2nd edn. Berlin: Mouton de Gruyter.

Laurence, Stephen and Eric Margolis (1999) 'Concepts and cognitive science', in E. Margolis and S. Laurence (eds), *Concepts: Core Readings*. Cambridge, MA: MIT Press, pp. 3–81.

Lee, David (2001) *Cognitive Linguistics: An Introduction*. Oxford: Oxford University Press.

Lee, Penny (1996) *The Whorf Theory Complex: A Critical Reconstruction*. Amsterdam: John Benjamins.

Leezenberg, Michel (2001) *Contexts of Metaphor*. Oxford: Elsevier.

Lehiste, Ilse (1969) ' "Being" and "having" in Estonian', *Foundations of Language*, 5, 324–41.

LePage, Robert and Andrée Tabouret-Keller (1985) *Acts of Identity*. Cambridge: Cambridge University Press.

Levinson, Stephen (1983) *Pragmatics*. Cambridge: Cambridge University Press.

Levinson, Stephen (1996) 'Introduction to part II', in *Rethinking Linguistic Relativity*, eds J. Gumperz and S. Levinson. Cambridge: Cambridge University Press, pp. 133–144.

Levinson, Stephen (1997) 'Language and cognition: the cognitive consequences of spatial descriptions in Guugu Yimithirr', *Journal of Linguistic Anthropology*, 7, 1, 98–131.

Levinson, Stephen (2003) *Space in Language and Cognition: Explorations in Cognitive Diversity*. Cambridge: Cambridge University Press.

Li, Penny and Lila Gleitman (2002) 'Turning the tables: language and spatial reasoning', *Cognition*, 83, 265–94.

Lindner, Susan (1981) *A Lexico-semantic Analysis of English Verb Particle Constructions with 'out' and 'up'*. Doctoral thesis, Linguistics dept, University of California, San Diego (available from UMI Dissertation Services: www.il.proquest.com/umi/dissertations/).

Lindstromberg, Seth (1997) *English Prepositions Explained*. Amsterdam: John Benjamins.

Lord, Carol (1976) 'Evidence for syntactic reanalysis: from verb to complementizer in Kwa', in S. B. Steever, C. A. Walker and S. S. Mufwen (eds), *Papers from the Parasession on Diachronic Syntax, April 22, 1976*. Chicago: Chicago Linguistics Society, pp. 179–91.

Lyons, John (1967) 'A note on possessive, existential and locative sentences', *Foundations of Language*, 3, 390–96.

Mandelbilt, Nili (2000) 'The grammatical marking of conceptual integration: from syntax to morphology', *Cognitive Linguistics*, 11, 197–252.

Mandler, Jean (1992) 'How to build a baby II. Conceptual primitives', *Psychological Review*, 99, 567–604.

Mandler, Jean (1996) 'Preverbal representation and language', in P. Bloom, M. A. Peterson, L. Nadel and M. F. Garrett (eds), *Language and Space*. Cambridge, MA: MIT Press, pp. 365–384.

Mandler, Jean (2004) *The Foundations of Mind: Origins of Conceptual Thought*. Oxford: Oxford University Press.

Meltzoff, Andrew (1995) 'Understanding the intentions of others: re-enactment of intended acts by 18-month-old children', *Developmental Psychology*, 31, 838–50.

Mircale, Andrew and Juan de Dios Yapita Moya (1981) 'Time and space in Aymara', in M. J. Hardman (ed.), *The Aymara Language and Its Social and Cultural Context*. Gainsville, FL: University of Florida Press, pp. 33–56.

Montague, Richard (1970) 'Universal grammar', *Theoria*, 36, 373–98.

Montague, Richard (1973) 'The proper treatment of quantification in ordinary English', in K. Hintikka, E. Moravcsik and P. Suppes (eds), *Approaches to Natural Language*. Dordrecht: Reidel, pp. 221–42.

Moore, Kevin Ezra (2000) *Spatial Experience and Temporal Metaphors in Wolof: Point of View, Conceptual Mapping and Linguistic Practice*. Doctoral thesis, Linguistics dept, University of California, Berkeley (available from UMI Dissertation Services: www.il.proquest.com/umi/dissertations/).

Murphy, Gregory (1996) 'On metaphoric representation', *Cognition*, 60, 173–204.

Murphy, M. Lynne (2003) *Semantic Relations and the Lexicon*. Cambridge: Cambridge University Press.

Musolff, Andreas (2004) *Metaphor and Political Discourse: Analogical Reasoning in Debates about Europe*. London: Palgrave.

Nerlich, Brigitte, Susan Johnson and David D. Clarke (2003) 'The first "designer baby": the role of narratives, clichés and metaphors in the year 2000 media debate', *Science as Culture*, 12, 4, 471–98.

Nerlich, Brigitte, Zazie Todd, Vimala Herman and David D. Clarke (2003) *Polysemy: Flexible Patterns of Meaning in Mind and Language*. Berlin: Mouton de Gruyter.

Newmeyer, Frederick J. (1999) 'Bridges between generative and cognitive linguistics', in L. de Stadler and C. Eyrich (eds), *Issues in Cognitive Linguistics*. Berlin: Mouton de Gruyter.

Núñez, Rafael and Eve Sweetser (forthcoming) 'Aymara, where the future is behind

you: convergent evidence from language and gesture in the crosslinguistic comparison of spatial construals of time', *Cognitive Science*.
Nuyts, Jan and Eric Pederson (eds) (1997) *Language and Conceptualization*. Cambridge: Cambridge University Press.
Oakley, Todd (1998) 'Conceptual blending, narrative discourse, and rhetoric'. *Cognitive Linguistics*, 9, 321–60.
Ohala, John (1989) 'Sound change is drawn from a pool of synchronic variation', in L. E. Breivik and E. H. Jahr (eds), *Language Change: Contributions to the Study of Its Causes*. Berlin: Mouton de Gruyter, pp. 173–98.
Ortony, Andrew (1993) *Metaphor and Thought*, 2nd edn. Cambridge: Cambridge University Press.
Östman, Jan-Ola and Mirjam Fried (2005a) 'The cognitive grounding of constructional grammar', in J.-O. Östman and M. Fried (eds), *Construction Grammars*. Amsterdam: John Benjamins, pp. 1–13.
Östman, Jan-Ola and Mirjam Fried (2005b) *Construction Grammars: Cognitive Grounding and Theoretical Extensions*. Amsterdam: John Benjamins.
Panther, Klaus-Uwe and Linda Thornburg (2003) *Metonymy and Pragmatic Inferencing*. Amsterdam: John Benjamins.
Pinker, Steven (1994) *The Language Instinct*. Harmondsworth: Penguin.
Pollard, Carl and Ivan Sag (1994) *Head-driven Phrase Structure Grammar*. Chicago: University of Chicago Press.
Pöppel, Ernst (1994) 'Temporal mechanisms in perception', in O. Sporns and G. Tononi (eds), *Selectionism and the Brain: International Review of Neurobiology*, Vol. 37. San Diego, CA: Academic Press, pp. 185–202.
Portner, Paul (2005) *What Is Meaning? Fundamentals of Formal Semantics*. Oxford: Blackwell.
Portner, Paul and Barbara Partee (2002) *Formal Semantics: The Essential Readings*. Oxford: Blackwell.
Pustejovsky, James (1995) *The Generative Lexicon*. Cambridge, MA: MIT Press.
Radden, Günter (1992) 'The cognitive approach to natural language', in M. Pütz (ed.), *Thirty Years of Linguistic Evolution: Studies in Honour of René Dirven on the Occasion of his Sixtieth Birthday*. Amsterdam: Benjamins, pp. 513–41.
Radden, Günter (1997) 'Time is space', in B. Smieja and M. Tasch (eds), *Human Contact Through Language and Linguistics*. Frankfurt: Peter Lang, pp. 147–66.
Radden, Günter (2003a) 'The metaphor TIME AS SPACE across languages', in N. Baumgarten, C. Böttger, M. Motz and J. Probst (eds), Übersetzen, Interkulturelle Kommunikation, Spracherwerb und Sprach-vermittlung – das Leben mit mehreren Sprachen. Festschrift für Juliane House zum 60. Geburtstag. *Zeitschrift für Interkulturellen Fremdsprachenunterricht* [online], 8, 2/3, 1–14 (available on-line at: http://zif.spz.tu-darmstadt.de/jg-08-2-3/beitrag/Radden1.htm).
Radden, Günter (2003b) 'How metonymic are metaphors?', in A. Barcelona (ed.), *Metaphor and Metonymy at the Crossroads*. Berlin: Mouton de Gruyter, pp. 93–108.

Radden, Günter and René Dirven (2005) *Cognitive English Grammar*. Amsterdam: John Benjamins.

Radden, Günter and Klaus-Uwe Panther (1999) *Metonymy in Language and Thought*. Amsterdam: John Benjamins.

Radford, Andrew (1997a) *Syntax: A Minimalist Introduction*. Cambridge: Cambridge University Press.

Radford, Andrew (1997b) *Syntactic Theory and the Structure of English: A Minimalist Approach*. Cambridge: Cambridge University Press.

Radford, Andrew (2004) *Minimalist Syntax: Exploring the Structure of English*. Cambridge: Cambridge University Press.

Ravin, Yael and Claudia Leacock (eds) (2002) *Polysemy: Theoretical and Computational Approaches*. Oxford: Oxford University Press.

Reddy, Michael ([1979] 1993) 'The conduit metaphor: a case of frame conflict in our language about language', in A. Ortony (ed.), *Metaphor and Thought*, 2nd edn. Cambridge: Cambridge University Press, pp. 164–201.

Reichenbach, Hans (1947) *Elements of Symbolic Logic*. New York: Macmillan.

Rice, Sally, Dominiek Sandra and Mia Vanrespaille (1999) 'Prepositional semantics and the fragile link between space and time', in M. Hiraga, Chris Sinha and S. Wilcox (eds), *Cultural, Psychological and Typological Issues in Cognitive Linguistics*. Amsterdam: John Benjamins, pp. 108–27.

Rosch, Eleanor (1975) 'Cognitive representations of semantic categories'. *Journal of Experimental Psychology: General*, 104, 192–233.

Rosch, Eleanor (1977) 'Human categorization', in N. Warren (ed.), *Studies in Cross-linguistic Psychology*. London: Academic Press, pp. 1–49.

Rosch, Eleanor ([1978] 1999) 'Principles of categorization', in B. Lloyd and E. Rosch (eds), *Cognition and Categorization*. Hillsdale, NJ: Erlbaum, pp. 27–48; reprinted in E. Margolis and S. Laurence (eds) (1999) *Concepts: Core Readings*. Cambridge, MA: MIT Press, pp. 189–206.

Rosch, Eleanor and Caroline Mervis (1975) 'Family resemblances: studies in the internal structure of categories', *Cognitive Psychology*, 7, 573–605.

Rosch, Eleanor, Caroline Mervis, Wayne Gray, David Johnson and Penny Boyes-Braem (1976) 'Basic objects in natural categories', *Cognitive Psychology*, 8, 382–439.

Rudzka-Ostyn, Brygida (1988) *Topics in Cognitive Linguistics*. Amsterdam: John Benjamins.

Ruhl, Charles (1989) *On Monosemy: A Study in Linguistic Semantics*. New York: SUNY.

Saeed, John (2003) *Semantics*, 2nd edn. Oxford: Blackwell.

Saffran Jenny, Richard Aslin and Elissa Newport (1996) 'Statistical learning by 8-month old infants', *Science*, 274, 1926–8.

Sampson, Geoffrey (2005) *Educating Eve: The Language Instinct Debate* (revised edition). London and New York: Continuum International.

Sandra, Dominiek (1998) 'What linguists can and can't tell you about the human mind: a reply to Croft', *Cognitive Linguistics*, 9, 4, 361–478.

Sandra, Dominiek and Sally Rice (1995) 'Network analyses of prepositional meaning: mirroring whose mind – the linguist's or the language user's?', *Cognitive Linguistics*, 6, 1, 89–130.

Saussure, Ferdinand de (1916) *Cours de Linguistique Générale*, trans. *Course in General Linguistics* by Roy Harris (1983). London: Duckworth.

Schmid, Hans-Jörg (2000) *English Abstract Nouns as Conceptual Shells: From Corpus to Cognition*. Berlin: Mouton de Gruyter.

Searle, John (1969) *Speech Acts: An Essay in the Philosophy of Language*. Cambridge: Cambridge University Press.

Semino, Elena and Jonathan Culpeper (eds) (2003) *Cognitive Stylistics: Language and Cognition in Text Analysis*. Amsterdam: John Benjamins.

Shinohara, Kazuko (1999) *Epistemology of Space and Time*. Kwansei, Japan: Gakuin University Press.

Shinohara, Kazuko (2000) '*Up-down Orientation in Time Metaphors: Analysis of English and Japanese*'. Manuscript, Tokyo University of Agriculture and Technology.

Sinha, Chris (1999) 'Grounding, mapping and acts of meaning', in T. Janssen and G. Redeker (eds), *Cognitive Linguistics, Foundations, Scope and Methodology*. Berlin: Mouton de Gruyter, pp. 223–56.

Sinha, Chris and Tania Kuteva (1995) 'Distributed spatial semantics', *Nordic Journal of Linguistics*, 18, 167–99.

Skinner, B. F. (1957) *Verbal Behavior*. New York: Appleton-Century-Crofts.

Spencer, Andrew (1991) *Morphological Theory*. Oxford: Blackwell.

Sperber, Dan and Deirdre Wilson ([1986] 1995) *Relevance: Communication and Cognition*, 2nd edn. Oxford: Blackwell.

Stassen, Leon (1997) *Intransitive Predication*. Oxford: Oxford University Press.

Stern, Josef (2000) *Metaphor in Context*. Cambridge, MA: MIT Press.

Stockwell, Peter (2002) *Cognitive Poetics: An Introduction*. London: Routledge.

Svorou, Soteria (1994) *The Grammar of Space*. Amsterdam: John Benjamins.

Sweetser, Eve (1988) 'Grammaticalization and semantic bleaching', in S. Axmaker, A. Jaisser and H. Singmaster (eds), *Proceedings of the 14th Annual Meeting of the Berkeley Linguistics Society*. Berkeley, CA: Berkeley Linguistics Society, pp. 389–405.

Sweetser, Eve (1990) *From Etymology to Pragmatics: Metaphorical and Cultural Aspects of Semantic Structure*. Cambridge: Cambridge University Press.

Sweetser, Eve (1999) 'Compositionality and blending: semantic composition, in a cognitively realistic framework', in T. Janssen and G. Redeker (eds), *Cognitive Linguistics: Foundations, Scope and Methodology*. Berlin: Mouton de Gruyter, pp. 129–62.

Sweetser, Eve (2000) 'Blended spaces and performativity', *Cognitive Linguistics*, 11, 3/4, 305–34.

Tallerman, Maggie (1998) *Understanding Syntax*. London: Arnold.

Talmy, Leonard (1985) 'Force dynamics in language and thought', in W. Eilfort, P.

Kroeber and K. Peterson (eds), *Papers from the Parasession on Causatives and Agentivity*. Chicago: Chicago Linguistics Society, pp. 293–337.

Talmy, Leonard (2000) *Toward a Cognitive Semantics* (2 vols). Cambridge, MA: MIT Press.

Tarski, Alfred ([1944] 2004) 'The semantic conception of truth and the foundations of semantics', in F. Schmitt (ed.), *Theories of Truth*. Oxford: Blackwell, pp. 115–51.

Taylor, John (1995) *Linguistic Categorization*, 2nd edn. Oxford: Oxford University Press.

Taylor, John (2002) *Cognitive Grammar*. Oxford: Oxford University Press.

Taylor, John (2003) *Linguistic Categorization*, 3rd edn. Oxford: Oxford University Press.

Tomasello, Michael (1992) *First Verbs: A Case Study of Early Grammatical Development*. Cambridge: Cambridge University Press.

Tomasello, Michael (1995) 'Language is not an instinct', *Cognitive Development*, 10, 131–56.

Tomasello, Michael (1999) *The Cultural Origins of Human Cognition*. Cambridge, MA: Harvard University Press.

Tomasello, Michael (2000) 'First steps in a usage based theory of language acquisition', *Cognitive Linguistics*, 11, 61–82.

Tomasello, Michael (ed.) (2002) *The New Psychology of Language: Cognitive and Functional Approaches to Language Structure*, 2nd edn. Hillsdale, NJ: Lawrence Erlbaum.

Tomasello, Michael (2003) *Constructing a Language: A Usage-based Theory of Language Acquisition*. Cambridge, MA: Harvard University Press.

Tomasello, Michael and Patricia J. Brooks (1998) 'Young children's earliest transitive and intransitive constructions', *Cognitive Linguistics*, 9, 379–95.

Trask, R. L. (1993) *A Dictionary of Grammatical Terms in Linguistics*. London: Routledge.

Trask, R. L. (1996) *Historical Linguistics*. London: Arnold.

Trask, R. L. (1997) *A Student's Dictionary of Language and Linguistics*. London: Arnold.

Trask, R. L. (1999) *Language: The Basics*, 2nd edn. London: Routledge.

Trask, R. L. (2000a) *The Penguin Dictionary of English Grammar*. Harmondsworth: Penguin.

Trask, R. L. (2000b) *The Dictionary of Historical and Comparative Linguistics*. Edinburgh: Edinburgh University Press.

Trask, R. L. (2004) 'What is a word?', *University of Sussex Working Papers in Linguistics and English Language*, no. LxWP11/04 (available online at: www.sussex.ac.uk/linguistics/documents/essay_-_what_is_a_word.pdf).

Traugott, Elizabeth Closs (1989) 'On the rise of epistemic meanings in English: an example of subjectification in semantic change', *Language*, 65, 31–55.

Traugott, Elizabeth Closs and Richard Dasher (2002) *Regularity in Semantic Change*. Cambridge: Cambridge University Press.

Traugott, Elizabeth Closs and Bernd Heine (1991) *Approaches to Grammaticalization* (2 vols). Amsterdam: John Benjamins.

Traugott, Elizabeth Closs and Jean-Pierre König (1991) 'The semantics-pragmatics of grammaticalization revisted', in E. Traugott and B. Heine (eds), *Approaches to Grammaticalization*, Vol. I. Amsterdam: John Benjamins, pp. 189–218.

Trudgill, Peter (1986) *Sociolinguistics: An Introduction to Language and Society*. Harmondsworth: Penguin Books.

Tuggy, David (1993) 'Ambiguity, polysemy and vagueness', *Cognitive Linguistics*, 4, 3, 273–90.

Tuggy, David (1999) 'Linguistic evidence for polysemy in the mind: a response to William Croft and Dominiek Sandra', *Cognitive Linguistics*, 10, 4, 343–68.

Turner, Frederick and Ernst Pöppel (1983) 'The neural lyre: poetic meter, the brain and time', *Poetry*, 142, 5, 277–309.

Turner, Mark (1990) 'Aspects of the invariance hypothesis', *Cognitive Linguistics*, 1, 2, 247–55.

Turner, Mark (1991) *Reading Minds: The Study of English in the Age of Cognitive Science*. Princeton, NJ: Princeton University Press.

Turner, Mark (1992) 'Design for a theory of meaning', in W. Overton and D. Palermo (eds), *The Nature and Ontogenesis of Meaning*. Hillsdale, NJ: Lawrence Erlbaum, pp. 91–107.

Turner, Mark (1996) *The Literary Mind*. Oxford: Oxford University Press.

Turner, Mark (2001) *Cognitive Dimensions of Social Science*. Oxford: Oxford University Press.

Turner, Mark and Gilles Fauconnier (1995) 'Conceptual integration and formal expression', *Metaphor and Symbolic Activity*, 10, 183–203.

Turner, Mark and Gilles Fauconnier (2000) 'Metaphor, metonymy and binding', in A. Barcelona (ed.), *Metaphor and Metonymy at the Crossroads*. Berlin: Mouton de Gruyter, pp. 264–86.

Tyler, Andrea and Vyvyan Evans (2001a) 'The relation between experience, conceptual structure and meaning: non-temporal uses of tense and language teaching', in M. Pütz, S. Niemeier and R. Dirven (eds), *Applied Cognitive Linguistics I: Theory and Language Acquisition*. Berlin: Mouton de Gruyter, pp. 63–108.

Tyler, Andrea and Vyvyan Evans (2001b) 'Reconsidering prepositional polysemy networks: the case of *over*', *Language*, 77, 4, 724–65.

Tyler, Andrea and Vyvyan Evans (2003) *The Semantics of English Prepositions: Spatial Scenes, Embodied Meaning and Cognition*. Cambridge: Cambridge University Press.

Ungerer, Hans-Jorg and Friedrich Schmid (1996) *An Introduction to Cognitive Linguistics*. London: Longman.

Vandeloise, Claude (1991) *Spatial Prepositions: A Case Study from French*, trans. Anna R. K. Bosch. Chicago: University of Chicago Press.

Vandeloise, Claude (1994) 'Methodology and analyses of the preposition *in*'. *Cognitive Linguistics*, 5, 2, 157–184.

REFERENCES

Varela, Francisco, Evan Thompson and Eleanor Rosch (1991) *The Embodied Mind: Cognitive Science and Human Experience*. Cambridge, MA: MIT Press.

Vendler, Zeno (1967) *Linguistics in Philosophy*. Ithaca, NY: Cornell University Press.

Visser, F. Th. (1969) *An Historical Syntax of the English Language, Vol. III*. Leiden: Brill.

Warner, Anthony R. (1993) 'Reworking the history of the English auxiliaries', in S. Adamson, V. Law, N. Vincent and S. Wright (eds), *Papers from the 5th International Conference on English Historical Linguistics*. Amsterdam: John Benjamins, pp. 537–58.

Whorf, Benjamin Lee (1956) *Language, Thought and Reality: Selected Writings by Benjamin Lee Whorf*, ed. John Carroll. Cambridge, MA: MIT Press.

Wierzbicka, Anna (1996) *Semantics: Primes and Universals*. Oxford: Oxford University Press.

Wittgenstein, Ludwig (1958) *Philosophische Untersuchungen*, trans. G. E. M. Anscombe as *Philosophical Investigations*, 3rd edn. 1999. Harlow, London: Prentice Hall.

Yu, Ning (1998) *The Contemporary Theory of Metaphor: A Perspective from Chinese*. Amsterdam: John Benjamins.

Zelinsky-Wibbelt, Cornelia (1993) *The Semantics of Prepositions: From Mental Processing to Natural Language*. Berlin: Mouton de Gruyter.

Zinken, Jörg, Iina Hellsten and Brigitte Nerlich (forthcoming) 'Discourse metaphors', in J. Zlatev, T. Ziemke and R. Frank (eds), *Body, Language and Mind: Sociocultural Situatedness*. Berlin: Mouton de Gruyter.

Index

The arrangement of the index is word-by-word. Page numbers followed by an asterisk indicate topics cited for further reading and italicised page numbers indicate exercises.

absolutive, 607
abstract domains, 207, 231, 235, 236
abstract nouns, 560
abstract thought, 191, 192, 301
abstractions, 115, 262, 544, 560–1
Access Principle, 312, 315, 376–81
accommodation, 132
accomplishment, 636
accusative, 500, 607
Achard, M., 148
achievement, 636
acquisition *see* language acquisition
action, 515, 516
action chains, 42, 603
actionalisation, 517
activation, 111, 312, 322
active voice, 609
active zones, 238–40
activities, 515, 636
actor-character connector, 378
acts, 515
additive, 402
Adger, D., 771
adjacency, 71
adjectives, 488, 566–7
adpositions, 567
adverbial function, 489, 497–8
adverbs, 488–9, 566–7
affect, 11
AFFECTION, 14–15

affixes, 485, 591, *612–13*
African languages, 719
AGENT, 37–8, 39, 42, 197, 540, 675
agentive -*er* suffix, 36–7
agentive nominalisation, 31–2, 593
Agnoli, F., 105
agreement, 498–500, 593–4, 751–2
Aitchison, J., 356
Akhtar, N., 140
Allwood, J., 22
alpha rhythm, 778
altered replication, 129–32, 133
Alverson, H., 104
ambiguity, 113, 329, 379, 381, *398*
amodal view, 242
analogy, 98–9, 423; *see also* disanalogy
analysis (parsing), 698–9
anaphora, 376
Anderson, L. B., 773
Anglo-Saxon (Old English), 121–2
antecedent, 376
anthropomorphism, 417
antonymy, 209
aphasia, 745, 772–3*
apparent motion, 76–8
apparent simultaneity, 76–7
apposition, 588
arbitrariness, 123, 217
argument roles, 673, 674–6
argument structure, 225, 668, 669–71, 674

arguments, 225, 351, 596–7, 674
articles, 573
Aslin, R., 137
aspect, 387–9, 490–1, 519, 523–4, 615, 634–7, 767–9
assimilation, 131
asterisk, 13
asymmetry, 562
atemporal relations, 563, 565–70, *579–80*, 619, 620, 621, 629, 630
Atkins, B. T., 358
attentional system, 41–3, 198–9, 526–7, 535–44
attenuation, 712, 729
ATTRACTION, 189
attributes (cues), 223–4, 257–8, 267
attrition, 709
Austin, J. L., 417–18, 447
Australia, 90, 144
autonomy, 353, *613*
    conceptual, 585–7
    phonological, 590–1
    syntax, 752, 753
auxiliary verbs, 388, 490–2, 730–2
axiality, 525–6
Aymara (South America), 92–4

Bach, E., 466
BACHELOR, 160–1, 169, 208, 251, 270–1
background dependent framing, 355
backstage cognition, 193, 368
backward projection, 410, *442–3*
BANANA, 216, 217, 220
Barcelona, A., 320, 321, 325
Barlow, M., 147
Barsalou, L., 223–4, 240, 245, 270
Bartlett, F. C., 223
base, 166–7, 237, *246*, 374, 538
basic domains, 207, 231, 232–5, *247*
basic level categories, 248, 257, 260, 261, 262, 282*
    language system, 263
    universality, 263–4
be, 620–1, *639*
    -ing, 621
be going to, 730, 732–3
Bechtel, W., 22
behaviourism, 141, 743–4
benefit tourist, *443–4*
Bergen, B., 481, 508, 697–9, 703
Berger, H., 78
bilabial, 34
binary features, 34, 35
BIRD, 267

Bishop, D., 772
bleaching, 712
blended space, 163, 458
blending, 51*, 124, 440–1*, 458
    multiple, 431–2
    nature of, 407–18
    see also Conceptual Blending Theory
BLOCKAGE, 187, 188
Bloom, P., 103–4
Bloomfield, L., 212, 743–4, 793
Boas, F., 97
BOAT RACE, 411–12, 421
Body in the Mind, The (Johnson), 46, 178
Börjars, K., 509
Boroditsky, L., 100–1, 105, 780
Borsley, B., 663
bound morphemes, 18–19, 159
boundaries, 254
bounded landmark, 157
boundedness, 519–20
bounding, 557–9, 634
Bowerman, M., 87–90
brain activity, 76
brain damage, 242
brain, localisation of function in, 146
Broca's area, 146
Brown, D., 102
Brugman, C., 330, 331–3, 357
building blocks, 756–7
building instructions, 371
by, 361
Bybee, J., 119, 735

calibration, 236
can, 190
Cann, R., 466
Caplan, D., 772
CAR, 223, 224, 225
cardinal points, 73
Carnie, A., 772
Carpenter, M., 140
Carston, R., 466
case, 498–500, 606–9, 765–7
case assignment, 766, *774*
Case Grammar, 206
case system, 122
CAT, 221–2
categories, 494
categorisation, 28–35, 43, *52*, 68, 282–3*, 502
    and cognitive semantics, 168–9, 249–55
    conceptual, 328
    grammar, 545–8
    and idealised cognitive models (ICMs), 248–85

categorisation, (*continued*)
  phonology, 34–5
  principles, 255
  spatial scenes (English and Korean), 87–90
  system, 256
  typical examples, 273
causality, 722
cause-effect, 424
CAUSED MOTION, 687–9
Celtic languages, 121
central systems, 242
centrality, 29, 219–20, *245*, 254
chain shifts, 130
chaining, 333
chains of reference, 375
Chang, N., 481, 508, 697–9, 703
CHANGE, 422, 691
characteristic knowledge, 217, 218–20
Chierchia, G., 466
Chilton, P., 323
Choi, S., 87–90
Chomsky, N., 55, 60–1, 103, 141, 212, 742, 750, 772
Christmas, temporal aspects, 8
Cienki, A., 201
circularity, 670
Clark, A., 51
class schemas, 570–1
classical (definitional) theory, 249, 251–2, *283*
Claudi, U., 714–21, 725, 732, 735
clausal head, 617–24
clauses, 7, 494, 594–610, *613*, 768
Clausner, T. C., 231, 234
cleft constructions, 493
CLINTON AS FRENCH PRESIDENT, 406–7, 408, 409–10, 426
closed languages, 447
closed-class, 19, 23*, *25*, 159, 165
  quantity, 521–3
  semantic system, 193
closed-class subsystems, features, 502–4
closure principle, 66–7
cluster models, 271–2
coalescence, 710
coded meaning, 113, 211, 213, 216, 721–2
codes, 123
coding, 506, 539, 609–10, *614*
coding links, 479
cognition, 240
cognitive commitment, 40–4, 250, 501, 669
cognitive economy *see* economy
cognitive grammar, 48–50, 109, 114–19, 206, 250, 475–511, 480–1, 615–40, 741–74
  architecture, 479–80

characteristics, 744
constructions, 581–614
further reading, 51*, 508*, 611–12*, 638*
vs. construction grammar, 660–1, *665*
vs. generative model, 752–8, 761, 781–2
word classes, 553–80
Cognitive Grammar Theory (Langacker), 533–45, 550*, 578*
cognitive lexical semantics, 250, 328, 358*
  assumptions, 333
cognitive linguistics, 5, 22–3*, 50, 51*
  assessing the enterprise, 777–82
  assumptions and commitments, 27–53
  competing theories, 779–82
  nature of, 27–51
  universals in, 63–8
  vs. formal linguistics, *105*
Cognitive Poetics, 294
Cognitive Principle of Relevance, 460
cognitive psychology, 222–3
cognitive reference points, 272–3
Cognitive Representation, 18, 165, 192, *203*, 514
cognitive routine, 501
cognitive sciences, 16, 22*
cognitive semantics, 48–50, 156–75, 163–70, *174*, 445–74, *466–7*
  categorisation and, 168–9, 249–55
  guiding principles, 157–63
  meaning construction in, 365–70
  truth-conditional (formal) semantics, comparison with, 455–8
coining, 124
coinstantiation construction, 656
collocational expressions, 211
COLOUR, 45, 64, 97, *107*, 233
COMMERCIAL EVENTS, 225–8
communication, 9
communicative function, 759
communicative intention, 139, 140, 459
Communicative Principle of Relevance, 461
comparison, 293
competence, 108, 111, 145, 745
complements, 42, 143, 492, 586, 587, 597
completion, 409–10, 433
complex domains, 307–8
complex units, 501
complexity hierarchies, 232, 234, *246*
component structures, 583
componential analysis (semantic decomposition), 60, 61–3, 208
componential features, 251
componential view, 207
composite prototypes, 275–6

composite structures, 583, 591–2
composition, 409
compositional semantics, 209–10, 213–15
compositionality, 171, 213, 268–9, 365, 450, 642, 671
compound metaphors, 304–5, 307–10, *326*
compound nouns, *613*
comprehension, 112
compressions, 418–26, *442*
   taxonomy, 420–5
COMPULSION, 187, 188, 189
computational model, 751, 753
COMPUTER DESKTOP blend, 415–17, 430
Comrie, B., 102, 773
conative constructions, *705*
concept elaboration, 79, 168–9, 310, 349
concept-dependency, 562
concepts, 7, *24–5*, 223, 714–15
conceptual alternativity, 516–17
conceptual autonomy, 585–7
Conceptual Blending Theory (Conceptual Integration), 51*, 163, 368, 400–44, *441–2*, 781
   comparison with Conceptual Metaphor Theory, 435–9
   constitutive processes, 410, *441*
   constraints, 433–4
   governing principles (optimality principles), 433–4
   origins, 401–3
   theoretical development, 440*
conceptual categories, 328
conceptual content system, 192
conceptual conversion operations, 516
conceptual dependence, 585–7
conceptual domains, 14–15, *24–5*, 286
conceptual evolution, 310
Conceptual integration *see* Conceptual Blending Theory
conceptual knowledge, 207
conceptual mappings, 164
Conceptual Metaphor Theory, 164–5, 167, 250, 286, 295, 296–304, 313, 322–3*, *325*, 419, 779–80
   comparison with Conceptual Blending Theory, 435–9
conceptual metonymy, 287, 311, 313, 314–18, 325
conceptual ontology, 231
conceptual organisation, 282–3*
conceptual phenomena, 778
conceptual projection, 46, 286, 364
   mappings, 167, 367

conceptual semantics, 746
conceptual space, 59, 696
conceptual structure, 48, 156, *174*
   embodiment and, 157–8, 176–205, 191–200
   semantic structure as, 158–60, 165
Conceptual Structuring System Model, 192, 480, 514–33, 549–50*
conceptual systems, 56–7
conceptualisation, 5, 7, 48
   language as facilitator, 98–101
   meaning construction as, 162–3, *175*, 363
   space, 68–75, 87–92, 100, 103–4*
   time, 75–87
conceptualising capacity, 56
concord, 498
conditioning, 131
configurational domains, 236
Configurational Structure System, 195–6, 518–26, *551*
connectives, 451–2
connectors, 375–6, 377, 379, 380
   directionality, 378
   vital relations, 420
connotation, 210–11
constants, 451
constative sentence types, 447–8
constituency, 492–4, 588, 694–5, 763
constitutive processes, 410, *441*
constraints, 58, 101, 353
   blending theory, 433–4
   cognitive vs. generative, 757–8
construals, *467*, 556, 559, 576, 620
construction grammar, 653–61, 663*, 747
   model, 653–9, 659–60
   vs. cognitive grammar, 660–1, *665*
construction grammars, 49, 206, 227, 481–2, 483, 508*, 641–65
   architecture, 666–706
   assumptions, 667–9
   comparing approaches, 699–701, *705*
   Goldberg, 667–92, 703*
construction (symbolic assembly/symbolic unit) *see* symbolic units (construction/symbolic assembly)
constructional meaning, 672–3
constructional schemas, 592–3
constructions, 23*, *24*, 581–614, 650, *703–5*, 754
   relationships between, 680–4
   verbs and, 671–84: case studies, 684–92
   vs. words and rules, 642–3
contact, 71

815

CONTAINER, 46–7, 158, 179, 181–2, 186, 191, 214, 234
containment, 46–7, 157–8
content function, 19
content meaning, 478
content requirement, 502, 758
content verb, 618
context, 12, 112–13, 211, 212, 220
   deictic expressions, 212, 498–9
   discourse, 716–17
   importance for polysemy, 352–5
   interpersonal, 221
   metonymy, 716–17
   posodic, 221
   role in meaning, 340–1
   role in polysemy, 359*
   sentential, 221, 353, 354–5
   situational, 221
   types, 221
   usage, 353–4
context-free literality, 288, 292–3
contextual effects, 460
contextual modulation, 220, 355
contextual (situated) meaning, 220–1
contiguity, 311
continuity principle, 67
continuous aspect, 388, 491
continuous matter, 515
contractibility, 559–60
contradiction, 364
contrast sets, 345–6
conventional knowledge, 217
conventional literality, 287, 289–90
conventionality (norms), 98, 109, 110, 123, 124–5, 332, 356, 501
   vs. regularity, 755–6
converging evidence, 17–18, 23*, 170
coordination, 492
copula, 621
copular clauses, 597–600
corpus linguistics, 358*, 780
correlated case systems, 607
   case study, 607–9
correlation-based metaphors, 322, *326*
correlational structure, 255
correspondence principle, 677
correspondence theory, 446, 454, 584–5
Coulson, S., 440, 441, 779, 780
COUNT, 186–7, 519–20, 561
counter-expectational interpretation, 384
counterfactual conceptualisation, 162, 369–70, 395, *399*, 406, 407
COUNTERFORCE, 188
counterpart connectors, 409

counterparts, 375–6, 379
Coventry, K., 104, 357
Croft, W., 57, 58–9, 103, 109–10, 120–33, 148, 173, 231, 234, 315, 357, 359, 481, 507, 509, 662–3, 692–7, 700, 701, 702, 703, 708–13, 734, 742, 759, 760, 773, 780–1
cross-domain mappings, 286
cross-linguistic patterns, semantic systems, 68–87
cross-linguistic universals, 308
cross-linguistic variation, semantic systems, 87–95, 102–3*
Cruse, D. A., 173, 352–4, 359, 507, 662–3, 700, 701, 702
cue (attribute) validity, 261
cues (attributes), 223–4, 257–8, 267
cultural constructs, 79
cultural products, 276–7
culture, *246*
Cummins, D. D., 22
Cummins, R., 22
CUP, 29
current relevance, 622
Cutrer, L. M., 397
Cuyckens, H., 22, 358, 359, 780

Dani tribe (New Guinea), 97
Dasher, R., 482, 509, 721–8, 732
data collection, *361*
dative shift (double object) constructions, 606, *614*, 668
Deane, P., 357
DEATH, 301–2
DEATH-AS-BUSINESS-FAILURE, 430–1
debounding, 187
declarative, 165
decompositional approaches, 251–2
decompression, 425–6
default information, 378
definite/indefinite interpretation, 371–2
definitional (classical) theory, 249, 251–2, *283*
definitional problem, 252–3
definitional structure, 251
degree of extension, 195–6, *203–4*, 523
degree modifiers, 525
deictic centre (locus of experience), 83, 197
deixis (perspectival location), 197, 212, 498–9, 528, 713, 722–3
denotation, 45, 210, 453, 456
denotatum, 453
deontic mood, *639–70*
dependence, 229, 349, 355, 562, 585–7, 590–1, *613*, 657

dependents, 492, 763
derivational affixes, 485
derivational morphemes, 589
derivational rules, 60
derivational system, 404, 748, 753–4
descriptive adequacy, 15, 564
descriptive grammar, 484
designation, 237
determiners, 489, 572–3, 574
determinism *see* linguistic determinism
developmental psycholinguistics, 134
deverbal nominalisations, 560
deviation, 109
Dewell, R., 357
diachronicity, 124–5, 707
dialect, 121
dictionary view, 160, 207–15, *245*
    problems, 210–13
dimensionality, 236
diminutive, 30–1
DIRECTION, 117
Dirven, R., 22, 359, 549, 577, 611, 638
disanalogy, 407, 410, 423–4
discourse context, 716–17
discreteness, 35, 515
disintegration, 425–6
distal morpheme, 628
distance, 71
distinctive features, 34–5
distinctiveness, 261
distribution patterns, 524
distributional approach, 487
distributional potential, 609
ditransitive constructions, 37–8, 39–40, 496, 605, 678, 685–7
divalence, 226
DIVERSION, 188
dividedness, 520–1
Dixon, R. M. W., 612, 702
domain highlighting, 315
domain matrix, 231, *246*
domain part-part relationships, 316–17
domain-general, 137, 138
domains
    abstract, 207, 231, 235, 236
    basic, 537–8
    complex, 307–8
    conceptual, 14–15, 286
    configurational, 236
    exercises, *24–5, 245, 246*
    locational, 236
    space, 233
    theory, 14–15, 137, 138, 166, 206, 211, 230–40, 244–5*

double dissociation, 145, 745
double object (dative shift) constructions, 606, *614*, 668
double raising, 32–3
double-scope networks, 429–31
Dowty, D., 612, 702
duality of patterning, 35
Dunbar, G., 356
DURATION, 79–80, 81, 83, 349
Dutch, 100
Dyirbal, 144
dynamic processes, 565

EARLIER, 86, 87, 94–5
economy, 118, 129, 255, 257
    of representation, 749
    vs. detail, 260–1
    vs. redundancy, 755
EEG (electroencephalogram), 76, 77, 78
ego, 84, 85–6
ego-based models, 84, 86–7
egocentricity, 715
elaboration (site), 586, 587, 589, 591, 595, 596
electricity analogy, 98–9
elements, 371–2
embedded clauses, 600–1, *613*
Embodied Construction Grammar, 481, 508*, 697–9, 700, 703*
embodied experience, 45
embodiment cognition thesis, 46–7, 51–2*, 64, 157–8, 176
embodiment (experiential grounding), 44–8, 64
    and conceptual structure, 157–8, 176–205
emergentism, 136, 403, 404, 405, 718, 760
empiricist approach, 44, 745, 778, 780–1
ENABLEMENT, 188–9, 190
encoding, 6, 20–1, *23–4*
encyclopaedic information, 221
encyclopaedic knowledge, 206, 216–22
    dynamism, 221–2
    points of access, 221
encyclopaedic meaning, 220–1
encyclopaedic network, 217
encyclopaedic semantics, 207
encyclopaedic view, 160–2, 206–44, 215–22, 244*
endpoint focus, 337–8
entailments, 298–9
entrenchment, 114, 117, 118–19, 120, 340, 501
environment, 64
epistemic distance, 387, 394–6, 628
epistemic modality, 387
epistemic model, 627–31

epistemic mood, *639–70*
epistemic stance, 395
equative clauses, 599
ergative, 607
etymology, 127
Euclidean properties, 503
Evans, V., 23, 47, 51, 75–87, 79, 104, 244, 324, 341, 342–8, 348–50, 357, 358, 780
EVENT, 80–2, 83, 84, 302, 389, *398*, 720
event structure metaphor, 299–300
event-sequence potential, 227
evidentiality, 93
exceptions, 253
excerpting, 187
exchange, 759
exemplars, 225, 249
expansibility, 559–60
expansion space, 394–5, *398*
experience, 64–5, *174*
experiential correlation, 165, 305–6, 717
experiential grounding (embodiment), 44–8, 64
experiential realism, 47–8
experientialism, 365, 715
explicature, 461–2
expressivity, 10–11
extension, 453
extraction, 111
extrinsic knowledge, 218

Fairclough, N., 759
false dichotomies, *774*
family resemblance, 29, 31, *52*, 265–7
Fauconnier, G., 50, 162, 163, 167, 173, 364–97, 400–40, 779
features, 34, 223
field-based reference, 72–3, 75, 90–2
figurative language, 287–93
figure, 18, *25–6*, 222
figure-ground segregation, 65, 66, 69–70
filler-gap analysis, 657
Fillmore, C., 14, 23, 166, 222–30, 231, 244, 250, 358, 481, 482–3, 508, 612, 642–62, *663*, 700, 701, 702, 742
fly, 351–2, *362*
focus, 389, *398*
focus of attention pattern, 526
focus input, 428, 429
Fodor, J. A., 103, 242, 772
Foley, W., 91–2, 105
FORCE, 182–3, 187
FORCE DYNAMICS, 183–4, 199–200, 531–3, 544–5, *551*, 630
foregrounding, 321

form change, 710–12
form-meaning, 6–7, 127, 131–2, 214, 478
formal approaches, 28, 44, 103*, 108, 171–2, 466*, 659
formal blends, 414–15
formal linguistics
  universals in, 60–3
  vs. cognitive linguistics, *105*
formal (truth-conditional) semantics *see* truth-conditional (formal) semantics
formal universals, 60
formalism, 758, 778–9
forms, 6
foundation spaces, 394–5, *398*, *399*
Frame Semantics, 166, 206, 222–30, 244*, *245–6*, 358*, 671–2
FrameNet project, 244
frames, 11, 69, 139, 166, 211, *245*, *245–6*, 245*
  categories dependent on, 229
  COMMERCIAL EVENTS, 225–8
  perspective provided, 229
  scene-structuring, 229–30
  semantic, 207, 222
  single situations, alternatives, 230
  speech events, 228
  words dependent on, 229
  see also reference frames
framing input, 428, 429
free morphemes, 18
Frege, G., 209, 447
French, 711–12
frequency, 114, 117, 118–20
fricatives, 34
Fried, M., 663, 703
Fromkin, V., 22
full-specification approach, 333–9, *359*
  problems, 339–42
functional asymmetry, 134
functional typological approaches, 758–61, 773*
  characteristics, 761
functional typology, 57, 59
functionalist approach, 108, 778–9
functions, 5
FURNITURE, 259
fusion, 594, 677–80
future, 92–4, 387–8, *739–40*
fuzziness, 29, 43, 253–4

GAME, 252–3
Gärdenfors, P., 22
Garrett, M. F., 103–4
Garrod, S., 104, 357

Geeraerts, D., 22, 356, 357
gender, 499, 547–8
gene pool, 126
generalisation, 721
  typological, 57
generalisation commitment, 28–40, 114, 250, 501, 661–2
Generalised Theory of Selection, 125–7
generative model, 44, 55–6, 118, 140–6, 449–50, 464, 661, 743–52, 753, 771–2*
  characteristics, 747
  vs. cognitive model, 752–8, 761, 781–2
generators, 274, 278
generic construals, 559
generic information, 249
generic knowledge, 217, 218
generic space, 404
generic-level metaphor, 302
Gentner, D., 98–9, 105
Gentner, D. R., 98–9
Germanic languages, 121
Gestalt psychology, 65
Gibbs, R. W., 287 322, 358, 780
Givon, T., 773
Gleitman, L., 62
global insight, 418
Goldberg, A., 22, 37, 39, 481, 508, 667–92, 700, 702, 703
Goldberg's Construction Grammar, 481, 508*
Goldin-Meadow, S., 105
Gonzales-Marquez, M., 781
goodness-of-example ratings, 265–6
Goossens, L., 319–20
Government and Binding Theory, 749
gradability, 352
GRADED CATEGORIES, 254
graded grammaticality, 111, 506
Grady, J., 304–10, 323–4, 441, 779
Graham, G., 22
grammar, 109, 484
  blending, 440–1*
  categorisation, 545–8
  cognitive approaches, 48–50, 250, 475–511, 480–3, *509*, 741–74: characteristics, 500–6; guiding assumptions, 476–80
  cognitive model, 476
  conceptual basis, 512–52, *551*
  core issues, 761
  and language change, *148–9*
  polysemy, 545–8
  theories, 741–61
grammatical concepts, 714
grammatical constructions, 214–15
grammatical functions, 494–8, *510*, 601–6, 765–7
grammatical knowledge, 501–2
grammatical marking, 626
grammatical meaning, *510*
grammatical morphemes, 593–4
grammatical subsystems, 18, 19–20, 23*, 480, 513–14
grammatical terminology, 483–500, 509*
grammatical units, 484–6
grammatical word forms, 486
grammaticalisation, 20, 131–2, 707–40
  cognitive approaches, 482–3, 508–9*, 713, *736–7*
  common patterns, 710
  nature of, 708–13
grammaticalisation continuum, 716
grammaticalisation micro and macrostructure, 717–18
grammaticality, 111, 505–6
granularity, 337
gravity, 45
Great English Vowel Shift, 130
Green, D., 22
Greenberg, J., 57
Grice, H. P., 209, 459, 461
Grice's maxims, 723–4
Gries, S. Th., 358, 780
GRIM REAPER, 431–2
ground, 18, *25–6*, 69, 222, 605
ground-based reference, 72, 74
grounded verb, 618
grounding, 575–6, 598–9, 626, 722
grounding predications, 575, *580*, 584, 617, 624–31
guidepost-based reference, 73, 76
Gumperz, J., 105
Guugu Yimithirr (Australia), 90–2, 100

habit, 627
Haiman, J., 244, 773
Halliday, M., 759, 773
Hampe, B., 201–2
Haser, V., 324
Haspelmath, M., 773
Hausa, 528–9, 719, 794, 802
have, 621–4
Head Driven Phrase Structure Grammar (HPSG), 61, 651, 660, 663*, 746
head final languages, 144
head initial languages, 144
head-complement structure, 586
head-dependent relations, 587
head-modifier structure, 587

heads, 492, 763
Heim, I., 466
Heine, B., 482, 508, 714–21, 725, 732, 734, 735, 742
Herskovits, A., 357
heterogeneity, 559, 637
hiding, 303–4
hierarchies of complexity, 232, 234, *246*
highlighting, 303–4, 312, 315, 355
historical linguistics, 121–2
holophrases, 134, 135, *149–50*
homogeneity, 559, 637
homographs, 329
homologous categories, 516
homonymy, 36, 209, 329, 356*
homophones, 329
Hopper, P., 102, 509, 709, 712, 734, 760
horizontal dimension, 94, 264–7
Hull, D., 125
human scale, 418–19
Humboldt, W. von, 97
Hünnemeyer, F., 714–21, 725, 732, 735
Hunt, E., 105
Hyams, N., 22
hydraulic system model, 99
hypermaxim of linguistic interaction, 128
hypocorrection, 131
HYPOTENUSE, 237
hypotheticals, *399*

I-language, 745
iconicity, 197, 760
idealisation, 110, 113, 169
idealised cognitive models (ICMs), 169, 248–85, 269–81, 282*, *284–5*, 627
ideals, 274
identity, 132, 375
identity connector, 375, 376, 377
identity constraint, 353
idiomatic expressions, 643–53
  case studies, 648–53
  typology, 643
idiomatic meaning, 12, 13, 23*
idiomatic units, 643
idioms, *664*
  decoding, 643–4
  encoding, 643–4
  exgrammatical, 644
  familiarity/unfamiliarity, 645–8
  formal, 644–5
  grammatical, 644
  pragmatic functions, 645
  productive, 644–5
  substantive, 644–5

types, 645
typology, 646
ignorance and error problem, 254, 268
image content, 235
image metaphors, 293–4
image schemas, 46–7, 158, 176, 177–91, 201–2*
  and abstract thought, 191–2
  as analogue representations, 184–5
  clustering, 187–9
  and domains, 233–5
  emergent, 178
  exercises, *52–3, 202–3, 247*
  idealised cognitive models, 280
  inherent meaning, 183–4
  interactional derivation, 182–3
  internal complexity, 185
  and linguistic meaning, 189–90
  metaphorical extension, 714
  metaphors and, 300–1
  more specific contexts, 180–2
  multi-modality, 186
  pre-conceptual origin, 180
  prepositions and, *359–60*
  properties, 179–89
  provisional list, 190
  transformation, 186–7, 337–9
  vs. mental images, 185
imagistic experience (sensory experience), 64–5, 178
immanence, 756
imperfect, 388, 621
imperfective processes, 632–3, 636–7, *640*
implausibility, 669–70
implicational scale, 649
implicational universals, 58–9
implicature, 461–2
implicit elements, 502
inclusiveness, 256–7, 261
incongruity, 652
indirect object, 606
Indo-European, 121
inferences, 298, 376, 461–2, 699; *see also* Invited Inferencing Theory
inferencing strategies, 162
infinitives, 568–70
inflectional morphemes, 589
informational significance, 45
inheritance relations, 301, 654, 681
initial state, 745
inner-space relations, 420, 429
innovation, 116, 123, 126–7, 136
input spaces, 404
inputs, 403

INSTANCE, 81, 82, 682–3
instantiation, 115–16, 117, 118, 119, 331
INSTRUMENT, 540
integration, 502
integration networks, 403–7, 405, 429–31, 431, *442–3*
    taxonomy, 426–31
integration principle, 433
intention, 130, 139
intention-reading, 137, 138, 139
intentional agents, 138, 139
interactive function, 9–11, 23*
interactors, 125, 127
interconnection, 558
interlocutors, 386–7
internal structuring, 373
International Phonetic Alphabet (IPA), 504
interpersonal context, 221
interpretation, 371–2, 384, 452–5
interrogative, 165
intersubjectification, 723–4
intransitive clauses, 605
intrinsic knowledge, 217, 218
introspective (subjective) experience, 65, 179, 235
intuitions, 16–17
Invariance Principle, 301–3, 304
inventory-based approaches, 481–2, 661, 753–4
Invited Inferencing Theory, 482, 721–8, 736*, *738*
    case study, 725–8
Invited Reference Theory, 344
irony, 288
Italian, 30–1
item-based, 134

Jackendoff, R., 48, 62–3, 103, 172, 612, 702, 773
Jaeger, J., 34
Janssen, T., 22
Japanese, 83
Johnson, M., 23, 46, 48, 51, 104, 158, 164–5, 173–4, 177–8, 179, 181, 187, 190, 202, 294–301, 310–21, 322
joint attention frames, 139
JOURNEYS, 294–5, 296–7, 298–9, 299–300

Kay, P., 14, 23, 481, 483, 642–62, *663*, 700, 701, 742
Keller, R., 128
Kemmer, S., 147, 773
knowledge context, 355
knowledge of language, 108, 112; *see also* encyclopaedic knowledge
knowledge representation, 223
    dynamism, 249
    perceptual basis, 240–3
KNUCKLE, 232, 233
Komatsu, L., 282–3
Korean, 87, 89–90, 91, 144
Kövecses, Z., 297–8, 312, 315, 316–19, 322, 325
Kreitzer, A., 357
Kuteva, T., 357, 734

Labov, W., 130
Lakoff, G., 3, 13, 15, 23, 27, 31, 33, 46, 48, 51, 104, 158, 164, 165, 169, 173–4, 177–8, 181, 190, 202, 250, 269–79, 282, 294–301, 310–21, 322, 323, 330, 333–9, 341, 342, 357, 703
landmarks, 157, 181, 334–5, 541, 597, 605; *see also* trajectory-landmark organisation
LANDYACHT blend, 416
Langacker, R., 3, 5, 6, 23, 49, 51, 109, 114–19, 147–8, 166, 174, 181, 193, 217, 230–40, 244, 476–81, 482, 504, 506, 508, 533–48, 553–77, 578, 581–610, 611, 617–37, 638, 700, 728–33, 736, 738, 767
language, 367–8
language acquisition, 108, 262–3, 745–6
    empirical findings, 134–6, 145–6
    generative view, 140–6
    sociocognitive mechanisms, 136–40
    theories, *150*
    usage-based approach, 148*
Language (Bloomfield), 743
language change, 108, 120–33, *148–9*
    causal mechanisms, 127–33
    usage-based approaches, 148*
Language Instinct, The (Pinker), 96
language modules, 41
language processing, 112, 697
language in use, 108, 109–14
    principles, 209
language users, 110, 127
language variation, 121
LATER, 86, 87, 94–5
lattices, 372, 374, *398*
Laurence, S., 254, 268, 282
layering, 712
Leacock, C., 359
learnability, 749
Lee, D., 173, 507, 577, 611, 638, 702, 734
Lee, P., 105
Leezenberg, M., 324

left-isolation construction, 657
let alone construction, 648–51, *664–5*
Levinson, S., 55–6, 61, 90, 100, 104, 105
lexical concepts (senses), 78–9, 180, 330, 342–4
    TIME, 79–81, 82–4
lexical entries, 329
Lexical Functional Grammar, 61, 746
lexical items, 169, 221, 328, 486
lexical relations, 209
lexical semantics, 36, 208, 213–15, 357*, 358*
lexical subsystem, 18, 19–20, 23*, 480
lexical verbs, 388
lexicalisation, 722
lexicographers, 207
lexicologists, 207
lexicon, 208, 450
    mental, 209
    metaphor, 38–9
    polysemy, 36
lexicon-grammar continuum, 193–4, 478, 669, 753
Li, P., 62
licensing (sanctioning), 115–16, 340, 505–6, 593
like, *361*
limitations (of language), 7–9
Lindner, S., 357
Lindstromberg, S., 357
Line, 78
lineage, 126, 127
linguemes, 126
linguistic determinism, 55, 96
linguistic modularity hypothesis, 28, 41, 144–5
linguistic relativity, 55, 71, 95–101, 105*
    Whorf, Benjamin Lee, 96–8
linguistic savants, 145
linguistic strategies, 110
linguistic typology, 56, 102–3*
    functional, 57, 59
linguistic units, 12, 159
linguistics, 22*
literal language, 12–13, 13–14, 287–93
literary theory, 441*
LOCATION, 117, 691
locational domains, 236
locus of experience (deictic centre), 83, 197
loss, 710–11, 712

Mandarin, 82, 94–5
Mandelbilt, N., 440
Mandler, J., 46, 47, 178, 180, 202, 515
manner of articulation, 34

mappings, 23*, 162, 164, 167–8, *175*
    conceptual projection, 367
    identity connectors, 376
    and mental spaces, 368
    metaphor, 308–9, *325*
Margolis, E., 254, 268, 282
marked coding, 609–10, *614*
MARRIAGE, 169
MASS, 186–7, 515, 519–20, 561
matching, 452–5
Matlock, T., 358
matter, 515, 516
maximisation of vital relations principle, 433, 434
maxims, 128–30, 132, 133, 723–4
may, 189–90
meaning, 6, 50*, 446, 556, 671–3
    encyclopaedic view, 206–47
meaning chains, 332
meaning change, 712–13
meaning construction, 161–2
    in cognitive semantics, 365–70, 458
    as conceptualisation, 162–3, *175*, 214
    dynamic nature, 386–96
    and mental spaces, 363–99
meaning extension, 38, 332
meaning potential (purport), 221, 371
Meltzoff, A., 140
mental image, 7
mental lexicon (semantic memory), 209, 331–2
mental space construction
    architecture, 371–82
    illustration, 382–6
Mental Spaces Theory, 162, 279–80, 368, 369–70, 380
    applications, 397*
    assumptions, 3*97*
    lattices, 374, *398*
    mappings, 368
    meaning construction and, 363–99
    tense-aspect system, 389–94
mentalism, 141, 207
merge, 751
meronymic (part-whole) relations, 231, 316, 424
message, 759
metalanguage, 446–7, 451–2
metaphor, 38–40, 43–4, 286, 293–310, 313
    exercises, *52*, *203*, *326*, *467*
    further reading, 324*, 325*, 441*
    Invited Referencing Theory, 724–5
    reconstruction, 714

relevance theory, 463, 465
  vs. blend, 437–9, 441*
  see also Conceptual Metaphor Theory,
    conceptual metaphors
metaphor mapping gaps, 439
metaphor systems, 299–300
metaphor theory see Conceptual Metaphor
  Theory; Primary Metaphor Theory
metaphor-metonymy interaction, 318–21
metaphorical blends, *444*
metaphorical extension, 339, 683–4, 714–15,
  737, 738
  approaches, 482, 714–21, 735–6*
  case studies, 718–21
metaphorical ICMs, 280
metaphorical projection, 158, 407
Metaphors We Live By (Lakoff and Johnson),
  177–8, 294–6, 310–11, 778
metaphtonymy, 319–20
methodology, 170, 342
metonymic ICMs, 281
metonymy, 167, 272–5, 275, 286, 287,
  310–21, 325*, *326*
  and blending, 441*
micro-senses, 353
Middle English, 130
mind/body dualism, 44
Minimalist Program, 61, 750, 751
mirror networks, 426–7
missing prototypes problem, 268
modal auxiliaries, 189, 547, 629–30, 723
  must, 189, 725–8
model-theoretic semantics, 453
modelling, 15, 17
modifiers, 143, 492, 586–7, 596–7
modular theory, 28, 41, 144–5, 242–3, 464,
  745, 746, 772–3*
Modularity of Mind, The (Fodor), 242
modules (subsystems), 28, 41, 145
Mogford, K., 772
MOMENT, 80, 82, 83, 84
monkeys, 138
monosemy, 329–30
monotransitive constructions, 496
Montague, R., 44, 446, 449
mood, 490, 615, 625–6, *639–70*, 767–9
morphemes, 12, 18–19, 159, 484–6, *509*, 589,
  593–4, 628, 629
morphology, 602
  categorisation, 30–1
  polysemy, 36–7
  verb forms, 626
morphosyntax, 61
MOTHER, 271–2, 276–7

MOTION, 15, 76–8, 82, 167, 691
motor movements, 259
movement, 493
movement analysis, 657
moving crowd model, 99
moving ego model, 84, 85–6
moving time model, 84–5
multi-modality, 186, 241
multiple blending, 431–2
Murphy, G., 324
must, 189, 725–8
mutual cognitive environment, 459–60

Nadel, L., 103–4
Narayan, S., 703
narrowing, 122
nasalisation, 131
nasals, 34–5
native speaker intuition, 746
nativism, 60, 137, 141–4, 464
natural language, 447–8, 451–2
natural language philosophers, 212
necessary conditions, 249, 251
negative polarity item, 649
Nerlich, B., 359
networks, 502, 545–8, 680
  double-scope, 429–31
  integration, 403–4, 405, 429–31, 431,
    442–3
  mirror, 426–7
  semantic, 332, 347, *360–1*
  simplex, 426
  single-scope, 427–9
Neural Lyre, The (Pöppel), 78
neuroscience, 241–2
New Guinea, 97
Newport, E., 137
Niemeier, S., 148
nominal complements, 597
nominal expressions (noun phrases), 371–2,
  381, 516, 597
nominal grounding predications, 572–6
nominal predications, 534, *551*, 556–7,
  561–3, *578*
nominative, 6, 500, 607
non-configurational languages, 144
non-derivational approach, 61
non-linguistic strategies, 110
non-literal language, 288–9
non-past, 627
non-perceptual view, 242
non-present present, *639*
nonmetaphorical literality, 288, 290–1
normal replication, 128, 133

823

norms (conventionality), 98, 109, 110, 123, 124–5, 332, 356, 501
noun phrases (nominal expressions), 371–2, 381, 516, 597
nouns, 32, 34, 487, 516, 556–7
    abstract, 560
    compound, *613*
NUMBER, 268, 278, 499, 519
Núñez, R., 93, 104
Nuyts, J., 23

Oakley, T., 441, 779
object language, 446–7
object-to-space, 718–19
objective construal, 543
objectivist approach, 647–8
objectivist thesis, 365
objectivist world view, 156
OBJECTs, 191, 496–7, 515, 595, 606; *see also* oblique object
obligatorification, 711
oblique object, 226
O'Connor, M. K., 23, 642–51, 663
off-line processing, 240–1
Ohala, J., 34
Old English (Anglo-Saxon), 121–2
one-word stage, 134
ontological categories, 62
open-classes, 19, 23*, *25*, 165, 489
    grammatical subsystems, 513–14
    semantic system, 193
openness, 377
operators, 451–2
optimal relevance, 461
Optimisation Principle, 372, 384
orthographic words, 485
Ortony, A., 324
ostensive communication, 459, 460–1
Östman, J.-O., 663, 703
outer-space relations, 420
over, 36, 38–9, 329, 330, 331, 333–9, 341, 343–7, 780
    distinct senses, 348
overshooting, 130
overt elements, 502

Pagliuca, W., 735
Panther, K.-U., 325
paradigmaticisation, 711
paragons, 274
paraphrase, 364
parsing (analysis), 698–9
part-whole (meronymic) relations, 231, 316, 424

partial sanction, 116–17
participant roles, 225, *246*, 673, 676
participles, 388, 567–8
parts of speech, 31–4, 486
passive voice, 32, 491, 609–10, *614*, 619, 620–1, 769–71
past participle, 389
past tense, 92–4, 120, *149*
PATH, 185, 186
path windowing, 198–9
PATIENT, 37–8, 42, 540, 675
pattern completion principle, 433
pattern-finding, 137–8
Pederson, E., 23
perceived resemblance, 293
perceived world structure, 255
perception, 65–8, 240
percepts, 7
perceptual meaning analysis, 47
perceptual moments, 75, 78
perceptual salience, 262
perceptual states, 184
perceptual symbol systems, 245*
perceptual symbols, 206, 245*
perf, 620–1, *639*
perfect aspect, 388, 490
perfect constructions, 619, 621–4
perfective, 634–6, 637
perfective processes, 632–3, *640*
performance, 108, 111
performative function, 417
performative sentences, 448
performative verbs, 635
performativity, 417–18
Perkins, R., 735
person, 498
personification, 301–2
perspectival direction, 530–1
perspectival distance, 528–9
perspectival location (deixis), 197, 212, 498–9, 528, 713, 722–3
perspectival mode, 529
perspectival system, 196–8, 528–31
perspective, 541–4, 724, 728
perspective point, 197
Peterson, A., 103–4
phonological attrition, 709
phonological autonomy, 590–1
phonological dependence, 590–1
phonological pole, 476–7, 479
phonological space, 130, 479
phonological words, 485–6
phonology, 34–5, 450
phrase structure, 60, 143, 144, 582–9, *612*, 749

phrases, 12, 492, *510*
PHYSICAL PROXIMITY, 15
Pinker, S., 96, 772
pivot schemas, 135
place of articulation, 34
PLACE-FUNCTION, 63
Plato, 209
plexity, 519
plosives, 34
poetry, 78
Pollard, C., 663
polysemy, 36–8, *52*, 169–70
    and blending, 441*
    as conceptual phenomenon, 329–31
    constructional, *706*
    fallacy, 342
    grammar, 545–8
    importance of context, 352–5
    lexical, *706*
    meaning change, 712
    psycholinguistics of, 358*
    role of context, 359*
    spatial particles, 357–8*
    and vagueness, 340–1, 356*
    verbs and constructions, 681–2
    see also Principled Polysemy approach
Pöppel, E., 45, 78, 104
Portner, P., 466
possibility space, *398–9*
potential reality, 630–1
potential relevance, 622
poverty of the stimulus, 141–2, *773*
pragmatic function, 375
    mappings, 167
pragmatic meaning, 113, 211
pragmatic strengthening, 344, 717
pragmatics, 208, 215–16, 364, 464
pre-conceptual experience, 46
pre-determiners, 575
pre-linguistic infants, 137
predicate, 494, 495–6, 596, 674
predicate calculus, 447, 451–2
predicative clauses, 599
predicative complement, 497
prediction, 680–1
preferential looking technique, 137
prefixes, 485
preposition phrases, 523–4
prepositions, 345–6, 348–52, *359–60*, 489, *579–80*
presuppositional mode, 372
primary information-bearing units, 695
primary lexical concepts, 79

Primary Metaphor Theory, 304–5, 304–10, 306, 307–10, 323–4*, *326*
primary reference object, 69
primes, 100–1, 208, 268
primitives (semantic primes), 28, 61, 62, 208, 251
Principle of Compositionality, 450
Principle of Maximised Motivation, 680–1
principled distinction, 514
Principled Polysemy approach, 329, 342–52, 347–8, 357*, *359*
Principles and Parameters approach, 60–1, 143, 749, 750
PROCESS, 555, 563–4, 565, 596, 619, 620, 621, 623, 629, 632–3, 635
production, 112
profile determinacy, 585
profile-base organisation, 166–7, 236–8, *247*, 538, 551–2
profiling, 41–3, 237, 673, 676, 679
    abstraction, 544
    selection, 537–41
progression, 515
progressive aspect, 388, 491
progressive constructions, 619, 621
progressive participle, 389
projected reality, 7, 48, 630–1
projection mapping, 167, 367
projector-based reference, 74, 77
prominence, 562
prompts, 8–9, 23*, 214, 366
pronouns, 490, 500
propagation, 123, 133, 372
PROPERTY, 424–5
propositional ICMs, 280
propositions, 448–9, *467*
proprioception, 233
prosodic context, 221
proto-scenes, 346–7
prototype structure (representativeness), 248, 256, *283–4*
    grammatical categories, 546–8
prototype theory, 249, 254, 255–69, 278–9, *283*, 555
prototype (typicality) effects, 169, 250, 254, 269
prototypes, 68, 169, 275–6, 282*
prototypical action chain model, 545, 601, 603
prototypical affixes, 591
prototypical grammatical constructions, 588–9
prototypical stems, 591
protracted duration, 79–80, 81, 83

proximity principle, 65–6, 71
psycholinguistics, 134
psychological reality problem, 254, 255
punctual events, 635
purport (meaning potential), 221, 371

qualitative distinction, 514
quantifiers, 452, 572–3, 574–5
QUANTITY, 14, 164–5, 286, 296, 515, 724
   closed class elements, 521–3

Radden, G., 23, 104, 312, 315, 316–19, 325, 549, 577, 611, 638
Radford, A., 772
radial categories, 275–8, *285*, 328–62, 357*
   illustration, 347–8
   words as, 331–3
Radical Construction Grammar, 481, 509*, 692–7, 700, 703*
   key features, 693–7
rationalist approach, 44, 745
Ravin, Y., 359
raw data, 16
reality, 446
reanalysis, 132
RECIPIENT, 37–8
recruitment, 373
Redeker, G., 22
reduction, 710–11
redundancy, 118, 755
reference, 209, 311, 380, 381, *398*, 456; *see also* Invited Reference Theory
reference assignment, 462
reference chains, 375
reference frames, 69, 71–4, *105–6*
   field-based, 72–3, 75
   ground-based, 72, 74
   Guugu Yimithirr, 90–2
   projector-based, 74
   taxonomy, 72
reference objects, 69–70
referential value, 658
referents, 176, 191
region, 558
regional dialect, 121
registers, 121, 228
regularity, 643, 658–9
   vs. conventionality, 755–6
reification, 516
relational complements, 597
relational knowledge, 249
relational predications, 535, *551*, 561–3, *579*
relativity *see* linguistic relativity
relevance, 460–1, 724

relevance principle, 433
Relevance-Theory, 212, 458, 459–65, 466*
   comparison with cognitive semantics, 463–5
REMOVAL OF RESTRAINT, 188, 189, 190
renewal, 709
replicability, 559–60
replication, 123, 126–7, 128–32, 133
replicator, 125, 126
representation, 456, 650, 759
   categories, 29
   cognitive, 18, 165, 192, *203*, 514
   economy of, 749
   spatial, 68–9
   as vital relation, 421–2
   see also knowledge representation
representative structure, 276
representativeness (prototype structure) *see* prototype structure (representativeness)
resemblance-based metaphors, 293–4, 322, *326*
resultative constructions, 682–3, 689–92
reversibility, 599
rhetoric, 293
Rice, S., 342, 358, 780
rigid reference, 381
rigidification, 711
RITUALS, 417–18
Rodman, R., 22
role reversal limitation, 139–40
role-value, 422
roles, 381–2
root, 189, 485, 590
Rosch, E., 68, 168–9, 249, 250, 255–69, 282
Rudzka-Ostyn, B., 23
rules, 118, 134
   vs. schemas, 754–5

Saeed, J., 173, 212, 466
Saffran, J., 137
salience, 217, 262, 275, 383
sampling, 57–8
sanctioning (licensing), 115–16, 340, 505–6, 593
Sandra, D., 342, 357, 358, 780
Sapir, E., 95–101
Sapir-Whorf hypothesis, 95–101, *105*
Saussure, F. de, 214, 217, 476
scaffolding, 756–7
scene encoding hypothesis, 673
scene-structuring, 229–30
scenes, 176, 191, 518
schema induction, 280, 367
schema mappings, 167–8, 280

schema-instance relations, 504–5, *510–11*
schemas, 115–16, 118, 135, 223, 502, 720
   class, 570–1
   coded meaning, 216
   frequency, 118
   non-reduction, 117–18
   vs. rules, 754–5
   see also image schemas, pivot schemas, utterance schemas
schematic categories, 195, *203–5*, 518–19
schematic meaning, 19, 176, 191, 478, 503, 514
schematic systems, 177, 194–5, 202*, 241
   Conceptual Structuring System Model, 517–18
schematisation, 115
Schmid, H.-J., 173
scope, 237, 451
scope of predication, 538
secondary landmark, 597
secondary lexical concepts, 79
secondary reference object, 69–70
   encompassing, 70
   external, 71
selection, 125–7, 133–4, 349
   profiling, 537–41
selective impairment, 145, 745
selective projection, 409
semantic arguments, 351
semantic coherence principle, 677
semantic decomposition (componential analysis), 60, 61–3, 208
semantic frames, 207, 222
semantic interpretation, 452–5
semantic map model, 59, 695–6, 760
semantic memory (mental lexicon), 209, 331–2
semantic networks, 332, 347, *360–1*
semantic parsimony, 670–1
semantic pole, 476, 479
semantic primes (primitives), 28, 61, 62, 208, 251
semantic roles, 37, 225, 597, 612*, 675, 702*
semantic selection, 349
semantic space, 479
semantic structure, 158–60, 191, 192–4
   bifurcation, 515
   and grammatical subsystems, 513–14
semantic systems
   cross-linguistic patterns, 68–87
   cross-linguistic variation, 87–95
semantic universals, 61–3
semantics, 209, 215–16, 364, 466*, 746
Semantics and Cognition (Jackendoff), 62–3

Semantics of English Prepositions, The (Tyler and Evans), 342
semi-vowels, 35
sense relations, 78, 208–9, 373
senses (lexical concepts) *see* lexical concepts (senses)
sensory experience (imagistic experience), 64–5, 178
sensory-perceptual systems, 179
sentence initial position, 18
sentence meaning, 364–5
sentence semantics, 450
sentences, 110–11, 448–9, 494
sentential context, 221, 353
   facets, 354–5
sequential scanning, 563
SER (Speech-Event-Reference), 387
set theory, 453, 454
shapes, 259–60
Shinohara, K., 95
SHIP OF STATE blend, 438–9
signified, 476
signifier, 476
signs, 476
similarity principle, 66, 67, 259
simile, 293
simplex networks, 426
simplex units, 501
simulations, 206, 225, 240, 241, 249
simulators, 241
single unified representation, 650
single-scope networks, 427–9
Sinha, C., 174, 357
situated (contextual) meaning, 220–1
situation aspect, 631–7
situation types, 631–2
situational context, 221
Skinner, B. F., 141, 744
Slobin, D., 90, 119
slope model, 95
smallness principle, 67–8, 69
social dialect, 121
social functions, 10, 759
social stereotypes, 273
solidarity, 132
somesthesia, 199
sound change, 130–1
SOURCE, 167, 295, 297–8, 337, 714–15
source domain hierarchy, 715
SOURCE-PATH-GOAL, 185
South America, 92
SPACE, 103–4*
   blended, 163
   conceptualisation, 59, 68–75, 87–92, 100

SPACE, (*continued*)
    configuration (Talmy), 515–16
    domain, 233
    expansion, 394–5, *398*
    foundation, 394–5, *398*, *399*
    phonological, 130
    possibility, *398–9*
    reference frames, 105–6
    as vital relation, 421
space building, 371, *397*
space-to-possession, 719–21
Spanish, 83, 730
spatial particles, 357–8*
spatial reference point, 622
spatial representation, 68–9
spatial scene categorisation, 87–9
spatial scenes, 69, 87–90
species-specific, 138–9
Specific Language Impairment, 145, 772–3*
specifiers, 143, 584
speech acts, 10, 447
speech event frames, 228
Speech-Event-Reference (SER), 387
Sperber, D., 459–65, 466
Stassen, L., 773
STATES, 158, 302, 515, 569, 636, 691
stative processes, 565
stative verbs, 43
Steen, G., 322
Stefanowitsch, A., 780
stems, 591, *612–13*
Stern, J., 324
strategies *see* linguistic/non-linguistic strategies
structural dependencies, 349
structural invariants, 223, 224–5
structural meaning, 176, 191
structure, 5
Structure of Time, The (Evans), 75–87
structured inventory of conventional linguistic units, 476
structuring function, 19, 502
styles, 228
sub-modules, 49, 745
subject-verb agreement, 584
subjectification, 482, 722–3, 728–32, 736*, *738–9*
    case studies, 730–2
subjective construal, 543
subjective (introspective) experience, 65, 179, 235
subjective motion, 338
subjectivism, 160
subjectivity, 304

subjects, 494–5, 595
subordinate categories, 257
subordinate clauses, 600
subparts, 682
subsenses, 353–4, *362*
substantive universals, 60
substitution, 492, 493
subsystems (modules), 28, 41, 145
sufficient conditions, 249, 251
suffixes, 485
summary scanning, 563
superordinate categories, 257
SURGEON AS BUTCHER blend, 401–2, 404–7
Svorou, S., 735
Swahili, 719
Sweetser, E., 93, 104, 156, 174, 397, 417–18, 735–6
symbolic assembly (construction/symbolic unit), 6–7, 12, 13, *23*, 23*, 109, 115, 477, 753
symbolic function, 6–9, 23*, 98, 184
symbolic ICMs, 281
symbolic system, 366, 702
symbolic thesis, 136, 214, 476–8
symbolic units (construction/symbolic assembly), 6–7, 12, 13, *23*, 23*, 109, 115, 477, 753
    grammatical knowledge, 501–2
symbolisation, 502
symbols, 6
synaesthetic metaphors, 321
synchronic semantic network, 345
synchronicity, 124–5, 707
syncopation, 421
synonymy, 209, 210–11
syntactic relations, 694–5
Syntactic Structures (Chomsky), 744
syntactic transformations, 748
syntax, 450, 492–500
    autonomy of, 752, 753
    categorisation, 31–4
    metaphor, 39–40
    polysemy, 37–8
systematicity, 5, 11–15
Systemic Functional Grammar, 759, 773*

tag question, 33
TALKING ANIMALS blend, 417
Tallerman, M., 509
Talmy, L., 4, 18, 33, 51, 68–74, 104, 156, 159, 174, 191, 192, 194–200, 202, 480, 502, 508, 514–33, 549–50, 612, 638
TAM (tense, aspect and modality), 387

target, 167, 297–8, 312, 322
Tarski, A., 446–7
Taylor, J., 23, 282, 320, 359, 507, 549, 578, 611, 636–7, 638, 648, 663, 702, 771
temporal anteriority, 623
temporal compression, 80, 82, 83, 349–50
temporal experience, 78
temporal lexical concepts *see* lexical concepts, TIME
temporal relations, 563–71, 564, 619
 verbs, 564–5
temporal sequence model, 84, 86
tense, 387–98, 615, 626–7, 767–9
 and aspect, 387–9
terms, 451
texts, 759
textual analysis, *327*
thematic hierarchy, 603
thematic relations, 63
thematic roles, 675
THEME, 63, 197
THEORIES ARE BUILDINGS, 307–8, 309
THINGs, 555–6, 560
thinking for speaking, 90
thought, 23*
 abstract, 191, 192, 301
 language as shaper, 98–101
 systematic structure, 14–15
 universals in, 55–68
through, 360–1
TIME, 14–15, 23*, 104*, *106–7*, 235, 298
 cognitive models, 79, 84–6
 configuration (Talmy), 515–16
 lexical concepts, 79–81, 82–4, 100
 patterns in conceptualisation, 75–87
 Principled Polysemy approach, 348–50
 variation in conceptualisation, 92–5
 as vital relation, 421
time reference, 387
time-based models, 84, 86–7
token frequency, 118–19
Tomasello, M., 108, 109, 110, 133–46, 148
topology principle, 433, 434, 503
Toward a Cognitive Semantics (Talmy), 194
TR-LM reversal, 610, 620–1, 728–9
trajector, 181, 334, 541
trajectory-landmark organisation, 541, *551–2*, 562
transducers, 242
transformational model, 746, 747–52
transformations, 60, *773*
transitive clauses, 601
transitive verbs, 31–2
transitivity, 601–6

transmitting, 6
Trask, R. L., 5, 22, 509, 735
Traugott, E. C., 482, 509, 709, 712, 721–8, 732, 734, 736
tree diagrams, 763–5
triadic relationships, 139, 140
trigger, 377
trivalence, 226
tropes, 293
truth, 446
truth conditional literality, 288, 291–2
truth-conditional (formal) semantics, 156, 171, 172, 212, 243, 380, 446–58
 comparison with cognitive semantics, 455–8
 and the generative enterprise, 449–50
 readings, 466*
 sentence meaning in, 364–5, 457
truth-conditions, 454–5
Tuggy, D., 340, 356, 357
Turner, F., 104
Turner, M., 163, 173, 174, 202, 297, 301, 323, 366–7, 400–40, 779
two-word stage, 134
Tyler, A., 23, 47, 104, 244, 341, 342–8, 357, 358, 780
type frequency, 118–19
typicality (prototype) effects, 169, 250, 254, 269
 cluster models, 271–2
 sources, 270–5
typological classification, 57
typological generalisation, 57
typological universals, 57–9
typology *see* linguistic typology

UNCLE, 237–8, 239
uncorrelated case systems, 607
undershooting, 130
Ungerer, H.-J., 173
ungrammaticality, 13
unidirectionality, 708, 715
unifying, 13
Universal Grammar, 55, 59, 60–1, 137, 744, 750
universals, 54–107, 102–3*, 695–7, 746
 in cognitive linguistics, 63–8
 conceptual space, 695
 cross-linguistic, 308
 formal, 60
 in formal linguistics, 60–3, 103*
 implicational, 58–9
 language acquisition, 143
 semantic, 61–3

universals, (*continued*)
   substantive, 60
   in thought and language, 55–68
   typological, 57–9
   unrestricted, 58
unmarked forms (zero morphemes), 629
unpacking principle, 433
unrestricted universals, 58
usage context, 353–4
usage events, 109–11, 221
usage-based model, 108, 114, 133–46, 147–8*, *148*, 478–83, 701, 702, 717
   meaning construction, 211, 216, 221
utterance schemas, 135
Utterance Selection Theory, 109, 123–7
utterances, 109–10

vagueness, 340–1, 341, 356*, *362*
valence, 225–6, 228, 231, 583–4
   clause level, 595–601
values, 223–4, 381–2, 452–5, 453, 454–5
Vandeloise, C., 183, 358
variable embodiment, 45
variables, 452
VEHICLE, 260
vehicles for metonymy, 312, 317–18, 322
Vendler, Z., 631–2, 635–6
verb phrases, 516
verb strings, 478, 490, 637, 767–9
verb-island constructions, 135
Verbal Behaviour (Skinner), 744
verbal communication, 459
verbs, 32, 34, 487–8, 490, 516, 564–5
   argument structure, 669–71
   and constructions, 671–84, 704*: case studies, 684–92
   form and function, 616–17, *638–9*
   meaning, 671–2
   simple, 630
Verspoor, M., 22
vertical axis, 94
vertical dimension, 256–64
VERTICAL ELEVATION, 15, 164–5, 286, 296
viewpoint, 389, *398*
visual experience, 45
vital relations, 418–26, *442*
   maximisation, 433
   taxonomy, 420–5

voice, 34, 490, 491, 609, 615
voice continuum, 35

way construction, *705*
ways of seeing, 355
web principle, 433
welfare shopping, *443–4*
Wernicke's area, 146
wh- question words, 62–3
wh- dependency, 657
Whorf, B. L., 55, 95–101, 96–8, 105
Wierzbicka, A., 208
Williams Syndrome, 145
Wilson, D., 459–65, 466
window, 198
windowing of attention, 198
windowing pattern, 527–8
Wittgenstein, L., 252
Wolof, 83, *106–7*
Women, Fire and Dangerous Things (Lakoff), 250, 269
word classes, 31, 486–7, *510*, 553–80
   cognitive vs. generative, 761–2
   conceptual basis, 533–4
   Langacker's model, 536, 571
   linguistic classification, 554–61
word meaning, 169–70, *175*, 211
   and radial categories, 328–62, 357*, *359*
   vs. sentence meaning, 213–15
word order, 13, *23*, *24*
word structure, 589–94, *612–13*
words, 485–6, *509*
   and rules, 642–3
world knowledge, 208
world view, 48
WXDY (*What's X doing Y*) construction, 651–3, 654, 655–9

x-bar theory, 143–4, 749, 750
XYZ constructions, 412–14

Yoruba, 709
Yu, L., 104

Zawada, B., 359
Zelinsky-Wibbelt, C., 358
zero morphemes (unmarked forms), 629
zeugma, 288, 354
Zinken, J., 324